HEART AND NEUROLOGIC DISEASE

HANDBOOK OF CLINICAL NEUROLOGY

Series Editors

MICHAEL J. AMINOFF, FRANÇOIS BOLLER, AND DICK F. SWAAB

VOLUME 177

HEART AND NEUROLOGIC DISEASE

Series Editors

MICHAEL J. AMINOFF, FRANÇOIS BOLLER, AND DICK F. SWAAB

Volume Editor

JOSÉ BILLER

VOLUME 177

3rd Series

ELSEVIER
Radarweg 29, PO Box 211, 1000 AE Amsterdam, Netherlands
The Boulevard, Langford Lane, Kidlington, Oxford OX5 1GB, United Kingdom
50 Hampshire Street, 5th Floor, Cambridge, MA 02139, United States

British Library Cataloguing-in-Publication Data
A catalogue record for this book is available from the British Library

Library of Congress Cataloging-in-Publication Data
A catalog record for this book is available from the Library of Congress

ISBN: 978-0-12-819814-8

For information on all Elsevier publications
visit our website at https://www.elsevier.com/books-and-journals

Publisher: Nikki Levy
Editorial Project Manager: Kristi Anderson
Production Project Manager: Punithavathy Govindaradjane
Cover Designer: Alan Studholme

Typeset by SPi Global, India

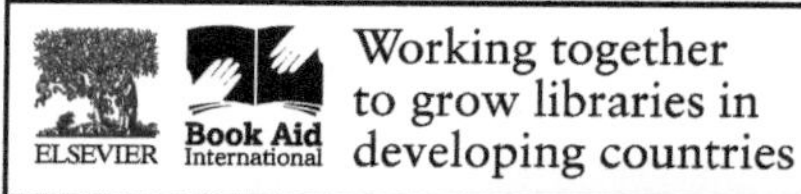

Handbook of Clinical Neurology 3rd Series

Available titles

Vol. 81, Pain, F. Cervero and T.S. Jensen, eds. ISBN 9780444519016
Vol. 82, Motor neurone disorders and related diseases, A.A. Eisen and P.J. Shaw, eds. ISBN 9780444518941
Vol. 83, Parkinson's disease and related disorders, Part I, W.C. Koller and E. Melamed, eds. ISBN 9780444519009
Vol. 84, Parkinson's disease and related disorders, Part II, W.C. Koller and E. Melamed, eds. ISBN 9780444528933
Vol. 85, HIV/AIDS and the nervous system, P. Portegies and J. Berger, eds. ISBN 9780444520104
Vol. 86, Myopathies, F.L. Mastaglia and D. Hilton Jones, eds. ISBN 9780444518996
Vol. 87, Malformations of the nervous system, H.B. Sarnat and P. Curatolo, eds. ISBN 9780444518965
Vol. 88, Neuropsychology and behavioural neurology, G. Goldenberg and B.C. Miller, eds. ISBN 9780444518972
Vol. 89, Dementias, C. Duyckaerts and I. Litvan, eds. ISBN 9780444518989
Vol. 90, Disorders of consciousness, G.B. Young and E.F.M. Wijdicks, eds. ISBN 9780444518958
Vol. 91, Neuromuscular junction disorders, A.G. Engel, ed. ISBN 9780444520081
Vol. 92, Stroke – Part I: Basic and epidemiological aspects, M. Fisher, ed. ISBN 9780444520036
Vol. 93, Stroke – Part II: Clinical manifestations and pathogenesis, M. Fisher, ed. ISBN 9780444520043
Vol. 94, Stroke – Part III: Investigations and management, M. Fisher, ed. ISBN 9780444520050
Vol. 95, History of neurology, S. Finger, F. Boller and K.L. Tyler, eds. ISBN 9780444520081
Vol. 96, Bacterial infections of the central nervous system, K.L. Roos and A.R. Tunkel, eds. ISBN 9780444520159
Vol. 97, Headache, G. Nappi and M.A. Moskowitz, eds. ISBN 9780444521392
Vol. 98, Sleep disorders Part I, P. Montagna and S. Chokroverty, eds. ISBN 9780444520067
Vol. 99, Sleep disorders Part II, P. Montagna and S. Chokroverty, eds. ISBN 9780444520074
Vol. 100, Hyperkinetic movement disorders, W.J. Weiner and E. Tolosa, eds. ISBN 9780444520142
Vol. 101, Muscular dystrophies, A. Amato and R.C. Griggs, eds. ISBN 9780080450315
Vol. 102, Neuro-ophthalmology, C. Kennard and R.J. Leigh, eds. ISBN 9780444529039
Vol. 103, Ataxic disorders, S.H. Subramony and A. Durr, eds. ISBN 9780444518927
Vol. 104, Neuro-oncology Part I, W. Grisold and R. Sofietti, eds. ISBN 9780444521385
Vol. 105, Neuro-oncology Part II, W. Grisold and R. Sofietti, eds. ISBN 9780444535023
Vol. 106, Neurobiology of psychiatric disorders, T. Schlaepfer and C.B. Nemeroff, eds. ISBN 9780444520029
Vol. 107, Epilepsy Part I, H. Stefan and W.H. Theodore, eds. ISBN 9780444528988
Vol. 108, Epilepsy Part II, H. Stefan and W.H. Theodore, eds. ISBN 9780444528995
Vol. 109, Spinal cord injury, J. Verhaagen and J.W. McDonald III, eds. ISBN 9780444521378
Vol. 110, Neurological rehabilitation, M. Barnes and D.C. Good, eds. ISBN 9780444529015
Vol. 111, Pediatric neurology Part I, O. Dulac, M. Lassonde and H.B. Sarnat, eds. ISBN 9780444528919
Vol. 112, Pediatric neurology Part II, O. Dulac, M. Lassonde and H.B. Sarnat, eds. ISBN 9780444529107
Vol. 113, Pediatric neurology Part III, O. Dulac, M. Lassonde and H.B. Sarnat, eds. ISBN 9780444595652
Vol. 114, Neuroparasitology and tropical neurology, H.H. Garcia, H.B. Tanowitz and O.H. Del Brutto, eds. ISBN 9780444534903
Vol. 115, Peripheral nerve disorders, G. Said and C. Krarup, eds. ISBN 9780444529022
Vol. 116, Brain stimulation, A.M. Lozano and M. Hallett, eds. ISBN 9780444534972
Vol. 117, Autonomic nervous system, R.M. Buijs and D.F. Swaab, eds. ISBN 9780444534910
Vol. 118, Ethical and legal issues in neurology, J.L. Bernat and H.R. Beresford, eds. ISBN 9780444535016
Vol. 119, Neurologic aspects of systemic disease Part I, J. Biller and J.M. Ferro, eds. ISBN 9780702040863
Vol. 120, Neurologic aspects of systemic disease Part II, J. Biller and J.M. Ferro, eds. ISBN 9780702040870
Vol. 121, Neurologic aspects of systemic disease Part III, J. Biller and J.M. Ferro, eds. ISBN 9780702040887
Vol. 122, Multiple sclerosis and related disorders, D.S. Goodin, ed. ISBN 9780444520012
Vol. 123, Neurovirology, A.C. Tselis and J. Booss, eds. ISBN 9780444534880
Vol. 124, Clinical neuroendocrinology, E. Fliers, M. Korbonits and J.A. Romijn, eds. ISBN 9780444596024
Vol. 125, Alcohol and the nervous system, E.V. Sullivan and A. Pfefferbaum, eds. ISBN 9780444626196
Vol. 126, Diabetes and the nervous system, D.W. Zochodne and R.A. Malik, eds. ISBN 9780444534804
Vol. 127, Traumatic brain injury Part I, J.H. Grafman and A.M. Salazar, eds. ISBN 9780444528926
Vol. 128, Traumatic brain injury Part II, J.H. Grafman and A.M. Salazar, eds. ISBN 9780444635211
Vol. 129, The human auditory system: Fundamental organization and clinical disorders, G.G. Celesia and G. Hickok, eds. ISBN 9780444626301

All volumes in the 3rd Series of the **Handbook of Clinical Neurology** are published electronically, on Science Direct: http://www.sciencedirect.com/science/handbooks/00729752.

Foreword

Dr. Joseph Foley used to say that the heart is nothing but a component of the nervous system: "a mass of neurons with a muscle." The humorous hyperbole of that illustrious neurologist well reflects the intimate relation of the cardiovascular and nervous systems. Acutely aware of this point, the Editors of previous series of the *Handbook of Clinical Neurology* published several volumes dedicated to the neurologic aspects of systemic diseases including those of the heart. In the current (third) series, the first part of the three volumes dealing with systemic diseases (Volumes 119–121) was a section dedicated to cardiovascular diseases, particularly their clinical aspects. Even though the 16 chapters of that section were prepared less than 10 years ago, we felt that a number of new developments justified a separate volume on "The heart and neurologic diseases." These new developments include current research on applied neuropsychology, genetics, pharmacology, imaging, and the relation with other disciplines such as sports neurology.

The present volume includes chapters dedicated to diseases that primarily affect the heart, such as congenital heart diseases, coronary artery diseases, bacterial endocarditis, and disorders of cardiac rhythm, as well as genetic syndromes. There is also emphasis on systemic effects of heart disease, strokes and pregnancy, and the neurologic complications of heart diseases in athletes. Special attention is given to the relationship between heart disease and cognitive impairment as well on the management of anxiety in patients with heart disease or undergoing cardiac surgery.

We have had the good fortune of enlisting the help of a highly experienced volume editor, Dr. Jose Biller of Loyola University in Chicago, who had previously coedited the earlier volumes devoted to the neurologic aspects of systemic diseases. He deserves our special thanks for assuring the right mix of continuity and highly updated information concerning the relationship between the heart and the nervous system.

As series editors, we reviewed all the chapters in the volume and made suggestions for improvement, but we are delighted that the volume editor and chapter authors produced such scholarly and comprehensive accounts of different aspects of the topic. We hope that the volume will appeal to clinicians as a state-of-the-art reference that summarizes the clinical features and management of the many neurologic manifestations of cardiovascular diseases. We also hope that basic researchers will find within it the foundations for new approaches to the study of the complex issues involved.

In addition to the print version, the volume is available electronically on Elsevier's Science Direct website, which is popular with readers and will improve the book's accessibility. Indeed, all of the volumes in the present series of the *Handbook* are available electronically on this website. This should make them even more accessible to readers and should facilitate searches for specific information.

As always, it is a pleasure to thank Elsevier, our publisher, and in particular Michael Parkinson in Scotland, Nikki Levy and Kristi Anderson in San Diego, and Punithavathy Govindaradjane at Elsevier Global Book Production in Chennai, for their assistance in the development and production of the *Handbook of Clinical Neurology*.

Michael J. Aminoff
François Boller
Dick F. Swaab

Preface

The interaction between the nervous and cardiovascular systems has been well established since the mid-19th century. At that time, physiologists were simultaneously describing electrical activity of the brain and heart and physiologists such as Walter Bradford Cannon (Boston, MA), J.N. Langley (Cambridge, England), and Claude Bernard (Paris, France) began to describe the links between the brain, autonomic nervous system, and cardiac function. Brain and heart diseases are linked by multimodal circuitries and several pathogenetic mechanisms.

Neurocardiology or cardioneurology refers to the pathophysiologic interplays of the nervous and cardiovascular systems. This overlapping multidisciplinary subspecialty has witnessed rapid groundbreaking scientific advances over the last several decades and continuous research innovation critically important to everyday clinical practice. The recognition of the role of complex neurohumoral mechanism in electrical disturbances of the heart, the complexity of the cardiac nervous system beyond a simple "flight or fight" sympathetic or parasympathetic response, and the complex overlay of cardiac and neurogenetic anomalies are just some of the translational research areas that have influenced our conceptualization of neurocardiology as an important clinical field. The current much deeper understanding of the complex multimodal and dynamic bidirectional dialogue between the heart and the brain has proved invaluable to this exciting and challenging interdisciplinary field.

Written primarily for physicians who care for patients with neurological and cardiac problems, and using a multidisciplinary approach, this comprehensive 33-chapter volume of the *Handbook of Clinical Neurology 3rd Series* devoted to the *Heart and Neurologic Disease* highlights critically important insights into how the brain and heart communicate with each other, how disorders of one system can impact the other and vice versa, the causal relationship between autonomic activity and sudden cardiac death, cardiac arrhythmias and myocardial function, as well as specific clinical applications of these brain and heart dynamics, and the value of neuroscience-based cardiovascular therapeutics.

I am greatly indebted to the many authors and coauthors for their invaluable contributions who made this volume possible. I greatly acknowledge the editorial assistance by Mike Parkinson, Development Editor (*Handbook of Clinical Neurology*), and to the rest of the editorial team at Elsevier, primarily Punithavathy Govindaradjane and Kristi L. Anderson. A special thanks to Linda Turner for her support during this project.

It is hoped that this volume with highly clinically relevant chapters will be an enjoyable and informative addition to standard textbooks and monographs.

José Biller

Contributors

H.P. Adams Jr
Division of Cerebrovascular Diseases, Department of Neurology, Carver College of Medicine, University of Iowa Hospitals and Clinics, University of Iowa, Iowa City, IA, United States

A.G. Almeida
Cardiology Service, Hospital Santa Maria, Centro Hospitalar Lisboa Norte and Faculty of Medicine, University of Lisbon, Lisbon, Portugal

S.F. Ameriso
Departmento de Neurología, Fleni, Buenos Aires, Argentina

A. Amireh
Department of Neurology, Ochsner/Louisiana State University Health Sciences Center, Shreveport, LA, United States

C. Arnold
Department of Neurology, Mayo Clinic, Rochester, MN, United States

A. Bhirud
Department of Cardiac Electrophysiology, Loyola University Medical Center, Maywood, IL, United States

J. Biller
Department of Neurology, Loyola University Chicago, Stritch School of Medicine, Maywood, IL, United States

H. Steven Block
SSM Health Dean Medical Group, Department of Neurology, St. Mary's Hospital, Madison, WI, United States

J.R. Brorson
Department of Neurology, University of Chicago, Chicago, IL, United States

T. Chakraborty
Department of Neurology, Mayo Clinic, Rochester, MN, United States

T. Cheng
Department of Neurology and Rehabilitation, University of Illinois at Chicago, Chicago, IL, United States

S.-M. Cho
Neurosciences Critical Care, Departments of Anesthesiology and Critical Care Medicine, Neurology, and Neurosurgery, Johns Hopkins University School of Medicine, Baltimore, MD, United States

S. Cruz-Flores
Department of Neurology, Paul L Foster School of Medicine, Texas Tech University Health Sciences Center, El Paso, TX, United States

R.M. Dafer
Department of Neurological Sciences, Rush University Medical Center, Chicago, IL, United States

M. Davis
Mental Health Service Line, Edward Hines Jr. VA Hospital, Hines, IL, United States

G. deVeber
Division of Neurology, Department of Pediatrics, Hospital for Sick Children, Toronto, ON, Canada

S.W. English
Department of Neurology, Mayo Clinic, Jacksonville, FL, United States

J.M. Ferro
Neurology Service, Hospital Santa Maria, Centro Hospitalar Lisboa Norte and Faculty of Medicine, University of Lisbon, Lisbon, Portugal

A.C. Fonseca
Neurology Service, Hospital Santa Maria, Centro Hospitalar Lisboa Norte and Faculty of Medicine, University of Lisbon, Lisbon, Portugal

W.H. Franklin
Department of Pediatrics, Loyola University Chicago, Stritch School of Medicine, Maywood, IL, United States

R. Gill
Department of Neurology, Loyola University Chicago, Stritch School of Medicine, Maywood, IL, United States

E. Gokcal
Department of Neurology, Massachusetts General Hospital, Boston, MA, United States

L.B. Goldstein
Department of Neurology, Kentucky Neuroscience Institute, University of Kentucky, Lexington, KY, United States

S.A. Goldstein
Department of Medicine, Division of Cardiology, Duke University, Durham, NC, United States

J.B. Gonzalez
Cardiology Department, Advocate Illinois Masonic Medical Center, Chicago, IL, United States

I.E. Green
Departments of Neurology and Public Health Sciences, University of Virginia, Charlottesville, VA, United States

M.E. Gurol
Department of Neurology, Massachusetts General Hospital, Boston, MA, United States

C.E. Hassett
Cerebrovascular Center, Neurological Institute, Cleveland Clinic, Cleveland, OH, United States

M.A. Hawkes
Departmento de Neurología, Fleni, Buenos Aires, Argentina

S. Hocker
Department of Neurology, Mayo Clinic, Rochester, MN, United States

M.J. Horn
Department of Neurology, Massachusetts General Hospital, Boston, MA, United States

V. Javalkar
Department of Neurology, Ochsner/Louisiana State University Health Sciences Center, Shreveport, LA, United States

R.E. Kelley
Department of Neurology, Ochsner/Louisiana State University Health Sciences Center, Shreveport, LA, United States

M.A. Kelly
Department of Neurology, Loyola University Chicago, Stritch School of Medicine, Maywood, IL, United States

J.P. Klaas
Department of Neurology, Mayo Clinic, Rochester, MN, United States

M. Laubham
Department of Medicine, Ohio State University, Columbus, OH, United States

E.C. Leira
Department of Neurology, Carver College of Medicine; Department of Epidemiology, College of Public Health, University of Iowa, Iowa City, IA, United States

R. Lichtenberg
Heart Care Centers of Illinois, Mokena, IL, United States

D.S. Liebeskind
Department of Neurology, Comprehensive Stroke Center, University of California Los Angeles, Los Angeles, CA, United States

D. Loewenstein
Department of Medicine, Division of Cardiology, Rush University Medical Center, Chicago, IL, United States

J.J. Lopez
Division of Cardiology, Loyola University Medical Center; Department of Medicine, Loyola University Chicago, Stritch School of Medicine, Maywood, IL, United States

R. Massaquoi
Department of Neurology, University Hospitals Cleveland Medical Center, Cleveland, OH, United States

A.R. Massaro
Department of Neurology, Hospital Sírio-Libanês, São Paulo, Brazil

K.L. Miller
Department of Neurology and Rehabilitation, University of Illinois at Chicago, Chicago, IL, United States

S. Morales-Vidal
Department of Neurology, Loyola University Chicago, Stritch School of Medicine, Maywood, IL, United States

P. Moshayedi
Department of Neurology, Comprehensive Stroke Center, University of California Los Angeles, Los Angeles, CA, United States

S. Nathan
Section of Cardiology, Department of Internal Medicine, University of Chicago, Chicago, IL, United States

A. Opaskar
Department of Neurology, University Hospitals Cleveland Medical Center, Cleveland, OH, United States

J.G. Ortiz Garcia
Department of Neurology, University of Oklahoma Health Sciences Center, Oklahoma City, OK, United States

N. Osteraas
Department of Neurologic Sciences, Rush University, Chicago, IL, United States

L. Pedelty
Department of Neurology and Rehabilitation, University of Illinois at Chicago, Chicago, IL, United States

W.J. Powers
Department of Neurology, University of North Carolina School of Medicine, Chapel Hill, NC, United States

E. Pulcine
Division of Neurology, Department of Pediatrics, Hospital for Sick Children, Toronto, ON, Canada

M. Rabbat
Department of Medicine, Division of Cardiology, Loyola University Medical Center, Maywood, IL, United States

A. Rabinstein
Department of Neurology, Mayo Clinic, Rochester, MN, United States

P. Riordan
Department of Neurology, Loyola University Chicago, Stritch School of Medicine, Maywood, IL, United States

S. Ruland
Department of Neurology, Loyola University Chicago, Stritch School of Medicine, Maywood, IL, United States

M.O. Santos
Neurology Service, Hospital Santa Maria, Centro Hospitalar Lisboa Norte and Faculty of Medicine, University of Lisbon, Lisbon, Portugal

M.J. Schneck
Department of Neurology, Loyola University Chicago, Stritch School of Medicine, Maywood, IL, United States

A. Shaban
Department of Neurology, Carver College of Medicine, University of Iowa, Iowa City, IA, United States

C. Sila
Department of Neurology, University Hospitals Cleveland Medical Center, Cleveland, OH, United States

A.M. Southerland
Departments of Neurology and Public Health Sciences, University of Virginia, Charlottesville, VA, United States

J.I. Suarez
Neurosciences Critical Care, Departments of Anesthesiology and Critical Care Medicine, Neurology, and Neurosurgery, Johns Hopkins University School of Medicine, Baltimore, MD, United States

M. Teitcher
Department of Neurology, Loyola University Chicago, Stritch School of Medicine, Maywood, IL, United States

F.D. Testai
Department of Neurology and Rehabilitation, University of Illinois at Chicago, Chicago, IL, United States

S. Vasaiwala
Department of Cardiac Electrophysiology, Loyola University Medical Center, Maywood, IL, United States

A. Waqar
Division of Cardiology, Loyola University Medical Center; Department of Medicine, Loyola University Chicago, Stritch School of Medicine, Maywood, IL, United States

E. Wijdicks
Department of Neurology, Mayo Clinic, Rochester, MN, United States

C. Woods
Department of Neurology, Loyola University Chicago, Stritch School of Medicine, Maywood, IL, United States

B.B. Worrall
Departments of Neurology and Public Health Sciences, University of Virginia, Charlottesville, VA, United States

Contents

Handbook of Clinical Neurology, Vol. 177 (3rd series)
Heart and Neurologic Disease
J. Biller, Editor
https://doi.org/10.1016/B978-0-12-819814-8.00010-X

Chapter 1

Neurologic complications of pediatric congenital heart disease

ELIZABETH PULCINE AND GABRIELLE DEVEBER*

Division of Neurology, Department of Pediatrics, Hospital for Sick Children, Toronto, ON, Canada

Abstract

Improved medical management and surgical outcomes have significantly decreased mortality in children with congenital heart disease; however, with increased survival, there is a greater lifetime exposure to neurologic complications with serious long-term neurodevelopmental consequences. Thus, recent focus has shifted to recognition and reduction of these extracardiac comorbidities. Vascular and infective complications, such as arterial ischemic stroke, infective endocarditis, and localization-related epilepsy are some of the most common neurologic comorbidities of congenital heart disease. In addition, it is now well recognized that congenital heart disease has an impact on overall brain development and contributes to adverse neurodevelopmental outcomes across multiple domains. The goal of this chapter is to summarize the most common neurologic comorbidities of congenital heart disease and its management.

VASCULAR COMPLICATIONS OF PEDIATRIC CONGENITAL HEART DISEASE

Incidence of arterial ischemic stroke in children with congenital heart disease

Neonates and children with congenital heart disease (CHD) have an increased risk of arterial ischemic stroke (AIS). The incidence of AIS is estimated to be 1 in 2500 to 1 in 4000 live births in neonates (deVeber et al., 2017; Ferriero et al., 2019) and 2–8 per 100,000 in children aged 28 days to 18 years (Mallick et al., 2014; Kirton and deVeber, 2015; deVeber et al., 2017). The incidence is even higher in neonates and children with cardiac disease and is reported to be as high as 132 in 100,000 or 1 in 757 (Hoffman et al., 2011), which translates into a 17-fold increased risk compared to that of the general pediatric population. Improved medical management and surgical outcomes have resulted in decreased mortality but contributed to a greater lifetime exposure of thromboembolic complications (Hoffman et al., 2011). Among different types of cardiac diagnosis, those with cyanotic congenital heart disease and single-ventricle physiology appear to be at greatest risk of AIS with a stroke incidence rate of 1380 in 100,000 or 1 in 72 (Hoffman et al., 2011). Table 1.1 summarizes the overall incidence of AIS and incidence of AIS associated with cardiac disorders in children from previous studies.

Pathophysiology of arterial ischemic stroke in children with congenital heart disease

Cardioembolic stroke accounts for 20%–30% of all ischemic strokes in adults (O'carroll and Barrett, 2017) and up to 30% of ischemic strokes in children (Sinclair et al., 2015; Chung et al., 2019). The ratio of spontaneous stroke to procedure-related stroke is approximately 3:1 (Chung et al., 2019). Cardioembolic stroke can arise via different pathophysiological mechanisms, including formation of a mural thrombus due to pooling of blood and nonlaminar flow in a dyskinetic ventricle or with cardiac arrhythmia, clot formation or vegetation on an abnormal heart valve, or paradoxical embolism of a venous thrombus in a right-to-left shunt due to congenital structural heart defects (Roach et al., 2008; Giglia et al., 2013; Sinclair et al., 2015; Ferriero et al., 2019). In children with cardioembolic stroke, the

*Correspondence to: Gabrielle deVeber, M.D., Division of Neurology, Department of Pediatrics, The Hospital for Sick Children, Toronto, ON, Canada. Tel: +1-905-845-8775, Fax: +1-416-813-6334, E-mail: gabrielle.deveber@sickkids.ca

Table 1.1

Incidence data for hemorrhagic and arterial ischemic stroke (AIS) and stroke associated with cardiac disease in neonates and children from different geographical regions

Study	Cohort	Overall incidence
Giroud et al. (1995)	Neonatal and childhood hemorrhagic stroke and AIS (France)	13/100,000 (7.9 per 100,000 for AIS only)
Agrawal et al. (2009)	Neonatal and childhood hemorrhagic stroke and AIS (USA)	4.6/100,000 (2.4 /100,000 for AIS alone)
Hoffman et al. (2011)	Neonatal and childhood cardiac AIS (USA)	132/100,000
Hoffman et al. (2011)	Neonatal and childhood single-ventricle AIS (USA)	1380/100,000
Mallick et al. (2014)	Childhood AIS (UK)	1.6/100,000
deVeber et al. (2017)	Childhood AIS (Canada)	1.72/100,000
deVeber et al. (2017)	Neonatal AIS (Canada)	10.2/100,000 live births
Surmava et al. (2019)	Neonatal and childhood hemorrhagic stroke and AIS (Ontario, Canada)	5.1/100,000 children (2.2 for AIS only)

presumed mechanism is often embolic; however, in situ arterial thrombosis may also play a role (Dowling et al., 2013). This is because children with cardiac disease often have underlying chronic conditions including anemia, polycythemia, hypoxemia, recurrent infections, procedural interventions, and mechanical circulatory support, predisposing them to acquired proinflammatory or prothrombotic states. Children with cardioembolic stroke also have a high frequency of acute systemic disorders at the time of stroke, seen in about one-third (Chung et al., 2019). Dowling et al. found that, compared to children with other stroke etiologies, children with cardiac disease are younger at presentation and are more likely to have a cardioembolic stroke pattern on imaging defined as multiple, bilateral, and involving both the anterior and posterior circulation supporting embolic mechanisms (Dowling et al., 2013).

Thrombosis risk factors in children with congenital heart disease

The incidence of congenital heart disease is 4–10 cases per 1000 live births in the United States (Go et al., 2014). Similarly, about 1 in 100 children in Canada are born with CHD (Irvine et al., 2015). Congenital heart defects are structural problems that result from abnormal formation of the heart or major blood vessels that arise from the heart. They range in severity from small connections between two chambers of the heart, which may spontaneously close over time, to complex malformations that require multiple corrective or palliative surgeries, which are life-limiting (Go et al., 2014). Children with CHD are at high risk of developing thrombosis and the risk can vary over the lifespan (Sinclair et al., 2015). This increased propensity to thrombosis, as illustrated in Fig. 1.1, is perhaps best explained by the interaction of three important variables historically described by Rudolph Virchow: alterations in blood flow; alterations in blood composition; and endothelial injury (Giglia et al., 2013; Sinclair et al., 2015).

Alterations in blood flow can result from the presence of a hypoplastic ventricle or a severely dilated atrium limiting the ability for inflow and outflow and resulting in a nonlaminar, turbulent circulation. Alterations in blood composition have long been noted in children with CHD with a greater frequency of genetic and acquired thrombophilias compared to healthy controls (Strater et al., 1999). Finally, endothelial injury can result from presence of a central venous catheter, mechanical circulatory support, or sutures from cardiac surgery, which expose the blood to artificial thrombogenic material (Sinclair et al., 2015).

Perhaps the most important risk factor in developing thrombosis depends on the type of structural cardiac defect. As early as 1961 unrepaired tetralogy of Fallot was recognized as one of the most important risk factors in patients with AIS due to the right-to-left intracardiac shunting needed to sustain adequate cardiac output (Martelle and Linde, 1961). Although stroke has been associated with most types of acquired and congenital cardiac disease, cyanotic and single-ventricle heart defects are at highest risk due to formation of intracardiac thrombi and paradoxical embolism due to right-to-left shunting (Bernson-Leung and Rivkin, 2016). In order to repair many types of single-ventricle heart defects including hypoplastic left heart syndrome, tricuspid atresia, and pulmonary atresia with intact ventricular septum, surgeons often perform a series of staged open-heart procedures over several years to allow the heart to function as a one-sided pump with two chambers (Lim, 2011).

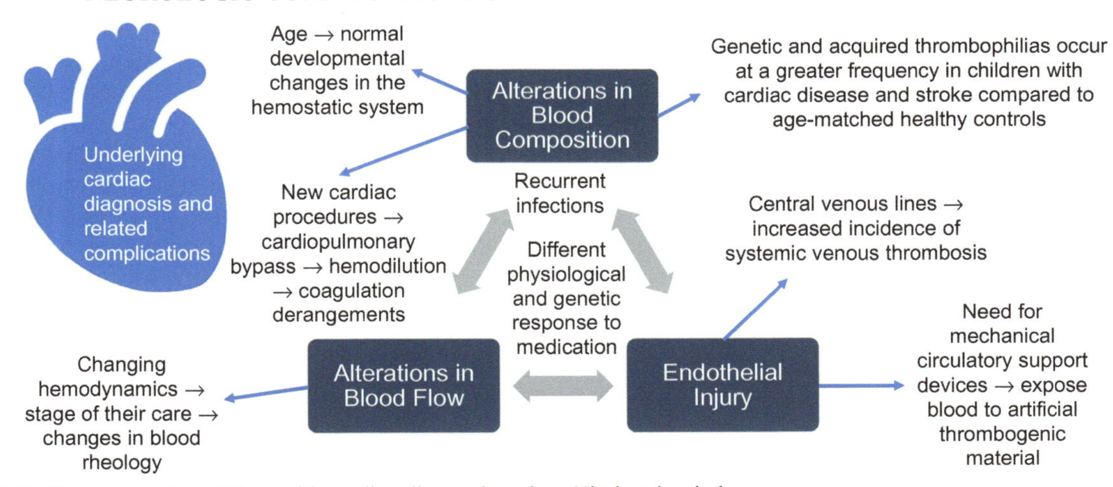

Fig. 1.1. Thrombosis in children with cardiac disease based on Virchow's triad.

Cardiac surgery itself poses a significant thromboembolic risk in addition to a risk of hypoxic–ischemic injury. During surgical repair, the use of cardiopulmonary bypass (CPB) is often necessary. CPB temporarily exposes the patient's blood to plastic tubing and other thrombogenic materials in the artificial circulatory system (Silvey and Brandao, 2017). This results in activation and aggregation of platelets and the fibrin-forming coagulation system with subsequent thrombus formation within the tubing or circuitry of the machine. A retrospective study from the Hospital for Sick Children, Toronto, Ontario, Canada examined 5526 children with congenital heart disease who underwent cardiac surgery from 1992 to 2001 and found that the incidence of ischemic stroke (28 with arterial ischemic stroke and 2 with cerebral sinus venous thrombosis) was 5.4 per 1000 children (Domi et al., 2008). Risk factors associated with procedural stroke included older age at the time of the surgery, longer duration of CPB, reoperation, and number of days hospitalized after the operation (Domi et al., 2008). The authors hypothesized that children who require reoperation likely have more severe underlying cardiac disease placing them at higher risk for procedure-related complications (Domi et al., 2008). More recently, Asakai et al. examined 76 children from Melbourne, Australia with cardiac disease and radiologically confirmed AIS and found that stroke occurred in 68% (95% CI: 58%–79%) of children following cardiac procedures (Asakai et al., 2015). This translated into 4.6 strokes per 1000 surgical procedures and 1.7 strokes per 1000 cardiac catheterizations from 1993 to 2010 (Asakai et al., 2015). The authors concluded that the prevalence of procedural stroke was the highest in patients with cyanotic CHD undergoing palliative surgery, which is consistent with previously published literature It remains unknown which individual patient characteristics further alter the risk of stroke in the procedural period. In infants and young children, normal developmental changes in the hemostatic system may also contribute to this risk and will be discussed in a later section.

Stroke recurrence in CHD is as high as 27% at 10 years and can occur even in children on antithrombotic therapy (Rodan et al., 2012). Rodan et al. showed that the recurrence risk was highest in the period immediately following the sentinel stroke and decreased with time (Rodan et al., 2012). Similarly, Asakai et al. found a 17% rate of stroke recurrence amongst children with cardiac disease at a median time of 21 days from the sentinel event (IQR 10.5–141 days) with a smaller follow-up interval (Asakai et al., 2015). For this reason, antithrombotic therapy is recommended by several consensus-based guidelines (Roach et al., 2008; Monagle et al., 2012; Giglia et al., 2013; Ferriero et al., 2019) for secondary stroke prevention. However, the duration of antithrombotic therapy remains institution dependent and ranges from at least 3 months poststroke and thereafter for 2–5 years to lifelong at our tertiary care center (The Hospital for Sick Children, Toronto, Ontario, Canada, unpublished observations). Studies have shown that stroke recurrence can occur many years following surgery, sometimes greater than 5 years after the most recent procedure (Rodan et al., 2012; Fox et al., 2015), resulting in controversy in management and knowing when it is safe to step-down, transition, or stop antithrombotic therapy altogether.

Thrombosis risk factors in adult survivors of congenital heart disease

Longer term studies of children surviving CHD show the increased risk for thromboembolism persists into adulthood. Studies have shown that the stroke risk continues to be elevated in adult CHD survivors many years after staged palliative cardiac repair with a prevalence of 0.05% per-patient year (Hoffmann et al., 2010). Although this prevalence may appear low, when compared to the general population of similar age, this is 10–100 times higher (Hoffmann et al., 2010). In addition to the risk factors commonly seen in neonates and children, including right-to-left shunting, adult-specific risk factors are unique and include atrial fibrillation, ventricular dysfunction and arrhythmias, Fontan circulation complicated by protein-losing enteropathy, pregnancy, and the use of estrogen containing oral contraception (Kirsh et al., 2002; Giglia et al., 2013).

Differences in the coagulation system of children with congenital heart disease

Hyperviscosity often occurs in cyanotic CHD due to a phenomenon called decompensated erythrocytosis (Tempe and Virmani, 2002). This occurs due to a cascade of physiological processes that become overactive in the presence of chronic tissue hypoxemia. In an attempt to increase tissue oxygenation, the kidneys release erythropoietin, a hormone that stimulates the bone marrow to increase the production of red blood cells (Tempe and Virmani, 2002). In the presence of a significant right-to-left shunt, erythropoietin continues to attempt to increase normal tissue oxygenation by increasing red cell mass and hemoglobin concentration thereby increasing blood viscosity and paradoxically reducing oxygen delivery (Tempe and Virmani, 2002). Normal systemic oxygen saturation is not usually achieved until after the Fontan procedure. Chronic hypoperfusion of the liver results in impaired metabolism of coagulation proteins and low-grade inflammation (Manlhiot et al., 2012). This leads to decreased levels of hepatically manufactured proteins that would normally protect against thrombosis: protein C, protein S and antithrombin (Silvey and Brandao, 2017). As CHD patients undergo corrective surgical procedures, their coagulation profile continues to be altered (Silvey and Brandao, 2017). A study by Odegard et al. showed that patients with single-ventricle physiology were more likely to have thrombophilic abnormalities compared with age-matched healthy controls at all three stages of their palliative surgical repair (Odegard et al., 2009). The study found that most coagulation proteins were significantly lower and that factor VIII levels increased after the Fontan (stage III palliation) procedure in children that were longitudinally followed from their first stage of repair (Odegard et al., 2009). Increased factor VIII levels have been demonstrated to be associated with an increased risk of thrombosis (Giglia et al., 2013). However, it is not known from this study if the coagulation profile of these children continues to change over the years post-Fontan or if this correlates with the risk of thrombotic events (Odegard et al., 2009). Not surprisingly, genetic and acquired thrombophilias have been reported to occur at a greater frequency in children with cardiac disease and AIS compared with age-matched healthy controls (Strater et al., 1999). The reason for this is thought to be multifactorial including those discussed previously as well as due to increased consumption, decreased production, and increased fibrinolysis of proteins involved in hemostasis (Silvey and Brandao, 2017). The increased prevalence of genetic thrombophilias in children with cardiac disease is only partly explained by the presence of syndromic disease such as Down syndrome and chromosome 8 deletion or duplication (Giglia et al., 2013), syndromes that are also associated with abnormalities in the function of coagulation factors. Additional reported abnormalities include elevated lipoprotein (a), protein C deficiency, presence of anticardiolipin antibodies and combined prothrombotic disorders (Strater et al., 1999).

Children with cyanotic CHD are also known to be at an increased risk of bleeding due to a variety of factors including polycythemia, hyperviscosity, thrombocytopenia, and platelet function abnormalities. CHD patients have been shown to have abnormal platelet numbers and function for several reasons including hypoxic inhibition of platelet production, increased platelet destruction, decreased platelet aggregation, and known genetic disorders such as Noonan's syndrome, which may present with both platelet dysfunction and cardiac disease (Silvey and Brandao, 2017). Platelet survival has also been demonstrated to be decreased in children with CHD: below 80 h compared to a normal survival time of 80–130 h (Waldman et al., 1975).

Acyanotic congenital heart disease and thrombosis risk

Acyanotic CHD includes atrial septal defect, ventricular septal defect, atrioventricular septal defect, and various aortic arch abnormalities including coarctation of the aorta. Children with acyanotic CHD are less likely to undergo repeated cardiac surgery with CPB, depending on the type of septal structural defect, as a number of them spontaneously close (Silvey and Brandao, 2017). For this reason, they are less likely to have thromboembolic complications. However, they are still at risk of paradoxical embolism due to transient increases in right atrial pressure with right-to-left shunting, should a septal defect remain open.

Patent foramen ovale (PFO) is a common anatomical variant found in 25% of the general population and should be distinguished from a congenital heart defect (Kent et al., 2013). Much like in adults who lack traditional atherosclerotic risk factors, it remains unclear whether isolated PFO plays a role in childhood stroke particularly given that the timing of normal physiological PFO closure is variable, remaining open in up to 35% of people between 1 and 29 years of age (Roach et al., 2008; Ferriero et al., 2019). It is also unclear if a PFO with right-to-left shunt is more prevalent in children with cryptogenic stroke and if the PFO should undergo interventional closure in order to prevent stroke recurrence (Ferriero et al., 2019). One small study suggested that PFO with right-to-left shunt is more prevalent in children with cryptogenic stroke than in healthy controls (Benedik et al., 2011). As in adults, interventional treatment of PFO remains controversial in pediatric stroke and there maybe anecdotal risk with the device itself including atrial fibrillation and risk of embolism during and after the procedure.

Complications of surgical management of pediatric congenital heart disease and its treatment

Paradoxically, improved surgical management of children with CHD may in fact facilitate the occurrence of neurologic complications, in some instances, by allowing individuals to survive who would have otherwise succumbed to their heart disease.

Cyanotic congenital heart disease and thrombosis risk according to different stages of cardiovascular surgery

In order to repair many types of cyanotic CHD, surgeons often perform a series of open-heart procedures over several years. This is known as staged reconstructive heart surgery (Lim, 2011). The ultimate goal is to have the heart function like a one-sided pump with two chambers (Lim, 2011). Thromboembolic complications are well recognized in patients undergoing cardiac surgery. One study by Manlhiot et al. from the Hospital for Sick Children and McMaster Children's Hospital found several predictors for thrombotic complications during surgery including age <31 days, baseline oxygen saturation <85%, previous thrombosis, heart transplantation, use of deep hypothermic circulatory arrest, longer cumulative time with central lines, and the postoperative use of extracorporeal membrane oxygenation (ECMO) (Manlhiot et al., 2011). Thrombotic complications at different stages of single-ventricle palliation will be discussed further below.

Norwood and Blalock–Taussig shunt (stage I palliation)

The period in and around the time of stage I palliation is generally considered to be the highest risk for thromboembolic complications (Manlhiot et al., 2012; Giglia et al., 2013; Silvey and Brandao, 2017). In most cases of univentricular physiology, stage I palliation will occur within several days of birth (Lim, 2011). Depending on the type of heart defect, different surgical procedures may be used, including the Norwood procedure (Lim, 2011). The purpose of this operation is to ensure that blood flow is controlled enough to prevent damage to the heart and lungs and that sufficient blood is reaching the lungs to keep the child adequately oxygenated until the second operation (Lim, 2011). One aspect of surgical palliation for many children with CHD is the placement of systemic-to-pulmonary artery shunts, which often vary in their diameter, flow characteristics, and composition (Giglia et al., 2013). The Blalock–Taussig shunt (BTS) is a common surgical procedure in neonates with single-ventricle physiology, where a shunt is created between the subclavian artery and the ipsilateral pulmonary artery to increase pulmonary blood flow (Lim, 2011). This shunt creates a low-flow area that increases the risk of thrombosis, and surgically removed shunts have been found to be thrombosed at a rate of 1%–17% (Manlhiot et al., 2012; Silvey and Brandao, 2017). Shunt thrombosis is a major cause of shunt failure and mortality in CHD patients. A 4% risk of death resulting from shunt failure has been reported (Giglia et al., 2013). For this reason, antithrombotic therapy is often required for BTS prophylaxis.

Bidirectional cavopulmonary anastomosis (stage II palliation)

The bidirectional cavopulmonary anastomosis (BCPS), also called the Glenn or hemi-Fontan, is the second-staged palliative procedure, which usually occurs within 6 months of birth (Lim, 2011). During this surgery, the superior vena cava, a large vein that carries deoxygenated blood from the upper body into the heart, is disconnected from the heart and attached to the pulmonary artery (Lim, 2011). After this operation, deoxygenated blood from the upper body goes to the lungs without passing through the heart (Lim, 2011). Although there are limited data, current experience suggests that the risk of thrombosis after bidirectional cavopulmonary anastomosis is low (Manlhiot et al., 2012; Giglia et al., 2013). After this surgery, patients are at an increased risk of developing pleural effusions and chylothoraxes (Giglia et al., 2013). This can result in a hypercoagulable state, especially if there is significant drainage, due to loss of important proteins, including protein C, protein S, and antithrombin. In addition, loss of antithrombin can limit the effectiveness of heparin. Chronic drainage from a pleural effusion can also result in dehydration and relative systemic hypotension (Giglia et al., 2013). With elevated superior vena cava pressures, there is slower drainage of the cerebral venous return, which may result in a cerebral sinovenous thrombosis and venous stroke (Giglia et al., 2013). The main concern regarding thrombosis at this palliative stage is the development of pulmonary embolism, with a subsequent increase in pulmonary vascular resistance, making patients unsuitable for further palliative surgeries (Manlhiot et al., 2012; Silvey and Brandao, 2017). The 2013 American Heart Association scientific statement on prevention and treatment of thrombosis in pediatric congenital heart disease recommends long-term prophylactic therapy with antiplatelet agents after BCPS (Giglia et al., 2013).

Fontan (stage III palliation)

The Fontan procedure is the last surgery in the staged repair of univentricular hearts. It occurs at approximately 2–3 years of age. During this surgery, the inferior vena cava, a large vein that carries deoxygenated blood from the lower body into the heart, is disconnected from the heart and attached to the pulmonary artery (Lim, 2011). After this operation, all of the deoxygenated blood from the body goes to the lungs without passing through the heart (Lim, 2011). The Fontan procedure was first performed in 1968. It has since undergone many modifications, although the mechanism of the anastomosis of the inferior vena cava to the pulmonary arteries remains unchanged (Gewillig and Brown, 2016). Patients undergoing the Fontan procedure are also at an increased risk of developing thromboembolism. The incidence of thrombosis after Fontan is reported to range from 17% to 33% in cross-sectional studies (Giglia et al., 2013), while the prevalence of ischemic stroke following a Fontan procedure is estimated to be between 1.4% and 19% (Manlhiot et al., 2012; Firdouse et al., 2014). However, this risk is not uniform and has been reported to vary over time. The highest risk is reported within the perioperative period extending 3–12 months and then again at 5–10 years after the procedure (Giglia et al., 2013). Previously identified post-Fontan thrombosis risk factors include passive blood flow, chronic venous hypertension, and atrial arrhythmias (Giglia et al., 2013; Sinclair et al., 2015; Silvey and Brandao, 2017). Liver congestion from chronic venous hypertension may result in decreased vitamin K-dependent proteins C and S (Giglia et al., 2013). As previously discussed, children post-Fontan have elevated factor VIII levels, which may further increase thrombosis risk (Odegard et al., 2009). Adding to this risk may be the type of Fontan modification performed. Initially, the patient may have a fenestration placed within the atria, called a fenestrated-Fontan, to provide a pop-off for venous blood to flow to the left side, if right-sided pressures are high (Lim, 2011). This in turn could lead to the development of paradoxical emboli if a thrombus travels to or originates within the right side of the heart (Lim, 2011; Giglia et al., 2013). It is still not clear, however, why the risk increases after 5–10 years. Several authors have postulated that at that time additional chronic risk factors come into play that are more commonly seen in adult survivors of CHD: ventricular dysfunction, atrial arrhythmia, prolonged immobilization, protein-losing enteropathy, and chronic pleural effusions (Giglia et al., 2013). Barker et al. reviewed 402 children who underwent the Fontan procedure between 1975 and 1998 for single-ventricle physiology and followed them for a median of 3.5 years postoperatively (Barker et al., 2005). The study found that risk of stroke was neither related to the type of Fontan nor the presence of fenestration (Barker et al., 2005). Not surprisingly, they found a significantly lower rate of stroke in patients on antithrombotic treatment with aspirin or warfarin compared to those not treated with antithrombotic therapy (2.4 per 1000 patient-years vs 13.4 per 1000 patient-years; $P=0.02$) (Barker et al., 2005). In order to determine the most effective antithrombotic therapy regimen, Monagle et al. performed a multicenter randomized trial comparing aspirin of 5 mg/kg/day to warfarin, with a target international normalized ratio of 2–3, for prevention of thrombosis following Fontan (Monagle et al., 2011). Children were screened at 3 months and 2 years with transthoracic and transesophageal echocardiograms (Monagle et al., 2011). Interestingly, the risk for both asymptomatic and symptomatic thrombosis was 19% at 2 years and was similar in both groups with no significant difference between those on aspirin or warfarin; however, the trial was underpowered with only 111 of the required 226 patients enrolled (Monagle et al., 2011). Of the 111 patients studied, 12 developed thrombosis in the aspirin group and 13 in the heparin/warfarin group with an increase in minor bleeding rate in the latter (Monagle et al., 2011). The study concluded that prophylaxis with either aspirin or warfarin is warranted, and clinical practice still varies both within and amongst centers.

Thromboprophylaxis across all stages of cardiovascular surgery

Because of the high risk of developing thromboembolic complications during cardiac surgery and cardiac procedures, a number of studies have looked at the efficacy of thromboprophylaxis in children with univentricular CHD (Monagle et al., 2011; Manlhiot et al., 2012). Antiplatelet and anticoagulant agents have both been successfully used to reduce thromboembolic complications (Monagle et al., 2011, Manlhiot et al., 2012).

Manlhiot et al. examined the association between thromboprophylaxis and thrombosis risk across all three stages of palliative cardiac repair (Manlhiot et al., 2011, 2012). The study found that after stage I palliation [HR 0.5; $P=0.05$] and after stage II palliation [HR 0.2; $P=0.04$] enoxaparin compared to no antithrombotic therapy was associated with a reduced risk of thromboembolic complications (Manlhiot et al., 2012). After stage III palliation, both warfarin [HR 0.27; $P=0.05$] and aspirin [HR 0.18; $P=0.02$] were associated with a reduced rate of thromboembolic complications compared to no antithrombotic therapy (Manlhiot et al., 2012). In terms of bleeding risk, 2 patients on enoxaparin and 1 patient on warfarin experienced major bleeding complications without any associated morbidity or mortality: two with subdural hematomas and one with an intrathoracic bleed (Manlhiot et al., 2012). Because thrombotic complications were associated with increased mortality after stage I palliation [HR 5.5; $P<0.001$] and stage II palliation [HR 12.5; $P<0.001$], the authors concluded that the risk–benefit ratio was in favor of thromboprophylaxis in children with CHD undergoing palliative repair, but the best type of antithrombotic therapy is not known (Manlhiot et al., 2012). Interestingly, while the study found that thromboprophylaxis across all stages of palliative cardiac repair had an overall reduction in thrombosis risk, it did not have the same reduction on thromboembolic complications directly associated with cardiovascular surgery (Manlhiot et al., 2012). This procedure-related thrombosis risk represents a significant proportion of overall thrombosis risk in children with CHD (Manlhiot et al., 2012). Multifactorial strategies are needed to specifically target this high-risk period such as use of peripheral venous lines instead of central venous lines and thrombophilia screening prior to surgery with more intense thromboprophylaxis for those at highest risk.

Cardiac catheterization

Diagnostic and interventional cardiac catheterization is performed in children with congenital heart disease for both diagnostic and therapeutic purposes. Known complications of these procedures include in situ thrombosis and distal embolus (Giglia et al., 2013; Silvey and Brandao, 2017). The prevalence of AIS in children due to cardiac catheterization is estimated to be 0.28%–1.3% (Weissman et al., 1985; Liu et al., 2001). When undergoing cardiac catheterization, the femoral artery or vein is accessed for catheter insertion immediately prior to a bolus of unfractionated heparin for prophylaxis (Lim, 2011; Giglia et al., 2013). The typical anticoagulation protocol for diagnostic or interventional cardiac catheterization includes a loading dose of 100U/kg (up to 5000U maximum) and an additional 50–100U/kg of heparin bolus to keep the activated clotting time (ACT) >200s (Giglia et al., 2013). ACT is a quantitative assay for monitoring heparin anticoagulation during various medical and surgical procedures (Giglia et al., 2013).

One common interventional catheterization procedure is called a balloon atrial septostomy (BAS), where a balloon catheter is used to create or enlarge a patent foramen ovale or atrial septal defect between the two upper chambers of the heart in order to increase oxygen saturation (Lim, 2011). In essence, this procedure is used to temporarily rescue the physiology of transposition of the great arteries (TGA), a life-threatening cyanotic CHD seen in infants, while awaiting definitive corrective surgery: the arterial switch operation. In one study, Block et al. found that BAS was significantly associated with preoperative AIS in infants with TGA [relative risk 4; 95% CI:1.5–9.3; $P=0.0015$] (Block et al., 2010). However, other studies have not found the same association (Petit et al., 2009). In practice, the use of prophylactic anticoagulation during BAS currently varies both within and amongst centers and is not evidence based.

Cardiopulmonary bypass

Cardiopulmonary bypass is defined as "the process of diverting venous blood from a patient's heart and lungs to a gas exchange system for the addition of oxygen, removal of carbon dioxide and subsequent reinfusion to the patient's arterial system" (Shann et al., 2008). Thereby CPB maintains the circulation of blood and oxygen to all organs while facilitating surgery on the open heart and its great vessels. Patients who undergo CPB are temporarily exposed to an artificial vascular surface, which can result in platelet activation and downstream effects that transiently alter hemostasis (Giglia et al., 2013; Silvey and Brandao, 2017). Secondary to the platelet activation, CPB in children is associated with an initial drop in the platelet count that subsequently recovers postsurgery (Giglia et al., 2013). Patients younger than 1 year of age are more likely to develop thrombocytopenia while undergoing CPB (Silvey and Brandao, 2017). In parallel, thrombi can originate in the bypass circuit due to tissue factor activation from contact with an artificial surface, which leads to increased thrombin generation and thereby fibrin clot formation. Thrombi can travel and enter the cerebral circulation directly, bypassing the pulmonary circulation, and embolizing in distal vessels (Sinclair et al., 2015). Anticoagulation during CPB aims to reduce clot formation within the CPB circuit and minimize consumption of coagulation factors while minimizing excessive intraoperative bleeding (Giglia et al., 2013). Heparin is the most common anticoagulant used for CPB. A loading dose of 300–400 U/kg is given IV and is also used to prime the pump and circuit. The optimal target ACT that will prevent clot formation within the CPB circuit is not precisely known, but clot formation is unlikely to occur with an ACT >300 s (Giglia et al., 2013). Commonly, a target of ACT >480 s is used in children (Giglia et al., 2013). Typically, ACT is measured 2–5 min after administration of heparin. Factors known to prolong ACT include hypothermia, hemodilution and decreased platelet function (Giglia et al., 2013). All these factors may be present during CPB making this a time both of increased risk of hemorrhage and thrombosis.

Postpump chorea

Postcardiopulmonary bypass chorea, frequently referred to as "postpump chorea" may be seen after cardiac surgery in a small number of children. The estimated frequency is 10% (0.6%–18%) per procedure, but the incidence has decreased with time, presumably as a result of advances in operative techniques (Nita, 2014). Manifestations are not limited to chorea and include orofacial dyskinesias, hypotonia, ballismus, supranuclear gaze palsy, affective changes, and pseudobulbar signs. Risk factors include increased duration on CPB, deeper hypothermia (<36°C), and circulatory arrest (Medlock et al., 1993). Chorea begins 3–12 days postoperatively and either remits with time or remains persistent (Nita, 2014). Neuroimaging studies and EEG do not show any focality (Bhidayasiri and Tarsy., 2012; Nita, 2014). The underlying pathophysiology of this syndrome is not entirely clear. One hypothesis is that chorea results from microembolic sequala due to platelet-fibrin aggregates, air, fat, or polyvinylchloride tubing (Bhidayasiri and Tarsy., 2012; Nita, 2014).

Extracorporeal membrane oxygenation

Extracorporeal membrane oxygenation is a closed CPB circuit designed to provide cardiorespiratory support for a short period of time, generally up to 14 days (Giglia et al., 2013). ECMO is used frequently for cardiac support in neonates and children with acquired and congenital heart disease. Indications include intractable low cardiac output, cardiac arrest, or as a bridge to heart transplant (Giglia et al., 2013). Like CPB, ECMO exposes the blood to foreign material resulting in thrombus formation and requires use of anticoagulation in order to prevent clot formation (Giglia et al., 2013). Despite anticoagulation, thromboembolic complications are not infrequent on ECMO. The prevalence of ischemic stroke in children treated with ECMO is reported to be 7%–11%, while the incidence rate on ECMO is reported to be 1–2 events per 100 days of support (Cengiz et al., 2005; Almond et al., 2011). The typical anticoagulation protocol for ECMO includes a loading dose of 100 U/kg heparin before cannulation and a continuous infusion of heparin to maintain ACT between 180 and 220 s (Giglia et al., 2013). Other suggested targets for anticoagulation on ECMO include prolongation of the partial thromboplastin time (PTT) to 1.5–2.5 times the control value and antifactor Xa level of 0.3–0.7 IU/mL (Giglia et al., 2013). In general, a lower level of anticoagulation is necessary compared with patients undergoing surgery using CPB, although a recent study suggested that higher heparin doses in pediatric patients on ECMO result in improved survival despite the potential increased risk of bleeding complications (Baird et al., 2007). When ECMO is used to support the circulation of patients after cardiac surgery, it is commonly referred to as postcardiotomy ECMO (Giglia et al., 2013). Postcardiotomy ECMO has been reported at a rate of 2%–5% of all postoperative patients in large tertiary care centers (Giglia et al., 2013). Indications include failure to wean from CPB, low cardiac output, or cardiac arrest postoperatively (Giglia et al., 2013). This subpopulation of patients is at a particularly increased risk of bleeding as they cannot be weaned from CPB and their heparin cannot be reversed (Giglia et al., 2013).

Ventricular assist devices

Ventricular assist devices (VADs) are primarily designed to support patients with terminal heart failure who are refractory to medical therapy while they await heart transplantation (Almond et al., 2011). Pediatric VADs, specifically the Berlin Heart EXCOR VAD, is superior to ECMO for bridging to heart transplant and has emerged as the new standard of care with device approval granted in 2011 by the U.S. Food and Drug Administration (Almond et al., 2011). Similarly to CPB and ECMO, the use of VAD also increases the risk of thrombosis by exposing blood to an artificial vascular surface, but the increased risk of thromboembolic complications occurs for a more prolonged period of time, thereby increasing the overall cumulative prevalence of these complications (Giglia et al., 2013). Children with a Berlin Heart EXCOR VAD have a combined hemorrhagic and ischemic stroke prevalence of 28%–47%, with a reported incidence rate of 0.5 events per 100 days of support (Fraser et al., 2012; Rohde et al., 2019). When compared to ECMO, the prevalence rate appears significantly higher while the incidence rate somewhat lower. This may be explained by the shorter exposure period for ECMO as the typical duration of support is less than 14 days compared to VAD duration, which often exceeds 30 days (Sinclair et al., 2015). In contrast, VADs typically used developed for adult use such as the HeartMate II appear to have a significantly lower risk of stroke in adolescents, approximately 6%–12% based on small pediatric studies (Cabrera et al., 2013). The reasons for this are not entirely known but may be a reflection of the differing underlying cardiac diseases or a shorter duration of use related to a greater availability of organ donation in older children and adults. In any case, multiple concurrent anticoagulant and antiplatelet agents are required with VADs to prevent thrombosis given the typical prolonged duration of use while awaiting heart transplant.

INFECTIVE COMPLICATIONS OF PEDIATRIC CONGENITAL HEART DISEASE

Brain abscess

Brain abscess is a serious and potentially fatal complication of cyanotic CHD, especially if there is diagnostic delay. A brain abscess begins as a localized area of brain infection and over time becomes surrounded by a highly vascularized capsule (Lumbiganon and Chaikitpinyo, 2013). It is estimated that the incidence of brain abscess in cyanotic heart disease is 4%–6% (Clark, 1966), although this number has steadily declined with time, perhaps as a result of more stringent antibiotic prophylaxis and advances in surgical correction of complex right-to-left intracardiac shunts (Brandt, 1981). Brain abscesses occur by hematogenous dissemination of infection from a primary infectious source (Takeshita et al., 1992). Children with cyanotic CHD are particularly susceptible because of the presence of an intracardiac right-to-left shunt, which allows venous blood to circulate through the arterial system without first being filtered by the pulmonary circulation (Brandt, 1981; Takeshita et al., 1992). In turn, the venous blood bypasses the phagocytic filtering action of the lungs, allowing direct entry into the cerebral circulation (Brandt, 1981; Takeshita et al., 1992). The most common bacteria found in brain abscesses of children with cyanotic congenital heart disease is reported to be *Streptococcus milleri* (52%) (Mehnaz et al., 2006). Risk factors for brain abscess in patients with CHD include significantly lower mean arterial oxygen saturation (Takeshita et al., 1992). Takeshita et al. reported that the mean arterial oxygen saturation, arterial partial pressure of oxygen, arterial blood oxygen content, and base excess in children with brain abscess were significantly lower compared to a control population who did not develop brain abscess but had cyanotic CHD. For this reason, a low threshold for neuroimaging is recommended for children with cyanotic CHD and early signs and symptoms of increased intracranial pressure or focality on neurological examination. Complications include hemiparesis and localization-related epilepsy. A recent Cochrane review found no robust evidence on the most effective antibiotic regimens for treating patients with cyanotic CHD and brain abscesses postsurgical aspiration (Lumbiganon and Chaikitpinyo, 2013).

Infective endocarditis

With the decline of rheumatic heart disease as a result of antibiotic treatment for group A streptococcus, infective endocarditis has emerged as a leading complication of congenital heart disease in children (Knirsch and Nadal, 2011). One risk factor for infective endocarditis is the use of prosthetic material to correct or palliate various acyanotic and cyanotic congenital heart lesions (Rushani et al., 2013). In one large population-based cohort of children with CHD, the cumulative incidence of infective endocarditis was reported to be 6.1 per 1000 or 4.1 cases per 10,000 patient-years (Rushani et al., 2013). Predictors of infective endocarditis included cyanotic CHD with unrepaired right-to-left shunts [adjusted relative risk 7.56; 95% CI 4.03–14.18], unrepaired left-sided lesions [adjusted relative risk 2.35; 95% CU 1.16–4.73], and unrepaired endocardial cushion defects [adjusted relative risk 3.00; 95% CU 1.06–8.51] (Rushani et al., 2013). In 2007, the American Heart Association guidelines for prevention of infective endocarditis revised and reduced the list of CHD lesions for which antibiotic prophylaxis was indicated prior to invasive dental, gastrointestinal, or genitourinary tract procedures (Wilson et al., 2007). The reason for the changes was due to several considerations including the relative importance of everyday bacteremia versus invasive procedures in causing infective endocarditis and comparison of absolute versus relative risk of acquisition of infective endocarditis and the cost-effectiveness of prophylaxis (Wilson et al., 2007). Rushani et al. showed that the relative risk of developing infective endocarditis was five times higher in the 6-month postoperative period. This is thought to be partly due to the endothelialization of prosthetic material, which takes approximately 6 months if implanted (Rushani et al., 2013). In a study by Fortun et al., nearly half of all cases of infective endocarditis (44%) occurred in children with CHD with native valves, which were uncorrected and of the 24 patients with prosthetic-valve endocarditis postoperative acquisition occurred in the first 6 months postsurgery, further supporting the aforementioned findings (Fortun et al., 2013). In addition, Rushani et al. showed that the highest risk of infective endocarditis was in children under 3 years of age [adjusted relative risk 3.53 95% CI 2.51–4.96], perhaps because most staged cardiac repairs are done in the first years of life. These findings help identify groups of children with CHD who are at the greatest risk of infective endocarditis and may benefit the most from antibiotic prophylaxis (Rushani et al., 2013). The most common etiological agents of infective endocarditis found in children with cyanotic CHD are reported to be *Streptococcus* spp. (33%) and *Staphylococcus* spp. (32%) (Fortun et al., 2013). Initial treatment is with empiric antibiotic therapy until the infective agent is identified and antibiotics can be further tailored. Neurological complications occur at a frequency of 20%–40% in children with infective endocarditis, especially those involving the left side of the heart (Roach et al., 2008). Stroke accounts for about half of those complications (Roach et al., 2008). Pathophysiological mechanisms include septic emboli at times complicated by infective aneurysmal formation (mycotic aneurysm) and vasculitis (Roach et al., 2008). Mycotic aneurysms can rupture causing hemorrhagic stroke including subarachnoid hemorrhage, which can be fatal, leading to a general recommendation to avoid anticoagulation in infective endocarditis with suspected septic embolization (Roach et al., 2008).

EPILEPSY AND PEDIATRIC CONGENITAL HEART DISEASE

Overall, the incidence of epilepsy in children born with CHD is reported to be as high as 3%–5%, and persists into adulthood (Leisner et al., 2016; Desnous et al., 2019). In a population-based observational study from Denmark, Leisner et al. examined 15, 222 children diagnosed with CHD between 1980 and 2010—a 30-year cohort. When compared to the general population, they found that children with CHD were more than three times more likely [hazard ratio 3.7; 95% CI 3.2–4.2] to be diagnosed with epilepsy before 5 years of age, and more than twice more likely [hazard ratio 2.3; 95% CI 2.1–2.7] from 5 to 32 years of age (Leisner et al., 2016). Overall, the cumulative incidence of epilepsy in children with CHD by age 15 was 5% (Leisner et al., 2016). The risk of epilepsy was significantly increased in those who underwent multiple-staged cardiac repairs [having greater than three surgeries was associated with a hazard ratio of 4.8; 95% CI 2.8–8.2], although the risk remained elevated [hazard ratio 1.8; 95% CI 1.5–2.3] amongst those children with relatively mild congenital heart defects, without need for surgical intervention, even when prematurity and extracardiac comorbities were excluded (Leisner et al., 2016). The authors concluded that multiple risk factors contribute to increased incidence of epilepsy in this cohort, such as chronic hypoxia in the fetus and genetic predisposition, which maybe more important than surgical factors alone (Leisner et al., 2016). In contrast, a more recent prospective study found a high incidence or perioperative clinical seizures in children with congenital heart disease (7.8%), and this was a predictor of future epilepsy with an incidence of 3% (Desnous et al., 2019). Predictors of perioperative clinical seizures included higher surgical complexity, delayed sternal closure, use of ECMO, longer CPB and aortic cross-clamp times, deep hypothermic circulatory arrest, and longer intensive care stay (Desnous et al., 2019). Almost all (90%) children with perioperative clinical seizures had some form of acquired brain injury including stroke, white matter, and hypoxic–ischemic injury (Desnous et al., 2019). Predictors of epilepsy included use of ECMO as well as length of hospital stay (Desnous et al., 2019). None of the patients without perioperative clinical seizures developed epilepsy (Desnous et al., 2019). As a result, the authors stressed the importan of systematic perioperative EEG monitoring in this cohort of children because the presence of clinical seizures during the vulnerable perioperative period was found to be one the main predictors of future epilepsy (Desnous et al., 2019). It remains to be seen if prophylactic antiepileptic medication can mitigate this latent epilepsy risk.

In children with CHD and focal brain injury, medically refractory epilepsy can develop. Treatment decisions may be challenging due to competing priorities between cardiac and neurologic considerations, and evidence-based algorithms are lacking to guide the treating physicians. Hemispherectomy is commonly performed in children with epilepsy secondary to neonatal or presumed perinatal arterial ischemic stroke (Scavarda et al., 2009) as there is growing evidence that early surgical treatment of localization-related medically refractory epilepsy can improve and even prevent adverse neurodevelopmental outcomes (Moosa et al., 2013). However, until recently, there were no reported cases in the literate of a child undergoing a hemispherectomy in the midst of staged cardiac repair and a potential vulnerable circulatory system (Jain et al., 2019). Jain et al. describe the first successful case of a functional hemispherectomy in a child with complex CHD, consisting of univentricular physiology, who developed medically refractory epilepsy secondary to a left middle cerebral artery infarction following a bidirectional cavopulmonary connection, at 11 months of age (Jain et al., 2019). The child had improvements in neurodevelopmental skills and remained seizure free following functional hemispherectomy post-Glenn and prior to Fontan. This case highlights beautifully a multidisciplinary approach to care, in the absence of any guidelines, which was able to mitigate adverse neurodevelopmental outcomes from acquired brain injury despite multiple competing risks in a child with complex CHD.

NEURODEVELOPMENTAL SEQUELAE OF PEDIATRIC CONGENITAL HEART DISEASE

Congenital heart disease has been shown to adversely influence the neurodevelopmental outcomes of children across multiple domains, independently of intraoperative management (Bird et al., 2008). Mechanisms include both acute acquired injury and abnormal brain development. In neonates with severe CHD, newly acquired brain injury has been frequently observed on routine MRI scanning both preoperatively (41%) and postoperatively (30%) (Dimitropoulos et al., 2013). Children with CHD also have been found to have preexisting deficits in cerebral development including high incidence of white matter injury despite being born full term, lower brain volumes, and more immature brain structure (Limperopoulos et al., 2001; Miller et al., 2007). The mechanism underlying this is hypothesized to be long-term exposure to hypoxia and resultant reduced fetal cerebral oxygen delivery and utilization (Donofrio and Massaro, 2010). There has been substantial research into both patient-specific factors and modifiable factors that influence neurological outcomes in children with congenital heart disease. Patient-specific factors include gender, underlying genetic predisposition, and in utero brain development (Ballweg et al., 2007). On the other hand, modifiable variables include intraoperative factors such as CPB time, deep hypothermic circulatory arrest, and nonoperative factors such as chronic hypoxemia and low cardiac output states (Ballweg et al., 2007). Multiple genetic disorders, such as Down syndrome, velocardiofacial syndrome, Noonan syndrome, and Turner syndrome, which cause congenital heart disease can lead to cognitive impairment independent of the underlying cardiac lesion (de Los Reyes and Roach, 2014). It is estimated that severe neurological comorbidity occurs in 5% of children undergoing surgery for CHD, while 28% are left with varying degrees of neurological impairments, the most common of which are hypotonia, fine and gross motor incoordination and developmental delay (Majnemer et al., 2006). The type and severity of cardiac lesion also has a strong impact on cognitive development. Children with hypoplastic left heart syndrome are particularly at risk of neurodevelopmental abnormalities including development delay and attention deficit hyperactivity disorder (Donofrio and Massaro, 2010; de Los Reyes and Roach, 2014). When compared to patients with transposition of the great arteries, patients with hypoplastic left heart syndrome have more problems with visual-motor skills, expressive language, attention, and externalizing behaviors (Brosig et al., 2007). Another study showed that neurodevelopment was better in children with transposition of the great arteries than in those with tetralogy of Fallot (Bellinger et al., 2001). The hypothesis is that children with tetralogy of Fallot are more likely to have an underlying genetic syndrome (Bellinger et al., 2001). As the rate of neurodevelopmental impairments continue to be substantial in this cohort, standardized neurodevelopmental evaluation in all children with CHD should be routinely performed so children who are most at risk could benefit from targeted interventions.

References

Agrawal N, Johnston SC, Wu YW et al. (2009). Imaging data reveal a higher pediatric stroke incidence than prior US estimates. Stroke 40: 3415–3421.

Almond CS, Singh TP, Gauvreau K et al. (2011). Extracorporeal membrane oxygenation for bridge to heart transplantation among children in the United States: analysis of data from the organ procurement and transplant network and extracorporeal life support organization registry. Circulation 123: 2975–2984.

Asakai H, Cardamone M, Hutchinson D et al. (2015). Arterial ischemic stroke in children with cardiac disease. Neurology 85: 2053–2059.

Baird CW, Zurakowski D, Robinson B et al. (2007). Anticoagulation and pediatric extracorporeal membrane oxygenation: impact of activated clotting time and heparin dose on survival. Ann Thorac Surg 83: 912–919. discussion 919-20.

Ballweg JA, Wernovsky G, Gaynor JW (2007). Neurodevelopmental outcomes following congenital heart surgery. Pediatr Cardiol 28: 126–133.

Barker PC, Nowak C, King K et al. (2005). Risk factors for cerebrovascular events following fontan palliation in patients with a functional single ventricle. Am J Cardiol 96: 587–591.

Bellinger DC, Wypij D, Du Plessis AJ et al. (2001). Developmental and neurologic effects of alpha-stat versus pH-stat strategies for deep hypothermic cardiopulmonary bypass in infants. J Thorac Cardiovasc Surg 121: 374–383.

Benedik MP, Zaletel M, Meglic NP et al. (2011). A right-to-left shunt in children with arterial ischaemic stroke. Arch Dis Child 96: 461–467.

Bernson-Leung ME, Rivkin MJ (2016). Stroke in neonates and children. Pediatr Rev 37: 463–477.

Bhidayasiri R, Tarsy D (2012). Postpump chorea. Movement disorders: A video atlas. Current clinical neurology, Humana Press, Totowa, NJ.

Bird GL, Jeffries HE, Licht DJ et al. (2008). Neurological complications associated with the treatment of patients with congenital cardiac disease: consensus definitions from the multi-societal database Committee for Pediatric and Congenital Heart Disease. Cardiol Young 18 (Suppl 2): 234–239.

Block AJ, Mcquillen PS, Chau V et al. (2010). Clinically silent preoperative brain injuries do not worsen with surgery in neonates with congenital heart disease. J Thorac Cardiovasc Surg 140: 550–557.

Brandt M (1981). Brain abscess in children with congenital heart disease. In: H Altenburg, P Bohm, W Schiefer, M Klinger, M Borck (Eds.), Brain abscess and meningitis. Advances in neurosurgery. Springer, Berlin, Heidelberg.

Brosig CL, Mussatto KA, Kuhn EM et al. (2007). Neurodevelopmental outcome in preschool survivors of complex congenital heart disease: implications for clinical practice. J Pediatr Health Care 21: 3–12.

Cabrera AG, Sundareswaran KS, Samayoa AX et al. (2013). Outcomes of pediatric patients supported by the HeartMate II left ventricular assist device in the United States. J Heart Lung Transplant 32: 1107–1113.

Cengiz P, Seidel K, Rycus PT et al. (2005). Central nervous system complications during pediatric extracorporeal life support: incidence and risk factors. Crit Care Med 33: 2817–2824.

Chung MG, Guilliams KP, Wilson JL et al. (2019). Arterial ischemic Stroke secondary to cardiac Disease in neonates and children. Pediatr Neurol 100: 35–41.

Clark DB (1966). Brain abscess and congenital heart disease. Clin Neurosurg 14: 274–287.

De Los Reyes E, Roach ES (2014). Neurologic complications of congenital heart disease and its treatment. Handb Clin Neurol 119: 49–59.

Desnous B, Lenoir M, Doussau A et al. (2019). Epilepsy and seizures in children with congenital heart disease: a prospective study. Seizure 64: 50–53.

Deveber GA, Kirton A, Booth FA et al. (2017). Epidemiology and outcomes of arterial ischemic stroke in children: the Canadian pediatric ischemic Stroke registry. Pediatr Neurol 69: 58–70.

Dimitropoulos A, Mcquillen PS, Sethi V et al. (2013). Brain injury and development in newborns with critical congenital heart disease. Neurology 81: 241–248.

Domi T, Edgell DS, Mccrindle BW et al. (2008). Frequency, predictors, and neurologic outcomes of vaso-occlusive strokes associated with cardiac surgery in children. Pediatrics 122: 1292–1298.

Donofrio MT, Massaro AN (2010). Impact of congenital heart disease on brain development and neurodevelopmental outcome. Int J Pediatr 2010.

Dowling MM, Hynan LS, Lo W et al. (2013). International Paediatric Stroke Study: stroke associated with cardiac disorders. Int J Stroke 8 (Suppl A100): 39–44.

Ferriero DM, Fullerton HJ, Bernard TJ et al. (2019). Management of Stroke in neonates and children: a scientific statement from the American Heart Association/ American Stroke Association. Stroke 50: e51–e96.

Firdouse M, Agarwal A, Chan AK et al. (2014). Thrombosis and thromboembolic complications in fontan patients: a literature review. Clin Appl Thromb Hemost 20: 484–492.

Fortun J, Centella T, Martin-Davila P et al. (2013). Infective endocarditis in congenital heart disease: a frequent community-acquired complication. Infection 41: 167–174.

Fox CK, Sidney S, Fullerton HJ (2015). Community-based case-control study of childhood stroke risk associated with congenital heart disease. Stroke 46: 336–340.

Fraser Jr CD, Jaquiss RD, Rosenthal DN et al. (2012). Prospective trial of a pediatric ventricular assist device. N Engl J Med 367: 532–541.

Gewillig M, Brown SC (2016). The Fontan circulation after 45 years: update in physiology. Heart 102: 1081–1086.

Giglia TM, Massicotte MP, Tweddell JS et al. (2013). Prevention and treatment of thrombosis in pediatric and congenital heart disease: a scientific statement from the American Heart Association. Circulation 128: 2622–2703.

Giroud M, Lemesle M, Gouyon JB et al. (1995). Cerebrovascular disease in children under 16 years of age in the city of Dijon, France: a study of incidence and clinical features from 1985 to 1993. J Clin Epidemiol 48: 1343–1348.

Go AS, Mozaffarian D, Roger VL et al. (2014). Heart disease and stroke statistics–2014 update: a report from the American Heart Association. Circulation 129: e28–e292.

Hoffman JL, Mack GK, Minich LL et al. (2011). Failure to impact prevalence of arterial ischemic stroke in pediatric cardiac patients over three decades. Congenit Heart Dis 6: 211–218.

Hoffmann A, Chockalingam P, Balint OH et al. (2010). Cerebrovascular accidents in adult patients with congenital heart disease. Heart 96: 1223–1226.

Irvine B, Luo W, Leon JA (2015). Congenital anomalies in Canada 2013: a perinatal health surveillance report by the Public Health Agency of Canada's Canadian perinatal surveillance system. Health Promot Chronic Dis Prev Can 35: 21–22.

Jain P, Haller C, Pulcine E et al. (2019). A child with a stroke, drug-refractory epilepsy and congenital heart disease: can a hemispherectomy be safely performed between staged cardiac procedures? Childs Nerv Syst 35: 1245–1249.

Kent DM, Ruthazer R, Weimar C et al. (2013). An index to identify stroke-related vs incidental patent foramen ovale in cryptogenic stroke. Neurology 81: 619–625.

Kirsh JA, Walsh EP, Triedman JK (2002). Prevalence of and risk factors for atrial fibrillation and intra-atrial reentrant tachycardia among patients with congenital heart disease. Am J Cardiol 90: 338–340.

Kirton A, Deveber G (2015). Paediatric stroke: pressing issues and promising directions. Lancet Neurol 14: 92–102.

Knirsch W, Nadal D (2011). Infective endocarditis in congenital heart disease. Eur J Pediatr 170: 1111–1127.

Leisner MZ, Madsen NL, Ostergaard JR et al. (2016). Congenital heart defects and risk of epilepsy: a population-based cohort study. Circulation 134: 1689–1691.

Lim ADEDS (2011). Illustrated field guide to congenital heart disease and repair, Scientific Software Solutions Inc, Charlottesville, VA.

Limperopoulos C, Majnemer A, Shevell MI et al. (2001). Functional limitations in young children with congenital heart defects after cardiac surgery. Pediatrics 108: 1325–1331.

Liu XY, Wong V, Leung M (2001). Neurologic complications due to catheterization. Pediatr Neurol 24: 270–275.

Lumbiganon P, Chaikitpinyo A (2013). Antibiotics for brain abscesses in people with cyanotic congenital heart disease. Cochrane Database Syst Rev CD004469.

Majnemer A, Limperopoulos C, Shevell M et al. (2006). Long-term neuromotor outcome at school entry of infants with congenital heart defects requiring open-heart surgery. J Pediatr 148: 72–77.

Mallick AA, Ganesan V, Kirkham FJ et al. (2014). Childhood arterial ischaemic stroke incidence, presenting features, and risk factors: a prospective population-based study. Lancet Neurol 13: 35–43.

Manlhiot C, Menjak IB, Brandao LR et al. (2011). Risk, clinical features, and outcomes of thrombosis associated with pediatric cardiac surgery. Circulation 124: 1511–1519.

Manlhiot C, Brandao LR, Kwok J et al. (2012). Thrombotic complications and thromboprophylaxis across all three stages of single ventricle heart palliation. J Pediatr 161 (513–519): e3.

Martelle RR, Linde LM (1961). Cerebrovascular accidents with tetralogy of Fallot. Am J Dis Child 101: 206–209.

Medlock MD, Cruse RS, Winek SJ et al. (1993). A 10-year experience with postpump chorea. Ann Neurol 34: 820–826.

Mehnaz A, Syed AU, Saleem AS et al. (2006). Clinical features and outcome of cerebral abscess in congenital heart disease. J Ayub Med Coll Abbottabad 18: 21–24.

Miller SP, Mcquillen PS, Hamrick S et al. (2007). Abnormal brain development in newborns with congenital heart disease. N Engl J Med 357: 1928–1938.

Monagle P, Cochrane A, Roberts R et al. (2011). A multicenter, randomized trial comparing heparin/warfarin and acetylsalicylic acid as primary thromboprophylaxis for 2 years after the Fontan procedure in children. J Am Coll Cardiol 58: 645–651.

Monagle P, Chan AKC, Goldenberg NA et al. (2012). Antithrombotic therapy in neonates and children: antithrombotic therapy and Prevention of thrombosis, 9th ed: American College of Chest Physicians Evidence-Based Clinical Practice Guidelines. Chest 141: e737S–e801S.

Moosa AN, Jehi L, Marashly A et al. (2013). Long-term functional outcomes and their predictors after hemispherectomy in 115 children. Epilepsia 54: 1771–1779.

Nita DA, Soman TB (2014). Movement disorders in the ICU. In: D Wheeler, H Wong, T Shanley (Eds.), Pediatric critical care medicine. Springer, London.

O'carroll CB, Barrett KM (2017). Cardioembolic Stroke. Continuum (Minneap Minn) 23: 111–132.

Odegard KC, Zurakowski D, Dinardo JA et al. (2009). Prospective longitudinal study of coagulation profiles in children with hypoplastic left heart syndrome from stage I through Fontan completion. J Thorac Cardiovasc Surg 137: 934–941.

Petit CJ, Rome JJ, Wernovsky G et al. (2009). Preoperative brain injury in transposition of the great arteries is associated with oxygenation and time to surgery, not balloon atrial septostomy. Circulation 119: 709–716.

Roach ES, Golomb MR, Adams R et al. (2008). Management of stroke in infants and children: a scientific statement from a special writing Group of the American Heart Association Stroke Council and the council on cardiovascular Disease in the young. Stroke 39: 2644–2691.

Rodan L, Mccrindle BW, Manlhiot C et al. (2012). Stroke recurrence in children with congenital heart disease. Ann Neurol 72: 103–111.

Rohde S, Antonides CFJ, Dalinghaus M et al. (2019). Clinical outcomes of paediatric patients supported by the Berlin heart EXCOR: a systematic review. Eur J Cardiothorac Surg 56: 830–839.

Rushani D, Kaufman JS, Ionescu-Ittu R et al. (2013). Infective endocarditis in children with congenital heart disease: cumulative incidence and predictors. Circulation 128: 1412–1419.

Scavarda D, Major P, Lortie A et al. (2009). Periinsular hemispherotomy in children with stroke-induced refractory epilepsy. J Neurosurg Pediatr 3: 115–120.

Shann KG, Giacomuzzi CR, Harness L et al. (2008). Complications relating to perfusion and extracorporeal circulation associated with the treatment of patients with congenital cardiac disease: consensus definitions from the multi-societal database Committee for Pediatric and Congenital Heart Disease. Cardiol Young 18 (Suppl 2): 206–214.

Silvey M, Brandao LR (2017). Risk factors, prophylaxis, and treatment of venous thromboembolism in congenital heart Disease patients. Front Pediatr 5: 146.

Sinclair AJ, Fox CK, Ichord RN et al. (2015). Stroke in children with cardiac disease: report from the international pediatric Stroke study group symposium. Pediatr Neurol 52: 5–15.

Strater R, Vielhaber H, Kassenbohmer R et al. (1999). Genetic risk factors of thrombophilia in ischaemic childhood stroke of cardiac origin. A prospective ESPED survey. Eur J Pediatr 158 (Suppl 3): S122–S125.

Surmava AM, Maclagan LC, Khan F et al. (2019). Incidence and current treatment gaps in pediatric Stroke and TIA: an Ontario-wide population-based study. Neuroepidemiology 52: 119–127.

Takeshita M, Kagawa M, Yonetani H et al. (1992). Risk factors for brain abscess in patients with congenital cyanotic heart disease. Neurol Med Chir (Tokyo) 32: 667–670.

Tempe DK, Virmani S (2002). Coagulation abnormalities in patients with cyanotic congenital heart disease. J Cardiothorac Vasc Anesth 16: 752–765.

Waldman JD, Czapek EE, Paul MH et al. (1975). Shortened platelet survival in cyanotic heart disease. J Pediatr 87: 77–79.

Weissman BM, Aram DM, Levinsohn MW et al. (1985). Neurologic sequelae of cardiac catheterization. Cathet Cardiovasc Diagn 11: 577–583.

Wilson W, Taubert KA, Gewitz M et al. (2007). Prevention of infective endocarditis: guidelines from the American Heart Association: a guideline from the American Heart Association Rheumatic Fever, Endocarditis and Kawasaki Disease Committee, Council on Cardiovascular Disease in the Young, and the Council on Clinical Cardiology, Council on Cardiovascular Surgery and Anesthesia, and the Quality of Care and Outcomes Research Interdisciplinary Working Group. J Am Dent Assoc 138 (739–45): 747–760.

Handbook of Clinical Neurology, Vol. 177 (3rd series)
Heart and Neurologic Disease
J. Biller, Editor
https://doi.org/10.1016/B978-0-12-819814-8.00011-1

Chapter 2

Neurologic complications of congenital heart disease in adults

SARAH A. GOLDSTEIN[1] AND LARRY B. GOLDSTEIN[2]*

[1]*Department of Medicine, Division of Cardiology, Duke University, Durham, NC, United States*

[2]*Department of Neurology, Kentucky Neuroscience Institute, University of Kentucky, Lexington, KY, United States*

Abstract

Congenital heart disease (CHD) is a heterogeneous group of structural abnormalities of the cardiovascular system that are present at birth. Advances in childhood medical and surgical treatment have led to increasing numbers of adults with CHD. Neurological complications of CHD in adults are varied and can include an increased risk of stroke not only related to the underlying congenital defect and its surgical management but also due to atherosclerotic disease associated with advancing age. In addition to cerebrovascular events, CHD in adults is also associated with an increased risk of neurodevelopmental disorders, cognitive impairment, psychiatric disease, and epilepsy. Collaborative multidisciplinary care with contributions from neurologists and cardiologists with expertise in adult CHD is necessary to provide optimal long-term care for this complex and rapidly evolving population.

INTRODUCTION

Congenital heart disease (CHD) is a heterogeneous group of embryologic structural abnormalities of the cardiovascular system that are present at birth. CHD affects 6–8/1000 live births each year and represents the most frequent cause of major birth defects (van der Linde et al., 2011). Although once primarily a pediatric disease due to high mortality rates in early childhood, the advent of cardiopulmonary bypass and novel surgical techniques have extended the life expectancy of affected individuals resulting in an aging cohort of patients with CHD (Tutarel et al., 2014). About 90% of children born with CHD now survive into adulthood (Marelli et al., 2007; Khairy et al., 2010; Benjamin et al., 2017). Currently, there are more adults than children living with some form of CHD (Marelli et al., 2007). As the life expectancy of patients with CHD continues to increase, it has become clear that childhood surgical and medical management strategies do not address all of the sequelae of these conditions, with late complications being common. Adults with CHD are at increased risk of cerebrovascular events and neurocognitive impairment compared to the age-matched general population (Lanz et al., 2015; Gilboa et al., 2016). This chapter will focus on the neurologic complications of CHD occurring during adulthood.

CLASSIFICATION, ANATOMY, AND CLINICAL PRESENTATION OF CONGENITAL HEART DISEASE

Historically, congenital heart lesions were classified anatomically by severity and further categorized based on the presence or absence of cyanosis. Due to the wide range of underlying structural defects and surgical management strategies (Tables 2.1 and 2.2), the classification system used to define congenital heart disease in adults (ACHD) is complex and has been repeatedly revised. Anatomic severity and the resulting hemodynamic physiology are not always correlated as they change over the course of a patient's lifetime. The ACHD Anatomic and Physiological classification system was, therefore,

*Correspondence to: Larry B. Goldstein, MD, FAAN, FANA, FAHA, Ruth L. Works Professor and Chairman, Department of Neurology, Co-Director, Kentucky Neuroscience Institute and UK Neuroscience Research Priority Area, Interim Director, UK-Norton Stroke Care Network, KY Clinic—University of Kentucky, 740 S. Limestone Street, J401, Lexington, KY 40536, United States. Tel: +1-859-218-5039, Fax: +1-859-323-5943, E-mail: larry.goldstein@uky.edu

Table 2.1

Congenital heart disease anatomy

I: Simple
Unrepaired lesions

- Isolated small ASD
- Isolated small VSD

Repaired lesions

- Repaired ASD without residual shunt
- Repaired VSD without residual shunt

II: Moderate complexity
Repaired or unrepaired lesions

- Congenital aortic valve disease
- Coarctation of the aorta
- Ebstein anomaly
- Moderate or large unrepaired ASD
- Moderate or large unrepaired VSD
- VSD with associated anomaly
- Repaired TOF

III: Great complexity
Repaired or unrepaired lesions

- Cyanotic congenital heart defect (unrepaired or palliated)
- Fontan circulation/single ventricle
- TGA

ASD, atrial septal defect; *TGA*, transposition of the great arteries; *TOF*, tetralogy of Fallot; *VSD*, ventricular septal defect.

developed as part of the 2018 AHA/ACC Guideline for the Management of Adults with Congenital Heart Disease and incorporates native anatomy, state of surgical repair, and the patient's current resulting physiology and functional status (Stout et al., 2018). A summary of the current anatomic classification system highlighting the congenital lesions that are most commonly associated with neurologic complications is presented in Table 2.1.

CHD represents a large and diverse group of anatomic abnormalities and this chapter focuses only on those that are the most prevalent and clinically relevant in adults. Simple shunt lesions comprise a large portion of the ACHD population. The atrial septal defect (ASD) is a persistent direct communication between the atrial chambers. Most children are initially asymptomatic, leading to the defect typically being identified later in life, either incidentally or because of the presence of a heart murmur, arrhythmia, or during an etiological evaluation after ischemic stroke (Hoffman and Kaplan, 2002). The ventricular septal defect (VSD) is another common shunt lesion and leads to direct and persistent communication between the ventricular chambers. Although large defects are commonly symptomatic early in life, small VSDs may go undiagnosed until adulthood when a murmur is detected or mild heart failure symptoms develop.

A bicuspid aortic valve (BAV) is the most common congenital heart defect, affecting 0.5%–2% of the general adult population (Hoffman and Kaplan, 2002). Coarctation of the aorta (CoA) is an embryologically derived narrowing of the aorta that is typically distal to the origin of the left subclavian artery. Patients most commonly present in childhood with unexplained hypertension measured in the upper extremities. Even after repair, those born with coarctation frequently develop significant hypertension by early adulthood either in the setting of restenosis or underlying endothelial dysfunction (Kenny et al., 2011). There is an association between these two congenital anomalies; approximately 50% of patients with CoA also have BAV (only 5% of the total population of patients with BAV have diagnosed CoA) (Lewin and Otto, 2005).

Ebstein anomaly is characterized by apical displacement of the tricuspid valve leaflets and results in right atrial enlargement and atrial arrhythmias. The need for surgical repair is dependent on the severity of tricuspid valve displacement and resulting structural valvular abnormalities.

Tetralogy of Fallot (TOF) affects 3%–10% of patients born with CHD and is the most common cause of cyanotic CHD (Apitz et al., 2009). This defect encompasses four distinct structural abnormalities: 1, right ventricular outflow obstruction; 2, ventricular septal defect; 3, right ventricular hypertrophy; and 4, an overriding aorta with displacement toward the right heart. Surgical repair is typically performed in early childhood. Adults with repaired TOF can have a normal lifespan, but some have cardiac arrhythmias and heart failure.

Transposition of the great arteries (TGA) is characterized by ventriculoarterial discordance. The left ventricle is in the subpulmonic position with the right ventricle being subaortic. Patients may either have congenitally corrected TGA in which atrioventricular discordance is also present or complete transposition in which there is preservation of the typical atrioventricular relationship. Congenitally corrected TGA does not typically require surgical repair in childhood. Complete transposition leads to severe cyanosis and requires surgical intervention early in life. The atrial switch procedure, involving atrial baffling to divert blood to the opposite ventricle, was common until the 1980s following which the arterial switch procedure became the standard repair.

Table 2.2

Common childhood surgical repairs for congenital heart disease in adults

Heart lesion	Repair	Details	Long-term complications
ASD/VSD	Surgical patch	Requires atriotomy/ventriculotomy	Scar-mediated arrhythmias
	Percutaneous device closure	Not possible for all types of intracardiac shunt lesions due to anatomic constraints	Erosion, residual shunt, infection
TOF	Complete repair	VSD patch and relief of right ventricular outflow tract obstruction	Arrhythmias, heart failure, infection
Complete TGA	Atrial switch procedure (Mustard/Senning)	Atrial baffling to divert blood to the opposite ventricle	Arrhythmias, baffle obstruction, infection, heart failure
	Arterial switch procedure	Aorta and pulmonary artery are detached from their native roots and reattached to the opposite root	Great vessel dilation or stenosis, coronary artery obstruction
Single ventricle	Fontan procedure	Connection of caval blood flow directly to pulmonary circulation	Arrhythmia, heart failure, thromboembolism, liver dysfunction with coagulopathy

ASD, atrial septal defect; *TOF*, tetralogy of Fallot; *VSD*, ventricular septal defect.

The most complicated and severe group of congenital heart defects are those that result in only one fully formed ventricle. These patients with a "single ventricle" require a series of palliative procedures early in life that result in Fontan circulation, which involves diversion of systemic caval blood return directly into the pulmonary circulation. Fontan circulation, therefore, bypasses the morphologic subpulmonary ventricle, resulting in blood flow to both the systemic and pulmonary beds relying on the single subaortic ventricle.

CEREBROVASCULAR COMPLICATIONS OF ACHD

Adults with CHD have both ischemic and hemorrhagic stroke at higher rates than the general population; 1 in 11 men and 1 in 15 women with CHD have a stroke between ages 18 and 64 years, with the majority being ischemic strokes (Lanz et al., 2015). This increased incidence is especially pronounced among younger patients with CHD, in whom stroke is 10–100 times more likely to occur than in the general population (Hoffmann et al., 2010). Adults with severe CHD and cyanosis are at highest risk (Lanz et al., 2015). As found in pediatric patients, embolic events are the most common cause of stroke in adults with CHD (Bokma et al., 2018). In adults with CHD, atherosclerotic disease related to traditional acquired cardiovascular risk factors has also emerged as an important cause of ischemic stroke as this population continues to live later into adulthood (Hoffmann et al., 2010).

Embolic stroke

Ischemic stroke caused by cerebral embolism can occur due to a multitude of etiologies in adults with CHD. A common cause of embolic stroke among patients with adults with CHD occurs when an atrial arrhythmia predisposes to intracardiac thrombus formation. This is the most common cause of embolic stroke in the general population (Ornello et al., 2018). Adults with CHD, however, are at higher risk of atrial arrhythmia compared to age-matched patients without CHD due to a combination of underlying structural abnormalities leading to cardiac remodeling, as well as the presence of atriotomy scars from prior surgical interventions. The overall prevalence of atrial arrhythmias among adults with CHD is approximately 15%, but the risk varies greatly based on the underlying congenital lesion, state of surgical repair, and patient age (Bouchardy et al., 2009). The most common atrial arrhythmias occurring in adults with CHD are intraatrial reentrant tachycardia (IART), atrial fibrillation, and atrial flutter, all of which are associated with a higher risk of intracardiac thrombus formation (Kaemmerer et al., 2003). The prevalence of atrial fibrillation in patients with CHD increases with age. As such, atrial fibrillation is becoming increasingly common as the population of patients with CHD continues to live longer into adulthood (Philip et al., 2012). The congenital defects with the highest prevalence of atrial arrhythmias include atrial septal defects, tetralogy of Fallot, Ebstein anomaly, complete transposition of the great arteries treated with an atrial switch procedure, and patients with Fontan circulation (Gelatt et al., 1997; Gatzoulis et al., 1999, 2000; Delacretaz et al., 2001; Giannakoulas et al., 2012).

Adult CHD patients with atrial arrhythmias are at higher risk of thromboembolic events compared to age-matched adults without CHD. Furthermore, commonly used atrial fibrillation-related stroke risk stratification scores such as $CHADS_2$ and $CHADS_2$Vasc do not include high-risk features specific to the adult CHD population, such as severity of CHD and ventricular dysfunction. These risk stratification systems, therefore, likely underestimate the risk of atrial arrhythmia-related thromboembolic events in some adult patients with CHD (Masuda et al., 2017). Adults with complex CHD (Class I, LOE B) or moderately complex CHD (Class II, LOE C) and sustained or recurrent IART, atrial fibrillation or atrial flutter should be treated with long-term oral anticoagulation, irrespective of $CHADS_2$/$CHADS_2$Vasc score. For patients with mildly complex CHD, decisions regarding systemic anticoagulation may be guided by the established risk scores used in the general population. Currently, vitamin K antagonists are the treatment of choice for arrhythmia-related stroke prevention in adult CHD patients, although evidence is emerging for the use of direct oral anticoagulants in those without mechanical heart valves or significant mitral stenosis (Khairy et al., 2014).

Although less common, certain patients with CHD are at risk for nonarrhythmogenic embolic stroke. Adults with residual shunt lesions such as ASDs and VSDs are at risk for paradoxical embolism, in which thrombus originating in the venous vasculature traverse through an intracardiac shunt into the systemic circulation leading to embolic ischemic events (Windecker et al., 2014). Patients with prosthetic vascular implants, especially those with mechanical heart valves, are at risk for associated thrombus formation and resulting embolic stroke. Guidelines indicate that all patients with a mechanical heart valve should receive systemic anticoagulation for prevention of thromboembolism, irrespective of arrhythmia history (Nishimura et al., 2014). All patients with congenital heart disease, but particularly those with prosthetic implants, are at higher risk for infective endocarditis, which can be complicated by cerebral septic emboli causing ischemic stroke, brain abscess, and potentially mycotic aneurysms with subarachnoid hemorrhage (Vincent and Otto, 2018).

Another relatively uncommon but important cause of embolic stroke can occur in patients with cyanotic heart disease. To deliver more oxygen to chronically hypoxic tissue in the setting of cyanotic heart disease, patients develop a compensatory erythrocytosis, which over time can lead to blood hyperviscosity. A major consequence of hyperviscosity can be thrombus formation due to sludging in the microvasculature. This process likely contributes significantly to the high prevalence of stroke among patients with cyanotic heart disease (DeFilippis et al., 2007).

Atherosclerotic cerebrovascular disease

Ischemic stroke due to atherosclerotic disease is less common among adults with CHD than embolic events. However, as long-term survival continues to improve, the incidence of ischemic stroke due to cerebrovascular atherosclerotic disease is expected to increase. Patients with CHD are now living long enough to develop age-related acquired stroke risk factors. Furthermore, patients with CHD tend to develop stroke risk factors at much younger ages than the general population (Moons et al., 2006). The risk of metabolic syndrome among adult CHD patients is at least twofold higher compared age-matched controls, most likely related to exercise restrictions and feeding strategies used to treat failure to thrive in infancy and early childhood (Deen et al., 2016).

Certain groups of patients with CHD have particularly high rates of acquired cardiovascular and stroke risk factors. Those with a history of CoA frequently develop hypertension in childhood or early adulthood, even after coarctation repair and in the absence of restenosis (Bocelli et al., 2013; Bhatt and Yeh, 2015). The prevalence of hypertension in patients with CoA over the age of 50 years is estimated to be approximately 90% (Nattel et al., 2017). The presence of CoA is associated with increased risk of ischemic stroke overall and at younger ages when compared to the general population (Pickard et al., 2018). Additionally, patients with moderate or severe congenital heart lesions are frequently asked to follow activity restrictions, which can result in a sedentary lifestyle thereby increasing their risk for hypertension, hyperlipidemia, and obesity (Barbiero et al., 2014). Patients with adult CHD are also at increased risk for developing diabetes mellitus compared to the general population. This risk is particularly profound among patients with cyanotic heart disease (Madsen et al., 2016).

Hemorrhagic stroke

Adult patients with CHD have higher rates of intracerebral hemorrhage and subarachnoid hemorrhage compared to the age-matched general population (Giang et al., 2018). Patients with highly complex congenital cardiac lesions are at highest risk for intracerebral hemorrhage (Giang et al., 2018). Those with CoA of the aorta are also at increased risk for hemorrhagic events, specifically subarachnoid hemorrhage, and develop hemorrhagic stroke at younger ages compared to the general population (Pickard et al., 2018). The etiology of hemorrhagic stroke among patients with CoA is multifactorial, due both to early onset chronic hypertension and cerebrovascular structural abnormalities (Kenny et al., 2011). Patients with CoA have an approximate fivefold increased frequency of intracranial aneurysm compared to the general population (Connolly et al., 2003). Patients with BAV also have an increased frequency of intracranial aneurysms, irrespective of whether CoA is also present (Schievink et al., 2010).

Management of acute stroke

There are no specific guidelines regarding the treatment of acute stroke in the setting of adult CHD. Management, therefore, should generally follow that of the general population with a few additional considerations. When thrombolysis is considered, careful attention to whether chronic anticoagulation is being used is important, as it is common for young patients and those without atrial arrhythmias to be receiving an anticoagulant. Specific coagulation tests might be needed to detect therapeutic levels of the newer direct oral anticoagulants. For endovascular interventions, it must be appreciated that patients with CHD may have atypical arterial anatomy that could make vascular access difficult. Adult CHD patients will frequently have had prior catheterizations or vascular imaging studies, all of which should be reviewed prior to attempting endovascular intervention. Expert cardiology and anesthesia consultation should be pursued urgently for procedural guidance in the setting of acute stroke.

COGNITIVE IMPAIRMENT AND PSYCHIATRIC DISEASE IN ACHD

The association of CHD with abnormal neurodevelopment in childhood is well documented (Marelli et al., 2016; Nattel et al., 2017). Children with CHD are also at higher risk of having autism spectrum disorder (Sigmon et al., 2019). Both may have cognitive and behavioral consequences that persist in adulthood. Although less thoroughly described, adults with CHD have higher rates of neurocognitive disorders including age-related cognitive impairment and psychiatric disease (Marelli et al., 2016). Although a large proportion of adult CHD patients with abnormal neurocognitive function first developed deficits in childhood, some chronic neurologic manifestations of CHD are more specific to the adult population.

Dementia and age-related cognitive impairment

Adults with CHD are at increased risk of developing dementia, irrespective of disease severity, compared to the general population. Furthermore, adult CHD patients are more than twice as likely to develop early onset dementia (Bagge et al., 2018). Although the underlying mechanism is not clear, it is hypothesized that adult patients with CHD are more susceptible to age-related cognitive decline due to abnormal brain and cognitive reserve. The concepts of brain and cognitive reserve were initially described in the setting of Alzheimer's disease in the general population. Brain reserve refers to differences in the brain structure that may increase tolerance to pathology, whereas cognitive reserve describes differences in how individuals perform tasks that affect the brain's resilience to changes over time (Stern, 2012). Potential factors unique to the adult CHD population that may reduce brain and cognitive reserve include neuronal migration abnormalities, chromosomal abnormalities, the long-term effects of abnormal hemodynamic physiology, complex medical, and surgical management interventions, particularly recurrent cardiopulmonary bypass, and acquired comorbidities (Bagge et al., 2018). The development of dementia, particularly early onset dementia, is a strong predictor of increased mortality among adults with CHD (Afilalo et al., 2011).

Psychosocial issues

Adults with CHD face unique medical and social challenges that are associated with an increased burden of psychiatric disease, particularly anxiety and depression (Kasmi et al., 2018). At least 1 in 3 adults with CHD have evidence of psychiatric disease, with some contemporary estimates indicating that up to 80% of adult CHD patients meet diagnostic criteria for a psychiatric diagnosis (Brandhagen et al., 1991; Horner et al., 2000; Bromberg et al., 2003). Potential contributing factors include, but are not limited to, concerns about mortality, treatment decision making, anxiety about heart health and the need for future procedures, difficult pediatric-adult transitions, the burden of managing chronic illness, and a feeling of injustice and isolation regarding the state of their personal health (Kovacs et al., 2006).

Certain patients with adult CHD are at higher risk for psychiatric comorbidities. Those with poorer functional status and physician-imposed activity restrictions are more likely to develop psychiatric disease, suggesting that severity of disease likely plays an important role (Popelova et al., 2001; van Rijen et al., 2004). Patients who do not receive regular specialized care throughout adulthood and are therefore confronted with sudden health declines or the need for unexpected intervention are particularly vulnerable (Bhatt et al., 2015). Another at risk group includes those adult CHD patients with age-related cognitive impairment in whom the prevalence of comorbid psychiatric disease is 3–4 times higher compared to the general population (Kovacs et al., 2009). Resulting at least in part from high rates of psychiatric comorbidities, adults with CHD are less likely to pursue advanced educational degrees and are more likely to be unemployed (Zomer et al., 2012).

EPILEPSY

Children with CHD are at higher risk of developing epilepsy (Leisner et al., 2016; Desnous et al., 2019). One population-based study found a 5% overall incidence of epilepsy in patients with CHD by age 15 years compared to 3% for those without CHD (Leisner et al., 2016). Even children with CHD who did not undergo surgical repair are at higher risk of epilepsy.

Epilepsy can further complicate neurodevelopment, which, as reviewed previously, may already be negatively affected in the setting of CHD (Desnous et al., 2019). Higher complexity of surgical interventions, surgical factors including the use of extracorporeal membrane oxygenation, prolonged need for intensive care, and longer hospitalizations are associated with clinical seizures and subsequent epilepsy in infants with CHD (Desnous et al., 2019). Although the mechanism of seizures and epilepsy in children with CHD is likely multifactorial, perioperative clinical seizures are frequently associated with new brain lesions (Desnous et al., 2019).

The possibility of seizures and epilepsy needs to be considered in adults with CHD who are having episodes of altered consciousness or syncope, which may commonly be attributed to cardiac arrhythmia or other cardiovascular causes. Epilepsy and the use of antiseizure medicines may further complicate pregnancies in women with CHD and have a detrimental effect on the quality of life of adults.

CONCLUSION

Improved medical and surgical interventions in childhood have significantly improved survival among patients with CHD, the majority of whom are now living into adulthood, with their life expectancy continually increasing. CHD confers increased risk among adult patients for stroke, cognitive impairment, psychosocial issues, and epilepsy. Understanding this risk in the setting of each patient's unique anatomy, dynamic physiology and conventional risk factors are important for surveillance, prevention, and management of neurologic sequelae of adult CHD. Collaborative multidisciplinary care with contributions from neurologists and adult cardiologist with expertise in adult CHD is necessary to provide optimal long-term care for this complex and rapidly evolving population.

REFERENCES

Afilalo J, Therrien J, Pilote L et al. (2011). Geriatric congenital heart disease: burden of disease and predictors of mortality. J Am Coll Cardiol 58: 1509–1515.

Apitz C, Webb GD, Redington AN (2009). Tetralogy of Fallot. Lancet 374: 1462–1471.

Bagge CN, Henderson VW, Laursen HB et al. (2018). Risk of dementia in adults with congenital heart disease: population-based cohort study. Circulation 137: 1912–1920.

Barbiero SM, Sica CDA, Schuh DS et al. (2014). Overweight and obesity in children with congenital heart disease: combination of risks for the future? BMC Pediatr 14: 271.

Benjamin EJ, Blaha MJ, Chiuve SE et al. (2017). Heart disease and stroke statistics-2017 update: a report from the American Heart Association. Circulation 135: e146–e603.

Bhatt AB, Yeh DD (2015). Long-term outcomes in coarctation of the aorta: an evolving story of success and new challenges. Heart 101: 1173–1175.

Bhatt AB, Foster E, Kuehl K et al. (2015). Congenital heart disease in the older adult: a scientific statement from the American Heart Association. Circulation 131: 1884–1931.

Bocelli A, Favilli S, Pollini I et al. (2013). Prevalence and long-term predictors of left ventricular hypertrophy, late hypertension, and hypertensive response to exercise after successful aortic coarctation repair. Pediatr Cardiol 34: 620–629.

Bokma JP, Zegstroo I, Kuijpers JM et al. (2018). Factors associated with coronary artery disease and stroke in adults with congenital heart disease. Heart 104: 574–580.

Bouchardy J, Therrien J, Pilote L et al. (2009). Atrial arrhythmias in adults with congenital heart disease. Circulation 120: 1679–1686.

Brandhagen DJ, Feldt RH, Williams DE (1991). Long-term psychologic implications of congenital heart disease: a 25-year follow-up. Mayo Clin Proc 66: 474–479.

Bromberg JI, Beasley PJ, D'angelo EJ et al. (2003). Depression and anxiety in adults with congenital heart disease: a pilot study. Heart Lung 32: 105–110.

Connolly HM, Huston Iii J, Brown Jr RD et al. (2003). Intracranial aneurysms in patients with coarctation of the aorta: a prospective magnetic resonance angiographic study of 100 patients. Mayo Clini Proc 78: 1491–1499 Elsevier.

Deen JF, Krieger EV, Slee AE et al. (2016). Metabolic syndrome in adults with congenital heart disease. J Am Heart Assoc 5: e001132.

Defilippis AP, Law K, Curtin S et al. (2007). Blood is thicker than water: the management of hyperviscosity in adults with cyanotic heart disease. Cardiol Rev 15: 31–33.

Delacretaz E, Ganz LI, Soejima K et al. (2001). Multi atrial maco-re-entry circuits in adults with repaired congenital heart disease: entrainment mapping combined with three-dimensional electroanatomic mapping. J Am Coll Cardiol 37: 1665–1676.

Desnous B, Lenoir M, Doussau A et al. (2019). Epilepsy and seizures in children with congenital heart disease: a prospective study. Seizure 64: 50–53.

Gatzoulis MA, Freeman MA, Siu SC et al. (1999). Atrial arrhythmia after surgical closure of atrial septal defects in adults. N Engl J Med 340: 839–846.

Gatzoulis MA, Balaji S, Webber SA et al. (2000). Risk factors for arrhythmia and sudden cardiac death late after repair of tetralogy of Fallot: a multicentre study. Lancet 356: 975–981.

Gelatt M, Hamilton RM, Mccrindle BW et al. (1997). Arrhythmia and mortality after the mustard procedure: a 30-year single-center experience. J Am Coll Cardiol 29: 194–201.

Giang KW, Mandalenakis Z, Dellborg M et al. (2018). Long-term risk of hemorrhagic stroke in young patients with congenital heart disease. Stroke 49: 1155–1162.

Giannakoulas G, Dimopoulos K, Yuksel S et al. (2012). Atrial tachyarrhythmias late after Fontan operation are related to increase in mortality and hospitalization. Int J Cardiol 157: 221–226.

Gilboa SM, Devine OJ, Kucik JE et al. (2016). Congenital heart defects in the United States: estimating the magnitude of the affected population in 2010. Circulation 134: 101–109.

Hoffman JI, Kaplan S (2002). The incidence of congenital heart disease. J Am Coll Cardiol 39: 1890–1900.

Hoffmann A, Chockalingam P, Balint O et al. (2010). Cerebrovascular accidents in adult patients with congenital heart disease. Heart 96: 1223–1226.

Horner T, Liberthson R, Jellinek MS (2000). Psychosocial profile of adults with complex congenital heart disease. Mayo Clin Proc 75: 31–36.

Kaemmerer H, Fratz S, Bauer U et al. (2003). Emergency hospital admissions and three-year survival of adults with and without cardiovascular surgery for congenital cardiac disease. J Thorac Cardiovasc Surg 126: 1048–1052.

Kasmi L, Calderon J, Montreuil M et al. (2018). Neurocognitive and psychological outcomes in adults with Dextro-transposition of the great arteries corrected by the arterial switch operation. Ann Thorac Surg 105: 830–836.

Kenny D, Polson JW, Martin RP et al. (2011). Hypertension and coarctation of the aorta: an inevitable consequence of developmental pathophysiology. Hypertens Res 34: 543–547.

Khairy P, Ionescu-Ittu R, Mackie AS et al. (2010). Changing mortality in congenital heart disease. J Am Coll Cardiol 56: 1149–1157.

Khairy P, Van Hare GF, Balaji S et al. (2014). PACES/HRS expert consensus statement on the recognition and management of arrhythmias in adult congenital heart disease: developed in partnership between the pediatric and congenital electrophysiology society (PACES) and the Heart Rhythm Society (HRS). Endorsed by the governing bodies of PACES, HRS, the American College of Cardiology (ACC), the American Heart Association (AHA), the European heart rhythm association (EHRA), the Canadian Heart Rhythm Society (CHRS), and the International Society for Adult Congenital Heart Disease (ISACHD). Can J Cardiol 30: e1–e63.

Kovacs AH, Silversides C, Saidi A et al. (2006). The role of the psychologist in adult congenital heart disease. Cardiol Clin 24: 607–618.

Kovacs AH, Saidi AS, Kuhl EA et al. (2009). Depression and anxiety in adult congenital heart disease: predictors and prevalence. Int J Cardiol 137: 158–164.

Lanz J, Brophy JM, Therrien J et al. (2015). Stroke in adults with congenital heart disease: incidence, cumulative risk, and predictors. Circulation 132: 2385–2394.

Leisner MZ, Madsen NL, Ostergaard JR et al. (2016). Congenital heart defects and risk of epilepsy: a population-based cohort study. Circulation 134: 1689–1691.

Lewin MB, Otto CM (2005). The bicuspid aortic valve: adverse outcomes from infancy to old age. Circulation 111: 832–834.

Madsen NL, Marino BS, Woo JG et al. (2016). Congenital heart disease with and without cyanotic potential and the long-term risk of diabetes mellitus: a population-based follow-up study. J Am Heart Assoc 5: e003076.

Marelli AJ, Mackie AS, Ionescu-Ittu R et al. (2007). Congenital heart disease in the general population: changing prevalence and age distribution. Circulation 115: 163–172.

Marelli A, Miller SP, Marino BS et al. (2016). Brain in congenital heart disease across the lifespan: the cumulative burden of injury. Circulation 133: 1951–1962.

Masuda K, Ishizu T, Niwa K et al. (2017). Increased risk of thromboembolic events in adult congenital heart disease patients with atrial tachyarrhythmias. Int J Cardiol 234: 69–75.

Moons P, Deyk KV, Dedroog D et al. (2006). Prevalence of cardiovascular risk factors in adults with congenital heart disease. Eur J Cardiovasc Prev Rehabil 13: 612–616.

Nattel SN, Adrianzen L, Kessler EC et al. (2017). Congenital heart disease and neurodevelopment: clinical manifestations, genetics, mechanisms, and implications. Can J Cardiol 33: 1543–1555.

Nishimura RA, Otto CM, Bonow RO et al. (2014). 2014 AHA/ACC guideline for the management of patients with valvular heart disease: a report of the American College of Cardiology/American Heart Association task force on practice guidelines. J Am Coll Cardiol 63: e57–185.

Ornello R, Degan D, Tiseo C et al. (2018). Distribution and temporal trends from 1993 to 2015 of ischemic stroke subtypes: a systematic review and meta-analysis. Stroke 49: 814–819.

Philip F, Muhammad KI, Agarwal S et al. (2012). Pulmonary vein isolation for the treatment of drug-refractory atrial fibrillation in adults with congenital heart disease. Congenit Heart Dis 7: 392–399.

Pickard SS, Gauvreau K, Gurvitz M et al. (2018). Stroke in adults with coarctation of the aorta: a national population-based study. J Am Heart Assoc 7: e009072.

Popelova J, Slavik Z, Skovranek J (2001). Are cyanosed adults with congenital cardiac malformations depressed? Cardiol Young 11: 379–384.

Schievink WI, Raissi SS, Maya MM et al. (2010). Screening for intracranial aneurysms in patients with bicuspid aortic valve. Neurology 74: 1430–1433.

Sigmon ER, Kelleman M, Susi A et al. (2019). Congenital heart disease and autism: a case-control study. Pediatrics 144: e20184114.

Stern Y (2012). Cognitive reserve in ageing and Alzheimer's disease. Lancet Neurol 11: 1006–1012.

Stout KK, Daniels CJ, Aboulhosn JA et al. (2018). 2018 AHA/ACC Guideline for the Management of Adults with Congenital Heart Disease. Circulation 139: e831–e832.

Tutarel O, Kempny A, Alonso-Gonzalez R et al. (2014). Congenital heart disease beyond the age of 60: emergence of a new population with high resource utilization, high morbidity, and high mortality. Eur Heart J 35: 725–732.

Van Der Linde D, Konings EE, Slager MA et al. (2011). Birth prevalence of congenital heart disease worldwide: a systematic review and meta-analysis. J Am Coll Cardiol 58: 2241–2247.

Van Rijen EH, Utens EM, Roos-Hesselink JW et al. (2004). Medical predictors for psychopathology in adults with operated congenital heart disease. Eur Heart J 25: 1605–1613.

Vincent LL, Otto CM (2018). Infective endocarditis: update on epidemiology, outcomes, and management. Curr Cardiol Rep 20: 86.

Windecker S, Stortecky S, Meier B (2014). Paradoxical embolism. J Am Coll Cardiol 64: 403–415.

Zomer AC, Vaartjes I, Uiterwaal CS et al. (2012). Social burden and lifestyle in adults with congenital heart disease. Am J Cardiol 109: 1657–1663.

Handbook of Clinical Neurology, Vol. 177 (3rd series)
Heart and Neurologic Disease
J. Biller, Editor
https://doi.org/10.1016/B978-0-12-819814-8.00002-0

Chapter 3

Neurologic complications of rheumatic fever

MAXIMILIANO A. HAWKES AND SEBASTIÁN F. AMERISO*

Departmento de Neurología, Fleni, Buenos Aires, Argentina

Abstract

Sydenham chorea, also known as St. Vitus dance, is a major clinical criterion for the diagnosis of acute rheumatic fever. Clinically, it results in a combination of movement disorders and complex neuropsychiatric symptoms. Cardiac damage due to rheumatic fever may also predispose to neurologic complications later in life. Rheumatic heart disease (RHD) is associated with heart remodeling, cardiac arrhythmias, and ischemic stroke. Furthermore, chronically damaged heart valves are predisposed to infection. Septic brain embolism, a known complication of infective endocarditis, may result in brain ischemia, hemorrhage, and spread of the infection to the brain.

INTRODUCTION

Rheumatic fever (RF) is a postinfectious, multisystem inflammatory disease typically occurring 2–4 weeks following group A β-hemolytic streptococcal (GABHS) infection (usually pharyngitis), due to cross-linked autoimmunity between streptococcal and host antigens. The main neurologic manifestation, Sydenham chorea (SC), results in movement disorders and complex neurophysiologic symptoms. Rheumatic heart disease (RHD) may also indirectly cause neurologic complications such as ischemic stroke and septic embolism from infective endocarditis.

In this chapter we provide a review of the neurologic complications of rheumatic fever.

HISTORIC VIGNETTE

Saint Vitus was a Christian saint, famous for healing neurologic disorders. He died as a martyr in boiling oil in 303 AD. Following his death, many believed his relics could cure "unsteady step, trembling limbs, limping knees, bent fingers and hands, paralyzed hands, lameness, crookedness, and withering body" (Eftychiadis, 2001). The link between St. Vitus dance and SC remains to this date unclear. Some believe that neurologic conditions cured by Vitus mimicked the movements of a dance, so St. Vitus dance became an umbrella term for various movement disorders. Others claim the link as an homage to the manic dancing that historically took place in front of his statue during the feast of Saint Vitus. Its association with outbreaks of dancing mania and other delirious behavior that occurred in Europe in the Middle Ages has also been proposed. In the 16th century, Paracelsus noticed the central role of emotional instability and lack of voluntary motor control of St. Vitus dance.

A century later, Thomas Sydenham provided more details on the kinetic disturbances of the disease: "… Saint Vitus' dance is a sort of convulsion which attacks boys and girls from the tenth year till they have done growing. At first it shows itself by a halting, or rather an unsteady movement of one of the legs, which the patient drags. Then it is seen in the hand of the same side. The patient cannot keep it a moment in its place, whether he lay it upon his breast or any other part of his body. Do what he may, it will be jerked elsewhere convulsively …" (Schechter, 1975; Eftychiadis, 2001). He also described the articular manifestations of RF. However, he failed to recognize its relationship with the movement disorders.

*Correspondence to: Sebastián F. Ameriso, MD, Montañeses 2325, Ciudad de Buenos Aires, 1428, Argentina. Tel: +541157773200 ext. 2462, Fax: +541157773209, E-mail: sameriso@fleni.org.ar

Richard Bright grouped together chorea and carditis as manifestations of rheumatism for the first time in the 19th century. The central role of an infectious process as the trigger of RF was elucidated in 1930s by Coburn et al. (Eftychiadis, 2001).

EPIDEMIOLOGY

RF is closely associated with socioeconomic disadvantage. During epidemics, approximately 3% of patients with untreated exudative pharyngitis or scarlet fever develop RF. The incidence falls to 1% with endemic infections. There are approximately half a million new cases of RF yearly worldwide. While RF can develop at any age, most cases occur among children between 5 and 15 years of age (Carapetis et al., 2005; GBD 2013 Mortality and Causes of Death Collaborators, 2015). The average incidence of acute RF is 19 per 100,000 school-aged children worldwide, with substantial geographic differences (Tibazarwa et al., 2008). The lowest incident rates have been reported in the developed countries of Western Europe and North America, where the incidence falls to below 2 cases per 100,000 school-aged children (Miyake et al., 2007). In these countries, the incidence of RF dramatically decreased after World War II as a result of improved health care. Nonetheless, outbreaks have occurred even though the disease was believed to be eradicated (Veasy et al., 1987). The highest incidences are found among school-aged indigenous Australian children (153–380 cases per 100,000) and in Brazil (700 per 100,000) (Parnaby and Carapetis, 2010; de Figueiredo et al., 2019). However, the true incidence, prevalence, and burden of RF is likely to be underestimated due to underreporting in many endemic areas (Lee et al., 2009). Poor sanitation, including overcrowding, which favors the transmission of streptococcal infections, and suboptimal access to medical care explain the higher incidence of RF in economically underprivileged areas of the world.

SC occurs in about one-fourth of patients with RF, ranging from 10% to 50% in different series (Cardoso et al., 1997). Female children (2:1 FM ratio) between 5 and 13 years of age are more frequently affected (Zomorrodi, 2006). A family history of RF has been found in about 30% of patients with SC, suggesting that a genetic predisposition plays a pathogenic role (Aron et al., 1965; Zomorrodi, 2006). In developed countries, the incidence of SC has declined along with RF. SC accounted for 0.9% of admissions to hospitals of children in Chicago before 1940, as compared to 0.2% in the period 1950–1980. Likewise, the incidence of RF and SC have steadily declined in Tel Aviv, Israel, from 99.3 and 3.47 per 100,000 in the decade 1960–1970 to 15.5 and 1.21 per 100,000 during 1980–1990. Nevertheless, SC is still the most common cause of acute acquired childhood onset chorea worldwide (Zomorrodi, 2006). For instance, in Brazil SC accounts for 64% of cases of chorea among children (Cardoso, 2011).

The prevalence of RHD is around 1–5 cases per 1000 school-age children in developing countries (Steer and Carapetis, 2009). Without proper medical care, these patients are at increased risk of embolic stroke later in life. Overall, between 3% and 7.5% of new ischemic strokes are due to RHD. It has been suggested that controlling RHD could prevent more than 300,000 stroke deaths yearly, worldwide (Steer and Carapetis, 2009; The Lancet Neurology, 2010). A hospital-based study conducted in Iran reported that RHD was present in 45% of patients with cardioembolic stroke. This suggests that the role of RHD as a cause of ischemic stroke may be underestimated in many areas, due to reporting bias (Ghandehari and Izadi-Mood, 2007).

PATHOPHYSIOLOGY

Rheumatic fever is an infection-triggered autoimmune disease

Both the humoral and cellular immune systems participate in the development of RF. Antibodies against the cell walls and membranes of GABHS antigens cross-react with self-antigens in genetically predisposed hosts. This process is called molecular mimicry. After cross-reaction, cytotoxicity, altered cellular signaling, and recruitment of T cells to the site of attack lead to disruption of host tissues (Cunningham, 2019). Three types of molecular mimicry have been recognized (Cunningham, 2019). The first type occurs due to identical amino acid sequences shared by molecules present in GABHS and human tissues (joints, heart, brain) (Fujinami et al., 1983; Ellis et al., 2005). The second type results from structural similarity (e.g., alpha-helical coiled-coil molecules like streptococcal M protein and myosin), despite less than 40% identity between antigens (Krisher and Cunningham, 1985; Adderson et al., 1998). Finally, the third type occurs between diverse molecules such as DNA and proteins or carbohydrates and peptides (Cunningham and Swerlick, 1986; Shikhman et al., 1993). The antibodies cannot penetrate into the brain unless the blood–brain barrier (BBB) is damaged by the infection (Cutforth et al., 2016; Cunningham, 2019).

Experimental studies have shown that intranasal GABHS infection allows activated Th17 cells to penetrate into the brain through the olfactory bulb and open the BBB to IgG (Dileepan et al., 2016). Monoclonal antibodies and T cell clones have allowed the identification of multiple host and streptococcal antigens responsible for immunological mimicry. Host brain antigens include lysoganglioside, located in the cell membrane, and tubulin, an abundant intracellular protein. Antibodies also increase calcium/calmodulin-dependent protein kinase II (CaMKII) activity, and can activate dopaminergic receptors D1 and D2 (Cunningham, 2019). For this reason, SC can be considered a dopamine receptor encephalitis (Cox et al., 2013), which in turn may explain the development of movement disorders.

Sydenham chorea affects mostly the basal ganglia

Different studies in patients with SC have shown selective basal ganglia involvement.

Postmortem brain analysis of patients with SC has shown perivascular infiltration with round cells, endothelial swelling, petechial hemorrhages, and gliosis affecting the basal ganglia and thalami (Loiselle and Singer, 2001). In a case control study, Giedd et al. compared brain MRIs of 24 patients with SC with 48 age, height, weight, gender, and handedness matched controls. The volume of the caudate, putamen, and globus pallidus was higher among patients with SC. No differences in total cerebral volume, prefrontal, midfrontal or thalamic areas were found between the SC patients and the healthy controls (Giedd et al., 1995). MRI signal changes within the striatum may resolve upon resolution of the neurologic manifestation (Traill et al., 2008). The amount of gamma-amino-butyric acid (GABA) and acetylcholine has been reported to decrease, along with the increasing dopaminergic activity in the basal ganglia in cases of SC (Genel et al., 2002). Absence of anti-basal ganglia antibodies (ABGA) in some cases of recurrent SC suggests that dopamine hypersensitivity of chronically damaged basal ganglia circuits could play an additional pathogenic role in some cases (Peña et al., 2002). Also, the occurrence of chorea with negative ABGA in elder patients with a history of SC suggests that coexistent pathology, such as dementia or simply cumulative neuronal loss associated with aging, may disrupt compensation of previously damaged circuits, leading to late reemergence of symptoms (Harrison, 2004). This theory could have therapeutic implications. Treating multiple relapses of SC could prevent additive damage of the basal ganglia, thus decreasing the odds of developing senile chorea later in life (Harrison, 2004).

Brain-heart interaction

Around 60% of patients with RF develop rheumatic heart disease (RHD), characterized by chronic damage to the heart valves eventually leading to cardiac remodeling, heart failure, and cardiac arrhythmias, particularly atrial fibrillation. These structural and electric abnormalities are associated with cardioembolic stroke (Negi et al., 2018). Moreover, chronic valvular heart disease predisposes to the potential neurologic complications of infective endocarditis (Holland et al., 2016).

CLINICAL PRESENTATION

Sydenham chorea

The clinical presentation of chorea and neuropsychiatric symptoms vary among people of differing ethnic backgrounds (Carapetis and Currie, 1999). Symptoms generally appear between 4–8 weeks following GABHS pharyngitis. More delayed forms have been reported up to 8 months after the inciting infection (Eshel et al., 1993). Chorea results in brief, involuntary, nonrhythmic, purposeless, "dance-like" movements, which characteristically occur in a continuous fashion while awake and improve with sleep. Onset is often insidious and the course progressive over hours or days. Although usually generalized, hemichorea can be seen in 20%–30% of cases (Aron et al., 1965; Cardoso et al., 1997; Zomorrodi, 2006).

About 60%–80% of patients with chorea develop carditis, while 30% develop arthritis (Cardoso et al., 1997). Chorea occurs in isolation in about 20% of cases (Dale, 2013). Other associated manifestations include motor impersistence, facial grimacing, tongue protrusion, and dysarthria (Cardoso, 2011). Hypometric saccades are found in 80% of patients. Oculogyric crisis is rare (Cardoso et al., 1997). Limb hypotonia is usually present; in a few severe cases, it can result in paralysis (chorea paralytica). A small series reported vocal tics in 80% of affected children, which may reflect dysfunction of cortical-thalamic-striatal-cortical circuits (Mercadante et al., 1997). Since SC patients lack a premonitory urge and vocalizations are unusually complex, other authors suggest avoiding the term "tic." They argue that vocalizations are rather associated with choreatic contractions of the upper respiratory tract muscles (de Teixeira et al., 2009; Cardoso, 2011). Patients may present decreased verbal fluency, suggesting dysfunction of the dorsolateral prefrontal-basal ganglia circuit (Cunningham et al., 2006). Prosody can be affected, usually with impaired modulation of fundamental frequency and longer duration of emission of sentences. This results in a monotone and slow speech, similar to what has been described in other basal ganglia disorders (Oliveira et al., 2010). A case control study found that patients with SC (21.8%) and patients with RF without neurologic symptoms (18.2%) had higher frequency of migraines compared to matched controls (8.1%). These findings are similar to what has been found in other basal ganglia disorders, such as essential tremor and Tourette syndrome (Teixeira et al., 2005b). A few case reports described peripheral neuropathy associated with streptococcal infections. However, a detailed neurophysiologic study of 26 patients with SC failed to find evidence of peripheral neuropathy in this population (Cardoso et al., 2005).

Neuropsychiatric symptoms are highly prevalent among individuals with SC. These symptoms include obsessive compulsive disorder (OCD), anxiety, mood disorders, psychosis, tics, attention deficit hyperactivity disorder (ADHD), and nonspecific symptoms such as emotional lability, irritability, and regressive behavior (Ridel et al., 2010; Punukollu et al., 2016).

Obsessive–compulsive symptoms may occur before (28.6%), concomitantly (33.3%), or after (38.1%) the onset of SC. A study suggests that OCD symptoms occur at a similar rate in patients with RF with and without chorea (Hounie et al., 2004). Aggressive obsessions (63%) and fear of contamination (34%) are the most common obsessional symptoms, while checking (53%), cleaning (42%), and repeating (36%) are the most common compulsions (Asbahr et al., 2005).

Depression and anxiety occur more frequently in patients with SC than in the general population, and account for the most frequent psychiatric disorders in individuals with SC. Anxiety has been found during and after the onset of neurologic symptoms, but depression appears more commonly after them (Ridel et al., 2010). In case–control studies, ADHD (30%–45%) and tics (72%) were more frequent among patients with SC than in those with RF without chorea and healthy controls (Hounie et al., 2004; Beato et al., 2010).

Executive dysfunction has been reported in adults with either remitted or persistent SC (Beato et al., 2010; Cavalcanti et al., 2010). Finally, available data indicate that patients with SC are between 9 and 13 times more likely to develop schizophrenia than healthy individuals (Wilcox and Nasrallah, 1986, 1988).

Post-GABHS infection autoimmune disorders spectrum

Until the 1990s, SC was considered the only neurologic sequela of GABHS infection (van Toorn et al., 2004). Currently, the spectrum of poststreptococcal autoimmune disorders has expanded to include Tourette syndrome, chronic tic disorder, pediatric autoimmune neuropsychiatric disorder associated with streptococcal infection (PANDAS), paroxysmal dyskinesias, acute disseminated encephalomyelitis (ADEM), dystonia, myoclonus, and anorexia nervosa (DiFazio et al., 1998; Sokol, 2000; Dale et al., 2001, 2002).

The acronym PANDAS was originally proposed in 1998 to describe a subset of children with OCD or tic disorders following GABHS infection who did not meet criteria for SC. Whether PANDAS is different from typical OCD or tic disorder and the nature of its pathogenic relationship with GABHS infection remain controversial. Dedicated reviews on this topic have been published elsewhere (Williams and Swedo, 2015). Briefly, the abrupt onset and episodic course of OCD and/or tic disorder, occurring between 3 years and onset of puberty, accompanied by neurologic symptoms (motor hyperactivity, choreiform movements, or tics), showing a temporal relation between GABHS infection at onset or exacerbation have been proposed as diagnostic criteria (Swedo et al., 1998, 2004).

Poststreptococcal ADEM (PSADEM) is characterized by dystonic movement disorders (50%) and behavioral symptoms (70%). Remarkably, the clinical presentation with dystonia is unusual in classic ADEM. The median age of onset (4.8 years) is lower than in PANDAS (6.8 years) and SC (8.4 years). While usually monophasic, relapses are possible. Autoreactive antibodies against the basal ganglia are positive in all patients, while basal ganglia involvement can be evidenced by brain MRI in most cases (80%) (Dale et al., 2001).

The importance of differentiating the different poststreptococcal autoimmune disorders from SC relies on their different response to treatments (van Toorn et al., 2004). For example, the efficacy of penicillin prophylaxis to prevent symptom exacerbations in children with PANDAS is unclear. Conversely, poststreptococcal disorders such as PANDAS, PSADEM, and dystonia may respond well to immunomodulatory therapies, such as plasma exchange and intravenous immunoglobulin (IVIG) (Dale et al., 2001; Edwards et al., 2004; van Toorn et al., 2004).

Rheumatic heart disease and stroke

Stroke is a leading cause of mortality and acquired disability worldwide. Most of the stroke burden occurs in low-income and middle-income countries (Krishnamurthi et al., 2013). RHD, the most common form of cardiac disease in low-income populations, is the link between RF and stroke. The median age of RHD at diagnosis is 17 years. This is relevant because 80%–85% of children younger than 15 years live in economically disadvantaged areas where RHD is endemic. Progression to RHD appears to be highest within the first year following acute RF diagnosis. Also, the development of complications is highest in the first year after the diagnosis of RHD. In order of frequency, these complications include heart failure, atrial fibrillation, infectious endocarditis, and stroke (He et al., 2016). The presence of spontaneous echo contrast on the left atrium and atrial fibrillation are significantly associated with stroke occurrence in patients with RHD. Subclinical atrial fibrillation is common in patients with RHD (Gupta et al., 2015).

DIAGNOSIS

SC should be suspected in children with acute chorea and behavioral symptoms, particularly in endemic areas for RF. The initial evaluation is oriented on diagnosing acute RF according to the revised Jones criteria (Gewitz et al., 2015). Patients and/or caregivers should be questioned about recent symptoms of pharyngitis. The physical examination should focus on identifying further signs such as erythema marginatum, subcutaneous nodules, arthritis, and carditis. Required ancillary tests are shown in Table 3.1. Brain imaging (MRI) and cerebrospinal fluid (CSF) analysis may be useful in excluding other causes of chorea (Table 3.2), particularly in atypical cases.

Functional neuroimaging studies, such as single photon emission computed tomography (SPECT) and positron emission tomography (PET), have consistently shown a pattern of basal ganglia hyperperfusion in patients with SC, which may persist even after the remission of abnormal movements (Beato et al., 2014). However, similar findings have been described in other infectious or other autoimmune causes of chorea, and they are unlikely to change the medical management (Nordal et al., 1999; Krakauer and Law, 2009). Hence, these studies are not routinely performed in the clinical practice.

In cases of ischemic stroke, echocardiographic studies provide hints of RHD. Once the diagnosis of RHD has been made, covert atrial fibrillation should be strongly suspected, if not evident upon initial evaluation. Serial Holter electrocardiogram monitoring, implantable loop recording devices, and even empiric anticoagulation may be considered in patients with embolic strokes of undetermined source and findings suggestive of RHD.

Table 3.1

Diagnostic workup

Group A streptococcal infection
• Current infection
○ Throat culture
• Preceding infection
○ Antistreptolysin O (ASO)
○ Antideoxyribonuclease (antiDNAase)
Cardiac evaluation
• Electrocardiogram
• Echocardiogram
Inflammatory markers
• C-reactive protein
• Erythrocyte sedimentation rate

Table 3.2

Differential diagnoses of chorea

Encephalitis
Medication side effect[a]
Cerebral palsy
Huntington disease
Wilson disease
Lesch–Nyhan syndrome
Hyperalaninemia
Ataxia telangiectasia
Tic disorder
Intracranial tumor
Lyme disease
Antiphospholipid antibody syndrome
Systemic lupus erythematosus
Vasculitis
Sarcoidosis
Hyperthyroidism

[a]Levodopa, amantadine, anticholinergics, neuroleptics, reserpine, tetrabenazine, calcium channel blockers, lamotrigine, phenytoin, carbamazepine, gabapentin, valproic acid, amphetamines, cocaine, cyproheptadine, methylphenidate, aminophylline, theophylline, antihistamines, baclofen, benzodiazepines, cimetidine, cyclosporine, digoxin, oral contraceptives, glucocorticoids, isoniazid, levofloxacin, lithium, opioids, selective serotonin reuptake inhibitors, tricyclic antidepressants.

TREATMENT

Comprehensive reviews on current treatment options for SC have been recently published (Dean and Singer, 2017; Vasconcelos et al., 2019). All patients should receive antibiotic therapy for eradication of GABHS (up to one-third of RF cases can occur after subclinical GABHS infections) (Dajani, 1991), and later on, for prophylactic purposes. This is important because prevention of recurrent GABHS infections lowers the risk of RHD and prevents recurrent chorea. Penicillin is the drug of choice, while cephalosporins, macrolides, and clindamycin are accepted options.

There is general agreement that corticosteroids shorten the course of symptoms with an acceptable safety profile (Green, 1978; Cardoso et al., 2003; Barash et al., 2005; Teixeira et al., 2005a; Fusco et al., 2012; Fusco and Spagnoli, 2018). A randomized, double-blind, parallel study compared the clinical course of 22 children treated with prednisone (2 mg/kg/day during 4 weeks, followed by a gradual discontinuation) and 15 children treated with placebo. Initial chorea intensity was similar in both groups. At 1 week the intensity of symptoms had significantly improved with prednisone. Also, complete resolution occurred more rapidly (mean duration of 54 days vs 120 days). No severe adverse events were observed (Paz et al., 2006). According to available evidence, oral corticosteroids can be used both as sole therapy in moderate to severe SC, and as maintenance therapy after IV corticosteroids in more severe cases (Vasconcelos et al., 2019).

Compared with haloperidol alone, IVIG plus haloperidol showed improvement in the severity of chorea at 1 month, without difference at 3 and 6 months of follow-up. The IVIG group also required haloperidol for significantly less time (mean time of use 51 days vs 136.7 days). Relapses were similar in both groups, and no adverse events were adjudicated to IVIG (Walker et al., 2012). It has also been suggested that IVIG may improve cognitive and psychiatric symptoms (Gregorowski et al., 2016). There are no data to support the superiority of IVIG over plasmapheresis and oral prednisone (Garvey et al., 2005).

Current evidence for most medications used for symptomatic treatment of chorea is weak, and the literature is diverse regarding recommendations. In this context, drug selection depends on the severity of symptoms, drug side effects, and patient/caregiver and physician preferences.

Valproic acid (Axley, 1972; Appleton and Jan, 1988; Daoud et al., 1990; al-Eissa, 1993; Genel et al., 2002; Pena et al., 2002; Oosterveer et al., 2010), antipsychotics (haloperidol, pimozide, risperidone, and olanzapine) (Peña et al., 2002), and carbamazepine (Harel et al., 2000; Genel et al., 2002; Peña et al., 2002) are broadly accepted options for treatment of moderate symptoms, and moderate–severe forms in combination with steroids or IVIG. Levetiracetam, tetrabenazine, and deutetrabenazine are less commonly reported treatments (Sahin and Cansu, 2015; Vasconcelos et al., 2019). Fig. 3.1 provides guidance for the treatment of SC.

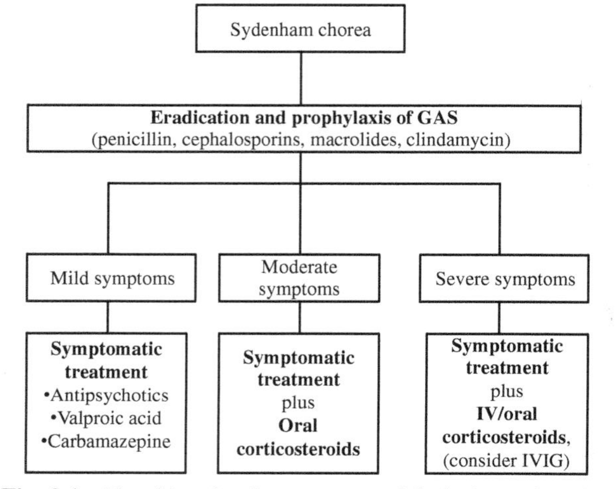

Fig. 3.1. Algorithm for the treatment of Sydenham chorea.

PROGNOSIS

SC is often a self-limited condition. Most patients present a gradual yet complete improvement of their symptoms within 12–15 weeks (Lessof and Bywaters, 1956; Hitchens, 1958). However, there are reports of persistent symptoms for 2 years (Carapetis and Currie, 1999; Cardoso et al., 1999; Gurkas et al., 2016). Up to one-third of patients present recurrences, mostly within 3 years of the initial presentation (range 3 months to 10 years) (Carapetis and Currie, 1999). Most recurrent episodes occur in patients receiving no antibiotic prophylaxis (al-Eissa, 1993; Gurkas et al., 2016). Other insults, such as autoimmunity, pregnancy (chorea gravidarum), oral contraceptive pills, dementia, or simply aging, acting upon chronically damaged basal ganglia, are other reported triggers (Berrios et al., 1985; Church et al., 2002, 2003; Korn-Lubetzki et al., 2004).

A similar clinical course can be anticipated with neuropsychiatric symptoms, although persistence of psychiatric disorders is possible. While some studies showed that patients with SC had a good functional, educational, socioeconomic, and employment status in the long term, others suggested lower college graduation, professional or executive positions, and marital stability compared to healthy controls (Punukollu et al., 2016).

Prognosis of patients with stroke will depend on the severity of the index event and use of reperfusion treatments. In recent years, the development of combined intravenous thrombolysis and endovascular reperfusion treatments, and advanced imaging to detect the best candidates for reperfusion beyond the classic time window, have dramatically changed the prognosis of stroke patients. However, these options may be limited in economically disadvantaged settings, where RHD-related strokes are more likely to occur. The early diagnosis of cardioembolic sources before the development of stroke and the appropriate workup to detect RHD in stroke survivors are essential for the early institution of proper primary or secondary prevention strategies.

CONCLUSIONS

RF is a postinfectious, inflammatory, multisystem disease affecting the brain through autoimmune mechanisms. Although many of these mechanisms have already been elucidated, more research is needed to completely understand this complex disease and to design more effective treatments. SC, the classical neurologic complication of RF, is characterized by movement disorders and complex neuropsychiatric symptoms. The disease is usually monophasic but can recur. Treatment is aimed at eradicating and preventing GABHS recurrences and to control symptoms of chorea. Steroids may shorten the course of the disease and should be considered in moderate–severe cases. RHD is associated with valvopathies, heart remodeling, and arrhythmias with the consequent risk of ischemic stroke. Abnormal heart valves also predispose to infectious endocarditis, which in turn increases the risk of septic embolism to the brain, causing both ischemia and hemorrhage. SC and RHD should be suspected when evaluating children and adolescents with chorea and young people with stroke, particularly in endemic regions.

References

Adderson EE, Shikhman AR, Ward KE et al. (1998). Molecular analysis of polyreactive monoclonal antibodies from rheumatic carditis: human anti-N-acetylglucosamine/anti-myosin antibody V region genes. J Immunol 161: 2020–2031.

al-Eissa A (1993). Sydenham's chorea: a new look at an old disease. Br J Clin Pract 47: 14–16.

Appleton RE, Jan JE (1988). Efficacy of valproic acid in the treatment of Sydenham's chorea. J Child Neurol 3: 147.

Aron AM, Freeman JM, Carter S (1965). The natural history of Sydenham's chorea. Am J Med 38: 83–95.

Asbahr FR, Garvey MA, Snider LA et al. (2005). Obsessive-compulsive symptoms among patients with Sydenham chorea. Biol Psychiatry 57: 1073–1076.

Axley J (1972). Rheumatic chorea controlled with haloperidol. J Pediatr 81: 1216–1217.

Barash J, Margalith D, Matitiau A (2005). Corticosteroid treatment in patients with Sydenham's chorea. Pediatr Neurol 32: 205–207.

Beato R, Maia DP, Teixeira AL et al. (2010). Executive functioning in adult patients with Sydenham's chorea. Mov Disord 25: 853–857.

Beato R, Siqueira CF, Marroni BJ et al. (2014). Brain SPECT in Sydenham's chorea in remission. Mov Disord 29: 256–258.

Berrios X, Quesney F, Morales A et al. (1985). Are all recurrences of "pure" Sydenham chorea true recurrences of acute rheumatic fever? J Pediatr 107: 867–872.

Carapetis JR, Currie BJ (1999). Rheumatic chorea in northern Australia: a clinical and epidemiological study. Arch Dis Child 80: 353–358.

Carapetis JR, Steer AC, Mulholland EK et al. (2005). The global burden of group A streptococcal diseases. Lancet Infect Dis 5: 685–694.

Cardoso F (2011). Sydenham's chorea. Handb Clin Neurol 100: 221–229. https://doi.org/10.1016/B978-0-444-52014-2.00014-8.

Cardoso F, Dornas L, Cunningham M et al. (2005). Nerve conduction study in Sydenham's chorea. Mov Disord 20: 360–363.

Cardoso F, Eduardo C, Silva AP et al. (1997). Chorea in fifty consecutive patients with rheumatic fever. Mov Disord 12: 701–703.

Cardoso F, Maia D, Cunningham MCQS et al. (2003). Treatment of Sydenham chorea with corticosteroids. Mov Disord 18: 1374–1377.

Cardoso F, Vargas AP, Oliveira LD et al. (1999). Persistent Sydenham's chorea. Mov Disord 14: 805–807.

Cavalcanti A, Hilário MOE, dos Santos FH et al. (2010). Subtle cognitive deficits in adults with a previous history of Sydenham's chorea during childhood. Arthritis Care Res 62: 1065–1071.

Church AJ, Cardoso F, Dale RC et al. (2002). Anti-basal ganglia antibodies in acute and persistent Sydenham's chorea. Neurology 59: 227–231.

Church AJ, Dale RC, Cardoso F et al. (2003). CSF and serum immune parameters in Sydenham's chorea: evidence of an autoimmune syndrome? J Neuroimmunol 136: 149–153.

Cox CJ, Sharma M, Leckman JF et al. (2013). Brain human monoclonal autoantibody from sydenham chorea targets dopaminergic neurons in transgenic mice and signals dopamine D2 receptor: implications in human disease. J Immunol 191: 5524–5541.

Cunningham MW (2019). Molecular mimicry, autoimmunity, and infection: the cross-reactive antigens of group A streptococci and their sequelae. Microbiol Spectr 7.

Cunningham MCQS, Maia DP, Teixeira AL et al. (2006). Sydenham's chorea is associated with decreased verbal fluency. Parkinsonism Relat Disord 12: 165–167.

Cunningham MW, Swerlick RA (1986). Polyspecificity of antistreptococcal murine monoclonal antibodies and their implications in autoimmunity. J Exp Med 164: 998–1012.

Cutforth T, DeMille MM, Agalliu I et al. (2016). CNS autoimmune disease after *Streptococcus pyogenes* infections: animal models, cellular mechanisms and genetic factors. Future Neurol 11: 63–76.

Dajani AS (1991). Current status of nonsuppurative complications of group A streptococci. Pediatr Infect Dis J 10: S25–S27.

Dale RC (2013). Immune-mediated extrapyramidal movement disorders, including Sydenham chorea. Handb Clin Neurol 112: 1235–1241.

Dale RC, Church AJ, Cardoso F et al. (2001). Poststreptococcal acute disseminated encephalomyelitis with basal ganglia involvement and auto-reactive antibasal ganglia antibodies. Ann Neurol 50: 588–595.

Dale RC, Church AJ, Surtees RAH et al. (2002). Post-Streptococcal autoimmune neuropsychiatric disease presenting as paroxysmal dystonic choreoathetosis. Mov Disord 17: 817–820.

Daoud AS, Zaki M, Shakir R et al. (1990). Effectiveness of sodium valproate in the treatment of Sydenham's chorea. Neurology 40: 1140–1141.

Dean SL, Singer HS (2017). Treatment of Sydenham's chorea: a review of the current evidence. Tremor Other Hyperkinet Mov (N Y) 7: 456.

de Figueiredo ET, Azevedo L, Rezende ML et al. (2019). Rheumatic fever: a disease without color. Arq Bras Cardiol 113: 345–354.

de Teixeira AL, Cardoso F, Maia DP et al. (2009). Frequency and significance of vocalizations in Sydenham's chorea. Parkinsonism Relat Disord 15: 62–63.

DiFazio MP, Morales J, Davis R (1998). Acute myoclonus secondary to group A β-hemolytic streptococcus infection: a PANDAS variant. J Child Neurol 13: 516–518.

Dileepan T, Smith ED, Knowland D et al. (2016). Group A streptococcus intranasal infection promotes CNS infiltration by streptococcal-specific Th17 cells. J Clin Invest 126: 303–317.

Edwards MJ, Dale RC, Church AJ et al. (2004). A dystonic syndrome associated with anti-basal ganglia antibodies. J Neurol Neurosurg Psychiatry 75: 914–916.

Eftychiadis AC (2001). Saint Vitus and his dance. J Neurol Neurosurg Psychiatry 70: 14.

Ellis NMJ, Li Y, Hildebrand W et al. (2005). T cell mimicry and epitope specificity of cross-reactive T cell clones from rheumatic heart disease. J Immunol 175: 5448–5456.

Eshel G, Lahat E, Azizi E et al. (1993). Chorea as a manifestation of rheumatic fever—a 30-year survey (1960–1990). Eur J Pediatr 152: 645–646.

Fujinami RS, Oldstone MB, Wroblewska Z et al. (1983). Molecular mimicry in virus infection: crossreaction of measles virus phosphoprotein or of herpes simplex virus protein with human intermediate filaments. Proc Natl Acad Sci 80: 2346–2350.

Fusco C, Spagnoli C (2018). Corticosteroid treatment in Sydenham's chorea. Eur J Paediatr Neurol 22: 327–331.

Fusco C, Ucchino V, Frattini D et al. (2012). Acute and chronic corticosteroid treatment of ten patients with paralytic form of Sydenham's chorea. Eur J Paediatr Neurol 16: 373–378.

Garvey MA, Snider LA, Leitman SF et al. (2005). Treatment of Sydenham's chorea with intravenous immunoglobulin, plasma exchange, or prednisone. J Child Neurol 20: 424–429.

GBD 2013 Mortality and Causes of Death Collaborators (2015). Global, regional, and national age–sex specific all-cause and cause-specific mortality for 240 causes of death, 1990–2013: a systematic analysis for the Global Burden of Disease Study 2013. Lancet 385: 117–171.

Genel F, Arslanoglu S, Uran N et al. (2002). Sydenham's chorea: clinical findings and comparison of the efficacies of sodium valproate and carbamazepine regimens. Brain Dev 24: 73–76.

Gewitz MH, Baltimore RS, Tani LY et al. (2015). Revision of the Jones Criteria for the diagnosis of acute rheumatic fever in the era of Doppler echocardiography. Circulation 131: 1806–1818.

Ghandehari K, Izadi-Mood Z (2007). Khorasan stroke registry: analysis of 1392 stroke patients. Arch Iran Med 10: 327–334.

Giedd JN, Rapoport JL, Kruesi MJP et al. (1995). Sydenham's chorea: magnetic resonance imaging of the basal ganglia. Neurology 45: 2199–2202.

Green LN (1978). Corticosteroids in the treatment of Sydenham's chorea. Arch Neurol 35: 53–54.

Gregorowski C, Lochner C, Martin L et al. (2016). Neuropsychological manifestations in children with Sydenham's chorea after adjunct intravenous immunoglobulin and standard treatment. Metab Brain Dis 31: 205–212.

Gupta A, Bhatia R, Sharma G et al. (2015). Predictors of ischemic stroke in rheumatic heart disease. J Stroke Cerebrovasc Dis 24: 2810–2815.

Gurkas E, Karalok ZS, Taskin BD et al. (2016). Predictors of recurrence in Sydenham's chorea: clinical observation from a single center. Brain Dev 38: 827–834.

Harel L, Zecharia A, Straussberg R et al. (2000). Successful treatment of rheumatic chorea with carbamazepine. Pediatr Neurol 23: 147–151.

Harrison NA (2004). Late recurrences of Sydenham's chorea are not associated with anti-basal ganglia antibodies. J Neurol Neurosurg Psychiatry 75: 1478–1479.

He VYF, Condon JR, Ralph AP et al. (2016). Long-term outcomes from acute rheumatic fever and rheumatic heart disease. Circulation 134: 222–232.

Hitchens RA (1958). Recurrent attacks of acute rheumatism in school-children. Ann Rheum Dis 17: 293–302.

Holland TL, Baddour LM, Bayer AS et al. (2016). Infective endocarditis. Nat Rev Dis Primers 2: 16059.

Hounie AG, Pauls DL, Mercadante MT et al. (2004). Obsessive-compulsive spectrum disorders in rheumatic fever with and without Sydenham's chorea. J Clin Psychiatry 65: 994–999.

Korn-Lubetzki I, Brand A, Steiner I (2004). Recurrence of Sydenham chorea: implications for pathogenesis. Arch Neurol 61: 1261–1264.

Krakauer M, Law I (2009). FDG PET brain imaging in neuropsychiatric systemic lupus erythematosis with choreic symptoms. Clin Nucl Med 34: 122–123.

Krisher K, Cunningham M (1985). Myosin: a link between streptococci and heart. Science 227: 413–415 80.

Krishnamurthi RV, Feigin VL, Forouzanfar MH et al. (2013). Global and regional burden of first-ever ischaemic and haemorrhagic stroke during 1990-2010: findings from the Global Burden of Disease Study 2010. Lancet Glob Health 1: e259–e281.

Lee JL, Naguwa SM, Cheema GS et al. (2009). Acute rheumatic fever and its consequences: a persistent threat to developing nations in the 21st century. Autoimmun Rev 9: 117–123.

Lessof MH, Bywaters EG (1956). The duration of chorea. Br Med J 1: 1520–1523.

Loiselle CR, Singer HS (2001). Genetics of childhood disorders: XXXI. Autoimmune disorders, part 4: is Sydenham chorea an autoimmune disorder? J Am Acad Child Adolesc Psychiatry 40: 1234–1236.

Mercadante MT, do Rosario Campos MC, Marques-Dias MJ et al. (1997). Vocal tics in Sydenham's chorea. J Am Acad Child Adolesc Psychiatry 36: 305–306.

Miyake CY, Gauvreau K, Tani LY et al. (2007). Characteristics of children discharged from hospitals in the United States in 2000 with the diagnosis of acute rheumatic fever. Pediatrics 120: 503–508.

Negi PC, Sondhi S, Rana V et al. (2018). Prevalence, risk determinants and consequences of atrial fibrillation in rheumatic heart disease: 6 years hospital based-Himachal Pradesh- Rheumatic Fever/Rheumatic Heart Disease (HP-RF/RHD) registry. Indian Heart J 70: S68–S73.

Nordal B, Nielsen J, Gudmu E (1999). Chorea in juvenile primary antiphospholipid syndrome: CASE REPORT. Scand J Rheumatol 28: 324–327.

Oliveira PM, Cardoso F, Maia DP et al. (2010). Acoustic analysis of prosody in Sydenham's chorea. Arq Neuropsiquiatr 68: 744–748.

Oosterveer DM, Overweg-Plandsoen WCT, Roos RAC (2010). Sydenham's chorea: a practical overview of the current literature. Pediatr Neurol 43: 1–6.

Parnaby MG, Carapetis JR (2010). Rheumatic fever in indigenous Australian children. J Paediatr Child Health 46: 527–533.

Paz JA, Silva CAA, Marques-Dias MJ (2006). Randomized double-blind study with prednisone in Sydenham's chorea. Pediatr Neurol 34: 264–269.

Pena J, Mora E, Cardozo J et al. (2002). Comparison of the efficacy of carbamazepine, haloperidol and valproic acid in the treatment of children with Sydenham's chorea: clinical follow-up of 18 patients. Arq Neuropsiquiatr 60: 374–377.

Punukollu M, Mushet N, Linney M et al. (2016). Neuropsychiatric manifestations of Sydenham's chorea: a systematic review. Dev Med Child Neurol 58: 16–28.

Ridel KR, Lipps TD, Gilbert DL (2010). The prevalence of neuropsychiatric disorders in Sydenham's chorea. Pediatr Neurol 42: 243–248.

Sahin S, Cansu A (2015). A new alternative drug with fewer adverse effects in the treatment of Sydenham chorea: Levetiracetam efficacy in a child. Clin Neuropharmacol 38: 144–146.

Schechter DC (1975). St. Vitus' dance and rheumatic disease. N Y State J Med 75: 1091–1102.

Shikhman AR, Greenspan NS, Cunningham MW (1993). A subset of mouse monoclonal antibodies cross-reactive with cytoskeletal proteins and group A streptococcal M proteins recognizes N-acetyl-beta-D-glucosamine. J Immunol 151: 3902–3913.

Sokol MS (2000). Infection-triggered anorexia nervosa in children: clinical description of four cases. J Child Adolesc Psychopharmacol 10: 133–145.

Steer AC, Carapetis JR (2009). Prevention and treatment of rheumatic heart disease in the developing world. Nat Rev Cardiol 6: 689–698.

Swedo SE, Leonard HL, Garvey M et al. (1998). Pediatric autoimmune neuropsychiatric disorders associated with streptococcal infections: clinical description of the first 50 cases. Am J Psychiatry 155: 264–271.

Swedo SE, Leonard HL, Rapoport JL (2004). The pediatric autoimmune neuropsychiatric disorders associated with streptococcal infection (PANDAS) subgroup: separating fact from fiction. Pediatrics 113: 907–911.

Teixeira ALJ, Maia DP, Cardoso F (2005a). Treatment of acute Sydenham's chorea with methyl-prednisolone pulse-therapy. Parkinsonism Relat Disord 11: 327–330.

Teixeira A, Meira F, Maia D et al. (2005b). Migraine headache in patients with Sydenham's chorea. Cephalalgia 25: 542–544.

The Lancet Neurology (2010). Neurological burden of acute rheumatic fever and rheumatic heart disease: the case for action. Lancet Neurol 9: 447.

Tibazarwa KB, Volmink JA, Mayosi BM (2008). Incidence of acute rheumatic fever in the world: a systematic review of population-based studies. Heart 94: 1534–1540.

Traill Z, Pike M, Byrne J (2008). Sydenham's chorea: a case showing reversible striatal abnormalities on CT and MRI. Dev Med Child Neurol 37: 270–273.

van Toorn R, Weyers HH, Schoeman JF (2004). Distinguishing PANDAS from Sydenham's chorea: case report and review of the literature. Eur J Paediatr Neurol 8: 211–216.

Vasconcelos LPB, Vasconcelos MC, Nunes MDCP et al. (2019). Sydenham's chorea: an update on pathophysiology, clinical features and management. Expert Opin Orphan Drugs 7: 501–511. https://doi.org/10.1080/21678707.2019.1684259.

Veasy LG, Wiedmeier SE, Orsmond GS et al. (1987). Resurgence of acute rheumatic fever in the intermountain area of the United States. N Engl J Med 316: 421–427.

Walker K, Brink A, Lawrenson J et al. (2012). Treatment of Sydenham chorea with intravenous immunoglobulin. J Child Neurol 27: 147–155.

Wilcox JA, Nasrallah HA (1986). Sydenham's chorea and psychosis. Neuropsychobiology 15: 13–14.

Wilcox A, Nasrallah H (1988). Sydenham's chorea and psychopathology. Neuropsychobiology 19: 6–8.

Williams KA, Swedo SE (2015). Post-infectious autoimmune disorders: Sydenham's chorea, PANDAS and beyond. Brain Res 1617: 144–154.

Zomorrodi A (2006). Sydenham's chorea in Western Pennsylvania. Pediatrics 117: e675–e679.

Handbook of Clinical Neurology, Vol. 177 (3rd series)
Heart and Neurologic Disease
J. Biller, Editor
https://doi.org/10.1016/B978-0-12-819814-8.00003-2

Chapter 4

Neurologic complications of nonrheumatic valvular heart disease

SALVADOR CRUZ-FLORES*

Department of Neurology, Paul L Foster School of Medicine, Texas Tech University Health Sciences Center, El Paso, TX, United States

Abstract

Valvular heart disease (VHD) is frequently associated with neurologic complications. Cerebral embolism is the most common, since thrombus formation results from the abnormalities in the valvular surfaces and the anatomic and physiologic changes associated with valve dysfunction, including atrial or ventricular enlargement, intracardiac thrombi, and cardiac dysrhythmias. Prosthetic heart valves, particularly mechanical valves, are very thrombogenic, which explains the high risk of thromboembolism and the need for long-term anticoagulation. Transcatheter aortic valve replacement (TAVR) has emerged as a nonoperative alternative to surgical aortic valve replacement for patients with intermediate or high surgical risk, and the procedure also has a risk of cerebral ischemia. In addition, anticoagulation, the mainstay of treatment to prevent cerebral embolism, has known potential for hemorrhagic complications. The emergence of new oral anticoagulants with similar effectiveness to warfarin and a better safety profile has facilitated the management of patients with atrial fibrillation. However, their application in patients with mechanical heart valves is still evolving. The prevention and management of these complications requires an understanding of their natural history to balance the risks posed by valvular heart disease, as well as the risks and benefits associated with the treatment.

Valvular heart disease is a term indicating the presence of an anatomic or functional disruption of the heart valves resulting in valvular stenosis and/or valvular insufficiency. In addition, shearing forces on the surface of the valves can lead to platelet aggregation and the formation of thrombi or the seeding of bacteria or other microorganisms, resulting in infective or noninfective endocarditis. Finally, severe myocardial dysfunction with dilated cardiomyopathy can also result in stretching and distortion of the valves or the subvalvular structures sufficient to cause valvular insufficiency or regurgitation (Boudoulas et al., 1994, 2013).

Valvular heart disease was frequently caused by infection conditions such as rheumatic fever and syphilis during the first half of the 20th century. The advent of antibiotics and extended life expectancy have resulted in a change in the epidemiology regarding the causes of valvular heart disease. In fact, longer life expectancy has resulted in an increasing prevalence of ischemic heart disease with cardiomyopathy, atherosclerosis resulting in calcific deposits leading to valvular calcification, and more effective cardiac surgery with the use of prosthetic heart valves, both bioprosthetic and mechanical, among other disease processes (Boudoulas et al., 1994, 2013; Supino et al., 2006).

Some studies have found that the prevalence of valvular heart disease in adults increases with age such that the prevalence is <1% in subjects under 45 years of age and >15% among those older than 75 years of age (Singh et al., 1999; Nkomo et al., 2006). In contrast, in the Framingham study the prevalence among those 50–60 years of age was 19% for mitral regurgitation and 13%–14.8% for aortic regurgitation (Singh et al., 1999).

The causes of valvular heart disease are multiple and include heritable connective tissue disorders, such as

*Correspondence to: Salvador Cruz-Flores, MD, MPH, Professor and Founding Chair, Department of Neurology, Paul L Foster School of Medicine, Texas Tech University Health Sciences Center El Paso, 4100 Alberta Ave, El Paso TX, 79912, United States. Tel: +1-915-215-5904, Fax: +1-915-215-5969, E-mail: salvador.cruz-flores@ttuhsc.edu

Marfan syndrome and Ehlers–Danlos syndrome; congenital valvular heart disease, such as bicuspid aortic valve disease and endocardial cushion defects; inflammatory/immunologic disorders, such as rheumatic fever and systemic lupus erythematosus; infective and noninfective endocarditis; myocardial disorders; neoplastic disorders, such as carcinoid, myxoma, and fibroelastoma; iatrogenic causes, as in surgical valve repair or valve replacement; drugs and physical agents, such as radiation; infiltrative disorders, such as hypereosinophilic syndrome; and finally, idiopathic disease (Boudoulas et al., 1994, 2013). Some of these categories are explored in other chapters of this volume. This chapter describes the most common causes of valvular heart disease in the present era, namely calcific aortic stenosis and mitral valve prolapse.

The most common neurologic complication of valvular heart disease is cerebral embolism and ischemic stroke, with its built-in complications and the complications related to acute treatment and/or stroke prevention strategies. In fact, as many as 10% of all patients with valvular heart disease have cardioembolic strokes (Cerebral Embolism Task Force, 1986). Platelet thrombi and red thrombi that result from multiple mechanisms including valvular rough surfaces, anatomic and physiologic changes related to heart valve dysfunction secondary to atrial or ventricular enlargement, intracardiac thrombi, and cardiac dysrhythmias in addition to prosthetic heart valves, explain the high frequency of embolism from valvular heart disease. Other complications arise from medical measures required to prevent stroke in patients with valvular heart disease, particularly anticoagulation, with its risk of brain hemorrhage.

CALCIFIC VALVE DISEASE

Calcific aortic stenosis and mitral annular calcification (MAC) are primarily seen in the elderly. An early study showed an incidence of MAC of 8.5% (Pomerance, 1970), although in later autopsy studies, MAC was seen in about 30% of 100 elderly subjects (McKeoun, 1975; Lauzier, 1987). The primary mechanisms for its development were initially thought related to shear injury with degeneration. However, recent evidence suggests that the process is similar to the pathogenesis of atherosclerosis in that common risk factors such as arterial hypertension and atherosclerosis in association with inflammatory endothelial processes lead to a chain of events that culminate in atherosclerosis, fibrosis, and calcification of the valves and the annulus (Boudoulas et al., 2013).

More importantly, MAC, aortic annular calcification (AAC), and aortic valve sclerosis (AVS) have been associated with the presence of cerebral infarctions and white matter disease through magnetic resonance imaging (MRI). In the Cardiovascular Health Study (CHS) 2680 participants without history of stroke or transient ischemic attack had MRI and echocardiography. Their mean age was 74.5 years and they were 39% men. MAC was found in 40%, AAC in 44.3%, and AVS in 53.3%. The prevalence of any calcification was 77%. Age, creatinine, male sex, white race, history of coronary disease, N-terminal pro B-type natriuretic peptide (NT-proBNP), and cystatin C were all associated with the presence of calcifications, although total cholesterol did not show an association with the presence of calcifications. Of note, 26.6% of the participants had one or more cerebral infarctions and 6% of them had severe white matter disease. Whether the presence of calcifications is a surrogate marker for cerebral ischemia or is involved as a causal agent requires further study, to decide if intervention in this process as a method of stroke prevention is needed (Rodriguez et al., 2011).

Aortic stenosis

Calcific aortic stenosis of a normal or congenitally bicuspid aortic valve is the most common cause of aortic valve disease in developed countries. The disease process resembles the natural course of atherosclerosis with lipid accumulation, inflammatory response, and calcification (Boudoulas et al., 1994, 2013; Rajamannan et al., 2007; Bonow et al., 2008). Valve calcification develops by the fourth or fifth decade in patients with bicuspid aortic valves, while calcific aortic stenosis usually occurs in normal tricuspid valves in the sixth through eighth decades (Carabello and Crawford, 1997). Clinical pathologic studies show that embolism from calcific aortic valve disease is not uncommon. In a series of 81 subjects with calcific aortic stenosis, 33% had emboli (Soulie et al., 1969). In another autopsy study of 165 subjects with calcific valve disease, 22% had emboli and 32 individuals had emboli in the coronary arteries, while emboli were found in the renal arteries in 11, the central retinal artery in 1, and in the middle cerebral artery in another (Holley et al., 1963a,b). Calcium emboli have frequently been found in the retinal vessels, where their appearance is as small white densities. Retinal calcium embolism is uncommon but is often associated with calcific aortic valve disease. In a small series of 24 patients with retinal calcium emboli, 9 (38%) had calcific aortic valve stenosis (Ramakrishna et al., 2005). In another series of 103 patients with retinal artery occlusions, 11 patients had aortic stenosis (Wilson et al., 1979). Aortic valve surgery, in particular percutaneous endovascular intervention, is associated with an incidence of embolism as high as 61% (Holley et al., 1963a). Despite the frequency of calcific emboli in the brain and retina found in autopsy studies, the frequency of symptomatic cerebral ischemia resulting from this type of emboli is rather low; the reason for this discrepancy is unclear, although some authors have hypothesized that the small size of the embolic particles is responsible. Therefore, in the absence of other indications such as AF or prosthetic heart valves, antithrombotic therapy is not recommended for stroke prevention in patients with calcified aortic valve disease. Antiplatelet agents should be used for secondary stroke prevention, as is recommended for patients with noncardiogenic ischemic stroke (Whitlock and Eikelboom, 2012).

Mitral annulus calcification

MAC occurs most commonly among the elderly and is a degenerative disorder with calcification of the support system and mitral valve annulus. MAC was found in 27% of 100 elderly patients in an autopsy series (McKeoun, 1975; Lauzier, 1987). MAC is associated with an increased risk of ischemic stroke. In a case control study of 151 patients with brain and retinal ischemia, 8 had MAC, compared with none of the matched age and gender controls (de Bono and Warlow, 1979). In the cohort of 1159 individuals in the Framingham study, 160 subjects had MAC; the incidence of ischemic stroke was 13.8% among those with MAC compared with 5.1% among those without MAC, for a relative risk of 2.10 (CI 95%, 1.24, 3.57) (Benjamin et al., 1992). Moreover, the frequency of stroke increased in correlation with the severity of MAC, with each millimeter of thickness as seen on the echocardiogram increasing the relative risk of stroke by 1.24. The embolic material causing stroke in MAC can be composed of calcium or can be thrombotic. The current recommendations for stroke prevention are to treat patients with MAC and embolism with antiplatelet agents. Long-term anticoagulation does not play a role in the secondary prevention of stroke in this condition. Mitral valve replacement should be considered in patients with recurrent embolism despite antiplatelet therapy (Fulkerson et al., 1979; Nestico et al., 1984; Kizer et al., 2005; Lansberg et al., 2012).

MITRAL VALVE PROLAPSE

Mitral valve prolapse (MVP) is a condition defined by echocardiography as classic MVP when there is a superior displacement of the mitral leaflets of more than 2 mm during systole and a maximal leaflet thickness of at least 5 mm during diastasis; in comparison, nonclassic MVP is defined by a leaflet thickness <5 mm. In the Framingham study, which included 1845 women and 1646 men followed for over 10 years, the overall prevalence of MVP was 2.4% (1.3% classic and 1.1% nonclassic) (Freed et al., 1999). Patients with MVP often seek consultation for a variety of symptoms that occur as part of this syndrome, including atypical chest pain, dyspnea and/or fatigue, and neuropsychiatric complaints that include panic attacks (Fontana et al., 1991; Bonow et al., 2008). Although early series suggested a causal association between MVP and stroke, more recently that association has not been supported. In a study of 213 patients 45 years or younger with cerebral ischemia and 263 matched controls, MVP was present in 1.9% of patients compared to 2.7% of controls, OR 0.70 (CI 95% 0.12, 2.5) (Gilon et al., 1999). Nevertheless, the Framingham study showed that MVP confers an excess risk of cerebral embolism with ischemia, which is 7% at 10 years compared with the expected rate of 3.2% (Avierinos et al., 2003). Thus it is still unclear what the causal role of MVP is among patients with ischemic stroke. Considering this uncertainty and the lack of evidence of benefit of anticoagulation for the prevention of embolic events among patients with MVP, antiplatelet agents are the only recommended therapy for secondary stroke prevention (Whitlock and Eikelboom, 2012).

PROSTHETIC HEART VALVES

Cardiac valve prostheses were developed in the early 1960s. There are many different designs but they can be categorized into two groups: bioprosthetic valves and mechanical valves. Bioprosthetic heart valves are usually heterografts from pig or cow pericardial or heart valve tissue, which are then mounted on a mechanical frame and have low thrombogenesis; therefore they have the advantage that they do not require anticoagulation with warfarin, but they tend to develop time-related structural failure. In contrast, mechanical heart valves are made of metal and carbon alloy, which provides structural stability, but these valves are particularly thrombogenic and require long-term anticoagulation with warfarin (Vongpatanasin et al., 1996; Chikwe and Filsoufi, 2011).

Prosthetic heart valves are associated with complications, some of which are considered of critical importance, including: structural valve deterioration, nonstructural dysfunction, valve thrombosis, embolism, bleeding events, and infection (endocarditis) (Chikwe and Filsoufi, 2011). Although many of these complications can have an impact on the nervous system, the most common neurologic complications are thrombosis with cerebral embolism, hemorrhage related to anticoagulation, and infective endocarditis. Prosthetic heart valves can develop thrombosis leading to valve dysfunction, but more importantly to systemic or cerebral thromboembolism. Thrombosis is more common in mechanical heart valves than in bioprosthetic heart valves, and is more common in the mitral valve than in the aortic valve. The estimated risk of valve thrombosis has been previously reported as high as 5.7% per year (Metzdorff et al., 1984). However, more recent estimates are 0.2% per year for mechanical heart valves and less than 0.1% for bioprosthetic heart valves (Hammermeister et al., 2000; Chikwe and Filsoufi, 2011). In a study of 575 patients undergoing aortic and mitral valve replacement in which patients were randomized to receive a Bjork-Shiley mechanical heart valve or a Hancock porcine heart valve, they noted a valve thrombosis rate of 1%–2%. More importantly, they did not find differences in the rate between valve type or location. No difference has been found in the rate of thrombosis between bileaflet and tilting disc valves (Bonow et al., 2008; Chikwe and Filsoufi, 2011). The average risk of thromboembolism resulting from mechanical heart valves is estimated at 4% a year in the absence of anticoagulation treatment. The risk is decreased to 2% a year by antiplatelet agents and to 1% by warfarin anticoagulation (Cannegieter et al., 1994a; Chikwe and Filsoufi, 2011).

Thromboembolism rates are associated with anticoagulation practices. In a study of 541 patients with a follow-up of 20 years, there were 158 embolic events in 121 patients (23%); the mean occurrence of embolism at 20 years was 24% for mechanical heart valves as compared to 39.2% for bioprosthetic heart valves in the aortic position, in contrast to 53.4% for mechanical versus 32% for bioprosthetic heart valves in the mitral position (Oxenham et al., 2003). In the Veterans Affairs Cooperative Study on Valvular Heart Disease, the embolism rates were comparable at 18% for either type of valve in the aortic position as compared to 18% for mechanical and 22% for bioprosthetic valves in the mitral position (Hammermeister et al., 2000). However, the thromboembolism rate in patients fully anticoagulated was in the range of 0.5%–1% per year (Cannegieter et al., 1994a; Grunkemeier et al., 2000).

Bioprosthetic heart valves

The risk of thromboembolism in patients with bioprosthetic heart valves and normal sinus rhythm is on average 0.7% per year. The risk is greater in patients with a valve in the mitral position than in the aortic position. The risk is generally greater early in the course after the implantation (first 3 months) before the valve is fully endothelialized (Bonow et al., 2008). Because of the high early risk, anticoagulation with unfractionated heparin is often used in the first few days overlapping with warfarin until the INR is within therapeutic range. After the first 3 months, warfarin can be discontinued in as many as 60% of patients; the rest often have to stay on warfarin due to high risk factors such as AF or previous thromboembolism (Bonow et al., 2008).

Aortic bioprosthetic valves

Evidence to date suggests that anticoagulation does not decrease the risk of embolism and increases nonsignificantly the risk of hemorrhage. In a retrospective study of 185 patients (109 on anticoagulation and 76 with no anticoagulation) and followed for the first 3 months post-valve implant, the risk of stroke was 7.4% among those anticoagulated compared to 6.5% among those not anticoagulated (RR 0.99 CI 95% 0.92, 1.06). The bleeding complication rate was the same in both groups (Moinuddeen et al., 1998). In another retrospective study of patients with aortic bioprosthetic heart valve implant, 103 patients were treated with warfarin, 509 with aspirin, and 136 received no antithrombotic therapy, and the rate of hemorrhage was 16.7%, 3.4%, and 3.1%, respectively. The risk of thromboembolism was 0.8% among those treated with aspirin compared to 2.9% and 1.5% in those treated with warfarin and on no antithrombotics, respectively (Blair et al., 1994). Two clinical trials compared antiplatelet therapy vs warfarin anticoagulation and showed no therapeutic effect of warfarin over trifusal in one, RR 1.98 (CI 95% 0.51, 7.68) (Aramendi et al., 1998, 2003; Colli et al., 2007). Although the quality of these studies is low, the current recommendation for patients with an aortic bioprosthetic heart valve replacement who have no other indication for anticoagulation is aspirin during the first 3 months postreplacement (Whitlock and Eikelboom, 2012; Whitlock et al., 2012). With regards to endovascular aortic valve replacement with a bioprosthesis, the evidence is very limited; however, since this procedure is considered an extension of coronary artery stenting, the current antithrombotic prophylaxis for embolism is a combination of aspirin and clopidogrel (Whitlock and Eikelboom, 2012; Whitlock et al., 2012).

Mitral bioprosthetic valves

The risk of thromboembolism after mitral valve bioprosthetic replacement is very high in the early postoperative period. The overall risk is 55% between days 1 and 10; 10% between days 11 and 90; and 2.4%/year thereafter. The risk is significantly decreased by anticoagulation. In fact, the risk on and off anticoagulation is 50% vs 60% within the first 10 days, 10% vs 13% between days 11 and 90; and 2.5% vs 3.9% at more than 90 days (Heras et al., 1995). There is no randomized controlled trial (RCT) supporting the use of anticoagulation in the first 3 months postmitral valve replacement. Despite the low quality of the evidence, the current recommendation is for patients with bioprosthetic mitral valve replacement to be anticoagulated during the first 3 months postvalve replacement (Whitlock and Eikelboom, 2012; Whitlock et al., 2012).

The long-term risk of thromboembolism and stroke in patients with bioprosthetic valves is 0.2%–2.6% per year, with the lowest risk among patients with aortic valve replacement; thus it is currently recommended to administer aspirin for patients with a bioprosthetic valve replacement (Cohn et al., 1981; Whitlock and Eikelboom, 2012; Whitlock et al., 2012). It is important to note that when atrial fibrillation (AF) coexists with the valve replacement, the risk of embolism is as high as 16% at 36 months and, therefore, anticoagulation with warfarin is indicated when atrial fibrillation coexists. Other associated factors potentially increasing the risk of thromboembolism include a low ejection fraction, enlarged atrium, hypercoagulable state, and history of thromboembolism. Patients with any of these conditions, even in the absence of AF, should receive warfarin in addition to aspirin therapy (Cohn et al., 1981; Goldsmith et al., 1998; Gonzalez-Lavin et al., 1984; Nunez et al., 1984; Whitlock and Eikelboom, 2012; Whitlock et al., 2012).

Mechanical heart valves

The risk of thrombosis and embolism increases from the time of valve implantation. Prosthetic materials and injured perivalvular tissue lead to platelet aggregation. Dacron sewing rings are prime material for platelet activation. In a large systematic review including 46 studies between 1970 and 1992, studying 13,088 patients with a total 53,647 patient-years of follow-up, the incidence of major embolism including stroke was 4 per 100 patient-years among patients not receiving antithrombotic therapy, compared to 2.2 per 100 patient-years among those receiving antiplatelet therapy and 1 per 100 patient-years in patients receiving anticoagulation with warfarin (Cannegieter et al., 1994a,b). In the same study, the risk was twice as high among patients with mitral valve prosthesis as compared to aortic valve prosthesis. In addition, bileaflet and disc tilting valves had a lower incidence than caged ball valves.

Generally, all patients with mechanical heart valves require warfarin anticoagulation. Aspirin in addition to warfarin is recommended in all patients with mechanical prosthetic heart valves. For heart valves in the aortic position, the target INR is 2–3, although for disc valves and Starr-Edwards valves the recommended INR is 2.5–3.5. A target INR of 2.5–3.5 is also recommended for patients with mechanical heart valves in the aortic position and at high risk defined by the presence of AF, low ejection fraction, hypercoagulable state, and history of thromboembolism (Bonow et al., 2008; Whitlock and Eikelboom, 2012; Whitlock et al., 2012). For mechanical prostheses in the mitral position the recommended target INR is 2.5–3.5 for all valve types given the higher risk of thromboembolism (Bonow et al., 2008; Whitlock and Eikelboom, 2012; Whitlock et al., 2012).

The use of bridging therapy with either unfractionated heparin or low molecular weight heparin is rather controversial and based on observational studies. Although the thromboembolic risk is certainly increased early after the implantation of the prosthetic heart valve, the reported risk of thromboembolism with bridging therapy with either agent is about 0.6%–1.1% with a bleeding risk of 3.3%–7.2%. Therefore bridging therapy is recommended after valve implantation until INR becomes therapeutic (Bonow et al., 2008;Whitlock and Eikelboom, 2012; Whitlock et al., 2012).

Recently, a phase 2 dose-validation randomized clinical trial testing several doses of dabigatran vs warfarin in patients with mechanical valves in the aortic or mitral position or both was stopped prematurely after the enrollment of 252 patients due to the excess thromboembolic (5% vs 0%) and bleeding (27% vs 12%) events among patients receiving dabigatran (Eikelboom et al., 2013). Therefore, for the long-term prevention of thrombosis and thromboembolic events, it is recommended to treat patients with warfarin anticoagulation in addition to aspirin. For individuals that are allergic to aspirin, the addition of clopidogrel is appropriate (Bonow et al., 2008; Whitlock and Eikelboom, 2012; Whitlock et al., 2012).

Transcatheter aortic valve replacement

In recent years transcatheter aortic valve replacement (TAVR) has emerged as a less-invasive therapy to replace aortic valves. This therapy was initially reserved for patients with severe aortic stenosis (AS) who were nonoperative candidates, yet TAVR is now a Class I recommendation for patients with symptomatic AS for patients that are high surgical risk and a Class IIa recommendation for patients at intermediate surgical risk. As with other cardiac surgery, the main neurologic complications pertain to ischemic stroke. Comparatively, the 30-day risk of stroke from surgical aortic valve replacement has been reported at 0.7%–6.5% while the risks for TAVR have been reported at 3%–6% with most complications occurring procedurally (Ng et al., 2018). Evidence also suggests that as many as 84% of patients may have silent ischemic embolic lesions after the procedure. While there are options for embolic protection devices, a meta-analysis of 16 studies with 1170 patients did not find differences in the rate of clinical stroke (Tay et al., 2011; Alassar et al., 2015; Bagur et al., 2017; Reardon et al., 2017; Ng et al., 2018).

ANTICOAGULATION AND ITS COMPLICATIONS

Since valvular thrombosis with systemic or cerebral embolism represents an important complication of valvular heart disease, antithrombotic therapy is an important aspect of the treatment. To summarize, the current recommendations for antithrombotic therapy are (Whitlock and Eikelboom, 2012; Whitlock et al., 2012):

1. Warfarin anticoagulation:
   a. Rheumatic mitral valve disease with normal sinus rhythm and a left atrial diameter >55 mm. Target INR 2–3
   b. Rheumatic mitral valve disease associated with left atrial thrombus. Target INR 2–3.
   c. Rheumatic valve disease associated with AF
   d. Bioprosthetic mitral valve in the first 3 months from implantation
   e. Mechanical heart valves
2. Unfractionated heparin full dose or subcutaneous low molecular weight heparin
   a. Nonbacterial thrombotic endocarditis and systemic or cerebral emboli
   b. Mechanical valves and normal sinus rhythm until INR is therapeutic with warfarin therapy
3. Antiplatelet agents
   a. Rheumatic mitral valve disease and normal sinus rhythm with normal sized left atrium (<55 mm)
   b. Aortic bioprosthetic valves and normal sinus rhythm (aspirin 50–100 mg/day) plus clopidogrel
   c. Bioprosthetic valves and normal sinus rhythm beyond 3 months postoperatively
   d. Mitral or aortic valve repair

Anticoagulation is contraindicated in patients with native valve infective endocarditis. In patients with mechanical heart valve endocarditis, anticoagulation should be stopped for the first 1 or 2 weeks after diagnosis and early antibiotic treatment. There are a few special circumstances that are worth mentioning.

Recurrent embolic events during optimal anticoagulation

When there is convincing evidence of recurrent embolic events despite optimal anticoagulation, it has been recommended to increase the dose of warfarin to a higher INR target. Another consideration is the addition of aspirin or increase of the aspirin dose when it is already being used in combination with warfarin. These recommendations should be put in place as long as it is considered safe (Bonow et al., 2008; Whitlock and Eikelboom, 2012; Whitlock et al., 2012).

Interruption of warfarin therapy for invasive procedures

A frequent clinical scenario is the need for invasive procedures among patients on long-term anticoagulation, from dental procedures to noncardiac surgery. The risks of bleeding during the procedure and the immediate perioperative period have to be weighed against the risk of thromboembolism.

In general, antithrombotic therapy should not be stopped for procedures during which the risk of bleeding is negligible, such as skin surgery, dental cleaning, or simple caries procedures. For those procedures with potential for significant bleeding complications, the antithrombotic treatment will have to be modified (Bonow et al., 2008). Since the risk of thromboembolic events in patients with mechanical heart valves not taking warfarin is 10%–20% annually, the inference is that the risk for stopping anticoagulation for 3 days is about 0.08%–0.16% (Bonow et al., 2008). In patients with mechanical aortic valves and no added risk factors for thromboembolism (AF, low ejection fraction, history of thromboembolism), it is possible to stop warfarin for 2 or 3 days and restart it 24 h after the procedure (Kearon and Hirsh, 1997; Chikwe and Filsoufi, 2011). In contrast, patients at high risk of thromboembolic events such as those patients with mechanical heart valves in the mitral position or patients with aortic valve prosthesis and risk factors such as AF, low ejection fraction, history of thromboembolism, or a hypercoagulable state, bridging therapy with unfractionated heparin at therapeutic doses is recommended and should be started when the INR is <2. Heparin should be stopped 4–6 h before the procedure and should be restarted once it is deemed safe after the procedure. While alternatives include therapeutic subcutaneous doses of unfractionated heparin or low molecular weight heparin, the evidence in their favor is not as strong (Kovacs et al., 2004; Bonow et al., 2008).

Emergent use of anticoagulation during acute ischemic stroke

Multiple studies to date have demonstrated that anticoagulation in acute ischemic stroke does not improve the outcome, and although it decreases the rate of deep vein thrombosis and pulmonary embolism, it increases the rate of hemorrhagic complications, including intracranial hemorrhage (Paciaroni et al., 2007; Sandercock et al., 2009). Prevention of early recurrent stroke was considered the primary indication for emergent anticoagulation, as the risk of early recurrence was initially estimated to be 1% per day and as high as 14% in the first 2 weeks after a stroke. However, randomized clinical trials with anticoagulation showed that the risk of stroke early recurrence is much lower, at 1.1%–4.9%, among patients not anticoagulated, with a cardiac source of embolism (Adams, 2002).

An important issue in the emergent use of anticoagulation during ischemic stroke refers to its safety. Hemorrhagic transformation of an ischemic infarct is a known potential complication, and all antithrombotics can be associated with it. Clinical trials testing unfractionated heparin and a variety of low molecular weight heparin have shown that the risk of hemorrhagic transformation ranges from 0.6% to 6.1%. Furthermore, the risk of symptomatic hemorrhagic transformation of ischemic stroke is associated with stroke severity (NIHSS >15) and high doses of anticoagulation.

While acute anticoagulation to improve the outcome of stroke is not indicated, early anticoagulation for secondary stroke prevention may be needed in situations of high risk of reembolism, as is the case with mechanical valves. In atrial fibrillation, the stroke guidelines suggest that it is reasonable to start anticoagulation within 14 days from stroke onset, although the optimal time is not established (Kernan et al., 2014). Although there is general agreement that large strokes are associated with a higher risk of bleeding, there are no studies looking specifically at CT findings as predictive of hemorrhagic complications (Adams, 2002). Although in the past decade, new oral anticoagulants (NOACs) have emerged as an alternative to warfarin for anticoagulation with a similar or better profile for stroke prevention, while having a lower rate of hemorrhagic complications, the optimal time for anticoagulation after a cardioembolism from AF or mechanical valves is still not known (Ruff et al., 2014). Patients with mechanical heart valves do have a higher risk of embolic events and therefore early anticoagulation after an ischemic stroke may be necessary. In these circumstances, it is recommended to repeat a CT scan in the first 2–3 days and start anticoagulation, provided there is no evidence of hemorrhagic transformation. After intracranial hemorrhage, anticoagulation may need to be withheld for as long as 2 weeks (Adams, 2002; Ferro, 2003). Although there are no randomized controlled trials that can answer the question of timing in cases of mechanical heart valves with large ischemic strokes, and in particular with intracranial hemorrhage, it is suggested that the window to start anticoagulation after the ischemic or hemorrhagic event may be around the 14-day mark, since the number of hemorrhagic events tends to be higher earlier than 10–14 days and ischemic events tend to increase beyond the second week (Flint et al., 2015; Kuramatsu et al., 2018). Unfortunately, as of the writing of this chapter, NOACs are not a viable or approved option for anticoagulation in patients with mechanical valves; while the door is still open to prove their benefit in stroke prevention for patients with prosthetic valves, more research is needed (Eikelboom et al., 2013; Aimo et al., 2018).

LAMBL'S EXCRESCENCES

Lambl's excrescences (LEs) are filiform fronds that appear at sites of valvular closure. They occur usually at sites of minor endothelial damage caused by wear and tear. Pathologically, these fronds are acellular and are fibrous covered by layering of acid mucopolysaccharide matrix. Their confluence and adherence can lead to their increase in size into larger masses. They were first described in 1856 by Vilem Dusan Lambl, a Bohemian physician. Given their location, their association as a potential cause of stroke is possible, and suspect. A 2018 review of the literature found 27 cases reported to date. In that series the mean age was 51% and 55% of the patients were younger than 55 years. The LEs were found most commonly by transesophageal echo and they were located in the aortic valve (73%) (Kariyanna et al., 2018). In a case-controlled study of patients with embolic strokes of undetermined source (ESUS), LEs were found in 8.9% of 393 patients. Most of the cases were identified by transesophageal echocardiogram. Interestingly enough, LEs did not have association with ESUS (Salehi Omran et al., 2020). At present, the actual prevalence of LEs among patients with stroke is not known; although they may be a potential embolic source, the evidence to date is not convincing and thus no specific aggressive stroke prevention strategy such as anticoagulation can be advocated.

References

Adams Jr. HP (2002). Emergent use of anticoagulation for treatment of patients with ischemic stroke. Stroke 33: 856–861.

Aimo A, Giugliano RP, De Caterina R (2018). Non-vitamin K antagonist oral anticoagulants for mechanical heart valves: is the door still open? Circulation 138: 1356–1365.

Alassar A, Soppa G, Edsell M et al. (2015). Incidence and mechanisms of cerebral ischemia after transcatheter aortic valve implantation compared with surgical aortic valve replacement. Ann Thorac Surg 99: 802–808.

Aramendi JL, Agredo J, Llorente A et al. (1998). Prevention of thromboembolism with ticlopidine shortly after valve repair or replacement with a bioprosthesis. J Heart Valve Dis 7: 610–614.

Aramendi JI, Mestres CA, Campos V et al. (2003). Triflusal versus oral anticoagulation for primary prevention of thromboembolism after bioprosthetic valve replacement (TRAC): rationale and design for a prospective, randomized, co-operative trial. Interact Cardiovasc Thorac Surg 2: 170–174.

Avierinos JF, Brown RD, Foley DA et al. (2003). Cerebral ischemic events after diagnosis of mitral valve prolapse: a community-based study of incidence and predictive factors. Stroke 34: 1339–1344.

Bagur R, Solo K, Alghofaili S et al. (2017). Cerebral embolic protection devices during ttranscatheter aortic valve implantation: systematic review and meta-analysis. Stroke 48: 1306–1315.

Benjamin EJ, Plehn JF, D'agostino RB et al. (1992). Mitral annular calcification and the risk of stroke in an elderly cohort. N Engl J Med 327: 374–379.

Blair KL, Hatton AC, White WD et al. (1994). Comparison of anticoagulation regimens after Carpentier-Edwards aortic or mitral valve replacement. Circulation 90: II214–9.

Bonow RO, Carabello BA, Chatterjee K et al. (2008). 2008 Focused update incorporated into the ACC/AHA 2006 guidelines for the management of patients with valvular heart disease: a report of the American College of Cardiology/American Heart Association Task Force on Practice Guidelines (Writing Committee to Revise the 1998 Guidelines for the Management of Patients With Valvular Heart Disease): endorsed by the Society of Cardiovascular Anesthesiologists, Society for Cardiovascular Angiography and Interventions, and Society of Thoracic Surgeons. Circulation 118: e523–e661.

Boudoulas H, Vavuranakis M, Wooley CF (1994). Valvular heart disease: the influence of changing etiology on nosology. J Heart Valve Dis 3: 516–526.

Boudoulas KD, Borer JS, Boudoulas H (2013). Etiology of valvular heart disease in the 21st century. Cardiology 126: 139–152.

Cannegieter SC, Rosendaal FR, Briet E (1994a). Thrombo embolic and bleeding complications in patients with mech anical heart valve prostheses. Circulation 89: 635–641.

Cannegieter SC, Van Der Meer FJ, Briet E et al. (1994b). Warfarin and aspirin after heart-valve replacement. N Engl J Med 330: 507–508; author reply 508-9.

Carabello BA, Crawford Jr FA (1997). Valvular heart disease. N Engl J Med 337: 32–41.

Cerebral Embolism Task Force (1986). Cardiogenic brain embolism. Cerebral Embolism Task Force. Arch Neurol 43: 71–84.

Chikwe J, Filsoufi F (2011). Durability of tissue valves. Semin Thorac Cardiovasc Surg 23: 18–23.

Cohn LH, Mudge GH, Pratter F et al. (1981). Five to eight-year follow-up of patients undergoing porcine heart-valve replacement. N Engl J Med 304: 258–262.

Colli A, Mestres CA, Castella M et al. (2007). Comparing warfarin to aspirin (WoA) after aortic valve replacement with the St. Jude Medical Epic heart valve bioprosthesis: results of the WoA Epic pilot trial. J Heart Valve Dis 16: 667–671.

De Bono DP, Warlow CP (1979). Mitral-annulus calcification and cerebral or retinal ischaemia. Lancet 2: 383–385.

Eikelboom JW, Connolly SJ, Brueckmann M et al. (2013). Dabigatran versus warfarin in patients with mechanical heart valves. N Engl J Med 369: 1206–1214.

Ferro JM (2003). Cardioembolic stroke: an update. Lancet Neurol 2: 177–188.

Flint AC, Lingamneni R, Rao VA et al. (2015). Risks of thrombosis and rehemorrhage during early management of intracranial hemorrhage in patients with mechanical heart valves. J Am Coll Cardiol 66: 1738–1739.

Fontana ME, Sparks EA, Boudoulas H et al. (1991). Mitral valve prolapse and the mitral valve prolapse syndrome. Curr Probl Cardiol 16: 309–375.

Freed LA, Levy D, Levine RA et al. (1999). Prevalence and clinical outcome of mitral-valve prolapse. N Engl J Med 341: 1–7.

Fulkerson PK, Beaver BM, Auseon JC et al. (1979). Calcification of the mitral annulus: etiology, clinical associations, complications and therapy. Am J Med 66: 967–977.

Gilon D, Buonanno FS, Joffe MM et al. (1999). Lack of evidence of an association between mitral-valve prolapse and stroke in young patients. N Engl J Med 341: 8–13.

Goldsmith I, Lip GY, Mukundan S et al. (1998). Experience with low-dose aspirin as thromboprophylaxis for the Tissuemed porcine aortic bioprosthesis: a survey of five years' experience. J Heart Valve Dis 7: 574–579.

Gonzalez-Lavin L, Chi S, Blair TC et al. (1984). Thromboembolism and bleeding after mitral valve replacement with porcine valves: influence of thromboembolic risk factors. J Surg Res 36: 508–515.

Grunkemeier GL, Li HH, Naftel DC et al. (2000). Long-term performance of heart valve prostheses. Curr Probl Cardiol 25: 73–154.

Hammermeister K, Sethi GK, Henderson WG et al. (2000). Outcomes 15 years after valve replacement with a mechanical versus a bioprosthetic valve: final report of the Veterans Affairs randomized trial. J Am Coll Cardiol 36: 1152–1158.

Heras M, Chesebro JH, Fuster V et al. (1995). High risk of thromboemboli early after bioprosthetic cardiac valve replacement. J Am Coll Cardiol 25: 1111–1119.

Holley KE, Bahn RC, Mcgoon DC et al. (1963a). Calcific embolization associated W with valvotomy for calcific aortic stenosis. Circulation 28: 175–181.

Holley KE, Bahn RC, Mcgoon DC et al. (1963b). Spontaneous calcific embolization associated with calcific aortic stenosis. Circulation 27: 197–202.

Kariyanna PT, Jayarangaiah A, Rednam C et al. (2018). Lambl's excrescences and stroke: a scoping study. Int J Clin Res Trials 3 (2): 127. https://doi.org/10.15344/2456-8007/2018/127.

Kearon C, Hirsh J (1997). Management of anticoagulation before and after elective surgery. N Engl J Med 336: 1506–1511.

Kernan WN, Ovbiagele B, Black HR et al. (2014). Guidelines for the prevention of stroke in patients with stroke and transient ischemic attack: a guideline for healthcare professionals from the American Heart Association/American Stroke Association. Stroke 45: 2160–2236.

Kizer JR, Wiebers DO, Whisnant JP et al. (2005). Mitral annular calcification, aortic valve sclerosis, and incident stroke in adults free of clinical cardiovascular disease: the Strong Heart Study. Stroke 36: 2533–2537.

Kovacs MJ, Kearon C, Rodger M et al. (2004). Single-arm study of bridging therapy with low-molecular-weight heparin for patients at risk of arterial embolism who require temporary interruption of warfarin. Circulation 110: 1658–1663.

Kuramatsu JB, Sembill JA, Gerner ST et al. (2018). Management of therapeutic anticoagulation in patients with intracerebral haemorrhage and mechanical heart valves. Eur Heart J 39: 1709–1723.

Lansberg MG, O'donnell MJ, Khatri P et al. (2012). Antithrombotic and thrombolytic therapy for ischemic stroke: antithrombotic therapy and prevention of thrombosis, 9th ed: American College of Chest Physicians Evidence-Based Clinical Practice Guidelines. Chest 141: e601S–e636S.

Lauzier S (1987). Cerebral ischemia with mitral valve prolapse and mitral annular calcification, Springer Verlag, London.

Mckeoun EF (1975). De Senectu: the FE Williams lecture. JR Coll Physicians Lond 10: 79.

Metzdorff MT, Grunkemeier GL, Pinson CW et al. (1984). Thrombosis of mechanical cardiac valves: a qualitative comparison of the silastic ball valve and the tilting disc valve. J Am Coll Cardiol 4: 50–53.

Moinuddeen K, Quin J, Shaw R et al. (1998). Anticoagulation is unnecessary after biological aortic valve replacement. Circulation 98: II95–8; discussion II98–9.

Nestico PF, Depace NL, Morganroth J et al. (1984). Mitral annular calcification: clinical, pathophysiology, and echocardiographic review. Am Heart J 107: 989–996.

Ng VG, Kodali S, George I (2018). Transcatheter trans-septal mitral valve-in-valve implantation. Ann Cardiothorac Surg 7: 821–823.

Nkomo VT, Gardin JM, Skelton TN et al. (2006). Burden of valvular heart diseases: a population-based study. Lancet 368: 1005–1011.

Nunez L, Gil Aguado M, Larrea JL et al. (1984). Prevention of thromboembolism using aspirin after mitral valve replacement with porcine bioprosthesis. Ann Thorac Surg 37: 84–87.

Oxenham H, Bloomfield P, Wheatley DJ et al. (2003). Twenty year comparison of a Bjork-Shiley mechanical heart valve with porcine bioprostheses. Heart 89: 715–721.

Paciaroni M, Agnelli G, Micheli S et al. (2007). Efficacy and safety of anticoagulant treatment in acute cardioembolic stroke: a meta-analysis of randomized controlled trials. Stroke 38: 423–430.

Pomerance A (1970). Pathological and clinical study of calcification of the mitral valve ring. J Clin Pathol 23: 354–361.

Rajamannan NM, Bonow RO, Rahimtoola SH (2007). Calcific aortic stenosis: an update. Nat Clin Pract Cardiovasc Med 4: 254–262.

Ramakrishna G, Malouf JF, Younge BR et al. (2005). Calcific retinal embolism as an indicator of severe unrecognised cardiovascular disease. Heart 91: 1154–1157.

Reardon MJ, Van Mieghem NM, Popma JJ et al. (2017). Surgical or transcatheter aortic-valve replacement in intermediate-risk patients. N Engl J Med 376: 1321–1331.

Rodriguez CJ, Bartz TM, Longstreth Jr. WT et al. (2011). Association of annular calcification and aortic valve sclerosis with brain findings on magnetic resonance imaging in community dwelling older adults: the cardiovascular health study. J Am Coll Cardiol 57: 2172–2180.

Ruff CT, Giugliano RP, Braunwald E et al. (2014). Comparison of the efficacy and safety of new oral anticoagulants with warfarin in patients with atrial fibrillation: a meta-analysis of randomised trials. Lancet 383: 955–962.

Salehi Omran S, Chaker S, Lerario MP et al. (2020). Relationship between Lambl's excrescences and embolic strokes of undetermined source. Eur Stroke J 5: 169–173.

Sandercock PA, Gibson LM, Liu M (2009). Anticoagulants for preventing recurrence following presumed non-cardioembolic ischaemic stroke or transient ischaemic attack. Cochrane Database Syst Rev CD000248.

Singh JP, Evans JC, Levy D et al. (1999). Prevalence and clinical determinants of mitral, tricuspid, and aortic regurgitation (the Framingham Heart Study). Am J Cardiol 83: 897–902.

Soulie P, Caramanian M, Soulie J (1969). Calcified aortic stenosis; pathological anatomy. Arch Mal Coeur Vaiss 62: 1096–1118.

Supino PG, Borer JS, Preibisz J et al. (2006). The epidemiology of valvular heart disease: a growing public health problem. Heart Fail Clin 2: 379–393.

Tay EL, Gurvitch R, Wijesinghe N et al. (2011). A high-risk period for cerebrovascular events exists after transcatheter aortic valve implantation. JACC Cardiovasc Interv 4: 1290–1297.

Vongpatanasin W, Hillis LD, Lange RA (1996). Prosthetic heart valves. N Engl J Med 335: 407–416.

Whitlock RP, Eikelboom JW (2012). Prevention of thromboembolic events after bioprosthetic aortic valve replacement: what is the optimal antithrombotic strategy? J Am Coll Cardiol 60: 978–980.

Whitlock RP, Sun JC, Fremes SE et al. (2012). Antithrombotic and thrombolytic therapy for valvular disease: antithrombotic therapy and prevention of thrombosis, 9th ed: American College of Chest Physicians Evidence-Based Clinical Practice Guidelines. Chest 141: e576S–e600S.

Wilson LA, Warlow CP, Russell RW (1979). Cardiovascular disease in patients with retinal arterial occlusion. Lancet 1: 292–294.

Handbook of Clinical Neurology, Vol. 177 (3rd series)
Heart and Neurologic Disease
J. Biller, Editor
https://doi.org/10.1016/B978-0-12-819814-8.00009-3

Chapter 5

Advances and ongoing controversies in PFO closure and cryptogenic stroke

TIFFANY CHENG[1], JOAQUIN B. GONZALEZ[2], AND FERNANDO D. TESTAI[1]*

[1]*Department of Neurology and Rehabilitation, University of Illinois at Chicago, Chicago, IL, United States*

[2]*Cardiology Department, Advocate Illinois Masonic Medical Center, Chicago, IL, United States*

Abstract

Approximately one-third of strokes are cryptogenic in origin. These patients have a higher prevalence of patent foramen ovale (PFO) compared to individuals with stroke of known origin. It has been proposed that some cryptogenic strokes (CSs) can be caused by paradoxical embolism across a PFO. PFOs can be treated medically with antithrombotic agents and percutaneously with occluder devices. Large randomized clinical trials have found transcatheter PFO closure to be superior to medical treatment for the prevention of recurrent stroke in young patients with CS. However, the superiority of PFO closure over medical treatment in unselected populations has not been demonstrated. In this chapter, we review the evidence supporting PFO closure and the selection of patients for such intervention.

INTRODUCTION

In fetal development, a foramen ovale shunts placental oxygenated blood from the right atrium to the left atrium allowing bypass of the high-resistance pulmonary vasculature. After birth, the lungs expand, resulting in a drop of resistance in the pulmonary circulation and right atrium. Blood then begins to circulate through the pulmonary vasculature, cardiac shunting decreases, and the septum primum and septum secundum fuse. This causes the interatrial septum to close, leaving a remnant called the foramen ovale. Occasionally, this foramen ovale fails to close completely, resulting in an atrial septal abnormality called a patent foramen ovale (PFO). Based on data obtained from autopsy studies, nearly 25% of the general population has a PFO. This PFO can function as a valve, permitting shunting of blood from the right-to-left atrium. Most PFOs in the general population are found incidentally. These are typically small and only become apparent after provocative maneuvers that increase the intrathoracic pressure, such as Valsalva or cough. However, it has been suggested that a type of stroke, called a cryptogenic stroke (CS), can be caused by paradoxical embolism through a PFO.

CS refers to an ischemic stroke that is not attributable to lacunar infarction, large-artery atherosclerosis, cardioembolism, or other identifiable etiology after vascular, serologic, and cardiac workup (Adams Jr et al., 1993). According to the Trial of Org10172 in Acute Stroke Treatment (TOAST) statement, CS also encompasses strokes where the evaluation is incomplete or more than one mechanism has contributed (Adams Jr et al., 1993). However, within these definitions, there are no universally accepted criteria as to what constitutes a complete evaluation for stroke. With this limitation in mind, it is estimated that approximately 30% of all strokes are cryptogenic and that 63% of the recurrent events remain cryptogenic after extensive workup (Hart et al., 2014). Compared to individuals with strokes of known etiologies, the Oxford Vascular Study found that individuals with CS have lower prevalence of traditional atherosclerotic risk factors or comorbid atherosclerotic diseases,

*Correspondence to: Fernando D. Testai, Department of Neurology and Rehabilitation, University of Illinois at Chicago, 912 S Wood St, Chicago, IL 60647, United States. Tel: +1-312-996-1047; Fax: +1-312-413-8215, E-mail: testai@uic.edu

characterized by myocardial infarction, peripheral vascular disease, and carotid artery occlusive disease (Li et al., 2015). This suggests that CS is a distinct stroke subtype, which could have unique pathogenic mechanisms and require targeted interventions. This concept is further supported by information obtained in observational studies. In patients <55 years old with CS, the prevalence of PFO was higher compared to patients with stroke of known etiology (40% vs 15%; $P < 0.001$), and a similar observation has been made for patients >55 years old (28% for CS vs 12% for non-CS: $P < 0.001$) (Handke et al., 2007). These data not only suggest that CS could be due to paradoxical embolism through a PFO but they also support the notion that stroke recurrence may be prevented by targeted strategies, such as percutaneous PFO closure or medical therapy. Studies examining these two preventative approaches have found discordant results. An observational study of 581 patients with stroke of unknown origin demonstrated that PFO alone does not affect the rate of recurrent stroke (Mas et al., 2001). In comparison, a meta-analysis of 48 observational studies showed that the rate of recurrent neurologic events after CS or transient ischemic attack (TIA) is almost six times higher with medical therapy compared to percutaneous PFO closure (Agarwal et al., 2012). These discrepant results may be explained by differences in both patient selection and cardiac anatomy. To this end, several cardioanatomic characteristics, such as atrial septum hypermobility, shunt severity, and associated atrial septal aneurysm (ASA), have been found to increase the stroke recurrence risk in patients with PFO. ASA, in particular, is a >10–15 mm bulging of the atrial septum into the right atrium, left atrium, or both due to redundant atrial septal tissue. Although it is far less common than PFO, it is often found incidentally in healthy individuals. Observational studies have shown that the association of PFO with ASA increases the odds of initial stroke 5 times and the risk of recurrent stroke almost 24 times (Mas and Zuber, 1995; Mas et al., 2001; Piechowski-Jozwiak and Bogousslavsky, 2013) The pathogenesis of stroke in ASA is not completely understood and different hypotheses have been considered, including paradoxical embolism through a coexistent PFO, associated atrial arrhythmias, and/or thrombus formation in the aneurysmal sac.

Six randomized clinical trials have compared the efficacy of percutaneous PFO closure to medical treatment for the prevention of recurrent stroke patients with cryptogenic infarction or TIA (Table 5.1). In this chapter, we will review these six trials, the approaches used to diagnose PFO and cryptogenic embolism, the evidence supporting percutaneous PFO closure in CS, and the selection of patients that may benefit from such intervention.

DIAGNOSIS AND CHARACTERIZATION OF PFO

PFO is typically an incidental finding that does not result in left-to-right shunt-related complications. However, PFO can also been found in association with stroke, migraine, platypnea-orthodeoxia, and provoked exercise desaturation. The likelihood of a PFO to be associated with an embolic event is influenced by volume status and hemodynamic fluctuations that occur with respiration and intrathoracic pressure changes. In addition, several anatomic factors, such as a hypermobile septum primum, known as ASA, increased shunt size, prominent eustachian valve or Chiari network, increase the risk of stroke in patients with PFO. Thus, it is important to capture these structural details on imaging.

Cardiac catheterization is the most accurate method for the diagnosis of PFO. The crossing of guidewire from the right to the left atrium through the septum confirms unequivocally the presence of an interatrial communication. In addition, cardiac catheterization is necessary for the treatment of PFO with percutaneous occluder devices. Despite its accuracy though, additional noninvasive studies, such as transthoracic echocardiogram (TTE), transesophageal echocardiogram (TEE), cardiac CT, or cardiac MR, are commonly required to plan the procedure (Hara et al., 2005). These supplemental imaging techniques provide essential details that are not captured by cardiac angiography, such as the type of the defect and its anatomic characteristics.

Echocardiography is the most commonly ancillary study used for the diagnosis of PFO. Studies comparing right heart catheterization to TEE have found that TEE has a sensitivity of 91%–100% and an accuracy of 88%–97%. TEE with color Doppler and agitated saline injection, or TEE bubble study, is the gold standard for the diagnosis of PFO. This modality allows the evaluation of the anatomy, size, and specific location of the defect as well as the identification of coexisting atrial septal defects or pulmonary shunts. In particular, the shunt size can be assessed by direct measurement of the PFO length or by counting the number of microbubbles that cross the interatrial septum after 3–6 cardiac cycles. Valsalva maneuver, cough, or other provocative maneuvers that transiently increase the pressure in the right atrium can augment the crossing of microbubbles and make the PFO more apparent. Different criteria have been developed to quantify the PFO size. In general, PFOs are considered large when they measure more than 2 mm in length and/or are associated with more than 20 microbubbles in the left atrium after opacification of the right atrium (Horlick et al., 2019). Large PFOs, ASAs, and other anatomic findings, such as prominent eustachian valves or Chiari networks, are easily confirmed with TEE.

Table 5.1

Randomized clinical trials comparing PFO closure to medical treatment for the prevention of recurrent stroke in patients with cryptogenic cerebral ischemia

Study	*n*	Major inclusion criteria	Interventions	Primary outcome	Average follow up (years)	Results hazard ratio or relative risk (95% Confidence interval)	Safety
CLOSURE (Furlan et al., 2012)	909	– 18–60 years – Cryptogenic stroke or TIA – PFO	– PFO closure with STARFlex device – Antithrombotics (antiplatelet agents and/or anticoagulation)	Stroke or TIA, death from any cause during 30 days of the procedure or neurologic death	2	*PFO closure*: 23 events (5.5%) *Antithrombotics*: 29 events (6.8%) 0.78 (0.45–1.35); $P=0.37$[a]	Atrial fibrillation *PFO closure*: 5.7% *Antithrombotics*: 0.7% ($P<0.001$)
PC-Trial (Meier et al., 2013)	414	– <60 years – Cryptogenic stroke or peripheral thromboembolism – PFO	– PFO closure with Amplatzer PFO occluder – Antithrombotics (antiplatelet agents or anticoagulation)	Death, nonfatal stroke, TIA, or peripheral embolism	4.1	*PFO closure*: 9 events (3.4%) *Antithrombotics*: 18 events (5.2%) 0.63 (0.24–1.62); $P=0.34$[a]	Atrial fibrillation *PFO closure*: 2.9% *Antithrombotics*: 1.0% (P 0.16)
RESPECT (Carroll et al., 2013)	980	– 18–60 years – Cryptogenic ischemic stroke – PFO identified by transesophageal echocardiography	– PFO closure with Amplatzer PFO occluder – Antithrombotics (antiplatelet agents or anticoagulation)	Nonfatal ischemic stroke, fatal ischemic stroke, or early death after randomization	2.6	*PFO closure*: 9 events (0.66 events per 100 patients-years) *Antithrombotics*: 16 events (1.38 events per 100 patient-years) 0.49 (0.22–1.11); $P=0.08$[a]	Atrial fibrillation *PFO closure*: 3.0% *Antiplatelet*: 1.5% (P 0.13)
RESPECT Long-term outcomes (Saver et al., 2017)					5.9	*PFO closure*: 18 events (0.58 events per 100 patient-years) *Antiplatelet*: 28 events (1.07 events per 100 patient-years) 0.55 (0.31–0.999); $P=0.046$[a]	
REDUCE (Sondergaard et al., 2017)	664	– 18–59 years – Cryptogenic ischemic stroke within 180 days – PFO with a right-to-left shunt	– PFO closure with Helex Septal occluder or Cardioform Septal occluder – Antiplatelet alone	– Clinical ischemic stroke – New brain infarction (clinical or silent brain infarction on brain imaging)	3.2	*Clinical ischemic stroke* *PFO closure*: 6 strokes (1.4%) *Antiplatelet*: 12 strokes (5.4) 0.23 (0.09–0.62); $P=0.002$[a] *New brain infarction* *PFO closure*: 22 events (5.7%) *Antiplatelet*: 20 events (11.3%) 0.51 (0.29–0.91); $P=0.04$[b]	Atrial fibrillation or flutter *PFO closure*: 6.6% *Antiplatelet*: 0.4% $P<0.001$

Continued

Table 5.1

Continued

Study	*n*	Major inclusion criteria	Interventions	Primary outcome	Average follow up (years)	Results hazard ratio or relative risk (95% Confidence interval)	Safety
CLOSE (Mas et al., 2017)	663	– 16–60 years – Cryptogenic stroke – PFO with associated atrial septal aneurysm or large interatrial shunt[c]	– PFO closure with any device approved by the Interventional Cardiology Committee – Antiplatelet alone – Anticoagulation (warfarin [INR 2–3] or direct oral anticoagulants)	Fatal or nonfatal stroke	5.3	*Closure vs antiplatelet* *PFO closure*: no strokes (0%) *Antiplatelet*: 14 strokes (4.9%) 0.03 (0–0.26); $P<0.001$[a] *Anticoagulation vs antiplatelet* *Anticoagulation*: 3 strokes (1.5%) *Antiplatelet*: 7 strokes (3.8%) 0.44 (0.11–1.48); $P=0.17$[a]	Atrial fibrillation or flutter *PFO closure*: 4.6% *Antiplatelet*: 0.9% *P* 0.02 Major of fatal bleeding *Anticoagulation*: 5.3% *Antiplatelet*: 2.3% *P* 0.18
DEFENSE-PFO (Lee et al., 2018)	120	– No age limit (mean age: 52 years) – Cryptogenic stroke – High-risk PFO (atrial septal aneurysm, hypermobility of the atrial septum, or large PFO)[d]	– PFO closure with Amplatzer PFO occluder – Antithrombotics (antiplatelet agents or anticoagulation)	Stroke, vascular death, or major bleeding	2.8	*Primary outcome* *PFO closure*: no events (0%) *Antithrombotics*: 6 events (12.9%) $P<0.013$[e] Ischemic stroke *PFO closure*: no events (0%) *Antithrombotics*: 5 events (10.5%) $P<0.023$[e]	Atrial fibrillation or flutter *PFO closure*: 2 event (2.2%) *Antiplatelet*: none *P*=not provided Major of fatal bleeding *Anticoagulation*: 0% *Antiplatelet*: 2 events (2.5%) *P* 0.15

Reproduced with permission from Gonzalez JB, Testai FD (2021). Advances and ongoing controversies in patent foramen ovale closure and cryptogenic stroke. Neurol Clin 39: 51–69.
[a]Large PFO defined as ≥2 mm.
[b]Results expressed as hazard ratio (95% confidence interval).
[c]HR not provided due to early termination resulting in small sample size than originally planned.
[d]Large shunt defined as >30 microbubbles on TTE or TEE detected either spontaneously or during provocation maneuvers.
[e]Results expressed as relative risk (95% confidence interval).
Abbreviations: *CLOSE*, The PFO Closure or Anticoagulants vs Antiplatelet Therapy to Prevent Stroke Recurrence; *CLOSURE*, Evaluation of the STARFlex Septal Closure System in Patients with a Stroke or TIA Due to the Possible Passage of a Clot of Unknown Origin Through a PFO; *DEFENSE-PFO*, Device Closure vs Medical Therapy for CS Patients With High-Risk PFO; *HR*, hazard ratio; *PC-trial*, PFO and Cryptogenic Embolism; *REDUCE*, The GORE Septal Occluder Device for PFO Closure in Stroke Patients; *RESPECT*, Randomized Evaluation of Recurrent Stroke Comparing PFO Closure to Established Current Standard of Care Treatment.

Characterizing the PFO and any relevant anatomic details contribute to risk stratification and the planning of the closure procedure, including the selection of the type, size, and number of devices. In comparison to TEE, TTE with bubble study has a sensitivity for the diagnosis of PFO of approximately 50%. Furthermore, TTE lacks the ability to provide the detailed anatomic characteristics that can be obtained from a properly done TEE (Mojadidi et al., 2014).

Transcranial Doppler (TCD) with injection of agitated saline, or TCD bubble study, is another noninvasive imaging method that can be used for the diagnosis of right-to-left shunts. The benefits of TCD include high sensitivity and low cost. However, TCD cannot determine the precise location of the shunt (intracardiac vs extracardiac) and does not provide anatomic details that can help guide definite treatment (Mojadidi et al., 2014; Mahmoud et al., 2017). Other potential noninvasive imaging options include cardiac CT and cardiac MRI. Although both have excellent accuracy, they are not routinely used due to their high costs. In addition, cardiac CT exposes patients to radiation and IV contrast with the associated risk of nephrotoxicity (Horlick et al., 2019).

Different groups and institutions have proposed various algorithms for the diagnosis of PFOs and some of these recommend the initial screening with noninvasive and less expensive tests, such as TTE or TCD. However, due to the limitations aforementioned, a second imaging study, such as TEE, is commonly required (Mojadidi et al., 2014; Horlick et al., 2019). Thus, from the practical standpoint, it has been proposed that TEE with color Doppler and bubble study is the best first study for the workup of cardioembolism as it is safe, highly accurate, and provides essential information required in planning for PFO closure.

DIAGNOSIS OF CRYPTOGENIC STROKE

CS is a nonlacunar embolic stroke that typically occurs in individuals who lack traditional risk factors for atherosclerotic disease and have a negative workup for defined causes of stroke, including procoagulability, artery-to-artery thromboembolism, and cardioembolism. As mentioned previously, in its most traditional definition, the term CS is also used for cases with incomplete evaluation and those with competing stroke mechanisms. CS often affects patients younger than 60. However, there is evidence that CS, particularly in the setting of PFO, can also occur in older individuals (Mazzucco et al., 2018). CS is a diagnosis of exclusion that requires discarding other defined causes of cerebral ischemia. The shortcoming of this definition is that there is lack of consensus on what constitutes a reasonable stroke workup. A three-tier diagnostic algorithm has been proposed (Fig. 5.1) (Saver and Clinical Practice, 2016). However, this may be expensive and uncertainties exist in relation to the yield and cost effectiveness of pursuing such evaluations in all the patients. More recently, the concept of embolic stroke of undetermined source (ESUS) has emerged (Hart et al., 2014). In comparison to CS, the diagnosis of ESUS requires completing a minimum number of investigations. These are: (1) brain CT or brain MR showing a nonlacunar infarct, (2) precordial echocardiography to exclude the possibility of a defined cardioembolic source such as valvulopathy or intracardiac tumor, (3) cardiac monitoring for at least 24h to exclude AFib, and (4) imaging of the extracranial and intracranial cerebral arteries looking for evidence of large-artery pathology that could be responsible for the stroke. TEE is not necessary for the diagnosis ESUS. Thus, a shortcoming of ESUS is that the left atrium is partially characterized and the aortic arch is not assessed. Despite this limitation, ESUS constitutes a subset of patients with CS who underwent a standardized minimal workup to exclude major causes of stroke.

PFO CLOSURE VS MEDICAL TREATMENT

Until recently, the treatment of CS in patients with PFO relied on the use of antithrombotic agents. However, six randomized clinical trials have shed light in the role of PFO closure in selected populations. CLOSURE (Evaluation of the STARFlex® Septal Closure System in Patients With a Stroke or TIA Due to the Possible Passage of a Clot of Unknown Origin Through a Patent Foramen Ovale) was a multicenter, randomized, and open-label trial that examined the effect of percutaneous closure of PFO plus medical therapy vs medical therapy alone for the prevention of recurrent stroke in patients with CS (Furlan et al., 2012). A total of 909 patients aged 18–60 with recent stroke of unknown origin or TIA were randomized to one of these two groups and followed for up to 2 years. In the closure group, patients were treated with the STARFlex Septal occluder device plus clopidogrel 75mg daily for 6 months and aspirin (81 or 325mg daily) for 2 years. Patients in the medical group were treated with aspirin (81 or 352mg daily), warfarin (INR goal of 2.0–3.0), or both, depending on the preference of the treating physician. The primary endpoint of stroke or TIA occurred in 5.5% of patients in the closure group and 6.8% of those in the medical group, a difference that was not statistically different (adjusted hazard ratio [HR] 0.78, 95% confidence interval [CI] 0.45–1.35; $P=0.37$). Based on these findings, the authors concluded that PFO closure plus aspirin did not have a significant benefit over medical therapy alone for the prevention of recurrent events in CS patients. A major concern of this trial was that almost a quarter of strokes in the percutaneous closure group occurred within the first 30 days of the procedure. This suggests that the device placement itself could be

Tier 1	
Brain imaging	MRI brain CT of the brain
Cerebrovascular imaging	MRA of head and neck CTA of head and neck Carotid duplex Transcranial Doppler ultrasonography
Cardiac assessment	Transthoracic echocardiography Transesophageal echocardiography 12-Lead electrocardiogram Inpatient cardiac telemetry 24 h Holter monitoring
Hematologic workup	Complete blood count Prothrombin time/INR Activated partial thromboplastin time

↓ Negative workup

Tier 2	
Cerebrovascular imaging	Catheter angiography Transcranial Doppler with microemboli detection Workup for vasculitis
Cardiac assessment	Prolonged (2–4 weeks) cardiac monitoring
Hematologic workup	Hypercoagulable assessment

↓ Negative workup

Tier 3	
Genetic testing	Mitochondrial diseases Fabry disease Cerebral autosomal dominant arteriopathy with subcortical infarcts and leukoencephalopathy (CADASIL) Other genetics causes of stroke
Cardiac assessment	Cardiac CT Cardiac MR Prolonged (1–3 years) loop recording
Hematologic workup	Workup for occult malignancy

↓ Negative workup

Cryptogenic stroke

Fig. 5.1. Proposed workup for the diagnosis of cryptogenic stroke (Saver and Clinical Practice, 2016). Reproduced from Gonzalez JB, Testai FD (2021). Advances and ongoing controversies in patent foramen ovale closure and cryptogenic stroke. Neurol Clin 39: 51–69.

related to the development of recurrent events. In addition, although PFO closure was effective in 86% of patients in this group, 1.1% of patients developed a left atrial thrombus. Other limitations to this trial included a lack of standardization in the workup of CS and hindrance in subject enrollment due to patient and physician preference for PFO closure. Furthermore, the rates of atrial fibrillation (AF) significantly differed between the two groups, with rates of 5.7% in the closure group and 0.7% in the medical treatment group ($P < 0.0001$). In the closure group, more than half of the AF cases were periprocedural, suggesting that the procedure itself could predispose to AF.

PC-Trial (*Patent Foramen Ovale and Cryptogenic Embolism*) was a randomized, open-label study that compared percutaneous closure of PFO to medical therapy for the prevention of recurrent stroke in patients with ischemic stroke, TIA, or peripheral thromboembolism of unknown origin (Meier et al., 2013). The primary outcome was death, nonfatal stroke, TIA, or peripheral embolism after randomization. Patients in the closure group were treated with the Amplatzer PFO occluder device in addition to acetylsalicylic acid (100–325 mg daily) plus ticlopidine (250–500 mg daily) or clopidogrel (75–150 mg daily) for 6 months. In the medical group, patients received antiplatelet treatment or anticoagulants at the discretion of the treatment team. A total of 414 subjects less than 60 years old were randomized to the closure group or medical group, and all patients were followed for approximately 4 years. Death, nonfatal stroke, TIA, or peripheral embolism occurred in 3.4% of the closure group and 5.2% of the medical group (HR 0.63, 95% CI 0.24–1.62; $P=0.34$). The rate of stroke in the closure group (1.5%) and the medical group (2.4%) did not statistically differ among the groups ($P=0.14$). Thus, the trial concluded that PFO closure was not superior to medical therapy alone for the prevention of recurrent embolism or death after CS. When examining complications, the rate of new-onset AF did not statistically differ among the two groups, with a rate of 2.9% in the closure group and 1.0% in the medical group ($P=0.16$). Limitations of this study included elevated attrition and slow recruitment. In addition, there may have been selective reporting bias, as there was indirect evidence that mild TIAs were more likely to be reported in the medical group.

RESPECT (*Patent Foramen Ovale Closure or Medical Therapy After Stroke*) was a randomized, open-label study with blinded adjudication that compared PFO closure to medical treatment alone for the prevention of recurrent stroke in CS patients (Carroll et al., 2013). The primary endpoint was a composite of nonfatal ischemic stroke, fatal ischemic stroke, or early death after randomization. A total of 980 subjects aged 18–60 were randomized to PFO closure or medical treatment, and the mean follow up for patients was 2.6 years. In the medical group, patients could be treated with antiplatelet agents or warfarin, with no protocol for the selection of the antithrombotic. Patients in the closure group underwent treatment with the Amplatzer PFO occluder device and aspirin (81–325 mg daily) plus clopidogrel for 1 month followed by aspirin alone for 5 months. In the intention-to-treat analysis, the rate of recurrent stroke was 0.66 per 100 patient-year in the closure arm and 1.38 per 100 patient-year in the medical arm (HR $=0.49$, 95% CI 0.22–1.11; $P=0.08$). However, it is important to note that three of nine strokes in the closure arm occurred without device placement. Thus, based on the prespecified per-protocol analysis, the number of strokes was 6 in the closure group and 16 in the medical group (HR $=0.37$, 95% CI 0.14–0.96; $P=0.03$). In addition, the as-treated analysis found that five strokes occurred in the closure group and 16 strokes occurred in the medical group (HR $=0.27$, 95% CI 0.10–0.75; $P=0.007$). Although the results of the intention-to-treat analysis were not statistically significant, the overall findings of the trial suggested that PFO closure was superior to medical treatment alone for the prevention of recurrent stroke. Furthermore, the control and the interventional arms had asymmetric event rates, trending toward superiority of the PFO closure group. As a result, the US Food and Drug Administration (FDA) requested long-term outcomes in order to approve the Amplatzer occluder for PFO closure in the prevention of recurrent stroke in CS patients. In 2017, the results of the long-term data showed a recurrent stroke rate of 0.58 per 100 patient-year in the closure group and 1.07 per 100 patient-year in the medical group (HR $=0.55$, 95% CI 0.31–0.99; $P=0.046$) (Saver et al., 2017). When adjusting the rate of recurrent CS based on the TOAST criteria, the rate of recurrent stroke was 0.03 per 100 patient-year in the closure group and 0.41 per 100 patient-year in the medical group (HR $=0.08$, 95% CI 0.01–0.58; $P=0.01$). In terms of safety, both groups had similar rates of AF and other serious adverse events. RESPECT became a landmark study, demonstrating that PFO closure with the Amplatzer PFO occluder device was superior to medical treatment alone for the prevention of recurrent stroke in CS patients <60 years of age. This ultimately led to FDA approval of the Amplatzer occluder for PFO closure in young patients with CS and PFO (FDA, 2016). However, it is worth mentioning that in a post hoc analysis, about one-third of the recurrent strokes were unrelated to PFOs. This finding emphasizes the importance of extensive neurologic and cardiac workup to exclude other mechanisms of stroke and determine appropriate patient selection for closure.

REDUCE (GORE® Septal Occluder Device for Patent Foramen Ovale) Closure in Stroke Patients was an open-label trial that randomized patients with blinded adjudication to either transcatheter PFO closure plus medical therapy or medical therapy alone for the prevention of recurrent stroke after cryptogenic cerebral ischemia (Sondergaard et al., 2017). A total of 664 patients 60 years or younger participated in this study. Exclusion criteria included uncontrolled vascular risk factors or a specific indication for anticoagulation therapy. The coprimary outcomes were recurrent clinical ischemic stroke and incidence of new brain infarction, as defined by clinical ischemic stroke or silent brain infarction on imaging. In contrast to previous studies, the medical group restricted antithrombotic treatment to antiplatelet agents only. The closure group received treatment with the Helex Septal Occluder, which mid-trial, had to be replaced by the Cardioform septal occluder due to design refinements. These patients also received clopidogrel 75 mg daily for 3 days postprocedure, and subsequent antiplatelet treatment was not dictated by protocol. All patients were followed for 2–5 years. During the median follow up of 3.2 years, the rate of ischemic stroke was 1.4% in the closure group and 5.4% in the medical group (effect size [ES] $=0.23$, 95% CI 0.09–0.62; $P=0.002$). The other primary outcome of new brain infarction occurred in 5.7% of the closure group and 11.3% of the medical group (ES $=0.51$, 95% CI 0.29–0.91; $P=0.04$). Within the closure group, 1.4% of patients had serious device-associated complications, including device dislocation and thrombosis. In addition, the rate of AF or atrial flutter was higher in the PFO group than in the medical group (6.6% vs 0.4%; $P < 0.001$). Almost 80% of these adverse events occurred within 45 days of PFO closure, with 60% resolution within 2 weeks of the adverse event. Based on these findings, the Gore Caridoform Septal Occluder device was approved by the FDA for prevention of recurrent stroke in patients aged 18–60 years with PFO and CS presumed to be caused by paradoxical embolism (GORE® CARDIOFORM Septal Occluder, 2018).

The previously described studies were followed by CLOSE (PFO closure or anticoagulants vs antiplatelet therapy to prevent stroke recurrence).This was a three-arm open-label study that randomized 664 CS patients aged 16–60 years old to receive PFO closure, antiplatelet treatment alone, or oral anticoagulation for the prevention of recurrent stroke (Mas et al., 2017). A unique feature of this study was that, by design, it only allowed the inclusion of patients with PFO and associated ASA or large interatrial shunt. The primary outcome examined at the average follow up of 5.3 years was fatal or nonfatal stroke. ASA was defined as aneurysm $\geq$15 mm and excursion $>$10 mm on TEE, and large interatrial shunt was defined as more than 30 microbubbles seen on TTE or TEE detected in the left atrium. These microbubbles were either spontaneously seen or induced by provocative maneuvers. Within the closure group, the medical device used for PFO closure was not dictated by protocol. However, all devices had to be approved by the Interventional Cardiology Committee. In addition to closure, these patients were treated with ASA 75 mg daily plus clopidogrel 75 mg daily for 3 months followed by single antiplatelet treatment. The antiplatelet group received aspirin, clopidogrel, or aspirin plus extended-release dipyridamole. The oral anticoagulant group received either a vitamin K antagonist (target INR of 2–3) or direct oral anticoagulants. In the antiplatelet only group, 14 subjects experienced recurrent stroke resulting in a 5-year cumulative stroke risk of 4.9%. In comparison, no subjects in the closure group experienced stroke (HR = 0.03, 95% CI 0.00–0.26; $P < 0.001$). In the intention-to-treat analysis, the anticoagulation group had three strokes with a 5-year cumulative stroke risk of 1.5%, while the antiplatelet group had seven strokes and a 5-year cumulative stroke risk of 3.8%. These findings suggested a trend of anticoagulation superiority over antiplatelet agents for prevention of recurrent stroke in young CS patients. However, the study was underpowered to further compare these groups. As a result, additional statistical analysis was not done. In terms of safety, the rate of AF was 4.6% in the closure group and 0.9% in the antiplatelet group. Limitations of this study included slow recruitment, the open-label nature, and the relatively low number of primary events. Despite this, the results confirmed that in young patients with CS and an associated ASA or large interatrial shunt, PFO closure is superior to antiplatelet treatment for prevention of recurrent stroke. With regards to anticoagulation therapy, questions remain as to whether anticoagulation is superior to antiplatelet agents for prevention of stroke in patients with CS.

The Device Closure vs Medical Therapy for Cryptogenic Stroke Patients With High-Risk Patent Foramen Ovale (DEFENSE-PFO) study compared percutaneous PFO closure with the Amplatzer PFO occluder to antithrombotic treatment in CS patients with high-risk PFO (Lee et al., 2018). High-risk PFO was defined as PFO with associated ASA, hypermobile interatrial septum with excursion into either atrium of at least 10 mm or interatrial shunt of at least 2 mm on TEE. The authors initially planned to include 210 patients based on power analysis, but the positive results of the previously discussed studies led to the early termination of the trial. As a result, the final sample size was 120 patients. In comparison to other trials, older age was not an exclusion criterion. Nonetheless, the mean age was 51.8 years, indicating that patients were generally young. In addition, patients in the closure group were 5 years younger on average compared to the medical group, and smoking and hypercholesterolemia were more frequent in the medical group. In relation to antithrombotic treatment in the medical group, patients received either antiplatelet agents or warfarin at the discretion of the treatment team. Primary endpoint events of stroke, vascular death, or major bleeding at 2-year follow-up occurred in six patients in the medical group (2-year event rate of 12.9%). There were no events in the closure group (P 0.013). In subgroup analysis, five ischemic strokes occurred in the medical group, (2-year event rate of 10.5%) while no strokes were reported in the closure group (P 0.023). In terms of complications, two events of AF were identified in the closure group while none were reported in the medical group. Limitations of this study included slow recruitment, early termination which undermined power, low number of participating sites, and differences in stroke risk factors among the two groups, including age, smoking, and hypercholesterolemia. Despite these limitations, the results of DEFENSE-PFO support PFO closure in young patients with CS, particularly in those with high-risk features.

Different meta-analyses compared the benefits and risks associated with PFO closure. In one of them ($n = 3440$), the rate of recurrent stroke was 2.0% after closure and 4.5% after medical therapy alone (RR = 0.43, 95% CI 0.21–0.90; $P = 0.024$) (Lattanzi et al., 2018). The two groups had similar rates of TIA and mortality.

There was a substantial reduction in the relative risk of stroke of 51.5% in favor of PFO closure, but the absolute risk reduction was modest at 2.1% (Ntaios et al., 2018). In addition, the number of procedures necessary to prevent one stroke in 3.7 years was 47 (Ntaios et al., 2018). In comparison, another study reported a summary rate difference for recurrent stroke of −0.67% per year (95% CI, −0.39 to −0.94%) in favor of PFO closure with a number needed to treat of 29 to reduce one stroke at 5 years (Messé et al., 2020). In terms of serious adverse events, PFO closure was associated with a risk of major procedural complications 3.9% (95% CI 2.3%–5.7%) (Messé et al., 2020). New-onset AF or atrial flutter was reported more often in the in the closure group (4.3%) than in the medical group (0.7%; $P<0.001$) (Lattanzi et al., 2018). Similarly, pulmonary embolism was more commonly seen in patients undergoing PFO closure (0.41% per year vs 0.11% per year, P 0.04) (Messé et al., 2020). In subgroup analysis, the effect of PFO closure over medical treatment was more robust in patients with substantial shunt (RR = 0.23, 95% CI 0.11–0.48; $P<0.001$) than in patients without substantial shunt (RR = 0.94, 95% CI 0.46–1.95; P 0.874). Similarly, the superiority of PFO closure over medical treatment was also greater in patients with associated ASA (RR = 0.17, 95% CI 0.06–0.53; $P=0.002$) than in those without ASA (RR = 0.85, 95% CI 0.43–1.68; $P=0.65$). A few trials examined the role of anticoagulation for the prevention of recurrent stroke in patients with CS and PFO. However, the majority of these studies were underpowered. Three studies compared anticoagulation to antiplatelet agents ($n=503$), one study compared anticoagulation to PFO closure ($n=353$), and three studies compared mixed antiplatelet and anticoagulation treatment to PFO closure ($n=2303$). A meta-analysis of these trials did not conclusively demonstrate that anticoagulation is more effective than antiplatelet agents for secondary stroke prevention (HR 0.73, 95% CI 0.45–1.17) (Messé et al., 2020). The wide CI, however, prevent drawing solid conclusions. Importantly, the benefit of PFO closure in patients treated with anticoagulants has not been established either. However, a higher risk of hemorrhage must be considered with anticoagulant use (Mir et al., 2018).

PATIENT SELECTION

The results of the studies discussed in the previous section suggest that PFO closure should be considered for patients with CS and PFO. However, several methodological factors limit the generalizability of the lessons learned. First, the studies have several methodologic shortcomings, including open-label design, slow recruitment, low number of primary outcomes, and asymmetric dropout rates in the experimental and the control groups. In addition, the use of different inclusion and exclusion criteria prevents a head-to-head comparison of trials. Second, CLOSURE I and RESPECT had short follow-up periods and DEFENSE-PFO was terminated early. Third, two large studies were negative for the primary outcome (CLOSURE and PC) and those that provided positive results in favor of PFO closure either did not allow the use of anticoagulants (REDUCE) or were enriched with high-risk PFO patients (CLOSE and DEFENSE-PFO). Fourth, a meta-analysis of these studies showed that the beneficial effect of PFO closure is relatively modest with a number needed to treat between 29 and 47 (Ntaios et al., 2018; Messé et al., 2020). These observations have raised questions about the benefits of transcatheter interventions in unselected populations. There is also the question of procedure-associated complications that occur in approximately 4% of the cases (Messé et al., 2020). Some of these, although rare, are associated with significant mortality and morbidity. This is the case with device migration and erosion, cardiac thrombosis, hematoma at the puncture site, periprocedural embolism, and incomplete occlusion. The most commonly reported complication, however, is new-onset AF. The information available suggests that this is commonly transient and resolves within weeks of the intervention. However, properly powered long-term studies are still necessary to determine the rate of AF recurrence and the factors influencing in it. In this scenario, those that oppose transcatheter interventions claim that, even though PFO closure is beneficial in the short-term, in the long-term it may replace a stroke risk factor for another that may develop silently and be more severe.

The appropriate selection of candidates for closure is of critical importance. The first step in this process resides on establishing the diagnosis of CS with a reasonable level of certainty. After the diagnosis of CS has been determined, the next step is addressing if the PFO constitutes a pathogenic or an incidental finding. Some elements of the medical history and presentation provide clues about the stroke mechanism. For example, history of deep venous thrombosis, pulmonary embolism, or onset of symptoms after extended immobilization or during Valsalva suggests the possibility of paradoxical embolism through a PFO. In addition, the presence of a left common iliac vein compressed by the right common iliac artery may predispose to the occurrence of deep venous thrombosis and thromboembolism, particularly in young women (May–Thurner syndrome or iliac vein compression syndrome) (Ho et al., 2019). In contrast, the finding of electrocardiographic, echocardiographic, or serologic markers of left atrial dysfunction in a patient with a negative stroke workup suggests the possibility of paroxysmal AF. The presence of a CS, or ESUS, in a young patient lacking atherosclerotic risk factors increases the probability of the PFO being associated with the pathogenesis of ischemic insult. From the practical standpoint, the strength of such association can be estimated using the Risk of Paradoxical Embolism (RoPE) score. The RoPE score is composed by six variables and ranges from 0 to 10 points. The number of points correlates with the likelihood of the stroke being caused by the PFO (Table 5.2). In a study including 3674 patients with CS, the PFO-attributable fraction for the index stroke increased from 0% for scores 0–3 to 88% for scores 9–10 (Fig. 5.2) (Kent et al., 2013). It should be noted that there is no universally accepted cutoff value to identify CS patients that could benefit from PFO closure. Data obtained in a small cohort, however, suggests that the benefit of transcatheter closure is modest in patients with RoPE scores <7 (Prefasi et al., 2016). The final factor to consider is the risk of recurrent stroke. Data obtained in a single-center registry of patients considered for PFO closure show that a RoPE score of ≤6 identifies patients with a higher risk of mortality and recurrent ischemic events (Morais et al., 2018). Other factors associated with recurrent stroke in PFO patients are shown in Table 5.3 (Pristipino et al., 2019).

The 2019 European position statement on the management of PFO in CS patients emphasizes the importance of estimating the probability that the PFO was associated with the occurrence of stroke and the risk of stroke recurrence prior to deciding whether the PFO should be closed or not. It also highlights the importance of pursuing a complete evaluation looking for defined causes of stroke, in particular paroxysmal AF (Table 5.4) (Pristipino et al., 2019).

Table 5.2

Risk of paradoxical embolism (RoPE) score

Patient characteristic	Points
No history of hypertension	1
No history of diabetes mellitus	1
No history of previous stroke or TIA	1
Nonsmoker	1
Cortical infarct on imaging	1
Age (years)	
18–29	5
30–39	4
40–49	3
50–59	2
60–69	1
≥70	0

Table 5.3

Predictors of recurrent cryptogenic stroke in patients with PFO (Pristipino et al., 2019)

Variable	*n*	Number of studies	Odds or hazard ratio (95% confidence interval)
Older age	2171	4	1.47 (1.2–1.8)
Atrial septal aneurysm	630	5	3.0 (1.8–4.8)
Aspirin vs oral anticoagulants	1235	5	2.5 (1.1–6.1)
Coagulation disorders	258	2	2.75 (1.2–6.5)
Stroke at index	367	2	3.0 (1.4–6.5)
Large PFO	334	2	3.0 (1.9–4.6)

Reproduced with permission from Gonzalez JB, Testai FD (2021). Advances and ongoing controversies in patent foramen ovale closure and cryptogenic stroke. Neurol Clin 39: 51–69.

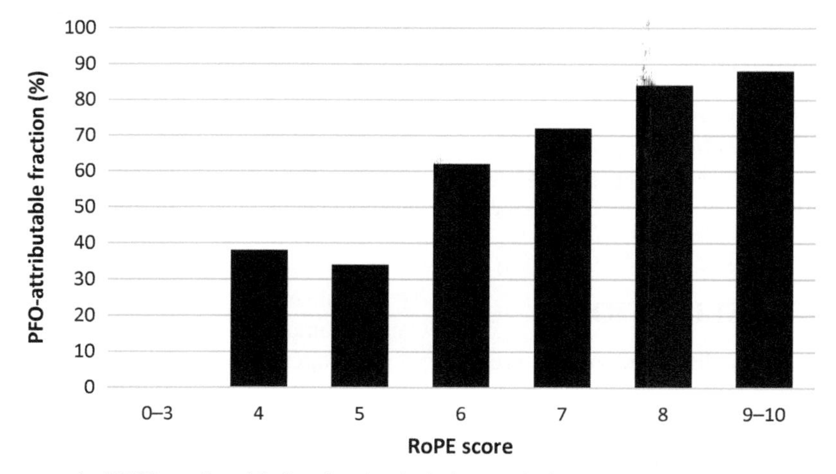

Fig. 5.2. Patent foramen ovale (PFO)-attributable fraction for the index stroke based on the Risk of Paradoxical Embolism (RoPE) score (Kent et al., 2013). Reproduced with permission from Gonzalez JB, Testai FD (2021). Advances and ongoing controversies in patent foramen ovale closure and cryptogenic stroke. Neurol Clin 39: 51–69.

Table 5.4

Highlights of the 2019 European position statement on the management of patients with patent foramen ovale (PFO) (Pristipino et al., 2019)

General management	1. Interdisciplinary assessment and decision making are recommended 2. Active involvement of the patient in the decision-making process is mandatory 3. The decision making should take into account a. Probability of a causal role of the PFO in the stroke b. Risk of stroke recurrence 4. Risk stratification should take into account clinical, anatomic, and imaging characteristics
PFO diagnosis	1. The most sensitive technique should be used as a first line for the diagnosis of PFO 2. Transcranial Doppler ultrasonography has a higher sensitivity than transthoracic echocardiography to detect right-to-left shunting 3. Transesophageal echocardiography should be performed to stratify risk
Estimating the probability of a causal role of the PFO in the stroke	1. No single clinical, anatomic, or imaging characteristics are sufficient to quantify the probability of a PFO causal role 2. PFO is more likely to have a causal role when patients are young and lack other stroke risk factors; however, the presence of vascular risk factors does not exclude a causative role of PFO 3. Atrial septal aneurysm, shunt severity, and atrial septal hypermobility are associated with a causal role of PFO 4. Simultaneous pulmonary embolism and/or deep vein thrombosis suggest a causal role of PFO 5. The RoPE score should only be part of a comprehensive individual evaluation
Estimating stroke recurrence	1. The risk of recurrent embolism in unselected patients with PFO is low 2. No single variable is sufficient to quantify the probability of stroke recurrence 3. Some variables are associated with higher recurrent stroke in PFO patients (Table 5.2) a. Older age b. Atrial septal aneurysm and/or PFO diameter c. Aspirin vs oral anticoagulants d. Coagulation disorders e. Stroke at index f. D-dimer > 1000 at admission
Atrial fibrillation rule-out strategy	1. All patients should undergo routine 12-lead ECG and either in-patient cardiac telemetry or 24-h Holter monitoring 2. In patients >65 years old with negative routine monitoring, it is reasonable to consider implantable cardiac monitoring before deciding on PFO closure or permanent oral anticoagulation 3. In patients 55–64 years old at risk for atrial fibrillation with negative routine monitoring, it is reasonable to consider implantable cardiac monitoring before deciding on PFO closure or permanent oral anticoagulation 4. In patients <55 years old with ≥2 high-risk factors for atrial fibrillation with negative routine monitoring, it is reasonable to consider implantable cardiac monitoring before deciding on PFO closure or permanent oral anticoagulation[a]
Position statements	1. Perform percutaneous closure of a PFO in carefully selected patients aged from 18 to 65 years with a confirmed cryptogenic stroke, TIA, or systemic embolism and an estimated high probability of a causal role of the PFO as assessed by clinical, anatomical, and imaging features 2. Percutaneous closure of a PFO must be proposed to each patient evaluating the individual probability of benefit based on an assessment of both the role of the PFO in the thromboembolic event and the expected results and risks of lifelong medical therapy 3. The choice of device should take into consideration that most available evidence has been obtained with the AMPLATZER™ PFO occluder or the GORE® CARDIOFORM septal occluder. The use of the latter should be balanced against a lower complete closure rate and a higher risk of atrial fibrillation as compared to medical therapy
Drug therapy after PFO closure	1. The choice of the type of antiplatelet drug in the follow-up is currently empiric 2. It is reasonable to use dual antiplatelet therapy for 1–6 months after PFO closure 3. We suggest a single antiplatelet therapy be continued for at least 5 years

Reproduced with permission from Gonzalez JB, Testai FD (2021). Advances and ongoing controversies in patent foramen ovale closure and cryptogenic stroke. Neurol Clin 39: 51–69.

[a]High-risk factors include uncontrollable hypertension, structural heart abnormalities (left ventricular hypertrophy, left atrial enlargement) uncontrollable diabetes, and congestive heart failure.

In 2020, the American Academy of Neurology (AAN) released a practice advisory report for patients with stroke and PFO (Messé et al., 2020). In agreement with the European position statement, the AAN report recommends pursuing a thorough evaluation looking for defined mechanisms of stroke. This statement also provides additional guidance on the workup to be done prior to considering PFO closure (Table 5.5). The 2019 European position statement on the management of PFO, the 2019 consensus statement of the Society for Cardiovascular Angiography and Interventions, and the 2020 AAN practice advisory report also emphasize the importance of using a multidisciplinary approach for the optimal selection of candidates for PFO closure. The team should include, at minimum, an interventional cardiologist and a neurologist with experience in the care of patients with stroke (Horlick et al., 2019). In addition, they support involving patients in the decision-making process and addressing with them risks and benefits of percutaneous interventions, areas of uncertainty, and alternative approaches. Fig. 5.3 depicts a recommended algorithm for the selection of candidates for PFO closure based on the randomized studies discussed in this chapter.

Table 5.5

Recommendations from the 2020 American Academy of Neurology practice advisory report for patients with stroke and PFO who are being considered for closure (Messé et al., 2020)

Suggested evaluations	1. Ensure that an appropriately thorough evaluation has been performed to rule-out alternative mechanisms of stroke 2. Obtain brain imaging to confirm stroke size and distribution, assessing for an embolic pattern or a lacunar infarct 3. Obtain complete vascular imaging (MRA or CTA) of the cervical and intracranial vessels to look for dissection, vasculopathy, and atherosclerosis 4. Perform a baseline ECG to look for atrial fibrillation. In patients thought to be at risk of atrial fibrillation should receive prolonged cardiac monitoring for at least 28 days 5. Assess for cardioembolic sources using TTE followed by TEE assessment if the first study does not identify a high-risk stroke mechanism 6. Perform hypercoagulable studies that would be considered a plausible high-risk stroke mechanism 7. Clinicians may use TCD-agitated saline contrast as a screening evaluation for right-to-left shunt. This does not obviate the need for TTE and TEE to rule-out alternative mechanisms of cardio embolism and confirm that right-to-left shunting is intracardiac and transseptal
Pre-operative evaluation	1. Patients should be assessed by a clinician with expertise in assessing the degree of shunting and anatomic features of a PFO, and performing PFO closure 2. Patients should be assessed by a clinician with expertise in stroke to ensure that the PFO is the most plausible mechanism of stroke 3. A shared decision-making approach between clinicians and the patient should be used
Counseling	1. Clinicians should counsel that having a PFO is common, that it is difficult to determine with certainty whether their PFO caused their stroke and that PFO closure probably reduces recurrent stroke risk in select patients 2. Clinicians may recommend closure following a discussion of potential benefits 3. Inform patients of the association of PFO size and stroke risk. Large stroke probably confer an elevated risk of stroke. Conversely, there probably is less likelihood of benefit of closure in patients with a small shunt 4. In patients who require long-term anticoagulation because of other medical condition, clinicians should counsel the patient that the efficacy of PFO closure in addition to anticoagulation cannot be confirmed or refuted
Presentations outside of the classic trial inclusion criteria	1. PFO closure may be offered in other populations, such as for a patient who is aged 60–65 years with a very limited degree of traditional vascular risk factors 2. PFO closure may be offered to younger patients (e.g., <30 years) with a single, small, deep stroke (<1.5 cm), a large shunt, and absence of any vascular risk factors
Medical management	1. In patients who opt to receive medical therapy alone without PFO closure, clinicians may recommend either an antiplatelet medication such as aspirin or anticoagulation

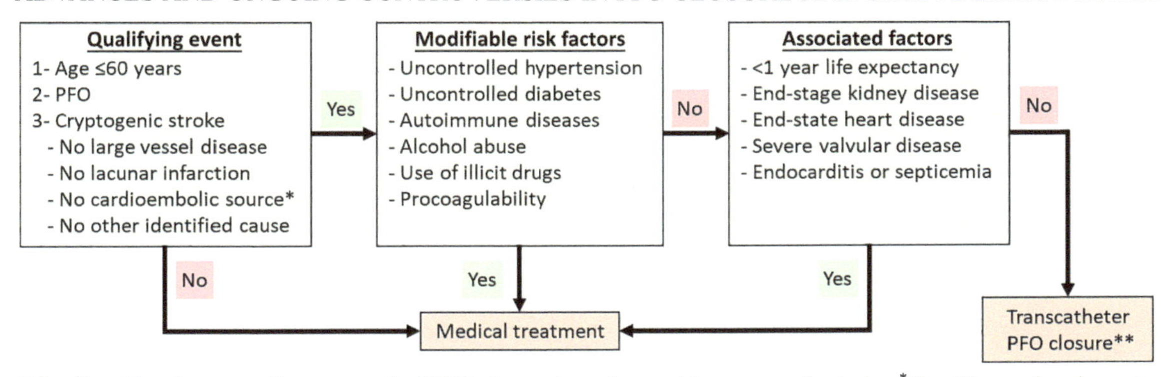

Fig. 5.3. Algorithm for patent foramen ovale (PFO) closure in patients with cryptogenic stroke. *Consider performing extended cardiac monitoring to evaluate for paroxysmal atrial fibrillation or atrial flutter. **Enhanced reasons for PFO closure include large PFO with substantial shunt and associated atrial septal aneurysm. Reproduced with permission from Gonzalez JB, Testai FD (2021). Advances and ongoing controversies in patent foramen ovale closure and cryptogenic stroke. Neurol Clin 39: 51–69.

CONCLUSIONS

PFO is commonly found in healthy adults, and its sole presence does not increase the risk of incident stroke. However, PFOs have a higher prevalence in patients with CS. Although there is insufficient evidence to support PFO closure in unselected cases, closure of the heart defect should be considered over antiplatelet treatment for the prevention of recurrent stroke in young individuals with cryptogenic ischemia, particularly in those with high-risk PFOs (i.e., atrial septal aneurysm, hypermobility of the atrial septum, or large interatrial shunt).

References

Adams Jr. H.P, Bendixen BH, Kappelle LJ et al. (1993). Classification of subtype of acute ischemic stroke. Definitions for use in a multicenter clinical trial. TOAST. Trial of Org 10172 in acute stroke treatment. Stroke 24: 35–41.

Agarwal S, Bajaj NS, Kumbhani DJ et al. (2012). Meta-analysis of transcatheter closure versus medical therapy for patent foramen ovale in prevention of recurrent neurological events after presumed paradoxical embolism. JACC Cardiovasc Interv 5: 777–789.

Carroll JD, Saver JL, Thaler DE et al. (2013). Closure of patent foramen ovale versus medical therapy after cryptogenic stroke. N Engl J Med 368: 1092–1100.

FDA 2016; FDA approves new device for prevention of recurrent strokes in certain patients. Available at: https://www.fda.gov/news-events/press-announcements/fda-approves-new-device-prevention-recurrent-strokes-certain-patients. Accessed July 02, 2019.

Furlan AJ, Reisman M, Massaro J et al. (2012). Closure or medical therapy for cryptogenic stroke with patent foramen ovale. N Engl J Med 366: 991–999.

GORE® CARDIOFORM Septal Occluder 2018; P050006/S060. Available at: https://www.fda.gov/medical-devices/recently-approved-devices/gorer-cardioform-septal-occluder-p050006s060. Accessed July 02, 2019.

Handke M, Harloff A, Olschewski M et al. (2007). Patent foramen ovale and cryptogenic stroke in older patients. N Engl J Med 357: 2262–2268.

Hara H, Virmani R, Ladich E et al. (2005). Patent foramen ovale: current pathology, pathophysiology and clinical status. J Am Coll Cardiol 46: 1768–1776.

Hart RG, Diener HC, Coutts SB et al. (2014). Embolic strokes of undetermined source: the case for a new clinical construct. Lancet Neurol 13: 429–438.

Ho A, Chung AD, Mizubuti GB (2019). A hairdresser's painful swollen left leg: artery compresses vein in May–Thurner syndrome. Lancet 394: e33.

Horlick E, Kavinsky CJ, Amin Z et al. (2019). SCAI expert consensus statement on operator and institutional requirements for PFO closure for secondary prevention of paradoxical embolic stroke: the American Academy of Neurology affirms the value of this statement as an educational tool for neurologists. Catheter Cardiovasc Interv 93: 859–874.

Kent DM, Ruthazer R, Weimar C et al. (2013). An index to identify stroke-related vs incidental patent foramen ovale in cryptogenic stroke. Neurology 81: 619–625.

Lattanzi S, Brigo F, Cagnetti C et al. (2018). Patent foramen ovale and cryptogenic stroke or transient ischemic attack: to close or not to close? A systematic review and meta-analysis. Cerebrovasc Dis 45: 193–203.

Lee PH, Song JK, Kim JS et al. (2018). Cryptogenic stroke and high-risk patent foramen ovale: the DEFENSE-PFO trial. J Am Coll Cardiol 71: 2335–2342.

Li L, Yiin GS, Geraghty OC et al. (2015). Incidence, outcome, risk factors, and long-term prognosis of cryptogenic transient ischaemic attack and ischaemic stroke: a population-based study. Lancet Neurol 14: 903–913.

Mahmoud AN, Elgendy IY, Agarwal N et al. (2017). Quantification of patent foramen ovale-mediated shunts: echocardiography and transcranial Doppler. Interv Cardiol Clin 6: 495–504.

Mas JL, Zuber M (1995). Recurrent cerebrovascular events in patients with patent foramen ovale, atrial septal aneurysm, or both and cryptogenic stroke or transient ischemic attack.

French Study Group on Patent Foramen Ovale and Atrial Septal Aneurysm. Am Heart J 130: 1083–1088.

Mas JL, Arquizan C, Lamy C et al. (2001). Recurrent cerebrovascular events associated with patent foramen ovale, atrial septal aneurysm or both. N Engl J Med 345: 1740–1746.

Mas JL, Derumeaux G, Guillon B et al. (2017). Patent foramen ovale closure or anticoagulation vs antiplatelets after stroke. N Engl J Med 377: 1011–1021.

Mazzucco S, Li L, Binney L et al. (2018). Prevalence of patent foramen ovale in cryptogenic transient ischaemic attack and non-disabling stroke at older ages: a population-based study, systematic review, and meta-analysis. Oxford vascular study phenotyped cohort. Lancet Neurol 17: 609–617.

Meier B, Kalesan B, Mattle HP et al. (2013). Percutaneous closure of patent foramen ovale in cryptogenic embolism. N Engl J Med 368: 1083–1091.

Messé SR, Gronseth GS, Kent DM et al. (2020). Practice advisory update summary: patent foramen ovale and secondary stroke prevention report of the guideline Subcommittee of the American Academy of Neurology. Neurology 94: 876–885.

Mir H, Siemieniuk RAC, Ge LC et al. (2018). Patent foramen ovale closure, antiplatelet therapy or anticoagulation in patients with patent foramen ovale and cryptogenic stroke: a systematic review and network meta-analysis incorporating complementary external evidence. BMJ Open 8: e023761-2018-023761.

Mojadidi MK, Winoker JS, Roberts SC et al. (2014). Accuracy of conventional transthoracic echocardiography for the diagnosis of intracardiac right-to-left shunt: a meta-analysis of prospective studies. Echocardiography 31: 1036–1048.

Morais LA, Sousa L, Fiarresga A et al. (2018). RoPE score as a predictor of recurrent ischemic events after percutaneous patent foramen ovale closure. Int Heart J 59: 1327–1332.

Ntaios G, Papavasileiou V, Sagris D et al. (2018). Closure of patent foramen ovale versus medical therapy in patients with cryptogenic stroke or transient ischemic attack: updated systematic review and meta-analysis. Stroke 49: 412–418.

Piechowski-Jozwiak B, Bogousslavsky J (2013). Stroke and patent foramen ovale in young individuals. Eur Neurol 69: 108–117.

Prefasi D, Martinez-Sanchez P, Fuentes B et al. (2016). The utility of the RoPE score in cryptogenic stroke patients ≤50 years in predicting a stroke-related patent foramen ovale. Int J Stroke 11: NP7–8. https://pubmed.ncbi.nlm.nih.gov/26763040/.

Pristipino C, Sievert H, D'Ascenzo F et al. (2019). European position paper on the management of patients with patent foramen ovale. General approach and left circulation thromboembolism. EuroIntervention 14: 1389–1402. PMID: 30141306. https://doi.org/10.4244/EIJ-D-18-00622.

Saver JL, Clinical Practice (2016). Cryptogenic stroke. N Engl J Med 374: 2065–2074.

Saver JL, Carroll JD, Thaler DE et al. (2017). Long-term outcomes of patent foramen ovale closure or medical therapy after stroke. N Engl J Med 377: 1022–1032.

Sondergaard L, Kasner SE, Rhodes JF et al. (2017). Patent foramen ovale closure or antiplatelet therapy for cryptogenic stroke. N Engl J Med 377: 1033–1042.

Handbook of Clinical Neurology, Vol. 177 (3rd series)
Heart and Neurologic Disease
J. Biller, Editor
https://doi.org/10.1016/B978-0-12-819814-8.00024-X

Chapter 6

Neurological complications of coronary heart disease and their management

ANEEQ WAQAR[1,2] AND JOHN J. LOPEZ[1,2]*

[1]*Division of Cardiology, Loyola University Medical Center, Maywood, IL, United States*

[2]*Department of Medicine, Loyola University Chicago, Stritch School of Medicine, Maywood, IL, United States*

Abstract

While risk factors for the development of neurovascular and coronary heart disease (CHD) are similar, it is important to consider neurologic complications of CHD separately, as many of these complications are a direct result of the underlying condition or procedures performed to treat atherosclerotic coronary disease. Stroke after myocardial infarction (MI) and acute coronary syndromes (ACSs) is not infrequent, occurring in 0.7%–2.5% of patients within 6 months of the coronary event. The etiology of these events can be frequently traced to the development of left ventricular thrombus (LVT) formation after large MI episodes. Often, however, these events are directly related to catheter-based procedures or anticoagulation strategies utilized to treat the ACS. Ischemic strokes outnumber hemorrhagic strokes in this population. While there is a modest evidence base for use of anticoagulation to treat LVT, catheterization-related ischemic stroke and anticoagulation-related hemorrhagic stroke are typically managed via standard approaches.

INTRODUCTION

In our current era where there is an epidemic of diabetes, hypertension, and obesity, there has been a concomitant rise in the prevalence of vascular disease including neurovascular and cardiovascular disease. Although the risk factors resulting in neurovascular and cardiovascular diseases are similar, it is also true that isolated cardiovascular and coronary heart disease (CHD) can result in devastating neurological complications.

This chapter will focus on the neurological complications of CHD and acute coronary syndrome (ACS) including the risk of embolic strokes in association with left heart catheterization (LHC); risk of developing left ventricular thrombi and embolic strokes; risk of hemorrhagic strokes secondary to initial management of ACS with anticoagulation; risk of long-term antiplatelet therapy; and finally, effects of hemodynamic instability during ACS on neurologic outcomes.

EPIDEMIOLOGY

The 2020 *Heart Disease and Stroke Statistics update of the American Heart Association* reports that 2 million persons ≥20 years of age in the United States have CHD (Virani et al., 2020). The prevalence of CHD was higher for males than females ≥60 years of age and this equates to a prevalence of 6.7% in US adults ≥20 years of age, with the overall prevalence for myocardial infarction (MI) at 3.0% in US adults ≥20 years of age. Per the *National Health and Nutrition Examination Survey* (NHANES) by the National Center for Health Statistics, approximately every 40 s, an American will have an MI. In 2020, approximately 720,000 Americans will have a new coronary event and approximately 335,000 will have a recurrent event per NHANES.

Data from 2002 showed that among 111,023 medicare patients discharged with a principal diagnosis of acute MI during an 8-month period between 1994 and 1995,

*Correspondence to: John J. Lopez, M.D., Division of Cardiology, Loyola University Medical Center; Department of Medicine, Loyola University of Chicago, Stritch School of Medicine, 2160 S. First Ave, Maywood, IL 60153, United States. Tel: +1-708-216-4720, Fax: +1-708-327-2771, E-mail: jlopez7@lumc.edu

2.5% had an ischemic stroke within 6 months of hospital discharge (Lichtman et al., 2002). In this study, independent predictors of ischemic stroke included age >75 years, African American ethnicity, no aspirin at discharge, frailty, prior stroke, atrial fibrillation, diabetes, hypertension, and history of peripheral vascular disease (Lichtman et al., 2002). In another study, among 15,904 stabilized patients with ACS, 113 (0.71%) had a stroke over a median follow-up of 90 days (Kassem-Moussa et al., 2004). It has been clearly demonstrated that the significant burden of CHD results in the downstream effect of an increased risk of neurological events.

ETIOLOGY

Strokes in the setting of CHD and ACS have several possible etiologies. Each etiology has its own complex and specific pathophysiology.

Cardiac catheterization and the associated risk of stroke

Cardiac catheterization remains the gold standard for the diagnosis of CHD. Strokes related to diagnostic LHC and interventional procedures remain a relatively common complication due to the large number of procedures that are being performed worldwide in the setting of an increasing burden of CHD.

Patients undergoing cardiac catheterization have an elevated risk of experiencing either an ischemic or a hemorrhagic stroke. In a majority of the cases, the underlying pathology of ischemic strokes is directly related to the catheterization procedure itself. Cardiac catheterization involves advancing single or multiple catheters over guide wires into the aorta via either that trans-femoral or trans-radial access. Most of the patients undergoing catheterization have risk factors for atherosclerotic plaque formation. During catheterization, manipulation of guide wires or catheter can possibly cause fragmentation of debris made up of thrombus, calcific materials, and cholesterol particles from these plaques with consequent central nervous system embolization (Keeley and Grines, 1998; Eggebrecht et al., 2000). The overall mechanism of ischemic strokes is essentially the same between diagnostic and interventional procedures. However, given interventional catheters are larger and the procedures are longer involving aggressive manipulation of the high-pressure arterial system, the overall risk of strokes is higher during these procedures. Other, uncommon causes of stroke after cardiac catheterization include procedural hypotension, air embolism, or arterial dissection.

A large retrospective study by the British Cardiovascular Intervention Society analyzed 426,046 patients who had percutaneous coronary intervention (PCI) in England and Wales between 2007 and 2012. Four hundred and thirty-six patients (0.1%) sustained an ischemic stroke/TIA complication and 107 patients (0.03%) had a hemorrhagic stroke (Kwok et al., 2015). Review of retrospective data shows that the highest incidence of such insults is in the initial 36 h after cardiac catheterization. However, this may be due to the fact that most of the patients undergoing elective catheterization are not hospitalized and of those who are hospitalized, the mean hospital admission time is less than 48 h (Ammann et al., 2003; Korn-Lubetzki et al., 2013).

Hemorrhagic strokes are less common and have been reported to account for 8%–46% of strokes following cardiac catheterization in registries, where strokes were categorized as ischemic or hemorrhagic (Brown and Topol, 1993; Myint et al., 2016). The risk of hemorrhagic strokes was typically attributable to acute coronary interventions, where enhanced intravenous and oral antithrombotic therapy is employed during these procedures (Brown and Topol, 1993; Oh et al., 2012). These patients also remain at an elevated risk for hemorrhagic strokes postprocedure and during follow-up, related to the continued need for an enhanced oral anticoagulation (OAC) and antiplatelet therapies postprocedure.

Left ventricular thrombus formation and risk of stroke

Left ventricular thrombus (LVT) formation is a serious but rare complication of MI, mostly occurring with ST-elevation myocardial infarction (STEMI). The feared complication related to the development of LVT is an embolic stroke, which can result in devastating disability and death. The risk of developing a LVT following an acute MI depends on the location and size of the infarct, ejection fraction postinfarction, presence of wall-motion abnormalities, and left ventricular aneurysm formation (Asinger et al., 1981; Chiarella et al., 1998). LVT are most commonly seen in patients with large anterior STEMI with resulting anteroapical wall aneurysm formation. Most commonly, these infarcts occur in the territory of the left anterior descending artery in patients with delayed or unsuccessful revascularization of this vessel as the infarct-related artery. A large area of anterior wall with hypokineses or akinesis combined with a reduced ejection fraction of ≤40% results in stasis of blood within the ventricular cavity. This stasis is associated with formation of LVT (Fig. 6.1).

In a prior era, when revascularization with primary PCI or fibrinolysis was not readily available, the incidence of LVT formation was reported as high as 40% (Asinger et al., 1981). Most of these thrombi are believed to develop within the first 2 weeks of infarction. More specifically, in a study of 30 patients who developed LVT after an acute anterior MI in the prethrombolytic/prerevascularization

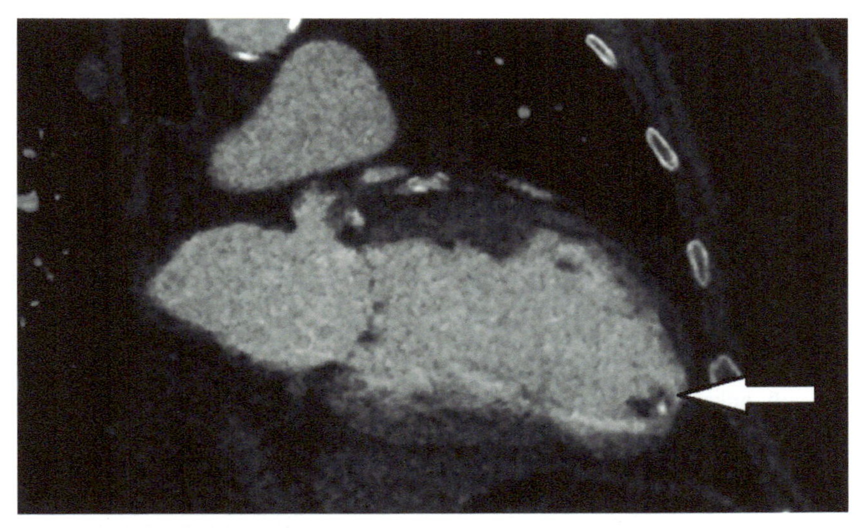

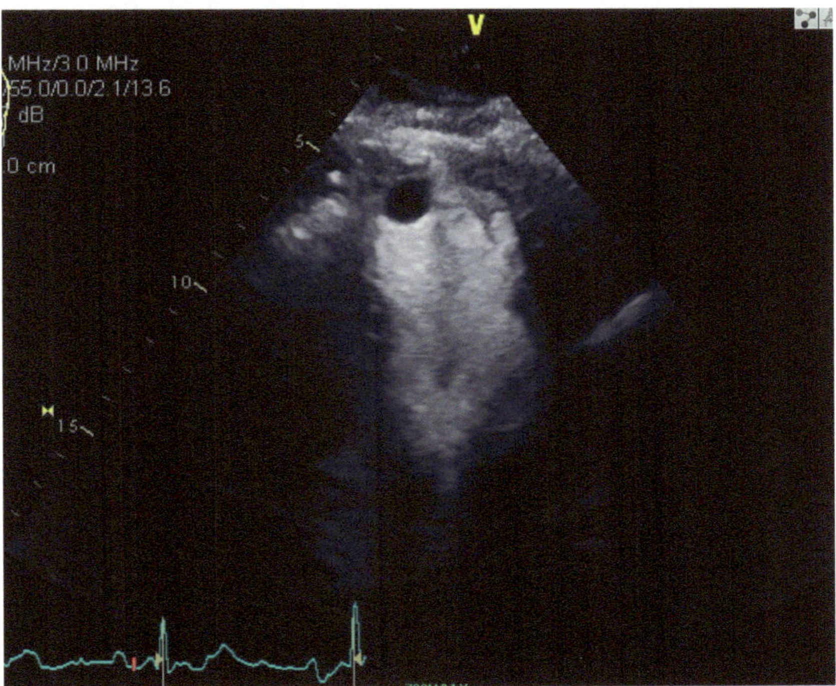

Fig. 6.1. Thrombus in left ventricular apex in a patient post-STEMI identified on MRI and echocardiography. Courtesy of Dr. Naeem Moulki.

era, LV thrombus was identified in 27% at 24 h, 57% at 48–72 h, 75% at 1 week, and 96% at 2 weeks (Weinrich et al., 1984).

With our current approach to STEMI, where PCI and fibrinolysis are the mainstays of therapy, the incidence of developing LVT is far lower as compared to the prerevascularization era. Data is limited, but the contemporary incidence of LVT is now reported to approximate 4%–8% (Rehan et al., 2006). However, even today these patients with LVT remain at a significantly elevated risk for embolic stroke. Thrombus size, thrombus mobility, and thrombus protrusion increase the risk for embolization. In patients with LVT not treated with anticoagulation, the risk of embolic stroke is as high as 10%–15% (Vaitkus and Barnathan, 1993). Diagnosis and management of LVT will be discussed later in this chapter.

Risk of hemorrhagic stroke secondary to initial management of acute coronary syndromes with anticoagulation

ACS is a broad term that includes unstable angina, non-STEMI (NSTEMI) and STEMI. Unstable angina describes a presentation with ischemic symptoms concerning for a cardiac etiology without the elevation of biomarkers of myocardial injury, i.e., troponins. These patients can also have subtle EKG changes including ST-depressions and T-wave inversions. NSTEMI describes a syndrome with a similar presentation but with an elevation of biomarkers of indicating myocardial injury. When patients present with a constellation of symptoms suggesting acute ischemia and an EKG demonstrating ST-segment elevation, this constellation of findings is required for the diagnosis of STEMI. Management of these various types of ACS is beyond the scope of this chapter, and we will only focus on the role of anticoagulation in the initial management of ACS and its associated risk with hemorrhagic strokes.

Per the 2014 *AHA/ACC Guidelines for the Management of Patients With Non-ST-Elevation Acute Coronary Syndromes*, there is a class I indication for patients with NSTEMI to receive initial treatment with parenteral anticoagulation in addition to antiplatelet therapy irrespective of future treatment strategy. Treatment options can include unfractionated heparin, enoxaparin, bivalirudin, and fondaparinux (Amsterdam et al., 2014). While the use of anticoagulation prevents ischemic myocardial complications, this occurs at an increased risk of bleeding complications; however, it is widely accepted that clinical benefit far outweighs this risk. Each year, approximately 2% of patients on OAC experience major bleeding, of which a small percentage are cerebral vascular accidents (Healey et al., 2008).

Risk of long-term dual antiplatelet therapy and stroke

Dual antiplatelet therapy (DAPT) is a widely used term in the cardiovascular literature that describes the administration of aspirin along with the use of an additional antiplatelet agent, typically consisting of one of the P2Y12 inhibitors clopidogrel, prasugrel, or ticagrelor. This approach of DAPT retains an integral role in the management of stable CHD and ACS (Table 6.1).

The recommended duration of DAPT varies based on the particular clinical scenario. Currently, the 2016 *ACC/AHA Guidelines on Duration of Dual Antiplatelet Therapy in Patients With Coronary Artery Disease* recommend that patients with stable CHD who have a PCI with a bare metal Stent have a class I indication to be on DAPT for a minimum duration of 1 month. In contrast, patients treated with a drug-eluting stent have a class I indication for treatment with DAPT for a minimum of 6 months. Finally, in patients with stable CHD who have undergone elective coronary artery bypass

Table 6.1

Classes of antiplatelet agents commonly used in the setting of acute coronary syndrome

Antiplatelet agent	Mode of action	Initial therapy and dose	Recommended long therapy length and dosage
Aspirin	Inhibits cyclooxygenase (prostaglandin H synthase), the enzyme that mediates the first step in the biosynthesis of prostaglandins and thromboxanes (including TXA2) from arachidonic acid	325 mg typically followed by 81 mg/day	After acute coronary syndrome or PCI, patients should continue aspirin indefinitely
Clopidogrel, Prasugrel, Ticagrelor	Platelet P2Y12 receptor blockers inhibit the binding of ADP to a specific platelet receptor P2Y12, thereby inhibiting activation of the GP IIb/IIIa complex and platelet aggregation	Loading dose of clopidogrel is 300 or 600 mg if patient going to catheterization the same day Loading dose for prasugrel 60 mg Loading dose for Ticagrelor 180 mg	Clopidogrel 75 mg/day with aspirin for at least 12 months Prasugrel 10 mg/day or Ticagrelor 90 mg BID with aspirin for 15 months for patients with drug-eluting stent and low-bleeding risk ASA dose for Ticagrelor is 81 mg/day
Eptifibatide, Abciximab	Blocks the final common pathway of platelet aggregation (the cross-bridging of platelets by fibrinogen binding to the GP IIb/IIIa receptor) and may also prevent adhesion to the vessel wall	Initiated at presentation for high-risk patients undergoing invasive approach and continued peri- and post-PCI	Eptifibatide is continued for 18 h post-PCI but shorter duration protocols are common Abciximab is continued for 12 h post-PCI

surgery (CABG) have a Class IIb indication to be on DAPT for at least 12 months (Helft, 2016). Fortunately, guidelines are much more straightforward for patients who have had an ACS event. Regardless of the initial management strategy with medical therapy, thrombolytics, PCI with bare metal or a drug-eluting stent or emergent CABG, patients diagnosed with ACS (NSTEMI or STEMI) have a class I indication to be on DAPT for a minimum of 12 months.

DAPT is also commonly used in the management of ischemic stroke; however, prolonged DAPT therapy does not appear to increase the risk of developing a hemorrhagic stroke. Large observational cohort studies have demonstrated that the use of aspirin and clopidogrel does not increase stroke risk above baseline, but the combination of aspirin and warfarin increases the risk of stroke to 0.9% per year compared to 0.2% per year for aspirin alone (Buresly et al., 2005). Triple therapy with aspirin, clopidogrel, and warfarin places the patient at three to five times the risk of major bleeding, which includes an increased stroke risk (Patti and Di Sciascio, 2010). One of the major predictor of bleeding with OAC (not noted with antiplatelet agents) is advanced patient's age. Risk of major bleeding on OAC combined with antiplatelet therapy increases by 3% per year in patients over the age of 75 years (Patti and Di Sciascio, 2010).

Effects of hemodynamic instability during ACS

Patients with ACS can have hemodynamically variable initial presentations. These can range from a patient with a large anterior infarct presenting with normal vitals to patients with small territorial infarct presenting with a ventricular fibrillation arrest or the full-blown development of cardiogenic shock. The wide spectrum of presentation of ACS makes it difficult to explain every possible scenario. The Initial goal for patients presenting with ACS is to maintain adequate end organ, including cerebral perfusion and function by focusing on hemodynamic stability including normal blood pressure, normal sinus rhythm, and heart rate to ensure an adequate cardiac output. Advanced heart failure or development of cardiogenic shock may necessitate the use of a pulmonary artery catheter monitoring to assess cardiac output, index, and cardiac power in an effort to assess the response to various therapeutic interventions. While decreased cardiac output for prolonged durations can lead to global cerebral ischemia, and increase the risk of ischemic strokes in patients with underlying cerebrovascular disease, there is no clear systolic blood pressure or mean arterial pressure target associated with improved neurologic outcomes, although a target value of 65 mmHg has been felt to be a reasonable target to maintain tissue perfusion (Nieminen et al., 2016; Van Diepen et al., 2017). Finally, CHD remains as one of the most common causes of sudden cardiac arrest and the current role of therapeutic hypothermia or targeted temperature monitoring is beyond the scope of this chapter.

CLINICAL PRESENTATION, DIAGNOSES, AND MANAGEMENT OF NEUROLOGICAL COMPLICATIONS OF CORONARY HEART DISEASE

Cardiac catheterization-related stroke

Stroke after cardiac catheterization is typically diagnosed within the first 24 h postprocedure. There are no set guidelines regarding the prevention of strokes secondary to cardiac catheterization, but limited data from a propensity matched analysis of 8726 patients reported an odds ratio of 0.33 for trans-radial approach compared with a femoral approach in the incidence of stroke (Shoji et al., 2018).

Strokes after cardiac catheterization have similar presentation to stroke secondary to any other etiology. Some of the frequent manifestations include visual disturbances, aphasia, dysarthria, and altered mental status. Patients with suspicion for an acute stroke after cardiac catheterization should be evaluated emergently with rapid activation of the hospital's stroke team. Given risk for both ischemic and hemorrhagic strokes in the setting of cardiac catheterization, patients should have urgent imaging with head CT or brain magnetic resonance imaging (MRI) and should be managed by the acute stroke team similarly to other ischemic and hemorrhagic strokes, with individualized decisions regarding management with conservative efforts, thrombolytics or acute endovascular intervention. From a cardiology perspective, there are no additional guidelines for management of postcardiac catheterization strokes.

Diagnoses and management of left ventricular thrombus

LVT remains a serious but rare complication of acute MI and is associated with systemic thromboembolism including embolic strokes. It is recommended that most of the patients diagnosed with an acute MI should have assessment of left ventricular function including evaluation of left ventricular apical function and presence of a LVT. The most common diagnostic modality that is used is the standard transthoracic echocardiography (TTE). In situations where left ventricular apex is poorly visualized on a TTE due to patients' body habitus and where there are anterior or apical wall-motion abnormalities with a high apical wall-motion score (≥ 5 on noncontrast

TTE), contrast TTE or cardiac MRI should be considered based on local availability and resources.

In patients who are diagnosed with an LVT, OAC therapy should be started immediately. Due to the lack of randomized control data in this particular area, guidelines are still new and frequently changing (Ibanez et al., 2018). The 2013 *American College of Cardiology Foundation/American Heart Association STEMI Guidelines* advise that OAC may be considered in patients with STEMI with anterior apical akinesis or dyskinesis to prevent LV thrombus formation (O'Gara et al., 2013). Similarly, the *AHA/American Stroke Association 2014 Guidelines on Stroke Prevention* advise that anticoagulation may be considered for 3 months in patients with acute anterior STEMI and ischemic stroke or transient ischemic attack who have anterior apical akinesis or dyskinesis (Kernan et al., 2015). The *European Society of Cardiology 2017 STEMI Guidelines* recommend once an LV thrombus is diagnosed, OAC should be considered for up to 6 months, guided by repeated echocardiography and with consideration of bleeding risk and need for concomitant antiplatelet therapy. Thus, the optimal duration of OAC in these patients is unclear, and decisions regarding continuation of OAC should be made on a case-by-case basis (Ibanez et al., 2018).

Most commonly in addition to DAPT, it is recommended that patients with evidence of LVT after MI be treated with warfarin. Given the high risk of thromboembolism and stroke, patients should be bridged with a parenteral anticoagulant until a therapeutic INR (2.0–3.0) is achieved for at least 24 h (McCarthy et al., 2018). While warfarin remains the most commonly used OAC and the agent with the greatest degree of historical use, there are substantial limitations in using warfarin including the need for frequent monitoring, a narrow therapeutic window, the need for dietary restrictions and multiple drug–drug interactions. Given these concerns, there has been a growing experience in the use of direct oral anticoagulants for this disorder (Patel et al., 2015). Finally, in 2014, the *AHA/American Stroke Association Guidelines on Stroke Prevention* introduced a new recommendation advising that low-molecular-weight heparin, dabigatran, rivaroxaban, or apixaban may be considered as an alternative to vitamin K antagonists for post-MI LV thrombus or anterior or apical wall-motion abnormalities with an LV ejection fraction less than 40%, who are intolerant of vitamin K antagonists because of nonhemorrhagic adverse events (Class IIb; Level of Evidence: C) (Kernan et al., 2015).

CONCLUSION

Given the high burden of vascular disease in society, the concomitant development of neurovascular and CHD will be noted with increasing frequency. While patients with CHD may independently develop cerebrovascular disease, it is important to understand the specific etiologies of neurovascular events that can occur in patients suffering from ACSs and acute MI. Understanding the etiology of these events, whether related to the development of LVT, due to procedural complications or related to anticoagulation regimens is critical to providing accurate recommendations in regards to prognosis and the most effective approach to management.

References

Ammann P, Brunner-La Rocca H, Angehrn W et al. (2003). Procedural complications following diagnostic coronary angiography are related to the operator's experience and the catheter size. Catheter Cardiovasc Interv 59: 13–18.

Amsterdam EA, Wenger NK, Brindis RG et al. (2014). 2014 AHA/ACC guideline for the management of patients with non–ST-elevation acute coronary syndromes: a report of the American College of Cardiology/American Heart Association task force on practice guidelines. J Am Coll Cardiol 64: e139–e228.

Asinger R, Mikell F, Elsperger J et al. (1981). Incidence of left-ventricular thrombosis after acute transmural myocardial infarction. N Engl J Med 305: 297–302.

Brown D, Topol E (1993). Stroke complicating percutaneous coronary revascularization. Am J Cardiol 72: 1207–1209.

Buresly K, Eisenberg M, Zhang X et al. (2005). Bleeding complications associated with combinations of aspirin, thienopyridine derivatives, and warfarin in elderly patients following acute myocardial infarction. Arch Intern Med 165: 784.

Chiarella F, Santoro E, Domenicucci S et al. (1998). Predischarge two-dimensional echocardiographic evaluation of left ventricular thrombosis after acute myocardial infarction in the GISSI-3 study 11GISSI-3 is supported by the Associazione Nazionale Cardiologi Ospedalieri (ANMCO) and Istituto di Ricerche Farmacologiche "Mario Negri". Am J Cardiol 81: 822–827.

Eggebrecht H, Oldenburg O, Dirsch O et al. (2000). Potential embolization by atherosclerotic debris dislodged from aortic wall during cardiac catheterization: histological and clinical findings in 7,621 patients. Catheter Cardiovasc Interv 49: 389–394.

Healey JS, Hart RG, Pogue J et al. (2008). Risks and benefits of oral anticoagulation compared with clopidogrel plus aspirin in patients with atrial fibrillation according to stroke risk: the atrial fibrillation clopidogrel trial with irbesartan for prevention of vascular events (ACTIVE-W). Stroke 39: 1482–1486.

Helft G (2016). New guidelines on duration of dual antiplatelet therapy in patients with coronary artery disease: what's the novelty? J Thorac Dis 8: E1301.

Ibanez B, James S, Agewall S et al. (2018). 2017 ESC guidelines for the management of acute myocardial infarction in patients presenting with ST-segment elevation: the task force for the management of acute myocardial infarction in patients presenting with ST-segment elevation of the European Society of Cardiology (ESC). Eur Heart J 39: 119–177.

Kassem-Moussa H, Mahaffey K, Graffagnino C et al. (2004). Incidence and characteristics of stroke during 90-day follow-up in patients stabilized after an acute coronary syndrome. Am Heart J 148: 439–446.

Keeley E, Grines C (1998). Scraping of aortic debris by coronary guiding catheters. J Am Coll Cardiol 32: 1861–1865.

Kernan W, Ovbiagele B, Kittner S (2015). Response to letter regarding article, "guidelines for the prevention of stroke in patients with stroke and transient ischemic attack: a guideline for healthcare professionals from the American Heart Association/American Stroke Association". Stroke 46: e87–e89.

Korn-Lubetzki I, Farkash R, Pachino R et al. (2013). Incidence and risk factors of cerebrovascular events following cardiac catheterization. J Am Heart Assoc 2: e000413.

Kwok C, Kontopantelis E, Myint P et al. (2015). Stroke following percutaneous coronary intervention: type-specific incidence, outcomes and determinants seen by the British Cardiovascular Intervention Society 2007–12. Eur Heart J 36: 1618–1628.

Lichtman J, Krumholz H, Wang Y et al. (2002). Risk and predictors of stroke after myocardial infarction among the elderly. Circulation 105: 1082–1087.

McCarthy C, Vaduganathan M, McCarthy K et al. (2018). Left ventricular thrombus after acute myocardial infarction. JAMA Cardiol 3: 642.

Myint P, Kwok C, Roffe C et al. (2016). Determinants and outcomes of stroke following percutaneous coronary intervention by indication. Stroke 47: 1500–1507.

Nieminen MS, Buerke M, Cohen-Solál A et al. (2016). The role of levosimendan in acute heart failure complicating acute coronary syndrome: a review and expert consensus opinion. Int J Cardiol 218: 150–157.

O'Gara PT, Kushner FG, Ascheim DD et al. (2013). 2013 ACCF/AHA guideline for the management of ST-elevation myocardial infarction: executive summary: a report of the American College of Cardiology Foundation/American Heart Association task force on practice guidelines. J Am Coll Cardiol 61: 485–510.

Oh M, Kwon J, Kim K et al. (2012). Subarachnoid hemorrhage mimicking leakage of contrast media after coronary angiography. Korean Circ J 42: 197.

Patel P, Zhao X, Fonarow G et al. (2015). Novel oral anticoagulant use among patients with atrial fibrillation hospitalized with ischemic stroke or transient ischemic attack. Circ Cardiovasc Qual Outcomes 8: 383–392.

Patti G, Di Sciascio G (2010). Antithrombotic strategies in patients on oral anticoagulant therapy undergoing percutaneous coronary intervention: a proposed algorithm based on individual risk stratification. Catheter Cardiovasc Interv 75: 128–134.

Rehan A, Kanwar M, Rosman H et al. (2006). Incidence of post myocardial infarction left ventricular thrombus formation in the era of primary percutaneous intervention and glycoprotein IIb/IIIa inhibitors. A prospective observational study. Cardiovasc Ultrasound 4: 20. https://doi.org/10.1186/1476-7120-4-20.

Shoji S, Kohsaka S, Kumamaru H et al. (2018). Stroke after percutaneous coronary intervention in the era of transradial intervention: report from a japanese multicenter registry. Circ Cardiovasc Interv 11: e006761.

Vaitkus P, Barnathan E (1993). Embolic potential, prevention and management of mural thrombus complicating anterior myocardial infarction: a meta-analysis. J Am Coll Cardiol 22: 1004–1009.

Van Diepen S, Katz JN, Albert NM et al. (2017). Contemporary management of cardiogenic shock: a scientific statement from the American Heart Association. Circulation 136: e232–e268.

Virani SS, Alonso A, Benjamin EJ et al. (2020). Heart disease and stroke statistics—2020 update: a report from the American Heart Association. Circulation 141: e139–e596.

Weinrich DJ, Burke JF, Pauletto FJ (1984). Left ventricular mural thrombi complicating acute myocardial infarction: long-term follow-up with serial echocardiography. Ann Intern Med 100: 789–794.

Handbook of Clinical Neurology, Vol. 177 (3rd series)
Heart and Neurologic Disease
J. Biller, Editor
https://doi.org/10.1016/B978-0-12-819814-8.00007-X

Chapter 7

Neurologic complications of heart surgery

AMIR SHABAN[1]* AND ENRIQUE C. LEIRA[1,2]

[1]*Department of Neurology, Carver College of Medicine, University of Iowa, Iowa City, IA, United States*

[2]*Department of Epidemiology, College of Public Health, University of Iowa, Iowa City, IA, United States*

Abstract

Cardiac surgeries are commonly associated with neurologic complications. The type and complexity of the surgery, as well as patients' comorbidities, determine the risk for these complications. Awareness and swift recognition of these complications may have significant implications on management and prognosis.

Recent trials resulted in an expansion of the time window to treat patients with acute ischemic stroke with intravenous thrombolysis and/or mechanical thrombectomy using advanced neuroimaging for screening. The expanded time window increases the reperfusion treatment options for patients that suffer a periprocedural ischemic stroke. Moreover, there is now limited data available to help guide management of intracerebral hemorrhage in patients undergoing treatment with anticoagulation for highly thrombogenic conditions, such as left ventricular assist devices and mechanical valves. In addition to cerebrovascular complications patients undergoing heart surgery are at increased risk for seizures, contrast toxicity, cognitive changes, psychological complications, and peripheral nerve injuries.

We review the neurological complications associated with the most common cardiac surgeries and discuss clinical presentation, diagnosis and management strategies.

INTRODUCTION

Neurological complications are some of the most feared events following heart surgery. The frequency and type of neurologic complications vary based on surgical factors and patients' comorbidities (Shaban and Leira, 2019). These complications significantly increase the length of hospital stay, morbidity, and mortality from heart surgery. Rapid recognition of these complications is essential, as it influences the prompt implementation of management strategies to improve outcomes. Treatment of these complications requires familiarity with the most recent guidelines of stroke management, as well as understanding risks and benefits of each treatment modality following that particular cardiac surgery. A treatment modality indicated for a neurological complication after one type surgery may be contraindicated for a similar complication after another type surgery.

Neurologic complications may involve both the central nervous system (CNS) and peripheral nervous system and can vary in severity from asymptomatic complications only detected on neuroimaging to serious complications leading to significant morbidity and mortality.

In this chapter, we discuss the possible neurologic complications following heart surgeries. We review their diagnosis, prevention, and treatment strategies (Table 7.1).

ACUTE ISCHEMIC STROKE

Acute ischemic stroke (AIS) is the most common complication following cardiac surgery. This feared complication is a major source of morbidity and mortality. The risk for AIS ranges from 0 and 18% based on the type of the surgery and the premorbid condition of the patient (Shaban and Leira, 2019). The rates of incidental

*Correspondence to: Amir Shaban, M.D., Department of Neurology, College of Public Health, University of Iowa, 200 Hawkins drive, Iowa City, IA 52242, United States. Tel: +1-319-384-9662, Fax: +1-319-356-4505, E-mail: amir-shaban@uiowa.edu

Table 7.1

Cardiac surgeries and their most common neurologic complications

Procedure	Complication	Frequency	Mechanism	Management
CABG	Ischemic stroke	1%–5%	Embolic or decreased flow due to arterial hypotension during surgery	IV-tPA should be weighed against risk for bleeding into the surgical site. Patients are often outside the window for IV-tPA Mechanical thrombectomy is the treatment of choice for large vessel occlusion and could be offered in the early window 0–6 h or in a delayed window 6–24 h
	ICH	<0.5%	Anticoagulation use during surgery	SBP < 140 mmHg and supportive care Consider anticoagulation reversal if needed. Surgical evaluation in some cases
	Retinal artery occlusion and ischemic optic neuropathy	<0.5%	More commonly embolic but could be due to arterial hypotension	Supportive care. IA-tPA has been tried with no evidence for benefit Ocular massage to reduce intraocular pressure and dislodgement of emboli
	RLN injury	<1%	Surgical injury, traction, or ice slush	Supportive care for unilateral injury. Respiratory support may be needed for bilateral RLN injury
	Seizures	Rare	Tranexamic acid or other medications that can decrease seizure threshold	Antiseizure medications; discontinuation of any medication that can decrease seizure threshold
Cardiac catheterization	Ischemic stroke	<1%	Embolic	IV-tPA and/or mechanical thrombectomy
	ICH	Rare	Anticoagulation use	SBP < 140 mmHg and supportive care Anticoagulation reversal Surgical evacuation may be needed
	Contrast neurotoxicity	1%–2%	BBB disruption	Rule out other etiologies and supportive care
	Lumbosacral plexus and femoral nerve injury (femoral approach)	<0.5%	Mechanical pressure due to hematoma formation	Supportive care
	Radial nerve injury (radial approach)	Rare	Direct surgical injury or pressure from a hematoma	Supportive care

Valve replacement	ICH	Varies based on anticoagulation regimen used	due to anticoagulation and antiplatelet use	SBP < 140 mmHg Anticoagulation reversal in the acute settings; duration of anticoagulation withholding are decided on a case-by-case basis
	Ischemic stroke	1.5%–17% depending on the procedure, location, type of prosthetic	Most commonly embolic	Despite increased risk for hemorrhagic transformation anticoagulation often continued
LVAD	ICH	8%–11%	Anticoagulation and acquired von Willebrand disease	SBP < 140 mmHg Anticoagulation reversal, duration of anticoagulation withholding is decided on a case-by-case basis
	Ischemic stroke	6%–18%	Hypercoagulable state and thrombosis	Continue anticoagulation
Heart transplant	Ischemic stroke	10%	Embolic from manipulation of the heart and aorta or hypoperfusion	IV-tPA is contraindicated but mechanical thrombectomy is an option
	Seizures	13%	Medication side effect, infections, malignancy	Antiseizure medications and evaluation for a secondary cause
Congenital heart surgery	Seizures	1.3%–7.8%	Medication toxicity and sequelae of cardiac arrest	Antiseizure medications
Aortic aneurysm repair	Spinal cord ischemia	0.5%	Hypotension or embolic	IV-tPA often contraindicated due to risk for bleeding into surgical site. CSF drain or steroids are used experimentally for prevention

ischemic strokes found on neuroimaging studies with no clinical symptoms are much higher (Kahlert et al., 2010).

Embolization from the heart and major arteries is the predominant mechanism of periprocedural ischemic strokes, while ischemia due to decreased cerebral blood flow as a result of a combination of arterial hypotension and stenosis of a cerebral vessel is a less common mechanism.

In general, two treatments are available for the management of AIS including intravenous tissue plasminogen activator (IV-tPA) and mechanical thrombectomy for proximal large vessel occlusion (LVO) strokes of the anterior circulation.

The type of cardiac surgery itself could affect treatment options. For example, IV-tPA will likely be contraindicated following a major open-heart surgery, but cardiac catheter procedures in general are not a contraindication. Ultimately, the decision to give IV-tPA should be individualized, and a careful review of benefits and potential risks for bleeding discussed thoroughly with members of the surgical team, patient, and appropriate patient's family members.

The duration of cardiac surgery is another important factor to determine the last known normal (LKN) time, as symptoms often go unnoticed until the procedure is completed, when the anesthetic effects have subsided. Patients with LKN time less than 4.5 h are eligible for IV-tPA if no other contraindications. However, cardiac surgeries are often prolonged procedures and symptoms are often discovered beyond the 4.5 h window. New data from several clinical trials suggest that IV-tPA could be administered beyond 4.5 h for selected patients. The selection is based on an estimation of the age of the ischemic lesion or the presence of viable penumbra based on advanced neuroimaging such as computed tomography perfusion (CTP) or magnetic resonance imaging (MRI) beyond 4.5 h or when the LKN is unknown (Thomalla et al., 2018; Ma et al., 2019). The recently updated American Heart Association guidelines recommend considering the use of IV-tPA for patients with unknown symptom onset if their MRI shows diffusion-positive FLAIR-negative lesion and IV-tPA could be administered within 4.5 h from symptoms discovery (Powers et al., 2019).

In addition to the LKN time and the risk for bleeding into the surgical bed, several other patient-related factors should be taken into consideration when deciding whether to treat postsurgical patients with IV-tPA. For example, use of anticoagulation during the surgery and recent myocardial infarction both of which could be contraindications for IV-tPA (Powers et al., 2018).

Patients with suspected LVO should be evaluated for possible mechanical thrombectomy. Patients within 6 h from LKN time with vessel imaging showing LVO and favorable head CT scan of the head and no contraindications should be treated. Patients presenting after 6 h and up to 24 h should be evaluated using CTP or magnetic resonance perfusion and treated with mechanical thrombectomy if specific mismatch criteria are met (Albers et al., 2018; Nogueira et al., 2018).

The recent extension of the time window for mechanical thrombectomy and the possible extension of the time window for IV-tPA will have a great impact on postsurgical stroke treatment and may greatly improve the outcomes of stroke after cardiac surgery.

Coronary artery bypass surgery

AIS occurs in 1%–5% of patients following coronary artery bypass surgery (CABG) (Shaban and Leira, 2019). This is higher when compared to percutaneous coronary intervention (PCI) (<1%). However, patients having PCI have higher chances for requiring a repeat procedure, thus increasing the risk for ischemic stroke in the long term (Serruys et al., 2009). The risk for embolization following CABG is associated with several factors related to the procedure itself such as its complexity, manipulation of the aorta, occurrence of perioperative atrial fibrillation (POAF), and use of the on-pump technique (Devgun et al., 2018). Premorbid condition of the patient and presence of established vascular risk factors are also known to be associated with increased risk for ischemic complications (Devgun et al., 2018).

Several strategies have been attempted to prevent ischemic strokes during CABG. Avoiding aortic manipulation with the "aortic no touch" strategy may help decrease the risk for perioperative ischemic stroke (Moss et al., 2015). Aortic clamping strategies did not appear to reduce stroke risk (Chen et al., 2018). Hypothermia during CABG to prevent stroke has been tried with no benefit (Rees et al., 2001). Remote ischemic conditioning has been tested in a clinical trial and did not improve outcomes among patients having elective on-pump CABG (Hausenloy et al., 2015). Erythropoietin administration and use of intraoperative TEE or epiaortic ultrasound to detect aortic atheroma has been tried with limited benefit (Tasanarong et al., 2013; Biancari et al., 2020).

It should be noted that carotid artery atherosclerosis can coexist with coronary artery disease in 6%–8% of patients having CABG (Weimar et al., 2017). Screening for carotid artery disease is recommended for patients with vascular risk factors (Hillis et al., 2012). However, carotid revascularization prior to the cardiac procedure is recommended only for those patients with history of stroke or transient ischemic attack (TIA) and hemodynamically consequential ipsilateral carotid artery stenosis (Hillis et al., 2012). Revascularization is also recommended prior to cardiac procedure for patients with bilateral severe stenosis (Hillis et al., 2012). In general, the role for carotid revascularization of asymptomatic carotid artery stenosis remains controversial. A randomized clinical trial failed to show any benefit for synchronous CABG and carotid endarterectomy for patients with asymptomatic high-grade carotid stenosis when compared to CABG alone (Weimar et al., 2017).

Postoperatively, dual antiplatelet (DAP) therapy is recommended for patients having CABG as DAP decreases risk for cardiovascular death, MI, or stroke compared to a single antiplatelet agent (Yusuf et al., 2001).

Of note, POAF occurred in about 11% of patients in the SYNTAX trial and was associated with increased risk for stroke and mortality (Serruys et al., 2009). A meta-analysis showed an increased risk for ischemic stroke and mortality in the short and long terms in patient who had POAF (Lin et al., 2019). Treatment of short-lived POAF that resolves prior to discharge remains uncertain.

Percutaneous coronary intervention

As described above, PCI is associated with less than 1% risk for perioperative ischemic stroke; this is lower than with CABG. Several procedural aspects are thought to increase ischemic stroke risk following PCI including complexity of cardiac lesions, multiple catheter exchanges, caliber of the guidewires used, and use of intraaortic balloon pump (Devgun et al., 2018).

DAP treatment with aspirin and clopidogrel is recommended for patients undergoing PCI and is thought to lower the risk for ischemic stroke and stent thrombosis (Levine et al., 2011).

Valvular surgeries

Ischemic stroke is a known complication of heart valve surgeries. Risk for ischemic stroke following cardiac valvular procedures varies between 1.5% and 17% depending on the type of surgery, valve involved, and type of prosthetic valve chosen. Many ischemic strokes occur in the periprocedural phase; the rate of strokes subsides thereafter. Strokes that occur during the first 24 h are usually driven by procedural factors, while cardiac arrhythmias explain later strokes (Nombela-Franco et al., 2012).

Surgical and transcatheter approaches are currently used for cardiac valvular replacement. The risk for ischemic stroke or a TIA after transcatheter aortic valve replacement (TAVR) is estimated to be between 1.5% and 5.1% (Kahlert et al., 2010). The reported risk for ischemic stroke is higher in transcatheter procedures compared to surgical heart valve replacement (Jones et al., 2015).

Several procedural aspects increase ischemic stroke risk following TAVR including need for balloon valvuloplasty, crossing a severely calcified aortic valve, and position of the rigid device across the native valve (Hahn et al., 2014). No differences in stroke rates between the transfemoral and transapical approaches have been observed.

For prevention, patients having TAVR are placed on DAP therapy for 3–6 months per current guidelines. However, two meta-analyses did not show lower rates of strokes among patients taking DAP compared to a single antiplatelet agent (Aryal et al., 2015; Maes et al., 2018). Embolic protection devices have shown safety during TAVR procedures. However, to date, there is no clear evidence for their efficacy (Bagur et al., 2017). Embolic protection devices did not reduce the number of clinically relevant ischemic strokes or mortality in a recent meta-analysis (Bagur et al., 2017).

Surgical valve replacement risks for perioperative stroke increase with older age, prior strokes, emergency surgery, longer cardiopulmonary bypass (CPB) times, and high blood transfusion requirements (Hauville et al., 2012).

Following the procedure, risk for ischemic stroke depends on the heart valve involved and type of prosthetic used, which in turn help determine the prevention strategy. For example, a mechanical mitral valve requires treatment with warfarin with a target INR of 3 and aspirin. Conversely, a bioprosthetic TAVR requires DAP for 6 months and aspirin alone thereafter. The presence of atrial fibrillation increases risk for ischemic events, and this may require further adjustment in the prevention strategy (Baber et al., 2010).

Heart transplantation

Heart transplantation surgery is associated with a high risk for neurological complications. Ischemic stroke and nonvascular complications occur in up to 10% of patients having heart transplantation (Cemillan et al., 2004). Premorbid vascular risk factors, which many of heart transplantation patients have, increase risk for periprocedural stroke (Sakamoto, 2016). Factors related to surgery itself are associated with higher rates for ischemic stroke such as use of left ventricular assist device (LVAD), intraaortic balloon pump, prolonged CPB time, and postoperative hepatic failure (Zierer et al., 2007). It is possible that the excision and manipulation of the diseased heart may also a play role in the intraoperative stroke (Acampa et al., 2016).

Left ventricular assist devices

LVADs are used in patients with heart failure as a bridge to heart transplantation. Ischemic stroke, a major complication in patients with LVADs, occurs in about 6%–18% of LVAD patients (Shaban and Leira, 2019). LVADs cause a hypercoagulable state by exposing blood to the artificial surfaces and high shear forces inside the device (Blitz, 2014). Patients with LVAD are usually placed on anticoagulation to prevent LVAD thrombosis and to reduce risk for embolic events. Management of ischemic stroke occurring despite anticoagulation in the setting of LVAD is challenging. The risk for hemorrhagic transformation is weighed against the risk for LVAD thrombosis. Not infrequently, anticoagulation is continued without interruptions.

Congenital heart surgery

Most congenital heart defects require surgical correction early in childhood. Neurological complications following these complex surgeries have become more recognized due to higher rates of survival. Periprocedural ischemic stroke are seen in 1%–2% of children following congenital heart surgery (Menache et al., 2002; Sakamoto, 2016). In addition to the challenges for treatment of ischemic stroke following heart surgery described previously, the current data available for managing children with ischemic stroke is scarce, as clinical trials of intravenous thrombolysis and mechanical thrombectomy excluded children under age 18.

OTHER TYPES OF ISCHEMIC COMPLICATIONS

Retinal artery occlusion

Affected patients present with acute painless vision loss. Both central and peripheral visions are lost when the central retinal artery is occluded, whereas segmental peripheral vision loss occurs in branch retinal artery occlusions (Limaye et al., 2018). On funduscopy, a cherry-red spot is a classic sign for a central artery occlusion, whereas peripheral wedge whitening is seen in branch retinal artery occlusions. Retinal artery occlusion can rarely occur following heart surgery. The mechanism of retinal ischemia is thought to be embolic rather than hypoperfusion (Raphael et al., 2019). Risk factors associated with higher rates of retinal artery occlusion included diabetic retinopathy, aortic insufficiency, atrial myxoma, hypercoagulability, and carotid artery stenosis (Calway et al., 2017). Acute treatment of retinal artery occlusion remains controversial. Intraarterial alteplase showed no benefit in the EAGLE trial that enrolled patients up to 20h from symptoms onset (Schumacher et al., 2010). Ocular massage may be used to decrease intraocular pressure and attempt to dislodge emboli to more distal locations. Similarly, intravenous acetazolamide may be administered to decrease intraocular pressure and thus increase the retinal blood flow.

Ischemic optic neuropathy

Ischemic optic neuropathy occurs in less than 0.5% of patients having cardiac surgery (Holy et al., 2009). It presents with central and peripheral painless vision loss. On examination, an afferent pupillary defect is often noted. This condition could be categorized into anterior and posterior ischemic optic neuropathy based on the location of the injury. While anterior ischemic optic neuropathy could have cotton wool spots and pallor around the optic nerve head on funduscopic examination, posterior ischemic optic neuropathy often has no associated findings. Anterior ischemic optic neuropathy is the most common type following cardiac surgery (Raphael et al., 2019). The mechanism of injury remains unknown but thought to result from impaired perfusion in a watershed territory and is associated with anatomical variation in the blood supply (Raphael et al., 2019).

Spinal cord ischemia

Spinal cord ischemia has been reported in 0.5%–10% of patients following surgical repair of aortic aneurysm (Szilagyi et al., 1978; Awad et al., 2017; Moulakakis et al., 2018). Spinal cord ischemia could result from low flow due to arterial hypotension, surgical injury to spinal arteries or embolic events. Diagnosis is often suspected after the anesthetic effects have subsided. Presenting symptoms vary based on the location of spinal cord injury. Most commonly, spinal cord ischemia involves the anterior spinal artery territory, resulting in bilateral lower extremity weakness and sensory loss to pain and temperature modalities while preserving vibration and position sense. The Adamkiewicz artery territory is often involved (Moulakakis et al., 2018). MRI diffusion-weighted imaging can help make a definitive diagnosis, but the usefulness of this technique in the spinal cord is limited.

Intraoperative somatosensory evoked potentials monitoring can detect development of spinal cord ischemia during surgery. Corticosteroids and CSF drainage devices have been tried for prevention of spinal cord ischemia during surgery, but this approach remains experimental (Moulakakis et al., 2018).

The use of IV-tPA in spinal cord ischemia has not been studied thoroughly. Several anecdotal case reports have used IV-tPA in patients with spinal cord ischemia with no complications (Jankovic et al., 2019). However, following major vascular surgery like aortic aneurysm repair, thrombolysis will likely be contraindicated due to the high risk for bleeding.

HEMORRHAGIC STROKE

Patients having cardiac surgeries have higher risk for intracerebral hemorrhage (ICH) due to anticoagulation use during these procedures. Patients with ICH often present with symptoms clinically undistinguishable from AIS. Head CT is indicated to differentiate between the two conditions.

Cardiac valvular surgeries specifically require prolonged anticoagulation and antiplatelet therapy to prevent ischemic events. The risk for ischemic events is highest during the immediate periprocedural period, while the risk for intracranial hemorrhage tends to increase with prolonged anticoagulation. The anticoagulation regimen determines the ICH risk. The duration and type of such regimen depends on the valve replaced, type of prosthetic used, and presence of atrial fibrillation.

Similarly, patients need to continue anticoagulation following LVAD placement to prevent device thrombosis. About 8%–11% of patients with LVAD develop ICH (Slaughter et al., 2009). In addition to the increased risk of ICH related to anticoagulation, patients with LVAD can develop acquired von Willebrand disease, which can further increase the risk for hemorrhage (Suarez et al., 2011). A case report showed that only one of five patients with ICH had supratherapeutic INR at the time of the hemorrhage (Ramey et al., 2017). Even though continuous-flow LVAD devices are associated with higher survival benefits, they are thought to have higher risk for ICH compared to pulsatile-flow devices (Elder et al., 2019).

The risk for ICH following CABG is lower than that for ischemic stroke. About 0.5% of patients having CABG develop ICH; higher rates have been observed among patients with renal dysfunction (Holzmann et al., 2013).

The risk for ICH among patients having left atrial appendage closure procedures is lower when compared to patients on prolonged treatment with warfarin. This procedure is often recommended for patients who are at high risk for long-term oral anticoagulation therapy (Reddy et al., 2017).

Management of ICH following cardiac surgeries represents a major challenge. Care starts with stabilizing the patients who may require endotracheal intubation and resuscitation. Once patients are stable, a head CT scan should be obtained in order to promptly differentiate ischemic from hemorrhagic stroke, to determine size and severity of ICH, and need for neurosurgical evaluation for possible decompression. In the acute phase, systolic blood pressure should be maintained below 140 mmHg, and patients' head position should be at least 30 degrees to avoid potentially exacerbate an elevated intracranial pressure.

CT angiography (CTA) should be considered if aneurysms or CNS vascular malformations are suspected. CTA is also helpful to detect contrast extravasation into the hematoma (spot sign), which is associated with increased risk of hematoma expansion (Hemphill 3rd et al., 2015).

For patients on oral anticoagulants, rapid reversal of the INR to less than 1.4 in the acute setting of warfarin-related ICH is advised, as achieving adequate reversal is associated with lower rates of hematoma expansion. Direct oral anticoagulant-related ICH should be reversed with specific antidote if available. Reversal of anticoagulation should be decided on a case-by-case basis taking into consideration the size and severity of the hemorrhage and the reason patient is on anticoagulation.

For patients with LVADs receiving anticoagulation, the decision regarding reversal may be more challenging given the high risk for LVAD thrombosis and ischemic stroke. Data from case series studies showed that while LVAD patients with small volume ICH have been successfully managed without anticoagulant reversal, there was no increase in the risk for thrombotic events for those receiving anticoagulation reversal (Wong et al., 2016; Tahir et al., 2018). In our institution, we make the decision about anticoagulation reversal symptom severity, size, and location of the hemorrhage.

The ideal time to resume anticoagulation remains uncertain. Restarting anticoagulation in the first 2–10 weeks has been associated with higher risks for hematoma expansion (Majeed et al., 2010; Kuramatsu et al., 2018). Patients with mechanical hearts valve or LVADs have high risk for ischemic stroke; thus, anticoagulation should be restarted as soon as it is safe. In a study on 11,000 INRs drawn from LVAD patients, it was noted that the risk for thrombosis was inversely related to the INR values with higher risk associated with INRs less than 1.5 (Nassif et al., 2016). There are currently no trials to guide when and how to restart anticoagulation for those patients. In a study of 36 LVAD patients, the authors recommended holding anticoagulation for at least 10 days following ICH. Commonly, patients with LVAD are also receiving antiplatelet therapy along with the anticoagulation. Whether to hold antiplatelet therapy or not is another controversial decision that should be individualized. The RESTART trial showed that restarting antiplatelet therapy following ICH was not associated with higher recurrence. However, only 4% of the patients received antiplatelet in the first 7 days following the hemorrhage (Al-Shahi Salman et al., 2019).

SEIZURES

Seizures are reported in less than 1% of patients following cardiac surgeries. They are thought to result from multifactorial etiologies such as strokes, global hypoperfusion during surgery, metabolic derangements, and drug side effects (Pataraia et al., 2018). Generalized seizures are the most commonly encountered postoperatively (Pataraia et al., 2018). Convulsive focal or generalized seizures can be recognized and treated readily; however, nonconvulsive seizures and nonconvulsive status epilepticus are more challenging and the diagnosis is often delayed.

Some of the medications used during or after cardiac surgeries are known to decrease the seizure threshold. For example, tranexamic acid is an antifibrinolytic agent that is used during cardiac surgery to decrease blood loss and is associated with about fourfold increase in the risk of seizures (Takagi et al., 2017a). Other examples of medications that can decrease seizure threshold include cephalosporins, antihistamines, and corticosteroids.

Heart transplantation has a 13% risk for seizures (Cemillan et al., 2004). New onset seizures following heart transplant should trigger a full neurological work up. Medication side effects, infections, and malignancies are some of the possible etiologies. Patients undergoing aortic surgery have 5% risk for seizures (Pataraia et al., 2018).

Approximately, 1.3%–7.8% of children undergoing congenital heart surgery developed seizures, which have been attributed to medication toxicity or sequelae of post-cardiac arrest (Menache et al., 2002; Desnous et al., 2019). Some procedural factors are thought to be associated with seizures including high complexity and duration surgeries and extracorporeal membrane oxygenation (ECMO) use (Desnous et al., 2019). The incidence of seizures in adult patients on ECMO was estimated at 1%–2% (Pataraia et al., 2018).

Perioperative seizures can recur. A study followed 7280 patients for a median of 21 months and showed that the incidence of seizures following cardiac surgery was 0.8%, of those 59% of patients had a recurrent seizure (Manji et al., 2015).

COGNITIVE DECLINE

The risk for cognitive decline following cardiac surgeries remains controversial. It was thought that patients undergoing CABG surgery especially those using CPB are more likely to have cognitive decline from cerebral microembolization and hypoperfusion (Newman et al., 2001). Later studies have shown that the cognitive decline rates were similar in patients undergoing CABG or medical treatment and that the cognitive decline was related to the cardiac disease itself rather than the cardiac surgery (Cormack et al., 2012; Selnes et al., 2012).

Several neuroprotective strategies, such as hypothermia and ischemic preconditioning, have failed to show any benefit to prevent cognitive decline after cardiac surgery (Nathan et al., 2007; Meybohm et al., 2013). Studies suggested that patients who had a single aortic cross-clamp had better cognitive outcomes compared to those who had multiple cross clamps (Bhamidipati et al., 2017).

Other techniques used to reduce cognitive decline postcardiac surgery with variables degrees of success include intraoperative cerebral oxygen monitoring, higher blood pressure targets, antiinflammatory medications, and neuroprotective agents (Bhamidipati et al., 2017).

PSYCHOLOGICAL COMPLICATIONS

Depression and anxiety are commonly seen in patients undergoing heart surgery and are diagnosed in about 14% of the patients (Takagi et al., 2017b). This is particularly notable in the waiting period prior to CABG (Takagi et al., 2017b). Furthermore, depression is associated with higher mortality following CABG with severe depression increasing mortality by almost twofolds (Blumenthal et al., 2003).

The relationship between depression and cardiac events could be related to behavioral mechanisms such as sedentarism and poor adherence to medication and diet and biological mechanisms including alteration in the autonomic nervous system activity and elevated catecholamines and inflammatory activity, in addition to endothelial and platelet dysfunction (Carney and Freedland, 2017).

Cognitive behavioral therapy and supportive stress management are efficacious and recommended for treating depression after CABG (Freedland et al., 2009).

CONTRAST NEUROTOXICITY

This complication has been reported in 1%–2% of patients receiving iodine contrast for the catheterization procedure (Shaban and Leira, 2019). The mechanism of injury is thought to be related to blood–brain barrier (BBB) disruption and dysfunction in the cerebral autoregulation.

Diagnosis is made based on clinical symptoms including headache, seizures, encephalopathy, and sometimes cortical blindness following catheterization. Brain MRI can reveal cerebral edema or contrast leak especially in the posterior circulation, which has less capacity for autoregulation (Kocabay et al., 2014). Most common symptoms resolve with supportive care within a few days (Kocabay et al., 2014).

Previous studies have associated arterial hypertension and renal failure with contrast neurotoxicity, while there was no association with the amount of contrast or duration of the procedure (Kocabay et al., 2014; Shaban and Leira, 2019).

PERIPHERAL NERVE INJURIES

Peripheral nerve injuries are common during cardiac surgeries and often caused by mechanical injury.

Femoral artery access for catheterization is associated with low risk for femoral nerve or lumbosacral plexus injury (Kent et al., 1994). The most common mechanism of injury involves retroperitoneal hematoma formation causing pressure on the femoral nerve or the lumbar plexus. Of patients undergoing femoral catheterization, about 0.5% developed retroperitoneal hematoma (Kent et al., 1994). The risk for hematoma formation is exacerbated by doing multiple procedures, high femoral puncture point, use of large bore catheters, coagulopathies, older age, renal insufficiency, and peripheral vascular disease (Kent et al., 1994). The retroperitoneal hematoma can present with thigh or pelvic pain, decline in hemoglobin levels, arterial hypotension, and symptoms of femoral nerve injury or lumbar plexopathy. Diagnosis is confirmed by abdominal and pelvic imaging. While some may require urgent surgical decompression, most of femoral nerve and lumbosacral plexus injuries can be managed with supportive care.

Another nerve that can be injured during cardiac surgeries is the recurrent laryngeal nerve (RLN). About 1% of patients undergoing CABG can have RLN injury (Dimarakis and Protopapas, 2004). The injury results from direct surgical trauma, during traction, or can be caused by the ice slush used during surgery. The left RLN is more prone to injury than the right. Unilateral injury is the most common. Bilateral injury may occur following bilateral internal thoracic artery harvesting and can be life threatening and cause respiratory compromise (Dimarakis and Protopapas, 2004).

Radial nerve injury may result from direct injury or hematoma pressure after arterial line placement or transradial approach for catheterization (Sila, 2014). Similarly, the median nerve may be injured following brachial artery catheterization or hematoma compression.

References

Acampa M, Lazzerini PE, Guideri F et al. (2016). Ischemic stroke after heart transplantation. J Stroke 18: 157–168.

Albers GW, Marks MP, Kemp S et al. (2018). Thrombectomy for stroke at 6 to 16 hours with selection by perfusion imaging. N Engl J Med 378: 708–718.

Al-Shahi Salman R, Minks DP, Mitra D et al. (2019). Effects of antiplatelet therapy on stroke risk by brain imaging features of intracerebral haemorrhage and cerebral small vessel diseases: subgroup analyses of the RESTART randomised, open-label trial. Lancet Neurol 18: 643–652.

Aryal MR, Karmacharya P, Pandit A et al. (2015). Dual versus single antiplatelet therapy in patients undergoing transcatheter aortic valve replacement: a systematic review and meta-analysis. Heart Lung Circ 24: 185–192.

Awad H, Ramadan ME, El Sayed HF et al. (2017). Spinal cord injury after thoracic endovascular aortic aneurysm repair. Can J Anaesth 64: 1218–1235.

Baber U, van der Zee S, Fuster V (2010). Anticoagulation for mechanical heart valves in patients with and without atrial fibrillation. Curr Cardiol Rep 12: 133–139.

Bagur R, Solo K, Alghofaili S et al. (2017). Cerebral embolic protection devices during transcatheter aortic valve implantation: systematic review and meta-analysis. Stroke 48: 1306–1315.

Bhamidipati D, Goldhammer JE, Sperling MR et al. (2017). Cognitive outcomes after coronary artery bypass grafting. J Cardiothorac Vasc Anesth 31: 707–718.

Biancari F, Santini F, Tauriainen T et al. (2020). Epiaortic ultrasound to prevent stroke in coronary artery bypass grafting. Ann Thorac Surg 109: 294–301.

Blitz A (2014). Pump thrombosis—a riddle wrapped in a mystery inside an enigma. Ann Cardiothorac Surg 3: 450–471.

Blumenthal JA, Lett HS, Babyak MA et al. (2003). Depression as a risk factor for mortality after coronary artery bypass surgery. Lancet 362: 604–609.

Calway T, Rubin DS, Moss HE et al. (2017). Perioperative retinal artery occlusion: risk factors in cardiac surgery from the United States National Inpatient Sample 1998-2013. Ophthalmology 124: 189–196.

Carney RM, Freedland KE (2017). Depression and coronary heart disease. Nat Rev Cardiol 14: 145–155.

Cemillan CA, Alonso-Pulpon L, Burgos-Lazaro R et al. (2004). Neurological complications in a series of 205 orthotopic heart transplant patients. Rev Neurol 38: 906–912.

Chen L, Hua X, Song J et al. (2018). Which aortic clamp strategy is better to reduce postoperative stroke and death: single center report and a meta-analysis. Medicine (Baltimore) 97: e0221.

Cormack F, Shipolini A, Awad WI et al. (2012). A meta-analysis of cognitive outcome following coronary artery bypass graft surgery. Neurosci Biobehav Rev 36: 2118–2129.

Desnous B, Lenoir M, Doussau A et al. (2019). Epilepsy and seizures in children with congenital heart disease: a prospective study. Seizure 64: 50–53.

Devgun JK, Gul S, Mohananey D et al. (2018). Cerebrovascular events after cardiovascular procedures: risk factors, recognition, and prevention strategies. J Am Coll Cardiol 71: 1910–1920.

Dimarakis I, Protopapas AD (2004). Vocal cord palsy as a complication of adult cardiac surgery: surgical correlations and analysis. Eur J Cardiothorac Surg 26: 773–775.

Elder T, Raghavan A, Smith A et al. (2019). Outcomes after intracranial hemorrhage in patients with left ventricular assist devices: a systematic review of literature. World Neurosurg 132: 265–272.

Freedland KE, Skala JA, Carney RM et al. (2009). Treatment of depression after coronary artery bypass surgery: a randomized controlled trial. Arch Gen Psychiatry 66: 387–396.

Hahn RT, Pibarot P, Webb J et al. (2014). Outcomes with post-dilation following transcatheter aortic valve replacement: the PARTNER I trial (placement of aortic transcatheter valve). JACC Cardiovasc Interv 7: 781–789.

Hausenloy DJ, Candilio L, Evans R et al. (2015). Remote ischemic preconditioning and outcomes of cardiac surgery. N Engl J Med 373: 1408–1417.

Hauville C, Ben-Dor I, Lindsay J et al. (2012). Clinical and silent stroke following aortic valve surgery and transcatheter aortic valve implantation. Cardiovasc Revasc Med 13: 133–140.

Hemphill 3rd JC, Greenberg SM, Anderson CS et al. (2015). Guidelines for the management of spontaneous intracerebral hemorrhage: a guideline for healthcare professionals from the American Heart Association/American Stroke Association. Stroke 46: 2032–2060.

Hillis LD, Smith PK, Anderson JL et al. (2012). 2011 ACCF/AHA guideline for coronary artery bypass graft surgery: executive summary: a report of the American College of Cardiology Foundation/American Heart Association Task Force on Practice guidelines. J Thorac Cardiovasc Surg 143: 4–34.

Holy SE, Tsai JH, McAllister RK et al. (2009). Perioperative ischemic optic neuropathy: a case control analysis of 126,666 surgical procedures at a single institution. Anesthesiology 110: 246–253.

Holzmann MJ, Ahlback E, Jeppsson A et al. (2013). Renal dysfunction and long-term risk of ischemic and hemorrhagic stroke following coronary artery bypass grafting. Int J Cardiol 168: 1137–1142.

Jankovic J, Rey Bataillard V, Mercier N et al. (2019). Acute ischemic myelopathy treated with intravenous thrombolysis: four new cases and literature review. Int J Stroke 14: 893–897.

Jones BM, Tuzcu EM, Krishnaswamy A et al. (2015). Neurologic events after transcatheter aortic valve replacement. Interv Cardiol Clin 4: 83–93.

Kahlert P, Knipp SC, Schlamann M et al. (2010). Silent and apparent cerebral ischemia after percutaneous transfemoral aortic valve implantation: a diffusion-weighted magnetic resonance imaging study. Circulation 121: 870–878.

Kent KC, Moscucci M, Gallagher SG et al. (1994). Neuropathy after cardiac catheterization: incidence, clinical patterns, and long-term outcome. J Vasc Surg 19: 1008–1013. discussion 1013-4.

Kocabay G, Karabay CY, Kalayci A et al. (2014). Contrast-induced neurotoxicity after coronary angiography. Herz 39: 522–527.

Kuramatsu JB, Sembill JA, Gerner ST et al. (2018). Management of therapeutic anticoagulation in patients with intracerebral haemorrhage and mechanical heart valves. Eur Heart J 39: 1709–1723.

Levine GN, Bates ER, Blankenship JC et al. (2011). 2011 ACCF/AHA/SCAI guideline for percutaneous coronary intervention: executive summary: a report of the American College of Cardiology Foundation/American Heart Association Task Force on Practice guidelines and the Society for Cardiovascular Angiography and Interventions. Circulation 124: 2574–2609.

Limaye K, Wall M, Uwaydat S et al. (2018). Is management of central retinal artery occlusion the next frontier in cerebrovascular diseases? J Stroke Cerebrovasc Dis 27: 2781–2791.

Lin MH, Kamel H, Singer DE et al. (2019). Perioperative/postoperative atrial fibrillation and risk of subsequent stroke and/or mortality. Stroke 50: 1364–1371.

Ma H, Campbell BCV, Parsons MW et al. (2019). Thrombolysis guided by perfusion imaging up to 9 hours after onset of stroke. N Engl J Med 380: 1795–1803.

Maes F, Stabile E, Ussia GP et al. (2018). Meta-analysis comparing single versus dual antiplatelet therapy following transcatheter aortic valve implantation. Am J Cardiol 122: 310–315.

Majeed A, Kim YK, Roberts RS et al. (2010). Optimal timing of resumption of warfarin after intracranial hemorrhage. Stroke 41: 2860–2866.

Manji RA, Grocott HP, Manji JS et al. (2015). Recurrent seizures following cardiac surgery: risk factors and outcomes in a historical cohort study. J Cardiothorac Vasc Anesth 29: 1206–1211.

Menache CC, Du Plessis AJ, Wessel DL et al. (2002). Current incidence of acute neurologic complications after open-heart operations in children. Ann Thorac Surg 73: 1752–1758.

Meybohm P, Renner J, Broch O et al. (2013). Postoperative neurocognitive dysfunction in patients undergoing cardiac surgery after remote ischemic preconditioning: a double-blind randomized controlled pilot study. PLoS One 8: e64743.

Moss E, Puskas JD, Thourani VH et al. (2015). Avoiding aortic clamping during coronary artery bypass grafting reduces postoperative stroke. J Thorac Cardiovasc Surg 149: 175–180.

Moulakakis KG, Alexiou VG, Karaolanis G et al. (2018). Spinal cord ischemia following elective endovascular repair of infrarenal aortic aneurysms: a systematic review. Ann Vasc Surg 52: 280–291.

Nassif ME, Larue SJ, Raymer DS et al. (2016). Relationship between anticoagulation intensity and thrombotic or bleeding outcomes among outpatients with continuous-flow left ventricular assist devices. Circ Heart Fail 9.

Nathan HJ, Rodriguez R, Wozny D et al. (2007). Neuroprotective effect of mild hypothermia in patients undergoing coronary artery surgery with cardiopulmonary bypass: five-year follow-up of a randomized trial. J Thorac Cardiovasc Surg 133: 1206–1211.

Newman MF, Kirchner JL, Phillips-Bute B et al. (2001). Longitudinal assessment of neurocognitive function after coronary-artery bypass surgery. N Engl J Med 344: 395–402.

Nogueira RG, Jadhav AP, Haussen DC et al. (2018). Thrombectomy 6 to 24 hours after stroke with a mismatch between deficit and infarct. N Engl J Med 378: 11–21.

Nombela-Franco L, Webb JG, DE Jaegere PP et al. (2012). Timing, predictive factors, and prognostic value of cerebrovascular events in a large cohort of patients undergoing transcatheter aortic valve implantation. Circulation 126: 3041–3053.

Pataraia E, Jung R, Aull-Watschinger S et al. (2018). Seizures after adult cardiac surgery and interventional cardiac procedures. J Cardiothorac Vasc Anesth 32: 2323–2329.

Powers WJ, Rabinstein AA, Ackerson T et al. (2018). 2018 guidelines for the early management of patients with acute ischemic stroke: a guideline for healthcare professionals from the American Heart Association/American Stroke Association. Stroke 49: e46–e110.

Powers WJ, Rabinstein AA, Ackerson T et al. (2019). Guidelines for the early management of patients with acute ischemic stroke: 2019 update to the 2018 guidelines for the early management of acute ischemic stroke: a guideline for healthcare professionals from the American Heart Association/American Stroke Association. Stroke 50: e344–e418.

Ramey WL, Basken RL, Walter CM et al. (2017). Intracranial hemorrhage in patients with durable mechanical circulatory support devices: institutional review and proposed treatment algorithm. World Neurosurg 108: 826–835.

Raphael J, Moss HE, Roth S (2019). Perioperative visual loss in cardiac surgery. J Cardiothorac Vasc Anesth 33: 1420–1429.

Reddy VY, Doshi SK, Kar S et al. (2017). 5-year outcomes after left atrial appendage closure: from the PREVAIL and PROTECT AF trials. J Am Coll Cardiol 70: 2964–2975.

Rees K, Beranek-Stanley M, Burke M et al. (2001). Hypothermia to reduce neurological damage following coronary artery bypass surgery. Cochrane Database Syst Rev CD002138.

Sakamoto T (2016). Current status of brain protection during surgery for congenital cardiac defect. Gen Thorac Cardiovasc Surg 64: 72–81.

Schumacher M, Schmidt D, Jurklies B et al. (2010). Central retinal artery occlusion: local intra-arterial fibrinolysis versus conservative treatment, a multicenter randomized trial. Ophthalmology 117: 1367–1375.e1.

Selnes OA, Gottesman RF, Grega MA et al. (2012). Cognitive and neurologic outcomes after coronary-artery bypass surgery. N Engl J Med 366: 250–257.

Serruys PW, Morice MC, Kappetein AP et al. (2009). Percutaneous coronary intervention versus coronary-artery bypass grafting for severe coronary artery disease. N Engl J Med 360: 961–972.

Shaban A, Leira EC (2019). Neurological complications of Cardiological interventions. Curr Neurol Neurosci Rep 19: 6.

Sila C (2014). Neurologic complications of cardiac tests and procedures. Handb Clin Neurol 119: 41–47.

Slaughter MS, Rogers JG, Milano CA et al. (2009). Advanced heart failure treated with continuous-flow left ventricular assist device. N Engl J Med 361: 2241–2251.

Suarez J, Patel CB, Felker GM et al. (2011). Mechanisms of bleeding and approach to patients with axial-flow left ventricular assist devices. Circ Heart Fail 4: 779–784.

Szilagyi DE, Hageman JH, Smith RF et al. (1978). Spinal cord damage in surgery of the abdominal aorta. Surgery 83: 38–56.

Tahir RA, Rotman LE, Davis MC et al. (2018). Intracranial hemorrhage in patients with a left ventricular assist device. World Neurosurg 113: e714–e721.

Takagi H, Ando T, Umemoto T et al. (2017a). Seizures associated with tranexamic acid for cardiac surgery: a meta-analysis of randomized and non-randomized studies. J Cardiovasc Surg (Torino) 58: 633–641.

Takagi H, Ando T, Umemoto T et al. (2017b). Perioperative depression or anxiety and postoperative mortality in cardiac surgery: a systematic review and meta-analysis. Heart Vessels 32: 1458–1468.

Tasanarong A, Duangchana S, Sumransurp S et al. (2013). Prophylaxis with erythropoietin versus placebo reduces acute kidney injury and neutrophil gelatinase-associated lipocalin in patients undergoing cardiac surgery: a randomized, double-blind controlled trial. BMC Nephrol 14: 136.

Thomalla G, Simonsen CZ, Boutitie F et al. (2018). MRI-guided thrombolysis for stroke with unknown time of onset. N Engl J Med 379: 611–622.

Weimar C, Bilbilis K, Rekowski J et al. (2017). Safety of simultaneous coronary artery bypass grafting and carotid endarterectomy versus isolated coronary artery bypass grafting: a randomized clinical trial. Stroke 48: 2769–2775.

Wong JK, Chen PC, Falvey J et al. (2016). Anticoagulation reversal strategies for left ventricular assist device patients presenting with acute intracranial hemorrhage. ASAIO J 62: 552–557.

Yusuf S, Zhao F, Mehta SR et al. (2001). Effects of clopidogrel in addition to aspirin in patients with acute coronary syndromes without ST-segment elevation. N Engl J Med 345: 494–502.

Zierer A, Melby SJ, Voeller RK et al. (2007). Significance of neurologic complications in the modern era of cardiac transplantation. Ann Thorac Surg 83: 1684–1690.

Handbook of Clinical Neurology, Vol. 177 (3rd series)
Heart and Neurologic Disease
J. Biller, Editor
https://doi.org/10.1016/B978-0-12-819814-8.00005-6

Chapter 8

Neurological complications of heart failure

AYRTON ROBERTO MASSARO*

Department of Neurology, Hospital Sírio-Libanês, São Paulo, Brazil

Abstract

Heart failure (HF) is a major global cause of death with increasing absolute worldwide numbers of HF patients. HF results from the interaction between cardiovascular aging with specific risk factors, comorbidities, and disease modifiers. The failing heart and neuronal injury have a bidirectional interaction requiring specific management strategies. Decreased cardiac output has been associated with lower brain volumes. Cerebral blood flow (CBF) may normalize following heart transplantation among severe HF patients. Stroke and cognitive impairment remain the main neurologic conditions associated with HF. However, HF patients may also suffer from chronic cerebral hypoperfusion. It seems likely that HF-related ischemic strokes are primarily the result of cardiac embolism. Atrial fibrillation (AF) is present in half of stroke patient with HF. The increased risk of hemorrhagic strokes is less well characterized and likely multifactorial, but may in part reflect a higher use of long-term antithrombotic therapy. The steady improvement of neuroimaging techniques has demonstrated an increased prevalence of silent ischemic lesions among HF patients. The populations most likely to benefit from long-term anticoagulant therapy are HF patients with AF. Cognitive impairment in HF can have a variety of clinical manifestations from mild memory problems to dementia.

INTRODUCTION

Heart failure (HF) is a major global cause of death (Gronewegen et al., 2020). This is a reflection of its chronic nature as well as growth and aging of the world population (Gronewegen et al., 2020). The rate of HF hospitalization represents 1%–2% of all hospital admissions and is more common among patients aged >65 years (Berry et al., 2001; Alla et al., 2007; Epstein et al., 2011; Braunwald, 2015; Benjamin et al., 2018). In addition, HF has been associated with the highest 30-day hospital readmission rate (Dharmarajan et al., 2013; Patil et al., 2019). Approximately, half of HF patients are admitted at least once within 1 year following diagnosis (Nichols et al., 2015). Recent meta-analysis including over 1.5 million HF patients estimated the 1-, 2-, 5-, and 10-year survival rate to be 87%, 73%, 57%, and 35%, respectively (Jones et al., 2019).

An estimated 64.3 million people are currently living with HF worldwide (GBD 2017 Disease and Injury Incidence and Prevalence Collaborators, 2018). The prevalence of HF has been estimated at 1%–2% of the general adult population (Van Riet et al., 2016; Conrad et al., 2018). Variations in the diagnostic criteria may further increase these numbers (Van Riet et al., 2016).

Patients with HF typically show evidence of a structural or functional abnormality of the heart resulting in reduced cardiac output and/or elevated intracardiac pressures either at rest or during stress (Ponikowski et al., 2016). Typical symptoms or signs, early measurement of plasma natriuretic peptide level, and echocardiography are routinely used to diagnose and classify patients with HF (Ponikowski et al., 2016). Enhanced awareness of mild HF as well as appropriate and timely use of diagnostic ancillary tools remains of permanent importance (Ponikowski et al., 2016).

*Correspondence to: Ayrton Roberto Massaro, M.D., Department of Neurology, Hospital Sírio-Libanês, São Paulo, SP, Brazil. Tel: +55-11-31298574, E-mail: ayrton.massaro@gmail.com

Whereas left ventricular ejection fraction (LVEF) is decreased in HF patients with systolic dysfunction, the ejection fraction is normal in patients with diastolic dysfunction characterized by elevated end-diastolic ventricular pressure (Ponikowski et al., 2016). HF patients are most often categorized as having reduced LVEF <40% (HFrEF), mid-range (HFmrEF; LVEF = 40%–49%), or preserved ejection fraction (HFpEF; LVEF >50%) (Ponikowski et al., 2016). Many classification systems have been used, including those based on symptom severity (New York Heart Association Functional Classification System—NYHA) or on disease progression (Metra and Teerlink, 2017).

HF is associated with decreased systemic perfusion and increased organ congestion, leading to organ dysfunction. Important multiorgan interactions with the failing heart result in the cardio-renal and cardio-hepatic syndromes (Laribi and Mebazaa, 2014; Rangaswami et al., 2019), whereas the clinical interface between the fainting heart and brain injury has been less well defined (Havakuk et al., 2017; Doehner et al., 2018).

ETIOLOGY OF HEART FAILURE

The heart and brain share a set of common vascular risk factors associated with HF. The most remarkably modifiable risk factors predisposing to HF are coronary artery disease, arterial hypertension, diabetes mellitus, obesity, valvular heart disease, physical inactivity, and cigarette smoking (He et al., 2001).

HF is a life-threatening syndrome with distinct geographic variations in mortality and morbidity rates depending on different etiologies. A wide range of cardiac conditions, hereditary defects, and systemic diseases result in HF. Multiple causes may coexist in the same patient, and they should be thoroughly evaluated to ensure better treatment. Ischemic causes of HF are prevalent in high-income countries, whereas infectious diseases including rheumatic fever and Chagas cardiomyopathy remain important causes of HF in low-middle income countries across the globe (Khatibzadeh et al., 2013; Agbor et al., 2020).

Although many comorbidities exist with HF, coronary artery disease, arterial hypertension, obesity and diabetes mellitus are robustly associated with the highest risk for HF in the Western world (Komanduri et al., 2017). Almost a third of patients with chronic coronary artery disease have HF (Parma et al., 2020). The rate of myocardial infarctions as a surrogate for ischemic heart disease has recently been modified due to the epidemiological transitions among both high- and low-middle income countries (Reynolds et al., 2017). An increase in blood pressure enhances mechanical stress increasing myocardial mass resulting in left ventricular hypertrophy and HF (Yussuf et al., 1989). However, although arterial hypertension is a major modifiable risk factor for HF, gaps in the treatment of high blood pressure still remain, and longstanding uncontrolled arterial hypertension ultimately leads to HF (Messerli et al., 2017).

Thus, early identification and optimal control of modifiable vascular risk factors are crucial to improved outcomes among HF patients. Traditionally, diabetes mellitus and heart failure have been treated as two distinct disease entities. However, recent therapeutic advances demonstrate remarkable concomitant improvements in outcomes for both diabetes mellitus and cardiovascular disease (Metha et al., 2020). Diabetes mellitus-related heart disease occurs in the form of coronary artery disease, cardiac autonomic neuropathy, or diabetic cardiomyopathy (Tarquini et al., 2018). In early stages, diastolic dysfunction is the only abnormality, but systolic dysfunction may follow in later stages associated with impaired LVEF (Tarquini et al., 2018). Microvascular diabetic complications have the strongest association with diabetes cardiomyopathy (Tarquini et al., 2018).

Maximizing adherence to evidence-based practices remains the best strategy to improve the management of HF patients. It is also of the utmost importance to include among modifiable risk factors related to HF, sedentary lifestyle, and obesity, as physical activity may lower the cardiovascular risk (Florido et al., 2020).

HF mechanisms differ between young and elderly patients who represent the vast majority of patients. HF results from the interaction between cardiovascular aging with specific risk factors, comorbidities, and disease modifiers (Triposkiadis et al., 2020).

The prevalence of amyloidosis among elderly individuals is high (Porcari et al., 2020). Amyloidosis is a systemic disease due to the buildup of protein material in the extracellular space, mainly in its light chain and transthyretin forms (Porcari et al., 2020). Most frequent presentation includes HF associated with biventricular hypertrophy (Porcari et al., 2020). Other symptoms include orthostatic hypotension due to autonomic dysfunction or infiltration of the sinus node. Cardiac magnetic resonance imaging (MRI) may demonstrate diffuse late gadolinium subendocardial enhancement (Porcari et al., 2020).

Degenerative abnormalities associated with severe aortic stenosis and mitral regurgitation are found in population aged ≥75 years and are related to increasing HF risk with advancing age (Rostagno, 2019).

HEART FAILURE AND ATRIAL FIBRILLATION

The association between atrial fibrillation (AF) and HF has been established almost a century ago (Anter et al., 2009). Paul Dudley White noted *"since auricular fibrillation so often complicates very serious heart disease, its occurrence may precipitate heart failure or even death, unless successful therapy is quickly instituted"* (Anter et al., 2009). The coexistence of these two conditions is best explained by shared vascular risk factors (Prabhu et al., 2017). HF precipitates AF by contributing to atrial remodeling due to: (1) increased filling pressures; (2) alterations in calcium handling; and (3); alterations of the electrical properties of atrial tissue (Prabhu et al., 2017).

The prevalence of AF in HF series ranges from 13% to 27% (Middlekauff et al., 1991; Carson et al., 1993; Senni et al., 1998; Mahoney et al., 1999). The prevalence further increases in parallel with disease severity, ranging from 5% among patients with mild HF to 10%–26% among patients with moderate, up to 50% in patients with severe HF (Maisel and Stevenson, 2003). In a prospective study of HF patients in sinus rhythm, onset of AF was associated with highly significant worsening or the NYHA functional classification, peak oxygen consumption, and cardiac index, along with increased mitral and tricuspid regurgitation (Pozzoli et al., 1998). Interestingly, restoration of sinus rhythm improved cardiac output, exercise capacity, and maximal oxygen consumption (Gosselink et al., 1994).

PHYSIOLOGICAL BRAIN CHANGES IN HF PATIENTS

The failing heart and neuronal injury have a bidirectional interaction and require specific management (Havakuk et al., 2017). Prior reviews on the physiological brain changes associated with HF are crucial before approaching a discussion of major neurological complications (Scherbakov and Doehner, 2018).

Cerebral blood flow and heart failure

Given its high metabolic demands, the brain receives 12% of the cardiac output despite accounting for only 2% of total body weight (Williams and Leggett, 1989). Brain perfusion is protected by cerebral autoregulation such that brain perfusion remains constant over a wide range of perfusion pressures (Lassen, 1959). Clinical evidence of disturbed cerebral function in HF has been long recognized. However, difficulties in adequately measuring the cerebral blood flow limited our understanding. In the early 1950s, a reduction of the CBF and an increase in cerebrovascular resistance among HF patients have been observed using the nitrous oxide method (Scheinberg, 1950).

Subclinical reductions in cardiac output disrupt CBF homeostasis (Erkelens et al., 2017). The relationship between CBF and cardiac output has been unequivocally documented among patients with severe HF (Massaro et al., 2006). Abnormalities in regional and global CBF as well as lower blood flow velocities in the middle cerebral artery have been described in patients with severe HF (Massaro et al., 2006). During exercise, cerebral autoregulation is also impaired in HF patients (Caldas et al., 2018). This sharp decline in cerebral perfusion may be associated with cognitive impairment (Eisenberg et al., 1960).

Furthermore, cerebrovascular reactivity is related to the NYHA classification and degree of LVEF. HF patients with low cardiac output and impaired cerebrovascular reactivity seem more susceptible to ischemic brain damage (Georgiadis et al., 2000). CBF may normalize following heart transplantation (Massaro et al., 2006), suggesting that mechanisms underlying cerebral autoregulation may be reversible (Caldas et al., 2018).

Hypoxia and heart failure

Intermittent periods of hypoxia are observed in HF patients and may have direct effect on the heart and peripheral blood vessels (Mansur et al., 2018). Specialized chemoreceptors located in the aorta and carotid arteries play a key role in detecting changes in blood oxygen concentration (Mansur et al., 2018). Response to hypoxia resuting in improved organ perfusion, increased cardiac output, and peripheral vasodialtaion may differ among health individuals and HF patients (Ainslie and Poulin, 2004). There is an intensification of the compensatory mechanisms for preserving cerebral blood flow in HF patients during hypoxia (Mansur et al., 2018).

Increased chemosensitivity to both hypoxia and hypercapnia, eliciting neuro-hormonal derangements, ventilatory instability, and ventricular arrhythmias are serious adverse prognostic markers in HF (Giannoni et al., 2009).

Other physiological parameters and heart failure

HF may also induce a hypercoagulable state associated with decreased CBF and a variety of rheological alterations, including increased platelet aggregation, reduced fibrinolysis, endothelial dysfunction, and activation of the inflammatory response (Kim et al., 2016).

Excess sympathetic activation occurs in HF patients and is neutrally mediated (Kishi, 2016). The central renin–angiotensin system is the major mechanism regulating the sympathetic nervous system in HF (Leenen, 2007).

NEUROLOGICAL COMPLICATIONS ASSOCIATED WITH HEART FAILURE

Neurological complications are common among patients with HF. Central nervous system manifestations are well established as part of the HF syndrome (Yancy et al., 2017). Stroke and cognitive impairment are the main neurologic conditions associated with HF. Furthermore, HF patients may also have a chronic cerebral hypoperfusion state (Pullicino and Hart, 2001; Pullicino et al., 2001).

Stroke and heart failure

Strokes are common neurologic complications of HF; ischemic strokes are the most frequently observed subtype (Pogmoragot et al., 2016; Adelborg et al., 2017). A US cohort study of 630 HF patients reported a 17-fold increased risk at 30 days compared with the general population. The excess risk persisted over 5 years (Witt et al., 2006). Although less frequent, only one study demonstrated an increased risk of hemorrhagic strokes in HF patients (Lip et al., 2012).

HF was reported in 9.1% ($n=1124$) among 12,396 patients with ischemic stroke in the Registry of the Canadian Stroke Network (Pogmoragot et al., 2016). The frequency of HF in ischemic stroke patients increased among 632,225 stroke patients evaluated between 1995 and 2005 in the United States (Divani et al., 2009). The stroke risk seems higher within 1 month following diagnosis of HF but returns to baseline within 6 months (Alberts et al., 2010). HF patients with ischemic stroke were frequently old, female, African-Americans and more likely to have additional vascular risk factors (Divani et al., 2009; Alberts et al., 2010; Lip et al., 2012).

Mechanisms of stroke in heart failure

Several mechanisms underlie the increased risk of ischemic stroke in HF patients including the presence of intracavitary thrombi in the dilated and hypokinetic left ventricle due to wall motion abnormalities, and in the left atrium due to coexistent AF. In addition to shared stroke risk factors, as previously discussed, HF is associated with endothelial dysfunction and hypercoagulability. The increased risk of hemorrhagic stroke is less well characterized and likely related to a higher use of antithrombotic drugs.

HF-related strokes are primarily the result of cardiac embolism (Table 8.1) (Falk et al., 1992; Pullicino et al., 2000; Vemmos et al., 2012; Pogmoragot et al., 2016). Therefore, HF patients with decreased LVEF are at increased risk of cardioembolic stroke (Dries et al., 1997; Freudenberger et al., 2007). An annual thromboembolic rate of 1.0% has been observed in HF patients without AF; the risk increases with further reductions of the LVEF, particularly among those with an LVEF of <30% (Dries et al., 1997; Freudenberger et al., 2007).

Table 8.1

Differences in the frequency of ischemic stroke subtypes between patients with and without HF

Stroke type	Patients without HF ($n=2621$)	Patients with HF ($n=144$)
Large artery atherosclerotic	16.7% (437)	8.3% (12)
Cardioembolic	26.8% (703)	81.9% (118)
Lacunar	20.6% (539)	2.1% (3)
Systemic hypoperfusion	0.2% (6)	2.8% (4)
Underdiagnosed etiology	17.6% (461)	2.1% (3)
Intracerebral hemorrhage	15.9% (417)	5.3% (4)

Vemmos, K, Ntaios G, Savvari P, et al. (2012). Stroke aetiology and predictors of outcome in patients with heart failure and acute stroke: a 10-year follow-up study. Eur J Heart Fail 14: 211–218.

AF is present in about half of stroke patients with HF (Vemmos et al., 2012; Pogmoragot et al., 2016). The CHA_2DS_2-VASc score, originally developed to predict stroke risk in AF patients, has also been reported to be predictive for strokes among HF patients regardless of the presence of AF (Melgaard et al., 2015). Although the HF without AF group had a slightly lower overall stroke rate than the AF with HF group, stroke risk was comparable in both groups when patients were stratified by CHA_2DS_2-VASc scores (Kang et al., 2017). Not uncommonly, silent AF is often diagnosed following stroke among HF patients having routine electrocardiographic monitoring (Dobreanu et al., 2013).

Valvular heart disease and dilated cardiomyopathy have been mainly associated with cardioembolic strokes, whereas HF due to coronary artery disease or arterial hypertension has been mainly associated with atherosclerotic and lacunar strokes, respectively (Vemmos et al., 2012).

Patients with HF are also at increased risk of cerebral hypoperfusion, as a result of reduced cardiac output (Meng et al., 2015). Among patients with severe carotid artery stenosis, volume of ipsilateral brain infarctions may be greater among HF patients with critical impairment of the ejection fraction (<35%) (Fig. 8.1) (Pullicino et al., 2001).

The hypoperfusion state in HF patients may also result in nonfocal transient neurological attacks (TNAs) (Meng et al., 2015; Oudeman et al., 2019). Transient neurological attacks are characterized by acute onset of neurological signs or symptoms, such as unsteadiness, confusion, or

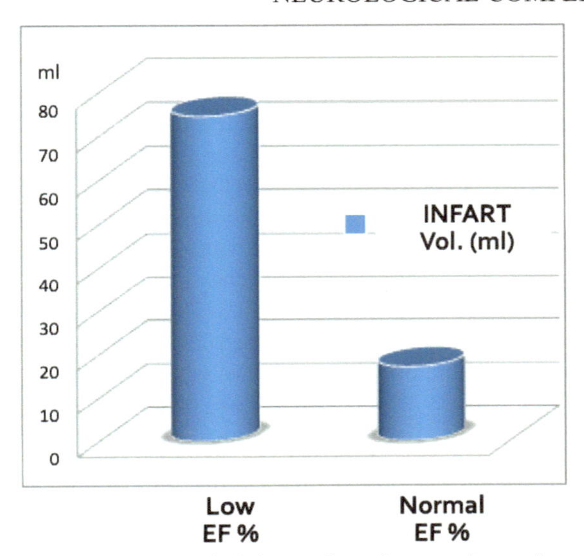

Fig. 8.1. Differences in infarct volume between low and normal cardiac ejection fraction in patients with ipsilateral severe carotid stenosis. From Pullicino P, Mifsud V, Wong E, et al. (2001). Hypoperfusion-related cerebral ischemia and cardiac left ventricular systolic dysfunction. J Stroke Cerebrovasc Dis 10: 178–182.

Table 8.2

Acute signs and symptoms characterized as transient neurological attacks (TNAs)

Nonfocal transient neurological attack	Number (%)
Unconsciousness	4 (14)
Confusion	1 (3)
Amnesia	1 (3)
Unsteadiness	4 (14)
Bilateral leg weakness	2 (7)
Blurred vision	3 (10)
Nonrotatory dizziness	22 (76)
Paresthesias	4(14)

Oudeman EA, Bron EE, van den Berg-Vos RM, et al. (2019). Cerebral perfusion and the occurrence of nonfocal transient neurological attacks. Cerebrovasc Dis 47: 303–308.

bilateral weakness (Table 8.2) (Bots et al., 1997). In contrast to transient ischemic attacks (TIAs), manifestations of TNAs cannot be attributed to one specific arterial territory of the brain (Koudstaal et al., 1992; Bots et al., 1997; Oudeman et al., 2019). TNAs have been associated with an increased risk of cardiac events, stroke, and cognitive impairment (Koudstaal et al., 1992). Transient and chronic global cerebral hypoperfusion are considered an important cause of TNAs (Koudstaal et al., 1992; Bots et al., 1997; Meng et al., 2015; Oudeman et al., 2019).

Heart failure during acute ischemic stroke

Early studies raised concern that HF might modify the response to administration of intravenous recombinant tissue-type plasminogen activator (rtPA) in patients with acute ischemic stroke due to low cardiac output potentially generating longstanding cardiac thrombi (Ois et al., 2008; Palumbo et al., 2012). Recent findings demonstrated that the efficacy and safety of intravenous rtPA and mechanical thrombectomy did not differ when compared to patients without HF (Abdul-Rahim et al., 2015; Siedler et al., 2019). In addition, the recanalization rate with endovascular treatment remains unaffected by the presence of HF (Siedler et al., 2019). Therefore, the unfavorable outcome of HF patients during acute ischemic stroke treatment may reflect long-range complications of HF, including thrombotic and embolic events (Siedler et al., 2019).

"Silent" stroke in heart failure

The improvement of advanced neuroimaging has increased the prevalence of silent ischemic lesions detected by MRI (Haeusler et al., 2011). Several studies have previously demonstrated silent strokes in patients with HF (Siachos et al., 2005; Kozdag et al., 2006). The severity of cardiac dysfunction is relevant to structural brain alterations. Left ventricular ejection fractions $<20\%$ have been associated with higher prevalence of ischemic brain lesions detected by MRI (Siachos et al., 2005). MRI performed in 20 asymptomatic HF patients demonstrated higher rate of silent cortical infarctions (20%) when compared to controls (Schmidlt et al., 1991).

Neuron specific enolase levels are elevated in HF patients with "silent" brain infarctions, suggesting silent neuronal injury (Ozyuncu et al., 2019). These findings further support the evidence for the cognitive impairment observed among HF patients.

Stroke outcome and heart failure

HF is a strong predictor of unfavorable functional outcome following acute ischemic stroke (Vemmos et al., 2012). Increasing age, diabetes mellitus, low systolic blood pressure, higher blood urea nitrogen levels, ischemic heart disease, stroke severity, and HF functional status have been related to worse outcome (Witt et al., 2006; Haeusler et al., 2011; Vemmos et al., 2012; Pogmoragot et al., 2016). There is also a relationship between LVEF with functional outcome among HF patients with acute ischemic stroke (Rahmayani et al., 2018; Li et al., 2019). The majority of deaths among HF patients is related to cardiac embolism or related to other cardiovascular causes (Vemmos et al., 2012). Stroke patients with HF have a worse neurologic deficits and longer hospital stay (Appleros et al., 2003; Ois et al., 2008; Divani et al., 2009; Milionis et al., 2013).

Stroke prevention in heart failure

HF mortality has decreased due to improvements in medical management, which includes the use of angiotensin converting enzyme inhibitors (ACEIs). Stroke-related HF patients with AF are more frequently discharged on oral anticoagulant therapy, whereas non-AF patients are mainly discharged on antiplatelet therapy (Vemmos et al., 2012). Optimal selection of antithrombotic therapy remains a paramount question in the care of HF patients (Schumacher et al., 2018).

A recent meta-analysis failed to demonstrate significant differences in major cardiovascular events comparing different antithrombotic strategies in HF patients in sinus rhythm (Ueyama et al., 2020).

The first trial to assess the effect of antithrombotic therapy among patients with HF was the unblinded, randomized WASH trial (Warfarin/Aspirin Study in Heart failure); WASH enrolled 279 HF patients requiring diuretic therapy were assigned to warfarin, aspirin (300 mg), or no antithrombotic treatment. The WASH trial failed to prove benefit with either aspirin or warfarin (Cleland et al., 2004).

The multicenter, randomized, double-blind and placebo-controlled Heart Failure Long-term Anti-Thrombotic Study trial was designed to demonstrate the potential benefit of long-term antithrombotic treatment in HF patients. However, very low enrollment precluded definitive conclusions about efficacy, and treatment did not seem to affect outcome (Cokkinos et al., 2006).

The Warfarin vs Aspirin in Reduced Cardiac Ejection Fraction trial that enrolled 2305 patients randomized to either aspirin 325 mg or warfarin (target international normalized ratio [INR] of 2.0–3.5) in patients in sinus rhythm who had a reduced LVEF. The initial design target 2860 participants, but recruitment difficulties also occurred in this trial like previous studies. After a mean follow-up of 3.5 years, warfarin did not reduce the primary composite endpoint of ischemic stroke, intracerebral hemorrhage, and all-cause death. A reduced risk of ischemic stroke with warfarin was offset by an increased risk of major hemorrhage (Homma et al., 2012). Patients with systolic dysfunction, NYHA class III/IV, poor health-related quality of life and higher rate of HF hospitalization were independently associated with suboptimal quality of warfarin anticoagulation control (Lee et al., 2018).

Therefore, it appears that the most likely group of patients to benefit from long-term anticoagulant therapy are HF patients who also have AF. Although anticoagulation is generally not recommended in patients with HF or sinus rhythm, it may be initiated in specific conditions closely related to the disease. The presence of intracardiac thrombi could be considered an indicator for high-embolic risk and stroke (Stratton and Resnick, 1987).

The direct oral anticoagulants (DOACs) are approved for the prevention of thromboembolism in patients with nonvalvular AF. As opposed to vitamin K antagonists, DOACs have a predictable anticoagulant effect, which is independent of dietary habits. Therefore, DOACs are administered at standard doses and do not require frequent laboratory measurements. In addition, patients are not exposed to either suboptimal or supratherapeutic anticoagulant effects, which frequently occurs with the vitamin K antagonists. DOACs in patients with HF and AF are more or at least equally effective compared with warfarin in the prevention of thromboembolism. In addition, DOACs use is associated with lower rates of major bleeding including intracranial bleeding (Savarese et al., 2016).

Despite these encouraging results observed with the use of DOACs in patients with HF and concomitant AF, as of today, data are still lacking regarding the impact of DOACs in HF patients in sinus rhythm. Toward this direction, the Cardiovascular Outcomes for People using Anticoagulation Strategies trial suggested that the addition of rivaroxaban 2.5 mg twice daily to aspirin in the subgroup of patients with HF and vascular disease significantly reduced the composite of cardiovascular death, myocardial infarction, and stroke. Major bleeding events were similar between the two treatment groups in HF patients (Branch et al., 2019). However, these positive results were not reproduced in the COMMANDER HF trial, which assessed the effect of rivaroxaban 2.5 mg twice daily on patients with a recent episode of HF decompensation, concomitant coronary disease and no AF (Zannad et al., 2018). In an exploratory analysis of this large global trial, it was demonstrated that low-dose rivaroxaban appeared to safely attenuate risk of stroke or TIA in the vulnerable early and late phase after a recent episode of worsening HF. Further investigations are need to evaluate selected at-risk populations of patients with HF and sinus rhythm with DOACs (Mehra et al., 2019).

Cognitive impairment in heart failure

Like King Lear, some have been made mad by their social circumstances, more commonly, a mild dementing state has become much more severe in the presence of physical illness such as HF (Anon, 1977).

Cardiovascular diseases and cardiovascular risk factors increase the incidence of cognitive impairment and dementia (Hooghiemstra et al., 2019). Approximately 25%–75% of all patients with HF experience neuropsychological changes, including impaired attention span and concentration, memory loss, reduced psychomotor speed, and decreased executive function (Hadjuk et al., 2013; Ampadu and Morley, 2015). Among elderly patients with HF, cognitive impairment is present in approximately 40% (Cannon et al., 2017; Brunen et al., 2020). Frailty, a syndrome characterized by an exaggerated decline in function and reserve of multiple physiological systems, is common among older patients with HF and is associated with worse outcomes (Pandey et al., 2019). Unfortunately, many HF patients with cognitive impairment are not recognized during their life (Hanon et al., 2014).

Cognitive impairment in HF can have a variety of clinical manifestations from mild memory problems through to full dementia. Delirium occurring during an admission for acute HF is associated with increased mortality and length of hospital stay.

Pathophysiology of cognitive impairment in heart failure

A report in The Lancet first described the term cardiogenic dementia and emphasized the importance of searching for a cardiac mechanism in conditions of brain functions (Anon, 1977).

Although the pathophysiological mechanisms involved in cognitive impairment associated with HF are not fully understood, chronic hypoperfusion, microembolic events, and other ischemic syndromes may play a role, with possible effects of disruptions of the blood–brain barrier, cerebral inflammation, and endothelial dysfunction. There are structural brain changes in areas related to cognition like the hippocampus. These findings may facilitate the development of cognitive impairment in HF patients (Mueller et al., 2020).

If the heart cannot pump out adequate amount of blood to the brain, there is increasing risk of cognitive impairment due to the intensification of chronic brain hypoperfusion. The hemodynamic instability will damage vulnerable brain and intensify age-dependent cognitive decline (De la Torre, 2020). Cerebral blood flow declines with increasing age, and gradual disruptions in cerebral hemodynamics appear to play a critical role in the pathogenesis of CI in HF patients (De la Torre, 2002; Alosco and Hayes, 2015).

Decreased cardiac output has also been shown to be associated with lower brain volumes (Jefferson et al., 2010, 2017). The Framingham Study demonstrated that cardiac index was positively correlated with total brain volume and information processing speed (Jefferson et al., 2010, 2017). These findings can be detected even in the absence of heart failure or severe cardiomyopathy (Jefferson, 2010). The degree of memory and CI in presence of a reduced LVEF seems to be variable with age (Festa et al., 2011).

There is a U-shaped association between left ventricular ejection fraction and the brain aging biological markers. Lower LVEF is related to abnormal brain aging, but also noted among those with the highest ejection fraction values (Jefferson et al., 2011).

Cardiac dysfunction associated with decreased CBF places aging brain at risk for Alzheimer's disease (Jefferson et al., 2015). Among cognitively normal adults, lower left ventricular blood flow was related to greater cerebral amyloid accumulation and greater evidence of tau phosphorylation and neurodegeneration. Therefore, modest age-related changes in cardiovascular function may have implications for pathophysiological changes in the brain later in life (Zuccala et al., 2003; Kresge et al., 2020).

Other possible causes include the interplay between decreased cardiac output and previously described comorbidities including microvascular damage from diabetes mellitus, vascular damage from hypertension, and arterial disease. The proximal aorta acts as a coupling device between heart and brain perfusion, modulating the amount of pressure and flow pulsatility transmitted into the cerebral microcirculation. The previous described multiple risk factors may enhance aortic stiffening as an independent and key determinant of brain structure and cognitive function (Mitchell et al., 2011). The detrimental effects of aortic stiffening may result in brain damage as well as heart failure. The resulting cerebral small vessel disease and heart failure may contribute to early cognitive decline and vascular dementia (de Roos et al., 2017).

AF may exacerbate cognitive impairment in HF patients due to decreased cerebral hypoperfusion associated with possible chronic microembolism leading to silent cerebral infarctions (Zito et al., 1996; Alosco et al., 2015).

Cerebral hypoperfusion associated with HF patients may be related to the breakdown of the neurovascular unit enabling progression into cognitive impairment and Alzheimer's disease (Zlokovic, 2011). Cognitive decline may start before cerebral hypoperfusion involving platelet hyperreactivity and activation of the brain endothelium inflammatory response (Adamski et al., 2018).

Depression and heart failure

Depression is a frequent and debilitating comorbidity affecting HF patients (Sbolli et al., 2020a). Moreover, depression carries a risk for HF and is associated with worse quality of life and clinical outcomes. Therapy for depression consists of nonpharmacological interventions and pharmacological interventions. Exercise training and cognitive therapy have beneficial effects on depressive symptoms. Selective serotonin reuptake inhibitors appear safe in HF patients. New therapies may offer advantages like repetitive transcranial magnetic stimulation.

HF patients with more severe disease showed whole-brain and regional brain hypometabolism in ^{18}F-FDG PET/CT. Depressed HF patients exhibited different metabolic patterns that could be used for differential depression diagnosis (Lyra et al., 2020).

Diagnosis of cognitive impairment

Cognitive impairment (CI) in HF patients is predicted by baseline cognitive status, demographic variables, and NYHA functional class. Therefore, there is a need to frequent assessment of cognitive function in patients with HF (Lee et al., 2019).

Multiple validated questionnaires exist to screen for CI in HF patients. These include the Mini-Mental Status Examination (MMSE), the Montreal Cognitive Assessment, and more recently the Mini-Cog (Athilingam et al., 2011; Saito et al., 2020). The MMSE seems not to be very sensitive in detecting mild CI. There is a growing need for a cognitive assessment tool, which is not time consuming and can provide clinically relevant information. Although the MMSE has been used in most studies and has been proposed as a cognitive function assessment tool for patients with HF, the Mini-Cog has been recently demonstrated a cognitive assessment tool that may be suitable in HF patients. The CI defined by Mini-Cog provided incremental prognostic information than MMSE in HF patients.

Therapeutic considerations in HF patients with cognitive impairment

Cerebral autoregulation may be impaired during exercises in patients with heart failure. Exercise capacity is a modifiable factor that could be improved in HF patients with the potential to improve cognition and other outcomes (Vellone et al., 2020).

There are supportive reasons for ACEIs that may be useful in reducing the risk of cognitive decline in HF patients. In addition to the antiinflammatory actions of ACEIs, there is a role modulation of the increased concentration of angiotensin converting enzymes, angiotensin II, and angiotensin I receptors in the cerebral cortex (Savioli Neto et al., 2009).

One of the biggest challenges in treating elderly HF patients with cognitive decline is the association of comorbidities, sarcopenia, cachexia, and polypharmacy. Therefore, before considering any specific treatment for the cognitive impairment, it is important to optimize the cardiac output through the management of the medications avoiding interactions or potential hypotension. Transcranial Doppler (TCD) is a useful tool to monitoring indirectly CBF changes in HF patients (Massaro et al., 2006).

Stroke-heart syndrome and heart failure

The concept of cardiac manifestations induced by ischemic strokes as a direct consequence of brain ischemia implies that cardiac disturbances occur after the onset of neurological deficits. Most of these stroke-associated cardiac disturbances are transient and peak within the first 3 days after the event. Takotsubo syndrome is a particular type of left ventricular dysfunction showing a characteristic pattern of left ventricular dysfunction that resembles a Japanese octopus trap called takotsubo (El-Battrawy et al., 2016; Pelliccia et al., 2017). There is a need to rule out acute coronary syndrome, especially in patients with ST segment elevation. Echocardiography should be considered to identify regional wall motion abnormalities in patients with stroke. In addition, there is a need to differentiate the apical ballooning type, mid-ventricular, basal, and focal types of Takotsubo syndrome (El-Battrawy et al., 2016; Pelliccia et al., 2017). The focal type of Takotsubo syndrome can resemble focal wall motion abnormalities seen in acute coronary syndrome and cardiac MRI can be useful to the differential diagnosis. This syndrome predominantly affects postmenopausal women and is often preceded by stressful physical or emotional triggers (El-Battrawy et al., 2016; Pelliccia et al., 2017). Acute neurological disorders like acute stroke are also common triggers of Takotsubo syndrome (El-Battrawy et al., 2016). The syndrome has been reported in 0.5%–1.2% of patients after acute stroke and when secondary to acute stroke can occur without the presence of emotional of psychological stress (El-Battrawy et al., 2016; Pelliccia et al., 2017).

Catecholamine levels are considerable higher in patients with Takotsubo syndrome than in those with myocardial infarction (Wittstein et al., 2005). In addition, the amount of endothelin in plasma is increased, which further supports the notion that endothelial dysfunction and microvascular constriction play an important part in the pathophysiology of Takotsubo syndrome. Patients with Takotsubo syndrome suffering from cancer might be considered as a high-risk group due to the increased incidence of stroke at long-term follow-up (Moller et al., 2018).

Sleep disorders and heart failure

Patients with HF often complain of sleep fragmentation or nonrestorative sleep, difficulties in initiating sleep, or waking up too early in the morning. Insomnia is also prevalent in patients with HF and is often suggested as a cause of poor sleep quality together occurring in an estimated 33% of patients with HF (Hayes et al., 2009). It may be caused by adverse effects of commonly prescribed HF medications, such as diuretics and β-blockers (Jimenez et al., 2011). Nocturia is also commonly associated with orthopnea (Redeker et al., 2012). Sleep-disordered breathing is highly prevalent in patients with HF. Both central and obstructive sleep apneas are frequently observed in these patients and were shown to have an important prognostic value. One relevant risk factor associated with sleep apnea in HF patients is obesity (Parati et al., 2016).

The negative intrathoracic pressure during obstructive sleep apnea is responsible for an increased left ventricular transmural pressure, which increases the afterload impairing the already failing left ventricular function. These recurrent events determine further increase of the elevated sympathetic activity in HF patients (Parati et al., 2016). This situation increased myocardial oxygen demand during sleep reducing cardiac output supporting myocardial dysfunction and ischemia. Nocturnal arrhythmias may also be present contributing to further cardiovascular and cerebrovascular diseases (Parati et al., 2016).

References

Abdul-Rahim AH, Fulton RL, Frank B et al. (2015). Association of chronic heart failure with outcome in acute ischemic stroke patients who received systemic thrombolysis: analysis from VISTA. Eur J Neurol 22: 163–169.

Adamski MG, Sternak M, Mohaissen T et al. (2018). Vascular cognitive impairment linked to brain endothelium inflammation n early stages of heart failure in mice. J Am Heart Assoc 7: e007694.

Adelborg K, Szepligeti S, Sundboll J et al. (2017). Risk of stroke in patients with heart failure: a population-based 30-year cohort study. Stroke 48: 1161–1168.

Agbor VM, Ntusi NAB, Noubiap JJ (2020). An overview of heart failure in low- and middle-income countries. Cardiovasc Diagn Ther 10: 244–251.

Ainslie PN, Poulin MJ (2004). Ventilatory cerebrovascular, and cardiovascular interactions in acute hypoxia: regulation by carbon dioxide. J Appl Physiol 97: 149–159.

Alberts VP, Bos MJ, Koudstaal PJ et al. (2010). Heart failure and the risk of stroke: the Rotterdam Study. Eur J Epidemiol 25: 807–812.

Alla F, Zannad F, Filippatos G (2007). Epidemiology of acute heart failure syndromes. Heart Fail Res 12: 91–95.

Alosco ML, Hayes SM (2015). Structural brain alterations in heart failure: a review of the literature and implications for risk of Alzheimer's disease. Heart Fail Rev 20: 561–571.

Alosco ML, Spitznagel MB, Sweet LH et al. (2015). Atrial fibrillation exacerbates cognitive dysfunction and cerebral perfusion in heart failure. Pacing Clin Electrophysiol 38: 178–186.

Ampadu J, Morley JE (2015). Heart and cognitive dysfunction. Int J Cardiol 178: 12–23.

Anon (1977). Cardiogenic dementia. Lancet 1: 27–28.

Anter E, Jessup M, Callans DJ (2009). Atrial fibrillation and heart failure: treatment considerations for a dual epidemic. Circulation 119: 2516–2525.

Appleros P, Nydevik I, Vitanen M (2003). Poor outcome after first-ever stroke: predictors for death, dependence, and recurrent stroke within the first year. Stroke 34: 122–126.

Athilingam P, King KB, Burgin SW et al. (2011). Montreal cognitive assessment and mini-mental status examination compared as cognitive screening tools in heart failure. Heart Lung 40: 521–529.

Benjamin EJ, Virani SS, Callaway CM et al. (2018). American Heart Association Council on Epidemiology and Prevention Statistics Committee and Stroke Statistics Subcommittee. Heart disease and stroke statistics-2018 update: a report from the American Heart Association. Circulation 137: 67–492.

Berry C, Murdoch DR, Mc Murray JJ (2001). Economics of chronic heart failure. Eur J Heart Fail 3: 283–291.

Bots ML, van der Wilk EC, Koudstaal PJ et al. (1997). Transient neurological attacks in the general population. Prevalence, risk factors, and clinical relevance. Stroke 28: 768–773.

Branch KR, Probstfield JL, Eikelboom JW et al. (2019). Rivaroxaban with or without aspirin in patients with heart failure and chronic coronary or peripheral artery disease. Circulation 140: 529–537.

Braunwald E (2015). The war against heart failure: the lancet lecture. Lancet 385: 812–824.

Brunen JMC, Echeverria MP, Diez-Manglano J et al. (2020). Cognitive impairment in patients hospitalized for congestive heart failure: data from the RICA Registry. Inter Emerg Med ahead of print.

Caldas JR, Panerai RB, Salinet AM et al. (2018). Dynamic cerebral autoregulation is impaired during submaximal isometric handgrip in patients with heart failure. Am J Physiol Heart Circ Physiol 315: H254–H261.

Cannon JA, Moffitt P, Perez-Moreno AC et al. (2017). Cognitive impairment and heart failure: systematic review and meta-analysis. J Card Fail 23: 464–475.

Carson PE, Johnson GR, Dunkman WB et al. (1993). The influence of atrial fibrillation on prognosis in mild to moderate heart failure: the V-HeFT Studies: the V-HeFT VA Cooperative Studies Group. Circulation 87: VI-102–VI-110.

Cleland JG, Findlay I, Jafri S et al. (2004). The warfarin/aspirin study in heart failure (WASH): a randomized trial comparing antithrombotic strategies for patients with heart failure. Am Heart J 148: 157–164.

Cokkinos DV, Haralabopoulos GC, Kostis JB et al. (2006). HELAS investigators. Efficacy of antithrombotic therapy in chronic heart failure: the HELAS study. Eur J Heart Fail 8: 428–432.

Conrad N, Judge A, Tran J et al. (2018). Temporal trends and patterns in heart failure incidence: a population-based study of 4 million individuals. Lancet 391: 572–580.

De la Torre JC (2002). Alzheimer disease as a vascular disorder: Nosological evidence. Stroke 33: 1152–1162.

De la Torre JC (2020). Hemodynamic instability in heart failure intensifies age-dependent cognitive decline. J Alzheimers Dis 76: 63–84.

de Roos A, van der Grond J, Mitchell G et al. (2017). Magnetic resonance imaging of cardiovascular function and the brain: is dementia a cardiovascular-driven disease? Circulation 135: 2178–2195.

Dharmarajan K, Hsieh AF, Lin Z et al. (2013). Diagnoses and timing of 30-day readmissions after hospitalization for heart failure, acute myocardial infarction, or pneumonia. JAMA 309: 355–363.

Divani AA, Vazquez G, Barrett AM et al. (2009). Risk factors associated with injury attributable to falling among elderly population with history of stroke. Stroke 40: 3286–3292.

Dobreanu D, Svendsen JH, Lewalter T et al. (2013). Current practice for diagnosis and management of silent atrial fibrillation: results of the European Heart Rhythm Association survey. Europace 15: 1223–1225.

Doehner W, Ural D, Haeusler KG et al. (2018). Heart and brain interaction in patients with heart failure: overview and proposal for a taxonomy. A position paper from the Study Group on Heart and Brain Interaction of the Heart Failure Foundation. Eur J Heart Fail 20: 199–215.

Dries DL, Rosenberg YD, Waclawiw MA et al. (1997). Ejection fraction and risk of thromboembolic events in patients with systolic dysfunction and sinus rhythm: evidence for gender differences in the studies of left ventricular dysfunction trials. J Am Coll Cardiol 29: 1074–1080.

Eisenberg S, Madison L, Sensenbach W (1960). Cerebral hemodynamic and metabolic studies in patients with congestive heart failure. II. Observations in confused subjects. Circulation 21: 704–709.

El-Battrawy I, Borggrefe M, Akin I (2016). Takotsubo syndrome and embolic events. Heart Fail Clin 12: 543–550.

Epstein AM, Jha AK, Orav EJ (2011). The relationship between hospital admission rates and rehospitalizations. N Engl J Med 365: 2287–2295.

Erkelens CD, van der Wal HH, de Jong BM et al. (2017). Dynamics of cerebral blood flow in patients with mild non-ischaemic heart failure. Eur J Heart Fail 19: 261–268.

Falk RH, Foster E, Coats MH (1992). Ventricular thrombi and thromboembolism in dilated cardiomyopathy: a prospective follow-up study. Am Heart J 123: 136–142.

Festa AR, Jia X, Cheung K et al. (2011). Association of low ejection fraction with impaired verbal memory in older patients with heart failure. Arch Neurol 68: 1021–1026.

Florido R, Kwak L, Lazo M et al. (2020). Physical activity and incident heart failure in high-risk subgroups: the ARIC Study. J Am Heart Assoc 9: e014885.

Freudenberger RS, Hellkamp AS, Halperin JL et al. (2007). Risk of thromboembolism in heart failure: an analysis from the Sudden Cardiac Death in Heart Failure Trial (SCD-HeFT). Circulation 115: 2637–2641.

GBD 2017 Disease and Injury Incidence and Prevalence Collaborators (2018). Global, regional, and national incidence, prevalence, and years lived with disability for 354 diseases and injuries for 195 countries and territories, 1990–2017: a systematic analysis for the Global Burden of Disease Study 2017. Lancet 392: 1789–1858.

Georgiadis D, Sievert M, Cencentti S et al. (2000). Cerebrovascular reactivity is impaired in patients with cardiac failure. Eur Heart J 21: 407–413.

Giannoni A, Emdin M, Bramanti F et al. (2009). Combined increased chemosensitivity to hypoxia and hypercapnia as a prognosticator in heart failure. J Am Coll Cardiol 53: 1975–1980.

Gosselink AT, Crijns HJ, van den Berg MP et al. (1994). Functional capacity before and after cardioversion of atrial fibrillation: a controlled study. Br Heart J 72: 161–166.

Gronewegen A, Rutten FH, Mosterd A et al. (2020). Epidemiology of heart failure. Eur J Heart Fail 22: 1342–1356.

Hadjuk AM, Lemon SC, McManus DD et al. (2013). Cognitive impairment and self-care in heart failure. Clin Epidemiol 5: 407–416.

Haeusler KG, Laufs U, Endres M (2011). Chronic heart failure and ischemic stroke. Stroke 42: 2977–2982.

Hanon O, Vidal JS, de Groote P et al. (2014). Prevalence of memory disorders in ambulatory patients aged ≥70 years with chronic heart failure (from the EFICARE study). Am J Cardiol 113: 1205–1210.

Havakuk O, King KS, Grazette L et al. (2017). Heart failure-induced brain injury. J Am Coll Cardiol 69: 1609–1616.

Hayes Jr D, Anstead MI, Ho J et al. (2009). Insomnia and chronic heart failure. Heart Fail Rev 14: 171–182.

He J, Ogden LG, Bazzano LA et al. (2001). Risk factors for congestive heart failure in US men and women. NHANES I epidemiologic follow-up study. Arch Intern Med 161: 996–1002.

Homma S, Thrompson JL, Pullicino PM et al. (2012). Warfarin and aspirin in patients with heart failure and sinus rhythm. N Engl J Med 366: 1859–1869.

Hooghiemstra A, Leeuwis AE, Bertens AS et al. (2019). Frequent cognitive impairment in patients with disorders along the heart-brain axis. Stroke 50: 3369–3375.

Jefferson AL (2010). Cardiac output as a potential risk factor for abnormal brain. J Alzheimers Dis 20: 813–821.

Jefferson AL, Himali JJ, Beiser AS et al. (2010). Cardiac index is associated with brain aging: the Framingham Heart Study. Circulation 122: 690–697.

Jefferson AL, Himali JJ, Au R et al. (2011). Relation of LVEF to cognitive aging. Framingham Heart Study. Am J Cardiol 108: 1346–1351.

Jefferson AL, Beiser AS, Himali JJ et al. (2015). Low cardiac index is associated with incident dementia and Alzheimer disease: the Framingham Heart Study. Circulation 131: 1333–1339.

Jefferson AL, Liu D, Gupta DK et al. (2017). Lower cardiac index levels relate to lower blood flow in older adults. Neurology 89: 2327–2334.

Jimenez JA, Greenberg BH, Mills PJ (2011). Effects of heart failure and its pharmacological management on sleep. Drug Discov Today Dis 8: 161–166.

Jones NN, Roalfe AK, Adoki I et al. (2019). Survival of patients with chronic heart failure in the community: a systematic review and meta-analysis. Eur J Heart Fail 21: 1306–1325.

Kang SH, Kim J, Park JJ et al. (2017). Risk of stroke in congestive heart failure with and without atrial fibrillation. Int J Cardiol 248: 182–187.

Khatibzadeh S, Farzadfar F, Oliver J et al. (2013). Worldwide risk factors for heart failure: a systematic review and pooled analysis. Int J Cardiol 168: 1186–1194.

Kim JU, Shah P, Tantry US et al. (2016). Coagulation abnormalities in heart failure: pathophysiology and therapeutic implications. Curr Heart Fail Rep 13: 319–328.

Kishi T (2016). Heart failure as a disruption of dynamic circulatory homeostasis mediated by the brain. Int Heart J 57: 145–149.

Komanduri S, Jadhao Y, Gudur SS et al. (2017). Prevalence and risk factors of heart failure in the USA: NHANES 2013–2014 epidemiological follow-up study. J Community Hosp Intern Med Perspect 7: 15–20.

Koudstaal PJ, Algra A, Pop GA et al. (1992). The Dutch TIA Study Group. Risk of cardiac events in atypical transient ischaemic attack or minor stroke. Lancet 340: 630–633.

Kozdag C, CIftci E, Vural A et al. (2006). Silent cerebral infarction in patients with dilated cardiomyopathy: echocardiographic correlates. Int J Cardiol 107: 376–381.

Kresge HA, Liu D, Gupta DK et al. (2020). Lower LVEF related to cerebrospinal fluid biomarker evidence of neurodegeneration in older adults. J Alzheimers Dis 74: 965–974.

Laribi S, Mebazaa A (2014). Cardiohepatic syndrome: liver injury in decompensated heart failure. Curr Heart Fail Rep 11: 236–240.

Lassen NA (1959). Cerebral blood flow and oxygen consumption in man. Physiol Rev 39: 183–238.

Lee TC, Qian M, Lip GYH et al. (2018). Heart failure severity and quality of warfarin anticoagulation control (From the WARCEF Trial). Am J Cardiol 122: 821–827.

Lee TC, Qian M, Liu Y et al. (2019). Cognitive decline over time in patients with systolic heart failure: insights from WARCEF. JACC Heart Fail 7: 1042–1053.

Leenen FH (2007). Brain mechanisms contributing to sympathetic hyperactivity and heart failure. Circ Res 101: 221–223.

Li Y, Fitzgibbons TP, McManus DD et al. (2019). LVEF and clinical defined heart failure to predict 90-day functional outcome after ischemic stroke. J Stroke Cerebrovasc Dis 28: 371–380.

Lip GY, Rasmussen LH, Skjoth F et al. (2012). Stroke and mortality in patients with incident heart failure: the diet, Cancer and health (DCH) cohort. BMJ Open 2: e000975.

Lyra V, Parissis J, Kallergi M et al. (2020). ^{18}F-FDG PET/CT brain glucose metabolism as marker of different types of depression comorbidity in chronic heart failure patients with impaired systolic function. Eur J Heart Fail 22: 2138–2146.

Mahoney P, Kimmel S, DeNofrio D et al. (1999). Prognostic significance of atrial fibrillation in patients at a tertiary medical center referred for heart transplantation because of severe heart failure. Am J Cardiol 83: 1544–1547.

Maisel WH, Stevenson LW (2003). Atrial fibrillation and heart failure: epidemiology, pathophysiology, and rationale therapy. Am J Cardiol 91: 2D–8D.

Mansur AP, Alvarenga GS, Kopel L et al. (2018). Cerebral blood flow changes during intermittent acute hypoxia in patients with heart failure. J Int Med Res 46: 4214–4225.

Massaro AR, Dutra AP, Almeida DR et al. (2006). Transcranial Doppler assessment of cerebral blood flow-effect of cardiac transplantation. Neurology 66: 124–126.

Mehra M, Vaduganathan M, Fu M et al. (2019). A comprehensive analysis of the effects or rivaroxaban on stroke or transient ischaemic attack in patients with heart failure, coronary artery disease, and sinus rhythm: the COMMANDER HF trial. Eur Heart J 40: 3593–3602.

Melgaard L, Gorst-Rassmussen A, Lane DA et al. (2015). Assessment of the CHA_2DS_2-VASc score in predicting ischemic stroke, thromboembolism, and death in patients with heart failure with and without atrial fibrillation. JAMA 314: 1030–1038.

Meng L, Hou W, Chui J et al. (2015). Cardiac output and cerebral blood flow. Anesthesiology 123: 1198–1208.

Messerli FH, Rimoldi SF, Bangalore S (2017). The transition from hypertension to heart failure: contemporary update. JACC Heart Fail 5: 543–551.

Metha A, Bhattacharya S, Estep J et al. (2020). Diabetes and heart failure: a marriage of inconvenience. Clin Geriatr Med 36: 447–455.

Metra M, Teerlink JR (2017). Heart failure. Lancet 390: 1981–1995.

Middlekauff HR, Stevenson WG, Stevenson LW (1991). Prognostic significance of atrial fibrillation in advanced heart failure: a study of 390 patients. Circulation 84: 40–48.

Milionis H, Faouizi M, Cordier M et al. (2013). Characteristics and early and long-term outcome in patients with acute ischemic stroke and low ejection fraction. Int J Cardiol 168: 1082–1087.

Mitchell GF, van Buchem MA, Sigurdsson S et al. (2011). Arterial stiffness, pressure and flow pulsatility and brain structure and function: the age, Gene/Environment Susceptibility–Reykjavik study. Brain 134: 3398–3407.

Moller C, Stiermaier T, Graf T et al. (2018). Prevalence and long-term prognostic impact of malignancy in patients with Takotsubo syndrome. Eur J Heart Fail 20: 816–818.

Mueller K, Thiel F, Beutner F et al. (2020). Brain damage with heart failure: cardiac biomarker alterations and gray matter decline. Circ Res 126: 750–764.

Nichols GA, Reynolds K, Kimes TM et al. (2015). Comparison of risk of re-hospitalization, all-cause mortality, and medical care resource utilization in patients with heart failure and preserved versus reduced ejection fraction. Am J Cardiol 116: 1088–1092.

Ois A, Gomis M, Cuadrado-Godia E et al. (2008). Heart failure in acute ischemic stroke. J Neurol 255: 385–389.

Oudeman EA, Bron EE, van den Berg-Vos RM et al. (2019). Cerebral perfusion and the occurrence of nonfocal transient neurological attacks. Cerebrovasc Dis 47: 303–308.

Ozyuncu N, Gulec S, Kaya CT et al. (2019). Relation of acute decompensated heart failure to silent cerebral infarcts in patients with reduced LVEF. Am J Cardiol 123: 1835–1839.

Palumbo V, Baldasseroni S, Nencini P et al. (2012). The coexistence of heart failure predicts short term mortality, but not disability, in patients with acute ischemic stroke treated with thrombolysis: the Florence Area Registry. Eur J Intern Med 23: 552–557.

Pandey A, Kitzman D, Reeves G (2019). Frailty is intertwined with heart failure: mechanisms, prevalence, prognosis, assessment, and management. JACC Heart Fail 7: 1001–1011.

Parati G, Lombardi C, Castagna F et al. (2016). Italian Society of Cardiology. Group on heart failure members. Heart failure and sleep disorders. Nat Rev Cardiol 13: 389–403.

Parma Z, Jasilek A, Greenlaw N et al. (2020). Incident heart failure in outpatients with chronic coronary syndrome: results from the international prospective CLARIFY registry. Eur J Heart Fail 22: 804–812.

Patil S, Shah M, Patel B et al. (2019). Readmissions among patients admitted with acute decompensated heart failure based on income quartiles. Mayo Clin Proc 94: 1939–1950.

Pelliccia F, Kaski JC, Crea F et al. (2017). Pathophysiology of Takotsubo syndrome. Circulation 135: 2426–2441.

Pogmoragot J, Lee DS, Park TH et al. (2016). Stroke and heart failure: clinical features, access to care, and outcomes. J Stroke Cerebrovasc Dis 25: 1048–1056.

Ponikowski P, Voors AA, Anker SD et al. (2016). 2016 ESC guidelines for the diagnosis and treatment of acute and chronic heart failure: the task force for the diagnosis and treatment of acute and chronic heart failure of the European Society of Cardiology (ESC) developed with the special contribution of the Heart Failure Association (HFA) of the ESC. Eur Heart J 37: 2129–2200.

Porcari A, Falco L, Lio V et al. (2020). Cardiac amyloidosis: do not forget to look for it. Eur Heart J 22: E142–E147.

Pozzoli M, Cioffi G, Traversi E et al. (1998). Predictors of primary atrial fibrillation and concomitant clinical and hemodynamic changes in patients with chronic heart failure: a prospective study in 344 patients with baseline sinus rhythm. J Am Coll Cardiol 32: 197–204.

Prabhu S, Voskobonik A, Kaye DM et al. (2017). Atrial fibrillation and heart failure: cause of effect? Heart Lung Circ 26: 967–974.

Pullicino PM, Hart J (2001). Cognitive impairment in congestive heart failure? Embolism vs hypoperfusion. Neurology 57: 945–946.

Pullicino PM, Halperin JL, Thompson JL (2000). Stroke in patients with heart failure and reduced LVEF. Neurology 54: 288–294.

Pullicino P, Mifsud V, Wong E et al. (2001). Hypoperfusion-related cerebral ischemia and cardiac left ventricular systolic dysfunction. J Stroke Cerebrovasc Dis 10: 178–182.

Rahmayani F, Paryono, Setyopranoto I (2018). The role of ejection fraction to clinical outcome of acute ischemic stroke patients. J Neurosc Rural Pract 9: 197–202.

Rangaswami J, Bhalla V, Blair JEA et al. (2019). American Heart Association Council on the kidney in cardiovascular disease and council on clinical cardiology. Cardiorenal syndrome: classification, pathophysiology, diagnosis and treatment strategies: a scientific statement from the American Heart Association. Circulation 139: e840–e878.

Redeker NS, Adams L, Berkowitz R et al. (2012). Nocturia, sleep and daytime function in stable heart failure. J Card Fail 18: 569–575.

Reynolds K, Go AS, Leong TK et al. (2017). Myocardial infarction in the cardiovascular research network. Am J Med 130: 317–327.

Rostagno C (2019). Heart valve disease in elderly. World J Cardiol 11: 71–83.

Saito H, Yamashita M, Endo Y et al. (2020). Cognitive impairment measured by Mini-Cog provides additive prognostic information in elderly patients with heart failure. J Cardiol 2: S0914.

Savarese G, Giugliano RP, Rosano GM et al. (2016). Efficacy and safety of novel oral anticoagulants in patients with atrial fibrillation and heart failure. A meta-analysis. JACC Heart Fail 4: 870–880.

Savioli Neto F, Magalhaes HM, Batlouni M et al. (2009). ACE inhibitors and plasma B-type natriuretic peptide levels in elderly patients with heart failure. Arq Bras Cardiol 92: 320–326.

Sbolli M, Fiuzat M, Cani D et al. (2020). Depression and heart failure: the lonely comorbidity. Eur J Heart Fail 22: 2007–2017.

Scheinberg P (1950). Cerebral circulation in heart failure. Am J Med 8: 148–152.

Scherbakov N, Doehner W (2018). Heart-brain interactions in heart failure. Card Fail Rev 4: 87–91.

Schmidlt R, Fazekas F, Offenbacher H et al. (1991). Brain magnetic resonance imaging and neuropsychologic evaluation of patients with idiopathic dilated cardiomyopathy. Stroke 22: 195–199.

Schumacher K, Kornei J, Shantsila E et al. (2018). Heart failure and stroke. Curr Heart Fail Rep 15: 287–296.

Senni M, Tribouilloy CM, Rodeheffer RJ et al. (1998). Congestive heart failure in the community: a study of all incident cases in Olmsted County, Minnesota, in 1991. Circulation 98: 2282–2289.

Siachos T, Vabakel A, Feldman DS et al. (2005). Silent strokes in patients with heart failure. J Card Fail 11: 485–489.

Siedler G, Sommer K, Macha K et al. (2019). Heart failure in ischemic stroke: relevance for acute care and outcome. Stroke 50: 3051–3056.

Stratton JR, Resnick AD (1987). Increased embolic risk in patients with left ventricular thrombi. Circulation 75: 1004–1011.

Tarquini R, Pala L, Brancati S et al. (2018). Clinical approach to diabetic cardiomyopathy: a review of human studies. Curr Med Chem 25: 1510–1524.

Triposkiadis F, Xanthopoulos A, Parissis J et al. (2020). Pathogenesis of chronic heart failure: cardiovascular aging, risk factors, comorbidities, and disease modifiers. Heart Fail Rev. https://doi.org/10.1007/s10741-020-09987-z (online ahead of print).

Ueyama H, Takagi H, Briasoulis A et al. (2020). Meta-analysis of antithrombotic strategies in patients with heart failure with reduced ejection fraction and sinus rhythm. Am J Cardiol 127: 92–98.

Van Riet EE, Hoes AW, Wagennar KP et al. (2016). Epidemiology of heart failure: the prevalence of heart failure and ventricular dysfunction in older adults over time. A systematic review. Eur J Heart Fail 18: 242–252.

Vellone E, Chiala O, Boyne J et al. (2020). Cognitive impairment in patients with heart failure: an international study. ESC Heart Fail 7: 46–53.

Vemmos K, Ntaios G, Savvari P et al. (2012). Stroke aetiology and predictors of outcome in patients with heart failure and acute stroke: a 10-year follow-up study. Eur J Heart Fail 14: 211–218.

Williams LR, Leggett RW (1989). Reference values for resting blood flow to organs of man. Clin Phys Physiol Meas 10: 187–217.

Witt BJ, Brown RD, Jacobsen SJ et al. (2006). Ischemic stroke after heart failure: a community-based study. Am Heart J 152: 102–109.

Wittstein IS, Thiemann D, Lima JA et al. (2005). Neurohumoral features of myocardial stunning due to sudden emotional stress. N Engl J Med 352: 530–548.

Yancy CW, Jessup M, Bozkurt B et al. (2017). 2017 ACC/AHA/HFSA focused update of the 2013 ACCF/AHA Guideline for the management of heart failure: a report of the American College of Cardiology/American Heart Association Task Force on Clinical Practice Guidelines and the Heart Failure Society of America. Circulation 136: e137–e161.

Yussuf S, Thom T, Abbott RD (1989). Changes in hypertension treatment and in congestive heart failure mortality in the United States. Hypertension 13: 174–179.

Zannad F, Anker SD, Byra WM et al. (2018). Rivaroxaban in patients with heart failure, sinus rhythm, and coronary disease. N Engl J Med 379: 1332–1342.

Zito M, Muscari A, Marini E et al. (1996). Silent lacunar infarcts in elderly patient with chronic non-valvular atrial fibrillation. Aging 8: 341–346.

Zlokovic BV (2011). Neurovascular pathways to neurodegeneration in Alzheimer's disease and other disorders. Nat Rev Neurosci 12: 723–738.

Zuccala G, Pedone C, Cesari M et al. (2003). The effects of cognitive impairment on mortality among hospitalized patients with heart failure. Am J Med 115: 97–103.

Handbook of Clinical Neurology, Vol. 177 (3rd series)
Heart and Neurologic Disease
J. Biller, Editor
https://doi.org/10.1016/B978-0-12-819814-8.00001-9

Chapter 9

Neurological complications of cardiomyopathies

ANA CATARINA FONSECA[1], ANA G. ALMEIDA[2], MIGUEL OLIVEIRA SANTOS[1], AND JOSÉ M. FERRO[1]*

[1]*Neurology Service, Hospital Santa Maria, Centro Hospitalar Lisboa Norte and Faculty of Medicine, University of Lisbon, Lisbon, Portugal*

[2]*Cardiology Service, Hospital Santa Maria, Centro Hospitalar Lisboa Norte and Faculty of Medicine, University of Lisbon, Lisbon, Portugal*

Abstract

There is a multifaceted relationship between the cardiomyopathies and a wide spectrum of neurological disorders. Severe acute neurological events, such as a status epilepticus and aneurysmal subarachnoid hemorrhage, may result in an acute cardiomyopathy the likes of Takotsubo cardiomyopathy. Conversely, the cardiomyopathies may result in a wide array of neurological disorders. Diagnosis of a cardiomyopathy may have already been established at the time of the index neurological event, or the neurological event may have prompted subsequent cardiac investigations, which ultimately lead to the diagnosis of a cardiomyopathy. The cardiomyopathies belong to one of the many phenotypes of complex genetic diseases or syndromes, which may also involve the central or peripheral nervous systems. A number of exogenous agents or risk factors such as diphtheria, alcohol, and several viruses may result in secondary cardiomyopathies accompanied by several neurological manifestations. A variety of neuromuscular disorders, such as myotonic dystrophy or amyloidosis, may demonstrate cardiac involvement during their clinical course. Furthermore, a number of genetic cardiomyopathies phenotypically incorporate during their clinical evolution, a gamut of neurological manifestations, usually neuromuscular in nature. Likewise, neurological complications may be the result of diagnostic procedures or medications for the cardiomyopathies and vice versa. Neurological manifestations of the cardiomyopathies are broad and include, among others, transient ischemic attacks, ischemic strokes, intracranial hemorrhages, syncope, muscle weakness and atrophy, myotonia, cramps, ataxia, seizures, intellectual developmental disorder, cognitive impairment, dementia, oculomotor palsies, deafness, retinal involvement, and headaches.

DEFINITION, CLASSIFICATION AND CLINICAL UPDATE

Cardiomyopathies refer to myocardial disorders whereby the heart muscle is structurally and functionally abnormal, in the absence of underlying coronary artery disease, arterial hypertension, valvular heart disease or congenital heart disease sufficient to cause the observed myocardial abnormality. Cardiomyopathies are grouped into specific morphological and functional phenotypes. Each phenotype is subclassified into familial and nonfamilial forms; the former is associated with specific genetic mutations and the latter as acquired conditions, influenced by genetic polymorphisms.

In 2006, the American Heart Association classified cardiomyopathies as genetic, acquired, or mixed. In 2008, the European Society of Cardiology classified the cardiomyopathies into familial/genetic and nonfamilial/nongenetic forms, with the following subtypes: hypertrophic, dilated, restrictive, arrhythmogenic right ventricular cardiomyopathy, unclassified – left ventricle

*Correspondence to: José M. Ferro, M.D., Ph.D., Faculdade de Medicina, Universidade de Lisboa, Hospital Santa Maria, Neurology, 6th floor, Avenida Professor Egas Moniz s/n 1649-035 Lisbon, Portugal. Tel: +351-217-957-474, Fax: +351-217-957-474, E-mail: jmferro@medicina.ulisboa.pt

noncompaction, and Takotsubo cardiomyopathy (Elliott et al., 2008). More recently, the World Heart Federation introduced the MOGE(S) classification, describing the morphological-functional (M) phenotype involvement of additional organs (O), familial/genetic (G) origin, and etiology including genetic mutations if applicable (E). Functional status (S) was considered optional (Arbustini et al., 2013).

Cardiomyopathies may result in heart failure, chest pain, cardiac arrhythmias, and sudden cardiac death. However, cardiomyopathies are often asymptomatic and diagnosis is made incidentally or during the screening of family members. Morphological diagnosis is essential and accomplished by the use of advanced cardiac imaging. An integrated approach is required for appropriate risk stratification and for further defining therapeutic strategies, preventive measures, and follow-up (Rapezzi et al., 2013).

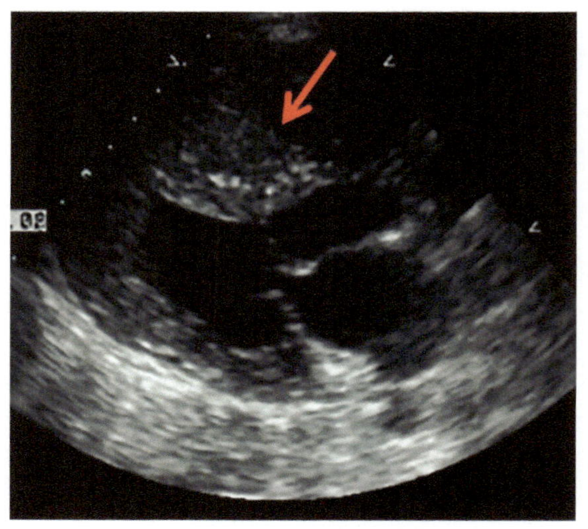

Fig. 9.1. Echocardiogram still image at parasternal view from a patient with hypertrophic cardiomyopathy, depicting a massive septal hypertrophy (*arrow*).

Hypertrophic cardiomyopathy

With an estimated prevalence of 1:500 of the general population, hypertrophic cardiomyopathy is the most frequent heritable cardiomyopathy. Mutations in genes encoding sarcomeric and associated proteins, cytoskeletal, and calcium-handling proteins have been pathogenically implicated. However, in 50% of cases of hypertrophic cardiomyopathy, the possible mutation remains unidentified (Maron et al., 2012).

The phenotypic hallmark of hypertrophic cardiomyopathy is characterized by left ventricular (LV) hypertrophy that is most often asymmetric involving a variable number of ventricular segments and with a predominance for the interventricular septum and the anterior wall (Fig. 9.1). Cardiomyocytes show both hypertrophy and disarray, with associated abnormal intramyocardial microvasculature, leading to ischemia and ultimately interstitial fibrosis and replacement fibrosis due to necrosis. Both hypertrophy and fibrosis are the basis for diastolic dysfunction that is a typical feature of hypertrophic cardiomyopathy. In contrast, LV systolic function as assessed by the ejection fraction is normal or supranormal until late progression to end-stage phases of the disease, where LV dilatation and reduced ejection fraction are present. Additionally, in two-thirds of patients with hypertrophic cardiomyopathy, a rest or exercise systolic LV outflow obstruction is present due to the abnormal septal morphology and mitral valve abnormalities namely regurgitation (Elliott et al., 2014).

Patients' presentation is variable, ranging from asymptomatic to heart failure, angina or mortality due to sudden cardiac death. Cardiac arrhythmias, including atrial fibrillation, are frequent events. Atrial fibrillation results from diastolic dysfunction and mitral regurgitation and is a risk factor for stroke and peripheral embolism with a prevalence and annual incidence of 27.1% and 3.8%, respectively (Guttmann et al., 2014). Sudden cardiac death due to ventricular arrhythmias is the most frequent cause of mortality. Although the incidence of sudden cardiac death ranges from 0.6% to 0.9% (O'Mahony et al., 2014), risk stratification remains challenging. Currently, it is estimated the 5-year risk of sudden cardiac death is assessed using a statistical risk prediction model (HCM Risk – SCD) that considers seven parameters: age, maximal left ventricular wall thickness, left atrial diameter, left ventricular outflow tract gradient, nonsustained ventricular tachycardia, unexplained syncope, and family history of sudden cardiac death (O'Mahony et al., 2014). Implantable cardioverter defibrillators (ICDs) prevent sudden cardiac death and are indicated for patients with higher risk (>6%) and intermediate risk with additional risk factors, such as magnetic resonance imaging (MRI), detection of fibrosis, apical aneurysms, or LV dysfunction (O'Hanlon et al., 2010).

The diagnosis of hypertrophic cardiomyopathy is most often accomplished by echocardiography. Cardiac MRI (CMRI) is another valuable ancillary diagnostic method, allowing the detection of myocardial fibrosis and the quantification and precise measurements of LV size and wall thickness, as well as the diagnosis of ischemia. ECG and Holter exercise test are commonly used additional methods (Elliott et al., 2014).

Treatment ranges from conventional heart failure medications to septal surgical or alcohol myectomy for drug-refractory symptoms. Antiarrhythmic therapy and ICD deployment are used as appropriate according to current recommendations (Elliott et al., 2014).

An important issue pertains to the concept of hypertrophic phenotype. In fact, classical hypertrophic cardiomyopathy shares morphological features with other conditions associated with LV hypertrophy, although mostly with a symmetrical pattern, unlike the asymmetrical pattern most typically found in hypertrophic cardiomyopathy. Examples of these phenocopies include amyloidosis, hemochromatosis, and other nonsarcomeric hypertrophic cardiomyopathies such as Anderson–Fabry disease, Danon disease, Pompe disease, mitochondrial disorders, and Friedreich's ataxia (Rapezzi et al., 2013).

Dilated cardiomyopathy

Dilated cardiomyopathies are defined by the presence of LV or biventricular dilatation and systolic dysfunction in the absence of abnormal loading conditions such as arterial hypertension, valvular heart disease, or coronary artery disease sufficient to cause global systolic impairment (Elliott et al., 2008). Prevalence of dilated cardiomyopathy has been estimated as 1:250 of population, often affecting young patients. The histopathology of dilated cardiomyopathies is variable, depending on the underlying cause, but late stages, as a rule show poverty of cardiomyocytes due to necrosis and apoptosis, and increased replacement fibrosis (Pinto et al., 2016).

Genetic and nongenetic mechanisms underlie dilated cardiomyopathy, but genetic predisposition often interacts with extrinsic factors. About 20%–35% of patients have a positive family history. More than 50 disease-related genes have been reported, with the *TTN* gene encoding for titin as the most frequent (Morales and Hershberger, 2013; Pugh et al., 2014). Regarding nongenetic causes for the dilated cardiomyopathy phenotype, myocarditis and systemic autoimmune disorders may result in chronic/irreversible evolutions, while others result in potentially reversible presentations like tachyarrhythmias, catecholamine storm, and toxic agents such as alcohol, chemotherapy, or cocaine use. In 20%–30% of cases, the disorder is considered idiopathic (Anzini et al., 2013).

Dilated cardiomyopathies have a variable phenotypic expression and may overlap with other cardiomyopathies. Arrhythmogenic cardiomyopathy has recently been described as an entity with a clinical phenotype between the classical arrhythmogenic right ventricular cardiomyopathy and dilated cardiomyopathy (Sen-Chowdhry et al., 2008).

The clinical spectrum of dilated cardiomyopathy ranges from asymptomatic to heart failure, due to systolic and diastolic dysfunction and cardiac arrhythmias. Functional mitral regurgitation is frequent, and it may aggravate the LV dysfunction. Sudden cardiac death is a possible presentation, occurring up to 0.5%/year, and often in association to reentry ventricular arrhythmias (Merlo et al., 2018). Strokes and peripheral thromboembolism are major adverse events, due to ventricular or atrial thrombi formation as a result of intracardiac stasis and atrial fibrillation, which may induce further LV dysfunction (Santhanakrishnan et al., 2016).

Determining the etiology is crucial for detecting possible reversible causes, amenable to timely appropriate interventions. Family screening is required to search for genetic causes and should be pursued in order to define further guidance strategies. Echocardiography remains the initial diagnostic method, capable of depicting the morphological and functional phenotypic features of dilated cardiomyopathies. CMRI is a useful complementary method to identify myocardial patterns pertaining to specific etiologies and for detecting myocardial interstitial and replacement fibrosis, which are associated with outcome (Halliday et al., 2019). Natriuretic peptides and Holter monitoring are additional tests, and, when appropriate, electrophysiological testing and endomyocardial biopsy are indicated (Santhanakrishnan et al., 2016).

Current therapeutic strategies have shown higher rates of subsequent LV reverse remodeling and improved long-term prognosis with current therapeutic approaches (Serdoz et al., 2016; Merlo et al., 2018). However, in a proportion of patients without a reversible cause, prognosis remains poor with reduced survival due to progressive heart failure and risk of sudden cardiac death. Resynchronization therapy (CRT) and ultimately heart transplantation may be additional therapeutic options (Serdoz et al., 2016). For sudden cardiac death prevention, ICD deployment has specific indications, namely involving specific genetic mutations and in the arrhythmogenic forms (Al-Khatib et al., 2018).

Restrictive cardiomyopathy

Restrictive cardiomyopathy is a myocardial disease characterized by the presence of restrictive ventricular physiology (impaired ventricular filling) in the presence of normal or reduced diastolic volumes and normal ventricular wall thickness in the absence of ischemic heart disease, arterial hypertension, valvular heart, or congenital heart disease (Elliott et al., 2008). In most cases, both the left and right ventricles are involved and typically there is biatrial enlargement.

Restrictive cardiomyopathies are caused by a number of heterogeneous etiologies, which may be limited to the heart or involve other organs. Additionally the restrictive cardiomyopathies may be due to familial/genetic or non-familial/nongenetic causes. Familial restrictive cardiomyopathies, due to sarcomeric gene mutations (also found in other cardiomyopathies like hypertrophic and dilated cardiomyopathies) or to desmin or alpha-B-crystallin gene mutations, are possible causes (Arbustini et al., 2006; Gambarin et al., 2008). Other familial/genetic causes include amyloidosis, hemochromatosis, and Anderson–Fabry disease (Seward and Casaclang-Verzosa, 2020). Also to be considered are nonfamilial causes like endocardial fibrosis due to hypereosinophilic syndromes (Mankad et al., 2016) and the myocardial lesions associated with radiotherapy.

Patients with restrictive cardiomyopathies frequently present with heart failure and cardiac arrhythmias. Thromboembolic events are also common originating from the dilated atria, particularly when atrial fibrillation is present (Muchtar et al., 2017).

The diagnosis is initially based on echocardiography that demonstrates the typical restrictive cardiomyopathy features with a normal or slightly impaired LV systolic function, severe diastolic dysfunction, and biatrial dilatation. Among other complementary diagnostic methods, CMRI is most useful for the purpose of etiological assessment, showing typical myocardial patterns of late gadolinium enhancement, T1 mapping, and T2* (Habib et al., 2017). Other diagnostic investigations may include DPD-scintigraphy (for transthyretin amyloidosis), endomyocardial biopsy, and familial/genetic investigations (Habib et al., 2017).

Prognosis depends on etiology and early treatment but may be poor (Muchtar et al., 2017).

Arrhythmogenic cardiomyopathy

Arrhythmogenic cardiomyopathy is currently named so, from the classical form of right ventricular cardiomyopathy (ARVC), which is a myocardial disease characterized by life-threatening ventricular arrhythmias, and histopathologically by a progressive dystrophy of the ventricular myocardium with necrosis of cardiomyocytes and fibro-fatty replacement (Basso et al., 2012). The right ventricle is primarily affected showing chamber dilatation, and systolic dysfunction, with regional wall thinning and motion abnormalities such as akinesis, diskinesis, and aneurysms (Fig. 9.2). The LV may also be affected showing regional wall fibro-fatty replacement and, more recently, a form limited to the LV has been recognized (Pilichou et al., 2016). These recent findings led to the current naming of the disease.

This type of cardiomyopathy is inherited mainly as an autosomal dominant inheritance pattern and is familial in more than 50% of cases. A recessive mode of inheritance has also been reported in syndromic cases. In the classic form, ARVC is a disease of desmosomes with pathogenic variants present in genes encoding desmosomal proteins (plakoglobin, desmoplakin, plakophilin-2, desmoglein-2, and desmocollin-2) that account for 30%–50% of cases. Furthermore, there is genetic heterogeneity with other genes found to be associated with the ARVC phenotype. Likewise, the relationship genotype–phenotype is not straightforward, with variable clinical expression for the same mutations (Campuzano et al., 2013).

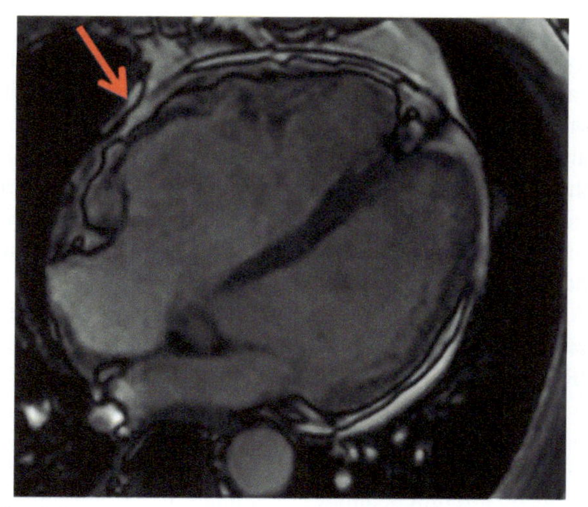

Fig. 9.2. Cardiac magnetic resonance cine frame of a four-chamber plane from a patient with arrhythmogenic right ventricular cardiomyopathy, showing a dilated right heart with wall aneurysms (*arrow*).

The clinical manifestations of ARVC are variable, depending on the phenotype phase of evolution. The diagnosis may be challenging. Patients with ARVC may be asymptomatic and only be diagnosed during a familial screening of an affected proband. Symptoms are mainly related to ventricular arrhythmias, such as syncope or sudden cardiac death, the most feared complication. Heart failure occurs late in the natural history of ARVC; symptoms are frequently severe and often unresponsive to treatment leading to heart transplantation. Athletic activities are contraindicated due to risk of worsening of ventricular arrhythmias and associated sudden death (Corrado et al., 2015). Patients with ARVC have an annual thromboembolic risk of 0.5%, including embolic events from right ventricular thrombi or atrial fibrillation (Wlodarska et al., 2006).

Diagnosis is based on right ventricular morphological and functional abnormalities detected by echocardiography, angiography and/or CMRI, endomyocardial biopsy, ECG, cardiac arrhythmias, and family history/genetics (Marcus et al., 2010). Differential diagnosis includes a number of conditions that are associated with right ventricular dilatation. CMRI is particularly useful for the morphologic/functional assessment allowing for a precise characterization of right and left ventricular volumes, ejection fraction, and wall abnormalities, with an important role for detecting right and left ventricular free wall fibrosis, using late-gadolinium enhancement (Aquaro et al., 2016).

Restriction of athletic activities is recommended for patients with ARVC even when asymptomatic. Due to the risk of sudden cardiac death, indications for ICD deployment should be considered following onset of syncope, episodes of sustained or nonsustained ventricular tachycardia, ventricular fibrillation, or in the presence of ventricular dysfunction (Corrado et al., 2015). Antiarrhythmic medications as adjunctive therapy and catheter ablation are valuable additional therapeutic measures. Heart failure is managed with standard recommendations. Heart transplantation is indicated for severe ventricular dysfunction or on rare occasions for uncontrollable ventricular arrhythmias (Tedford et al., 2012).

Left ventricular noncompaction

Left ventricular noncompaction is characterized by a morphological appearance of the myocardium with an inner noncompacted hypertrabeculated layer and deep recesses communicating with the left ventricular cavity and an outer compacted myocardial layer. The pathophysiological mechanism for left ventricular noncompaction has been attributed to the arrest of LV compaction during embryogenesis, but other nonembryogenic hypotheses have been proposed (Jenni et al., 2007).

Left ventricular noncompaction is a specific morphological pattern that may occur in isolation or be associated with other cardiac or systemic conditions including the X-linked Barth syndrome, cardiomyopathies, and certain congenital heart diseases (Arbustini et al., 2016). Several genetic mutations have been associated with isolated left ventricular noncompaction. Recently, several cases overlapping with dilated and hypertrophic cardiomyopathies were reported (Arbustini et al., 2016). Screening of relatives should be proposed when familial disease is suspected or in isolated cases of left ventricular noncompaction.

Patients with left ventricular noncompaction may be asymptomatic or present with LV dysfunction and heart failure, supraventricular and ventricular arrhythmias, heart block or thromboembolic events (Stöllberger et al., 2011a,b). Diagnosis is made by echocardiography or CMRI. For both diagnostic modalities, several diagnostic criteria have been proposed (Almeida and Pinto, 2013). Pharmacological therapy is recommended according to standard treatment of heart failure and arrhythmias, as well as ICD deployment. Anticoagulation should be considered when left ventricular noncompaction appears in patients with impaired LV systolic function.

Amyloidosis

Cardiac amyloidosis is in most cases associated with systemic amyloidosis, where the misfolded amyloid proteins form fibrils deposits in the heart and other organs. Heart failure is the most important clinical feature, with impaired ventricular filling as the predominant mechanism.

Several types of amyloidosis have been recognized. From the more than 30 types of amyloidosis, only a few affect the heart: (a) AL—light chain amyloidosis where the precursor protein is derived from a clonal plasma cell bone marrow disorder; (b) ATTR—transthyretin amyloidosis, where ATTR amyloid is derived from transthyretin produced by the liver, with two types, hereditary (familial, autosomal dominance) and wild-type (senile systemic amyloidosis); (c) unusual forms: AA (serum amyloid A), beta-2 microglobulin (ABM), gelsolin (AGel), and other even more unusual types (Gertz et al., 2015; Lousada et al., 2015).

Myocardial infiltration occurs in almost all cases, with associated LV hypertrophy. There is also evidence of myocardial toxicity from circulating light chains in AL amyloidosis (Falk et al., 2016). The pathophysiology of cardiac amyloidosis is classically characterized by left and right ventricular hypertrophy without cavity dilatation. There is preserved systolic function until late phases of the disease process, where ventricular diastolic dysfunction and atrial dilatation also occur (Siddiqi and Ruberg, 2018).

The extent of cardiac involvement is the primary determinant of clinical outcome. Patients often present with heart failure and arrhythmias. Diagnosis is often challenging and quite frequently delayed. Thromboembolic events are frequent (Rapezzi et al., 2009). In cardiac AL, median survival of untreated patients is about 6 months after onset of heart failure. In the wild-type ATTR, the clinical course is more indolent, but survival is only 3–5 years from diagnosis (Feng et al., 2009).

Echocardiography and CMRI are essential for the diagnosis. Echocardiography depicts the characteristic features of increased ventricular wall thickness, apical sparing in strain imaging, diastolic dysfunction, and pericardial effusion. However, these features are not enough for the diagnosis; CMRI should strongly suggest the diagnosis by the presence of abnormal delayed gadolinium enhancement and typical T1 mapping. Nevertheless, the amyloid type is not diagnosed. A biopsy for definitive diagnosis is required in AL. For ATTR, diagnosis may be performed by nuclear scintigraphy, 99mTc-DP in the absence of a monoclonal protein (Maurer et al., 2017; Chacko et al., 2019).

Treatment of cardiac amyloidosis depends on the type of amyloid and is directed at the precursor protein or aims at disrupting existing deposits. Standard heart failure medications are not always indicated and may cause hemodynamic failure. New pharmacological therapies are ongoing, under several clinical trials.

Myocarditis

Myocarditis is histologically defined by the presence of inflammatory infiltrates in the myocardium in conjunction with degenerative and/or necrotic changes of adjacent cardiomyocytes not typical of ischemic lesions (Caforio et al., 2015). A complex interplay of environmental and genetic factors causes inflammatory myocardial injury. It is likely that cardiac clinical phenotype in patients with myocarditis is influenced by concomitant functional autoantibodies or toxic compounds that directly impair metabolic and mechanical functions of the cardiomyocytes (Nagatomo and Tang, 2014).

A variety of exogenous physical, chemical, and microbiological agents may damage the myocardium, inducing inflammatory reactions, with viral etiologies being the most frequent (Schultheiss et al., 2011). Chagas disease is a specific cause of myocarditis with acute and chronic forms that frequently evolves with apical aneurysms.

Myocarditis may be asymptomatic or have different clinical presentations, ranging from sudden or rapidly evolving signs of acute cardiac disease to slowly and inadvertently progressing cardiac dysfunction (Chacko et al., 2019). Chest pain, palpitations, syncope, and dyspnea are commonly present. Cardiogenic shock may be the presenting feature in cases of advanced disease (Nagatomo and Tang, 2014; Caforio et al., 2015).

Histopathology remains the gold standard for diagnosis. According to the 1986 Dallas Criteria, myocarditis is defined by the presence of inflammatory infiltrates in the myocardium accompanied by degenerative and/or necrotic changes of adjacent cardiomyocytes not typical of ischemic damage associated with myocardial infarction. The addition of immunohistochemistry has updated the diagnostic criteria for myocarditis using biopsy data (Basso et al., 2013).

For the clinical diagnosis, beyond typical ECG and biomarkers of cardiac necrosis, cardiac imaging remains essential. Echocardiography is the initial bedside diagnostic method although relatively insensitive when LV function is preserved. CMRI may serve as an in vivo histological method by depicting typical myocardial changes with high diagnostic sensitivity (Ferreira et al., 2018). Moreover, CMRI provides target guidance for biopsy due to the typical focal distribution of lesions.

Prognosis is variable. Isolated myocarditis may lead to a chronic inflammatory cardiomyopathy, with a dilated cardiomyopathy phenotype, in about 30% of cases (Caforio et al., 2007).

Other myocardial diseases

Several other important myocardial diseases have to be considered (Elliott et al., 2008; Arbustini et al., 2013). Takotsubo cardiomyopathy is characterized by acute ventricular dysfunction with specific wall motion patterns, usually precipitated by physical or emotional stress, with catecholamines and microvascular mechanisms involved. Hemodynamic impairment, cardiac arrhythmias, and thromboembolism are common. A potentially life-threatening condition, emerging toward the end of pregnancy, or in the months following delivery, peripartum cardiomyopathy is an idiopathic form of cardiomyopathy, presenting with heart failure secondary to LV dysfunction. Risk factors and prognosis are still poorly characterized. Lysosomal and glycogen storage diseases are rare and heterogeneous diseases with accumulation of macromolecules in the heart and manifesting as hypertrophic or dilated cardiomyopathy phenotypes. Heart involvement in neuromuscular diseases is discussed further below.

NEUROLOGICAL MANIFESTATIONS OF CARDIOMYOPATHIES

Although the relationship between the cardiomyopathies and neurological diseases is usually thought to be bidirectional, such relationship is in fact more complex (Finsterer et al., 2013; Finsterer and Wahbi, 2014) (Fig. 9.3). Severe acute neurological events, such as status epilepticus and aneurysmal subarachnoid hemorrhage, may result in an acute cardiomyopathy like of Takotsubo cardiomyopathy. Conversely, cardiomyopathies may result in a wide array of neurological disorders. Diagnosis of a cardiomyopathy may have been already established at the time of the index neurological event or the neurological event may have prompted subsequent cardiac investigations, which ultimately lead to the diagnosis of a cardiomyopathy. The cardiomyopathies

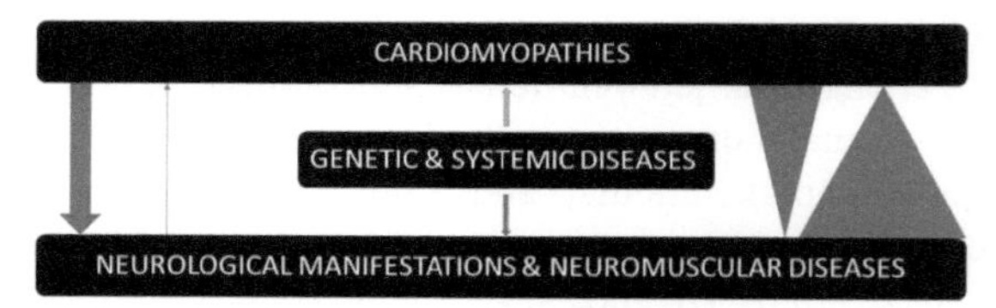

Fig. 9.3. Schematic representation of the relationship between cardiomyopathies and neurological disorders and clinical manifestation. *Left*—Complications of cardiomyopathies cause neurological events (e.g., stroke), less often the reverse occurs (e.g., seizure causes Takotsubo). *Center*—Genetic and systemic diseases cause both neurological phenotypes and cardiomyopathy. *Right*—Neuromuscular disease (predominant phenotype) with cardiomyopathy and vice versa (less common).

belong to one of the many phenotypes of complex genetic diseases or syndromes, which may also involve the central or peripheral nervous systems. A number of exogenous agents or risk factors such as diphtheria, alcohol, and several viruses may result in secondary cardiomyopathies accompanied by several neurological manifestations. A variety of neuromuscular disorders (NMD) such as myotonic dystrophy or amyloidosis may demonstrate cardiac involvement during their clinical course. Furthermore, a number of genetic cardiomyopathies phenotypically incorporate during their clinical evolution, a gamut of neurological manifestations, usually neuromuscular in nature. Likewise, neurological complications may be the result of diagnostic procedures or medications for the cardiomyopathies and vice versa.

Neurological manifestations of the cardiomyopathies are broad and include, among others, transient ischemic attacks, ischemic strokes, intracranial hemorrhages, syncope, muscle weakness and atrophy, myotonia, cramps, ataxia, seizures, intellectual developmental disorder, cognitive impairment, dementia, oculomotor palsies, deafness, retinal involvement, and headaches.

Axial and appendicular ataxia and nystagmus are key neurological manifestations of Friedreich's ataxia, ataxia due to vitamin E deficiency, and of the unusual PMM2-CGD congenital disorder of glycosylation (Schiff et al., 2017). Oculomotor palsies, deafness, or seizures point toward mitochondrial diseases. Orofacial dyskinesia, in particular lip-biting, suggests neuro-acanthocytosis. Static or progressive cognitive impairment is a feature of several neuromuscular and genetic disorders producing also resulting in cardiomyopathy. Less often, cognitive impairment and dementia may result from the cumulative effect of multiple brain infarctions.

Seizures are commonly observed and are often generalized tonic–clonic or myoclonic. However, other severe types can be observed such as focal myoclonic, focal atonic, generalized atonic, generalized myoclonic-atonic and typical absences, or tonic–clonic seizures. Besides mitochondrial diseases, several other unusual genetic and metabolic diseases can cause both epilepsy, generally of the myoclonic type, and cardiomyopathies. Examples are CoQ10 deficiency, euKariocytic elongation factor (EEF1A) mutation, action myoclonus renal failure syndrome (SCARB2 mutations), hereditary moyamoya syndrome, congenital generalized lipodystrophy type 2 (Bernardelli-Seip syndrome), leukodystrophy related to AARS2 mutations, KARS-related diseases, organic acidurias, urea cycle disorders, and 1p36 deletion syndrome.

Conversely, seizures can also rarely cause asystole, Takotsubo cardiomyopathy, and mortality due to sudden death [SUDEP-sudden death in epilepsy] (Belcour et al., 2015; Krishnamoorthy et al., 2016).

There are also a few cases of cardiomyopathies associated with prolonged use of antiseizure medications (lamotrigine and gabapentin) (Stöllberger et al., 2011b; Tellor et al., 2019). In patients with mitochondrial disorders and epilepsy, antiseizure drugs of choice are levetiracetam, topiramate, zonisamide, piracetam, and benzodiazepines. Mitochondrion-toxic agents, such as valproate, carbamazepine, phenytoin, and barbiturates, should be avoided as well those potentially enhancing the frequency of myoclonus, such as phenytoin, carbamazepine, lamotrigine, vigabatrin, tiagabine, gabapentin, pregabalin, and oxcarbazepine (Finsterer and Zarrouk-Mahjoub, 2017).

CEREBROVASCULAR COMPLICATIONS

Through different physiopathological mechanisms, the cardiomyopathies have been associated with a higher risk of ischemic stroke and transient ischemic attacks. Most of the information regarding the association between cardiomyopathies and stroke derives from single case reports or case series.

Incidence

Patients with hypertrophic cardiomyopathy have a stroke frequency of 6% and an incidence of 0.8% per year (Maron et al., 2002). Adult patients with hypertrophic cardiomyopathy with an established pathogenic mutation have increased risk for the combined endpoints of nonfatal stroke, cardiovascular death, or progression to New York Heart Association (NYHA) functional class III/IV compared with patients with hypertrophic cardiomyopathy in whom no mutation is identified (Olivotto et al., 2008). In cases of dilated cardiomyopathy embolic events have an incidence ranging from 1.5 to 3.5 per 100 patient-years (Fuster et al., 1981). The event rate of stroke in patients with left ventricular noncompaction cardiomyopathy (LVNC) is 1%–2% per year with a total risk of thromboembolism of 21%–38% (Stöllberger et al., 2011a,b; Cevik et al., 2012) (Fig. 9.4). During long-term follow-up, thromboembolic events occurred in 24% of patients in one series (Oechslin et al., 2000).

The incidence of stroke in cardiac amyloidosis remains unknown, as there are only isolated case reports and case series describing this association (Zhang et al., 2017).

The cardiomyopathies can be previously known at the time of index stroke or be the first clinical manifestation that leads to their diagnosis (Fonseca et al., 2020).

In a prospective study that included 132 patients, CMRI identified a cardiomyopathy in seven patients (5.3%, 95% confidence interval 2.59%–10.54%) without a previous diagnosis of cardiomyopathy and with an unremarkable transthoracic echocardiogram (Fonseca

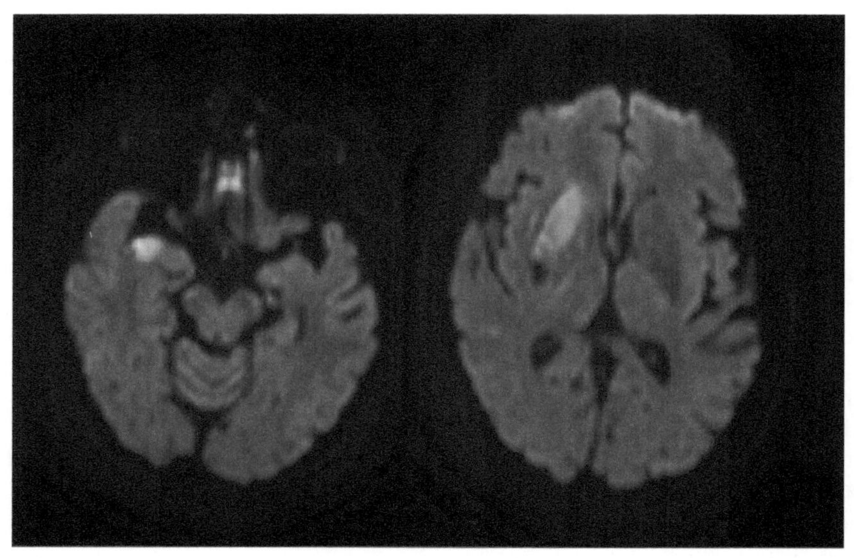

Fig. 9.4. Brain magnetic resonance imaging with diffusion-weighted sequence from a 77-year-old patient with a left ventricular noncompaction cardiomyopathy showing a right middle cerebral artery territory ischemic stroke.

et al., 2020). Four patients had hypertrophic cardiomyopathy, two had restrictive cardiomyopathy, and one had LVNC. Six of these patients had been classified after standard evaluation as having undetermined stroke, and one patient as having cardioembolic stroke due to atrial fibrillation.

Stroke mechanisms

Cardiomyopathies can be the cause of cardioembolic strokes. Cardiomyopathies can be associated with an increased risk of dysrhythmias like atrial fibrillation or atrial flutter. Furthermore, cardiomyopathies are associated with an increased risk of atrial or ventricular intracavitary thrombi.

The prevalence of atrial fibrillation in hypertrophic cardiomyopathy is about 20%–30%, compared to 1% in the general population (Debonnaire et al., 2017). Diagnosing atrial fibrillation can be challenging, since most episodes of paroxysmal atrial fibrillation are asymptomatic. Predictors for stroke in patients with hypertrophic cardiomyopathy include the presence of heart failure (NYHA class III/IV), age >60 years of age, left ventricular outflow track obstruction and atrial fibrillation/left atrium size >45 mm (Maron et al., 2002). Left ventricle late gadolinium enhancement extent (>14.4%) in CMRI has also been shown to be an independent predictor for thromboembolic complications (Hohneck et al., 2020).

Patients with the mid-cavity variant of hypertrophic obstructive cardiomyopathy may have other complications like apical left ventricular aneurysms with the formation of intramural thrombi than can lead to ischemic strokes (Basavarajaiah et al., 2009).

Dilated cardiomyopathy in sinus rhythm and with a left ventricular ejection fraction <40% has been associated with thrombus formation in the left heart chambers. In one study, left ventricular thrombi were found in 13% of patients, and left atrial appendage thrombi were found in 68% of patients (Mischie et al., 2013).

In LVNC, there is an increased probability of intraventricular thrombi formation that is due not only to systolic dysfunction or to ventricular dilation but also to thrombis formation within the recesses of the myocardial meshwork. The intertrabecular recesses in the cardiac apex and/or inferior and lateral left ventricular walls are prone to thrombus formation with a potential risk of secondary systemic embolism (Lestienne et al., 2017). Patients with LVNC also have an increased probability of atrial fibrillation. In a review of 144 patients diagnosed with LVNC, 22 patients had strokes or embolism (Stöllberger et al., 2011a,b). Sixty-four percent of these patients had reduced left ventricular function on echocardiography. It was not clarified the percentage of patients with reduced ejection fraction who also had atrial fibrillation. In total, 27% had atrial fibrillation. Features of LVNC that increase stroke risk include: left ventricular dysfunction (fractional shortening <25% or ejection fraction <40%), atrial fibrillation, history of thromboembolic complications, known ventricular thrombi, chamber dilatation, and evidence of spontaneous contrast (Bhatia et al., 2011).

Patients with amyloid cardiomyopathy also have an increased prevalence of atrial fibrillation due to deposition and infiltration of the amyloid protein in the atria. The prevalence of atrial fibrillation in cardiac amyloidosis is approximately 20% (Zhang et al., 2017). Atrial infiltration causes structural dissociation and increases the probability of atrial thrombi formation, even in patients in sinus rhythm (Stables and Ormerod, 1996). Deposition of amyloid in the endocardial tissue leads to fibrous thickening and impaired wall kinetics that contribute to the formation of endocardial thrombi (Rice et al., 1981). Intracardiac thrombi are common in patients with cardiac amyloidosis. In an analysis of 54 necropsy patients with cardiac amyloidosis, 14 (26%) had intracardiac thrombi (Robinson et al., 2020). A retrospective investigated the association between primary systemic amyloidosis and stroke (Zubkov et al., 2007). The study included 40 patients with biopsy proven amyloidosis and ischemic stroke. Thirteen patients (32.5%) had ischemic strokes as a primary presentation before been diagnosed with systemic amyloidosis. Patients with an initial stroke presentation had the worst outcome, with an average survival of 6.9 months following the diagnosis of amyloidosis; strokes developed 9.6 months before the diagnosis with primary amyloidosis. Thirty-seven percent experienced recurrent ischemic strokes.

A study examining the autopsy results of 116 patients with cardiac amyloidosis (55 of immunoglobulin light chain (AL) and 61 of other subtypes) discovered intracardiac thrombi in 33 patients. The AL subgroup had more intracardiac thrombi found at autopsy (51% vs 16%) and additional fatal embolic events (26% vs 8%) when compared with the other amyloidosis subgroups (Feng et al., 2017).

Strokes may be the cause but also the consequence of Takotsubo cardiomyopathy. The associated acute left ventricular dysfunction may lead to the formation of intracardiac thrombi. Most common thromboembolic events are cerebral, renal, or peripheral. Approximately, one-third of these patients with left ventricular thrombi develop embolic complications, although cardioembolism may also occur without detectable ventricular thrombi (Y-Hassan et al., 2019).

Specific systemic diseases that may cause cardiomyopathy may later result in ischemic strokes such as Fabry disease, Chagas disease, alcohol cardiomyopathy, and peripartum cardiomyopathy.

Patients with Fabry disease may have a restrictive cardiomyopathy and have a higher probability of atrial fibrillation than the general population due to glycosphingolipid deposition and fibrosis as well as atrial dilatation (Schiffmann et al., 2009). In a case series, dysrhythmias were 1.5 times more likely among males. This was attributed to the more advanced cardiomyopathy and renal dysfunction in homozygotes compared to heterozygotes (Schiffmann et al., 2009).

Chronic cardiomyopathy is the most common clinical form of cardiomyopathy in Chagas disease (Carod-Artal et al., 2005). The apical region of the LV is affected and aneurysms, thrombi, or both occur with high frequency. Chagas cardiomyopathy is independently associated with ischemic strokes, whereas hypercoagulable states do not appear to be major contributors to the excess stroke risk observed in these patients (Carod-Artal et al., 2005).

Chronic and heavy alcohol consumption may cause an acquired dilated cardiomyopathy that may be one of several possible etiological mechanisms linking alcohol consumption with ischemic strokes (Djoussé and Gaziano, 2008). Patients history of binge drinking are also prone to developing atrial fibrillation, which may result in cardioembolic strokes.

Peripartum cardiomyopathy refers to a rare dilated cardiomyopathy presenting in the last month of pregnancy or within 5 months following delivery (Dyken and Biller, 1994). This pathology is associated with a high incidence of mural thrombi and subsequent systemic and pulmonary embolism. The risk of cerebral embolism is relatively low (Dyken and Biller, 1994). A single center study with 45 patients reported a frequency of stroke of 9% (Laghari et al., 2013).

Stroke clinical profile

Few studies have described the clinical profile of ischemic strokes associated with cardiomyopathies. A literature review of amyloid cardiomyopathy with stroke reported nonspecific symptoms, including manifestations of heart failure and ischemic strokes (Zhang et al., 2017).

Etiological investigation

Although, in patients with a cardiomyopathy, stroke etiology is probably cardioembolic, a complete etiological investigation with laboratory analyses including blood cell counts and evaluation of the intracranial and extracranial cervical vessels should be performed to exclude other concomitant stroke causes.

Regarding cardiac imaging, transthoracic echocardiogram is a key noninvasive examination that assists in the diagnosis of a wide range of cardiomyopathies. Transthoracic echocardiography may help detect left atrial and ventricular thrombi and other features predisposing to embolism including the presence of spontaneous echocontrast, regional wall motion abnormalities, and large left ventricular end-diastolic dimensions in dilated cardiomyopathy (Innelli and Galderisi, 2004). Transoesophageal echocardiography is better suited for the investigation of atrial thrombi. In some cases, echocardiography may not allow the diagnosis of a cardiomyopathy for reasons like difficult visualization of certain heart structures such as the apex. CMRI has major diagnostic utility in the assessment of several cardiomyopathies

(Captur et al., 2016). In contrast to echocardiography, CMRI is not dependent on good acoustic windows, provides tissue characterization information, and has higher spatial resolution (Semsarian et al., 2015). The search for cardiomyopathies as a cause of stroke should include CMRI when echocardiography is normal, but there is a suspicion based on clinical grounds or known echocardiographic diagnostic limitations. CMRI is the most sensitive examination for the detection of left ventricular thrombus (Weinsaft et al., 2011).

Heart rhythm investigation is also important as the cardiomyopathies are associated with an increased risk of atrial fibrillation. In a cohort study of patients with hypertrophic cardiomyopathy who had stroke or embolic events, atrial fibrillation had not been documented before the index event in more than half of patients (Haruki et al., 2016). Ambulatory ECG monitoring has been recommended by the European Society of Cardiology at the initial clinical assessment of patients with hypertrophic cardiomyopathy to assess the stroke risk (Elliott et al., 2014). It has also been recommended that patients with hypertrophic cardiomyopathy and left atrial diameter $\geq$45 mms should undergo 6–12 monthly 48-h ambulatory ECG monitoring to search for occult atrial fibrillation (Elliott et al., 2014). The ACCF/AHA recommends that a 24 ambulatory (Holter) electrocardiographic monitoring be considered in adults with HCM to assess for asymptomatic paroxysmal atrial fibrillation/atrial flutter (Class IIb, Level of Evidence C) (Gersh et al., 2011).

Current recommendations, in general, for ischemic stroke etiological investigation include a 12-lead ECG and a 24-Holter ECG. A longer time of Holter-ECGs monitoring or the use of implantable cardiac monitors could be considered as the diagnosis of atrial fibrillation has major therapeutic implications.

Treatment

Beyond atrial fibrillation or the presence of intracardiac thrombi, there are no clear indications for ischemic stroke primary prevention in patients with cardiomyopathies.

In patients with hypertrophic cardiomyopathy and atrial fibrillation, long-term oral anticoagulation is recommended. For primary stroke prevention, as patients with hypertrophic cardiomyopathy tend to be younger than other high-risk groups and have not been included in clinical trials of thromboprophylaxis, the use of the CHA2DS2-VASc score to calculate stroke risk is not recommended (Elliott et al., 2014). According to the European Society of Cardiology and the ACCF/AHA guidelines given the high incidence of stroke in patients with HCM and paroxysmal, persistent or permanent atrial fibrillation, it is recommended that all patients with atrial fibrillation should receive treatment with vitamin-K antagonists. Lifelong therapy with oral anticoagulants is recommended, even when sinus rhythm is restored (Gersh et al., 2011; Elliott et al., 2014). Currently, there is no evidence from randomized controlled trials on the effectiveness of direct oral anticoagulants (DOACs) in reducing thromboembolic risk in this population. The four major prospective trials assessing the efficacy of DOACs vs warfarin in atrial fibrillation did not include patients with hypertrophic cardiomyopathy in their analyses (Camm and Camm, 2017).

It remains controversial whether patients with LVNC should receive oral anticoagulation in the absence of atrial fibrillation or intracardiac thrombi. Some investigators recommend long-term prophylactic anticoagulation for all patients with LVNC independently of previous thromboembolic complications or the degree of left ventricular dysfunction. Although oral anticoagulation has been frequently proposed for these patients, these recommendations rely only on single case reports or small case series.

Randomized clinical trials on anticoagulation have been performed in patients with dilated cardiomyopathy with low left ventricular ejection fraction but no cardiac arrhythmias. The WATCH trial (Warfarin and Antiplatelet Therapy in Chronic Heart Failure) studied antithrombotic therapy (aspirin, clopidogrel, and warfarin) in patients with a left ventricular ejection fraction $\leq$35% in sinus rhythm and found no significant differences between the antithrombotic agents in the primary composite endpoints of all-cause mortality, nonfatal myocardial infarction, and nonfatal stroke (Massie et al., 2009). In a subanalysis, warfarin was associated with fewer nonfatal strokes than aspirin or clopidogrel, but major bleeding episodes were more frequent in warfarin patients compared with clopidogrel ($P<0.01$) but not aspirin ($P=0.22$) (Massie et al., 2009).

For patients with left ventricular thrombi, anticoagulation with warfarin continues to have more evidence in favor than with the DOACs. In a cohort study that included a total of 514 patients with left ventricular thrombi, on multivariable analysis, anticoagulation with DOAC vs warfarin (HR, 2.64; 95% CI, 1.28–5.43; $P=0.01$) was significantly associated with stroke or systemic embolism (Stables and Ormerod, 1996).

In Fabry disease, the increased risk of stroke may be reduced by combined therapy with agalsidase alfa and renin–angiotensin system inhibitors.

Due to the high risk of thromboembolism, anticoagulation is advisable in cases of peripartum cardiomyopathy at least during pregnancy and the first 2 months postpartum. Unfractionated heparin is safe during pregnancy (Arany and Elkayam, 2016).

Treatment of alcoholic cardiomyopathy consists of alcohol abstinence and heart failure medications.

NEUROMUSCULAR DISORDERS

Heart involvement is a well-known feature of some NMD, mainly in a large number of inherited myopathies and the hereditary variant transthyretin (ATTRv) amyloidosis. Heart involvement may develop during the course of the disease or, more rarely, at its onset. The cardiac phenotype (cardiomyopathy and/or heart rhythm disorders) and its severity are variable, depending on the underlying neuromuscular condition. However, in the same inherited NMD, inter- and intrafamilial clinical heterogeneity may be seen. Cardiac complications could be responsible for the morbidity and even mortality (sudden cardiac death due to a life-threatening arrhythmias and end-stage congestive heart failure) observed in certain patients with neuromuscular disorders. Neurologists and cardiologists must remain involved in multidisciplinary teams, in order to offer patients and their affected relatives the optimal appropriate clinical monitoring and treatment.

Cardiologists may recognize an inherited cardiac disease, which could be allelic at the same locus of a neuromuscular disorder. However, in the setting of a cardiomyopathy of unknown cause, the question "When to suspect about an underlying neuromuscular disease?" could be a challenging issue. Apart from evident and classical signs and symptoms related to a specific neuromuscular disease (muscle weakness and atrophy, myotonia, cramps, myoglobinuria, calf pseudohypertrophy, sensory deficits, diminished/abolished muscle stretch reflexes, scapular wining, joint contractures, short stature, cataracts, vitreous opacities and/or retinal abnormalities), cardiac imaging findings have an important role, especially when the cardiac phenotype is the predominant feature.

A normal human left ventricle has a maximum of three prominent trabeculations and is thus less trabeculated than the right ventricle (Boyd et al., 1987; Sedmera et al., 2000). LVNC, also known as noncompaction, is a rare cardiac abnormality characterized by trabeculations aligned with the inner part of the endocardium, being the apex and lateral wall of the left ventricle the most commonly affected regions (Stollberger and Finsterer, 2004; Stollberger et al., 2019). LVNC has been associated almost always with other cardiac malformations (Ichida et al., 1999; Oechslin et al., 2000; Stollberger et al., 2002). Furthermore, an association with NMD has also been reported (Stollberger et al., 1999, 2001, 2002, 2006, 2015; Finsterer and Stollberger, 2013). Actually, if a neuromuscular disease is systematically investigated, an association with LVNC has been found in up to 80% (Stollberger et al., 2002, 2015; Stollberger and Finsterer, 2019). Mitochondrial disorders, Barth syndrome, zaspopathy (myofibrillar myopathy), and myotonic dystrophy type 1 were the most commonly related disorders (Stollberger et al., 2002, 2015; Stollberger and Finsterer, 2019), but more recently, other more unusual neuromuscular diseases have also been documented (Finsterer and Stollberger, 2019). In light of these findings, it has been suggested that patients with LVNC should be referred to a neurologist for a screening of underlying neuromuscular diseases. However, there are several limitations. Precise and widely accepted diagnostic criteria for LVNC are still lacking, which is a major concern regarding both clinical and research application. Conversely, LVNC is not specific of neuromuscular disease, as it has been described before among athletes (Gati et al., 2013), pregnant women (Gati et al., 2014), black Africans (Peters et al., 2012), and chromosomal defects (Finsterer, 2009), although the clinical significance in these conditions is thus far undefined.

Cardiac imaging findings may raise the suspicion of hereditary ATTRv amyloidosis, which is particularly significant when the heart involvement is the prominent manifestation (Ando et al., 2013). Searching for features of cardiac amyloid infiltration with echocardiography, a major difference between the assessment of the left ventricle wall mass with ECG and echocardiography, and the uptake of ^{99m}Tc-DPD by the infiltrated amyloid TTR myocardium are quite different but helpful methods for the diagnosis of cardiac amyloidosis (Falk and Dubrey, 2010; Rapezzi et al., 2010, 2011). CMR offers additional tolls for cardiac amyloid diagnosis (Captur et al., 2016).

Since the neuromuscular field is quite broad, our review will focus on the most prevalent and preferentially adult-onset inherited neuromuscular disorders associated with cardiac manifestations.

Hereditary myopathies

Inherited myopathies are a broad and complex spectrum of clinical entities, including muscular dystrophies, mitochondrial disorders, metabolic myopathies, myofibrillar myopathies, and muscle channelopathies. Heart involvement occurs in many of them with variable degrees of severity and clinical patterns. The clinical presentation of heart failure and cardiac conduction defects in hereditary myopathies are similar with other cardiac diseases. Besides Duchenne muscular dystrophy, there are no current specific guidelines for the management of the cardiac complications associated with other hereditary myopathies. Heart failure symptomatic treatment and devices are used as for other general cardiomyopathies. Pacemaker and implantable cardioverter-defibrillation may also be necessary for cardiac conduction abnormalities treatment and preclusion of sudden cardiac death.

Muscular dystrophies

Muscular dystrophies are a group of inherited muscle disorders characterized by progressive muscle weakness and dystrophic pathological pattern on muscle biopsy (Mercuri et al., 2019). Cardiac complications occur in a number of these disorders, in particular those associated with abnormalities in the dystrophin-associated glycoprotein complex (Blake et al., 2002). This complex molecular structure is expressed in both skeletal and heart muscle and links the subsarcolemmal cytoskeleton to the extracellular matrix (Blake et al., 2002).

The dystrophinopathies are a group of X-linked progressive muscular dystrophies caused by mutations in the *DMD* gene that result in total or partial absence of functional dystrophin protein (Brandsema and Darras, 2015; Thangarajh, 2019). They include Duchenne and Becker muscular dystrophies, X-linked dilated cardiomyopathy, muscle cramps, and myalgia (Brandsema and Darras, 2015; Thangarajh, 2019). Duchenne muscular dystrophy is the most common NMD, being the prevalence at birth 2.9 per 10,000 live male births for Duchenne muscular dystrophy and 0.5 per 10,000 for Duchenne muscular dystrophy worldwide (Emery, 1991). The more severe phenotype, Duchenne muscular dystrophy, usually presents in early childhood with leg and pelvic muscles weakness and delayed motor milestones, while Becker muscular dystrophies is its milder allelic form (Thangarajh, 2019). Cardiac phenotype in dystrophinopathies is progressive and frequently includes sinus tachycardia and dilated cardiomyopathy (Perloff, 1984; Mazur et al., 2012; Thomas et al., 2012). Almost all Duchenne muscular dystrophy patients have dilated cardiomyopathy in the beginning of the third decade of life, but due to lack of inactivity, signs and symptoms of heart failure are usually mild or even absent (McNally, 2007; Townsend et al., 2007; Meyers and Townsend, 2019). However, congestive heart failure and rhythm disorders ultimately develop in the end stage of the disease and are frequently a major cause of death (Mavrogeni et al., 2015). Becker muscular dystrophy patients usually have a more expressive and severe cardiomyopathy than Duchenne muscular dystrophy, probably because they can still perform active muscle exercise, inducing an additional physical stress to the partial dystrophin-deficient cardiac muscle cells (Melacini et al., 1996; Thangarajh, 2019).

X-linked dilated cardiomyopathy is a rare dystrophinopathy form with a prevalence raging between 3% and 7% (Arbustini et al., 2000; Nakamura, 2015). Specific *DMD* gene mutations localized in both first promoter and exon are the main cause of this clinical phenotype (Feng et al., 2002). Skeletal muscle sparing in this entity is the result of alternative promoters activation (the brain and Purkinje cells promoters), which are responsible to keep enough levels of functional dystrophin in the skeletal muscle cells (Feng et al., 2002).

Although female carriers are mostly asymptomatic, some of them may actually develop muscle and particularly cardiac symptoms (Ishizaki et al., 2018). Dilated cardiomyopathy risk seems to increase with aging in this peculiar population (Ishizaki et al., 2018).

Routine cardiac surveillance is recommended for Duchenne muscular dystrophy patients, with a baseline echocardiogram at the time of diagnosis or at least at 6–7 years of age (Birnkrant et al., 2018a,b; Thangarajh, 2019). Afterwards, an annual echocardiogram is the rule for boys older than 10 years (Birnkrant et al., 2018a,b; Thangarajh, 2019). Angiotensin-converting enzyme inhibitors or angiotensin-receptor blockers should be started in the presymptomatic cardiac stage in boys older than 10 years (McNally et al., 2015; Birnkrant et al., 2018a,b). Eplerenone seems to have a beneficial adding effect to the conventional symptomatic treatment (Raman et al., 2015). If clinical, laboratory, and/or imaging signs of cardiac dysfunction emerged at an earlier age, it is recommended to start treatment as soon as possible (Birnkrant et al., 2018a,b). Advanced heart failure therapies in Duchenne muscular dystrophy, including implantable cardioverter-defibrillator, left ventricular assist device, and heart transplantation, have shown promising results regarding cardiac outcomes (Wittlieb-Weber et al., 2019).

Myotonic dystrophy type 1 (DM1) and 2 (DM2) are both autosomal dominant progressive multisystem disorders caused by CTG or CCTG repeats expansions in the corresponding *DMPK* or *CNBP* genes, respectively (Thornton, 2014; Turner and Hilton-Jones, 2014; Johnson, 2019). A classical triad of symptoms characterizes them, namely progressive muscle weakness, myotonia, and early-onset cataracts (Thornton, 2014; Turner and Hilton-Jones, 2014; Johnson, 2019). Distal limb muscle weakness is a prominent feature of DM1, while proximal muscle weakness is the hallmark of DM2 (Thornton, 2014; Johnson, 2019). Other clinical manifestations may include cardiac disease, multiple endocrinopathies, frontal balding (in males), hypersomnia, cognitive impairment and gastrointestinal disorders (Thornton, 2014; Turner and Hilton-Jones, 2014; Johnson, 2019). Cardiac phenotype in DM1 is characterized by conduction defects and/or supraventricular or ventricular arrhythmias (Cudia et al., 2009; Lau et al., 2015). Progressive atrioventricular block is considered the leading cause of death (Groh et al., 2008; Johnson et al., 2015). The most common tachyarrhythmias are atrial, and the predisposing is probably due to regions of atrial fibrosis (Lau et al., 2015). Ventricular tachyarrhythmias occur due to classic reentry circuits and are promoted by fibrotic foci, fatty infiltration as well as trigged activity (Merino et al., 1998; Ballo et al., 2013). Conduction defects and a past history of atrial fibrillation are predictors of sudden cardiac death (Duboc and Wahbi, 2012; Lallemand et al., 2012). Ventricular tachycardia or fibrillation is usually the tachyarrhythmias involved in sudden cardiac death mechanisms (Hermans et al., 2012). Rhythm disorders are not so common in DM2, but a risk of sudden cardiac death is still considerable (Schoser et al., 2004; Peric et al., 2019). A clinical meaningful cardiomyopathy is not the rule in both types of myotonic dystrophy (Thornton, 2014; Turner and Hilton-Jones, 2014; Johnson, 2019). Cardiac evaluation and management have the objective to avoid complications leading to sudden cardiac death. Annual ECG monitoring and referral to the cardiologist when a prolonged PQ interval is detected and/or cardiac symptoms emerged are crucial for appropriate treatment with a pacemaker or implantable cardioverter-defibrillator (Ashizawa et al., 2018; Schoser et al., 2019).

Limb-girdle muscular dystrophies are a clinically, genetically, and biochemical heterogeneous group of disorders that share clinical and dystrophic pathological feature in muscle biopsy (Mercuri et al., 2019). Dilated cardiomyopathy has been described in at least four autosomal dominant limb-girdle muscular dystrophies associated with *MYOT* (Olive et al., 2005), *LMNA* (Captur et al., 2018), *CAV3* (Gazzerro et al., 2011), and *DES* (Arbustini et al., 2006) gene mutations, respectively. Rhythm disorders, namely conduction defects, are universally present in laminopathies (*LMNA* gene mutations) in patients older than 30 years (Captur et al., 2018).

Sarcoglycanopathies (especially *SGCB*, *SGCG*, *SGCD* genes mutations) (Fayssoil, 2010; Fayssoil et al., 2012) and fukutinopathies (*FKTN* gene mutations) (Poppe et al., 2003; Gaul et al., 2006) are the autosomal recessive limb-girdle muscular dystrophies forms mostly associated with cardiac manifestations, being dilated cardiomyopathies sometimes quite severe.

Emery-Dreifuss muscular dystrophy is a group of clinically and genetically heterogeneous disorder characterized by progressive muscle weakness and atrophy, early contractures, and cardiac complications (Mercuri et al., 2019). Cardiac manifestations present commonly at the second decade of life, may precede the onset of significant muscle weakness, and are usually associated with life-threatening rhythm disorders and cardiomyopathy (Wang and Peng, 2019; Heller et al., 2020).

Mitochondrial myopathies

Mitochondrial myopathies are a heterogeneous group of inherited conditions resulting from a primary defect in the mitochondrial respiratory chain with consecutively impaired cellular energy metabolism (Calvo et al., 2016; Cohen, 2019). Clinical manifestations are quite variable, and virtually any organ or system may be affected (Cohen, 2019). The overall prevalence of heart involvement in mitochondrial myopathies is still unknown. However, cardiac dysfunction is common as part of MM and is manifested by cardiomyopathy (hypertrophic, dilated, and noncompaction) and/or rhythm disorders (Meyers et al., 2013; Cohen, 2019). In a systematic review and meta-analysis of cardiac disease in mitochondrial myopathies, mitochondrial encephalopathy, lactic acidosis, and stroke-like episodes (MELAS), especially associated with the m.3243A > G mutation, was found to have higher ECG and echocardiography abnormalities than patients with other clinical syndromes (Quadir et al., 2019). The same was true for myoclonic epilepsy and ragged fibers, associated with the m.8344A > G mutation but only regarding echocardiography findings (Quadir et al., 2019). Kearns–Sayre syndrome may be complicated by complete atrioventricular block during their clinical course (Kenny and Wetherbee, 1990), while Wolff–Parkinson–White syndrome may occur in MELAS (Vydt et al., 2007). Cardiomyopathy, mainly hypertrophic, is a common feature to all mitochondrial myopathies in general (Cohen, 2019).

Metabolic myopathies

Other rare metabolic myopathies may also have heart involvement, namely glycogen storage disorders and fatty acid oxidation disorders (Kanungo et al., 2018; Cohen, 2019). In general, pediatric forms have a severe cardiac phenotype (hypertrophic cardiomyopathy with or without rhythm disorders) with significant prognostic implications, while in adult entities, they are usually milder or even absent (Arbustini et al., 2018; Kanungo et al., 2018). As such, their description is beyond the scope of this chapter.

Myofibrillar myopathies

Myofibrillar myopathies are clinically and genetically heterogeneous disorders pathologically characterized by abnormal accumulation of intrasarcoplasmic proteins with myofibril disorganization at the level of the *Z*-disks (Kley et al., 2016; Fichna et al., 2018). Classically, slowly progressive muscle weakness (distal, proximal or both) is the main hallmark of the disease, and some gene mutations are associated with other features (Kley et al., 2016; Fichna et al., 2018). A predominant dilated or restrictive cardiomyopathy is usually observed in one-third of myofibrillar myopathies, in general associated with *DES*, *CRYAB*, *MYOT*, *LDB3/ZASP*, *FLNC*, *BAG3*, *TTN*, and *LMNA* gene mutations (Kley et al., 2016; Fichna et al., 2018).

Muscle channelopathies

Hereditary muscle channelopathies are episodic muscle disorders caused by mutations in muscle ion channels (sodium, chloride, calcium, or potassium) related genes, and they are divided into nondystrophic myotonias and primary periodic paralysis (Veerapandiyan et al., 1993; Sansone, 2019). Andersen–Tawil syndrome, an autosomal dominant periodic paralysis caused by *KCNJ2* gene mutations, deserves especial attention for the associated cardiac manifestations (Statland et al., 2018; Sansone, 2019). The presence of dysmorphic features and heart involvement differentiate this clinical entity from the other episodic muscle disorders (Statland et al., 2018; Sansone, 2019). Rhythm disorders including sustained ventricular tachyarrhythmias, torsade de pointes, and prolonged QT interval have been described (Kokubun et al., 2019; Sansone, 2019). There is no data regarding sudden cardiac death, but a close monitoring annually or when clinical appropriate has been suggested (Sansone, 2019).

Hereditary ATTRv amyloidosis

Hereditary ATTRv amyloidosis is a rare multisystem disorder characterized by the extracellular deposition of amyloid fibrils composed by transthyretin (TTR) (Benson and Kincaid, 2007; Ando et al., 2013). It is an autosomal dominant disorder caused by mutations in *TTR* gene with an age of onset ranging from the second to the ninth decade of life (Ando et al., 2013; Conceicao et al., 2016). More than 130 *TTR* gene mutations have been reported worldwide (Sekijima et al., 2018). Phenotypic heterogeneity is a hallmark of this disorder, and it may include a progressive sensorimotor and autonomic neuropathy, infiltrative cardiomyopathy, nephropathy, and ocular involvement (Ando et al., 2013; Conceicao et al., 2016; Sekijima et al., 2018). Some *TTR* gene mutations are more likely associated with neuropathy (Val50Met) as the predominant feature, while others are associated particularly with cardiomyopathy (Val122Ile, Ile68Leu, Thr60Ala, Leu111Met), but both clinical manifestations may coexist in different proportions (Ando et al., 2013). Cardiac manifestations include bundle branch block, atrioventricular, and sinoatrial block (Ando et al., 2013; Maurer et al., 2019; Ruberg et al., 2019). Myocardial amyloid infiltration is a progressive process, in which a restrictive cardiomyopathy is the end spectrum of the cardiac disease (Ando et al., 2013; Maurer et al., 2019; Ruberg et al., 2019). To date, treatment of hereditary ATTRv amyloidosis has been directed to the progression of the neuropathy. Tafamidis, an already European Medicines Agency (European Medicines Agency, 2020) approved TTR tetramer stabilizer for the neuropathy, was demonstrated in the Transthyretin Amyloidosis Cardiomyopathy Clinical trial to reduce all-cause mortality and cardiovascular hospitalizations as well as to slow decline in quality of life and functional capacity in patients with ATTR amyloid cardiomyopathy and heart failure compared with patients who receive placebo with a comparable safety profile (Maurer et al., 2018). Recently, it was approved both by the FDA (US Food and Drug Administration, 2020) and the European Medicines Agency (2020) for the treatment of ATTR amyloid cardiomyopathy.

References

Al-Khatib SM, Stevenson WG, Ackerman MJ et al. (2018). 2017 AHA/ACC/HRS guideline for management of patients with ventricular arrhythmias and the prevention of sudden cardiac death: a report of the American College of Cardiology/American Heart Association task force on clinical practice guidelines and the Heart Rhythm Society. Heart Rhythm 15: e73–189.

Almeida AG, Pinto FJ (2013). Non-compaction cardiomyopathy. Heart 99: 1535–1542.

Ando Y, Coelho T, Berk JL et al. (2013). Guideline of transthyretin-related hereditary amyloidosis for clinicians. Orphanet J Rare Dis 8: 31.

Anzini M, Merlo M, Sabbadini G et al. (2013). Long-term evolution and prognostic stratification of biopsy-proven active myocarditis. Circulation 128: 2384–2394.

Aquaro GD, Barison A, Todiere G et al. (2016). Usefulness of combined functional assessment by cardiac cardiac magnetic resonance and tissue characterization versus task force criteria for diagnosis of arrhythmogenic right ventricular cardiomyopathy. Am J Cardiol 118: 1730–1736.

Arany Z, Elkayam U (2016). Peripartum cardiomyopathy. Circulation 133: 1397–1409.

Arbustini E, Diegoli M, Morbini P et al. (2000). Prevalence and characteristics of dystrophin defects in adult male patients with dilated cardiomyopathy. J Am Coll Cardiol 35: 1760–1768.

Arbustini E, Pasotti M, Pilotto A et al. (2006). Desmin accumulation restrictive cardiomyopathy and atrioventricular block associated with desmin gene defects. Eur J Heart Fail 8: 477–483.

Arbustini E, Narula N, Dec GW et al. (2013). The MOGE(S) classification for a phenotype-genotype nomenclature of cardiomyopathy: endorsed by the World Heart Federation. J Am Coll Cardiol 62: 2046–2072.

Arbustini E, Favalli V, Narula N et al. (2016). Left ventricular noncompaction: a distinct genetic cardiomyopathy? J Am Coll Cardiol 68: 949–966.

Arbustini E, Di Toro A, Giuliani L et al. (2018). Cardiac phenotypes in hereditary muscle disorders: JACC state-of-the-art review. J Am Coll Cardiol 72: 2485–2506.

Ashizawa T, Gagnon C, Groh WJ et al. (2018). Consensus-based care recommendations for adults with myotonic dystrophy type 1. Neurol Clin Pract 8: 507–520.

Ballo P, Giaccardi M, Colella A et al. (2013). Mechanical and electrophysiological substrate for recurrent atrial flutter detected by right atrial speckle tracking echocardiography and electroanatomic mapping in myotonic dystrophy type 1. Circulation 127: 1422–1424.

Basavarajaiah S, Siam M, Sharma S (2009). Unusual cause of stroke in hypertrophic cardiomyopathy. Heart 95: 1592.

Basso C, Corrado D, Bauce B et al. (2012). Arrhythmogenic right ventricular cardiomyopathy. Circ Arrhytm Electrophysiol 5: 1233–1246.

Basso C, Calabrese F, Angelini A et al. (2013). Classification and histological, immunohistochemical, and molecular diagnosis of inflammatory myocardial disease. Heart Fail Rev 18: 673–681.

Belcour D, Jabot J, Grard B et al. (2015). Prevalence and risk factors of stress cardiomyopathy after convulsive status epilepticus in ICU patients. Crit Care Med 43: 2164–2170.

Benson MD, Kincaid JC (2007). The molecular biology and clinical features of amyloid neuropathy. Muscle Nerve 36: 411–423.

Bhatia NL, Tajik AJ, Wilansky S et al. (2011). Isolated noncompaction of the left ventricular myocardium in adults: a systematic overview. J Card Fail 17: 7771–7778.

Birnkrant DJ, Bushby K, Bann CM et al. (2018a). Diagnosis and management of Duchenne muscular dystrophy, part 2: respiratory, cardiac, bone health, and orthopaedic management. Lancet Neurol 17: 347–361.

Birnkrant DJ, Bushby K, Bann CM et al. (2018b). Diagnosis and management of Duchenne muscular dystrophy, part 1: diagnosis, and neuromuscular, rehabilitation, endocrine, and gastrointestinal and nutritional management. Lancet Neurol 17: 251–267.

Blake DJ, Weir A, Newey SE et al. (2002). Function and genetics of dystrophin and dystrophin-related proteins in muscle. Physiol Rev 82: 291–329.

Boyd MT, Seward JB, Tajik AJ et al. (1987). Frequency and location of prominent left ventricular trabeculations at autopsy in 474 normal human hearts: implications for evaluation of mural thrombi by two-dimensional echocardiography. J Am Coll Cardiol 9: 323–326.

Brandsema JF, Darras BT (2015). Dystrophinopathies. Semin Neurol 35: 369–384.

Caforio AL, Calabrese F, Angelini A et al. (2007). A prospective study of biopsy-proven myocarditis: prognostic relevance of clinical and aetiopathogenetic features at diagnosis. Eur Heart J 28: 1326–1333.

Caforio AL, Marcolongo R, Basso C et al. (2015). Clinical presentation and diagnosis of myocarditis. Heart 101: 1332–1344.

Calvo SE, Clauser KR, Mootha VK (2016). MitoCarta2.0: an updated inventory of mammalian mitochondrial proteins. Nucleic Acids Res 44: D1251–D1257.

Camm CF, Camm AJ (2017). Atrial fibrillation and anticoagulation in hypertrophic cardiomyopathy. Arrhythm Electrophysiol Rev 6: 63–68.

Campuzano O, Alcalde M, Allegue C et al. (2013). Genetics of arrhythmogenic right ventricular cardiomyopathy. J Med Genet 50: 280–289.

Captur G, Manisty C, Moon JC (2016). Cardiac MRI evaluation of myocardial disease. Heart 102: 1429–1435.

Captur G, Arbustini E, Bonne G et al. (2018). Lamin and the heart. Heart 104: 468–479.

Carod-Artal FJ, Vargas AP, Horan TA et al. (2005). Chagasic cardiomyopathy is independently associated with ischemic stroke in Chagas disease. Stroke 36: 965–970.

Cevik C, Shah N, Wilson JM et al. (2012). Multiple left ventricular: in a patient with left ventricular noncompaction. Tex Heart Inst J 39: 550–553.

Chacko L, Martone R, Cappelli F et al. (2019). Cardiac amyloidosis. Updates in imaging. Curr Cardiol Rep 21: 108.

Cohen BH (2019). Mitochondrial and metabolic myopathies. Continuum (Minneap Minn) 25: 1732–1766.

Conceicao I, Gonzalez-Duarte A, Obici L et al. (2016). Red-flag symptom clusters in transthyretin familial amyloid polyneuropathy. J Peripher Nerv Syst 21: 5–9.

Corrado D, Wichter T, Link MS et al. (2015). Treatment of arrhythmogenic right ventricular cardiomyopathy/dysplasia: an international task force consensus statement. Eur Heart J 36: 3227–3237.

Cudia P, Bernasconi P, Chiodelli R et al. (2009). Risk of arrhythmia in type I myotonic dystrophy: the role of clinical and genetic variables. J Neurol Neurosurg Psychiatry 80: 790–793.

Debonnaire P, Joyce E, Hiemstra Y et al. (2017). Left atrial size and function in hypertrophic cardiomyopathy patients and risk of new-onset atrial fibrillation. Circ Arrhythm Electrophysiol 10: e004052.

Djoussé L, Gaziano JM (2008). Alcohol consumption and heart failure: a systematic review. Curr Atheroscler Rep 10: 117–120.

Duboc D, Wahbi K (2012). What is the best way to detect infra-Hisian conduction abnormalities and prevent sudden cardiac death in myotonic dystrophy? Heart 98: 433–434.

Dyken ME, Biller J (1994). Peripartum cardiomyopathy and stroke. Cerebrovasc Dis 4: 325–328.

Elliott P, Andersson B, Arbustini E et al. (2008). Classification of the cardiomyopathies: a position statement from the European Society Of CardiologyWorking Group on Myocardial and Pericardial Diseases. Eur Heart J 29: 270–276.

Elliott PM, Anastasakis A, Borger MA et al. (2014). 2014 ESC guidelines on diagnosis and management of hypertrophic cardiomyopathy: the Task Force for the Diagnosis and Management of Hypertrophic Cardiomyopathy of the European Society of Cardiology (ESC). Eur Heart J 35: 2733–2779.

Emery AE (1991). Population frequencies of inherited neuromuscular diseases—a world survey. Neuromuscul Disord 1: 19–29.

European Medicines Agency. February 25, 2020. Available from: https://www.ema.europa.eu/en/medicines/human/EPAR/vyndaqel#authorisation-details-section. Accessed May 24, 2020.

Falk RH, Dubrey SW (2010). Amyloid heart disease. Prog Cardiovasc Dis 52: 347–361.

Falk RH, Alexander KM, Liao R et al. (2016). AL (light-chain) cardiac amyloidosis: a review of diagnosis and therapy. J Am Coll Cardiol 68: 1323–1341.

Fayssoil A (2010). Cardiac diseases in sarcoglycanopathies. Int J Cardiol 144: 67–68.

Fayssoil A, Nardi O, Orlikowski D et al. (2012). Heart involvement in sarcoglycanopathies. Rev Neurol (Paris) 168: 779–782.

Feng J, Yan J, Buzin CH et al. (2002). Mutations in the dystrophin gene are associated with sporadic dilated cardiomyopathy. Mol Genet Metab 77: 119–126.

Feng D, Syed IS, Martinez M et al. (2009). Intracardiac thrombosis and anticoagulation therapy in cardiac amyloidosis. Circulation 119: 2490–2497.

Feng D, Edwards WD, Oh JK et al. (2017). Intracardiac thrombosis and embolism in patients with cardiac amyloidosis. Circulation 116: 2420–2426.

Ferreira VM, Schulz-Menger J, Holmvang G et al. (2018). Cardiovascular magnetic resonance in nonischemic myocardial inflammation: expert recommendations. J Am Coll Cardiol 72: 3158–3176.

Fichna JP, Maruszak A, Zekanowski C (2018). Myofibrillar myopathy in the genomic context. J Appl Genet 59: 431–439.

Finsterer J (2009). Cardiogenetics, neurogenetics, and pathogenetics of left ventricular hypertrabeculation/noncompaction. Pediatr Cardiol 30: 659–681.

Finsterer J, Stollberger C (2013). Unclassified cardiomyopathies in neuromuscular disorders. Wien Med Wochenschr 163: 505–513.

Finsterer J, Stollberger C (2019). Mutations in genes associated with either myopathy or noncompaction. Herz 44: 756–758.

Finsterer J, Wahbi K (2014). CNS-disease affecting the heart: brain-heart disorders. J Neurol Sci 345: 8–14.

Finsterer J, Zarrouk-Mahjoub S (2017). Management of epilepsy in MERRF syndrome. Seizure 50: 166–170.

Finsterer J, Stolberger C, Wahbi K (2013). Cardiomyopathy in neurological disorders. Cardiovasc Pathol 22: 389–400.

Fonseca AC, Marto JP, Pimenta D et al. (2020). Undetermined stroke genesis and hidden cardiomyopathies determined by cardiac magnetic resonance. Neurology 94: e107–e113.

Fuster V, Gersh BJ, Giuliani ER et al. (1981). The natural history of idiopathic dilated cardiomyopathy. Am J Cardiol 47: 525–531.

Gambarin FI, Tagliani M, Arbustini E (2008). Pure restrictive cardiomyopathy associated with cardiac troponin I gene mutation: mismatch between the lack of hypertrophy and the presence of disarray. Heart 94: 1257.

Gati S, Chandra N, Bennett RL et al. (2013). Increased left ventricular trabeculation in highly trained athletes: do we need more stringent criteria for the diagnosis of left ventricular non-compaction in athletes? Heart 99: 401–408.

Gati S, Papadakis M, Papamichael ND et al. (2014). Reversible de novo left ventricular trabeculations in pregnant women: implications for the diagnosis of left ventricular noncompaction in low-risk populations. Circulation 130: 475–483.

Gaul C, Deschauer M, Tempelmann C et al. (2006). Cardiac involvement in limb-girdle muscular dystrophy 2I: conventional cardiac diagnostic and cardiovascular magnetic resonance. J Neurol 253: 1317–1322.

Gazzerro E, Bonetto A, Minetti C (2011). Caveolinopathies: translational implications of caveolin-3 in skeletal and cardiac muscle disorders. Handb Clin Neurol 101: 135–142.

Gersh BJ, Maron BJ, Bonow RO et al. (2011). 2011 ACCF/AHA guideline for the diagnosis and treatment of hypertrophic cardiomyopathy: a report of the American College of Cardiology Foundation/American Heart Association task force on practice guidelines. Circulation 124: e783–e831.

Gertz MA, Benson MD, Dyck PJ et al. (2015). Diagnosis, prognosis, and therapy of transthyretin amyloidosis. J Am Coll Cardiol 66: 2451–2466.

Groh WJ, Groh MR, Saha C et al. (2008). Electrocardiographic abnormalities and sudden death in myotonic dystrophy type 1. N Engl J Med 358: 2688–2697.

Guttmann OP, Rahman MS, O'Mahony C et al. (2014). Atrial fibrillation and thromboembolism in patients with hypertrophic cardiomyopathy: a systematic review. Heart 100: 465–472.

Habib G, Bucciarelli-Ducci C, Caforio ALP et al. (2017). EACVI scientific documents committee; Indian academy of echocardiography. Multimodality imaging in restrictive cardiomyopathies: an EACVI expert consensus document. Eur Heart J Cardiovasc Imaging 18: 1090–1121.

Halliday BP, Baksi AJ, Gulati A et al. (2019). Outcome in dilated cardiomyopathy related to the extent, location, and pattern of late gadolinium enhenacement. JACC Cardiovasc Imaging 12: 1645–1655.

Haruki S, Minami Y, Hagiwara N (2016). Stroke and embolic events in hypertrophic cardiomyopathy: risk stratification in patients without atrial fibrillation. Stroke 47: 936–942.

Heller SA, Shih R, Kalra R et al. (2020). Emery-Dreifuss muscular dystrophy. Muscle Nerve 61: 436–448.

Hermans MC, Faber CG, Bekkers SC et al. (2012). Structural and functional cardiac changes in myotonic dystrophy type 1: a cardiovascular magnetic resonance study. J Cardiovasc Magn Reson 14: 48.

Hohneck A, Overhoff D, Doesch C et al. (2020). Extent of late gadolinium enhancement predicts thromboembolic events in patients with hypertrophic cardiomyopathy. Circ J 84: 754–762.

Ichida F, Hamamichi Y, Miyawaki T et al. (1999). Clinical features of isolated noncompaction of the ventricular myocardium: long-term clinical course, hemodynamic properties, and genetic background. J Am Coll Cardiol 34: 233–240.

Innelli P, Galderisi M (2004). Which role of echocardiography to predict systemic embolism in dilated cardiomyopathy? Am J Cardiol 94: 1479–1480.

Ishizaki M, Kobayashi M, Adachi K et al. (2018). Female dystrophinopathy: review of current literature. Neuromuscul Disord 28: 572–581.

Jenni R, Oechslin EN, van der Loo B (2007). Isolated ventricular non-compaction of the myocardium in adults. Heart 93: 11–15.

Johnson NE (2019). Myotonic muscular dystrophies. Continuum (Minneap Minn) 25: 1682–1695.

Johnson NE, Abbott D, Cannon-Albright LA (2015). Relative risks for comorbidities associated with myotonic dystrophy: a population-based analysis. Muscle Nerve 52: 659–661.

Kanungo S, Wells K, Tribett T et al. (2018). Glycogen metabolism and glycogen storage disorders. Ann Transl Med 6: 474.

Kenny D, Wetherbee J (1990). Kearns-Sayre syndrome in the elderly: mitochondrial myopathy with advanced heart block. Am Heart J 120: 440–443.

Kley RA, Olive M, Schroder R (2016). New aspects of myofibrillar myopathies. Curr Opin Neurol 29: 628–634.

Kokubun N, Aoki R, Nagashima T et al. (2019). Clinical and neurophysiological variability in Andersen-Tawil syndrome. Muscle Nerve 60: 752–757.

Krishnamoorthy V, Mackensen GB, Gibbons EF et al. (2016). Cardiac dysfunction after neurologic injury: what do we know and where are we going? Chest 149: 1325–1331.

Laghari AH, Khan AH, Kazmi KA (2013). Peripartum cardiomyopathy: ten year experience at a tertiary care hospital in Pakistan. BMC Res Notes 6: 495.

Lallemand B, Clementy N, Bernard-Brunet A et al. (2012). The evolution of infrahissian conduction time in myotonic dystrophy patients: clinical implications. Heart 98: 291–296.

Lau JK, Sy RW, Corbett A et al. (2015). Myotonic dystrophy and the heart: a systematic review of evaluation and management. Int J Cardiol 184: 600–608.

Lestienne F, Bruno C, Bertora D et al. (2017). Ischemic stroke in a young patient heralding a left ventricular noncompaction cardiomyopathy. Case Rep Neurol 9: 204–209.

Lousada I, Comenzo RL, Landau H et al. (2015). Light chain amyloidosis: patient experience survey from the amyloidosis research consortium. Adv Ther 32: 920–928.

Mankad R, Bonnichsen C, Mankad S (2016). Hypereosinophilic syndrome: cardiac diagnosis and management. Heart 102: 100–106.

Marcus FI, McKenna WJ, Sherrill D et al. (2010). Diagnosis of arrhythmogenic right ventricular cardiomyopathy/dysplasia: proposed modification of the task force criteria. Eur Heart J 31: 806–814.

Maron BJ, Olivotto I, Bellone P et al. (2002). Clinical profile of stroke in 900 patients with hypertrophic cardiomyopathy. J Am Coll Cardiol 39: 301–307.

Maron BJ, Maron MS, Semsarian C (2012). Genetics of hypertrophic cardiomyopathy after 20 years: clinical perspectives. J Am Coll Cardiol 60: 705–715.

Massie BM, Collins JF, Ammon SE et al. (2009). Randomized trial of warfarin, aspirin, and clopidogrel in patients with chronic heart failure: the warfarin and antiplatelet therapy in chronic heart failure (WATCH) trial. Circulation 119: 1616–1624.

Maurer MS, Elliott P, Comenzo R et al. (2017). Addressing common questions encountered in the diagnosis and management of cardiac amyloidosis. Circulation 135: 1357–1377.

Maurer MS, Schwartz JH, Gundapaneni B et al. (2018). Tafamidis treatment for patients with transthyretin amyloid cardiomyopathy. N Engl J Med 379: 1007–1016.

Maurer MS, Bokhari S, Damy T et al. (2019). Expert consensus recommendations for the suspicion and diagnosis of transthyretin cardiac amyloidosis. Circ Heart Fail 12: e006075.

Mavrogeni S, Markousis-Mavrogenis G, Papavasiliou A et al. (2015). Cardiac involvement in Duchenne and Becker muscular dystrophy. World J Cardiol 7: 410–414.

Mazur W, Hor KN, Germann JT et al. (2012). Patterns of left ventricular remodeling in patients with Duchenne muscular dystrophy: a cardiac MRI study of ventricular geometry, global function, and strain. Int J Cardiovasc Imaging 28: 99–107.

McNally EM (2007). New approaches in the therapy of cardiomyopathy in muscular dystrophy. Annu Rev Med 58: 75–88.

McNally EM, Kaltman JR, Benson DW et al. (2015). Contemporary cardiac issues in Duchenne muscular dystrophy. Working Group of the National Heart, Lung, and Blood Institute in collaboration with Parent Project Muscular Dystrophy. Circulation 131: 1590–1598.

Melacini P, Fanin M, Danieli GA et al. (1996). Myocardial involvement is very frequent among patients affected with subclinical Becker's muscular dystrophy. Circulation 94: 3168–3175.

Mercuri E, Bonnemann CG, Muntoni F (2019). Muscular dystrophies. Lancet 394: 2025–2038.

Merino JL, Carmona JR, Fernandez-Lozano I et al. (1998). Mechanisms of sustained ventricular tachycardia in myotonic dystrophy: implications for catheter ablation. Circulation 98: 541–546.

Merlo M, Cannatá A, Gobbo A et al. (2018). Evolving concepts in dilated cardiomyopathy. Eur J Heart Fail 20: 228–239.

Meyers TA, Townsend D (2019). Cardiac pathophysiology and the future of cardiac therapies in Duchenne muscular dystrophy. Int J Mol Sci 20: 4098.

Meyers DE, Basha HI, Koenig MK (2013). Mitochondrial cardiomyopathy: pathophysiology, diagnosis, and management. Tex Heart Inst J 40: 385–394.

Mischie AN, Chioncel V, Droc I et al. (2013). Anticoagulation in patients with dilated cardiomyopathy, low ejection fraction, and sinus rhythm: back to the drawing board. Cardiovasc Ther 31: 298–302.

Morales A, Hershberger RE (2013). Genetic evaluation of dilated cardiomyopathy. Curr Cardiol Rep 15: 375.

Muchtar E, Blauwet LA, Gertz MA (2017). Restrictive cardiomyopathy: genetics, pathogenesis, clinical manifestations, diagnosis, and therapy. Circ Res 121: 819–837.

Nagatomo Y, Tang WH (2014). Autoantibodies and cardiovascular dysfunction: cause or consequence? Curr Heart Fail Rep 11: 500–508.

Nakamura A (2015). X-linked dilated cardiomyopathy: a cardiospecific phenotype of dystrophinopathy. Pharmaceuticals (Basel) 8: 303–320.

O'Hanlon R, Grasso A, Roughton M et al. (2010). Prognostic significance of myocardial fibrosis in hypertrophic cardiomyopathy. J Am Coll Cardiol 56: 867–874.

O'Mahony C, Jichi F, Pavlou M et al. (2014). A novel clinical risk prediction model for sudden cardiac death in hypertrophic cardiomyopathy (HCM risk-SCD). Eur Heart J 35: 2010–2020.

Oechslin EN, Attenhofer Jost CH, Rojas JR et al. (2000). Long-term follow-up of 34 adults with isolated left ventricular noncompaction: a distinct cardiomyopathy with poor prognosis. J Am Coll Cardiol 36: 493–500.

Olive M, Goldfarb LG, Shatunov A et al. (2005). Myotilinopathy: refining the clinical and myopathological phenotype. Brain 128: 2315–2326.

Olivotto I, Girolami F, Ackerman MJ et al. (2008). Myofilament protein gene mutation screening and outcome of patients with hypertrophic cardiomyopathy. Mayo Clin Proc 83: 630–638.

Peric S, Bjelica B, Aleksic K et al. (2019). Heart involvement in patients with myotonic dystrophy type 2. Acta Neurol Belg 119: 77–82.

Perloff JK (1984). Cardiac rhythm and conduction in Duchenne's muscular dystrophy: a prospective study of 20 patients. J Am Coll Cardiol 3: 1263–1268.

Peters F, Khandheria BK, dos Santos C et al. (2012). Isolated left ventricular noncompaction in sub-Saharan Africa: a clinical and echocardiographic perspective. Circ Cardiovasc Imaging 5: 187–193.

Pilichou K, Thiene G, Bauce B et al. (2016). Arrhythmogenic cardiomyopathy. Orphanet J Rare Dis 11: 33.

Pinto YM, Elliott PM, Arbustini E et al. (2016). Proposal for a revised definition of dilated cardiomyopathy, hypokinetic non-dilated cardiomyopathy and its implications for clinical practice: a position statement of the ESC working group on myocardial and pericardial disease. Eur Heart J 37: 1850–1858.

Poppe M, Cree L, Bourke J et al. (2003). The phenotype of limb-girdle muscular dystrophy type 2I. Neurology 60: 1246–1251.

Pugh TJ, Kelly MA, Gowrisankar S et al. (2014). The landscape of genetic variation in dilated cardiomyopathy as surveyed by clinical DNA sequencing. Genet Med 16: 601–608.

Quadir A, Pontifex CS, Lee Robertson H et al. (2019). Systematic review and meta-analysis of cardiac involvement in mitochondrial myopathy. Neurol Genet 5: e339.

Raman SV, Hor KN, Mazur W et al. (2015). Eplerenone for early cardiomyopathy in Duchenne muscular dystrophy: a randomised, double-blind, placebo-controlled trial. Lancet Neurol 14: 153–161.

Rapezzi C, Merlini G, Quarta CC et al. (2009). Systemic cardiac amyloidoses: disease profiles and clinical courses of the 3 main types. Circulation 120: 1203–1212.

Rapezzi C, Quarta CC, Riva L et al. (2010). Transthyretin-related amyloidoses and the heart: a clinical overview. Nat Rev Cardiol 7: 398–408.

Rapezzi C, Quarta CC, Guidalotti PL et al. (2011). Usefulness and limitations of 99mTc-3,3-diphosphono-1, 2-propanodicarboxylic acid scintigraphy in the aetiological diagnosis of amyloidotic cardiomyopathy. Eur J Nucl Med Mol Imaging 38: 470–478.

Rapezzi C, Arbustini E, Caforio AL et al. (2013). Diagnostic work-up in cardiomyopathies: bridging the gap between clinical phenotypes and final diagnosis. A position statement from the ESC Working Group on Myocardial and Pericardial Diseases. Eur Heart J 34: 1448–1458.

Rice GP, Ebers GC, Newland F et al. (1981). Recurrent cerebral embolism in cardiac amyloidosis. Neurology 31: 904.

Robinson AA, Trankle CR, Eubanks G et al. (2020). Off-label use of direct oral anticoagulants compared with warfarin for left ventricular thrombi. JAMA Cardio e200652.

Ruberg FL, Grogan M, Hanna M et al. (2019). Transthyretin amyloid cardiomyopathy: JACC state-of-the-art review. J Am Coll Cardiol 73: 2872–2891.

Sansone VA (2019). Episodic muscle disorders. Continuum (Minneap Minn) 25: 1696–1711.

Santhanakrishnan R, Wang N, Larson MG et al. (2016). Atrial fibrillation begets heart failure and vice versa: temporal associations and differences in preserved versus reduced ejection fraction. Circulation 133: 484–492.

Schiff M, Roda C, Monin ML et al. (2017). Clinical, laboratory and molecular findings and long-term follow-up data in 96 French patients with PMM2-CDG (phosphomannomutase 2-congenital disorder of glycosylation) and review of the literature. J Med Genet 54: 843–851.

Schiffmann R, Warnock DG, Banikazemi M et al. (2009). Fabry disease: progression of nephropathy, and prevalence of cardiac and cerebrovascular events before enzyme replacement therapy. Nephrol Dial Transplant 24: 2102–2111.

Schoser BG, Ricker K, Schneider-Gold C et al. (2004). Sudden cardiac death in myotonic dystrophy type 2. Neurology 63: 2402–2404.

Schoser B, Montagnese F, Bassez G et al. (2019). Consensus-based care recommendations for adults with myotonic dystrophy type 2. Neurol Clin Pract 9: 343–353.

Schultheiss H, Kühl U, Cooper LT (2011). The management of myocarditis. Eur Heart J 32: 2616–2625.

Sedmera D, Pexieder T, Vuillemin M et al. (2000). Developmental patterning of the myocardium. Anat Rec 258: 319–337.

Sekijima Y, Ueda M, Koike H et al. (2018). Diagnosis and management of transthyretin familial amyloid polyneuropathy in Japan: red-flag symptom clusters and treatment algorithm. Orphanet J Rare Dis 13: 6.

Semsarian C, Ingles J, Maron MS et al. (2015). New perspectives on the prevalence of hypertrophic cardiomyopathy. J Am Coll Cardiol 65: 1249–1254.

Sen-Chowdhry S, Syrris P, Prasad SK et al. (2008). Left-dominant arrhythmogenic cardiomyopathy: an under-recognized clinical entity. J Am Coll Cardiol 52: 2175–2178.

Serdoz LV, Daleffe E, Merlo M et al. (2016). Predictors for restoration of normal left ventricular function in response to cardiac resynchronization therapy measured at time of implantation. Am J Cardiol 108: 75–80.

Seward JB, Casaclang-Verzosa G (2020). Infiltrative cardiovascular diseases: cardiomyopathies that look alike. J Am Coll Cardiol 55: 1769–1779.

Siddiqi OK, Ruberg FL (2018). Cardiac amyloidosis: an update on pathophysiology, diagnosis and treatment. Trends Cardiovasc Med 28: 10–21.

Stables RH, Ormerod OJ (1996). Atrial thrombi occurring during sinus rhythm in cardiac amyloidosis: evidence for atrial electromechanical dissociation. Heart 75: 426.

Statland JM, Fontaine B, Hanna MG et al. (2018). Review of the diagnosis and treatment of periodic paralysis. Muscle Nerve 57: 522–530.

Stollberger C, Finsterer J (2004). Left ventricular hypertrabeculation/noncompaction. J Am Soc Echocardiogr 17: 91–100.

Stollberger C, Finsterer J (2019). Understanding left ventricular hypertrabeculation/noncompaction: pathomorphologic findings and prognostic impact of neuromuscular comorbidities. Expert Rev Cardiovasc Ther 17: 95–109.

Stollberger C, Finsterer J, Valentin A et al. (1999). Isolated left ventricular abnormal trabeculation in adults is associated with neuromuscular disorders. Clin Cardiol 22: 119–123.

Stollberger C, Finsterer J, Blazek G (2001). Isolated left ventricular abnormal trabeculation: follow-up and association with neuromuscular disorders. Can J Cardiol 17: 163–168.

Stollberger C, Finsterer J, Blazek G (2002). Left ventricular hypertrabeculation/noncompaction and association with additional cardiac abnormalities and neuromuscular disorders. Am J Cardiol 90: 899–902.

Stollberger C, Winkler-Dworak M, Blazek G et al. (2006). Age-dependency of cardiac and neuromuscular findings in left ventricular noncompaction. Int J Cardiol 111: 131–135.

Stöllberger C, Blazek G, Dobias C et al. (2011a). Frequency of stroke and embolism in left ventricular hypertrabeculation/noncompaction. Am J Cardiol 7: 1021–1023.

Stöllberger C, Höftberger R, Finsterer J (2011b). Lamotrigine-trigged obstructive hypertrophic cardiomyopathy, epilepsy and metabolic myopathy. Int J Cardiol 149: e103–e105.

Stollberger C, Blazek G, Gessner M et al. (2015). Age-dependency of cardiac and neuromuscular findings in adults with left ventricular hypertrabeculation/noncompaction. Am J Cardiol 115: 1287–1292.

Stollberger C, Wegner C, Finsterer J (2019). Left ventricular hypertrabeculation/noncompaction, cardiac phenotype, and neuromuscular disorders. Herz 44: 659–665.

Tedford RJ, James C, Judge DP et al. (2012). Cardiac transplantation in arrhythmogenic right ventricular dysplasia/cardiomyopathy. J Am Coll Cardiol 59: 289–290.

Tellor KB, Ngo-Lam R, Badran D et al. (2019). A rare case of a gabapentin-induced cardiomyopathy. J Clin Pharm Ther 44: 644–646.

Thangarajh M (2019). The dystrophinopathies. Continuum (Minneap Minn) 25: 1619–1639.

Thomas TO, Morgan TM, Burnette WB et al. (2012). Correlation of heart rate and cardiac dysfunction in Duchenne muscular dystrophy. Pediatr Cardiol 33: 1175–1179.

Thornton CA (2014). Myotonic dystrophy. Neurol Clin 32: 705–719, viii.

Townsend D, Yasuda S, Metzger J (2007). Cardiomyopathy of Duchenne muscular dystrophy: pathogenesis and prospect of membrane sealants as a new therapeutic approach. Expert Rev Cardiovasc Ther 5: 99–109.

Turner C, Hilton-Jones D (2014). Myotonic dystrophy: diagnosis, management and new therapies. Curr Opin Neurol 27: 599–606.

US Food and Drug Administration. Frequently asked questions: breakthrough therapies. January 24, 2020. Available from: https://www.fda.gov/regulatory-information/food-and-drug-administration-safety-and-innovation-act-fdasia/frequently-asked-questions-breakthrough-therapies. Accessed May 24, 2020.

Veerapandiyan A, Statland JM, Tawil R (1993). Andersen-Tawil syndrome. In: MP Adam, HH Ardinger, RA Pagon et al. (Eds.), GeneReviews((R)). University of Washington, Seattle (WA).

Vydt TC, de Coo RF, Soliman OI et al. (2007). Cardiac involvement in adults with m.3243A>G MELAS gene mutation. Am J Cardiol 99: 264–269.

Wang S, Peng D (2019). Cardiac involvement in Emery-Dreifuss muscular dystrophy and related management strategies. Int Heart J 60: 12–18.

Weinsaft JW, Kim HW, Crowley AL et al. (2011). LV thrombus detection by routine echocardiography: insights into performance characteristics using delayed enhancement CMR. JACC Cardiovasc Imaging 4: 702–712.

Wittlieb-Weber CA, Villa CR, Conway J et al. (2019). Use of advanced heart failure therapies in Duchenne muscular dystrophy. Prog Pediatr Cardiol 53: 11–14.

Wlodarska EK, Wozniak O, Konka M et al. (2006). Thromboembolic complications in patients with arrhythmogenic right ventricular dysplasia/cardiomyopathy. Europace 8: 596–600.

Y-Hassan S, Holmin S, Abdula G, Böhm F (2019). Thromboembolic complications in takotsubo syndrome: review and demonstration of an illustrative case. Clin Cardiol 42: 312–319.

Zhang XD, Liu YX, Yan XW (2017). Cerebral embolism secondary to cardiac amyloidosis: a case report and literature review. Exp Ther Med 14: 6077–6083.

Zubkov AY, Rabinstein AA, Dispenzieri A, Wijdicks EF (2007). Primary systemic amyloidosis with ischemic stroke as a presenting complication. Neurology 69: 1136–1141.

Handbook of Clinical Neurology, Vol. 177 (3rd series)
Heart and Neurologic Disease
J. Biller, Editor
https://doi.org/10.1016/B978-0-12-819814-8.00030-5

Chapter 10

Neurologic complications of myocarditis

H. STEVEN BLOCK*

SSM Health Dean Medical Group, Department of Neurology, St. Mary's Hospital, Madison, WI, United States

Abstract

Myocarditis, a nonischemic acquired cardiomyopathy, is an uncommon condition with multiple presentation patterns which may be initially difficult to recognize and may simulate other conditions such as acute myocardial infarction, pericarditis, septicemia, etc. There are four distinct clinical presentation patterns that include: (1) low-grade nonspecific symptoms such as fatigue; (2) symptoms that resemble an acute myocardial infarction, especially in younger individuals; (3) a heart failure presentation which may be acute, subacute, or chronic and may be associated with cardiac conduction system defects and arrhythmias; and (4) an arrhythmia presentation that may produce sudden cardiac death, especially in young athletes with minimal or no prodromal symptoms. This chapter will provide a brief overview of various myocarditis etiologies and diagnostic modalities. The ultimate focus will be directed toward neurologic manifestations of myocarditis and its subtypes, complications of specific therapies including extracorporeal membrane oxygenation (ECMO) for refractory heart failure, and review the current literature regarding the appropriate use of therapeutic anticoagulation in myocarditis and heart failure for stroke prevention. Covid-19 infection has been discovered to cause myocarditis. The emerging science will be discussed. Nuances of brain death (BD) determination in patients receiving venoarterial ECMO for heart failure refractory to standard medical therapies will be discussed.

NEUROLOGICAL COMPLICATIONS OF MYOCARDITIS

Myocarditis is an infrequently occurring acquired nonischemic cardiomyopathy that can be caused by infectious, inflammatory, autoimmune, and toxic etiologies. It is an uncommon cause of stroke that may be caused by heart failure with hypotension, hypercoagulability, arrhythmias including atrial fibrillation, and following specific therapies. The initial presentation of myocarditis is variable with symptoms that may be subacute and subtle, fulminant and severe, or infrequently cause sudden death. Rarely, stroke from myocarditis may also occur with a paucity of typical cardiac symptoms such as angina-like chest pain, heart failure, or perception of arrhythmias. Nonspecific presenting symptoms may include atypical or pleuritic chest pain, palpitations, exertional fatigue, and dyspnea. Only with heightened awareness will myocarditis be considered in the differential diagnosis of stroke in select cases.

A detailed discussion of all aspects of myocarditis is beyond the scope of this chapter. Instead, following a succinct overview of general features of myocarditis, unique aspects of special interest to the neurologist will be presented.

MYOCARDITIS, GENERAL PRINCIPLES

Myocarditis affects all ages, most frequently in the young. There are several clinical presentation patterns: (1) subclinical symptoms such as fatigue or flulike symptoms; (2) an infarct-like presentation with chest pain, elevated cardiac injury enzymes, and EKG changes; (3) a heart failure presentation; and (4) an arrhythmia presentation characterized by ventricular and atrial arrhythmias

*Correspondence to: H. Steven Block, M.D., SSM Health Dean Medical Group, Department of Neurology, St. Mary's Hospital, Madison, WI, United States. Tel: +1-608-260-3425; Fax: +1-608-260-3444, E-mail: h.steven.block@ssmhealth.com

that may be life threatening as a first presentation, especially in young athletes.

Potential initial symptoms of acute myocarditis, variable based upon underlying etiology

1. Viral prodrome fever, rash, myalgias, arthralgias, fatigue, respiratory, or GI symptoms.
2. Chest pain mimicking MI or pericarditis.
3. Troponin elevation, especially in young people without coronary artery disease.
4. Tachycardia and bradycardia arrhythmias, ventricular tachycardia, atrial fibrillation, new unexplained heart block, and aborted sudden cardiac death.
5. Heart failure.
6. Cardiogenic shock.

Based upon the initial history and physical examination findings, other etiologies need to be rapidly sought and excluded such as occult coronary ischemia, stress cardiomyopathy, pericarditis, valvular disease, congenital heart disease, etc.

Etiologies of myocarditis

A broad range of infectious etiologies include multiple viruses, bacteria, spirochete, fungi, rickettsia, protozoa, and helminth organisms. Noninfectious etiologies include cardiac toxins including alcohol, cocaine, various antineoplastic agents, various autoimmune disorders, medication hypersensitivity reactions, various psychiatric medications, antibiotics, amphetamines, catecholamines from pheochromocytoma, insect stings, heavy metals, etc. (Leone et al., 2019).

Incidence of infectious etiologies vary by region

1. *Western Europe and North America:* Predominately viral: Human Herpesvirus 6, parvovirus B19, adenovirus, coxsackie B virus.
2. *Asia:* diphtheria, typhoid fever, rubella, scorpion bite, coxsackie B virus, chikungunya, H1N1 influenza in 2009 and hepatitis C.
3. *Australasia:* coxsackie B virus and enterovirus.
4. *Mexico, Central, and South America:* Chagas disease, Measles, meningococcal meningitis, AIDS, dengue fever, diphtheria, Kawasaki disease, coxsackie B virus.
5. *Sub-Saharan Africa and Middle East:* HIV both directly and through lowered immunity to opportunistic infection such as toxoplasmosis, *cryptococcus neoformans* and *mycobacterium avium* intracellulare. Trypanosomiasis and shigellosis have also been reported (Mayosi, 2007; Cooper et al., 2014).

Myocarditis is not a single entity but a heterogeneous disorder, with variable presentations and neurological manifestations based upon the underlying etiology. Lymphocytic myocarditis is the most common, often viral. Giant cell myocarditis, autoimmune mediated, should be considered in unexplained new-onset fulminant cardiac failure of 2 weeks–3 months duration, often in the presence of a dilated left ventricle and ventricular arrhythmias and high degree heart block. Histologically, this produces a mixed inflammatory infiltrate of macrophages, lymphocytes, and multinucleated giant cells without well-formed granulomas and necrosis. It is one of the causes of fulminant myocarditis, which is potentially fatal or requires cardiac transplantation. Cardiac sarcoidosis is generally observed in multiorgan disease albeit it has been described in isolation. Other infectious diseases such as tuberculosis, Chagas disease, various parasitic, bacterial, and fungal infections can produce caseating, necrotizing, or suppurative granulomas. Eosinophilic myocarditis can occur from a hypersensitivity reaction from medications, parasitic organisms, Churg–Strauss syndrome or from idiopathic hypereosinophilia (HE). This may produce severe myocardial injury due to release of eosinophilic acidic protein, which can be necrotizing. Neutrophilic myocarditis can be seen following bacteremia, especially in an immunocompromised patient, producing microabscesses and myocardial necrosis (Leone et al., 2019).

Fulminant myocarditis may occur with a myriad of disorders including giant cell myocarditis, hypersensitivity, and eosinophilic myocarditis, from cancer chemotherapies such as immune checkpoint inhibitors and from viral mediated lymphocytic myocarditis. Carrying a poor prognosis, this may manifest as rapidly progressive heart failure producing cardiogenic shock and abnormalities of the cardiac conduction system, which include tachyarrhythmias, bradyarrhythmias, heart block, and sudden death. It may mimic septicemia and should be considered in the differential diagnosis. Endomyocardial biopsy may help establish the diagnosis. General heart failure measures including inotropic and mechanical support may be needed (Kociol et al., 2020).

DIAGNOSIS OF MYOCARDITIS

The diagnosis of myocarditis was initially based upon clinical findings, later enhanced with the development of histologic criteria obtained from endomyocardial biopsy (Dallas criteria) (Aretz et al., 1987; Friedrich et al., 2009), further strengthened with the addition of immunohistochemistry and PCR of the biopsy material, and later by cardiac MRI imaging (Lake Louise criteria).

Cardiac imaging

Transthoracic echocardiography can assess bilateral ventricular function, chamber volumes, pulmonary artery pressure, regional and global wall motion abnormalities, and valvular function. Additionally, echocardiography can detect mural cardiac thrombi that may produce embolic stroke. Echocardiography of myocarditis may be unremarkable or demonstrate only nonspecific findings. Fulminant myocarditis may present with global left ventricular dysfunction and a thickened but not dilated left ventricle. Cardiac sarcoidosis may demonstrate a normal, dilated, or restrictive cardiomyopathy pattern.

Cardiac MR imaging (CMR) can assess the stage of cardiac inflammation, which includes the following: (1) an edema phase characterized by myocardial interstitial edema that may be detected by cardiac MRI T2w-STIR imaging; (2) inflammation-associated hyperemic phase caused by hyperemia and capillary leakage, which may be detected by early gadolinium enhancement; and (3) myocardial necrosis may be detected by late gadolinium enhancement. CMR may also be used when echocardiography is nondiagnostic as well as differentiating ischemic vs nonischemic myocardial fibrosis based upon whether the pattern of late gadolinium enhancement conforms to a coronary artery vascular territory. It is useful in detecting the stage of cardiac injury, as in eosinophilic myocarditis further described later. Additionally, it may be useful in identifying several etiologies of myocarditis including amyloidosis, sarcoidosis, Chagas disease, etc. Future imaging modalities may include CT and PET cardiac MRI (Agac et al., 2011; Ponikowski et al., 2016; Biere et al., 2019).

When combined with endomyocardial biopsy, cardiac MR is highly sensitive for an infarct-like myocarditis presentation manifesting as chest pain and troponin rise. CMR has a low sensitivity for a myocarditis heart failure presentation and very low sensitivity for a myocarditis arrhythmia presentation. When myocarditis manifestations are severe such as left ventricular deterioration, life-threatening arrhythmias, and/or unexplained syncope with clinical features suggestive of acute myocarditis, absence of cardiac MRI abnormalities should not preclude further testing with endomyocardial biopsy (Francone et al., 2014).

Neurological complications of myocarditis can occur from several etiologies: (1) cardioembolic stroke from mural thrombi and atrial fibrillation; (2) ischemic cerebral infarction from hypotension related to heart failure; (3) unique presentations and manifestations of specific myocarditis types; (4) unique treatment clinical considerations to prevent neurological manifestations; (5) neurological complications of therapy; and (6) unusual neurological associations of myocarditis and its treatments. The following will expand on these considerations.

Intracardiac thrombus from myocarditis

There are notable differences in the pathogenesis of mural thrombus development following myocarditis in comparison to myocardial infarction. Following myocardial infarction, regional wall motion abnormalities represent a major factor in mural thrombus formation that increases with infarction area, along with blood stasis and local hypercoagulability. Mural thrombus may provide structural integrity to an area of infarction (Delewi et al., 2012). In contrast, myocarditis may produce local endocardial inflammatory factors without significant wall motion abnormality. In experimental infection of mice with the enterovirus coxsackie B3 virus that also affects humans, thrombus appears early in the course of myocarditis, before significant myocardial necrosis and inflammatory cellular infiltration became prominent. Additionally, in animal models of experimental viral myocarditis, the presence of congestive heart failure increased the incidence of mural thrombus formation (Tomioka et al., 1986; Kishimoto et al., 1992).

Intracardiac thrombus may occur in myocarditis in one or multiple chambers. Agac and colleagues reported a 22-year-old with cardiac failure from fulminant myocarditis 2 weeks after a flulike illness with severely hypokinetic LV and mural thrombi in all four chambers. Despite therapeutic anticoagulation with unfractionated heparin, the patient succumbed to a fatal cardiogenic embolus the evening before implantation of a left ventricular assist device (Ramanan et al., 1993; Agac et al., 2011).

Can mural thrombus develop in the absence of akinetic myocardium? The following represents two clinical examples of mural cardiac thrombus formation from local endocardial inflammatory factors in the absence of akinetic myocardium. An immunocompromised patient with staphylococcal aureus septicemia and myocarditis was described to have transthoracic echocardiogram evidence of normal LV function and an apical LV thrombus. Lateral T wave inversion, troponin elevation, and cardiac MRI late gadolinium enhancement in the apex were consistent with myocarditis. The apical thrombus resolved with antibiotics and therapeutic anticoagulation (McGee et al., 2018). Similarly, recurrent LV mural thrombi developed in a patient with biopsy-proven myocarditis with only initial LV apical dyskinesia but not hypokinesia or akinesia. After initiation of therapeutic anticoagulation, additional LV mural thrombi formed in the absence of regional wall motion abnormalities, again consistent with an endocardial inflammatory process rather than focal stasis (Kojima et al., 1988).

Anticoagulation considerations for stroke prevention in myocarditis

Fulminant myocarditis may produce heart failure. Atrial fibrillation is one of most common arrhythmias in heart failure and can also occur from myocarditis without heart failure. Heart failure increases thromboembolic stroke risk as a result of both atrial fibrillation and Virchow's triad of stasis, hypercoagulability due to prothrombotic and proinflammatory cytokines and endothelial injury. Fourteen percent of all hospital-associated new onset atrial fibrillation occurred in the context of severe sepsis, representing a sixfold stroke increase in comparison to patients without severe sepsis. In one series, 11% of in-hospital strokes occurred in patients with severe sepsis. Additionally, atrial thrombi may form as little as 2 days following the onset of atrial fibrillation (Walkey et al., 2011).

Oral anticoagulation for chronic heart failure with permanent or paroxysmal atrial fibrillation is recommended. Anticoagulation is not recommended in patients with chronic heart failure with reduced ejection fraction without atrial fibrillation. Anticoagulation is also not recommended in heart failure without any prior thromboembolic event or a cardioembolic source (Yancy et al., 2013). In heart failure patients with atrial fibrillation, prophylactic oral anticoagulation is recommended after weighing risk vs benefit using the CHA2DS2 VASc and HAS-BLED scores and periodic reassessment for safety of continued use (Ponikowski et al., 2016).

Notably, heart failure patients are also at increased risk of stroke even without atrial fibrillation, with stroke rates almost as high as patients with atrial fibrillation (1.6/100 patient-years vs 2/100 patient-years, respectively) (Melgaard et al., 2015; Ferreira et al., 2018). In one series, almost half of first strokes during heart failure with reduced ejection fraction were disabling or fatal (Mehra et al., 2019).

Routine use of warfarin therapy in heart failure *without* atrial fibrillation is not supported by the current literature. The WARCEF trial demonstrated that in comparison to antiplatelet therapy, reduced risk of ischemic stroke provided by oral anticoagulation was offset by the increased risk of major hemorrhage (Homma et al., 2012). A meta-analysis of four randomized trials (WASH, HELAS, WATCH and WARCEF) also revealed dosage-adjusted warfarin in heart failure patients with sinus rhythm reduced the relative ischemic stroke rate by 51% (1.87% vs 3.53%, RR 0.49, $P = 0.006$) in comparison to antiplatelet therapy (aspirin 81–325 mg vs clopidogrel 0.75 mg daily) with an absolute risk reduction of 0.74% per year, but doubled the risk of major hemorrhage (5.59% vs 2.60%, $P < 0.00001$) with absolute risk increase of 0.99% per year. Warfarin did not significantly reduce the risk of death, myocardial infarction, hospitalization due to heart failure, or intracranial hemorrhage in comparison to antiplatelet therapy (Liew et al., 2014). The findings of the more recent Commander HF trial did not lower the composite endpoints of death, myocardial infarction, or stroke utilizing rivaroxaban 2.5 mg twice daily compared to placebo in addition to standard heart failure care in patients without atrial fibrillation, and there was no anticoagulation benefit for this composite endpoint. Notably, heart failure was the major cause of death rather than thrombin-mediated stroke (Zannad et al., 2018).

Certain exceptions to the aforementioned may apply. In an American Heart Association scientific statement regarding recognition and initial management of fulminant myocarditis, Kociol and colleagues have suggested a role for prophylactic anticoagulation in patients with acute NEM, a rare and frequently fatal hypersensitivity eosinophilic myocarditis that has a high likelihood of ventricular thrombus with embolization (Kociol et al., 2020).

Further study is needed to develop more targeted approach to patient selection and timing of anticoagulation to reduce heart failure-related stroke in patients without atrial fibrillation based upon a validated risk factor scale. A subsequent post hoc analysis of COMMANDER HF data following either a first HF event or worsening HF demonstrated a 32% reduction in stroke alone (of which almost half were disabling or fatal) and TIA in the low-dose rivaroxaban group compared to placebo, without an increased risk of bleeding. The most vulnerable time for the index stroke was within the first 6 months after the heart failure event and remained elevated thereafter. The overall number needed to treat (NNT) with rivaroxaban to prevent one stroke or TIA without fatal bleeding or bleeding into a critical space was 164. However, after targeted patient selection using the CHA2DS2 VASC score > 4, the NNT fell to 96 (Mehra et al., 2019).

Other studies have proposed anticoagulation selection parameters based upon the composite score utilizing age, prior stroke, hypertension, GFR, and Killip class (Ferreira et al., 2018). Siliste and colleagues highlight additional evidence gaps for stroke prevention in heart failure patients that deserve further investigation (Siliste et al., 2018).

Hypereosinophilia and myocarditis

Hypereosinophilia (HE) is defined as either 1.5×10^9/L eosinophils at least 1 month apart, at least 20% eosinophils on bone marrow biopsy, extensive eosinophilic tissue infiltration, or extensive deposition of eosinophilic granular protein. In 25% of eosinophilic myocarditis patients, initially blood hypereosinophilia was absent and later rose, especially in hypersensitivity reactions. Hypereosinophilic syndrome (HES) includes both the HE criteria and end-organ damage or dysfunction after excluding other causes (Valent et al., 2012). HE producing myocarditis may be caused by multiple etiologies including hypersensitivity reactions (34%; medications, allergic), autoimmune disorders (13%), lymphoproliferative and myeloproliferative states (9.5%), and infections (5%; especially parasitic and protozoal). Drugs most commonly producing hypersensitivity eosinophilic myocarditis included antibiotics such as minocycline and beta-lactam class, clozapine, carbamazepine, vaccines, and antituberculous agents. Idiopathic hypereosinophilia is the most common (36%) after exclusion of the aforementioned.

Acute NEM is a rare eosinophilic myocarditis most commonly triggered by drug hypersensitivity causing fulminant heart failure or sudden cardiac arrest. Death and cardiac transplantation occur up to 50% of those affected (Brambatti et al., 2017; Chan and Gibson, 2019; Kociol et al., 2020).

Cardiac involvement in hypereosinophilic syndrome (Loeffler's endocarditis) has significant potential to produce stroke from mural thrombus at a time when the left ventricular systolic function is still preserved. Cardiac involvement occurs in more than 50% of patients with hypereosinophilic syndrome and carries up to a 50% 2-year mortality due to difficulty recognizing the condition in the early stages, while it is potentially treatable. There is three-stage cardiac injury associated with HES. The first, acute necrotic stage occurs after release of eosinophil granules in the endomyocardium. These granules contain eosinophilic cationic protein, eosinophilic peroxidase, major basic protein, and cytokines, which impair tissue hemostasis and integrity producing endocardial inflammation. The second, intermediate thrombotic stage occurs in response to stage I. In general, patients may be asymptomatic or minimally symptomatic during stage I and stage II, demonstrating only nonspecific constitutional symptoms such as fever, weight loss, fatigue, or inattentiveness. Some patients may experience symptoms of pericarditis, systemic or pulmonary edema, and abdominal pain. Splinter hemorrhages in HES have a unique characteristic appearance, being multiple and situated side-by-side in a crescentic distribution parallel to the curve of the fingernails. This may represent an early clue to the diagnosis (Mulroy et al., 2018). Troponin can rise during the first or second essentially asymptomatic stages. During the second stage, mural thrombi (sometimes with thin attachment points, "floating thrombi") can propagate producing a stroke as one of the first presenting symptoms of HES. In the third stage, mural fibrosis occurs, which may produce restrictive cardiomyopathy, diastolic dysfunction, mitral, or tricuspid valve regurgitation. The opportunity to mitigate serious cardiac sequela by identifying the cause and treating hypereosinophilia generally occurs during the first or second stages, at a time when the condition may be difficult to clinically recognize. Endomyocardial biopsy or cardiac MR imaging with late gadolinium enhancement demonstrating features of early myocardial edema and later fibrotic stage may help establish the diagnosis (Fauci et al., 1982; Masaki et al., 2016; Salih et al., 2020).

Cardiogenic embolization from HES may not be the only mechanism of stroke. Eosinophilic granule protein has a procoagulation effect, influencing protein XII function and inhibition of endothelial anticoagulant activities by binding to thrombomodulin. It can also trigger disseminated intravascular coagulopathy. It can neutralize both endogenous and administered heparin, requiring increased dosages of the latter (Venge et al., 1979; Gambacorti Passerini et al., 1989; Slungaard et al., 1993; Mukai et al., 1995; Todd et al., 2014).

Stroke location in HES is also unique. In one series, multifocal infarctions in a hemispheric watershed distribution occurred in almost half of the cases and without large vessel occlusion. This infarction pattern is suspected to be caused by multifocal thrombosis of small terminal arterioles or delayed/incomplete microemboli clearance. Hemorrhage and cerebral venous thrombosis, while uncommon, may also occur. Altered metal status was a predictor of poor outcome (Lee and Ahn, 2014).

Eosinophilic granulomatosis with polyangiitis (EGPA, formerly Churg–Strauss syndrome) is one the most infrequent of ANCA vasculidities and should be considered when at least four of the following are present: Asthma, hypereosinophilia >10%, allergic rhinitis, pulmonary infiltrates, paranasal sinus abnormalities, peripheral neuropathy, and extravascular eosinophil biopsy findings. It can also produce multiorgan involvement affecting heart, lungs, kidney, and vascular systems. Asthma almost always present and represents a distinguishing factor from other hypereosinophilic syndromes. EGPA should be suspected when asthma is difficult to treat, steroid dependent or late onset. Cardiac involvement is common, previously considered to be present in up to 40% of patients but later determined to be present in 90% of cases manifest as myocarditis, endomyocardial fibrosis, myocardial edema, and mural thrombi. Cardiac failure refractory immunosuppressive therapy may lead to cardiac transplantation (Rosenberg et al., 2006; Szczeklik et al., 2011; André et al., 2017).

Cerebrovascular injury by vasculitis-associated eosinophilic acidic protein may produce ischemic stroke by microvascular inflammation with thrombosis as well as produce intracerebral, intraventricular, and/or subarachnoid hemorrhage. Endomyocardial involvement may produce mural thrombi with embolic stroke. In one reported case, multiple posterior circulation ischemic and hemorrhagic infarctions were associated with echocardiographic evidence of mobile structures arising from the left ventricular septum. Antibiotic therapy for the empiric treatment of presumptive infectious endocarditis did not produce any change in these lesions. Myocardial biopsy demonstrated eosinophilic endocarditis with myocarditis. The lesions resolved 10 days after initiating high-dose pulse corticosteroid therapy (Sakuta et al., 2019).

While the potential for clinical remission is favorable with immunosuppressive therapy, serious clinically silent cardiac findings may persist, with the potential for prolonged cardioembolic stroke risk. A study of 20 EGPA patients felt to be in full clinical remission, and 20 age and sex-matched healthy controls were evaluated with cardiac MRI, transthoracic echocardiogram, 24 h Holter cardiac rhythm monitoring, and cardiac stress test. In addition to cardiac MRI demonstrating features of late gadolinium enhancement in 17 of 19 patients consistent with irreversible myocardial scar formation and reduced left ventricular ejection fraction <50% consistent with residual heart failure, it also detected early gadolinium enhancement in 6 of 19 patients revealing ongoing active cardiac inflammation. Five patients demonstrated ventricular ectopy consisting of bigeminy, trigeminy, pairs, and ventricular tachycardia. Four of the 5 patients had supraventricular ectopy as well. Cardiac exercise stress test was abnormal in 4 patients without symptoms of angina. As all patients had normal epicardial coronary artery patency, this finding was suspected to represent myocardial small vessel disease from the granulomatous vasculitis or eosinophilic injury. These findings suggest the need for increased clinical surveillance for potentially treatable symptoms and potential refinement of EGPA therapeutic protocols (Szczeklik et al., 2011).

COVID-19 MYOCARDITIS

Our knowledge of the mechanism causing myocarditis in severe acute respiratory syndrome coronavirus 2 (SARS-CoV-2) infection is evolving and may occur from more than one discrete mechanism including direct viral myocarditis producing myocyte inflammation, cytokine-related myocardial dysfunction with necrosis, and myocardial edema. Underlying stroke incidence and risk factors related to SARS-CoV-2, myocarditis also awaits further study.

There are seven strains of human coronavirus. Four generally produce self-limited disease. Three strains of the virus have produced severe illness with cardiovascular manifestations and increased mortality. These include severe acute respiratory syndrome coronavirus (SARS-CoV) first reported in 2002, Middle East respiratory syndrome coronavirus (MERS-CoV) in 2012 and most recently severe acute respiratory syndrome coronavirus 2 first reported in December 2019, producing the global pandemic coronavirus disease 2019 (Covid-19). While the death rate of SARS-CoV and MERS-CoV was greater than SARS-CoV-2, the high infectivity rate of SARS-CoV-2 to date has produced more deaths than the first two combined (Driggin et al., 2020). As this chapter was completed during the early stages of the Covid-19 pandemic in late spring 2020, understanding of the pathophysiology of myocarditis from coronavirus is evolving. The following overview represents the current state of knowledge.

Presenting symptoms of SARS-CoV-2 may be variable with low-grade constitutional symptoms typical of many viral syndromes such as fever, cough, dyspnea, myalgias, nasal congestion, hypoosmia, diarrhea, and headache. Patients with minimal or no symptoms in the early stage of illness pose the greatest degree of infectivity. Symptoms may progress to pneumonia, acute respiratory distress syndrome (ARDS), and cardiovascular dysfunction producing hemodynamic compromise with a high rate of mortality. Multiorgan failure may occur due to hypotension, intravascular microthrombosis, and septicemia. In one series, the mortality in the ICU setting was 61.5% by 28 days (Majid et al., 2020; Yang et al., 2020). However, acute myocarditis may also be the first manifestation of SARS-CoV-2 infection (Hu et al., 2020).

Viral presence in noninvasive macrophages with low-grade cardiac inflammatory response, and low grade or focal myocyte necrosis may not conclusively prove a direct viral injury and might be explained by viral migration (Tavazzi et al., 2020). Additional study is needed to define the mechanism of myocardial injury, especially with regard to the contribution of inflammatory activation with cytokine release and possible autoimmune response (Hendren et al., 2020).

The stress of SARS-CoV-2 infection superimposed upon preexisting cardiovascular disease adds further complexity by triggering coronary ischemic symptoms producing type I or type II myocardial infarction. Type I MI is caused by typical epicardial coronary artery occlusion. Type II MI may be caused by heterogeneous factors such as demand ischemia from tachycardia and arrhythmias, aortic stenosis, heart failure, respiratory failure, hypoxemia, vasospasm, increased circulatory catecholamines, hypertension with or without left ventricular hypertrophy, microvascular thrombosis, and/or endothelial dysfunction without coronary artery occlusive disease (Lopez-Cuenca et al., 2016).

The early literature has a paucity of cardiac myocardial biopsy and postmortem pathologic confirmation of clinically suspected SARS-CoV-2 myocarditis. One case of clinically suspected combined fulminant viral myocarditis and reverse Takotsubo cardiomyopathy (defined as inferolateral wall hypokinesia with preserved or hyperkinetic apical function) autopsy demonstrated only rare foci of scattered CD45+ interstitial myocytes and myocardial edema without myocyte injury (Yan et al., 2020). Another similar case presenting with chest pain, elevated troponin, transthoracic echocardiogram features of a reverse Takotsubo cardiomyopathy pattern and cardiac MR findings of myocardial edema without late gadolinium enhancement. Cardiac myocardial biopsy demonstrated diffuse T-lymphocytic infiltrates, significant interstitial edema, and limited focal necrosis. Virus-negative lymphocytic myocarditis was suspected but not proven as there was no evidence of SARS-CoV-2 genome in the myocardium nor were there microvascular abnormalities (Sala et al., 2020; Zhou, 2020). This highlights the occasional complexity distinguishing myocarditis from stress cardiomyopathy as both can be present and potentially impact therapy. Kawai and Shimada highlight the clinical, radiographic, and pathologic distinction between the myocarditis and stress cardiomyopathy (Kawai and Shimada, 2014).

As with other causes of myocarditis, SARS-CoV-2-related myocarditis may trigger atrial and ventricular cardiac conduction system abnormalities including atrial fibrillation, ventricular tachycardia, fascicular block, and QT prolongation, which can be exacerbated by various treatments such as hydroxychloroquine (Kochi et al., 2020). Of note, 14% of all hospital associated new onset atrial fibrillation events and 11% of in-hospital strokes occurred in patients with severe sepsis. Patients with severe sepsis have a sixfold increase of in-hospital stroke compared to hospital patients without severe sepsis (Walkey et al., 2011).

A meta-analysis of 4189 SARS-CoV-2 patients demonstrated increased risk of cardiac injury with severe infection, which carries a 14 times greater mortality risk compared to mild infection. Acute cardiac injury detected by elevated biomarker levels is associated with a fourfold increase in mortality risk. High levels of cardiac injury biomarkers such as troponin, CK-MB, and NT-proBNP rise above normal by the midpoint of hospitalization and rise shortly before death. As such, longitudinal measurement of cardiac injury biomarkers during hospitalization may help predict poor outcome and guide resource allocation (Li et al., 2020).

While treatment of SARS-CoV-2 is beyond the scope of this chapter, general measures include supportive care, immune suppressants, pressors, and occasionally mechanical hemodynamic support. Venoarterial extracorporeal membrane oxygenation (VA ECMO) should be considered in refractory cardiogenic shock that has failed adequate fluid resuscitation, inotropes, and vasopressors support. Vasodilatory shock is unlikely to benefit from ECMO. As SARS-CoV-2 infection may produce a hypercoagulable state, significant attention should be directed toward thrombosis in the ECMO circuit. Highest priority for ECMO use should be based upon resource availability favoring younger patients with minor or no medical comorbidities (Chow et al., 2020).

As the understanding of SARS-CoV-2 clinical manifestations and treatments evolve, the Extracorporeal Life Support Organization has created a living consensus document regarding the indications and applications of ECMO in this patient population under various clinical scenarios (Shekar et al., 2020). A more detailed discussion of ECMO can be found later.

The recently initiated CASCADE study, an international, hospital-based multicenter study of stroke incidence and outcomes, will further the understanding of stroke associated with the Covid-19 pandemic (Abootalebi et al., 2020).

Extracorporeal membrane oxygenation and neurological complications

Two types of ECMO have been developed to provide temporary external cardiopulmonary or pulmonary bridge to recovery or bridge to cardiac or pulmonary transplantation. Venoarterial ECMO can be used to treat refractory cardiogenic shock. VA ECMO may be a therapeutic option when cardiogenic shock fails to respond to volume administration, inotropes, vasoconstrictors, and intraaortic balloon counterpulsation measures. Fulminant myocarditis producing heart failure with reduced ejection fraction represents one of several indications for VA ECMO as a bridge to myocardial recovery or potential bridge to cardiac transplantation in refractory cases (ELSO, 2013). Venovenous ECMO (VV ECMO) is employed for oxygenation in severe respiratory failure and right ventricular support refractory to conventional measures, utilizing a membrane oxygenator circuit bypassing the right heart and lungs. Indications for VV ECMO include ventilatory failure from massive pulmonary embolization, ARDS, severe pneumonia, interstitial lung disease, etc. This section will focus on VA ECMO.

Temporary external cardiopulmonary bypass is achieved by shunting blood through a circuit with a venous cannula placed in femoral vein, inferior vena cava, or right atrium. Blood is transported through a membrane oxygenator and returned to the arterial circulation via a large bore catheter, generally in the right femoral artery. Alternatively, arterial return can be achieved via a cannula in the ascending aorta, right common carotid artery, or subclavian artery. Therapeutic anticoagulation is necessary to mitigate thromboembolic complications originating from the ECMO circuit and stagnant areas of blood flow in the left ventricle and aortic root. While overall survival for all VA ECMO indications is low (approximately 50%–60% in-hospital mortality following VA ECMO following myocardial infarction), patients with potentially reversible conditions such as fulminant myocarditis appear to have better outcomes, as high as 71% (Asaumi et al., 2015; Keebler et al., 2018).

Survival with ECMO

Various prediction models have been created to estimate in-hospital VA ECMO survival. The SAVE score calculator (Survival After Veno-Arterial ECMO) was created utilizing data from a 10-year retrospect review of patients enrolled in the ECMO database ending in December 2013. Specific clinical features can be entered into an online calculator to determine estimated percentage of survival (www.save-score.com) (Schmidt et al., 2015). The Extracorporeal Life Support Organization Registry reported CNS complications occurred in 15.1%. Patients experiencing ECMO-related neurological complications had an 11% survival in comparison to 57% without. They also found in-hospital mortality 95% in patients with two neurological complications and 100% with three or more (Lorusso et al., 2016).

Neurological complications from ECMO

There is substantial variability in standardization and reporting methodology of ECMO-related complications. As such, the following statistics from database and retrospective studies should be considered as general trends. Additionally, in some circumstances it may be very difficult to distinguish hypoxic–ischemic changes caused by the underlying condition that required hemodynamic ECMO support from a brain injury induced by ECMO. In a systematic review of 44 studies of ECMO-related complications from 1990 to 2017, neurological complications following VA ECMO have been reported up to 13.3%, with 5.9%–7.8% for ischemic and/or hemorrhagic stroke. Ischemic stroke is more common in VA ECMO than VV ECMO due to propagation of thromboemboli into the systemic circulation that would otherwise be trapped in the lungs in a VV ECMO circuit. Hypoxic ischemic encephalopathy has been reported to be greater than 50% from low cardiac output. Intracranial hemorrhage (ICH) produces substantially increased mortality risk reported as high as 100% in one series. Risk factors for ICH include female sex, prolonged ventilation, decreased serum fibrinogen, anticoagulant use, serum creatinine greater than 2.6 mg/dL, hemodialysis, thrombocytopenia, and hypocapnia. Of specific mention, patients experiencing a rapid fall in pCO_2 at the initiation of ECMO were at substantial increased risk of ICH, necessitating careful monitoring of the pCO_2 rate of change (Sutter et al., 2018).

An 8-year retrospective review of 878 VA ECMO patients revealed ischemic stroke generally occurred in the second week of ECMO and was not associated with increased mortality. In contrast, intracranial hemorrhage, while occurring less frequently, occurred in the first week of ECMO and was associated with a significant increased risk of mortality. While no specific risk factors were identified for ischemic stroke, ICH was associated with initial rapid decline in pCO_2, which can alter cerebrovascular reactivity, thrombocytopenia under 100,000, and female sex (Le Guennec et al., 2018).

Stroke caused by a proximal cerebral large vessel occlusion can potentially be treated with mechanical thrombectomy. Anticoagulant use during ECMO contraindicates intravenous thrombolytics. Cerebral infarction caused by cerebral venous thrombosis can continue to receive anticoagulant therapy for ECMO. Intracerebral hemorrhage during ECMO may necessitate discontinuation of anticoagulation. There is limited data to guide duration of anticoagulation abstinence during VA ECMO, but in one case, the thromboembolic stroke risk did not increase for the 2 days that the patient received ECMO (Cho et al., 2019).

Additional ECMO complications include ischemic encephalopathy, subclinical cognitive impairment, seizures, paraplegia, limb ischemia related to catheter placement with associated peripheral nerve injuries, compartment syndrome, and death from various causes including solid or gaseous microemboli, hypoxemia, hyperoxia, anticoagulation complications. ECMO-related absence of pulsatile blood flow may alter cerebral vascular reactivity impairing autoregulation creating the risk of cerebral edema (Xie et al., 2017). The Extracorporeal Life Support Organization Registry reported multiple CNS complications in 70 of 4552 patients (Lorusso et al., 2016).

Daily neurological assessments temporarily off of sedating medication and paralytic agents are necessary for early recognition, investigation, and treatment of ECMO-neurological complications (Sutter et al., 2018).

Differential hypoxia from ECMO producing cerebral and coronary ischemia

As oxygenated blood is returned to either the femoral artery or aorta, there can be adequate oxygenation to the abdominal viscera and lower extremities with relative hypoxic blood perfusing the upper body. This scenario is especially evident if there is both cardiac and pulmonary failure and cardiac function improves prior to the lung's ability to oxygenate. As such, deoxygenated blood may be pumped from the recovering left heart. Called the "North–South Syndrome" or "Harlequin Syndrome," returned oxygenated VA ECMO blood flow produces a mixing zone with deoxygenated blood flow from the heart at various aortic levels dependent upon the relative degree of LV contractility and VA ECMO return flow. If the mixing zone occurs proximally at the level of the aortic arch, there is potential for ischemia of the coronary and cerebral circulation. Arterial blood saturation should be monitored in the right upper extremity rather than a lower extremity to remain vigilant for this scenario. Differential hypoxemia can be mitigated by either repositioning of the arterial return cannula to the upper body in the right subclavian artery, increasing ECMO flow until it supports nearly 100% of the total circulatory needs or placing an additional return catheter in the superior vena cava (veno-arterial-veno ECMO) (Rupprecht et al., 2015; Stevens et al., 2017).

Carotid artery cannulation for VA ECMO return flow poses a substantially increased risk in young patients and neonates as this site of cannulation may be used more frequently than in adults due to perceived increased risk in the latter. A retrospective review of the Extracorporeal Life Support Organization registry database from 2007 to 2008 identified 2977 patients that underwent VA ECMO. Carotid artery cannulation occurred in 64% (aorta 32%, femoral artery 4%) of patients. Twenty-one percent (611 patients) of patients experienced neurological injury with seizures, infarction, and hemorrhage. Contributory factors include embolization of ECMO circuit clots, air, small particulate matter, and anticoagulant use (Teele et al., 2014). Other carotid artery complications include carotid artery dissection and vocal cord paralysis following have been reported in pediatric patients (Schumacher et al., 1989; Ryerson et al., 2017).

Brain death determination complexities in ECMO patients

There is a high risk for mortality in ECMO patients. Some may progress to BD. The Extracorporeal Life Support Organization Registry reported BD in 358 of 4522 patients (7.9%) (Lorusso et al., 2016). BD determination by standard criteria in ECMO patients is complicated by difficulties with conventional apnea testing, as nonpulmonary gas exchange during ECMO continuously removes carbon dioxide in addition to providing oxygenation. BD criteria require apnea with proof of $pCO_2 \geq 60$ mmHg elevation or a 20 mm pCO_2 Hg rise in cases of preexisting hypercapnia. This can be controlled by adjusting the membrane oxygenator sweep gas flow down to 0.5–1 L per min, increasing be fraction of inspired membrane oxygenation to 100% and addition of positive end-expiratory pressure while monitoring for apnea. Further details of BD determination criteria have been documented by Greer et al., Bein et al., and Ihle and Burrell (Muralidharan et al., 2011; Saucha et al., 2015; Bein et al., 2019; Ihle and Burrell, 2019; Greer et al., 2020).

Posterior reversible encephalopathy syndrome, myocarditis, and ECMO

Posterior reversible encephalopathy syndrome (PRES) was described in an adenovirus-related myocarditis with severe biventricular failure with left ventricular ejection fraction 10% treated with ECMO, intraaortic balloon pump, and pressor support with gradual hemodynamic improvement. Following discontinuation of sedation, seizures, visual disturbances, lethargy, and slow motor responses were observed. Cerebral MRI demonstrated left hemispheric subcortical vasogenic edema in the frontal, parietal, and occipital lobes. MRA demonstrated vasospasm of the bilateral cavernous and supraclinoid ICAs, and middle and anterior cerebral arteries (Lorusso et al., 2014).

Posterior reversible encephalopathy syndrome and infectious myocarditis

In addition to the aforementioned VA ECMO case, there are several case reports of posterior reversible encephalopathy syndrome associated with leptospirosis myocarditis. Leptospirosis is caused by the spirochete Leptospira interrorgan, which causes illness in two phases. The first, septicemic phase, is characterized by high fever, myalgias, headache, cough, rash, nausea, vomiting conjunctival injection, and constitutional symptoms. After a brief period without fever, the second immune phase produces multiorgan dysfunction affecting kidney, liver, lung, coagulation, and heart, including myocarditis. Neurological manifestations described include aseptic meningitis, transverse myelitis, myeloradiculopathy, cerebellar syndrome, Guillain–Barré syndrome, and PRES. The reported cases of PRES were hypertensive during the course of the illness, developed seizures, and radiographic evidence biocipital cortical vasogenic edema (Bal, 2005; Aram et al., 2010; Priyankara and Manoj, 2019).

MYOCARDITIS AND MYASTHENIA GRAVIS ASSOCIATIONS

Myasthenia gravis (MG) patients, especially with thymoma, may have coexisting immunologic neurological disorders in addition to skeletal muscular weakness from antibodies directed toward the postsynaptic neuromuscular junction. These include neuromyotonia and meningoencephalitis related to voltage-gated potassium channel antibodies. Other coexisting nonneurologic immune-mediated disorders include pure red cell aplasia associated with thymoma caused by abnormal CD8 T cell clones, alopecia areata, various immune deficiency states, disturbance of taste, Sjogren's syndrome, thyroiditis, rheumatoid arthritis, ulcerative colitis, pemphigus, systemic lupus erythematosus, and myocarditis.

Myocarditis associated with MG has been associated with thymoma, especially malignant thymomas. Fifteen to twenty five percent of patients with MG have thymoma, 30%–44% of patients with thymoma have MG, and 20%–60% of patients with MG were found to have myocarditis an autopsy. Heart muscle antibodies measured by ELISA have been found in 29/30 MG patients with thymoma, none in 30 young onset MG patients with thymic hyperplasia, and 7/15 late onset MG patients with thymic atrophy. Recognition of coexisting myocarditis in a patient suspected to be in myasthenic crisis may be difficult due to similar symptoms such as dyspnea, fatigue, and respiratory failure initially attributable to MG. Concerning symptoms of coexisting myocarditis include arrhythmias, elevated cardiac injury enzymes, and declining cardiac output. Due to the concurrent cardiac conduction system inflammatory involvement, significant and sometimes lethal cardiac arrhythmias may ensue. Coexisting small vessel cardiac vasculitis with patent coronary arteries in a myasthenic with pathologically proven giant cell myocarditis and skeletal myositis has been described. In another case, a postviral autoimmune trigger was suspected to precipitate fulminant myocarditis presenting as acute respiratory failure, fever, elevated cardiac injury enzymes with no significant coronary artery disease followed by arrhythmias, and fatal cardiogenic shock. Two years prior the patient underwent resection of a malignant thymoma with pleural metastases. Previously, he also experienced neuromyotonia and was found to have elevated acetylcholine receptor antibodies, anti-voltage-gated potassium channel antibodies and antistriational antibodies. While direct inflammation from Coxsackie virus may have caused viral myocarditis, in this case, substantial Coxsackievirus A IgG titer elevation without IgM titer elevation suggested a prior viral exposure followed by a postviral autoimmune trigger (De Jongste et al., 1986; Mygland et al., 1991; Cohle and Lie, 1996; Suzuli et al., 2013; Premier et al., 2018).

Immune checkpoint inhibitors, myocarditis, and myasthenia gravis

Immune checkpoint inhibitors (ICIs) represent a class of immunotherapeutic agents that have significantly improved clinical outcomes in cancers that can metastasize to the central nervous system. These include melanoma, renal cell carcinoma, bladder, lungs, gastrointestinal, and thymic cancers. They may also produce potentially life-threatening immune-mediated adverse effects including myasthenia gravis and cardiovascular toxicities such as myocarditis. Cellular surface receptor ligands help modulate a balance of T-cell activation, immune tolerance, and immune-mediated tissue damage. Inhibition of these ligands removes self-tolerance, which may be used against the aforementioned neoplasia but may also trigger autoimmune phenomenon as an unintended consequence. ICIs consist of agents that block Programmed Death-1 receptors (PD-1; nivolumab, pembrolizumab, cemiplimab), Programmed Death Ligand-1 (PD-L1; atezolizumab, avelumab, durvalumab), and Cytotoxic T Lymphocyte Antigen-4 (CTLA-4; ipilimumab). Analysis of Vigibase, the World Health Organization's pharmacovigilance database revealed most neurological events occurred within the first 3 months of ICI therapy. Myasthenia gravis had the highest fatality rate among neurological complications (19.3%) and occurred early with a median time of onset 29 days. Additionally, when myasthenia gravis presented with myocarditis and myositis, the death rate was 62.5%. Death rates for myasthenia gravis alone were 16.2%, myasthenia gravis with myositis 20.7% and myocarditis alone 33%. Myasthenia gravis associated with ICI frequently did not demonstrate acetylcholine receptor antibodies and often demonstrated concurrent myositis. Other neurologic autoimmune disorders associated with ICIs include noninfectious encephalitis and/or myelitis, cerebral artery vasculitis including giant cell arteritis, Guillain–Barré syndrome and other peripheral neuropathies, noninfectious meningitis, encephalitis, transverse myelitis, etc. Presenting symptoms of some of these adverse effects may initially simulate CNS metastatic disease (Zamani et al., 2016; Salem et al., 2018; Fishman et al., 2019; Johnson et al., 2020; Takai et al., 2020; Valenti-Azcarate et al., 2020).

CONCLUSIONS AND FUTURE DIRECTIONS

Myocarditis is an uncommon condition with multiple presentation patterns, which may be initially difficult to recognize and may simulate other conditions such as acute myocardial infarction, pericarditis, septicemia, etc. Awareness of specific myocarditis types that have a high incidence of stroke or death such as eosinophilic myocarditis, fulminant myocarditis, myocarditis from immune checkpoint inhibitors, etc., will lead to earlier discovery and treatment. Awareness of specific clinical associations such as myocarditis and myasthenia gravis will avoid confusing a myasthenic crisis with symptoms of acute myocarditis. The neurologist should be mindful of unique complications of mechanical hemodynamic support from extracorporeal membrane oxygenation techniques such as differential hypoxia to the upper body. Insofar as heart failure even without atrial fibrillation is a risk factor for stroke, an important area of future study will be to further define which subgroup of patients with heart failure under specific clinical conditions might benefit from anticoagulation.

REFERENCES

Abootalebi S, Aertker BM, Andalibi MS et al. (2020). Call to action: SARS-CoV-2 and CerebrovAscular DisordErs (CASCADE). J Stroke Cerebrovasc Dis 29: 104938.

Agac M, Akyuz AR, Acar Z et al. (2011). Massive multi-chamber heart thrombosis as a consequence of acute fulminant myocarditis complicated with fatal ischaemic stroke. Eur J Echocardiogr 12: 885.

André R, Cottin V, Saraux JL et al. (2017). Central nervous system involvement in eosinophilic granulomatosis with polyangiitis (Churg-Strauss): report of 26 patients and review of the literature. Autoimmun Rev 16: 963–969.

Aram J, Cockerell O, Evanson (2010). J POC21 posterior reversible encephalopathy syndrome in a case of leptospirosis. J Neurol Neurosurg Psychiatry 81: e40.

Aretz HT, Billingham ME, Edwards WD et al. (1987). Myocarditis. A histopathologic definition and classification. Am J Cardiovasc Pathol 1: 3–14.

Asaumi Y, Yasuda S, Morii I et al. (2015). Favourable clinical outcome in patients with cardiogenic shock due to fulminant myocarditis supported by percutaneous extracorporeal membrane oxygenation. Eur Heart J 26: 2185–2192.

Bal AM (2005). Unusual clinical manifestations of leptospirosis. J Postgrad Med 51: 179–183.

Bein T, Müller T, Citerio G (2019). Determination of brain death under extracorporeal life support. Intensive Care Med 45: 364–366.

Biere L, Piriou N, Ernande L et al. (2019). Imaging of myocarditis and inflammatory cardiomyopathies. Arch Cardiovasc Dis 112: 630–641.

Brambatti M, Matassini M, Adler E et al. (2017). Eosinophilic myocarditis: characteristics, treatment, and outcomes. J Am Coll Cardiol 70: 2363–2375.

Chan K, Gibson P (2019). Eosinophilic myocarditis presenting with hypoactive delirium and cardioembolic stroke. CMAJ 191: E1159–E1163.

Cho SM, Farrokh S, Whitman G et al. (2019). Neurocritical care for extracorporeal membrane oxygenation patients. Crit Care Med 47: 1773–1781.

Chow J, Alhussaini A, Calvillo-Arguelles O et al. (2020). Cardiovascular collapse in Covid-19 infection: the role of veno-arterial extracorporeal membrane oxygenation (VA-ECMO). CJS Open. 2: 273–277. https://doi.org/10.1016/j.cjco.2020.04.003.

Cohle S, Lie J (1996). Myasthenia gravis-associated systemic vasculitis and myocarditis with involvement of the cardiac conducting tissue. Cardiovasc Pathol 5: 159–162.

Cooper LT, Keren A, Sliwa K et al. (2014). The global burden of myocarditis. Part one: a systematic literature review for the global burden of disease is, injuries and risk factors 2010 study. Glob Heart 9: 121–129.

De Jongste M, Oosterhuis H, Lie K (1986). Intractable ventricular tachycardia in a patient with giant cell myocarditis, thymoma and myasthenia gravis. Int J Cardiol 13: 374–378.

Delewi R, Zijlstra F, Piek J (2012). Left ventricular thrombus formation after acute myocardial infarction. Heart 98: 1743–1749.

Driggin E, Madhavan M, Bikdeli B et al. (2020). Cardiovascular considerations for patients, healthcare workers and help systems erring the Covid 19 pandemic. J Am Coll Cardiol 75: 2352–2371.

ELSO (2013). ELSO adult cardiac failure supplement to ELSO general guidelines, version 1.3, Extracorporeal Life Support Organization, https://www.elso.org.

Fauci A, Harley J, Roberts W et al. (1982). NIH conference. The idiopathic hypereosinophilic syndrome. Clinical, pathophysiologic, and therapeutic considerations. Ann Intern Med 97: 78–92.

Ferreira JP, Girerd N, Gregson J et al. (2018). Stroke risk in patients with reduced ejection fraction after myocardial infarction without atrial fibrillation. J Am Coll Cardiol 71: 727–735.

Fishman J, Hogan J, Maus (2019). Inflammatory and infectious syndromes associated with cancer immunotherapies. Clin Infect Dis 69: 909–920.

Francone M, Chimenti C, Galea N et al. (2014). CMR sensitivity varies with clinical presentation and extent of cell necrosis in biopsy-proven acute myocarditis. JACC Cardiovasc Imaging 7: 254–263.

Friedrich MG, Sechtem U, Schulz-Menger J et al. (2009). Cardiovascular magnetic resonance in myocarditis: a JACC white paper. J Am Coll Cardiol 53: 1475–1487.

Gambacorti Passerini C, Cortellaro M, Cofrancesco E et al. (1989). Possible mechanisms of fibrin deposition in the hypereosinophilic syndrome. Haemostasis 19: 32–37.

Greer D, Shemie S, Lewis A et al. (2020). Determination of brain death/death by neurologic criteria: the world brain death project. (Published online ahead of print, 2020 August 3). JAMA. https://doi.org/10.1001/jama.2020.11586.

Hendren N, Drazner M, Bozkurt B et al. (2020). Description and proposed management of the acute Covid 19 cardiovascular syndrome. Circulation 141: 1903–1914.

Homma S, Thompson J, Pullicino P et al. (2012). Warfarin and aspirin in patients with heart failure and sinus rhythm. New Eng J Med 366: 1859–1869.

Hu H, Ma F, Wei X et al. (2020). Coronavirus fulminant myocarditis treated with glucocorticoid and human immunoglobulin. Eur Heart J ehaa190. https://doi.org/10.1093/eurheartj/ehaa190. Epub ahead of print. PMID: 32176300; PMCID: PMC7184348.

Ihle J, Burrell A (2019). Confirmation of brain death on VA-ECMO should mandate simultaneous distal arterial and post-oxygenator blood gas sampling. Intensive Care Med 45: 1165–1166.

Johnson D, Manouchehri A, Haugh A et al. (2020). Neurological toxicity associated with immune checkpoint inhibitors: a pharmacovigilance study. J ImmunoTherapy Cancer 7: 134–143.

Kawai S, Shimada T (2014). Inflammation in Takotsubo cardiomyopathy? Inquiry from "Guidelines for diagnosis and treatment of myocarditis (JCS 2009)". J Cardiol 63: 247–249.

Keebler ME, Haddad E, Choi C et al. (2018). Venoarterial extracorporal membrane oxygenation in cardiogenic shock. JACC Heart Fail 6: 503–516.

Kishimoto C, Ochiai H, Sasayama S (1992). Intracardiac thrombus in murine coxsackievirus B3 myocarditis. Heart Vessels 7: 76–81.

Kochi A, Tagliari A, Forleo G et al. (2020). Cardiac and arrhythmic complications in patients with Covid 19. J Cariovasc Electrophysiol 31: 1003–1008.

Kociol RD, Cooper LT, Fang JC et al. (2020). Recognition and initial management of fulminant myocarditis: a scientific statement from the American Heart Association. Circulation 141: e69–e92.

Kojima J, Miyazaki S, Fujiwara H (1988). Recurrent left ventricular mural thrombi in a patient with acute myocarditis. Heart Vessels 4: 120–122.

Le Guennec L, Cholet C, Huang F et al. (2018). Ischemic and hemorrhagic brain injury during venoarterial-extracorporeal membrane oxygenation. Ann Intensive Care 8: 129.

Lee D, Ahn T-B (2014). Central nervous system involvement of hypereosinophilic syndrome: a report 10 cases and a literature review. J Neurol Sci 347: 281–287.

Leone O, Pieroni C, Rapezzi C et al. (2019). The spectrum of myocarditis: from pathology to the clinics. Virchows Arch 475: 279–301.

Li J, Han T, Woodward M et al. (2020). The impact of 2019 novel coronavirus on heart injury: a systematic review and meta-analysis. (Published online ahead of print, 2020 April 16). Prog Cardiovasc Dis. https://doi.org/10.1016/j.pcad.2020.04.008. S0033-0620(20)30080-3.

Liew A, Eikelboom J, Connolly S et al. (2014). Efficacy and safety of warfarin vs. antiplatelet therapy in patients with systolic heart failure and sinus rhythm: a systematic review and meta-analysis of randomized controlled trials. Int J Stroke 9: 199–206.

Lopez-Cuenca A, Gomez-Molina M, Flores-Blanco P et al. (2016). Comparison between type II and type I myocardial infarction: clinical features, treatment strategies and outcomes. J Geriatric Cardiol 13: 15–22.

Lorusso R, Vizzardi E, Pinelli L et al. (2014). Posterior reversible encephalopathy syndrome in a patient submitted to extracorporeal membrane oxygenation for acute fulminant myocarditis. Int J Cardiol 172: e329–e330.

Lorusso R, Barili F, Mauro MD et al. (2016). In-hospital neurologic complications in adult patients undergoing venoarterial extracorporeal membrane oxygenation: results from the extracorporeal life support organization registry. Crit Care Med 44: e964–e972.

Majid M, Safavi-Naeini P, Solomon S et al. (2020). Potential effects of coronavirus is on the cardiovascular system. A review. JAMA Cardiol 5: 831–840.

Masaki N, Issiki A, Kirimura M et al. (2016). Echocardiographic changes in eosinophilic endocarditis induced by Churg-Strauss syndrome. Intern Med 55: 2819–2823.

Mayosi BM (2007). Contemporary trends in the epidemiology and management of cardiomyopathy and pericarditis in sub-Saharan Africa. Heart 93: 1176–1183.

McGee M, Shiel E, Brienesse S et al. (2018). *Staphylococcus aureus* myocarditis with associated left ventricular apical thrombus. Case Rep Cardiol 2018: 7017286. doi:10.1155/2018/7017286. Published 2018 May 23.

Mehra MR, Vaduganathan M, Fu M et al. (2019). A comprehensive analysis of the effects of rivaroxaban one stroke or transient ischemic attack in patients with heart failure, coronary artery disease and sinus rhythm: the COMMANDER HF trial. Eur Heart J 40: 3593–3602.

Melgaard L, Gorst-Rasmussen A, Lane DA et al. (2015). Assessment of the CHA_2DS_2-VASc score in predicting ischemic stroke, thromboembolism, and death in patients with heart failure with and without atrial fibrillation. JAMA 314: 1030–1038.

Mukai HY, Ninomiya H, Ohtani K et al. (1995). Major basic protein binding to thrombomodulin potentially contributes to the thrombosis in patients with eosinophilia. Br J Haematol 90: 892–899.

Mulroy E, Cleland J, Anderson N (2018). Crescentic splinter hemorrhages reflect stroke pathophysiology and hypereosinophilic syndrome. Aust J Dermatol 59: E211–E246.

Muralidharan R, Mateen F, Shinohara R et al. (2011). The challenges of brain death determination in adult patients on extracorporal membrane oxygenation. Neurocrit Care 14: 423–426.

Mygland A, Aarli JA, Hofstad H et al. (1991). Heart muscle antibodies in myasthenia gravis. Autoimmunity 10: 263–267.

Ponikowski P, Voors A, Anker S et al. (2016). 2016 ESC Guidelines for the diagnosis and treatment of acute and chronic heart failure. Eur J Heart Fail 891–975.

Premier D, Davidson D, Loehre P et al. (2018). Giant cell polymyositis and myocarditis in a patient with thymoma and myasthenia gravis: a post-viral autoimmune process? J Neuropathol Exp Neurol 77: 661–664.

Priyankara W, Manoj E (2019). Posterior reversible encephalopathy in a patient with severe leptospirosis complicated with pulmonary haemorrhage, myocarditis, and acute kidney injury. Case Rep Crit Care 2019: 6498315. Published 2019 December 17.

Ramanan AS, Pandit N, Yashwant M et al. (1993). Embolic stroke in myocarditis. Indian Pediatr 30: 531–533.

Rosenberg M, Lorenz H, Gassler N et al. (2006). Rapid progressive eosinophilic cardiomyopathy in a patient with Churg-Strauss syndrome (CSS). Clin Res Cardiol 95: 289–294.

Rupprecht L, Lunz D, Philipp A et al. (2015). Pitfalls in percutaneous ECMO cannulation. Heart Lung Vessel 7: 320–326.

Ryerson L, Sanchez-Glanville C, Huberdeau C et al. (2017). Carotid artery dissection following neck cannulation for extracorporeal life support. World J Pediatr Congenit Heart Surg 8: 414–416.

Sakuta K, Miyagawa S, Suzuki K et al. (2019). Rapid disappearance of intraventricular Mobile structures with steroids in eosinophilic granulomatosis with polyangiitis. J Stroke Cerebrovasc Dis 28: 104326.

Sala S, Peratto G, Gramenga M et al. (2020). Acute myocarditis presenting as a reverse Tako-Tsubo syndrome in a patient with SARS-CoV-2 respiratory infection. Eur Heart J 41: 1861–1862. https://doi.org/10.1093/eurheartj/ehaa286. Published online 2020 April 8.

Salem J, Moey M, Lebrun-Vignes M et al. (2018). Cardiovascular toxicities associated with immune checkpoint inhibitors: an observational, retrospective and pharmacovigilance study. Lancet Oncol 19: 1579–1589.

Salih M, Ibrahim R, Tirunagiri D et al. (2020). Loeffler's endocarditis and hypereosinophilic syndrome. (Published online ahead of print, 2020 June 9). Cardiol Rev 2020. https://doi.org/10.1097/CRD.0000e000000000324.

Saucha W, Sołek-Pastuszka J, Bohatyrewicz R et al. (2015). Apnea test in the determination of brain death in patients treated with extracorporeal membrane oxygenation (ECMO). Anaesthesiol Intensive Ther 47: 368–371.

Schmidt M, Burrell A, Roberts L et al. (2015). Predicting survival after ECMO for refractory cardiogenic shock: the survival after veno-arterial-ECMO (SAVE)-score. Eur Heart J 36: 2246–2256.

Schumacher R, Weinfeld I, Bartlett R (1989). Neonatal vocal cord paralysis following extracorporeal membrane oxygenation. Pediatrics 84: 793–796.

Shekar K, Badulak J, Peek G (2020). Extracorporeal life support organization coronavirus disease 2019 interim guidelines: a consensus document from an International Group of Interdisciplinary Extracorporeal Membrane Oxygenation Providers. ASAIO J 66: 707–721.

Siliste R, Antohi E, Pepoyan S et al. (2018). Anticoagulation in heart failure without atrial fibrillation: gaps and dilemmas in current clinical practice. Eur J Heart Fail 20: 978–988.

Slungaard A, Vercellotti G, Tran T et al. (1993). Eosinophil cationic granule proteins impair thrombomodulin function. A potential mechanism for thromboembolism in hypereosinophilic heart disease. J Clin Invest 91: 1721–1730.

Stevens C, Callaghan M, Forrest P et al. (2017). Flow mixing during peripheral veno-arterial extra corporeal membrane oxygenation—a simulation study. J Biomech 55: 64–70.

Sutter R, Tisljar K, Marsch S (2018). Acute neurologic complications during extracorporeal membrane oxygenation: a systematic review. Crit Care Med 46: 1506–1513.

Suzuli S, Utsugisawa K, Suzuki N (2013). Overlooked non-motor symptoms in myasthenia gravis. J Neurol Neurosurg Psychiatry 84: 989–994.

Szczeklik W, Miszalski-Jamka T, Mastalerz L et al. (2011). Multimodality assessment of cardiac involvement in Churg-Strauss syndrome patients in clinical remission. Circ J 75: 649–655.

Takai M, Kato D, Iinuma K et al. (2020). Simultaneous pembrolizumab-induced myasthenia gravis and myocarditis in a patient with metastatic bladder cancer: a case report. Urol Case Rep 31: 101145.

Tavazzi G, Pellegrini C, Maurelli M et al. (2020). Myocardial localization of coronavirus in COVID-19 cardiogenic shock. Eur J Heart Failure 22: 911–915.

Teele S, Salvin J, Barrett C et al. (2014). The association of carotid artery cannulation and neurologic injury in pediatric patients supported with venoarterial extracorporeal membrane oxygenation. Pediatr Crit Care Med 15: 355–361.

Todd S, Hemmaway C, Nagy Z (2014). Catastrophic thrombosis in idiopathic hypereosinophilic syndrome. Br J Haematol 165: 425.

Tomioka N, Kishimoto C, Matsumori A et al. (1986). Mural thrombus in experimental viral myocarditis in mice: relation between thrombosis and congestive heart failure. Cardiovasc Res 20: 665–671.

Valent P, Klion A, Horny H-P et al. (2012). Contemporary consensus proposal on criteria and classification of eosinophilic disorders and related syndromes. J Allergy Clin Immunol 130: 607–612.e9.

Valenti-Azcarate R, Vazquez I, Illan C et al. (2020). Nivolumab and Ipilimuab-induced myositis and myocarditis mimicking a myasthenia gravis presentation. Neuromuscul Disord 3: 67–69.

Venge P, Dahl R, Hallgren R (1979). Enhancement of factor XII dependent reactions by eosinophil cationic protein. Thromb Res 14: 641–649.

Walkey A, Wiener R, Ghobrial J et al. (2011). Incidence stroke and mortality associated with new onset atrial fibrillation patients hospitalized with severe sepsis. JAMA 306: 2248–2254.

Xie A, Lo P, Yan T et al. (2017). Neurologic complications of extracorporeal membrane oxygenation: a review. J Cardiothorac Vasc Anesth 31: 1836–1846.

Yan L, Mir MS et al. (2020). Covid-19 in a Hispanic woman. Autopsy report with clinical pathological correlation. Arch Pathol Lab Med 144: 1041–1047.

Yancy CW et al. (2013). 2013 ACCF/AHA Guideline for the management of heart failure. Circulation 128: e240–e327.

Yang X, Yu Y, Xu J et al. (2020). Clinical course and outcomes a critically ill patient SARS-CoV-2 pneumonia in Wuhan, China: a single-centered, retrospective, observational study. Lancet Respir Med 8: 475–481.

Zamani M, Aslani S, Salmaninejad A et al. (2016). PD-1/PD-L and autoimmunity: a growing relationship. Cell Immunol 310: 27–41.

Zannad F, Anker S, Byra W et al. (2018). Rivaroxaban in patients with heart failure, sinus rhythm in coronary disease (COMMANDER HF Trial). N Engl J Med 379: 1332–1342.

Zhou R (2020). Does SARS-CoV-2 cause viral myocarditis in Covid-19 patients? Eur Heart J 41: 2123. https://doi.org/10.1093/eurheartj/ehaa392. Published May 3, 2020.

Handbook of Clinical Neurology, Vol. 177 (3rd series)
Heart and Neurologic Disease
J. Biller, Editor
https://doi.org/10.1016/B978-0-12-819814-8.00008-1

Chapter 11

Neurologic complications of infective endocarditis

TIA CHAKRABORTY, ALEJANDRO RABINSTEIN*, AND EELCO WIJDICKS

Department of Neurology, Mayo Clinic, Rochester, MN, United States

Abstract

Infective endocarditis (IE) is an infection primarily affecting the endocardium of heart valves that can embolize systemically and to the brain. Neurologic manifestations include strokes, intracerebral hemorrhages, mycotic aneurysms, meningitis, cerebral abscesses, and infections of the spine. Neurologic involvement is associated with worse mortality, though it does not always portend a poor functional prognosis. Neuroimaging is indicated in patients who have neurologic symptoms, including cerebral vessel imaging in patients who have subarachnoid hemorrhage. In the case of acute ischemic stroke (IS), IV thrombolysis is contraindicated but endovascular thrombectomy may be a consideration. Neurologic findings understandably raise concern about valve surgery when indicated due to the risk of hemorrhage with perioperative anticoagulation. However, most neurologic complications do not preclude valve surgery and valve surgery may in fact be indispensable in some cases to prevent further neurologic problems. Management decisions in patients with IE and neurologic complications should therefore be multidisciplinary with a major contribution from the neurologist.

INTRODUCTION

Infective endocarditis (IE) is an infection of the cardiac endothelium that causes vegetations on native or mechanical valves, or on implanted cardiac devices, and in extreme circumstances may result in an aortic root abscess. The incidence of IE has increased to 15 per 100,000 in the United States and has a 30-day mortality rate as high as 20%. Nocosomial IE has become more common, accounting for 25%–30% of contemporary cohorts with increasing use of long-term intravenous (IV) lines and invasive procedures. A major risk factor for IE remains IV drug use. Other key risk factors include rheumatic and degenerative heart disease, advanced age, diabetes, chronic kidney disease, cancer, and congenital heart disease. The most common organisms are *Staphylococcus aureus* (40% of patients), followed by viridans group Streptococcus (17%), enterococci (11%), and less common organisms such as the HACEK group (Haemophilus species, *Aggregatibacter actinomycetemcomitans*, *Cardiobacterium hominis*, *Eikenella corrodens*, *Kingella* species), coagulase-negative Staphylococcus on prosthetic valves and cardiac devices, and rarely fungal organisms (Fedeli et al., 2011; Cahill and Prendergast, 2016; Vincent and Otto, 2018; Wang et al., 2018).

Systemic emboli from cardiac vegetations are common. Embolism to the brain is a frequent complication, and rates of detection have increased with improved neuroimaging and its frequent use in IE. Between 20% and 40% of patients with IE have neurologic complications, and 15%–30% of them have neurologic symptoms (Heiro et al., 2000, 2007; Mangoni et al., 2003; Habib et al., 2015; Yanagawa et al., 2016). Both hemorrhagic and ischemic infarcts can be seen in as many as 80% of patients with neuroimaging findings (Cooper et al., 2009; Hess et al., 2013). Other complications include microhemorrhages, mycotic aneurysms (MAs), infectious arteriopathy, meningitis, brain and spinal abscesses, and vertebral osteomyelitis. Strokes are an independent predictor of increased mortality; other predictors are

*Correspondence to: Alejandro Rabinstein, Department of Neurology, Mayo Clinic, 200 First Street SW, Rochester, MN, 55905, United States. Tel: +1-507-538-1038, Fax: +1-507-266-0178, E-mail: rabinstein.alejandro@mayo.edu

Staphylococcus aureus infection, older age, male sex, mechanical valve involvement, and atrioventricular block (Pericart et al., 2016).

Neurologic complications in IE can be overt (focal findings) or subtle (microhemorrhages on MRI) and can even be asymptomatic or hard to detect clinically in a patient with severe cardiac failure. Management in patients who have cerebrovascular complications of endocarditis, particularly with mycotic aneurysms or intracerebral hemorrhages (ICH), can be challenging given the complexity of weighing the benefit of potential valve surgery against the risk of bleeding in the setting of high doses of heparin during surgery.

DIAGNOSIS OF INFECTIVE ENDOCARDITIS

Infectious endocarditis is not easy to diagnose in the earlier stages and frequently surprises physicians. Insidious presentations without fever can be seen (10% of patients), but most patients present with constitutional symptoms, and severe cases manifest with septic shock, or with neurologic symptoms such as encephalopathy, seizures, or focal deficits. If the patient is seen first by a neurologist due to a presenting ischemic or hemorrhagic stroke, IE should be considered if fever or signs of cardiac failure are also present. The modified Duke criteria provide a systemic framework for clinically diagnosing patients with suspected IE (Li et al., 2000). Other exam findings include splinter or subungual hemorrhages, Janeway lesions (painless erythematous lesions on palms and soles), Osler nodes (painful violet lesions on fingers and toes), and Roth spots (retinal hemorrhages with central pallor) (Fig. 11.1.). While these findings are classically taught for the recognition of IE, they are far less common than progressive cardiac failure and pulmonary edema (Cahill and Prendergast, 2016). Blood cultures are paramount in making the diagnosis of IE and should be obtained in patients within 48 h of otherwise unexplained persistent fever (a key sign in many cases), a new or changed heart murmur (may be a difficult clinical judgment), or echocardiogram evidence of new valve dysfunction (depending on accurate imaging of the valves, preferentially by transesophageal echocardiography (TEE)) (Habib et al., 2015; Nishimura et al., 2017; Vincent and Otto, 2018). At least three sets of blood cultures should be obtained from different venipuncture sites throughout the first day of presentation (Baddour et al., 2015).

Echocardiography should also ideally be performed within 12 h of presentation in patients suspected of having IE. Transthoracic echocardiography is often the initial modality, but transesophageal echocardiography is much more sensitive to detect vegetations in valves or intracardiac device leads. The differentiation between thrombus and vegetation requires substantial cardiac expertise. A repeat TEE within 3–5 days should be pursued if the initial TEE is negative in cases with high clinical suspicion. Echocardiography can help predict embolic risk by assessing the number, size, shape, location, echogenicity, and mobility of vegetations. The greatest risk of embolic complications appears to occur with vegetations $\geq$10 mm on the anterior leaflet of the mitral valve. ^{18}F-FDG PET—CT imaging/angiography may be useful in identifying cardiac vegetations and systemic emboli in patients with negative TEEs and ongoing suspicion of IE and can visualize systemic emboli (Baddour et al., 2015; Sordelli et al., 2019).

GENERAL MANAGEMENT OF INFECTIVE ENDOCARDITIS

The treatment approach for patients with IE should be guided by a multidisciplinary team comprised of a cardiologist and infectious diseases specialist. In more severe cases, expert opinion from specialists in radiology, cardiothoracic surgery, neurology, interventional neuroradiology, neurosurgery, and nephrology may be needed

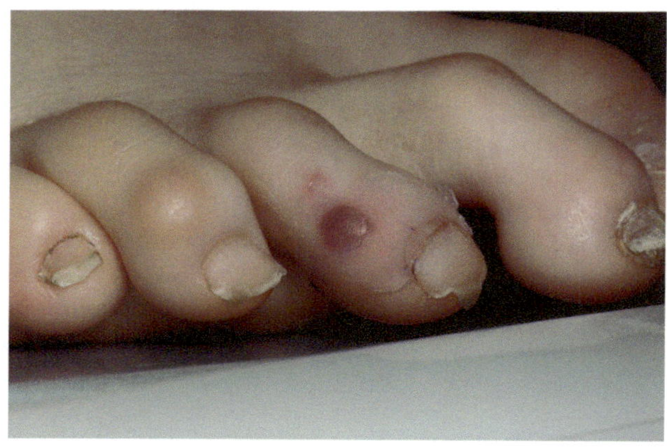

Fig. 11.1. Osler node in a patient with IE. These are painful lesions on fingers and toes.

(Cahill et al., 2017). Appropriate antibiotics are outlined in the ACC/AHA guidelines, and they should be administered under the guidance of infectious diseases specialists for 2–6 weeks depending on the location of the infection, whether the valve is native or prosthetic, and the bacterial strain and susceptibilities (Baddour et al., 2015; Vincent and Otto, 2018). Valve surgery is an early consideration in patients who present with heart failure, severe valve regurgitation with large mobile vegetations >10 mm, heart block, annular or aortic abscess, destructive penetrative lesions, fungal or highly resistant organisms, sepsis or persistent bacteremia, or multiple systemic emboli and multiorgan failure (Baddour et al., 2015).

When neurologic complications arise and are evidenced by neuroimaging, they could delay the timing of valve surgery for fear of perioperative ICH; these surgical delays are associated with a worse prognosis (Thuny et al., 2007; Baddour et al., 2015). On the other hand, removing the embolic source is a rational choice. Guidelines by the Society of Thoracic Surgeons advise to wait at least 4 weeks following an embolic ischemic or hemorrhagic stroke to replace an infected valve. This is due to the presumed increased risk of perioperative ICH in the setting of procedural anticoagulation and cardiopulmonary bypass (Byrne et al., 2011; Okazaki et al., 2013; Ballal et al., 2015). The only prospective study assessing timing of surgery found that the rate of systemic and cerebral emboli went down from 21% and 13%, respectively, in patients without early valve surgery to 0% in patients who underwent surgery within 48 h of presentation; however, this study excluded patients with large ischemic strokes (IS) (Kang et al., 2012).

Other retrospective studies have revealed varied results in patients with IS or ICH. More recent studies have shown favorable mortality rates and functional outcomes in patients with IS who have undergone valve surgery, even early (from within 2–7 days of presentation) (Okita et al., 2016; Chakraborty et al., 2017, 2019; Kim et al., 2018). Mortality and outcomes in the few reported patients who underwent valve surgery at any point in time seemed to be less favorable when there was evidence of ICH, particularly when the hematoma was large, or located in the posterior fossa (Okita et al., 2016; Chakraborty et al., 2019). Yet, patients with smaller ICH, including those with innumerable microhemorrhages and even some with subarachnoid hemorrhage, may still have good outcomes without enlargement of hemorrhages following valve surgery. A management algorithm for patients with neurologic symptoms has been proposed based on outcome analysis depending on specific neuroimaging findings (Fig. 11.2.) (Chakraborty et al., 2019). It is very important to highlight that clinically silent neuroimaging findings are very common and they frequently result in major delays or cancellation of valve surgery (Snygg-Martin et al., 2008; Cooper et al., 2009; Duval et al., 2010; Chakraborty et al., 2017, 2019). Thus, a measured approach to neuroimaging and interpretation of neuroimaging findings is essential. A head CT should be obtained in patients who have neurologic symptoms. A normal CT or one with small IS or small deep ICH should not preclude early valve surgery. If findings are equivocal or the patient is comatose, MRI is warranted for further investigation. A CT showing territorial IS or hemorrhage that is lobar or in the posterior fossa pose great concern for hemorrhagic risk. In these cases of major hemorrhage and in cases of subarachnoid hemorrhage, cerebral vessel imaging should be pursued to assess for mycotic aneurysms. Despite great recent improvement in the quality of noninvasive angiographic images, catheter cerebral angiography remains the modality of choice in these instances.

ISCHEMIC STROKE

Ischemic stroke is the most common neurologic manifestation of IE, with a reported incidence of 37%–83% (Klein et al., 2009; Morofuji et al., 2010; Hess et al., 2013; Iung et al., 2013; Malhotra et al., 2017). Most strokes are cardioembolic, and risk is influenced primarily by vegetation size, mobility, *Staphylococcus aureus* infection, and mitral valve involvement (Garcia-Cabrera et al., 2013). The embolic material is likely a fragile thrombus with an abundance of hyaline amidst the infectious organism (Fig. 11.3.). The majority of strokes appear embolic on CT or MRI (Fig. 11.4.). Another mechanism by which subcortical and cortical strokes may present in IE is an infectious pan vasculitis, which may also cause cerebral hemorrhages (Van de Beek et al., 2008).

When acute stroke is suspected or present, IV thrombolysis use is contraindicated given the high hemorrhagic risk in IE (Junna et al., 2007; Asaithambi et al., 2013; Ong et al., 2013; Marquardt et al., 2019; Powers et al., 2019). Endovascular mechanical thrombectomy for acute IS caused by large vessel occlusions may be considered for patients who meet other eligibility criteria (despite patients with suspected or presumed septic emboli having been excluded from thrombectomy trials) (Sveinsson et al., 2016; Scharf et al., 2017; Ambrosioni et al., 2018; Bolognese et al., 2018; Nogueira et al., 2018; Marquardt et al., 2019; Sloane et al., 2020).

Other acute stroke management guidelines should apply, including the maintenance of cerebral perfusion by avoiding hypotension, glucose control with a target of 140–180 mg/dL, maintaining normothermia, and early initiation of physical and/or speech rehabilitation

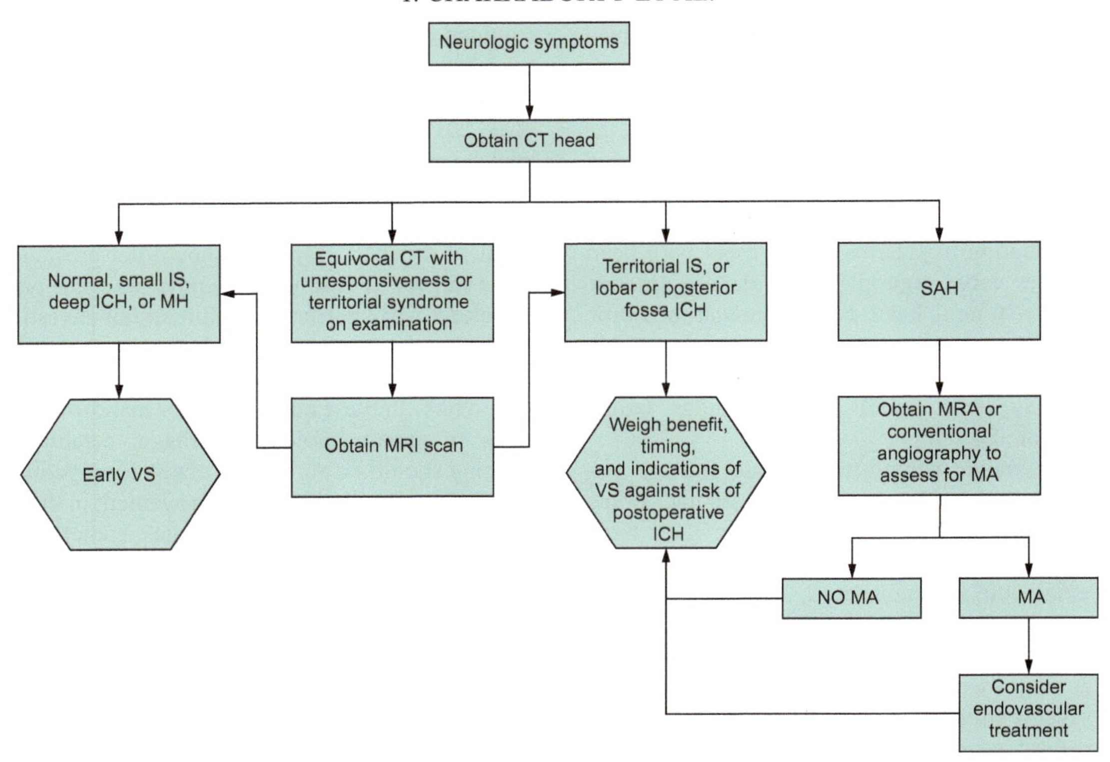

Fig. 11.2. Proposed management algorithm for patients with neurologic symptoms. Neuroimaging is recommended in patients only if they have neurologic symptoms. Depending on findings on CT and the severity of symptoms, further neuroimaging may be considered and the risk of intraoperative hemorrhage vs the benefit of valvular source control with surgery must be weighed.

(Powers et al., 2019). Data regarding the initiation of antiplatelet therapy for secondary stroke prevention in IE is lacking. Until such data is available, starting antiplatelet adjunctive therapy in IE is not recommended. However, the continuation of long-term antiplatelet therapy when IE develops without hemorrhagic complications may be reasonable (Powers et al., 2019). Anticoagulation is contraindicated until the IE is deemed controlled.

INTRACEREBRAL HEMORRHAGE

Intracerebral hemorrhages can account for up to 25% of neurologic complications in IE, not including cerebral microhemorrhages, which were the second most common complication in one recent cohort (35% of patients) (Chakraborty et al., 2017). ICH may often occur by way of hemorrhagic conversion of IS (15% of patients) (Murdoch et al., 2009). Other mechanisms of ICH are rupture of a mycotic aneurysm causing ICH, subarachnoid hemorrhage, subdural hemorrhage from parenchymal ICH, as well as hemorrhages from a pan vasculitis causing friable and fragile vessel walls (Van de Beek et al., 2008; Geisenberger et al., 2015; Sotero et al., 2019).

As in any ICH, abrupt discontinuation of antiplatelet or anticoagulant therapies, potential reversal of anticoagulation, urgent blood pressure control, and intracranial pressure management are of vital importance. ICH is not an absolute contraindication to valve surgery, though there is always a risk of perioperative hemorrhage in the setting of high doses of heparin during cardiopulmonary bypass. This risk varies depending on the size of the hemorrhage (small vs large), location (territorial and posterior fossa may present a higher risk of worsening or recurrent hemorrhage after surgery), and whether a mycotic aneurysm is present or has ruptured (Kume et al., 2018; Chakraborty et al., 2019).

A commonly fatal neurologic complication is a ruptured cerebral mycotic aneurysm. Mortality associated with MA is high but has declined from 80% after rupture and 30% when unruptured down to 12%–32% in more recent case series. MA may develop through endovascular or extravascular (meningitis or cerebral abscess formation) spread of septic emboli. The infectious organism causes marked proliferation of the intima and destruction of the internal elastic lamina (Molinari et al., 1973; Kannoth and Thomas, 2009). Compared to berry aneurysms, MAs are far less common overall, constituting

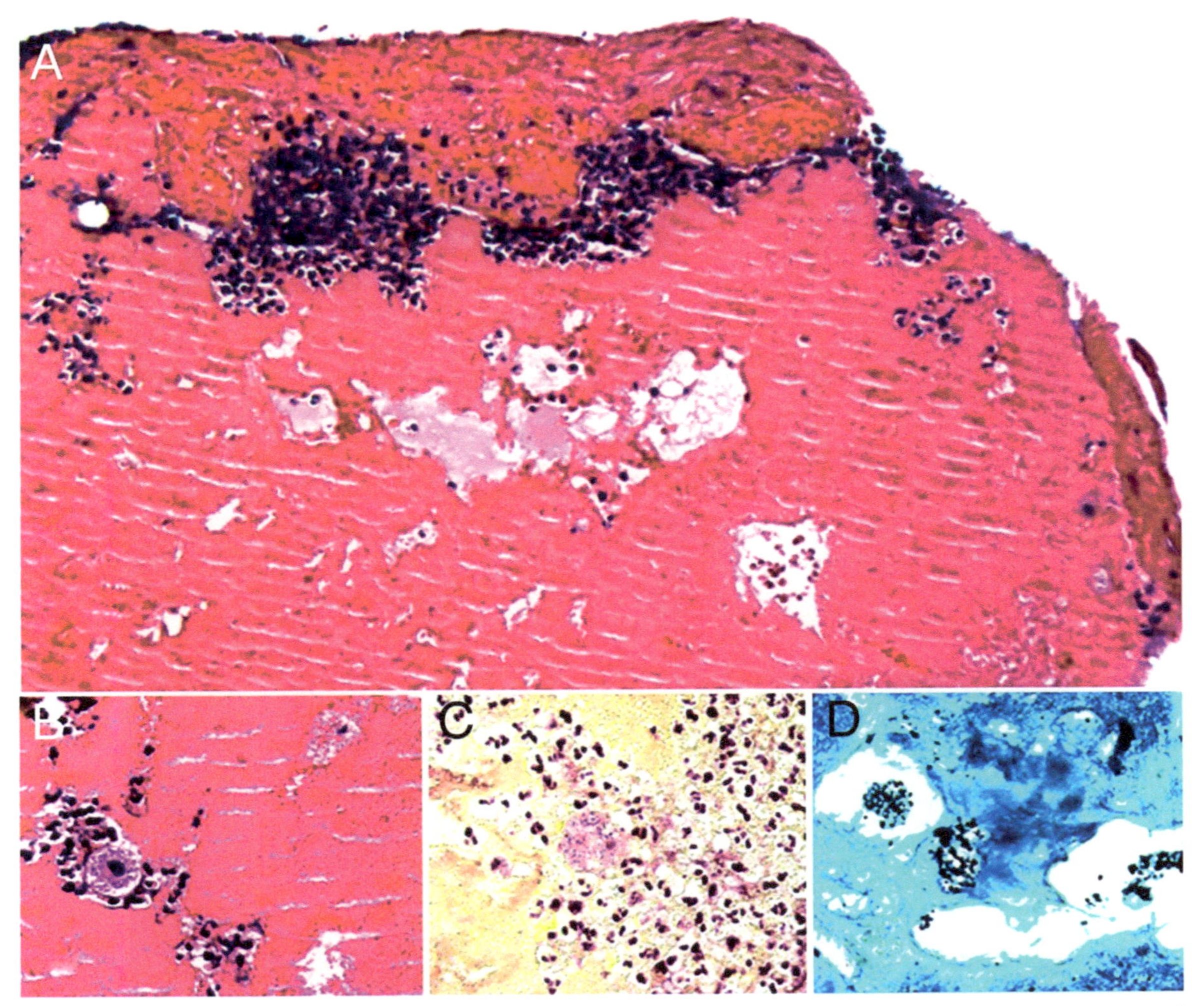

Fig. 11.3. Histopathology of intracranial septic embolus. H&E staining (A and B), gram stain (C), and GMS trichrome stain (D) show abundant hyaline and dead coccal forms. Adapted from Scharf, E.L., Chakraborty, T., Rabinstein, A., et al., 2017. Endovascular management of cerebral septic embolism: three recent cases and review of the literature. J Neurointerv Surg 9, 463–465.

0.7%–6.5% of intracranial aneurysms, more prevalent in younger patients, typically fusiform, can be multiple, and are generally located in more distal cerebral vessels (Kannoth and Thomas, 2009; Kume et al., 2018). Digital subtraction angiography is the gold standard for MA detection. CTA and MRA may detect aneurysms as small as 3 mm, but distal and smaller aneurysms may be missed by these noninvasive modalities. MAs are all the more challenging because they are dynamic and may develop rapidly, or rarely even form long after antibiotic completion (likely because of residual debility in the vessel wall), adding to the complexity of IE management and consideration of rupture if valve surgery is indicated (Fig. 11.5.) (Daneshmand et al., 2019).

There is no uniform way to risk stratify patients with unruptured MA when considering medical vs surgical treatment, and management should be individualized. Aneurysms up to 10 mm in diameter have been shown to resolve with antibiotics alone. Follow-up angiography is recommended in these cases in 7–14 days to assess antibiotic response or if new aneuryms have formed. If the MA is enlarging despite antibiotics or is large from the beginning, surgical management may be necessary (Peters et al., 2006). If a surgical approach is pursued or for ruptured aneurysms, endovascular treatment of the aneurysm (vs surgical clipping, which may be more technically difficult given lack of a more defined aneurysmal neck and friable aneurysmal wall) is often successful. Endovascular coiling can be attempted in more proximal aneurysms, but glue embolization or autologous clot injections is preferred to secure more distal aneurysms that may not be accessible through a microcatheter (Chapot et al., 2002; Phuong et al., 2002;

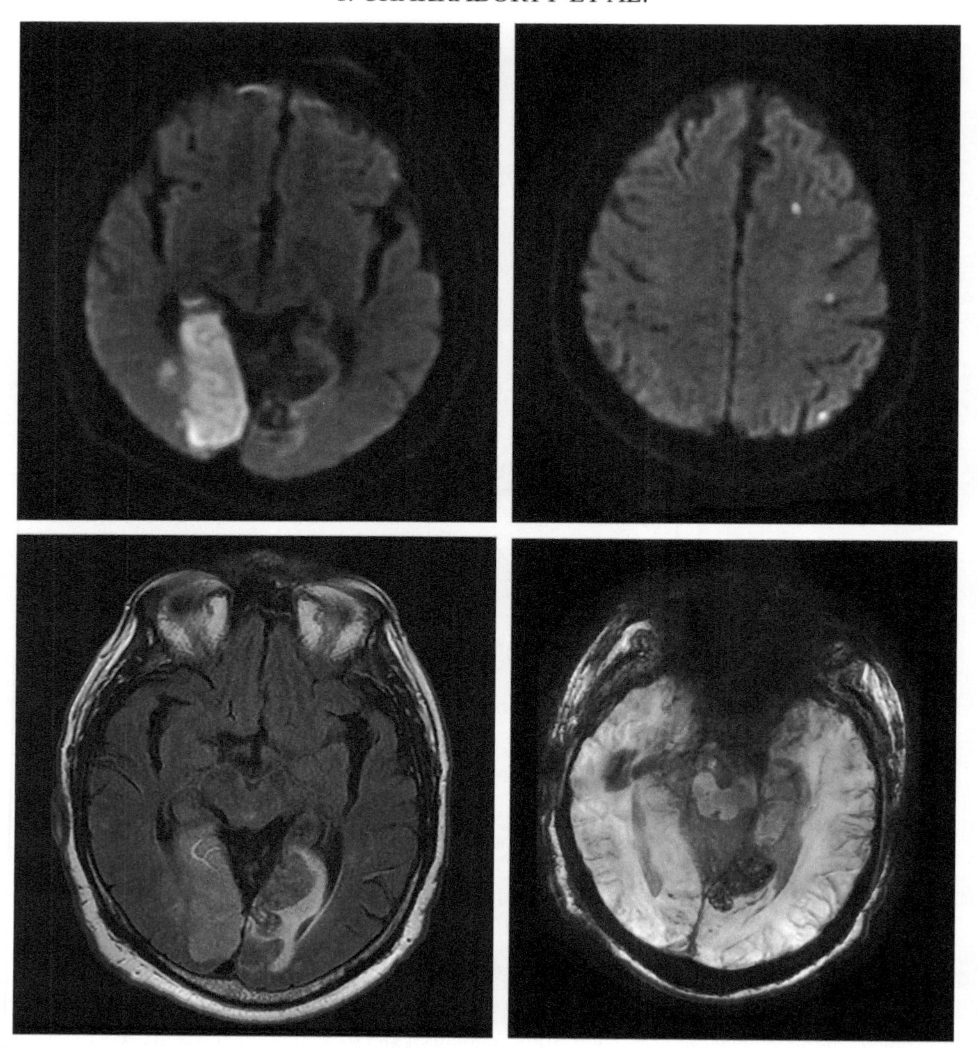

Fig. 11.4. Ischemic strokes caused by infective endocarditis. Acute ischemic strokes restrict on DWI-ADC sequences (*top right* and *left*). In conjunction with the FLAIR sequence (*bottom right*), the MRI appears to shows embolic strokes that are cortical, in different vascular territories, and of different ages. The susceptibility-weighted imaging (SWI) sequence also reveals hemorrhagic conversion of a chronic left occipital stroke (*bottom left*).

Kannoth and Thomas, 2009). Endovascular flow-diversion has also shown promise in terms of comparable efficacy and safety when compared to coiling. In fact in one study, coiling was associated with higher rates of MA recurrence and retreatment vs flow-diversion (Petr et al., 2016). The safety of valve surgery in patients with MA remains unknown and has yet to be investigated. Even if only one MA is seen and treated, these patients probably have a more diffuse septic arteritis and may have recurrent hemorrhages from different vessels with or without visible formation of new MAs.

CNS INFECTIONS

Infections of the central nervous system (CNS) are less common neurologic manifestations, present in 1%–6% of patients with IE. Complications include meningitis and cerebral abscesses. Septic emboli, mainly comprised of *Staphylococcus aureus*, are the most likely mechanistic cause (Garcia-Cabrera et al., 2013; Sotero et al., 2019). Cerebral abscesses may be uniform or multiple. They show the typical ring-enhancement, may have significant vasogenic edema causing mass effect, are located at the gray-white junction, and may cause hemorrhage (Fig. 11.6.). They may then also cause seizures aside from focal neurologic deficits. When patients present initially with cerebral abscess without known IE, it may be warranted to rule out IE as a potential cause, as well as obtaining dental panorex imaging. Intracranial pressure management as well as potential surgical resection and/or extraventricular drain placement for hydrocephalus may also be warranted.

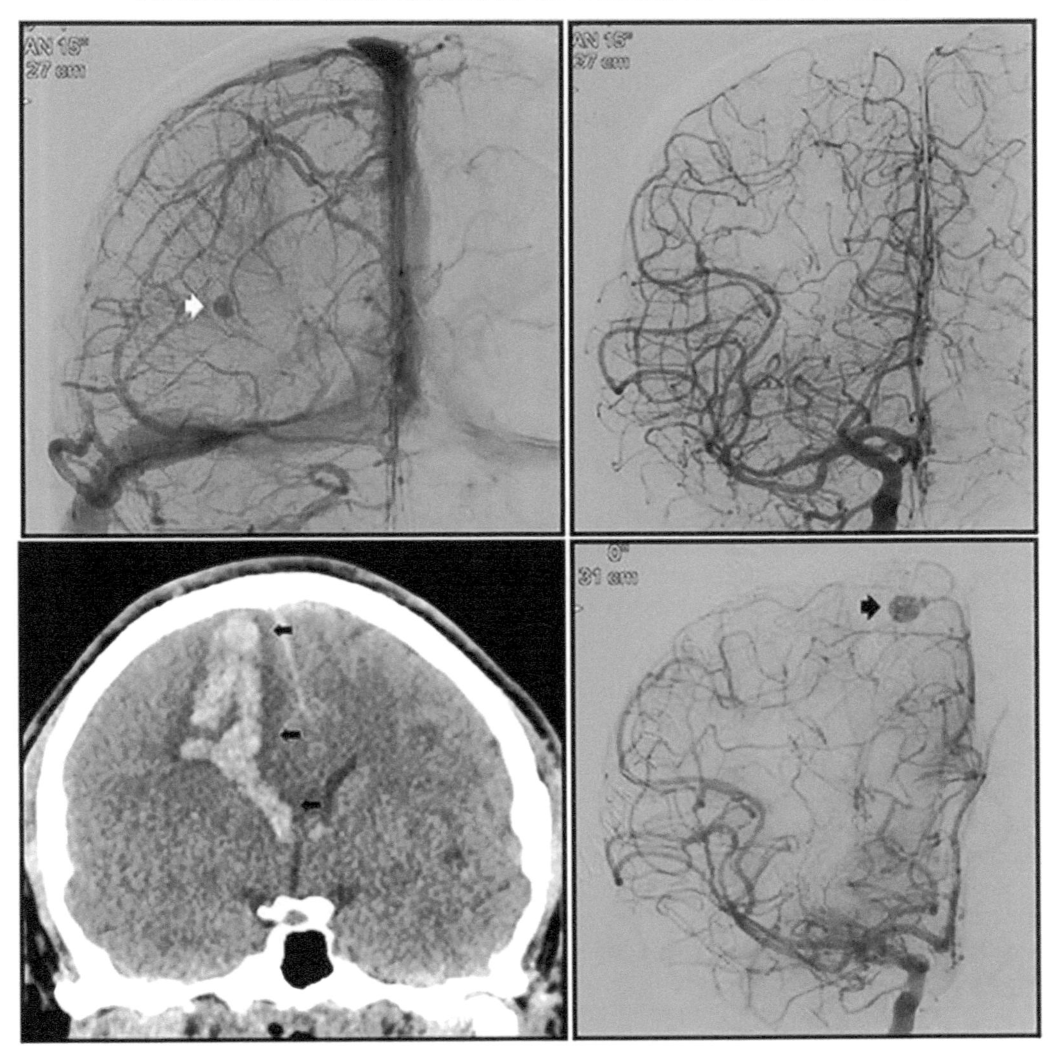

Fig. 11.5. Ultrarapid developing mycotic aneurysms. A 30-year-old male with cardiobacterium endocarditis had subarachnoid hemorrhage on MRI. Cerebral angiogram (*top left*) showed a 4 mm aneurysm in right MCA. He developed severe aortic regurgitation and had a large vegetation; thus, urgent valve surgery was planned. Preoperatively, a repeat angiogram (*top right*) showed stability of the aneurysm (same size, sluggish filling, with no feeding artery), and no intervention was pursued. He underwent valve surgery the next day. Postoperatively the same day, he became comatose. CT showed acute intracerebral hemorrhage (*bottom left*). A third angiogram (*bottom right*) showed a new aneurysm in the parasagittal area. Adapted from Daneshmand, A., Rangel-Castilla, L., Rydberg, C., Wijdicks, E., 2019. Ultra-rapid developing infectious aneurysms. Neurocritical Care 30, 487–489.

Spinal complications such as epidural abscess, spondylodiscitis, and vertebral osteomyelitis are additional potential neurologic complications through hematogenous spread of bacteria. Their incidences are low (<4%), and investigations should be prompted by back pain, which approximately one-third of patients with IE may have alongside musculoskeletal pain, arthralgias, radicular pain, and infectious symptoms. The lumbar region is most commonly affected (Cone et al., 2008; Sotero et al., 2019). Spondylodiscitis, which is infection of the intervertebral disc that may cause a secondary infection of the vertebrae starting at the endplates, may lead to osteomyelitis. Group D Streptococcus is the most common causative organism, and spondylodiscitis has not been found to worsen IE prognosis (Le Moal et al., 2002). In vertebral osteomyelitis, the lumbar spine is most commonly involved. MRI with contrast of pertinent spinal regions guided by the neurologic exam should be diagnostic (Oh et al., 2015). Antibiotics alone may help resolve spinal infections, but surgical management, particularly in the case of epidural abscess and spondylodiscitis, may be indicated when signs of cord compression are evident or when the infection fails to resolve with antibiotics.

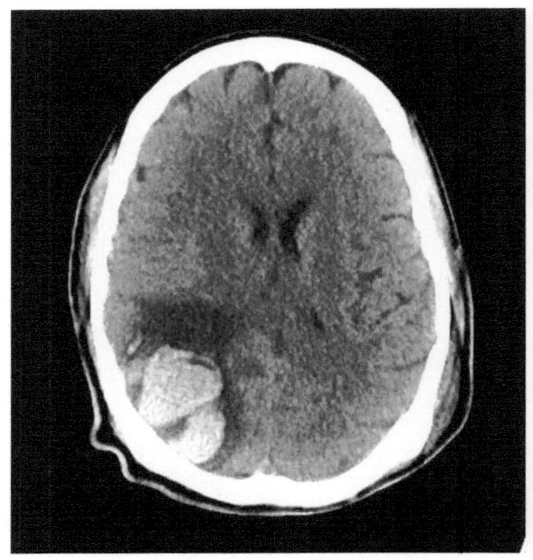
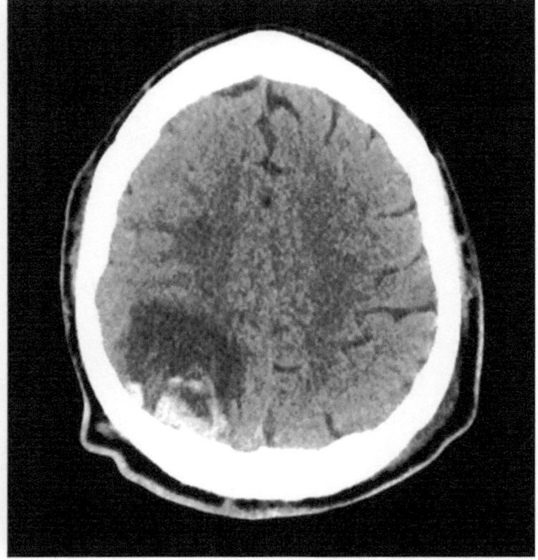

Fig. 11.6. Cerebral abscess in IE. A 22-year-old male with *Staphylococcus aureus* endocarditis on antibiotics for 1 week acutely became comatose, prompting this CT that shows acute intracerebral hemorrhage with vasogenic and cytotoxic edema that ultimately required resection for intracranial hypertension refractory to hyperosmolar therapy.

CONCLUSIONS

Neurologic complications are frequent and important to recognize in the overall management of patients with IE. The diagnosis of IE in patients presenting with neurologic deficits and fever should be considered. In addition to obtaining blood cultures urgently and initiating antibiotics, specific neurologic manifestations should be managed as they would be otherwise managed in the absence of a septic source, except for the avoidance of IV thrombolysis for acute ischemic stroke and the use of anticoagulants for stroke prevention. Neuroimaging should be reserved for patients with neurologic symptoms, and not every patient with IE and neurologic symptoms should have a brain MRI or cerebral angiography. MRI should be reserved for patients with neurologic symptoms unexplained by the CT scan, and MRI findings should be interpreted measuredly to avoid unnecessary surgical delays in patients who need valve replacement. Angiography is only indicated in patients with intracranial hemorrhage and particularly subarachnoid hemorrhage.. When an MA is found, neurosurgical or neuroradiologic intervention is warranted because it may grow rapidly. Treatment of a recently detected MA does not prevent future ICH from newly formed aneurysms or even in the absence of new visible aneurysms. The decision to pursue valve surgery should be made by a multidisciplinary team composed of cardiologists, infectious diseases specialists, cardiothoracic surgeons, and neurologists when neurologic problems have arisen.

REFERENCES

Ambrosioni J, Urra X, Hernandez-Meneses M et al. (2018). Mechanical thrombectomy for acute ischemic stroke secondary to infective endocarditis. Clin Infect Dis 66: 1286–1289.

Asaithambi G, Adil MM, Qureshi AI (2013). Thrombolysis for ischemic stroke associated with infective endocarditis: results from the nationwide inpatient sample. Stroke 44: 2917–2919.

Baddour LM, Wilson WR, Bayer AS et al. (2015). Infective endocarditis in adults: diagnosis, antimicrobial therapy, and Management of Complications: a scientific statement for healthcare professionals from the American Heart Association. Circulation 132: 1435–1486.

Ballal A, Mirza S, Raman J (2015). Neurological risk in urgent valve replacement cardiac surgery for infective endocarditis with cerebrovascular complications—a case series. J Cardiothorac Surg 10: A56.

Bolognese M, von Hessling A, Muller M (2018). Successful thrombectomy in endocarditis-related stroke: case report and review of the literature. Interv Neuroradiol 24: 529–532.

Byrne JG, Rezai K, Sanchez JA et al. (2011). Surgical management of endocarditis: the society of thoracic surgeons clinical practice guideline. Ann Thorac Surg 91: 2012–2019.

Cahill TJ, Prendergast BD (2016). Infective endocarditis. The Lancet 387: 882–893.

Cahill TJ, Baddour LM, Habib G et al. (2017). Challenges in infective endocarditis. J Am Coll Cardiol 69: 325–344.

Chakraborty T, Scharf E, Rabinstein AA et al. (2017). Utility of brain magnetic resonance imaging in the surgical Management of Infective Endocarditis. J Stroke Cerebrovasc Dis 26: 2527–2535.

Chakraborty T, Scharf E, DeSimone D et al. (2019). Variable significance of brain MRI findings in infective endocarditis and its effect on surgical decisions. Mayo Clin Proc 94: 1024–1032.

Chapot R, Houdart E, Saint-Maurice JP et al. (2002). Endovascular treatment of cerebral mycotic aneurysms. Radiology 222: 389–396.

Cone LA, Hirschberg J, Lopez C et al. (2008). Infective endocarditis associated with spondylodiscitis and frequent secondary epidural abscess. Surg Neurol 69: 121–125.

Cooper HA, Thompson EC, Laureno R et al. (2009). Subclinical brain embolization in left-sided infective endocarditis: results from the evaluation by MRI of the brains of patients with left-sided intracardiac solid masses (EMBOLISM) pilot study. Circulation 120: 585–591.

Daneshmand A, Rangel-Castilla L, Rydberg C et al. (2019). Ultra-rapid developing infectious aneurysms. Neurocrit Care 30: 487–489.

Duval X, Iung B, Klein I et al. (2010). Effect of early cerebral magnetic resonance imaging on clinical decisions in infective endocarditis: a prospective study. Ann Intern Med 152: W175.

Fedeli U, Schievano E, Buonfrate D et al. (2011). Increasing incidence and mortality of infective endocarditis: a population-based study through a record-linkage system. BMC Infect Dis 11: 48.

Garcia-Cabrera E, Fernandez-Hidalgo N, Almirante B et al. (2013). Neurological complications of infective endocarditis: risk factors, outcome, and impact of cardiac surgery: a multicenter observational study. Circulation 127: 2272–2284.

Geisenberger D, Huppertz LM, Buchsel M et al. (2015). Non-traumatic subdural hematoma secondary to septic brain embolism: a rare cause of unexpected death in a drug addict suffering from undiagnosed bacterial endocarditis. Forensic Sci Int 257: e1–e5.

Habib G, Lancellotti P, Antunes MJ et al. (2015). 2015 ESC guidelines for the management of infective endocarditis: the task force for the Management of Infective Endocarditis of the European Society of Cardiology (ESC). Endorsed by: European Association for Cardio-Thoracic Surgery (EACTS), the European Association of Nuclear Medicine (EANM). Eur Heart J 36: 3075–3128.

Heiro M, Nikoskelainen J, Engblom E et al. (2000). Neurologic manifestations of infective endocarditis: a 17-year experience in a teaching Hospital in Finland. Arch Intern Med 160: 2781–2787.

Heiro M, Helenius H, Hurme S et al. (2007). Short-term and one-year outcome of infective endocarditis in adult patients treated in a Finnish teaching hospital during 1980–2004. BMC Infect Dis 7: 78.

Hess A, Klein I, Iung B et al. (2013). Brain MRI findings in neurologically asymptomatic patients with infective endocarditis. Am J Neuroradiol 34: 1579–1584.

Iung B, Tubiana S, Klein I et al. (2013). Determinants of cerebral lesions in endocarditis on systematic cerebral magnetic resonance imaging: a prospective study. Stroke 44: 3056–3062.

Junna M, Lin C-CD, Espinosa RE et al. (2007). Successful intravenous thrombolysis in ischemic stroke caused by infective endocarditis. Neurocrit Care 6: 117–120.

Kang DH, Kim YJ, Kim SH et al. (2012). Early surgery versus conventional treatment for infective endocarditis. N Engl J Med 366: 2466–2473.

Kannoth S, Thomas SV (2009). Intracranial microbial aneurysm (infectious aneurysm): current options for diagnosis and management. Neurocrit Care 11: 120.

Kim YK, Choi CG, Jung J et al. (2018). Effect of cerebral embolus size on the timing of cardiac surgery for infective endocarditis in patients with neurological complications. Eur J Clin Microbiol Infect Dis 37: 545–553.

Klein I, Iung B, Labreuche J et al. (2009). Cerebral microbleeds are frequent in infective endocarditis: a case-control study. Stroke 40: 3461–3465.

Kume Y, Fujita T, Fukushima S et al. (2018). Intracranial mycotic aneurysm is associated with cerebral bleeding post-valve surgery for infective endocarditis. Interact Cardiovasc Thorac Surg 27: 635–641.

Le Moal G, Roblot F, Paccalin M et al. (2002). Clinical and laboratory characteristics of infective endocarditis when associated with spondylodiscitis. Eur J Clin Microbiol Infect Dis 21: 671–675.

Li JS, Sexton DJ, Mick N et al. (2000). Proposed modifications to the Duke criteria for the diagnosis of infective endocarditis. Clin Infect Dis 30: 633–638.

Malhotra A, Schindler J, Mac Grory B et al. (2017). Cerebral microhemorrhages and meningeal siderosis in infective endocarditis. Cerebrovasc Dis 43: 59–67.

Mangoni ED, Adinolfi LE, Tripodi M-F et al. (2003). Risk factors for "major" embolic events in hospitalized patients with infective endocarditis. Am Heart J 146: 311–316.

Marquardt RJ, Cho SM, Thatikunta P et al. (2019). Acute ischemic stroke therapy in infective endocarditis: case series and systematic review. J Stroke Cerebrovasc Dis 28: 2207–2212.

Molinari GF, Smith L, Goldstein MN et al. (1973). Pathogenesis of cerebral mycotic aneurysms. Neurology 23: 325.

Morofuji Y, Morikawa M, Yohei T et al. (2010). Significance of the T2*-weighted gradient echo brain imaging in patients with infective endocarditis. Clin Neurol Neurosurg 112: 436–440.

Murdoch DR, Corey GR, Hoen B et al. (2009). Clinical presentation, etiology, and outcome of infective endocarditis in the 21st century: the international collaboration on endocarditis-prospective cohort study. Arch Intern Med 169: 463–473.

Nishimura RA, Otto CM, Bonow RO et al. (2017). 2017 AHA/ACC focused update of the 2014 AHA/ACC guideline for the Management of Patients with Valvular Heart Disease: a report of the American College of Cardiology/ American Heart Association task force on clinical practice guidelines. Circulation 135: e1159–e1195.

Nogueira RG, Jadhav AP, Haussen DC et al. (2018). Thrombectomy 6 to 24 hours after stroke with a mismatch between deficit and infarct. N Engl J Med 378: 11–21.

Oh JS, Shim JJ, Lee KS et al. (2015). Cervical epidural abscess: rare complication of bacterial endocarditis with streptococcus viridans: a case report. Korean J Spine 12: 22–25.

Okazaki S, Yoshioka D, Sakaguchi M et al. (2013). Acute ischemic brain lesions in infective endocarditis: incidence, related factors, and postoperative outcome. Cerebrovasc Dis 35: 155–162.

Okita Y, Minakata K, Yasuno S et al. (2016). Optimal timing of surgery for active infective endocarditis with cerebral complications: a Japanese multicentre study. Eur J Cardiothorac Surg 50: 374–382.

Ong E, Mechtouff L, Bernard E et al. (2013). Thrombolysis for stroke caused by infective endocarditis: an illustrative case and review of the literature. J Neurol 260: 1339–1342.

Pericart L, Fauchier L, Bourguignon T et al. (2016). Long-term outcome and valve surgery for infective endocarditis in the systematic analysis of a community study. Ann Thorac Surg 102: 496–504.

Peters PJ, Harrison T, Lennox JL (2006). A dangerous dilemma: management of infectious intracranial aneurysms complicating endocarditis. Lancet Infect Dis 6: 742–748.

Petr O, Brinjikji W, Cloft H et al. (2016). Current trends and results of endovascular treatment of unruptured intracranial aneurysms at a single institution in the flow-diverter era. AJNR Am J Neuroradiol 37: 1106–1113.

Phuong LK, Link M, Wijdicks E (2002). Management of intracranial infectious aneurysms: a series of 16 cases. Neurosurgery 51: 1145–1151. discussion 1151-2.

Powers WJ, Rabinstein AA, Ackerson T et al. (2019). Guidelines for the early Management of Patients with Acute Ischemic Stroke: 2019 update to the 2018 guidelines for the early Management of Acute Ischemic Stroke: a guideline for healthcare professionals from the American Heart Association/American Stroke Association. Stroke 50: e344–e418.

Scharf EL, Chakraborty T, Rabinstein A et al. (2017). Endovascular management of cerebral septic embolism: three recent cases and review of the literature. J Neurointerv Surg 9: 463–465.

Sloane KL, Raymond SB, Rabinov JD et al. (2020). Mechanical Thrombectomy in stroke from infective endocarditis: case report and review. J Stroke Cerebrovasc Dis 29: 104501.

Snygg-Martin U, Gustafsson L, Rosengren L et al. (2008). Cerebrovascular complications in patients with left-sided infective endocarditis are common: a prospective study using magnetic resonance imaging and neurochemical brain damage markers. Clin Infect Dis 47: 23–30.

Sordelli C, Fele N, Mocerino R et al. (2019). Infective endocarditis: echocardiographic imaging and new imaging modalities. J Cardiovasc Echogr 29: 149–155.

Sotero FD, Rosário M, Fonseca AC et al. (2019). Neurological complications of infective endocarditis. Curr Neurol Neurosci Rep 19: 23.

Sveinsson O, Herrman L, Holmin S (2016). Intra-arterial mechanical Thrombectomy: an effective treatment for ischemic stroke caused by endocarditis. Case Rep Neurol 8: 229–233.

Thuny F, Avierinos JF, Tribouilloy C et al. (2007). Impact of cerebrovascular complications on mortality and neurologic outcome during infective endocarditis: a prospective multicentre study. Eur Heart J 28: 1155–1161.

Van de Beek D, Rabinstein AA, Peters SG et al. (2008). Staphylococcus endocarditis associated with infectious vasculitis and recurrent cerebral hemorrhages. Neurocrit Care 8: 48–52.

Vincent LL, Otto CM (2018). Infective endocarditis: update on epidemiology, outcomes, and management. Curr Cardiol Rep 20: 86.

Wang A, Gaca JG, Chu VH (2018). Management considerations in infective endocarditis: a review. JAMA 320: 72–83.

Yanagawa B, Pettersson GB, Habib G et al. (2016). Surgical Management of Infective Endocarditis Complicated by embolic stroke: practical recommendations for clinicians. Circulation 134: 1280–1292.

Handbook of Clinical Neurology, Vol. 177 (3rd series)
Heart and Neurologic Disease
J. Biller, Editor
https://doi.org/10.1016/B978-0-12-819814-8.00013-5

Chapter 12

Neurologic complications of nonbacterial thrombotic endocarditis

RIMA M. DAFER*

Department of Neurological Sciences, Rush University Medical Center, Chicago, IL, United States

Abstract

Endocarditis is an inflammatory or infective condition affecting the cardiac valves or endocardium, often associated with serious neurological sequelae. Nonbacterial thrombotic endocarditis (NBTE)—referred to as degenerative, Libman–Sachs, marantic, verrucous, or terminal endocarditis—is a serious but rare cause of valvular heart disease characterized by deposition of sterile vegetations of fibrin and platelet aggregates on the cardiac valves, eventually resulting in life-threatening embolization of these thrombi to the brain, limbs, or visceral organs. NBTE may complicate a heterogeneous group of chronic conditions, predominantly connective tissue and autoimmune disorders, malignancies, and diseases associated with hypercoagulability states. NBTE usually affects the native rather than prosthetic valves, and unlike infective endocarditis (IE), sparing the involved valve function without its destruction. Compared to those seen in IE, vegetations in NBTE are small and friable, thus may easily be dislodged leading to systemic thromboembolism with devastating morbidities and mortality. There are no diagnostic criteria for NBTE, and antemortem diagnosis is challenging. The condition should be suspected in patients with thromboembolic events and vegetations on the cardiac valves on echocardiographic or cardiac imaging studies, in the absence of underlying infection, especially in disorders predisposing to coagulopathy. Early recognition and prompt treatment of the primary underlying disorder is essential. Anticoagulation with heparin or heparinoid products is recommended to prevent recurrent embolism. Surgical intervention is not indicated except in selected patients with life-threatening recurrent embolism.

INTRODUCTION

Nonbacterial thrombotic endocarditis (NBTE) is a rare but serious noninfectious inflammatory condition of the native cardiac valves, where sterile noninfectious vegetations composed of a mixture of fibrin and platelets aggregate on the heart valves, leading to thromboembolism. NBTE was originally described by Ziegler as thromboendocarditis in 1888 (Ziegler, 1888). It was not until 1923 when the noninfective endocarditis, was described by Libman in a detailed review of patients with rheumatic fever, which showed the lack of infectious etiology in 50% of subjects at postmortem examination (Libman, 1923). He described noninfectious nonbacterial endocarditis as indeterminate or verrucous in type (Libman, 1923). The term NBTE was first introduced by Gross and Friedman, after the former described pathological changes in the heart of 11 patients with lupus-associated endocarditis (Gross, 1932; Gross and Friedberg, 1936).

Several heterogeneous group of conditions have been associated with NBTE, mainly connective tissue disorders, autoimmune diseases, malignancies, and maladies associated with prothrombotic states. Vegetations in NBTE commonly involve a single native, usually healthy, mitral, and/or aortic valve. Rarely, a superficially degenerated

*Correspondence to: Rima M. Dafer, MD, MPH, Professor of Neurology, Department of Neurological Sciences, Rush University Medical Center, Chicago IL 60091, United States. Tel: +1-(312)-942-4500, Fax: +1-(312)-563-2206, E-mail: rima_dafer@rush.edu

heart valve or a prosthetic heart valve may be involved. Compared with the lesions in IE, vegetations in NBTE are often small, more friable, and easily prone to dislodge, leading systemic embolization to the brain, visceral organs including the lung, spleen, and kidney, or extremities.

EPIDEMIOLOGY

NBTE is present in 0.3%–9.3% of autopsies (Kuramoto et al., 1984; Glass, 1993; Smeglin et al., 2008), with a reported incidence of around 1% in adults (Steiner, 1993). It most commonly affects individuals in the fourth and eight decades of life, although it has been reported in every age group, less commonly in children, with some predominance in women (Lopez et al., 1987b). The majority of cases of NBTE have been reported in patients with systemic lupus erythematosus (SLE), i.e., Libman Sacks endocarditis, or as marantic endocarditis as a paraneoplastic disorder in patients with neoplasms, usually advanced or late stage cancer. NBTE has also been associated with a wide spectrum of autoimmune conditions. These include rheumatoid arthritis (Choi et al., 2016; Morris et al., 2019), antiphospholipid syndrome (Eiken et al., 2001; Sirinvaravong et al., 2018), giant cell arteritis (Eftychiou et al., 2005), scleroderma (Kinney et al., 1979; De Langhe et al., 2016), systemic sclerosis (De Langhe et al., 2016), Sneddon syndrome (Berciano and Teran-Villagra, 2018), Behçet's disease (Demirelli et al., 2015; Nassenstein et al., 2015; Morris et al., 2019), adult-onset Still disease (Zenagui and De Coninck, 1995; Mimura et al., 2014; Quelven et al., 2018), and hypereosinophilic syndrome (i.e., Loeffler's endocarditis) (Mannelli et al., 2012; Lamba et al., 2016). Several other disorders leading to prothrombotic states have been associated with NBTE; these include pregnancy (George et al., 1984), disseminated intravascular coagulation (DIC) (Biller et al., 1982; Singh et al., 1998; Mimura et al., 2014), and infections such as with the human immunodeficiency virus (Currie et al., 1995), tuberculosis (Diaz Curiel et al., 1984), or sepsis. Thrombophilia is a common consequence of malignancy, occurring in 15% of patients with cancer (Deitcher, 2003). Cancer-associated thrombosis was first described by Armand Trousseau in 1865 (Trousseau, 1865a), himself dying from thrombotic complications of gastric cancer 2 years after publishing his observation (Khorana, 2003). Formerly known as marantic endocarditis (stemming from the Hellenistic Greek word marantikos for "wasting away"), malignancy-associated NBTE is a serious condition with significant morbidity and mortality. It usually accompanies disseminated or advanced end-stage cancer, with the highest occurrence in mucin-producing adenocarcinomas (Min et al., 1980; Edoute et al., 1997). Edoute et al. identified valvular vegetations consistent with NBTE in 19% of 200 cancer patients, with a prevalence of 4% of all end-stage cancer (Edoute et al., 1997). Postmortem studies showed a significantly high presence of malignancy in 60% of cases of NBTE, with an incidence of 1% in adults (Steiner, 1993). Tumors most frequently associated with NBTE include carcinomas of the pancreas, lung, ovaries, biliary system, colon, and stomach (Steiner, 1993; Cestari et al., 2004; el-Shami et al., 2007; Gundersen and Moynihan, 2016; Shoji et al., 2019). NBTE has also been reported in the setting of bladder cancer (del Villar Negro et al., 1983; Fujimoto et al., 2018), advanced leukemias (Ahmed et al., 2018; Jeong et al., 2018), and lymphomas (Edoute et al., 1997; Kraft et al., 2002; Ali et al., 2012), especially when DIC exists. Additionally, NBTE has been observed in 8% of patients undergoing allogenic bone marrow transplantation (Patchell et al., 1985).

PATHOGENESIS

The pathogenesis of NBTE is complex and not well understood. Acute cell injury in the setting of underlying predisposing autoimmune, paraneoplastic, or prothrombotic condition may lead to endothelial cell damage and deposition of sterile platelets and strands of fibrin on the heart valve leaflets. In malignancies, migrating inflammatory mononuclear cells interact with cancer cells, releasing cytokines, leading to further endothelial injury, and promoting platelets deposition (Eiken et al., 2001; el-Shami et al., 2007). The presence of chronic immune complex or antiphospholipid antibodies may accelerate endothelial injury and thus may promote thrombus formation (Moyssakis et al., 2007). These thrombotic vegetations are small and friable and tend to cluster in areas of high turbulence around the cardiac valves (el-Shami et al., 2007). Hence, these small vegetations with minimal cellular organization have a higher tendency to detach and to cause extensive systemic embolization, in particular to the brain. The mitral and aortic valves are the most commonly involved (Steiner, 1993), followed by the tricuspid and pulmonary valves. Vegetations are usually present on the atrial surface of the atrioventricular valves and the ventricular surface of the outflow semilunar valves. In over 80% of cases, the normal native valve is affected, usually remaining undamaged with preserved function. Less commonly, vegetations may cluster on a mildly degenerate or prosthetic valve (Steiner, 1993). The incidence of systemic embolization varies depending on the organ affected and usually ranges between 14% and 90%, with thromboembolism involving either the venous or the arterial trees. Ischemic stroke due to cerebral embolism is the most frequent and most devastating complication, occurring in over 40% of cases (Steiner, 1993; Amaral et al., 1997). Cerebral embolism may be the initial manifestations of the underlying primary disorder, thus making diagnosis difficult and challenging. DIC usually coexists in nearly half of the cases of malignancy-associated NBTE, playing a major role in thrombus deposition in the small and medial arteries and veins of major organs, with the splenic, pulmonary, and renal circulations being most commonly affected (Kim et al., 1977; Kuramoto et al., 1984). Limb embolization is also common (Joshi et al., 2009).

CLINICAL SYMPTOMATOLOGY

NBTE is typically silent with majority of cases remaining asymptomatic until embolization occurs, or are diagnosed postmortem (Smeglin et al., 2008). High index of suspicion for NBTE should be maintained in patients with systemic or cerebral embolism, especially in the setting of underlying predisposing conditions such as malignancy, autoimmune or connective tissue disorders, or other prothrombotic states (el-Shami et al., 2007), especially when infectious etiologies are ruled out. Petechial or splinter hemorrhages are absent. Fever, weight loss, lethargy, and night sweats are usually absent. When present in the absence of an underlying infections process, these constitutional symptoms should raise strong suspicion of an occult malignancy. Patients usually present with recurrent arterial and/or venous systemic embolism, with clinical manifestations depending on the site of embolization and on the organs involved. Arthritis and/or multiorgan involvement may be indicative of an underlying connective tissue disorder. A malar rash or livedo reticularis may raise the suspicion of SLE or Sneddon syndrome. Unexpected or migratory thrombophlebitis could also be a forewarning of an occult visceral malignancy (Trousseau, 1865b). Limb ischemia may present with skin and digits discoloration, pain, coldness, paresthesia, and pulselessness. Acute abdominal pain, nausea, and bloody stool may be the manifestations of mesenteric or splenic ischemia. Flank pain and hematuria may occur with renal ischemia. Shortness of breath and chest discomfort should raise suspicion of pulmonary embolism. Chest pain, dyspnea, orthopnea, and acute myocardial infarction may occur when the coronary arteries are affected. Unlike in IE, cardiac murmurs are rare and are detected in less than 30% of cases. Valvular lesions may result in rhythm anomalies and may lead to heart failure.

The most common neurological complication is acute onset neurological deficit due to cerebral ischemia. The incidence of cerebral embolism is considerably higher in NBTE (33%) when compared with IE (19%) (Joshi et al., 2009). Cerebral embolization is often a late complication of widespread cancer, although it may rarely be the initial presentation of an occult visceral malignancy (Rogers et al., 1987). Neurological dysfunction may affect the retina, spinal cord, or cerebrum. Patient may present with headaches, seizures, mental status changes, or focal neurological deficit. Retinal ischemia with acute monocular visual loss may occur. Unlike IE, primary intraparenchymal hemorrhage is an uncommon complication of NBTE (George et al., 1984; Fujishima et al., 1994; Wigger et al., 2016), with a higher risk in patients already on anticoagulation. When present, cerebral hemorrhage usually occurs due to hemorrhagic transformation within the infarcted tissue. Bleeding may also occur within a metastatic disease of the brain, or due to DIC. The presence of cerebral microhemorrhages or cerebral microbleeds on susceptibility-weighted MR brain imaging is uncommon.

DIAGNOSIS

Majority of patients with NBTE are asymptomatic, and the condition remains largely underdiagnosed antemortem until cerebral or systemic embolization occurs. When a primary predisposing etiology is lacking, diagnosis remains difficult. NBTE has no official diagnostic criteria as compared to Duke's criteria of IE. The condition should be suspected in patients with wide range of thromboembolism in particular acute cerebral embolism and cardiac vegetations when blood cultures are negative or in subjects with poor response to antibiotics. The presence of an underlying cancer or predisposing autoimmune or connective tissue disorders should raise high suspicion for NBTE. Clinical manifestations are nonspecific, and constitutional symptoms are usually absent. Microscopic testing and blood cultures are often negative. Connective tissue work-up, including testing for antiphospholipid antibodies, is indicated in patients with systemic manifestations, rash, arthritis, miscarriages, or multiorgan involvement. Echocardiogram is crucial to elucidate small vegetations or mobile echogenic mass on the cardiac valves. Although less sensitive for the detection of vegetations, two-dimensional transthoracic echocardiogram (TTE) may show valve regurgitation or destruction (Lopez et al., 1987a; Salem et al., 2004; Roldan et al., 2008). Three-dimensional transesophageal echocardiogram (TEE) should be considered when TTE is nondiagnostic, as it is superior to TTE in detecting small thrombi and in characterizing vegetations, with a sensitivity of 90% compared to 70% in TTE (Hojnik et al., 1996; Eiken et al., 2001) (Fig. 12.1). CT head and magnetic resonance imaging (MRI) of brain may show multiple cerebral emboli crossing a single arterial distribution. MRI is preferred to head CT as it may detect smaller disseminated emboli or may detect subtle micro

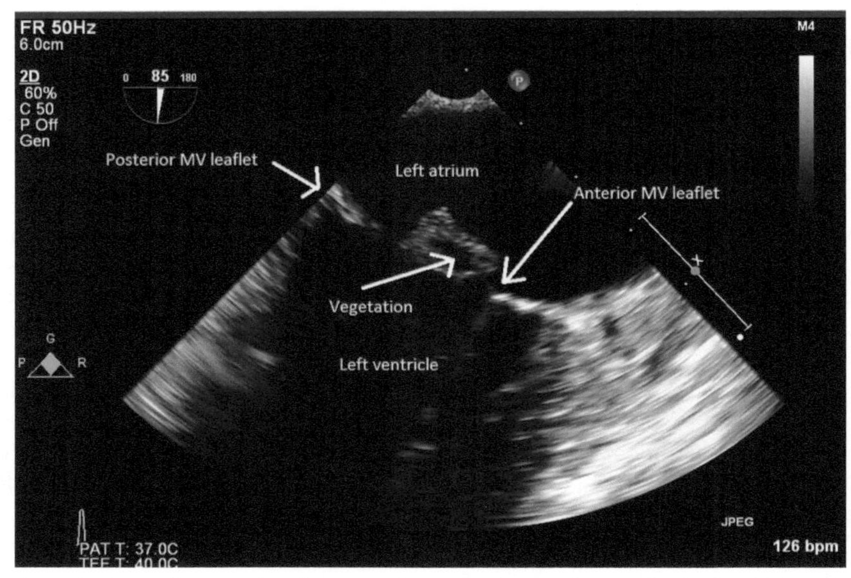

Fig. 12.1. Transesophageal echocardiogram showing a mobile vegetation on the anterior leaflet of the mitral valve. *MV*, mitral valve.

hemorrhages or hemorrhagic transformation within a lesion or within the infarct-bed (Fig. 12.2). Mycotic aneurysms are uncommon in NBTE as compared to IE. CT chest may be beneficial in patients with dyspnea as concomitant pulmonary embolisms may occur. Recurrent or multiple embolisms—especially when multiple organs are involved—should raise the suspicion of an underlying malignancy (Biller et al., 1982). When diagnosis remains uncertain, or when initial echocardiographic imaging is nonconfirmatory, multimodality diagnostic tools such as cardiac CT combined with molecular imaging with (18) F-Fluorodesoxyglucose positron emission tomography CT ((18)F-FDG-PET-CT) may help distinguishing between IE and NBTE, allowing for more accurate diagnosis (Dahl et al., 2015). In patients with no known risk factors, paraneoplastic disorders should be considered and prompt search for occult malignancy should be performed (Eiken et al., 2001).

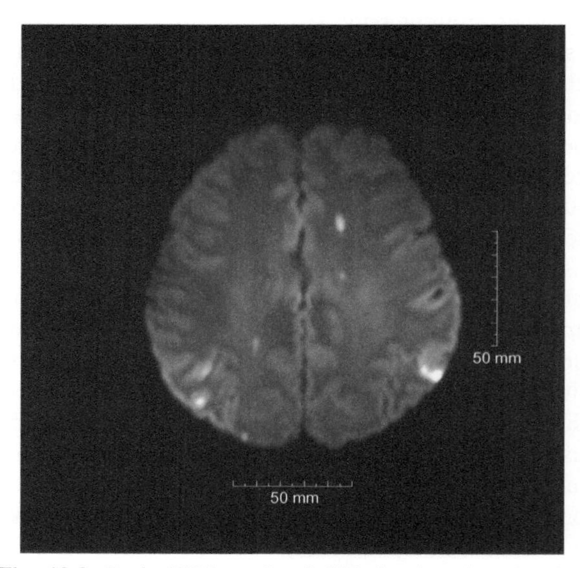

Fig. 12.2. Brain MRI-restricted diffusion imaging showing several scattered cerebral infarctions in various arterial distributions.

DIFFERENTIAL DIAGNOSIS

NBTE should always be differentiated from IE. NBTE should be suspected when fever, constitutional symptoms, or other signs of infection are absent, or when microscopic testing is negative and blood cultures are sterile. Similarly, there should be high suspicion for NBTE in patients who show no improvement, with persistent cardiac vegetations, despite an adequate course of antibiotics. When differentiation between NBTE and IE is difficult, advanced imaging studies should be obtained. Lambl's valve excrescences should also be excluded; these are small, thin filiform fronds which are usually seen near the valvular coaptation sites of the heart, often of the left heart valves, and are commonly present in up to 40% of normal subjects (Roldan et al., 2008). In rare cases, vegetations in NBTE may present as atrial tumor, posing diagnostic challenges, and the thrombus is only identified after cardiac surgery and resection (Abouarab et al., 2018).

MANAGEMENT

Diagnosis, management, and treatment of NBTE is challenging. Clinicians should focus on exclusion of an underlying infectious etiology, identification, and control of the concurrent primary underlying pathology and early initiation of anticoagulation therapy to prevent systemic embolization (Nishimura et al., 2014).

Autoimmune conditions should be treated with steroids and immunosuppressant agents (Sonsoz et al., 2019). In the case of neoplasms, debulking the primary tumor may be necessary and may help regressing the vegetations or may lead to suppression of further thrombi formation (Albright et al., 2016). Unfortunately, no specific therapy has consistently shown to reverse valvular vegetations.

The hyperacute ischemic stroke care in NBTE should not differ from that of any acute ischemic stroke situation. Despite the potentially higher risk of hemorrhagic complications, thrombolytic therapy with intravenous recombinant tissue plasminogen is standard of care and should be considered in patients presenting with acute focal ischemic neurological deficits (Yagi et al., 2014). Patients with cortical signs and symptoms and debilitating neurological deficits should undergo arterial imaging to identify intracranial large vessel arterial occlusive disease and, if appropriate, should undergo endovascular thrombectomy with aspiration and/or stent retriever techniques. The thrombectomy device should be carefully chosen as the usually floating thrombus is predominantly a white cardiac thrombus that has obliterated or occluded the intracranial arterial vasculature (Yasuda et al., 2019).

Once infectious etiologies are excluded, anticoagulation therapy should be initiated as soon as possible to prevent further thrombo-embolization. Heparin and heparinoid products are preferred over vitamin K antagonists and have been shown to be more effective in reducing the risk of recurrent embolism, especially in patients with cancer and venous thromboembolism, with no known increased risk of bleeding (Lee et al., 2003; Whitlock et al., 2012). Ideally, treatment with unfractionated heparin infusion to prevent recurrent thromboembolism should be initiated; then, patients should be transitioned to low molecular weight heparin for long-term anticoagulation treatment. Data on the use of direct oral anticoagulants (DOACs) in NBTE are lacking. Several prospective clinical trials with various DOACs in cancer-associated venous thromboembolism are ongoing. When antiphospholipid syndrome is suspected, rivaroxaban should be avoided, as its use has been associated with increased risk of recurrent thrombosis (Dufrost et al., 2016). Serial cardiac imaging preferably with transesophageal echocardiogram after completion of treatment with anticoagulation, or with cardiac CT, is necessary to insure resolution of the vegetations. Generally, surgical intervention is not recommended in NBTE. In selected patients with certain cardiac dysfunction or recurrent symptoms, cardiac surgery may be considered. These conditions include severe valvular dysfunction leading to congestive heart failure, recurrent embolism despite optimal anticoagulation, or in large mobile vegetations with recurrent embolism despite anticoagulation (Rabinstein et al., 2005), or when the appearance of the thrombotic vegetations mimics atrial mass on echocardiographic studies (Abouarab et al., 2018).

PROGNOSIS

NBTE carries a poor outcome with significant morbidity and mortality, especially when associated with cancer. Despite treating the primary predisposing underlying disorder, and despite appropriate anticoagulation, the risk of recurrent embolization remains high and is estimated at 50%–75% (Sack et al., 1977).

CONCLUSION

In summary, NBTE is a rare but serious cause of cerebral embolism with devastating outcome. The majority of patients have autoimmune conditions predominantly SLE, or advanced malignancies, in particular mucinous-secreting adenocarcinoma with associated prothrombotic states. Diagnosis should be made when the triad of cerebral or systemic embolization, small valvular vegetations, and systemic diseases known to be associated with NBTE is present. Treatment is directed toward controlling the primary underlying pathology, together with long-term anticoagulation to prevent recurrent thromboembolism. Prognosis of NBTE is often poor as the condition is usually diagnosed after systemic embolization has occurred, and in the setting of late stage and advanced malignancy.

References

Abouarab AA, Elmously A, Leonard JR et al. (2018). Nonbacterial thrombotic endocarditis presenting with leg pain and a left atrial mass lesion. Cardiology 139: 208–211.

Ahmed S, Jani P, Yamani MH et al. (2018). Marantic endocarditis associated with T-cell large granular lymphocytic leukemia: first report of its occurrence with a lymphoproliferative malignancy in adults. J Oncol Pract 14: 625–627.

Albright BB, Black JD, Vilardo N et al. (2016). Correction of coagulopathy associated with non-bacterial thrombotic endocarditis (NBTE) by surgical debulking in a case of ovarian clear cell carcinoma. Gynecol Oncol Rep 17: 13–15.

Ali N, Konstantinov I, Heath JA et al. (2012). Nonbacterial thrombotic endocarditis in a child with non-Hodgkin's lymphoma. Pediatr Cardiol 33: 843–845.

Amaral G, Santos Junior EH, De Azevedo LC et al. (1997). Nonbacterial thrombotic endocarditis. Arq Bras Cardiol 68: 373–375.

Berciano J, Teran-Villagra N (2018). Sneddon syndrome and non-bacterial thrombotic endocarditis: a clinicopathological study. J Neurol 265: 2143–2145.

Biller J, Challa VR, Toole JF et al. (1982). Nonbacterial thrombotic endocarditis. A neurologic perspective of clinicopathologic correlations of 99 patients. Arch Neurol 39: 95–98.

Cestari DM, Weine DM, Panageas KS et al. (2004). Stroke in patients with cancer: incidence and etiology. Neurology 62: 2025–2030.

Choi JH, Park JE, Kim JY et al. (2016). Non-bacterial thrombotic endocarditis in a patient with rheumatoid arthritis. Korean Circ J 46: 425–428.

Currie PF, Sutherland GR, Jacob AJ et al. (1995). A review of endocarditis in acquired immunodeficiency syndrome and human immunodeficiency virus infection. Eur Heart J 16: 15–18.

Dahl A, Schaadt BK, Santoni-Rugiu E et al. (2015). Molecular imaging in Libman-sacks endocarditis. Infect Dis (Lond) 47: 263–266.

De Langhe E, Seghers A, Demaerel P et al. (2016). Non-infective endocarditis with systemic embolization and recurrent stroke in systemic sclerosis. Rheumatology (Oxford) 55: 589–591.

Deitcher SR (2003). Cancer and thrombosis: mechanisms and treatment. J Thromb Thrombolysis 16: 21–31.

Del Villar Negro A, Merino Angulo J, Rivera Pomar JM (1983). Non-bacterial thrombotic endocarditis. Clinicopathological study. Rev Clin Esp 169: 343–346.

Demirelli S, Degirmenci H, Inci S et al. (2015). Cardiac manifestations in Behcet's disease. Intractable Rare Dis Res 4: 70–75.

Diaz Curiel M, Mata Lopez P, Fernandez Guerrero M et al. (1984). Non-bacterial thrombotic endocarditis associated with tuberculosis. Rev Clin Esp 174: 59–60.

Dufrost V, Risse J, Zuily S et al. (2016). Direct oral anticoagulants use in antiphospholipid syndrome: are these drugs an effective and safe alternative to warfarin? A systematic review of the literature. Curr Rheumatol Rep 18: 74.

Edoute Y, Haim N, Rinkevich D et al. (1997). Cardiac valvular vegetations in cancer patients: a prospective echocardiographic study of 200 patients. Am J Med 102: 252–258.

Eftychiou C, Fanourgiakis P, Vryonis E et al. (2005). Factors associated with non-bacterial thrombotic endocarditis: case report and literature review. J Heart Valve Dis 14: 859–862.

Eiken PW, Edwards WD, Tazelaar HD et al. (2001). Surgical pathology of nonbacterial thrombotic endocarditis in 30 patients, 1985–2000. Mayo Clin Proc 76: 1204–1212.

El-Shami K, Griffiths E, Streiff M (2007). Nonbacterial thrombotic endocarditis in cancer patients: pathogenesis, diagnosis, and treatment. Oncologist 12: 518–523.

Fujimoto D, Mochizuki Y, Nakagiri K et al. (2018). Unusual rapid progression of non-bacterial thrombotic endocarditis in a patient with bladder cancer despite undergoing intensification treatment with rivaroxaban for acute venous thromboembolism. Eur Heart J 39: 3907.

Fujishima S, Okada Y, Irie K et al. (1994). Multiple brain infarction and hemorrhage by nonbacterial thrombotic endocarditis in occult lung cancer—a case report. Angiology 45: 161–166.

George J, Lamb JT, Harriman DG (1984). Cerebral embolism due to non-bacterial thrombotic endocarditis following pregnancy. J Neurol Neurosurg Psychiatry 47: 79–80.

Glass JP (1993). The diagnosis and treatment of stroke in a patient with cancer: nonbacterial thrombotic endocarditis (NBTE): a case report and review. Clin Neurol Neurosurg 95: 315–318.

Gross L (1932). The heart in atypical verrucous endocarditis (Libman-Sacks). In: Contributions to the medical sciences in Honor of Dr. Emanuel Libman by his pupils, friends and colleagues, International Press, New York.

Gross L, Friedberg CK (1936). Nonbacterial thrombotic endocardities classificatin and general description. Arch Intern Med (Chic) 58: 620–640.

Gundersen H, Moynihan B (2016). An uncommon cause of stroke: non-bacterial thrombotic endocarditis. J Stroke Cerebrovasc Dis 25: e163–e164.

Hojnik M, George J, Ziporen L et al. (1996). Heart valve involvement (Libman-sacks endocarditis) in the antiphospholipid syndrome. Circulation 93: 1579–1587.

Jeong EG, Jung MH, Youn HJ (2018). Histologically confirmed nonbacterial thrombotic endocarditis in a febrile leukemic patient. Korean J Intern Med 33: 632–633.

Joshi SB, Richards MJ, Holt DQ et al. (2009). Marantic endocarditis presenting as recurrent arterial embolisation. Int J Cardiol 132: e14–e16.

Khorana AA (2003). Malignancy, thrombosis and Trousseau: the case for an eponym. J Thromb Haemost 1: 2463–2465.

Kim HS, Suzuki M, Lie JT et al. (1977). Nonbacterial thrombotic endocarditis (NBTE) and disseminated intravascular coagulation (DIC): autopsy study of 36 patients. Arch Pathol Lab Med 101: 65–68.

Kinney E, Reeves W, Zellis R (1979). The echocardiogram in scleroderma endocarditis of the mitral valve. Arch Intern Med 139: 1179–1180.

Kraft F, Torres Morales A, Giovannoni A et al. (2002). Nonbacterial thrombotic endocarditis as paraneoplastic manifestation of pulmonary adenocarcinoma. Arch Cardiol Mex 72: 303–305.

Kuramoto K, Matsushita S, Yamanouchi H (1984). Nonbacterial thrombotic endocarditis as a cause of cerebral and myocardial infarction. Jpn Circ J 48: 1000–1006.

Lamba H, Deo S, Altarabsheh S et al. (2016). Non-bacterial thrombotic endocarditis of aortic valve due to hypereosinophilic syndrome. J Heart Valve Dis 25: 760–763.

Lee AY, Levine MN, Baker RI et al. (2003). Low-molecular-weight heparin versus a coumarin for the prevention of recurrent venous thromboembolism in patients with cancer. N Engl J Med 349: 146–153.

Libman E (1923). Characterization of various forms of endocarditis. JAMA 80: 813–818.

Lopez JA, Fishbein MC, Siegel RJ (1987a). Echo-cardiographic features of nonbacterial thrombotic endocarditis. Am J Cardiol 59: 478–480.

Lopez JA, Ross RS, Fishbein MC et al. (1987b). Nonbacterial thrombotic endocarditis: a review. Am Heart J 113: 773–784.

Mannelli L, Cherian V, Nayar A et al. (2012). Loeffler's endocarditis in hypereosinophilic syndrome. Curr Probl Diagn Radiol 41: 146–148.

Mimura T, Shimodaira M, Kibata M et al. (2014). Adult-onset Still's disease with disseminated intravascular coagulation and hemophagocytic syndrome: a case report. BMC Res Notes 7: 940.

Min KW, Gyorkey F, Sato C (1980). Mucin-producing adenocarcinomas and nonbacterial thrombotic endocarditis: pathogenetic role of tumor mucin. Cancer 45: 2374–2382.

Morris A, Pal P, O'riordan E et al. (2019). Small vessel multi-organ vasculitis and marantic endocarditis complicating rheumatoid arthritis. Eur J Rheumatol 6: 223–225.

Moyssakis I, Tektonidou MG, Vasilliou VA et al. (2007). Libman-sacks endocarditis in systemic lupus erythematosus: prevalence, associations, and evolution. Am J Med 120: 636–642.

Nassenstein K, Deluigi CC, Afube T et al. (2015). Nonbacterial endocarditis presenting as a right ventricular tumor in assumed Behcet's disease. Herz 40: 225–227.

Nishimura RA, Otto CM, Bonow RO et al. (2014). 2014 AHA/ACC guideline for the management of patients with valvular heart disease: executive summary: a report of the American College of Cardiology/American Heart Association task force on practice guidelines. Circulation 129: 2440–2492.

Patchell RA, White 3rd CL, Clark AW et al. (1985). Nonbacterial thrombotic endocarditis in bone marrow transplant patients. Cancer 55: 631–635.

Quelven Q, Cador B, Poinot M et al. (2018). Intra-cardiac manifestation during adult-onset still's disease's, a tricuspid vegetation as a rare expression of systemic disease. Rev Med Interne 39: 816–819.

Rabinstein AA, Giovanelli C, Romano JG et al. (2005). Surgical treatment of nonbacterial thrombotic endocarditis presenting with stroke. J Neurol 252: 352–355.

Rogers LR, Cho ES, Kempin S et al. (1987). Cerebral infarction from non-bacterial thrombotic endocarditis. Clinical and pathological study including the effects of anticoagulation. Am J Med 83: 746–756.

Roldan CA, Qualls CR, Sopko KS et al. (2008). Transthoracic versus transesophageal echocardiography for detection of Libman-sacks endocarditis: a randomized controlled study. J Rheumatol 35: 224–229.

Sack GH, Levin J, Bell WR (1977). Trousseau's syndrome and other manifestations of chronic disseminated coagulopathy in patients with neoplasms: clinical, pathophysiologic, and therapeutic features. Medicine (Baltimore) 56: 1–37.

Salem DN, Stein PD, Al-Ahmad A et al. (2004). Antithrombotic therapy in valvular heart disease—native and prosthetic: the seventh ACCP conference on antithrombotic and thrombolytic therapy. Chest 126: 457S–482S.

Shoji MK, Kim JH, Bakshi S et al. (2019). Nonbacterial thrombotic endocarditis due to primary gallbladder malignancy with recurrent stroke despite anticoagulation: case report and literature review. J Gen Intern Med 34: 1934–1940.

Singh S, Dass A, Jain S et al. (1998). Fatal non-bacterial thrombotic endocarditis following viperine bite. Intern Med 37: 342–344.

Sirinvaravong N, Rodriguez Ziccardi MC, Patnaik S et al. (2018). Nonbacterial thrombotic endocarditis in a patient with primary antiphospholipid syndrome. Oxf Med Case Rep 2018: omy024.

Smeglin A, Ansari M, Skali H et al. (2008). Marantic endocarditis and disseminated intravascular coagulation with systemic emboli in presentation of pancreatic cancer. J Clin Oncol 26: 1383–1385.

Sonsoz MR, Tekin RD, Gul A et al. (2019). Treatment of Libman-sacks endocarditis by combination of warfarin and immunosuppressive therapy. Turk Kardiyol Dern Ars 47: 687–690.

Steiner I (1993). Nonbacterial thrombotic endocarditis—a study of 171 case reports. Cesk Patol 29: 58–60.

Trousseau A (1865a). Phlegmasia alba dolens. In: Clinique Médicale de l'Hôtel-Dieu de Paris, Ballière, Paris, pp. 654–712.

Trousseau A (1865b). Plegmasia alba dolens. Lectures on clinical medicine, delivered at the Hotel-Dieu, Paris. vol. 5, 281–332.

Whitlock RP, Sun JC, Fremes SE et al. (2012). Antithrombotic and thrombolytic therapy for valvular disease: antithrombotic therapy and prevention of thrombosis, 9th ed: American College of Chest Physicians evidence-based clinical practice guidelines. Chest 141: e576S–e600S.

Wigger O, Windecker S, Bloechlinger S (2016). Nonbacterial thrombotic endocarditis presenting as intracerebral hemorrhage. Wien Klin Wochenschr 128: 922–924.

Yagi T, Takahashi K, Tanikawa M et al. (2014). Fatal intracranial hemorrhage after intravenous thrombolytic therapy for acute ischemic stroke associated with cancer-related nonbacterial thrombotic endocarditis. J Stroke Cerebrovasc Dis 23: e413–e416.

Yasuda K, Ayaki T, Kawabata Y et al. (2019). An autopsy case after endovascular thrombectomy for cardioembolic stroke due to nonbacterial thrombotic endocarditis. Rinsho Shinkeigaku 59: 195–199.

Zenagui D, De Coninck JP (1995). Atypical presentation of adult still's disease mimicking acute bacterial endocarditis. Eur Heart J 16: 1448–1450.

Ziegler E (1888). Ueber den Bau und die Entstehung der endocarditis chen Efflorescenzen. Ver Kong Inn Med 7: 339–343.

Handbook of Clinical Neurology, Vol. 177 (3rd series)
Heart and Neurologic Disease
J. Biller, Editor
https://doi.org/10.1016/B978-0-12-819814-8.00012-3

Chapter 13

Neurologic complications of atrial fibrillation: Pharmacologic and interventional approaches to stroke prevention

ASHWIN BHIRUD AND SMIT VASAIWALA*

Department of Cardiac Electrophysiology, Loyola University Medical Center, Maywood, IL, United States

Abstract

Atrial fibrillation is a common cardiac arrhythmia that carries a risk of stroke. This is commonly stratified with the CHA_2DS_2-VASc score. Stroke risk can be reduced with anticoagulants or with interventions to close the left atrial appendage, the most common source of left atrial thrombi. While warfarin has been traditionally used as the only oral anticoagulant available, there are several direct oral anticoagulants that compare favorably with respect to both stroke and bleeding risk in randomized controlled trials. Multiple interventional options exist to close the left atrial appendage, but the Watchman device is the only one that compares favorably with warfarin in randomized controlled trials.

INTRODUCTION

Atrial fibrillation (AF), the most common cardiac arrhythmia that leads to one-sixth of all strokes, is a rising epidemic with a 2014 estimated US prevalence of 2.7–6.1 million people, which is projected to be greater than 12 million by 2030 (Misayaka et al., 2006; Go et al., 2014). AF is associated with significant risk of stroke, congestive heart failure, and mortality (Schnabel et al., 2015). Antithrombotics to reduce the risk of stroke or systemic embolism are an integral component of AF management, of which the vitamin K antagonist (VKA) warfarin has been the historic benchmark for decades. In the last decade, there has been explosive growth in the field, with several new pharmacologic and nonpharmacologic therapies. The evolution of stroke risk stratification and both old and novel therapies will be reviewed herein. For convenience, stroke and systemic embolism will be referred to simply as stroke.

RISK STRATIFICATION

The decision to anticoagulate a patient with AF is individualized based on a patient's relative stroke and bleeding risk. Risk factors for AF-related stroke were first elucidated after the first randomized controlled trials (RCT) comparing warfarin and aspirin to placebo were published in the early 1990s. Increasing age, history of arterial hypertension, prior TIA or stroke, and history of diabetes mellitus were identified as risk factors from pooled control groups from RCTs (Atrial Fibrillation Investigators, 1994). Female gender and recent congestive heart failure were additionally identified later (Hart et al., 1999). A study of echocardiographic factors identified left ventricular systolic dysfunction as a predictor, while left atrial diameter by M-mode echocardiography was not a predictor (Atrial Fibrillation Investigators, 1998).

Subsequently, various risk stratification schemes were suggested based on these associations (Feinberg et al., 1999; Hart et al., 1999; Wang et al., 2003). The $CHADS_2$ score (congestive heart failure, hypertension, age $\geq$75 years, diabetes mellitus, and stroke/transient ischemic attack) was validated in 2001 in the National Registry of Atrial Fibrillation as a convenient method to stratify risk without warfarin therapy, though 31% were taking aspirin (Gage et al., 2001). Patients with a $CHADS_2$ score of 0 had an annual stroke rate of 1.2%. The score became popular due to its simplicity.

*Correspondence to: Smit C. Vasaiwala, M.D., Loyola University Medical Center, 2160 South 1st Ave., Bldg 110, Room 6232, Maywood, IL, 60153, United States. Tel: +1-708-216-9449, Fax: +1-708-327-2377, E-mail: svasaiw@lumc.edu

The CHA_2DS_2-VASc score (congestive heart failure, hypertension, age ≥75 years, diabetes mellitus, stroke/transient ischemic attack, vascular disease, age 65–74 years, sex category) was devised in 2010 to better stratify intermediate risk patients by age and incorporate gender and vascular disease as risk factors (Lip et al., 2014). The CHA_2DS_2-VASc score has become pervasive as well for its combination of simplicity and discrimination. The ATRIA stroke risk score, devised in 2013, sacrifices simplicity to provide a more granular risk stratification tool (Singer et al., 2013). There are conflicting reports regarding its performance relative to CHA_2DS_2-VASc, which may depend on the population's absolute stroke rate as well as stroke definitions that are broader than the adjudicated strokes seen in RCTs (Chao et al., 2014; Lip et al., 2014; Friberg et al., 2015; Aspberg et al., 2016). Several other risk stratification tools have been developed (Alkhouli and Friedman, 2019).

A modeling study based on recent estimates of stroke and bleeding risk reported that warfarin was net beneficial when annual stroke risk was greater than 1.7%, while dabigatran was net beneficial when annual stroke risk was greater than 0.9% (Eckman et al., 2011). Risk stratification is an integral component of AF management when considering both pharmacologic and nonpharmacologic therapies.

PHARMACOLOGIC THERAPIES

Aspirin

Aspirin has been used for decades in AF patients to reduce the risk of stroke in patients deemed ineligible for warfarin, and RCT data started appearing in the early 1990s. The Stroke Prevention in Atrial Fibrillations (SPAF) trial compared aspirin 325 mg daily vs placebo in warfarin-ineligible patients and showed decreased stroke risk with aspirin; however, all patients above 75 years old were deemed warfarin ineligible (Stroke Prevention in Atrial Fibrillation Investigators, 1991). Subsequent evidence has been conflicting, and a more recent registry study showed no reduction in stroke associated with aspirin (Själander et al., 2013). A meta-analysis of RCT data demonstrated a significant attenuation of stroke benefit from aspirin as patients' age, becoming essentially ineffective beyond age 75 (van Walraven et al., 2009). Importantly, aspirin increases the risk of bleeding. Even low-dose aspirin is associated with a 70% relative increase in major bleeding (McQuaid and Laine, 2006).

Vitamin K antagonists (warfarin)

In 1991, the SPAF trial was the first RCT demonstrating the superiority of warfarin for stroke reduction against placebo in nonvalvular atrial fibrillation (NVAF) with similar bleeding risk (SPAF investigators, 1991). Subsequent meta-analyses of RCTs from the era have confirmed a relative risk reduction of 62%–68% with warfarin vs placebo and significant risk reduction vs aspirin (van Walraven et al., 2002; Hart et al., 2007).

Warfarin has since become the gold standard of stroke reduction in AF. In eight RCTs after 2000, the annual stroke rate for AF patients on warfarin was 1.66%, ranging from 0.89% in $CHADS_2$ score of 0–1 to 2.5% in $CHADS_2$ score of 3–6 (Agarwal et al., 2012).

However, significant issues relating to warfarin use have been discovered over the years. Oral anticoagulation (OAC) is historically underprescribed. OAC with warfarin requires ongoing bloodwork to assess levels of anticoagulation and monitoring for significant drug interactions. Between 2008 and 2012, before widespread use of direct oral anticoagulants (DOACs), OAC prescriptions even in patients with $CHADS_2$ score >3 did not exceed 50% (Hsu et al., 2016).

Warfarin is notoriously difficult to maintain a steady therapeutic level of anticoagulation, with higher levels increasing bleeding risk and lower levels exposing patients to stroke risk. In RCTs since the turn of the century, the time spent in therapeutic range (TTR) with warfarin varied from 55% to 68% (Agarwal et al., 2012). In nonrandomized studies of warfarin use, TTR was 55%, suggesting that real-world warfarin effectiveness is less than that in trials (Baker et al., 2009). Strokes are more likely, more severe, and more fatal when warfarin levels are subtherapeutic (Hylek et al., 2003; Reynolds et al., 2004). In a meta-analysis of phase 3 trials of DOACs vs warfarin, major bleeding was significantly less in DOAC vs warfarin users when TTR for warfarin was less than 66%, whereas no major difference was seen when TTR was greater than 66%, suggesting labile levels of anticoagulation mediate excess bleeding events in warfarin users (Ruff et al., 2014; Reiffel, 2017).

Dual antiplatelet therapy

After clopidogrel was discovered, there was uncertainty whether combination aspirin/clopidogrel therapy might improve outcomes upon warfarin. Two RCTs from the 2000s ended that notion. The ACTIVE W trial showed that low-dose aspirin and clopidogrel was inferior to warfarin in preventing strokes while still resulting in more bleeding (Connolly et al., 2006). In the subsequent ACTIVE A trial, patients with AF deemed unsuitable for warfarin were randomized to combination aspirin/clopidogrel or aspirin alone, and dual antiplatelet therapy was associated with reduced stroke but increased major bleeding without net benefit (Connolly et al., 2009b).

Direct oral anticoagulants

Since the first direct oral anticoagulants (DOAC) dabigatran was US Food and Drug Administration (FDA) approved in 2010 for the reduction of stroke risk in NVAF, there has been increasing utilization of these drugs. Dabigatran is a direct thrombin inhibitor that was studied in the RE-LY study. This was an RCT of high- and low-dose dabigatran compared to warfarin, which demonstrated reduced stroke risk with high-dose dabigatran (150 mg twice daily) compared to warfarin without increased major bleeding (Connolly et al., 2009a,b).

Rivaroxaban and apixaban are both factor Xa inhibitors and were subsequently FDA approved in 2011 and 2012, respectively, for the same indication after the publication of the ROCKET-AF and ARISTOTLE RCTs. ROCKET-AF demonstrated noninferiority between warfarin and rivaroxaban (20 mg daily) with respect to stroke with similar bleeding risk (Patel et al., 2011). ARISTOTLE showed superiority of apixaban (5 mg twice daily) compared to warfarin with respect to both stroke risk and major bleeding (Granger et al., 2011; Patel et al., 2011). The AVERROES RCT comparing apixaban with aspirin (81–325 mg daily) in warfarin-ineligible NVAF patients demonstrated decreased stroke risk with apixaban with similar risk of major bleeding (Connolly et al., 2011). The final DOAC and factor Xa inhibitor edoxaban was FDA approved in 2015 after the publication of the ENGAGE-TIMI-48 RCT. This compared high- (60 mg daily) and low-dose (30 mg daily) edoxaban with warfarin, and both doses of edoxaban were noninferior against warfarin with regard to stroke with less major bleeding (Giugliano et al., 2013).

In a meta-analysis from 2013 analyzing the four DOAC vs warfarin trials, DOACs significantly reduced stroke, mainly driven by a 50% reduction in hemorrhagic stroke (Ruff et al., 2014). DOACs were also associated with a 10% reduction in all-cause mortality against warfarin, though they were associated with greater GI bleeding. The meta-analysis showed a consistent relative benefit of DOACs across subgroups.

Given uncertainty as to the relative benefits with background aspirin therapy, a meta-analysis analyzing patients on aspirin in the DOAC vs warfarin trials demonstrated a pooled reduction in stroke risk and intracranial hemorrhage with DOACs without a difference in major bleeding risk (Bennaghmouch et al., 2018).

The availability of reversal agents has been a concern regarding increasing use of DOACs, as the anticoagulant effect of warfarin is able to readily reverse in the setting of major bleeding. As of now, two reversal agents for DOACs have been approved in the US: idarucizumab for dabigatran (Pollack et al., 2017) and andexanet alfa for apixaban and rivaroxaban (Connolly et al., 2019; Cuker et al., 2019). Nonspecific agents such as prothrombin complex concentrate have been used as well. The relative availability and cost of these agents remain a concern.

End stage renal disease

Patients with end stage renal disease (ESRD) have been largely excluded from RCTs of stroke prevention. The use of warfarin for stroke prevention in patients with ESRD is controversial. A meta-analysis of 12 observervational studies reported a nonsignificant reduction in ischemic stroke and a significant increase in bleeding risk (Van Der Meersch et al., 2017). While the other DOACs are not FDA approved in ESRD, apixaban has been approved based on limited pharmacokinetic data (Hylek, 2018). Retrospective data since then suggests benefit to apixaban relative to warfarin in this population (Siontis et al., 2018). The recent RENAL-AF trial showed equivalent stroke and major bleeding rates between apixaban and warfarin in AF patients with ESRD at 1-year follow-up, though TTR for warfarin was only 44% (Pokorney, 2019). The ongoing AXADIA trial will randomize patients with AF and ESRD on chronic hemodialysis to apixaban 2.5 mg twice daily or the VKA phenprocoumon.

Adherence

Despite the perceived benefits of DOACs relative to warfarin, clinician awareness and patient adherence is still an issue. Even with better risk stratification and increasing use of DOACs, the rate of anticoagulant prescriptions has only mildly improved (Katz et al., 2017). Regarding patient adherence, studies show the percent of patients covered by DOAC greater than 80% of days ranged from 47.5% to 56.8%, compared to 40.2% with warfarin (Brown et al., 2016; Yao et al., 2016). There was no observed difference in adherence between once-daily rivaroxaban and twice-daily apixaban.

NONPHARMACOLOGIC THERAPIES OR LEFT ATRIAL APPENDAGE CLOSURE

Surgical left atrial appendage closure

The left atrial appendage (LAA) has been identified as the origin of greater than 90% of thrombi in patients with NVAF (Blackshear and Odell, 1996). Surgical closure of the LAA by excision or exclusion has become a standard component of the MAZE procedure to treat AF during concomitant cardiac surgery. Observational and small randomized studies suggest significant improvement in stroke risk without OAC (Cox et al., 1999; Tsai et al., 2014; Caliskan et al., 2017). However, incomplete surgical closure is a persistent issue that complicates study results and practical management. Up to 60% of surgical left atrial appendage closure (LAAC) with excision or exclusion results in incomplete closure, which has been associated with undiminished risk of stroke (Kanderian et al., 2008). An RCT of LAA amputation or ligation in AF patients undergoing cardiac surgery, the left atrial appendage occlusion Study III (LAAOS III) trial, is underway.

Percutaneous endocardial LAAC

Percutaneous endocardial LAAC was initially introduced in 2001 (Sievert et al., 2002). The WATCHMAN device (Boston Scientific, Inc.) is a self-expanding LAAC device composed of a nitinol frame with fixation barbs and a permeable polyester fabric that carries a risk of thrombus formation while endothelialization of the device takes place. In the initial RCTs, warfarin and aspirin 81 mg were given for 45 days postprocedure, then switched to aspirin 81 mg and clopidogrel 75 mg if no significant peri-device leak was seen at follow up TEE. Finally, at 6 months clopidogrel was stopped and aspirin 325 mg started indefinitely. It is the only percutaneous device studied in RCTs. The PROTECT AF and PREVAIL trials tested the WATCHMAN device against warfarin in warfarin-eligible patients. In 2015, it received FDA approval after these studies demonstrated noninferiority with respect to postprocedure strokes, though there were significant periprocedural complications (Holmes et al., 2009, 2015).

Efficacy of the WATCHMAN device against warfarin proved durable in longer follow-up of these RCTs and their respective registries, exchanging a nonsignificant increase in ischemic strokes for a significant reduction in hemorrhagic strokes and major bleeding, which may mediate a mortality reduction (Reddy et al., 2017). Subsequent registry data suggests persistent efficacy and improvement in periprocedural safety (Holmes et al., 2019).

RCTs have not been published regarding warfarin-ineligible patients; however, there is an RCT in progress—the Assessment of the Watchman Device in Patients Unsuitable for Oral Anticoagulation trial (Holmes et al., 2017). Additionally, there is significant registry data suggesting comparable stroke rates using antiplatelet postimplant instead of OAC (Reddy et al., 2013; Boersma et al., 2019). In the PROTECT AF and PREVAIL trials, device-related thrombus was seen in ~3.7% of patients at follow up, of which 26% had strokes, and seemed to be independent of antithrombotic regimen (Dukkipati et al., 2018).

Given the rise in use of both LAAC devices and DOACs over the last several years, randomized comparisons between the two are anxiously awaited. Preliminary results from the PRAGUE-17 RCT comparing LAAC followed by predominantly antiplatelet therapy to DOACs with predominantly apixaban showed similar stroke and bleeding outcomes (Osmancik et al., 2019). This may suggest DOACs are equivalent to LAAC in minimizing both stroke and major bleeding risk.

Other endocardial LAAC devices are in development, of which the AMPLATZER Amulet device (Abbott Vascular, Inc.) and the WaveCrest device (Coherex Medical, Inc.) are being compared to the Watchman device in RCTs (Bergmann and Landmesser, 2014).

Percutaneous epicardial LAAC

The LARIAT device was introduced clinically in 2013. The LARIAT device is a snare with a pretied suture that is guided epicardially over the LAA base after which the suture is released (Bartus et al., 2013). The largest registry data showed a 98% rate of complete LAA closure with significant rates of complications that improved with advances in safety techniques (Lakkireddy et al., 2016). Though FDA approved for surgical soft tissue approximation, it is not specifically approved for stroke prevention in AF and has not been tested for stroke prevention in RCTs.

CONCLUSION

The development of new pharmacologic and nonpharmacologic therapies for stroke prevention in AF has introduced a new era in stroke prevention in AF. Though the historic benchmark has been warfarin, the DOACs and LAAC devices appear to be outperforming warfarin with respect to major bleeding outcomes. Persistent issues with DOACs include cost, adherence, and availability of reversal agents, while issues of postprocedural antithrombotic regimens and incomplete closure persist with LAAC interventions. Future studies will continue to provide further guidance in this arena.

References

Agarwal S, Hachamovitch R, Menon V (2012). Current trial-associated outcomes with warfarin in prevention of stroke in patients with nonvalvular atrial fibrillation: a meta-analysis. Arch Internal Med 172: 623–631.

Alkhouli M, Friedman PA (2019). Ischemic stroke risk in patients with Nonvalvular atrial fibrillation: JACC review topic of the week. J Am Coll Cardiol 74: 3050–3065.

Aspberg S, Chang Y, Atterman A (2016). Comparison of the ATRIA, CHADS 2 , and CHA 2 DS 2 -VASc stroke risk scores in predicting ischaemic stroke in a large Swedish cohort of patients with atrial fibrillation. Eur Heart J 37: 3203–3210.

Atrial Fibrillation Investigators (1994). Risk factors for stroke and efficacy of antithrombotic therapy in atrial fibrillation: analysis of pooled data from five randomized controlled trials. Arch Intern Med 154: 1449–1457.

Atrial Fibrillation Investigators (1998). Echocardiographic predictors of stroke in patients with atrial fibrillation: a prospective study of 1,066 patients from 3 clinical trials. Arch Intern Med 158: 1316–1320.

Baker WL, Cios DA, Sander SD et al. (2009). Meta-analysis to assess the quality of warfarin control in atrial fibrillation patients in the United States. J Manag Care Pharm 15: 244–252.

Bartus K, Han FT, Bednarek J et al. (2013). Percutaneous left atrial appendage suture ligation using the LARIAT device in patients with atrial fibrillation: initial clinical experience. J Am Coll Cardiol 62: 108–118.

Bennaghmouch N, De Veer AJ, Bode K et al. (2018). Efficacy and safety of the use of non–vitamin K antagonist Oral anticoagulants in patients with Nonvalvular atrial fibrillation and concomitant aspirin therapy. Circulation 137: 1117–1129.

Bergmann MW, Landmesser U (2014). Left atrial appendage closure for stroke prevention in non-valvular atrial fibrillation: rationale, devices in clinical development and insights into implantation techniques. EuroIntervention 10: 497–504.

Blackshear JL, Odell JA (1996). Appendage obliteration to reduce stroke in cardiac surgical patients with atrial fibrillation. The Ann Thorac Surg 61: 755–759.

Boersma LV, Ince H, Kische S et al. (2019). Evaluating real-world clinical outcomes in atrial fibrillation patients receiving the WATCHMAN left atrial appendage closure technology: final 2-year outcome data of the EWOLUTION trial focusing on history of stroke and hemorrhage. Circ Arrhythm Electrophysiol 12: e006841.

Brown JD, Shewale AR, Talbert JC (2016). Adherence to rivaroxaban, dabigatran, and apixaban for stroke prevention in incident, treatment-naïve nonvalvular atrial fibrillation. J Manag Care Pharm 22: 1319–1329.

Caliskan E, Sahin A, Yilmaz M et al. (2017). Epicardial left atrial appendage AtriClip occlusion reduces the incidence of stroke in patients with atrial fibrillation undergoing cardiac surgery. Europace 20: e105–e114.

Chao TF, Liu CJ, Wang KL et al. (2014). Using the CHA_2DS_2-VASc score for refining stroke risk stratification in 'low-risk' Asian patients with atrial fibrillation. J Am Coll Cardiol 64: 1658–1665.

Connolly S, Yusuf S, Camm J et al. (2006). Clopidogrel plus aspirin versus oral anticoagulation for atrial fibrillation in the atrial fibrillation Clopidogrel trial with Irbesartan for prevention of vascular events (ACTIVE W): a randomised controlled trial. Lancet 367: 1903.

Connolly SJ, Ezekowitz MD, Yusuf S et al. (2009a). Dabigatran versus warfarin in patients with atrial fibrillation. N Engl J Med 361: 1139–1151.

Connolly SJ, Pogue J, Hart RG et al. (2009b). Effect of clopidogrel added to aspirin in patients with atrial fibrillation. The N Engl J Med 360: 2066–2078.

Connolly SJ, Eikelboom J, Joyner C et al. (2011). Apixaban in patients with atrial fibrillation. N Engl J Med 364: 806–817.

Connolly SJ, Crowther M, Eikelboom JW et al. (2019). Full study report of andexanet alfa for bleeding associated with factor xa inhibitors. N Engl J Med 380: 1326–1335.

Cox JL, Ad N, Palazzo T (1999). Impact of the maze procedure on the stroke rate in patients with atrial fibrillation. J Thorac Cardiovasc Surg 118: 833–840.

Cuker A, Burnett A, Triller D et al. (2019). Reversal of direct oral anticoagulants: guidance from the anticoagulation forum. Am J Hematol 94: 697–709.

Dukkipati SR, Kar S, Holmes DR et al. (2018). Device-related thrombus after left atrial appendage closure: incidence, predictors, and outcomes. Circulation 138: 874–885.

Eckman MH, Singer DE, Rosand J et al. (2011). Moving the tipping point: the decision to anticoagulate patients with atrial fibrillation. Circ Cardiovasc Qual Outcomes 4: 14–21.

Feinberg WM, Kronmal RA, Newman AB et al. (1999). Stroke risk in an elderly population with atrial fibrillation. JGIM 14: 56–59.

Friberg L, Skeppholm M, Terént A (2015). Benefit of anticoagulation unlikely in patients with atrial fibrillation and a CHA_2DS_2-VASc score of 1. J Am Coll Cardiol 65: 225–232.

Gage BF, Waterman AD, Shannon W et al. (2001). Validation of clinical classification schemes for predicting stroke: results from the National Registry of Atrial Fibrillation. JAMA 285: 2864–2870.

Giugliano RP, Ruff CT, Braunwald E et al. (2013). Edoxaban versus warfarin in patients with atrial fibrillation. N Engl J Med 369: 2093–2104.

Go AS, Mozaffarian D, Roger VL et al. (2014). American Heart Association statistics committee and stroke statistics subcommittee. Heart disease and stroke statistics–2014 update: a report from the American Heart Association. Circulation 129: e28–e292.

Granger CB, Alexander JH, McMurray JJ et al. (2011). Apixaban versus warfarin in patients with atrial fibrillation. N Engl J Med 365: 981–992.

Hart RG, Pearce LA, McBride R et al. (1999). Factors associated with ischemic stroke during aspirin therapy in atrial fibrillation: analysis of 2012 participants in the SPAF I–III clinical trials. Stroke 30: 1223–1229.

Hart RG, Pearce LA, Aguilar MI (2007). Meta-analysis: antithrombotic therapy to prevent stroke in patients who have nonvalvular atrial fibrillation. Ann Intern Med 146: 857–867.

Holmes DR, Reddy VY, Turi ZG et al. (2009). Percutaneous closure of the left atrial appendage versus warfarin therapy for prevention of stroke in patients with atrial fibrillation: a randomised non-inferiority trial. Lancet 374: 534–542.

Holmes DR, Doshi SK, Kar S et al. (2015). Left atrial appendage closure as an alternative to warfarin for stroke prevention in atrial fibrillation: a patient-level meta-analysis. J Am Coll Cardiol 65: 2614–2623.

Holmes DR, Reddy VY, Buchbinder M et al. (2017). The assessment of the Watchman device in patients unsuitable for oral anticoagulation (ASAP-TOO) trial. Am Heart J 189: 68–74.

Holmes DR, Reddy VY, Gordon NT et al. (2019). Long-term safety and efficacy in continued access left atrial appendage closure registries. J Am Coll Cardiol 74: 2878–2889.

Hsu JC, Maddox TM, Kennedy KF et al. (2016). Oral anticoagulant therapy prescription in patients with atrial fibrillation across the spectrum of stroke risk: insights from the NCDR PINNACLE registry. JAMA Cardiol 1: 55–62.

Hylek EM (2018). Apixaban for end-stage kidney disease. Circulation 138: 1534.

Hylek EM, Go AS, Chang Y et al. (2003). Effect of intensity of oral anticoagulation on stroke severity and mortality in atrial fibrillation. N Engl J Med 349: 1019–1026.

Kanderian AS, Gillinov AM, Pettersson GB et al. (2008). Success of surgical left atrial appendage closure: assessment by transesophageal echocardiography. J Am Coll Cardiol 52: 924–929.

Katz DF, Maddox TM, Turakhia M et al. (2017). Contemporary trends in oral anticoagulant prescription in atrial fibrillation patients at low to moderate risk of stroke after guideline-recommended change in use of the $CHADS_2$ to the CHA_2DS_2-VASc score for thromboembolic risk assessment: analysis from the National Cardiovascular Data Registry's outpatient practice innovation and clinical excellence atrial fibrillation registry. Circ Cardiovasc Qual Outcomes 10: e003476.

Lakkireddy D, Afzal MR, Lee RJ et al. (2016). Short and long-term outcomes of percutaneous left atrial appendage suture ligation: results from a US multicenter evaluation. Heart Rhythm 13: 1030–1036.

Lip GYH, Nielsen PB, Skjøth F et al. (2014). The value of the European society of cardiology guidelines for refining stroke risk stratification in patients with atrial fibrillation categorized as low risk using the anticoagulation and risk factors in atrial fibrillation stroke score: a nationwide cohort study. Chest 146: 1337–1346.

McQuaid KR, Laine L (2006). Systematic review and meta-analysis of adverse events of low-dose aspirin and clopidogrel in randomized controlled trials. Am J Med 119: 624–638.

Misayaka Y, Barnes ME, Gersh BJ et al. (2006). Secular trends in incidence of atrial fibrillation in Olmsted county Minnesota, 1980 to 2000 and implications on the projection for future prevalence. Circulation 114: 119–125.

Osmancik P, Herman D, Neuzil P et al. (2019). Percutaneous left atrial appendage closure versus novel anticoagulation agents in high-risk atrial fibrillation patients (PRAGUE-17 study). Presented at the European Society of Cardiology Congress, France, Paris.

Patel MR, Mahaffey KW, Garg J et al. (2011). Rivaroxaban versus warfarin in nonvalvular atrial fibrillation. N Engl J Med 365: 883–891.

Pokorney S (2019). RENal hemodialysis patients ALlocated apixaban versus warfarin in atrial fibrillation—RENAL-AF. Presented at the American Heart Association annual scientific sessions, PA, Philadelphia.

Pollack Jr CV, Reilly PA, Van Ryn J et al. (2017). Idarucizumab for dabigatran reversal—full cohort analysis. N Engl J Med 377: 431–441.

Reddy VY, Möbius-Winkler S, Miller MA et al. (2013). Left atrial appendage closure with the Watchman device in patients with a contraindication for oral anticoagulation: the ASAP study (ASA Plavix feasibility study with Watchman left atrial appendage closure technology). J Am Coll Cardiol 61: 2551–2556.

Reddy VY, Doshi SK, Kar S et al. (2017). 5-year outcomes after left atrial appendage closure: from the PREVAIL and PROTECT AF trials. J Am Coll Cardiol 70: 2964–2975.

Reiffel JA (2017). Time in the therapeutic range for patients taking warfarin in clinical trials: useful, but also misleading, misused, and overinterpreted. Circulation 135: 1475–1477.

Reynolds MW, Fahrbach K, Hauch O et al. (2004). Warfarin anticoagulation and outcomes in patients with atrial fibrillation: a systematic review and metaanalysis. Chest 126: 1938–1945.

Ruff CT, Giugliano RP, Braunwald E et al. (2014). Comparison of the efficacy and safety of new oral anticoagulants with warfarin in patients with atrial fibrillation: a meta-analysis of randomised trials. Lancet 383: 955–962.

Schnabel RB, Yin X, Gona P et al. (2015). 50 year trends in atrial fibrillation prevalence, incidence, risk factors, and mortality in the Framingham heart study: a cohort study. Lancet 386: 154–162.

Sievert H, Lesh MD, Trepels T et al. (2002). Percutaneous left atrial appendage transcatheter occlusion to prevent stroke in high-risk patients with atrial fibrillation: early clinical experience. Circulation 105: 1887–1889.

Singer DE, Chang Y, Borowsky LH et al. (2013). A new risk scheme to predict ischemic stroke and other thromboembolism in atrial fibrillation: the ATRIA study stroke risk score. J Am Heart Assoc 2: e000250.

Siontis KC, Zhang X, Eckard A et al. (2018). Outcomes associated with apixaban use in patients with end-stage kidney disease and atrial fibrillation in the United States. Circulation 138: 1519–1529.

Själander S, Själander A, Svensson PJ et al. (2013). Atrial fibrillation patients do not benefit from acetylsalicylic acid. Europace 16: 631–638.

Stroke Prevention in Atrial Fibrillation Investigators (1991). Stroke prevention in atrial fibrillation study: final results. Circulation 84: 527–539.

Tsai YC, Phan K, Munkholm-Larsen S et al. (2014). Surgical left atrial appendage occlusion during cardiac surgery for patients with atrial fibrillation: a meta-analysis. Eur J Cardiothorac Surg 47: 847–854.

Van Der Meersch H, De Bacquer D, De Vriese AS (2017). Vitamin K antagonists for stroke prevention in hemodialysis patients with atrial fibrillation: a systematic review and meta-analysis. Am Heart J 184: 37–46.

van Walraven C, Hart RG, Singer DE et al. (2002). Oral anticoagulants vs aspirin in nonvalvular atrial fibrillation: an individual patient meta-analysis. JAMA 288: 2441–2448.

van Walraven C, Hart RG, Connolly S et al. (2009). Effect of age on stroke prevention therapy in patients with atrial fibrillation: the atrial fibrillation investigators. Stroke 40: 1410–1416.

Wang TJ, Massaro JM, Levy D et al. (2003). A risk score for predicting stroke or death in individuals with new-onset atrial fibrillation in the community: the Framingham Heart Study. JAMA 290 (8): 1049–1056.

Yao X, Abraham NS, Alexander GC et al. (2016). Effect of adherence to oral anticoagulants on risk of stroke and major bleeding among patients with atrial fibrillation. J Am Heart Assoc 5: e003074.

Handbook of Clinical Neurology, Vol. 177 (3rd series)
Heart and Neurologic Disease
J. Biller, Editor
https://doi.org/10.1016/B978-0-12-819814-8.00004-4

Chapter 14

Tachyarrhythmias and neurologic complications

CATHERINE E. HASSETT[1], SUNG-MIN CHO[2], AND JOSE I. SUAREZ[2]*

[1]*Cerebrovascular Center, Neurological Institute, Cleveland Clinic, Cleveland, OH, United States*

[2]*Neurosciences Critical Care, Departments of Anesthesiology and Critical Care Medicine, Neurology, and Neurosurgery, Johns Hopkins University School of Medicine, Baltimore, MD, United States*

Abstract

Tachyarrhythmias are abnormal heart rhythms with a ventricular rate of 100 or more beats per minute. These rhythms are classified as either narrow or wide-complex tachycardia with further subdivision into regular or irregular rhythm. Patients are frequently symptomatic presenting with palpitations, diaphoresis, dyspnea, chest pain, dizziness, and syncope. Sudden cardiac death may occur with certain arrhythmias. Recognizing tachyarrhythmia and understanding its management is important as a wide spectrum of neurologic complications have been associated with such arrhythmias. The purpose of this chapter is to provide a comprehensive overview on the neurologic complications of tachyarrhythmias, neurologic adverse events of antiarrhythmic interventions, and neurologic conditions that can precipitate tachyarrhythmia.

ATRIOVENTRICULAR ELECTROPHYSIOLOGY AND THE AUTONOMIC SYSTEM

Electrical activity begins when an action potential arises in the sinoatrial (SA) node within the right atrium and travels to the atrioventricular (AV) junction (Fernandez-Jimenez et al., 2017). The role of the AV node is to slow and limit the number of signals conducted from an atrial focus to the ventricles (Fernandez-Jimenez et al., 2017). After a brief delay at the AV node, the conduction stimulus travels from the bundle of His to the left and right bundle branches, the Purkinje fibers, apical endocardium, and culminates at the ventricular myocardium (Fernandez-Jimenez et al., 2017). This allows for a synchronous contraction between the atria and the ventricles. Disruptions in the circuit from the atrial tissue to the AV node will create supraventricular tachycardias (SVT), whereas a stimulus originating from ventricular myocardium will elicit ventricular tachycardias (VT) (Link, 2012).

The SA node is under the direct influence of sympathetic (SNS) and parasympathetic (PNS) divisions of the autonomic nervous system (Collins et al., 2006; Fernandez-Jimenez et al., 2017). The SNS will raise the heart rate by increasing the production of action potentials within the SA node, while the PNS functions to suppress the SA node's activity (Collins et al., 2006). In a healthy individual, an increase in heart rate involves reciprocal actions of increasing sympathetic and decreasing parasympathetic involvement (Collins et al., 2006). Therefore, hyperactivity of the SNS may elicit tachyarrhythmias.

TYPES OF TACHYARRHYTHMIA

Narrow complex tachycardia

Narrow complex tachycardia, or more commonly known as SVT, originate from or incorporate supraventricular tissue with a prevalence of 2.29 per 1000 persons (Link, 2012; Fernandez-Jimenez et al., 2017). SVT is defined by a QRS duration of less than 120 milliseconds (ms), which can be further divided into regular versus irregular rate (Fernandez-Jimenez et al., 2017). The most

*Correspondence to: Jose I. Suarez, M.D., Department of Neurology, Neurosurgery, Anesthesiology and Critical Care Medicine, Division of NCCU, Johns Hopkins Medical Institutions, 600 N. Wolfe Street, Phipps 455, Baltimore, MD 21287, United States. Tel: 410-955-7481; Fax: 410-614-7903, E-mail: jsuarez5@jhmi.edu

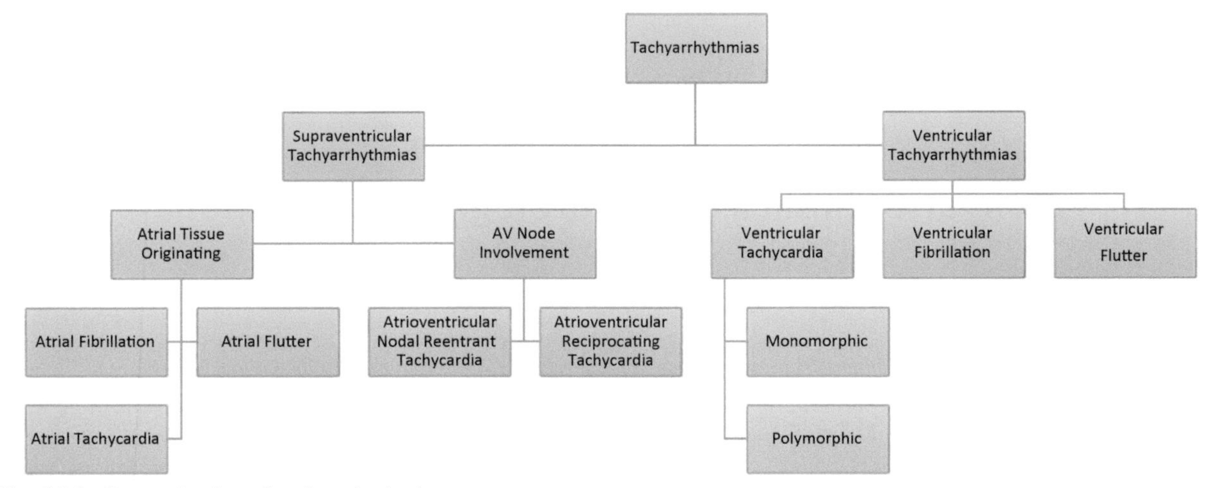

Fig. 14.1. Categorization of tachyarrhythmias.

common SVTs include atrial tachycardia (AT), atrial fibrillation (AF), atrial flutter, atrioventricular nodal reentrant tachycardia (AVNRT), and atrioventricular reciprocating tachycardia (AVRT) (Fig. 14.1). Paroxysmal SVT (PSVT) is a subset of SVT distinguished by a regular rhythm with an abrupt onset and termination, such as AT, AVNRT, and AVRT.

ATRIAL TACHYCARDIA

Atrial tachycardia (AT) is the least common PSVT accounting for only 10% of the total number of PSVT cases (Link, 2012). It is defined as either unifocal or multifocal atrial tachycardia depending on the number of focal circuits that originate from atrial tissue or the musculature of the great cardiac veins (Fernandez-Jimenez et al., 2017). Unifocal AT can be caused by reentry from underlying structural heart disease or enhanced automaticity leading to accelerated cardiac action potentials (Collins et al., 2006; Link, 2012). It is an important distinction as enhanced automaticity can precipitate incessant AT.

Once considered a benign arrhythmia that occasionally required symptomatic management, it is now widely accepted that incessant AT with rapid ventricular rate can cause a reversible tachycardia-induced cardiomyopathy (Belkin et al., 2007; Link, 2012; Fernandez-Jimenez et al., 2017). Reversible tachycardia-induced cardiomyopathy has consequences of increased blood stasis, subendothelial damage from adverse atrial remodeling, and induction of a prothrombotic, hypercoagulable state leading to increased risk of ischemic stroke (IS), heart failure, and sudden death (Belkin et al., 2007).

While atrial fibrillation and atrial flutter fall under the broad category of atrial tachycardia, these are often considered separate entities with their increased risk of IS necessitating anticoagulation (AC) consideration (Belkin et al., 2007; Yamada et al., 2017). The neurologic implications of these arrhythmias are discussed in greater detail in other chapters.

AVNRT AND AVRT

These tachycardias arise from the atrioventricular region secondary to reentry circuits (Michaud and Stevenson, 2013). Reentry circuits are found either within the AV node or in the presence of an accessory pathway in junctional tissue (Braunwald, 1997). Accessory pathways are extranodal connections from the epidural surface of the atrium to the ventricle along the atrioventricular groove (Braunwald, 1997; Michaud and Stevenson, 2013). These pathways are secondary to an in utero failure to separate the atrium and ventricle by fibrous AV rings (Braunwald, 1997; Collins et al., 2006). These bypass pathways can conduct antegrade or retrograde but most conduct in both directions (Goodacre and Irons, 2002).

AVNRT is a reentry supraventricular arrhythmia involving the two pathways of the AV node (Goodacre and Irons, 2002). In typical AVNRT, antegrade conduction is via the slow pathway and retrograde conduction via the fast pathway, ultimately causing simultaneous depolarization of the atria and ventricles (Goodacre and Irons, 2002). It is relatively common occurring in 10% of the population, usually being most symptomatic in women and those in their fourth decade of life or beyond (Braunwald, 1997; Link, 2012). Patients can experience an episode lasting seconds to several hours resulting in significant debility and decreased quality of life (Link, 2012).

AVRT is a reentrant tachycardia secondary to an aberrant pathway involving the atrium, the accessory pathway, the AV node, and the ventricle (Michaud and

Stevenson, 2013). The best known AVRT is Wolff–Parkinson–White (WPW) syndrome, where the aberrant pathway between the atria and ventricles is called the bundle of Kent (Braunwald, 1997, Link, 2012). This pathway creates an electrical circuit that bypasses the AV node in an orthodromic (or antidromic) transmission causing life-threatening arrhythmias (Braunwald, 1997; Goodacre and Irons, 2002; Link, 2012). Clinicians should ensure patients presenting with atrial fibrillation with underlying AVRT do not receive pharmacologic AV nodal blockade (American College of Cardiology Foundation, 2003). If the AV node is blocked in an AVRT patient, the accessory pathway will transmit fibrillation waves causing rapid ventricular rate or ventricular fibrillation (American College of Cardiology Foundation, 2003).

Wide complex tachycardias

Ventricular tachycardia, ventricular fibrillation, and Torsade de Pointes (TdP) are the most common wide complex tachyarrhythmias. These tachycardias can originate from either supraventricular tissue with aberrant conduction or the ventricular myocardium, such as the left ventricle, left ventricular outflow tract, right ventricle, and right ventricular outflow tract (Link, 2012). Wide complex tachycardia is defined by a cardiac rhythm of 100 beats per minute (bpm) or more with a QRS duration of 120 ms or more (John and Stevenson, 2003). It is subclassified into regular or irregular rhythm. The morphology of wide complex tachycardia can also suggest potential underlying etiologies. If a wide complex is monomorphic, it may be originating from a focus within the ventricles, whereas polymorphic wide complex raises concern for torsades de pointes tachycardia and QT interval prolongation (Collins et al., 2006; Link, 2012).

VENTRICULAR TACHYCARDIA

Ventricular tachycardia is defined as an arrhythmia with rates faster than 100 bpm originating from the ventricular myocardium or His-Purkinje system (John and Stevenson, 2003). Sustained ventricular tachycardia lasting greater than 30 s can lead to hemodynamic instability (John and Stevenson, 2003). The most common etiology of ventricular tachycardia is scar-related myocardium in patients with structural heart disease, such as a prior myocardial infarction. Other causes include dilated cardiomyopathy, hypertrophic cardiomyopathy, myocarditis, valvular heart disease, or congenital heart disease (Braunwald, 1997). Idiopathic ventricular tachycardia can be seen in patients without structural disease, such as epileptic patients that have a threefold increased risk of ventricular tachycardia (Borowicz, 2013).

VENTRICULAR FIBRILLATION AND FLUTTER

Physiologic triggers, like premature ventricular beats (PVC), under normal conditions do not degenerate into sustained ventricular arrhythmia (Collins et al., 2006). However, in the presence of a structurally abnormal heart, PVCs can induce reentry circuits. These two ventricular arrhythmias cause disorganized electrical activity that prohibits the ventricles from beating synchronously (Braunwald, 1997; John and Stevenson, 2003). Ventricular flutter and fibrillation are difficult to clinically distinguish; however, there is limited clinical significance as both can lead to cardiac arrest and sudden death if not terminated (John and Stevenson, 2003). Cardiac arrest can lead to increased risk of neurologic brain injury from reperfusion injury, oxidative damage, and abnormal hemodynamics (Uchino et al., 2016). This has led to the incorporation of several different postresuscitative approaches to cardiac arrest management, including therapeutic hypothermia and extracorporeal cardiopulmonary resuscitation (ECPR); however, mortality and morbidity remain high (Uchino et al., 2016).

TORSADE DE POINTES

Torsade de Pointes is defined by a prolonged QT interval evolving into a rapid ventricular tachycardia with a continuous alteration of the QRS morphology (Unger and Fassbender, 2013). Prolongation of the cardiac action potential creates a persistent depolarization effect that provokes this life-threatening arrhythmia (Unger and Fassbender, 2013).

Long QT syndrome (LQTS) is characterized by a prolonged QT interval that can cause syncope and sudden death (John and Stevenson, 2003). While congenital LQTS can be secondary to an inherited genetic cardiac channelopathy, an acquired cause is more common (John and Stevenson, 2003). The most common causes of acquired LQTS is linked to pharmacologic agents, such as class 1A and III antiarrhythmics, macrolides or fluoroquinolones antimicrobials, or metabolic derangements, such as hypokalemia and hypomagnesemia (Unger and Fassbender, 2013a). Several common neuropsychotropic medications, including dopamine receptor antagonists, selective serotonin reuptake inhibitors, and amantadine, have the ability to prolong the QT interval (Ozeki et al., 2010; Feldman and Gidal, 2013; Unger and Fassbender, 2013).

Neurologic complications from tachyarrhythmia

Ischemic stroke

Once thought to be noncausative, recent literature has questioned the relationship between PSVT and ischemic stroke. A recently published meta-analysis of studies on the risk of IS among patients with PSVT found a higher risk of IS compared to individuals without PSVT with a pooled relative risk of 2.03 (95% confidential interval (CI), 1.22–3.38, $I^2 = 89\%$) (Rujirachun et al., 2019). This meta-analysis is supportive of a large retrospective cohort analysis of 4.8 million patients from acute, federal care emergency departments in California (Kamel et al., 2013). Out of 14,121 patients diagnosed with PSVT, the rate of IS after PSVT diagnosis (0.94%; 95% CI, 0.76%–1.16%) significantly exceeded those without a diagnosis of PSVT (0.21%; 95% CI, 0.21%–0.22%, $P < 0.001$).

There are several mechanisms that can support an increased IS risk for patients with PSVT. First the elevated heart rate can create an atrial cardiomyopathy leading to stasis of blood flow and increased thrombus formation (Brembilla-Perrot and Blangy, 2006; Johnson et al., 2018; Rujirachun et al., 2019). Another explanation is that frequent episodes of PSVT may lead to the development of atrial fibrosis and left atrial enlargement creating reentrant circuits that may precipitate atrial fibrillation (Johnson et al., 2018). Even short runs of AT, defined as supraventricular ectopic beats less than 5 s captured on 24 h, outpatient cardiac monitoring, had significantly higher stroke rates in comparison to patients without short run AT (11.4% vs 8.3%, $P < 0.001$) (Yamada et al., 2017). However, older patients with PSVT have been found to have a higher burden of atherosclerotic diseases, which argues that perhaps PSVT is simply a marker of higher cardiovascular disease burden without causative effects on stroke risk (Csepe et al., 2017).

Unlike atrial fibrillation, it is unclear if initiation of anticoagulation for fast AT is beneficial. A post hoc analysis of the TRENDS study, which is a study assessing the relationship between low levels of paroxysmal AF and AT and the risk of stroke, found no significant difference in thromboembolic events between patients with AT/AF with the use of aspirin and warfarin vs those that did not (Glotzer et al., 2009). A meta-analysis of four randomized controlled trials evaluating prolonged cardiac monitoring greater than 7 days versus less than 48 h in patients with transient ischemic attack or cryptogenic stroke concluded that there was an increased use of AC in patients with prolonged cardiac monitoring (Dahal et al., 2016). However, there was no difference in rate of recurrent stroke or TIA between these two groups when followed up from 3 to 12 months. ACARDIA, a currently enrolling clinical trial, is attempting to answer if apixaban is superior to aspirin for the prevention of recurrent stroke in patients with cryptogenic ischemic stroke and atrial cardiopathy from undetected atrial fibrillation or other supraventricular arrhythmias (ACARDIA, n.d., unpublished work).

Hypoxic ischemic brain injury

Despite advances in postresuscitative care, cardiac arrest caused by ventricular arrhythmias remains a major cause of mortality and morbidity (Gräsner et al., 2016). Hypoxic ischemic brain injury (HIBI) is the primary cause of death in up to 68% of inpatient cardiac arrest and 23% of out-of-hospital cardiac arrest (Gräsner et al., 2016). The complex pathophysiology of HIBI focuses on the initial hypoxic injury followed a secondary reperfusion injury after return of spontaneous circulation (Sekhon et al., 2017). Large randomized trials for cardiac arrest patients receiving targeted temperature management (TTM), with target temperature of 33–34 or 36°C, have shown improved neurologic outcomes (Laver et al., 2004). TTM mitigates reperfusion injuries by reducing cerebral metabolism to avoid excessive intracellular anaerobic metabolism, hyperemic blood flows, and uncontrolled intracranial pressures (Nolan et al., 2018).

Management of cardiac arrest remains an active area of research with recent evidence and guidelines promoting the use of extracorporeal cardiopulmonary resuscitation in patients with refractory cardiac arrest (Brooks et al., 2015; Link et al., 2015, Cho et al., 2019). Initiation of ECPR allows for immediate cerebral perfusion during the postresuscitation period with recent studies showing benefit in survival and neurologic outcome at hospital discharge and at 3–6 month follow-up (Wang et al., 2014; Kim et al., 2016; Yukawa et al., 2017; Beyea et al., 2018; Cho et al., 2019).

Syncope

Syncope is defined as a transient loss of consciousness associated with an inability to maintain postural tone with rapid and spontaneous recovery (Sheldon et al., 2015). All tachyarrhythmias have the ability to produce syncope (Sheldon et al., 2015). Tachycardia leads to diminished left ventricular volume activating cardiac mechanoreceptors causing loss of sympathetic tone, enhanced vagal tone, and hypotension. Several studies have shown a prolonged QTc can predispose patients to syncope, which can be misdiagnosed as epileptic seizure (Unger and Fassbender, 2013; Van der Lende et al., 2016). In a study of 31 patients with genetic and electrographically confirmed LQTS, seizure disorder was the most common misdiagnosis and resulted in a significantly longer diagnostic delay compared to other misdiagnoses (MacCormick et al., 2009). QTc prolongation should be considered in patients who are not responding to antiepileptic medications as a delay in diagnosis can have consequences of sudden cardiac death (Van der Lende et al., 2016).

Postural orthostatic tachycardia syndrome

Postural orthostatic tachycardia syndrome (POTS) is a clinical syndrome defined by worsening syncope-like symptoms upon standing upright, an associated rise in heart rate of at least 30 bpm, and the absence of orthostatic hypotension (Sheldon et al., 2015). The prevalence of POTS is 0.2% with a strong female predominance within the ages of 15–25 years (Thieben et al., 2007; Sheldon et al., 2015). A number of proposed mechanisms include autonomic denervation, hypovolemia, hyperadrenergic stimulation, deconditioning, and hypervigilance (Benarroch, 2012). Analysis from tertiary care centers has reported that up to 50% of POTS patients have concomitant autonomic neuropathies of small and distal postganglionic sudomotor fibers, which can be detected by a thermoregulatory sweat test (Peltier et al., 2010; Benarroch, 2012). Treatment is primarily focused on conservative measures with pharmacologic interventions, such as low-dose beta-blockers, fludrocortisone, and pyridostigmine, reserved for refractory symptomatic management (Singer et al., 2004; Raj et al., 2009; Kanjwal et al., 2011). As far as we know, there are no epidemiologic studies associating POTS with IS.

Mitochondrial disorders

WPW syndrome can be a cardiac manifestation of several primary neurologic disorders, especially mitochondrial diseases. Both Leber's hereditary optic neuropathy and MELAS (mitochondrial encephalopathy, lactic acidosis, and stroke-like episodes) have associations with WPW syndrome (Hirano et al., 1992; Sproule et al., 2007; Finsterer et al., 2018). Patients presenting with WPW syndrome should be clinically screened for these syndromes and genetic testing should be pursued if appropriate. There is one isolated report of WPW and familial hypokalemic periodic paralysis; however, causality has not been established (Robinson et al., 2000).

Cognitive impairment

There is increasing evidence that atrial fibrillation has a negative impact on cognitive health (Kanmanthareddy et al., 2014). In addition to the silent infarct burden commonly seen in atrial fibrillation, altered cerebral hemodynamics, proinflammatory state, and cerebral microbleeds from chronic anticoagulation have been associated with a higher risk of cognitive impairment (Kanmanthareddy et al., 2014). Further investigation is needed to evaluate if these brain imaging findings are linked to other tachyarrhythmias. One study evaluated the cognitive function of 113 pediatric patients undergoing ablation for AVRT or AVNRT (Maryniak et al., 2013). These patients underwent standardized neuropsychological testing, which revealed a high prevalence of cognitive impairment in children and adolescents with an underlying AVRT and AVNRT rhythm. While the exact mechanism is unknown, the authors proposed low level or short-term anoxia may cause insults to the hippocampus, which is exquisitely sensitive to changes in oxygen levels (Allen et al., 2006). There is limited data on the impact of other tachyarrhythmias on cognitive outcomes; however, eliminating arrhythmias and maintaining sinus rhythm appears to be a modifiable risk factor for the development of dementia.

Management of tachyarrhythmia

Acute management of supraventricular arrhythmias

In the absence of hemodynamic instability, Valsalva maneuvers and carotid sinus massage can be used to increase vagal tone to slow conduction through the AV node and terminate the reentrant circuit (Al-Zaiti and Magdic, 2016; Page et al., 2016). If these maneuvers fail, administration of intravenous adenosine or non-dihydropyridine calcium channel blocker should be attempted (Page et al., 2016). If the arrhythmia resolves after administration of adenosine, it establishes AVNRT or AVRT as the likely PSVT. Ertipamil, an intranasal, short-acting calcium channel blocker, can also be used for acute termination of PSVT (Kashou and Noseworthy, 2020). Its high rate of conversion to sinus rhythm and low side-effect profile has made ertipamil an appealing alternative. Unstable SVT should proceed directly to synchronized cardioversion (Littmann et al., 2019).

Acute management of ventricular tachyarrhythmia

In hemodynamically unstable sustained VT, the priority is stabilization and electrical cardioversion (Littmann et al., 2019). Stable VT should be managed with antiarrhythmic medications and intravenous amiodarone is considered the most effective first pharmacologic choice. Other agents, such as procainamide and lidocaine, can also be considered for acute termination (Thomas and Behr, 2016; Ortiz et al., 2017; Littmann et al., 2019). Caution should be used with these agents as hypotension can cause hemodynamic instability in a stable VT patient (Littmann et al., 2019).

Long-term treatment of tachyarrhythmia

A referral to an electrophysiologist is necessary for patients with poorly tolerated or medically resistant tachyarrhythmia for antiarrhythmic management and catheter ablation consideration (Al-Zaiti and Magdic, 2016). Several agents can be used to slow or block AV nodal conduction for PSVT, whereas sotalol and amiodarone are the principal agents for ventricular arrhythmia medical management (Leary et al., 2014; Littmann et al., 2019). Implantable cardioverter defibrillator implantation may be necessary to prevent sudden cardiac death from ventricular arrhythmias for patients with underlying cardiac conditions, such as myocardial infarctions or heart failure (Littmann et al., 2019).

Given the high success rates and favorable safety profiles of catheter ablation, current practices are advocating for early catheter ablation of paroxysmal supraventricular, atrioventricular, and certain ventricular tachycardias (Mahtani and Nair, 2019). Despite its success, studies have reported a periprocedural stroke risk during ablation procedures ranging from 0.8% to 1.8%. A prospective study attempted to further investigate the incidence of major complications, defined as an adverse event requiring intervention, causing long-term disability, or resulting in prolonged hospitalization, among different catheter ablation indications at a high-volume center (Bohnen et al., 2011). Major catheter ablation-related complications were found to differ between procedure indications with 0.8% for SVT, 3.4% for idiopathic ventricular fibrillation, 5.2% atrial fibrillation, and 6% for ventricular tachycardia with structural heart disease. Out of 1676 ablations, there were 10 (0.6%) ischemic strokes events, which were found to occur at significantly higher rates in atrial fibrillation compared to SVT ablations ($P=0.017$).

Other studies have reported a silent embolic stroke risk detected by brain magnetic resonance imaging (MRI) after atrial fibrillation ablation to range from 6% to 38% (Bohnen et al., 2011). Another study performed brain MRIs in 18 patients immediately after ventricular ablation (Whitman et al., 2017). Seven patients (58%) undergoing LV ablation experienced a total of 16 cerebral emboli, in comparison with zero patients undergoing right ventricular ablation ($P=0.04$). Although this silent infarct burden was not clinically apparent, it is still clinically significant as the lesions may contribute to poor neurocognitive outcomes (Bohnen et al., 2011; Whitman et al., 2017).

Neurologic complications of antiarrhythmic medications

Prior to initiating chronic antiarrhythmic, the potential side effects of pharmacotherapy should be considered (Table 14.1). The most commonly used pharmacologic agents for tachyarrhythmias (excluding atrial fibrillation/flutter) and their associated neurologic complications are discussed below:

Table 14.1

Reported neurologic complications of antiarrhythmic medications

Neurologic complications	Drug
Ataxia	Amiodarone, Mexiletine, Propafenone
Autonomic neuropathy	Amiodarone
Dizziness	Diltiazem, Dofetilide, Dronedarone, Ecainide, Esmolol, Flecainide, Ibutilide, Isoprenaline
Dyskinesia	Amiodarone
Headache	Adenosine, Dofetilide, Dronedarone, encainide, Esmolol, Flecainide, Ibutilide, Isoprenaline, Metoprolol, Propafenone, Quinidine, Sotalol, Tocainide, Verapamil
Hemiballism	Amiodarone
Myoclonus	Amiodarone, Diltiazem, Propafenone, Verapamil
Myopathy	Propranolol
Optic neuropathy	Amiodarone
Nystagmus	Amiodarone, Tocainide
Parkinsonism	Amiodarone, Diltiazem, Verapamil
Pseudotumor cerebri	Amiodarone
Peripheral neuropathy	Amiodarone, Flecainide, Procainamide, Propafenone
Seizure	Diltiazem, Esmolol, Flecainide, Lidocaine, Propranolol, Propafenone, Quinidine, Tocainide, Verapamil
Stroke	Amiodarone, Metoprolol
Tinnitus	Propafenone, Quinidine
Tremor	Amiodarone, Flecainide, Isoprenaline, Mexiletine, Propafenone, Tocainide
Visual symptoms	Amiodarone, Digoxin, Flecainide, Metoprolol, Mexeltine, Quinidine

- *Adenosine* is a short-acting agent that is used to identify and convert an SVT rhythm by creating a transient heart block by slowing the atrioventricular node conduction (Leary et al., 2014). It lacks negative inotropic effects, is quickly metabolized, and can be given intravenously (Leary et al., 2014). Adenosine should be avoided in irregular wide-complex tachycardia, since it may cause unstable rhythms (Link, 2012). Side effects include flushing, dyspnea, chest pain, and severe headaches secondary to intracranial vessel dilation with superficial temporal artery dilatation noted on a transcranial Doppler study (Meschia and Biller, 1998; Leary et al., 2014).
- *Calcium channel blockers* can be used to disrupt a reentrant pathway by decreasing conduction velocity and prolonging repolarization at the AV node (Leary et al., 2014). Case reports of myoclonus and dystonia have been reported for nifedipine and verapamil (Vadlamudi and Wijdicks, 2002; Klein, 2003). Long-term use of diltiazem has been associated with movement disorders, such as myoclonus (Dick and Barold, 1989; Vadlamudi and Wijdicks, 2002; Klein, 2003; Swanoski et al., 2011) and drug-induced Parkinsonism, especially akathisia (Dick and Barold, 1989; Graham and Stewart-Wynne, 1994). As a CYP3A4 inhibitor, diltiazem inhibits the metabolism of 3-hydroxy-3-methyl-glutaryl coenzyme A reductase inhibitors (statins) creating increased circulating levels of statin leading to rhabdomyolysis and myopathies in certain patients (Peces and Pobes, 2001; Jasińska et al., 2006; Search Collaborative Group, 2008; Hu et al., 2011).
- *Esmolol* is a short-acting cardioselective beta-1 beta blocker (Leary et al., 2014). The most frequent neurologic complications include dizziness and somnolence similar to other agents in with beta blockade properties (Leary et al., 2014). A case report described an intravenous infusion of esmolol, which precipitated new-onset generalized convulsions, but otherwise neurologic complications are rare (Leary et al., 2014).
- *Amiodarone* is a class III antiarrhythmic drug that blocks potassium currents to prevent repolarization, prolong action potential duration, slow conduction, and reduce electrical impulses (Littmann et al., 2019; Leary et al., 2014). It has been implicated in several neurologic disorders including tremors, ataxia, and peripheral neuropathies (Leary et al., 2014). A large clinical trial found patients with prior myocardial infarctions receiving treatment with amiodarone and simvastatin are at an increased risk of myopathy secondary to its CYP3A inhibitor properties (Search Collaborative Group, 2008). In addition, amiodarone can cause thromboembolic strokes in a nonanticoagulated patient secondary to potential chemical cardioversion (Leary et al., 2014).
- *Flecainide* functions as a class 1c antiarrhythmic that blocks sodium channels to suppress intracardiac automaticity (Valentino et al., 2017). The most common neurologic adverse effect is dizziness, visual disturbance, and headaches (Page et al., 2016). However, case reports of rare central nervous system involvement have reported episodes of psychosis, dystonia, and hyperkinetic movements, such as myoclonus and tremors (Leary et al., 2014). A report of flecainide overdose caused predictable significant cardiac conduction disturbance, as well as generalized tonic–clonic seizure (Kennedy et al., 1989).

For the majority of antiarrhythmic-induced neurologic adverse effect, cessation of the offending drug will resolve the symptoms.

Tachyarrhythmia from acute brain injury

Epilepsy-related tachyarrhythmia

The incidence of arrhythmia in epilepsy is common. The limbic system, specifically the amygdala and insular cortex, is involved in control of the autonomic nervous system (Oppenheimer et al., 1992). The insula is balanced by inhibitory and excitatory pathways to the lateral hypothalamus. An animal study by Oppenheimer et al. showed significant dysrhythmias on electrocardiogram after electrical stimulation of the anterior insular cortex (Oppenheimer et al., 1991). Many studies have reported ventricular arrhythmias after insula-originating seizures (Oppenheimer et al., 1991, 1992, 1996).

Cardiac arrhythmias are common in focal and generalized epilepsy with the highest rate in patients with Lennox–Gastaut syndrome (Nei et al., 2012; Desai et al., 2017). In fact, for epileptic patients with a known cardiac disturbance, 90% of these patients will experience an acute ictal arrhythmia with the most common being sinus tachycardia (Nei et al., 2012). Postictal states can provoke life-threatening arrhythmias, such as ventricular fibrillation or flutter (Thomas and Behr, 2016).

Seizure-related cardiac arrhythmias are frequently reported and implicated in the pathophysiology of sudden unexpected death in epilepsy (SUDEP) (Nei et al., 2012). Ventricular arrhythmias are common during the postictal state and have been witnessed in association with SUDEP (Nei et al., 2012; Van der Lende et al., 2016). Several mechanisms during a seizure may contribute to postictal ventricular fibrillation, including a sudden rise in catecholamines, peri-ictal QT prolongation, and ST changes (Van der Lende et al., 2016). In addition, epilepsy appears to be an independent risk factor for the development of VT and VF arrhythmias with a threefold higher risk in comparison to the general population (Bardai et al., 2012).

Heart rate variability has been well studied in epilepsy. While high variability in heart rate correlates with healthy autonomic control mechanisms, a decreased variability is a risk factor for arrhythmias and an independent predictor of mortality in cardiac patients (Hallioglu et al., 2008; Feldman and Gidal, 2013). Several studies have shown decreased heart-rate variability in patients with focal and generalized epilepsy (Tomson and Kennebäck, 1997). Experimental and clinical data suggest that antiepileptic medications do not appear to increase the risk of heart rate variability or QT prolongation for most healthy patients (Tomson and Kennebäck, 1997). However, certain groups, such as the elderly with underlying structural heart disease, electrolyte imbalances, or polypharmacy, should have increased surveillance for malignant arrhythmias (Feldman and Gidal, 2013).

Tachyarrhythmia as cerebrovascular complication

Patients after an acute stroke have an increased risk for sudden cardiac death with ventricular arrhythmias being the leading cause (Kallmünzer et al., 2012). A higher incidence of QT interval prolongation and ventricular arrhythmias was reported in patients with insular infarcts in comparison to other locations (Oppenheimer et al., 1991, 1992, 1996; Sander and Klingelhofer, 1995). Lateralization of the insular cortex remains controversial but overall, left-sided anterior insular lesions decreased parasympathetic activity and amplified cardiac sympathetic response (Oppenheimer et al., 1996).

Sympathetic hyperactivity after an ischemic stroke or subarachnoid hemorrhage can lead to increased catecholamine levels and electrolyte derangements (Chen et al., 2014). Several studies have found an association between these electrolyte disturbances and QTc interval prolongation, TdP, and other ventricular arrhythmias (Behr and Mahida, 2011; Madias et al., 2011). This excess sympathetic stimulus can cause disruptions in the membrane sodium–potassium ATPase transporters causing a potassium influx resulting in hypokalemia (Chen et al., 2014). In addition, sympathetic hyperactivity can promote high intracellular calcium concentrations, which will prevent myocardial relaxation causing further metabolic imbalance and cell death (Behr and Mahida, 2011). This has been classically seen in Takotusbo cardiomyopathy, which can lead to life-threatening ventricular arrhythmias and TdP (Ghosh et al., 2009; Behr and Mahida, 2011; Madias et al., 2011).

Tachycardia after acute brain or spinal cord injury

Multiorgan dysfunction is common after acute brain injury (Lim and Smith, 2007). With a massive catecholamine rise, sympathetic hyperactivity, and proinflammatory response, the presence of nonneurologic injury has been associated with a higher mortality rate after ABI (Macmillan et al., 2002; Lim and Smith, 2007). A study of 74 traumatic brain injury patients revealed that the cardiovascular system had the highest incidence of organ dysfunction (Zygun et al., 2005).

Electrocardiogram (ECG) changes are common during ABI including prolonged QT interval, ST segment abnormalities, presence of Q waves, and flat or inverted T waves (Jachuck et al., 1975; Cheung and Hachinski, 2000). Prolonged QT syndrome has been associated with traumatic subarachnoid hemorrhage, and the degree of prolongation has been directly linked to the severity of the hemorrhage (Collier et al., 2004). While these ECG changes can precipitate life-threatening ventricular arrhythmias, this risk is transient resolving within 2 weeks (Lim and Smith, 2007). The current evidence from subarachnoid hemorrhage supports early sympathetic blockade to mitigate against a catecholamine surge, reduce myocardial damage, and improve neurologic outcomes (Neil-Dwyer et al., 1978).

Patients with cervical and upper thoracic (T1-T6) spinal cord injury (SCI) can also develop autonomic dysreflexia with increased resting heart rate, increased blood-pressure variability, and hypertension increasing their risk for cardiovascular complications (Collins et al., 2006). Patients with spinal cord injury above T6 remove control of the preganglionic cardiac sympathetic fibers leading to disruption of autonomic cardiac control (Bonica, 1968). This may cause increased susceptibility to tachyarrhythmias as noted in multiple cord-injured animal studies (Colachis III and Clinchot, 1997; Mayorov et al., 2001; Collins et al., 2006). In addition, life-threatening arrhythmias may more frequently occur as animal studies depict lower electrical stimulation thresholds needed for induction of ventricular arrhythmias (Colachis and Clinchot, 1997; Rodenbaugh et al., 2003a,b).

Structural-related tachyarrhythmia

There is a report of medically resistant AT as a presenting sign of a right temporoparietoccipital arachnoid cyst

(Palomer, 2003). After evaluating for common cardiac and extracardiac causes, the mass effect from the arachnoid cyst was thought to be the culprit. After fistulization of the cyst, the atrial tachycardia gradually normalized and antiarrhythmic medications were successfully discontinued.

CONCLUSION

There is a complex physiologic connection between the central and autonomic nervous system. It is important to understand the potential neurologic complications contributing to both the development and consequence of tachyarrhythmias. In addition, clinicians should be aware of the adverse implications of antiarrhythmic medications on the central and peripheral nervous system.

References

ACARDIA n.d. "ACARDIA: AtRial cardiopathy and antithrombotic drugs in prevention after cryptogenic stroke—full text view." ClinicalTrials.gov, clinicaltrials.gov/ct2/show/NCT03192215.

Allen JS, Tranel D, Bruss J et al. (2006). Correlations between regional brain volumes and memory performance in anoxia. J Clin Exp Neuropsychol 28: 457–476.

Al-Zaiti SS, Magdic KS (2016). Paroxysmal supraventricular tachycardia: pathophysiology, diagnosis, and management. Crit Care Nurs Clin North Am 28: 309–316. https://doi.org/10.1016/j.cnc.2016.04.005.

American College of Cardiology Foundation (2003). ACC/AHA/ESC guidelines for the management of patients with supraventricular arrhythmias. Circulation 108 (15): 1871–1909. (Accessed June 9, 2020).

Bardai A, Lamberts RJ, Blom MT et al. (2012). Epilepsy is a risk factor for sudden cardiac arrest in the general population. PLoS One 7: e42749.

Behr ER, Mahida S (2011). Stress cardiomyopathy and the acquired long QT syndrome. Heart Rhythm 8: e1–e2.

Belkin M, Hayes DL, Upadhyay GA (2007). Device-detected atrial tachycardia and risk of thromboembolism, American College of Cardiology. https://www.acc.org/latest-in-cardiology/articles/2017/02/03/09/44/device-detected-atrial-tachycardia-and-risk-of-thromboembolism. (Accessed June 9, 2020).

Benarroch EE (2012). Postural tachycardia syndrome: a heterogeneous and multi-factorial disorder. Mayo Clin Proc 87: 1214–1225.

Beyea MM, Tillmann BW, Iansavichene AE et al. (2018). Neurologic outcomes after extracorporeal membrane oxygenation assisted CPR for resuscitation of out-of-hospital cardiac arrest patients: a systematic review. Resuscitation 130: 146–158.

Bohnen M, Stevenson WG, Tedrow UB et al. (2011). Incidence and predictors of major complications from contemporary catheter ablation to treat cardiac arrhythmias. Heart Rhythm 8: 1661–1666.

Bonica JJ (1968). Autonomic innervation of the viscera in relation to nerve block. Anesthesiology 29: 793–813.

Borowicz K (2013). Antiarrhythmic drugs and epilepsy. Pharmacol Rep 65: 17–18.

Braunwald E (1997). Heart disease: a textbook of cardiovascular medicine, fifth edn. WB Saunders Co, Philadelphia, Pa, pp. 641–656.

Brembilla-Perrot B, Blangy H (2006). Prevalence of inducible paroxysmal supraventricular tachycardia during esophageal electrophysiologic study in patients with unexplained stroke. Int J Cardiol 109: 344–350.

Brooks SC, Anderson ML, Bruder E et al. (2015). Part 6: alternative techniques and ancillary devices for cardiopulmonary resuscitation: 2015 American Heart Association guidelines update for cardiopulmonary resuscitation and emergency cardiovascular care. Circulation 132: S436–S443.

Chen S, Li Q, Wu H et al. (2014). The harmful effects of subarachnoid hemorrhage on extracerebral organs. Biomed Res Int 2014: 858496.

Cheung RTF, Hachinski VC (2000). The insula and cerebrogenic sudden death. Arch Neurol 57: 1685–1688.

Cho SM, Farrokh S, Whitman G et al. (2019). Neurocritical care for extracorporeal membrane oxygenation patients. Crit Care Med 47: 1773–1781.

Colachis III SC, Clinchot DM (1997). Autonomic hyperreflexia associated with recurrent cardiac arrest: case report. Spinal Cord 35: 256–257.

Collier BR, Miller SL, Kramer GS et al. (2004). Traumatic subarachnoid hemorrhage and QTc prolongation. J Neurosurg Anesthesiol 16: 196–200.

Collins HL, Rodenbaugh DW, DiCarlo SE (2006). Spinal cord injury alters cardiac electrophysiology and increases the susceptibility to ventricular arrhythmias. Prog Brain Res 152: 275–288.

Csepe TA, Hansen BJ, Fedorov VV (2017). Atrial fibrillation driver mechanisms: insight from the isolated human heart. Trends Cardiovasc Med 27: 1–11.

Dahal K, Chapagain B, Maharjan R et al. (2016). Prolonged cardiac monitoring to detect atrial fibrillation after cryptogenic stroke or transient ischemic attack: a meta-analysis of randomized controlled trials. Ann Noninvasive Electrocardiol 21: 382–388.

Desai R, Rupareliya C, Patel U et al. (2017). Burden of arrhythmias in epilepsy patients: a Nationwide inpatient analysis of 1.4 million hospitalizations in the United States. Cureus 9: e1550.

Dick RS, Barold SS (1989). Diltiazem-induced parkinsonism. Am J Med 87: 95–96.

Feldman AE, Gidal BE (2013). QTc prolongation by antiepileptic drugs and the risk of torsade de pointes in patients with epilepsy. Epilepsy Behav 26: 421–426.

Fernandez-Jimenez R, Hoit BD, Walsh RA et al. (2017). Normal physiology of the cardiovascular system. In: Hurst's the heart, McGraw-Hill, New York, NY.

Finsterer J, Stollberger C, Gatterer E (2018). Wolff-Parkinson-White syndrome and noncompaction in Leber's hereditary optic neuropathy due to the variant m.3460GA. J Int Med Res 46: 2054–2060.

Ghosh S, Apte P, Maroz N et al. (2009). Takotsubo cardiomyopathy as a potential cause of long QT syndrome and torsades de pointes. Int J Cardiol 136: 225–227.

Glotzer TV, Daoud EG, Wyse DG et al. (2009). The relationship between daily atrial tachyarrhythmia burden from implantable device diagnostics and stroke risk: the TRENDS study. Circ Arrhythm Electrophysiol 2: 474–480. https://doi.org/10.1161/CIRCEP.109.849638.

Goodacre S, Irons R (2002). ABC of clinical electrocardiography: atrial arrhythmias. BMJ 324: 594–597.

Graham DF, Stewart-Wynne EG (1994). Diltiazem-induced acute parkinsonism. Aust NZ J Med 24: 70.

Gräsner JT, Lefering R, Koster RW et al. (2016). 27 Nations, ONE Europe, ONE Registry: a prospective one month analysis of out-of-hospital cardiac outcomes in 27 countries in Europe. Resuscitation 105: 188–195.

Hallioglu O, Okuyaz C, Mert E et al. (2008). Effects of antiepileptic drug therapy on heart rate variability in children with epilepsy. Epilepsy Res 79: 49–54.

Hirano M, Ricci E, Koenigsberger MR et al. (1992). MELAS: an original case and clinical criteria for diagnosis. Neuromuscul Disord 2: 125–135.

Hu M, Mak VW, Tomlinson B (2011). Simvastatin-induced myopathy, the role of interaction with diltiazem and genetic predisposition. J Clin Pharm Ther 36: 419–425.

Jachuck J, Ramani PS, Clark F et al. (1975). Electrocardiographic abnormalities associated with raised intracranial pressure. Br Med J 1: 242–244.

Jasińska M, Owczarek J, Orszulak-Michalak D (2006). The influence of simvastatin at high dose and diltiazem on myocardium in rabbits, the biochemical study. Acta Pol Pharm 63: 386–390.

John RM, Stevenson WG (2003). Ventricular arrythmias. In: Harrison's cardiovascular medicine, McGraw-Hill, New York, NY.

Johnson LSB, Persson AP, Wollmer P et al. (2018). Irregularity and lack of p waves in short tachycardia episodes predict atrial fibrillation and ischemic stroke. Heart Rhythm 15: 805–811.

Kallmünzer B, Breuer L, Kahl N et al. (2012). Serious cardiac arrhythmias after stroke: incidence, time course, and predictors—a systematic, prospective analysis. Stroke 43: 2892–2897.

Kamel H, Elkind MS, Bhave PD et al. (2013). Paroxysmal supraventricular tachycardia and the risk of ischemic stroke. Stroke 44: 1550–1554.

Kanjwal K, Karabin B, Sheikh M et al. (2011). Pyridostigmine in the treatment of postural orthostatic tachycardia: a single-center experience. Pacing Clin Electrophysiol 34: 750–755.

Kanmanthareddy A, Vallakati A, Sridhar A et al. (2014). The impact of atrial fibrillation and its treatment on dementia. Curr Cardiol Rep 16: 519.

Kashou AH, Noseworthy PA (2020). Etripamil nasal spray: an investigational agent for the rapid termination of paroxysmal supraventricular tachycardia (SVT). Expert Opin Investig Drugs 29: 1–4.

Kennedy A, Thomas P, Sheridan DJ (1989). Generalized seizures as the presentation of flecainide toxicity. Eur Heart J 10: 950–954.

Kim SJ, Kim HJ, Lee HY et al. (2016). Comparing extracorporeal cardiopulmonary resuscitation with conventional cardiopulmonary resuscitation: a meta-analysis. Resuscitation 103: 106–116.

Klein C (2003). Myoclonus and myoclonus-dystonias. In: Genetics of movement disorders, Academic Press. 451–471.

Laver S, Farrow C, Turner D et al. (2004). Mode of death after admission to an intensive care unit following cardiac arrest. Intensive Care Med 30: 2126–2128.

Leary MC, Veluz JS, Caplan LR (2014). Neurologic complications of arrhythmia treatment. In: Handbook of clinical neurology neurologic aspects of systemic disease part I, Elsevier 119: 129–150.

Lim HB, Smith M (2007). Systemic complications after head injury: a clinical review. Anaesthesia 62: 474–482.

Link MS (2012). Evaluation and initial treatment of supraventricular tachycardia. N Engl J Med 367: 1438–1448.

Link MS, Berkow LC, Kudenchuk PJ et al. (2015). Part 7: adult advanced cardiovascular life support: 2015 American Heart Association guidelines update for cardiopulmonary resuscitation and emergency cardiovascular care. Circulation 132: S444–S464.

Littmann L, Olson EG, Gibbs MA (2019). Initial evaluation and management of wide-complex tachycardia: a simplified and practical approach. Am J Emerg Med 37: 1340–1345.

MacCormick JM, McAlister H, Crawford J et al. (2009). Misdiagnosis of long QT syndrome as epilepsy at first presentation. Ann Emerg Med 54: 26–33.

Macmillan C, Grant I, Andrews P (2002). Pulmonary and cardiac sequelae of subarachnoid haemorrhage. Intensive Care Med 28: 1012–1023.

Madias C, Fitzgibbons TP, Alsheikh-Ali AA et al. (2011). Acquired long QT syndrome from stress cardiomyopathy is associated with ventricular arrhythmias and torsades de pointes. Heart Rhythm 8: 555–561.

Mahtani AU, Nair DG (2019). Supraventricular tachycardia. Med Clin North Am 103: 863–879.

Maryniak A, Bielawska A, Bieganowska K et al. (2013). Does atrioventricular reentry tachycardia (AVRT) or atrioventricular nodal reentry tachycardia (AVNRT) in children affect their cognitive and emotional development? Pediatr Cardiol 34: 893–897.

Mayorov DN, Adams MA, Krassioukov AV (2001). Telemetric blood pressure monitoring in conscious rats before and after compression injury of spinal cord. J Neurotrauma 18: 727–736.

Meschia JF, Biller J (1998). Complications of cardiovascular drugs. In: J Biller (Ed.), Iatrogenic neurology. Butterworth-Heinemann, Boston, pp. 379–396.

Michaud GF, Stevenson WG (2013). Supraventricular tachycarrythmias. In: Harrison's cardiovascular medicine, McGraw-Hill, New York, NY.

Nei M, Sperling MR, Mintzer S et al. (2012). Long-term cardiac rhythm and repolarization abnormalities in refractory focal and generalized epilepsy. Epilepsia 53: e137–e140.

Neil-Dwyer G, Walter P, Cruickshank JM (1978). Effect of propranolol and phentolamine on myocardial necrosis after subarachnoid haemorrhage. Br Med J 6143: 990–992.

Nolan JP, Neumar RW, Adrie C et al. (2018). Post-cardiac arrest syndrome: epidemiology, pathophysiology, treatment, and prognostication. A scientific statement from the International Liaison Committee on Resuscitation; the American Heart Association Emergency Cardiovascular Care Committee; the council on cardiovascular surgery and anesthesia; the council on cardiopulmonary, perioperative, and critical care; the council on stroke. Resuscitation 79: 350–379.

Oppenheimer SM, Wilson JX, Guiraudon C et al. (1991). Insular cortex stimulation produces lethal cardiac arrhythmias: a mechanism of sudden death? Brain Res 550: 115–121.

Oppenheimer SM, Gelb A, Girvin JP et al. (1992). Cardiovascular effects of human insular cortex stimulation. Neurology 42: 1727–1732.

Oppenheimer SM, Kedem G, Martin WM (1996). Left-insular cortex lesions perturb cardiac autonomic tone in humans. Clin Auton Res 6: 131–140. https://doi.org/10.1007/bf02281899.

Ortiz M, Martín A, Arribas F et al. (2017). Randomized comparison of intravenous procainamide vs. intravenous amiodarone for the acute treatment of tolerated wide QRS tachycardia: the PROCAMIO study. Eur Heart J 38: 1329–1335.

Ozeki Y, Fujii K, Kurimoto N et al. (2010). QTc prolongation and antipsychotic medications in a sample of 1017 patients with schizophrenia. Prog Neuro-Psychopharmacol Biol Psychiatry 34: 401–405.

Page RL, Joglar JA, Caldwell MA et al. (2016). 2015 ACC/AHA/HRS Guideline for the management of adult patients with supraventricular tachycardia: a report of the American College of Cardiology/American Heart Association Task Force on Clinical Practice Guidelines and the Heart Rhythm Society. Circulation 133: e506–e574.

Palomer J (2003). Incessant tachycardia and a cerebral tumor. Rev Esp Cardiol 56: 519–522.

Peces R, Pobes A (2001). Rhabdomyolysis associated with concurrent use of simvastatin and diltiazem. Nephron 89: 117–118.

Peltier AC, Garland E, Raj SR et al. (2010). Distal sudomotor findings in postural tachycardia syndrome. Clin Auton Res 20: 93–99.

Raj SR, Black BK, Biaggioni I et al. (2009). Propranolol decreases tachycardia and improves symptoms in the postural tachycardia syndrome: less is more. Circulation 120: 725–734.

Robinson JE, Morin VI, Douglas MJ et al. (2000). Familial hypokalemic periodic paralysis and Wolff-Parkinson-White syndrome in pregnancy. Can J Anaesth 47: 160–164.

Rodenbaugh DW, Collins HL, DiCarlo SE (2003a). Paraplegia differentially increases arterial blood pressure relatedcardiovascular disease risk factors in normotensive and hypertensive rats. Brain Res 980: 242–248.

Rodenbaugh DW, Collins HL, DiCarlo SE (2003b). Increased susceptibility to ventricular arrhythmias in hypertensive paraplegic rats. Clin Exp Hypertens 25: 349–358.

Rujirachun P, Wattanachayakul P, Winijkul A et al. (2019). Paroxysmal supraventricular tachycardia and risk of ischemic stroke: a systematic review and meta-analysis. J Arrhythm 35: 499–505.

Sander D, Klingelhofer J (1995). Stroke-associated pathologic sympathetic activation related to size of infarction and extent of insular damage. Cerebrovasc Dis 5: 381–385.

Sekhon MS, Ainslie PN, Griesdale DE (2017). Clinical pathophysiology of hypoxic ischemic brain injury after cardiac arrest: a "two-hit" model. Crit Care 21: 90.

Sheldon RS, Grubb 2nd BP, Olshansky B et al. (2015). 2015 Heart rhythm society expert consensus statement on the diagnosis and treatment of postural tachycardia syndrome, inappropriate sinus tachycardia, and vasovagal syncope. Heart Rhythm 12: e41–e63.

Singer W, Spies JM, McArthur J et al. (2004). Prospective evaluation of somatic and autonomic small fibers in selected autonomic neuropathies. Neurology 62: 612–618.

Sproule DM, Kaufmann P, Engelstad K et al. (2007). Wolff-Parkinson-White syndrome in patients with MELAS. Arch Neurol 64: 1625–1627.

Swanoski MT, Chen JS, Monson MH (2011). Myoclonus associated with long-term use of diltiazem. Am J Health Syst Pharm 68: 1707–1710.

The SEARCH Collaborative Group (2008). SLCO1B1 variants and statin-induced myopathy—a genome-wide study. N Engl J Med 359: 789–799.

Thieben MJ, Sandroni P, Sletten DM et al. (2007). Postural orthostatic tachycardia syndrome: the Mayo clinic experience. Mayo Clin Proc 82: 308–313.

Thomas SH, Behr ER (2016). Pharmacological treatment of acquired QT prolongation and torsades de pointes. Br J Clin Pharmacol 81 (3): 420–427.

Tomson T, Kennebäck G (1997). Arrhythmia, heart rate variability, and antiepileptic drugs. Epilepsia 38: S48–S51.

Uchino H, Ogihara Y, Fukui H et al. (2016). Brain injury following cardiac arrest: pathophysiology for neurocritical care. J Intensive Care 4: 31.

Unger MM, Fassbender K (2013). Relevance of long QT syndrome in clinical neurology. World J Neurol 3: 25.

Vadlamudi L, Wijdicks EF (2002). Multifocal myoclonus due to verapamil overdose. Neurology 58: 984–985.

Valentino MA, Panakos A, Ragupathi L et al. (2017). Flecainide toxicity: a case report and systematic review of its electrocardiographic patterns and management. Cardiovasc Toxicol 17: 260–266.

Van der Lende M, Surges R, Sander JW et al. (2016). Cardiac arrhythmias during or after epileptic seizures. J Neurol Neurosurg Psychiatry 87: 69–74.

Wang CH, Chou NK, Becker LB et al. (2014). Improved outcome of extracorporeal cardiopulmonary resuscitation for out-of-hospital cardiac arrest–a comparison with that for extracorporeal rescue for in-hospital cardiac arrest. Resuscitation 85: 1219–1224.

Whitman IR, Gladstone RA, Badhwar N et al. (2017). Brain emboli after left ventricular endocardial ablation. Circulation 135: 867–877.

Yamada S, Lin C-Y, Chang S-L (2017). Risk of stroke in patients with short-run atrial tachyarrhythmia. Stroke 48: 3232–3238.

Yukawa T, Kashiura M, Sugiyama K et al. (2017). Neurological outcomes and duration from cardiac arrest to the initiation of extracorporeal membrane oxygenation in patients with out-of-hospital cardiac arrest: a retrospective study. Scand J Trauma Resusc Emerg Med 25: 95.

Zygun DA, Kortbeek JB, Fick GH et al. (2005). Nonneurologic organ dysfunction in severe traumatic brain injury. Crit Care Med 33: 654–660.

Handbook of Clinical Neurology, Vol. 177 (3rd series)
Heart and Neurologic Disease
J. Biller, Editor
https://doi.org/10.1016/B978-0-12-819814-8.00006-8

Chapter 15

Neurologic complications of brady-arrhythmias

NICHOLAS OSTERAAS*

Department of Neurologic Sciences, Rush University, Chicago, IL, United States

Abstract

Brady-arrhythmias are responsible for both overt as well as subtle neurologic signs and symptoms, from the seemingly benign and nonspecific symptoms associated with presyncope, to sudden focal neurologic deficits. A brief background on nodal and infra-nodal brady-arrhythmias is provided, followed by extensive discussion regarding neurologic complications of brady-arrhythmias. The multiple mechanisms of and associations between Brady-arrhythmias and transient ischemic attacks and ischemic stroke are discussed. Controversial associations between brady-arrhythmias and neurologic disease are discussed as well, such as potential roles of brady-arrhythmias in cognitive impairment and sequelae of chronotropic incompetence; and the contribution of brady-arrhythmias to syncope and associated injuries to the nervous system. The chapter is written to stand on its own, with guidance toward other pertinent sections of this text where appropriate for further reading.

NEUROLOGIC COMPLICATIONS OF BRADY-ARRHYTHMIAS

The neurologic complications and clinical manifestations of brady-arrhythmias are varied, and generally acute in their associated presentations. The most clinically apparent neurologic manifestations include syncope and presyncopal states, transient ischemic attacks (TIAs) as well as ischemic strokes. This heterogeneous collection of arrhythmias may potentially (albeit controversially) contribute to more subtle and chronic neurologic presentations, such as cognitive impairment and neurologic complications of exercise intolerance in certain select cases.

This chapter will begin with a background overview of the various brady-arrhythmias, followed by an in-depth review of the literature on the associated neurologic complications of brady-arrhythmias, along with other associations between neurologic disease and brady-arrhythmias with direction to other closely related chapters in this volume as appropriate for further reading.

BACKGROUND PHYSIOLOGY

A brief review of normal cardiac conduction is in order prior to further discussion of brady-arrhythmia, which are defined as ventricular rates of under 60 beats per minute in adults, with varying parameters suggested for children based on age given the large variation in normal resting heart rate (Rijnbeek et al., 2001). The sinus node is located at the junction of the superior vena cava and right atrium. These specialized atrial pacemaker cells self-depolarize and subsequently transmit depolarization impulses through internodal tracts initially in the right atrium, with subsequent atrial contraction seen as the P wave on EKG. The impulse is subsequently transmitted to the atrioventricular node in the intraatrial septum, visually represented on EKG as the PR interval. The signal proceeds to the bundle of Hiss followed by left and right bundle branches, visible as the QRS complex, followed by ventricular repolarization visible as the T wave (Sova and Kolda, 1949; Mangrum and DiMarco, 2000).

Although the sinus node pacemaker cells self-depolarize (as the ventricular system will if normal

*Correspondence to: Nicholas Osteraas, M.D. M.S., Department of Neurologic Sciences, Rush University, Chicago, IL, United States. Tel: +1-312-942-4500, Fax: +1-312-942-2380, E-mail: nicholas_d_osteraas@rush.edu

physiologic cardiac conduction does not occur), the sinus node and AV node specifically, and the cardiac conduction system in general, are innervated by both the parasympathetic and sympathetic branches of the autonomic nervous system. This system allows for physiologic regulation of heart rate, with increase in parasympathetic tone resulting in lower heart rates, and inversely for increased sympathetic tone (Brubaker et al., 2006). Parasympathetic tone generally predominates in healthy individuals in the resting state (Mangrum and DiMarco, 2000) with decreasing parasympathetic tone with increasing age (Korczyn et al., 1976; Brubaker and Kitzman, 2011). Although physiology is useful in regulating heart rate and cardiac output in response to states of differing sympathetic and parasympathetic tone, this intervention can be susceptible to pathologic processes that affect both the peripheral nervous system and CNS. A primary CNS process can therefore potentially result in brady-arrhythmias, in addition to potentially suffering sequela from these same arrhythmias (Oppenheimer, 2006).

Our overview of the specific brady-arrhythmias begins with the sinus node. Dysfunction at the level of the sinus node is associated with a wide variety of arrhythmias involving both the generation and transmission of electrical impulses. This can manifest as a tachy-arrhythmia, a brady-arrthymia, or alterations between the two (so called "Tachy-brady" syndrome). Cardiac arrhythmias originating in the sinus node have historically been labeled sick sinus syndrome, or sinus node dysfunction. It has an estimated incidence of approximately 0.8 per 1000 years, or 1 in 600 patients, with mean age of diagnosis in the early 1970s (Jensen et al., 2014) and one of the most common indications for pacemaker implantation (Vintila et al., 1988; Brignole et al., 1990; Mangrum and DiMarco, 2000).

Brady-arrhythmias is associated with a disordered SA node commonly include sinus bradycardia, nodal exit block, and sinus arrest (Ferrer, 1968; Petermeyer, 1976; Mangrum and DiMarco, 2000). Sinus pauses (or sinus arrest) can be captured on EKG as the absence of P waves for a duration of greater than 2 s and in otherwise young, healthy individuals are not necessarily indicative of a disease state. SA nodal exit block occurs when SA nodal impulses have difficulty transmitting to the atrial myocardium. These blocks are classified as first, second, or third degree. First degree blocks describe slowed atrial conduction, the second degree blocks include increased prolongation until there is a failure of conduction (or "drop"), and the third degree is complete absence of transmission. The arrhythmias associated with sinus node dysfunction are commonly interspersed with long periods of normal cardiac conduction (Lien et al., 1977).

The etiology of sinus node dysfunction varies widely and can include medications, cardiac disease, systemic conditions, SA nodal fibrosis, and genetic/inherited causes. Regardless of underlying etiology (further discussion of which is beyond the scope of this chapter), progression of disease is common, with brady-arrhythmias predominating in the early course, followed by concurrent tachy-arrhythmias. The opposite pattern has been proposed as well, with paradoxical atrial fibrillation hypothesized to lead to atrial remodeling and subsequent SA node dysfunction (John and Kumar, 2016).

The majority of neurologic symptoms attributed to sinus node dysfunction are related to cerebral hypoperfusion, with frequent complaints including syncope or presyncope being more common than "cardiac" symptoms such as palpitations (Fairfax and Lambert, 1976). Vague and nonspecific symptoms may be present including malaise, vague digestive complaints, oliguria, fatigue (Bigger and Reiffel, 1979). Associated neurologic symptoms and complications will be discussed in further detail.

Sinus bradycardia is well known as a common finding in otherwise young and healthy adults, with normal pacemaker firing and cardiac conduction with origin in the SA node at the junction of the superior vena cava and right atrium with subsequent transmission through the AV node, bundle of Hiss, left and right branches followed by Purkinje fibers at a rate of fewer than 60 beats per minute (Viitasalo et al., 1982; Dreifus et al., 1983; Anderson et al., 2009). Nocturnal resting heart rate is on average even lower than this, and in healthy individuals AV block and sinus pauses have been demonstrated to occur while sleeping during 24 h monitoring (Brodsky et al., 1977; Sobotka et al., 1981). Those working in a hospital setting with patients on cardiac monitors may be alarmed at the low rates possible in otherwise young patients with high levels of cardiovascular fitness. The author has not infrequently rounded on otherwise healthy patients whom have had atropine placed ready at the bedside for benign sinus bradycardia occurring overnight.

In addition to associations with good physical conditioning, there are genetic predispositions to asymptomatic sinus bradycardia as well (Milanesi et al., 2006). In addition to genetic predispositions to symptomatic bradycardia (Lehmann and Klein, 1978), there have been case reports of transient second or third degree block in otherwise healthy athletes immediately after intense exercise (Meytes et al., 1975). Notably, some authors contend that the EKGs in these studies were not interpreted properly and instead are best classified as first degree blocks (Barold et al., 1997), and this continues to be debated (Barold and Herweg, 2012).

Careful assessment is needed, however, as sinus bradycardia is not always a benign sign of excellent conditioning in the young population. Symptomatic bradycardia (along with other arrhythmias) has been associated with severe weight loss in patients with anorexia (Yahalom et al., 2013; Sall and Timperley, 2015; Cotter et al., 2019) along with multiple drugs of abuse and disease states. Asymptomatic sinus bradycardia in healthy individuals is otherwise unlikely to be associated with neurologic sequela (although of course there are cardiac conditions of concern that can occur in young athletes, please see Chapter 25). In a similar vein, sinus arrhythmia in healthy individuals is representative of normal physiologic beat to beat variability and is likewise unconcerning in the majority of cases (Grossman and Taylor, 2007).

In the elderly, less healthy population, presumably asymptomatic sinus bradycardia can be a sign of preexisting sick sinus syndrome (Gann et al., 1979) myocardial injury, or a side effect (or intended therapeutic effect) of medications. When occurring in the setting of cardiac ischemia, sinus bradycardia has been postulated to be protective against malignant arrhythmias and assist with the optimization of myocardial oxygen utilization (Myers et al., 1974; Kjekshus et al., 1981).

AV block occurs with normal SA node origin but with a delay or even complete lack of subsequent signal transmission to the ventricular myocardium. It is classified primarily as first, second, or third degree and can occur at multiple points on the conduction system (Kosowsky et al., 1976). The first degree AV block is demonstrable on an EKG as a PR interval of greater than 200ms with no conduction failure. When failure intermittently occurs, second degree AV block is the most appropriate classification, with two main classifications (type 1 Wenckebach or type II, Mobitz) based on EKG patterns. This classification cannot be utilized to precisely localize the exact site of conduction disturbance (Barold and Herweg, 2012; Vogler et al., 2012). Complete conduction failure is third degree or complete heart block, with no synchrony between atrial and ventricular contraction, and although more frequently acquired, there are congenital cases as well (Pinsky et al., 1982; Michaelsson et al., 1997).

Paroxysmal AV block characterized by intermittent transitions between 1:1 AV conduction and second or third degree heart block (Burton, 1962; Lee et al., 2009; Aste and Brignole, 2017) with differing subtypes based on potential etiologies, high vagal tone, and disease of the AV conduction system and idiopathic (Aste and Brignole, 2017). It commonly presents with syncope and can be found incidentally and is not always clearly symptomatic (Brignole et al., 2011; Vogler et al., 2012).

The presence of a single brady-arrhythmia by no means excludes the presence of other arrhythmias, and associations with other arrhythmias may potentially have more serious consequences for the nervous system than the known brady-arrhythmia. Escape rhythms, extra nodal tachycardia along with atrial fibrillation, have been documented to occur in individuals with the second degree AV block (Rasmussen, 1971). One group followed 210 patients who had pacemakers implanted for a variety of brady-arrhythmias (sick sinus syndrome, heart block type II Mobitz, or complete (Mattioli et al., 1998)) over a 2-year period and detected new atrial fibrillation in 11% of the cohort, which is a potentially management changing finding.

A detailed discussion regarding the etiology of brady-arrhythmias is beyond the scope of this text but will be briefly touched upon. Although commonly cardiac in origin (Zoob and Smith, 1963), other etiologies abound (Vogler et al., 2012) and many neurologic disease states have been known to have effects on the cardiovascular system, including induction of brady- arrhythmias; this is worth a brief mention. The most well-known brady-arrhythmia in association with neurologic diseases is likely the sinus bradycardia associated with the Cushing reflex (Fodstad et al., 2006). The exact mechanism by which the Cushing reflex induces bradycardia has been theorized to be a component of the response to sustained systemic sympathetic activity by the baroreceptors, activation of the neuro-cardiac axis (Schmidt et al., 2018) and postulated in some cases to be, albeit with limited support, as a result of activation the vagus nerve as a result of direct compression in herniation syndromes (Tsai et al., 2018).

Brady-arrhythmias have been associated with brain tumors in the absence of significant mass effect that would trigger a Cushing response; seizures (Devinsky et al., 1997); postcarotid endarterectomy (Mlekusch et al., 2003); rare cases have been reported to occur remotely after vagal nerve stimulation use for the treatment of epilepsy (Amark et al., 2007); and it is frequent after spinal cord injury as well (Grigorean et al., 2009). It is also worth discussing that medications may be entirely causal of or contributory to brady-arrhythmias, as well as exacerbating an associated low volume state, in turn further reducing cerebral perfusion and increasing risk for neurologic complications (Jansen et al., 1986; Zeltser et al., 2004; Osmonov et al., 2012).

Calcium channel blockers (common culprits in pharmacologically induced heart block (Zeltser et al., 2004)) are sometimes utilized in the management of headache by neurologists. Symptomatic (or incidentally discovered) brady-arrhythmias can occur with agents not commonly known by prescribers to act on the cardiovascular system, such as the newer generation of antidepressants (Pacher and Kecskemeti, 2004), and combinations of medications for to treat memory impairment and assist with mood regulation. Substances better known to cause concerning side cardiovascular and neurologic effects can have less commonly known effects upon cardiac conduction, such as cocaine (Castro and Nacht, 2000) along with substances commonly thought to be much more benign from a cardiovascular standpoint, such as marijuana (Kariyanna et al., 2019).

NEUROLOGIC COMPLICATIONS

The most common neurologic presentations and symptoms associated with brady-arrhythmias are those associated with cerebral hypoperfusion. The more common arrhythmias predisposing to presyncope (as well as syncope) are Mobitz type II block and complete AV Block (Task Force et al., 2009). Cardiac output clearly has a key role in providing adequate cerebral perfusion, approximately 20% of the cardiac output in adults and up to 50% in children (van Dijk and Wieling, 2013). Cardiac output is frequently taught throughout medical schools and physiology classes as a simple product of the stroke volume and heart rate (Mangrum and DiMarco, 2000; Morris, 2011). With decreasing heart rates, the ventricles may fill to a greater extend between beats to maintain cardiac output, but only to an extent, as ventricular volume is generally fixed (Dreifus et al., 1983), and this compensation can be further limited in the presence of any coexisting structure heart disease (Samet, 1973).

The physiologic factors involved in developing symptoms related to a brady-arrhythmia are not as simple as determining a decreased cardiac output. A specific cardiac output will not necessarily lead to the same cerebral perfusion pressure among different individuals nor even the same individual during different physiologic states as the intracranial compartment in a baseline state offers some resistance to the systemic mean arterial pressure. Cerebral perfusion pressure is mathematically defined as the difference between mean arterial pressure and intracerebral pressure (Moraine et al., 2000; van Dijk and Wieling, 2013).

However, cerebral perfusion is not entirely dependent upon consistent cardiac output. The autoregulatory process involving vasodilation and constriction though multiple mechanisms affecting cerebral vascular resistance help ensure that the brain is optimally perfusion in the event of low mean arterial pressure (Armstead, 2016). The presence of these autoregulatory mechanisms allows for wide ranges of systemic arterial pressure while insulating the central nervous system from the effects of what would otherwise be large variations in cerebral perfusion. This autoregulatory system has both upper and lower limits; the lower point of autoregulatory failure is a matter of debate but has been reported as between 50 and 70 mmHg mean arterial pressure (Aaslid et al., 1989; Drummond, 1997).

Below a certain threshold of perfusion, no autoregulatory mechanism will allow for sufficient cerebral perfusion, and an individual in this case may suffer symptoms and sequela of decreased cerebral perfusion. Cerebral hypoperfusion can lead to a variety of subjective symptoms that can be difficult for patients to clearly articulate. Individuals with presyncope, regardless of the underlying etiology, can complain of a large variety of symptoms. They will frequently endorse lightheadedness or dizziness; can experience cardiac symptoms such as palpitations and chest pain depending on the etiology of the hypoperfusion; and will frequently complain of and present with additional neurologic symptoms such as confusion, visual disturbances with potentially superimposed so called “focal” neurologic deficits (Vogler et al., 2012; Semelka et al., 2013; Novak, 2016). A state of heightened emotional arousal such as fear or sense of impending doom is not common but has been the presenting complaint of a minority of those with presyncope (Kharod et al., 2014), and a case of transient global amnesia associated with bradycardia and hypotension has been reported (Dugan et al., 1981).

Investigators questioned patients regarding symptoms of presyncope in a cohort of individuals with autonomic insufficiency, 84% reported feelings of dizziness and 63% “giddiness” and “light headedness.” Visual disturbances were also present in 84% of the cohort, but the classic tunnel vision was not nearly as common (13%) as “blurring,” “graying out.” Neck pain in the suboccipital and shoulder area was common (81%) along with low back pain. Hearing changes, not infrequent in patients with cerebral hypoperfusion were uncommon in this cohort (Mathias et al., 1999). Other neurologic symptoms that have been attributed to chronically low perfusion have included lack of motivation along with subjective memory and attention problems (Duschek and Schandry, 2007).

The presyncopal state that can occur with brady arrhythmias is not merely a collection of subjectively disturbing phenomenon, especially when combined with an otherwise safe activity (such as working while on a ladder) or with other comorbid conditions common in the elderly that could predispose to falling (Nickens, 1985). Falls with associated injury and neurologic damage are certainly possible without loss of consciousness (although not common nor severe as that associated with completed syncope (Auer, 2008)). Even if a patient with a presyncopal state secondary to a brady-arrhythmia manages to avoid a fall, injury, and neurologic sequela is still possible to occur in the event of patient struggling to remain upright. As an example, severely painful lumbar muscular skeletal strain was the presenting complaint of an individual who suffered from pharmacologically induced pre syncope (Isbister et al., 2001).

At first glance, this may seem like a somewhat trivial neurologic sequela of brady-arrhythmias; however, up to one in three of older adults will fall each year, and in turn up to one in three of these falls can result in trauma (Alexander et al., 1992). Although hip fractures have a well-deserved reputation as associated with debilitating outcomes, injuries to the CNS are the most devastating. Traumatic injuries to the head occurring during falls are associated with the highest risk of death; the most costly and debilitating injuries otherwise are those to the spinal cord (Mahdian, 2013). Admittedly, it is difficult to know how many of these fall-related traumas are secondary, at least in part, to brady-arrhythmias although it has been estimated to be approximately 1 in 20 (D'Ascenzo et al., 2013).

Presyncope leading to head injury may lead to traumatic subdural or epidural hematomas. Even if there is no immediate neurologic sequelae from such pathology, the presence of blood in an extra axial compartment may potentially lead to withholding of therapeutic anticoagulation with subsequent consequences for the central nervous system. The author has personal experience with such a case; a patient with atrial fibrillation had felt light headed and fallen 3 months prior to presentation and was diagnosed with a traumatic subdural hematoma; no operation was required and anticoagulation for primary ischemic stroke prevention had been held with an antiplatelet utilized instead. The patient presented with acute onset aphasia and right-sided weakness, no evidence of any blood products on emergent CT scanning, and intravenous tissue plasminogen activator was administered for suspected ischemic stroke. Unfortunately, the middle cerebral artery did not recanalize after IVtPA administration, and the patient was not a candidate for endovascular therapy. Presyncope can of course progress to syncope; and brady-arrhythmias have the potential in some cases lead to cardiac death (Greene, 1990; Arsenos et al., 2013).

Brady-arrhythmias can directly contribute to focal neurologic symptoms (aside from as well as superimposed with the constellation of symptoms associated with global cerebral hypoperfusion) via the major mechanisms of hypoperfusion, and cardiac emboli may be associated in certain cases. Alternative indirect methods are theoretically possible as well, such as a syncopal or pre syncopal state induced by an arrhythmia leading to a traumatic carotid or vertebral artery dissection, with subsequent CNS ischemia.

From a pathophysiological standpoint, reduced blood flow to the CNS from either arterial occlusion or hypoperfusion results in loss of neuronal function; however, this need not result in permanent cell death. The exact amount of cerebral blood flow required in order to maintain normal function, results in impaired functioning but not death, and flow incompatible with continued neuronal survival has been and continues to remain an area of active investigation and research (Bandera et al., 2006; Fisher and Bastan, 2012), but there is general agreement in the concept along with associated terms such as ischemic "penumbra" to define areas of brain parenchyma with hypoperfused but non ischemic neurons, and "core" to label those neurons that have or will suffer irreversible cell death. The clinical manifestations are generally the loss of neuronal function that is normally present at baseline, with some notable exceptions. Distinguishing between "penumbra" and "core" cannot be done on a clinical basis; this requires the use of neuroimaging. Likewise, the clinical context can be helpful in postulating regarding possible mechanisms of the cerebral ischemia, but one cannot know for certain at first if symptoms all attributable to arterial occlusion or hypo-perfusion.

Traditional clinical labels utilized in describing CNS ischemia are transient ischemic attack and ischemic stroke. Historically, a TIA was defined as a return to neurologic baseline within an arbitrary 24 h time window. Should neuronal function be restored to normal, either spontaneously or through clinical intervention, it is possible that advanced imaging with MRI may demonstrate evidence of irreversible ischemia; the proper classification of this common clinical scenario is debatable, but may be labeled as an ischemic stroke or TIA with MRI diffusion positivity (Prabhakaran et al., 2007; Prabhakaran and Lee, 2009; Brazzelli et al., 2014).

As a general rule, ischemia to the left cerebral hemisphere will result in symptoms contra laterally. A notable exception is a variation of the anterior cerebral artery in which both branches arise from a unilateral origin; in this event should this common vascular origin be either occluded or hypoperfusion; the patient may demonstrated bilateral lower extremity weakness (LeMay and Gooding, 1966). The most commonly associated symptoms with cerebral ischemia include the absence of normal sensory and motor functioning and can include higher processing or so called "cortical" signs such as aphasia, homonymous hemianopia, and neglect depending on the arterial supply involved and the presence and robustness of an alternative collateral blood supply to the affected tissue.

Ischemia in the posterior circulation can result in less readily recognizable stroke syndromes, and many isolated symptoms potentially referable to brainstem syndromes are nonspecific for cerebral ischemia in isolation, such as vertigo (which can be described by patients as dizziness or lightheadedness), blurred vision, hearing changes, and dysarthria (Paul et al., 2013). Some of these may overlap with symptoms described by those with presenting with syncope and presyncope. If the symptoms are transient with return to baseline before a neurologic evaluation can be conducted, the detection of a brady-arrhythmia may obfuscate the clinical picture, with symptoms potentially attributed to the detected brady-arrhythmia even if mechanistically an alternative ischemic risk factor (such as atherosclerosis of a vertebral artery) may ultimately be responsible for the transient symptoms.

In a similar vein, posterior circulation strokes affecting the basilar artery can result in transient or persistent neurologic deficits affecting both sides of the body simultaneously, as all major motor and sensory pathways as well as brainstem tracts and nuclei dwell in critical territory in the brainstem. This is one of the few instances in which a patient complaining of entire body weakness or transient syncope/drop attacks can have a cerebrovascular etiology for these transient symptoms (Demel and Broderick, 2015). Another atypical but very pertinent vascular syndrome to be aware of that may present with focal deficits in the setting of brady-arrhythmia-related hypotension is that of bilateral anterior circulation hypoperfusion, significant enough to result in ischemia in territories susceptible to hypoperfusion (so called "border zone" territories). These are areas of brain parenchyma that lie in the border of vascular supply, for example between the middle cerebral artery and anterior cerebral arteries. Ischemia can affect both sides of the anterior circulation, resulting in bilateral symptoms (Mangla et al., 2011). This has been classically taught to present as a "man in a barrel syndrome," with the distal aspects of the upper extremities relatively spared compared to the proximal as though a barrel had been placed over the patient's torso given the relationship to the motor homunculus involving the upper extremities and the watershed distribution of vascular supply, but this specific pattern of weakness has been associated with a variety of clinical presentations with a variety of underlying etiologies (Orsini et al., 2009; Finsterer, 2010).

As in the case of minor isolated symptoms, the clinical picture can be confusing if the patient has returned to normal at the time of medical evaluation and could potentially be confused with a primary cardiac syndrome. Neuroimaging of the vasculature, utilizing magnetic resonance imaging or CT scan angiography, along with imaging of the brain parenchyma can be useful in such circumstances. Significant stenosis in arteries supplying brain parenchyma may suggest cerebral ischemia as a possible explanatory etiology for neurologic deficits, and likewise restricted diffusion on magnetic resonance imaging of the brain parenchyma in those same territories suggests ischemia as well. Although neuroimaging patterns involving infarction in border zone territories in the setting of hypoperfusion strongly support hypoperfusion as a mechanism, this is not always the case. There are reports of embolic phenomenon demonstrating similar patterns on neuroimaging studies (Mangla et al., 2011) and based on imaging alone, it is not necessarily possible to conclusively identify the etiology.

The sudden onset of "focal" neurologic deficits brings to mind embolic phenomena. However, based on presentation alone a specific mechanism cannot be determined with certainty. How would hypoperfusion lead to a focal deficit as opposed to only symptoms of global hypoperfusion? Hypoperfusion in the setting of focal vascular stenosis makes the most sense mechanistically. However, the presence of a severe stenosis is not required for a TIA or stroke to occur in the presence of a brady-arrhythmia-induced hypotensive episode. As a representative case series, 39 patients with patent large arteries presented with presyncopal or syncopal symptoms in association with the diagnosis of either ischemic stroke or TIA (implying that focal neurologic deficits were present). Despite the absence of what would be traditionally called symptomatic large artery atherosclerosis (individuals with narrowing up to 50% in the carotids were not excluded for this cohort), nearly three quarters of these patients had an infarct imagining pattern categorized as "border zone" (Ryan et al., 2015). The majority of patients in this cohort had alternative etiologies for the systemic hypotension other than brady-arrhythmias, although approximately 10% of this cohort was determined to have an etiology involving an arrhythmia for the hypotension and associated infarction.

An atypical and potentially confusing presentation of neurologic dysfunction secondary to hypoperfusion in the setting of focal stenosis is that of the "limb shaking TIA." Patients will complain of abnormal uncontrollable movements in a limb, with a description that calls to mind a focal motor seizure or intermittent tremor more so than cerebral ischemia (Ali et al., 2006; Das and Baheti, 2013). This is unlikely to occur as the result of a brady-arrhythmia alone in the setting of normal vasculature, diagnostic work up may potentially reveal critical or potentially complete occlusion of pertinent vasculature, most commonly the internal carotid artery on the contralateral side from the affected limb. Although much attention is appropriately given to surgical management of near occlusions in the internal carotid arteries, there are cases with alteration in blood pressure medication regimens with the aim of achieving greater than standard blood pressure ranges for those with a history of cerebral ischemia, with results in improvement and/or resolution of symptoms (Das and Baheti, 2013).

Although the underlying pathophysiology is not entirely clear, Junctional brady-arrhythmia has been associated with higher rates of thromboembolism and ischemic stroke. Junctional brady-arrhythmia can be associated with or without retrograde p waves. One study followed 69 patients with junctional bradycardia for an approximate 2 year mean duration; during this follow-up period, eight patients (17.8%) without retrograde p waves and in one of the patients (4.2%) with retrograde P waves suffered a stroke or TIA, a difference that was statistically significant in this study (Kim et al., 2016). Spontaneous echo contrast, a risk factor for embolic stroke, along with atrial enlargement (Vincelj et al., 2002), has been found in individuals with junctional bradycardia (Koller-Strametz et al., 1995). In the previously described study, left atrial volume was significantly larger in the junctional bradycardia group compared to controls. The possibility of concurrent atrial fibrillation (or other arrhythmias) was investigated with Holter monitors in some patients, but given the nature of paroxysmal atrial fibrillation and limited timeframe of monitoring, was unable definitively ruled out. Likewise, details regarding intracranial vasculature were not reported in detail, and so it is difficult to definitively rule out other etiologies for the stroke or TIA in this population.

Sick sinus syndrome is likewise known as a risk factor for cerebral ischemia with patient characteristics including concurrent atrial fibrillation (Fisher et al., 1988), hypertension, heart failure, prior stroke or embolism being predictive of future stroke risk (Greenspon et al., 2004) as are the other risk factors associated with the atrial fibrillation ischemic stroke scoring systems irrespective of confirmed atrial fibrillation diagnosis (Svendsen et al., 2013). As one may expect, pacemaker implantation is not entirely protective against cerebral ischemia (Fisher et al., 1988; Mattioli et al., 1998) and has been associated with neurologic complications (please see Chapter 20).

Brady-arrhythmias may be associated with focal neuronal deficits via other mechanisms than hypoperfusion and embolic phenomenon as well. Patients with prior strokes have been known to have a recrudescence, or "unmasking" of prior stroke symptoms in association with other conditions, commonly with infectious or metabolic etiologies or medications with sedative properties (Thal et al., 1996). This unmasking phenomenon has been reported with other neurologic conditions, such as multiple sclerosis (Ratchford et al., 2014). In general, the deficits are not as severe as the prior, initial neurologic deficits and there are no new accompanying changes on magnetic resonance imaging despite the symptoms lasting a long enough duration for these changes to occur if they were secondary to new or progressive pathology. Up to 13% of patients in one case series of patients with stroke symptom recrudescence have had hypotension as a proposed "trigger" for an unmasking episode (Topcuoglu et al., 2017). The etiology of hypotension was not discussed nor is detailed description of the vascular anatomy provided in these patients. Given this, potential criticisms could include that unmasking did not occur per say, but that a patient with a prior stroke secondary to focal arterial stenosis may have had a transient ischemic attack in the same territory. Traditional symptom/imaging correlates and clinical reasoning dictates that neurologic deficits lasting for over for several hours in duration should be associated with restricted diffusion on MRI on brain imaging (spine imaging is less sensitive) if they are to be attributable to ischemia; but longer durations of ischemia without evidence of new infarction have been reported (Sacco et al., 2013).

The author presents the following case in support of brady-arrthymia associated hypotension and unmasking of prior deficits. An elderly patient was admitted to the hospital with suspected tachy-brady syndrome and developed focal neurologic deficits similar to his prior stroke (in addition to a simultaneous global hypoperfusion syndrome) after administration of intravenous diltiazem in an attempt to control a tachy-arrhythmia resulted in a heart rate in the low 30s along with a drop of over 40% of mean arterial pressure; the patient gradually recovered to neurologic baseline over 12 h later after all nodal agents were held with corresponding improvement in heart rate along with mean arterial pressure and had no restricted diffusion on MRI and no visible focal stenosis of significance on magnetic resonance angiography (although the study was somewhat motion limited).

Although in general patients with focal stenosis and associated hypoperfusion tend to either infarct when symptomatic, resulting in long-lasting deficits and associated changes on subsequent imaging, or resolve with the patient returning to neurologic baseline (with or without imaging changes), gray areas exist in between this dichotomy, a patient with congestive heart failure as well as severe middle cerebral artery stenosis presented with aphasia of 1-week duration and had complete resolution of aphasia postangioplasty, with no infarcted territory on magnetic resonance imaging (Ehtisham and Chimowitz, 2005). Given the fact that the symptoms lasted far too long of a duration to be considered a transient ischemic attack, yet completely resolved clinically with no evidence of infarcts on imaging the syndrome was classified as a case of "limping brain."

The discussion regarding neurologic complications of brady-arrhythmias until this point has largely focused on acute neurologic presentations, but more subtle subacute or chronic presentations are possible. Brady-arrhythmias have been hypothesized to contribute to a prolonged state of baseline cerebral hypo-perfusion, and in some patients, a concurrent state of cognitive impairment. This is a subset of the fairly large and heterogeneous category of cognitive impairment secondary to vascular mechanisms known as vascular cognitive impairment

(Sulkava and Erkinjuntti, 1987; Lane, 1991; Ruitenberg et al., 2005; O'Brien, 2006; de la Torre, 2012). In brief, the pathophysiology has been debated and is most likely multifactorial; decreased energy substrate delivery along with impaired toxin clearance likely play a role. Impairment in cerebrovascular autoregulation (which can be protective against lower perfusion pressures as previously discussed) has been postulated to be a factor as well (Shapiro and Chawla, 1969; de la Torre, 2012) along with "traditional" ischemic risk factors (Meyer et al., 1988).

Although most likely chronic in nature, the development of cognitive impairment meeting criteria for dementia has been reported following isolated events of hypotension. Although this is a well-known occurrence after cardiac arrest, cases have been reported with hypotension secondary to brady-arrhythmias that did not require resuscitation (Sulkava and Erkinjuntti, 1987). The authors reported a case series of patients with a new diagnosis dementia in temporal connection to brady-arrhythmia's associated with systemic hypotension, excluding patients requiring resuscitation. Of note, this was an uncommon presentation, representing only 4.5% cases of "vascular dementia" that were reviewed.

It is likely that ischemia resulting in hypoperfusion mediated these events. Brady-arrhythmias may otherwise potentially contribute to cognitive impairment via cerebral ischemia, although the traditional risk factors for multiinfarct dementia include other disease states traditionally associated with small vessel ischemia (such as hypertension, diabetes, and smoking (Meyer et al., 1988)) with infarcts more frequently located in subcortical territories (Gorelick et al., 1992). For much more in-depth discussion on the relationship to the heart and dementia along with cardiac output, cerebral blood flow and the brain, please see other dedicated chapters in this text.

If there is a casual link between brady-arrhythmias and cognitive impairment (or dementia), is it possible that some of this impairment may be reversed or at least halted with intervention? It is known that pacemaker implantation in patients with heart block results in improved cardiac output; pacemaker implantation yielded linear improvement in cardiac output as well cerebral blood flow in patients with complete heart block up to rates 60 bpm, but the association did not hold past this point (Shapiro and Chawla, 1969; Samet, 1973). Can this corresponding increase in cerebral blood blow lead to subsequent improvement in cognitive impairment in this patient population or has the damage to neuronal networks involved in cognition of a permanent, ischemic nature and the best that can be hoped for is reduction of further impairment?

Induction and subsequent reversal of potential pathologic substrates to cognitive impairment associated with low cerebral blood volume have been demonstrated in an animal model (ElAli et al., 2013). Rodents underwent iatrogenic unilateral CCA occlusion, with subsequent imaging demonstrating increased update of radiolabeled AB protein (thought to contribute on a pathophysiologic basis to Alzheimer's dementia) and alterations in glucose metabolism in the affected hemisphere only, with levels returning to baseline state after reperfusion. In humans, despite common subjective cognitive complaints associational with some brady-arrhythmias (primarily in the domains of memory and attention (Duschek and Schandry, 2007)), pacemaker implantation has generally not resulted in a subjective sense of cognitive improvement in patients when postimplantation quality of life surveys are conducted, even though quality of life overall seems to be perceived as improved (Sulfg et al., 1969; Stofmeel et al., 2001; Chen and Chao, 2002). One research group examined electroencephalograms (EEG) and measures of cerebral blood flow in seven patients postpacemaker implantation for complete heart block. All patients had increases in measures of cerebral perfusion, and EEG results postpacemaker implantation demonstrated significantly more appropriate fast activity and less slow activity. Cognition was not formally tested; however, patients in his cohort reported a subjective sense of memory improvement (Sulfg et al., 1969).

Koide and colleagues evaluated 14 patients who underwent permanent pacemaker placement for either complete heart block or sick sinus syndrome, with measures of cardiac output, cerebral blood flow, and verbal and nonverbal memory compared to matched controls on brief measures of verbal and nonverbal memory. Cerebral blood flow improved post pacemaker placement. When tested 2 months postprocedure, the bradycardic patients improved 14.4% on the verbal metric, placing them in the same range as matched controls; no improvement was noted on the nonverbal cognitive testing (Koide et al., 1994). Rockwood and colleagues administered neuropsychological batteries evaluating abstract reasoning and memory, to 19 patients both pre- and postpacemaker implantation; and despite subjective reports of cognitive improvement, the pre- and posttesting results did not reach statistical significance (Rockwood et al., 1992). Another small trial involving 26 patients demonstrated some evidence of cognitive improvement over a 6-month timeframe on measures of verbal fluency and the mini-mental state examination (Barbe et al., 2002).

There has not yet been any evidence that pacing modes for symptomatic bradycardia (atrial or ventricular) are associated with different cognitive outcomes as measured by cognitive section of the Cambridge Examination for Mental Disorders of the Elderly and the Cambridge Neuropsychological Test Battery over a 2-year period in

74 patients (Gribbin et al., 2005). To date, the majority of studies examining the relationship between cognition and intervention for symptomatic brady arrhythmias have included a very small number of participants and heterogeneous out measures of cognitive functioning; the most accurate evaluation of the available evidence at this time is best summarized, as the relationship between brady-arrhythmia, cognitive impairment, and any improvement postpacemaker implantation is an area that requires further investigation.

Another "subacute" or potentially clinically less obvious source of neurologic sequelae of brady-arrhythmias are those associated with chronotropic incompetence. Brady-arrhythmias are in general a potential cause of chronotropic incompetence, defined as an inadequate heart rate response to the physiologic demands placed on the cardiovascular system (which is generally described as under 80% of the age-predicated maximal heart rate (Brubaker and Kitzman, 2011)). Although there are associations with cardiovascular mortality, neurologic sequelae are not in general directly associated with isolated chronotropic incompetence, but indirect sequelae are possible. Although it is debated to what extent chronotropic incompetence limits exercise capacity (Jamil et al., 2016), a potential limitation on exercise capacity may lead to decreased aerobic and daily activity levels, which in turn are well known risk factors for ischemic stroke (Knight, 2012) via various postulated mechanisms (Howard and McDonnell, 2015). Higher levels of physical activity may have a protective effect in patients with Parkinson's disease (Oliveira de Carvalho et al., 2018), are associated with improved quality of life and reduced fatigue in patients with multiple sclerosis (Halabchi et al., 2017), and is currently being examined in a randomized controlled trial in dementia prevention (Iuliano et al., 2019) among others.

In summation, brady-arrhythmias manifest acute neuronal dysfunction via hypoperfusion, which can lead to presyncope states in addition to syncope and associated trauma, focal neurologic deficits via a variety of mechanisms and may be associated with a variety of less subtle presentations (such as cognitive impairment and those associated with chronotropic incompetence).

References

Aaslid R, Lindegaard KF, Sorteberg W et al. (1989). Cerebral autoregulation dynamics in humans. Stroke 20: 45–52.

Alexander BH, Rivara FP, Wolf ME (1992). The cost and frequency of hospitalization for fall-related injuries in older adults. Am J Public Health 82: 1020–1023.

Ali S, Khan MA, Khealani B (2006). Limb-shaking transient ischemic attacks: case report and review of literature. BMC Neurol 6: 5.

Amark P, Stodberg T, Wallstedt L (2007). Late onset bradyarrhythmia during vagus nerve stimulation. Epilepsia 48: 1023–1024.

Anderson RH, Yanni J, Boyett MR et al. (2009). The anatomy of the cardiac conduction system. Clin Anat 22: 99–113.

Armstead WM (2016). Cerebral blood flow autoregulation and dysautoregulation. Anesthesiol Clin 34: 465–477.

Arsenos P, Gatzoulis K, Dilaveris P et al. (2013). Arrhythmic sudden cardiac death: substrate, mechanisms and current risk stratification strategies for the post-myocardial infarction patient. Hell J Cardiol 54: 301–315.

Aste M, Brignole M (2017). Syncope and paroxysmal atrioventricular block. J Arrhythm 33: 562–567.

Auer J (2008). Syncope and trauma. Are syncope-related traumatic injuries the key to find the specific cause of the symptom? Eur Heart J 29: 576–578.

Bandera E, Botteri M, Minelli C et al. (2006). Cerebral blood flow threshold of ischemic penumbra and infarct core in acute ischemic stroke: a systematic review. Stroke 37: 1334–1339.

Barbe C, Puisieux F, Jansen I et al. (2002). Improvement of cognitive function after pacemaker implantation in very old persons with bradycardia. J Am Geriatr Soc 50: 778–780.

Barold SS, Herweg B (2012). Second-degree atrioventricular block revisited. Herzschrittmacherther Elektrophysiol 23: 296–304.

Barold SS, Jais P, Shah DC et al. (1997). Exercise-induced second-degree AV block: is it type I or type II? J Cardiovasc Electrophysiol 8: 1084–1086.

Bigger Jr. JT, Reiffel JA (1979). Sick sinus syndrome. Annu Rev Med 30: 91–118.

Brazzelli M, Chappell FM, Miranda H et al. (2014). Diffusion-weighted imaging and diagnosis of transient ischemic attack. Ann Neurol 75: 67–76.

Brignole M, Menozzi C, Lolli G et al. (1990). Carotid sinus syndrome and sick sinus syndrome: 2 frequent and distinct indications for pacemaker implantation. G Ital Cardiol 20: 5–11.

Brignole M, Deharo JC, De Roy L et al. (2011). Syncope due to idiopathic paroxysmal atrioventricular block: long-term follow-up of a distinct form of atrioventricular block. J Am Coll Cardiol 58: 167–173.

Brodsky M, Wu D, Denes P et al. (1977). Arrhythmias documented by 24 hour continuous electrocardiographic monitoring in 50 male medical students without apparent heart disease. Am J Cardiol 39: 390–395.

Brubaker PH, Kitzman DW (2011). Chronotropic incompetence: causes, consequences, and management. Circulation 123: 1010–1020.

Brubaker PH, Joo KC, Stewart KP et al. (2006). Chronotropic incompetence and its contribution to exercise intolerance in older heart failure patients. J Cardiopulm Rehabil 26: 86–89.

Burton CR (1962). Paroxysmal atrial tachycardia with atrioventricular block. Can Med Assoc J 87: 114–120.

Castro VJ, Nacht R (2000). Cocaine-induced bradyarrhythmia: an unsuspected cause of syncope. Chest 117: 275–277.

Chen HM, Chao YF (2002). Change in quality of life in patients with permanent cardiac pacemakers: a six-month follow-up study. J Nurs Res 10: 143–150.

Cotter R, Lyden J, Mehler PS et al. (2019). A case series of profound bradycardia in patients with severe anorexia nervosa: thou shall not pace? HeartRhythm Case Rep 5: 511–515.

Das A, Baheti NN (2013). Limb-shaking transient ischemic attack. J Neurosci Rural Pract 4: 55–56.

D'Ascenzo F, Biondi-Zoccai G, Reed MJ et al. (2013). Incidence, etiology and predictors of adverse outcomes in 43,315 patients presenting to the emergency department with syncope: an international meta-analysis. Int J Cardiol 167: 57–62.

de la Torre JC (2012). Cardiovascular risk factors promote brain hypoperfusion leading to cognitive decline and dementia. Cardiovasc Psychiatry Neurol 2012: 367516.

Demel SL, Broderick JP (2015). Basilar occlusion syndromes: an update. Neurohospitalist 5: 142–150.

Devinsky O, Pacia S, Tatambhotla G (1997). Bradycardia and asystole induced by partial seizures: a case report and literature review. Neurology 48: 1712–1714.

Dreifus LS, Michelson EL, Kaplinsky E (1983). Bradyarrhythmias: clinical significance and management. J Am Coll Cardiol 1: 327–338.

Drummond JC (1997). The lower limit of autoregulation: time to revise our thinking? Anesthesiology 86: 1431–1433.

Dugan TM, Nordgren RE, O'Leary P (1981). Transient global amnesia associated with bradycardia and temporal lobe spikes. Cortex 17: 633–637.

Duschek S, Schandry R (2007). Reduced brain perfusion and cognitive performance due to constitutional hypotension. Clin Auton Res 17: 69–76.

Ehtisham A, Chimowitz MI (2005). A clinical example of the "limping brain" syndrome: a case report. AJNR Am J Neuroradiol 26: 1521–1524.

ElAli A, Theriault P, Prefontaine P et al. (2013). Mild chronic cerebral hypoperfusion induces neurovascular dysfunction, triggering peripheral beta-amyloid brain entry and aggregation. Acta Neuropathol Commun 1: 75.

Fairfax AJ, Lambert CD (1976). Neurological aspects of sinoatrial heart block. J Neurol Neurosurg Psychiatry 39: 576–580.

Ferrer MI (1968). The sick sinus syndrome in atrial disease. JAMA 206: 645–646.

Finsterer J (2010). Man-in-the-barrel syndrome and its mimics. South Med J 103: 9–10.

Fisher M, Bastan B (2012). Identifying and utilizing the ischemic penumbra. Neurology 79: S79–S85.

Fisher M, Kase CS, Stelle B et al. (1988). Ischemic stroke after cardiac pacemaker implantation in sick sinus syndrome. Stroke 19: 712–715.

Fodstad H, Kelly PJ, Buchfelder M (2006). History of the cushing reflex. Neurosurgery 59: 1132–1137, Discussion 1137.

Gann D, Tolentino A, Samet P (1979). Electrophysiologic evaluation of elderly patients with sinus bradycardia: a long-term follow-up study. Ann Intern Med 90: 24–29.

Gorelick PB, Chatterjee A, Patel D et al. (1992). Cranial computed tomographic observations in multi-infarct dementia. A controlled study. Stroke 23: 804–811.

Greene HL (1990). Sudden arrhythmic cardiac death–mechanisms, resuscitation and classification: the Seattle perspective. Am J Cardiol 65: 4B–12B.

Greenspon AJ, Hart RG, Dawson D et al. (2004). Predictors of stroke in patients paced for sick sinus syndrome. J Am Coll Cardiol 43: 1617–1622.

Gribbin GM, Gallagher P, Young AH et al. (2005). The effect of pacemaker mode on cognitive function. Heart 91: 1209–1210.

Grigorean VT, Sandu AM, Popescu M et al. (2009). Cardiac dysfunctions following spinal cord injury. J Med Life 2: 133–145.

Grossman P, Taylor EW (2007). Toward understanding respiratory sinus arrhythmia: relations to cardiac vagal tone, evolution and biobehavioral functions. Biol Psychol 74: 263–285.

Halabchi F, Alizadeh Z, Sahraian MA et al. (2017). Exercise prescription for patients with multiple sclerosis; potential benefits and practical recommendations. BMC Neurol 17: 185.

Howard VJ, McDonnell MN (2015). Physical activity in primary stroke prevention: just do it!. Stroke 46: 1735–1739.

Isbister GK, Prior FH, Foy A (2001). Citalopram-induced bradycardia and presyncope. Ann Pharmacother 35: 1552–1555.

Iuliano E, di Cagno A, Cristofano A et al. (2019). Physical exercise for prevention of dementia (EPD) study: background, design and methods. BMC Public Health 19: 659.

Jamil HA, Gierula J, Paton MF et al. (2016). Chronotropic incompetence does not limit exercise capacity in chronic heart failure. J Am Coll Cardiol 67: 1885–1896.

Jansen PA, Gribnau FW, Schulte BP et al. (1986). Contribution of inappropriate treatment for hypertension to pathogenesis of stroke in the elderly. Br Med J (Clin Res Ed) 293: 914–917.

Jensen PN, Gronroos NN, Chen LY et al. (2014). Incidence of and risk factors for sick sinus syndrome in the general population. J Am Coll Cardiol 64: 531–538.

John RM, Kumar S (2016). Sinus node and atrial arrhythmias. Circulation 133: 1892–1900.

Kariyanna PT, Wengrofsky P, Jayarangaiah A et al. (2019). Marijuana and cardiac arrhythmias: a scoping study. Int J Clin Res Trials 4: 132.

Kharod S, Norman C, Ryan M et al. (2014). Severe unexplained relative hypotension and bradycardia in the emergency department. Case Rep Emerg Med 2014: 969562.

Kim GS, Uhm JS, Kim TH et al. (2016). Junctional bradycardia is a potential risk factor of stroke. BMC Neurol 16: 113.

Kjekshus JK, Blix AS, Grottum P et al. (1981). Beneficial effects of vagal stimulation on the ischaemic myocardium during beta-receptor blockade. Scand J Clin Lab Invest 41: 383–389.

Knight JA (2012). Physical inactivity: associated diseases and disorders. Ann Clin Lab Sci 42: 320–337.

Koide H, Kobayashi S, Kitani M et al. (1994). Improvement of cerebral blood flow and cognitive function following pacemaker implantation in patients with bradycardia. Gerontology 40: 279–285.

Koller-Strametz J, Wieselthaler G, Kratochwill C et al. (1995). Thromboembolism associated with junctional escape rhythm and atrial standstill after orthotopic heart transplantation. J Heart Lung Transplant 14: 999–1002.

Korczyn AD, Laor N, Nemet P (1976). Sympathetic pupillary tone in old age. Arch Ophthalmol 94: 1905–1906.

Kosowsky BD, Latif P, Radoff AM (1976). Multilevel atrioventricular block. Circulation 54: 914–921.

Lane RJ (1991). 'Cardiogenic dementia' revisited. J R Soc Med 84: 577–579.

Lee S, Wellens HJ, Josephson ME (2009). Paroxysmal atrioventricular block. Heart Rhythm 6: 1229–1234.

Lehmann H, Klein UE (1978). Familial sinus node dysfunction with autosomal dominant inheritance. Br Heart J 40: 1314–1316.

LeMay M, Gooding CA (1966). The clinical significance of the azygos anterior cerebral artery (A.C.A.). Am J Roentgenol Radium Therapy, Nucl Med 98: 602–610.

Lien WP, Lee YS, Chang FZ et al. (1977). The sick sinus syndrome: natural history of dysfunction of the sinoatrial node. Chest 72: 628–634.

Mahdian M (2013). Fall injuries: an important preventable cause of trauma. Arch Trauma Res 2: 101–102.

Mangla R, Kolar B, Almast J et al. (2011). Border zone infarcts: pathophysiologic and imaging characteristics. Radiographics 31: 1201–1214.

Mangrum JM, DiMarco JP (2000). The evaluation and management of bradycardia. N Engl J Med 342: 703–709.

Mathias CJ, Mallipeddi R, Bleasdale-Barr K (1999). Symptoms associated with orthostatic hypotension in pure autonomic failure and multiple system atrophy. J Neurol 246: 893–898.

Mattioli AV, Castellani ET, Vivoli D et al. (1998). Prevalence of atrial fibrillation and stroke in paced patients without prior atrial fibrillation: a prospective study. Clin Cardiol 21: 117–122.

Meyer JS, McClintic KL, Rogers RL et al. (1988). Aetiological considerations and risk factors for multi-infarct dementia. J Neurol Neurosurg Psychiatry 51: 1489–1497.

Meytes I, Kaplinsky E, Yahini JH et al. (1975). Wenckebach A-V block: a frequent feature following heavy physical training. Am Heart J 90: 426–430.

Michaelsson M, Riesenfeld T, Jonzon A (1997). Natural history of congenital complete atrioventricular block. Pacing Clin Electrophysiol 20: 2098–2101.

Milanesi R, Baruscotti M, Gnecchi-Ruscone T et al. (2006). Familial sinus bradycardia associated with a mutation in the cardiac pacemaker channel. N Engl J Med 354: 151–157.

Mlekusch W, Schillinger M, Sabeti S et al. (2003). Hypotension and bradycardia after elective carotid stenting: frequency and risk factors. J Endovasc Ther 10: 851–859, discussion 860-851.

Moraine JJ, Berre J, Melot C (2000). Is cerebral perfusion pressure a major determinant of cerebral blood flow during head elevation in comatose patients with severe intracranial lesions? J Neurosurg 92: 606–614.

Morris SG (2011). A model for teaching the determinants of cardiac output. Can J Cardiovasc Nurs 21: 8–12.

Myers RW, Pearlman AS, Hyman RM et al. (1974). Beneficial effects of vagal stimulation and bradycardia during experimental acute myocardial ischemia. Circulation 49: 943–947.

Nickens H (1985). Intrinsic factors in falling among the elderly. Arch Intern Med 145: 1089–1093.

Novak P (2016). Orthostatic cerebral hypoperfusion syndrome. Front Aging Neurosci 8: 22.

O'Brien JT (2006). Vascular cognitive impairment. Am J Geriatr Psychiatry 14: 724–733.

Oliveira de Carvalho A, Filho ASS, Murillo-Rodriguez E et al. (2018). Physical exercise for Parkinson's disease: clinical and experimental evidence. Clin Pract Epidemiol Ment Health 14: 89–98.

Oppenheimer S (2006). Cerebrogenic cardiac arrhythmias: cortical lateralization and clinical significance. Clin Auton Res 16: 6–11.

Orsini M, Catharino AM, Catharino FM et al. (2009). Man-in-the-barrel syndrome, a symmetrical proximal brachial amyotrophic diplegia related to motor neuron diseases: a survey of nine cases. Rev Assoc Med Bras (1992) 55: 712–715.

Osmonov D, Erdinler I, Ozcan KS et al. (2012). Management of patients with drug-induced atrioventricular block. Pacing Clin Electrophysiol 35: 804–810.

Pacher P, Kecskemeti V (2004). Cardiovascular side effects of new antidepressants and antipsychotics: new drugs, old concerns? Curr Pharm Des 10: 2463–2475.

Paul NL, Simoni M, Rothwell PM et al. (2013). Transient isolated brainstem symptoms preceding posterior circulation stroke: a population-based study. Lancet Neurol 12: 65–71.

Petermeyer BE (1976). Sick sinus (brady cardia-tachycardia) syndrome. J Am Osteopath Assoc 75: 1037–1045.

Pinsky WW, Gillette PC, Garson Jr A et al. (1982). Diagnosis, management, and long-term results of patients with congenital complete atrioventricular block. Pediatrics 69: 728–733.

Prabhakaran S, Lee VH (2009). Does diffusion-weighted imaging in transient ischemic attack patients improve accuracy of diagnosis, prognosis, or both? Stroke 40: e408, author reply e409.

Prabhakaran S, Chong JY, Sacco RL (2007). Impact of abnormal diffusion-weighted imaging results on short-term outcome following transient ischemic attack. Arch Neurol 64: 1105–1109.

Rasmussen K (1971). Chronic sinoatrial heart block. Am Heart J 81: 39–47.

Ratchford JN, Brock-Simmons R, Augsburger A et al. (2014). Multiple sclerosis symptom recrudescence at the end of the natalizumab dosing cycle. Int J MS Care 16: 92–98.

Rijnbeek PR, Witsenburg M, Schrama E et al. (2001). New normal limits for the paediatric electrocardiogram. Eur Heart J 22: 702–711.

Rockwood K, Dobbs AR, Rule BG et al. (1992). The impact of pacemaker implantation on cognitive functioning in elderly patients. J Am Geriatr Soc 40: 142–146.

Ruitenberg A, den Heijer T, Bakker SL et al. (2005). Cerebral hypoperfusion and clinical onset of dementia: the Rotterdam study. Ann Neurol 57: 789–794.

Ryan DJ, Kenny RA, Christensen S et al. (2015). Ischaemic stroke or TIA in older subjects associated with impaired dynamic blood pressure control in the absence of severe large artery stenosis. Age Ageing 44: 655–661.

Sacco RL, Kasner SE, Broderick JP et al. (2013). An updated definition of stroke for the 21st century: a statement for healthcare professionals from the American Heart Association/American Stroke Association. Stroke 44: 2064–2089.

Sall H, Timperley J (2015). Bradycardia in anorexia nervosa. BMJ Case Rep 2015: bcr2015211273.

Samet P (1973). Hemodynamic sequelae of cardiac arrhythmias. Circulation 47: 399–407.

Schmidt EA, Despas F, Pavy-Le Traon A et al. (2018). Intracranial pressure is a determinant of sympathetic activity. Front Physiol 9: 11.

Semelka M, Gera J, Usman S (2013). Sick sinus syndrome: a review. Am Fam Physician 87: 691–696.

Shapiro W, Chawla NP (1969). Observations on the regulation of cerebral blood flow in complete heart block. Circulation 40: 863–870.

Sobotka PA, Mayer JH, Bauernfeind RA et al. (1981). Arrhythmias documented by 24-hour continuous ambulatory electrocardiographic monitoring in young women without apparent heart disease. Am Heart J 101: 753–759.

Sova J, Kolda M (1949). EKG in normal pregnancy. Ceska Gynekol 14: 282–291.

Stofmeel MA, Post MW, Kelder JC et al. (2001). Changes in quality-of-life after pacemaker implantation: responsiveness of the Aquarel questionnaire. Pacing Clin Electrophysiol 24: 288–295.

Sulfg IA, Cronqvist S, Schuller H et al. (1969). The effect of intracardial pacemaker therapy on cerebral blood flow and electroencephalogram in patients with complete atrioventricular block. Circulation 39: 487–494.

Sulkava R, Erkinjuntti T (1987). Vascular dementia due to cardiac arrhythmias and systemic hypotension. Acta Neurol Scand 76: 123–128.

Svendsen JH, Nielsen JC, Darkner S et al. (2013). CHADS2 and CHA2DS2-VASc score to assess risk of stroke and death in patients paced for sick sinus syndrome. Heart 99: 843–848.

Task Force for the Diagnosis and Management of Syncope of the European Society of Cardiology et al. (2009). Guidelines for the diagnosis and management of syncope (version 2009). Eur Heart J 30: 2631–2671.

Thal GD, Szabo MD, Lopez-Bresnahan M et al. (1996). Exacerbation or unmasking of focal neurologic deficits by sedatives. Anesthesiology 85: 21–25, discussion 29A-30A.

Topcuoglu MA, Saka E, Silverman SB et al. (2017). Recrudescence of deficits after stroke: clinical and imaging phenotype, triggers, and risk factors. JAMA Neurol 74: 1048–1055.

Tsai YH, Lin JY, Huang YY et al. (2018). Cushing response-based warning system for intensive care of brain-injured patients. Clin Neurophysiol 129: 2602–2612.

van Dijk JG, Wieling W (2013). Pathophysiological basis of syncope and neurological conditions that mimic syncope. Prog Cardiovasc Dis 55: 345–356.

Viitasalo MT, Kala R, Eisalo A (1982). Ambulatory electrocardiographic recording in endurance athletes. Br Heart J 47: 213–220.

Vincelj J, Sokol I, Jaksic O (2002). Prevalence and clinical significance of left atrial spontaneous echo contrast detected by transesophageal echocardiography. Echocardiography 19: 319–324.

Vintila M, Fagarasanu R, Luca R (1988). Pacemaker therapy in a group of 132 patients with sick sinus syndrome—indications and results. Med Int 26: 305–309.

Vogler J, Breithardt G, Eckardt L (2012). Bradyarrhythmias and conduction blocks. Rev Esp Cardiol (Engl Ed) 65: 656–667.

Yahalom M, Spitz M, Sandler L et al. (2013). The significance of bradycardia in anorexia nervosa. Int J Angiol 22: 83–94.

Zeltser D, Justo D, Halkin A et al. (2004). Drug-induced atrioventricular block: prognosis after discontinuation of the culprit drug. J Am Coll Cardiol 44: 105–108.

Zoob M, Smith KS (1963). The aetiology of complete heart-block. Br Med J 2: 1149–1153.

Handbook of Clinical Neurology, Vol. 177 (3rd series)
Heart and Neurologic Disease
J. Biller, Editor
https://doi.org/10.1016/B978-0-12-819814-8.00027-5

Chapter 16

Effects of acute neurologic disease on the heart

CATHERINE ARNOLD AND SARA HOCKER*

Department of Neurology, Mayo Clinic, Rochester, MN, United States

Abstract

Patients with acute neurologic disease often also have evidence of cardiac dysfunction. The cardiac dysfunction may result in a number of clinical signs including abnormal EKG changes, variations in blood pressure, development of cardiac arrhythmias, release of cardiac biomarkers, and reduced ventricular function. Although typically reversible, these cardiac complications are important to recognize as they are associated with increased morbidity and mortality. In this chapter, we discuss the suspected pathophysiology, clinical presentation, and management of the cardiac dysfunction that occur as a consequence of different types of acute neurologic illness.

INTRODUCTION

Through a variety of mechanisms, the nervous system modulates cardiac function. When the nervous system suffers injury, these pathways may be disrupted and cardiac dysfunction may occur. In this chapter, we discuss the normal physiologic interplay between the heart and nervous system as well as the pathophysiology and clinical manifestations of cardiac dysfunction following several types of acute neurologic injury.

NORMAL PHYSIOLOGY

The parasympathetic and sympathetic nervous systems make up the autonomic nervous system and play fundamental and antagonistic roles in the regulation of cardiac function (Gordan et al., 2015; Murthy et al., 2015). The parasympathetic output to the heart is primarily mediated by the vagus nerve that originates from nuclei within the medulla. Increased parasympathetic output to the heart results in decreased heart rate, atrioventricular (AV) conduction, and ventricular excitability. In contrast, sympathetic output originates in the thoracic and results in increased heart rate, AV conduction, and ventricular excitability and contractility. Higher cerebral structures including the frontal cortex, insula, amygdala, cingulate, hypothalamus, and periaqueductal gray matter also influence cardiac function by exhibiting descending modulation of the autonomic outflow (Scorza et al., 2009; Gordan et al., 2015; Murthy et al., 2015; Chen et al., 2017).

In addition to the autonomic nervous system, the cardiovascular system is also influenced by the neuroendocrine system, predominantly via the hypothalamic–pituitary–adrenal (HPA)-axis. The HPA axis is responsible for creating the heightened "stress–response" mediated by the adrenal glands via its production and release of cortisol and catecholamines (Gordan et al., 2015). Catecholamines, specifically, have direct impact on cardiovascular function via their interactions with adrenergic receptors resulting in the elevation of heart rate, increased force of cardiac contraction, and changes in blood pressure (Hinson and Sheth, 2012; Krishnamoorthy et al., 2016).

Thus, together, the autonomic nervous and neuroendocrine systems allow the cardiovascular system to appropriately respond to the environment. Injury to the nervous system may disrupt the multiple pathways involved in this process and result in a dysfunctional cardiac response. The severity of cardiac dysfunction can vary from mild and clinically insignificant to life-threatening

*Correspondence to: Sara Hocker, MD, Mayo Clinic, Department of Neurology, 200 1st Street SW, Rochester, MN 55905, United States. Tel: +1-507-284-4741, E-mail: hocker.sara@mayo.edu

and devastating and is largely determined by the structures affected and severity of injury to the nervous system (Gordan et al., 2015).

BRAIN

Increased intracranial pressure

The Cushing reflex is a well-recognized response to significantly elevated intracranial pressure (ICP). It has been described in severe cases of subarachnoid or intracerebral hemorrhage, CNS infection, tumor, and traumatic brain injury (TBI). It is estimated to occur in up to one-third of patients with increased ICP and may predict impending herniation (Fodstad et al., 2006). The classic Cushing response is comprised of a triad of clinical signs including increased systolic pressure/wide pulse pressure, bradycardia, and respiratory irregularity. Increased intracranial pressure leads to diminished perfusion pressure within the brain. This leads to a compensatory activation of the sympathetic nervous system to maintain cerebral perfusion. As a result, systolic blood pressure increases. Baroreceptor stimulation from the increased blood pressure leads to activation of the parasympathetic nervous system and in response, heart rate is slowed. Additionally, increased intracranial pressure may lead to increased pressure on the brainstem resulting in dysfunction of the respiratory center and irregular breathing (Wan et al., 2008).

Paroxysmal sympathetic hyperactivity

Paroxysmal sympathetic hyperactivity (PSH) is not uncommon following acute and severe injury to the central nervous system. Although classically described in patients with TBI, it may also be seen in severe ischemic injury, subarachnoid hemorrhage, intracerebral hemorrhage, tumors, CNS infection, or in cases of acute hydrocephalus (Rabinstein, 2007; Fernandez-Ortega et al., 2012; Meyfroidt et al., 2017). Although best described as paroxysmal sympathetic hyperactivity, this phenomenon has been variably named as autonomic storming, sympathetic storming, dysautonomia, and diencephalic seizures among others.

Clinical manifestations of PSH can be quite striking with rather abrupt increases in temperature, blood pressure, heart rate, and respiratory rate as well as diaphoresis and posturing. Often episodes are triggered by external stimulation (Meyfroidt et al., 2017). The exact pathophysiology behind these changes is not entirely clear, but it is thought that there is a disruption of descending inhibitory networks, which ultimately results in uncontrolled excitation of the spinal circuit and paroxysms of increased sympathetic output (Fernandez-Ortega et al., 2017; Meyfroidt et al., 2017).

Treatment of PSH is multifaceted. Triggers that provoke episodes such as physical stimulation (i.e., physical touch, suctioning) should be avoided as much as possible. Multiple pharmacological options are available to both prevent and halt paroxysms. Medications such as opioids (primarily morphine and fentanyl), nonselective beta-blockers like propranolol or labetalol, benzodiazepines, and neuromodulating agents such as gabapentin, baclofen, and bromocriptine have been used to treat PSH with success (Meyfroidt et al., 2017). Often a combination of these medications is required to adequately control the paroxysms (Rabinstein, 2007). Without appropriate treatment of PSH, secondary systemic harm, including cardiovascular injury, may occur. Close monitoring for signs/symptoms of such complications is critical so that appropriate management is not delayed (Fernandez-Ortega et al., 2012; Meyfroidt et al., 2017).

Acute cerebrovascular disorders

Much of what we know regarding the cardiac dysfunction that results as a consequence of neurologic injury stems from experimental animal studies and clinical data from patients who suffer acute subarachnoid hemorrhage and ischemic stroke (Krishnamoorthy et al., 2016). Signs of cardiac dysfunction following acute cerebrovascular disease can be broadly classified into EKG changes and arrhythmias, left ventricular dysfunction, and release of cardiac biomarkers. The presence and severity of these complications correlates with degree of the neurologic injury (Ripoll et al., 2017). Increased age, premorbid cardiac disease, and vascular risk factors are also associated with increased risk of cardiac complications following ischemic or hemorrhagic stroke (Rincon et al., 2008; Al-Qudah et al., 2015; Krishnamoorthy et al., 2016; Scheitz et al., 2018).

EKG changes are quite prevalent in patients with acute cerebrovascular disease with frequencies reported as high as 80% and 90% in subarachnoid hemorrhage and ischemic stroke, respectively (Manea et al., 2015; Krishnamoorthy et al., 2016; Chen et al., 2017; Scheitz et al., 2018). Repolarization changes, particularly QT prolongation, are the most common abnormalities seen (Chen et al., 2017; Kerro et al., 2017). Ischemic changes including ST-depression or elevation and T-wave inversion are also seen, particularly within the first 72h following the inciting event (Kerro et al., 2017). Changes are often best seen in the anterolateral or inferolateral leads (Samuels, 2007). Some of these changes may be reflective of demand ischemia. However, in most cases, the EKG changes do not represent true ischemic heart disease but rather a manifestation of autonomic dysregulation; this is especially true in cases of hemorrhagic stroke (Samuels, 2007). Most of the EKG changes are transient but have an associated risk of cardiac arrhythmia, and therefore, should be interpreted cautiously (Scheitz et al., 2018). Cardiac arrhythmias including bradycardia, atrial fibrillation, supraventricular tachycardia, as well as more malignant ventricular arrhythmias like ventricular tachycardia or ventricular fibrillation, may also occur as a result of autonomic dysregulation (Al-Qudah et al., 2015; Tahsili-Fahadan and Geocadin, 2017; Scheitz et al., 2018).

Impaired left ventricular function is also not uncommon following an acute cerebrovascular event. Signs of ventricular dysfunction are present in up to 30% of patients with ischemic stroke or subarachnoid hemorrhage (Murthy et al., 2015; Ripoll et al., 2017; Scheitz et al., 2018). The sites of dyskinesia may differ in patients with neurogenic disease compared to those with other forms of stress-related cardiomyopathy with the basal and mid-ventricular segments of the anteroseptal and anterior walls being more often affected than the more traditional left ventricular apical or mid-ventricular segments involvement seen in nonneurologenic stress cardiomyopathy (Murthy et al., 2015). Identifying ventricular dysfunction is important as it has been associated with poor functional outcome (Ripoll et al., 2017; Scheitz et al., 2018).

Takotsubo cardiomyopathy, an acute heart failure syndrome characterized by apical ballooning and severe impairment of ventricular function, may also be seen in acute cerebrovascular disease. It occurs in 1%–6% of SAH patients and 0.5%–1.2% of ischemic stroke patients (Murthy et al., 2015; Kerro et al., 2017; Ripoll et al., 2017; Scheitz et al., 2018). Typically, the cardiac impairment occurs early after the inciting event. Patients with Takotsubo cardiomyopathy require close monitoring of their hemodynamic status (Kerro et al., 2017; Scheitz et al., 2018). Euvolemia should be targeted and adrenergic blockade or inotropic agents may be necessary to achieve hemodynamic stability (Murthy et al., 2015; Chen et al., 2017; Kerro et al., 2017). Takotsubo cardiomyopathy is also associated with systemic complications including prolonged intubation and vasopressor support requirement, pulmonary edema, cerebral vasospasm, and cardiac arrhythmias. Although it is associated with overall increased morbidity and mortality (Murthy et al., 2015; Kerro et al., 2017), once beyond the acute phase, either full or partial cardiac recovery usually occurs within 3 weeks (Kerro et al., 2017). Consultation with a cardiologist is advised in these settings for initial recommendations and follow-up to ensure resolution of the abnormalities.

Elevated troponin levels have been documented in >30% of patients with subarachnoid hemorrhage (Krishnamoorthy et al., 2016; Ripoll et al., 2017) and roughly 10%–20% of patients with ischemic stroke (Scheitz et al., 2018). Notably, the degree of increase in troponin in either group is lower than that seen in myocardial infarction. In up to 50% of patients with SAH, creatine kinase myocardial isoenzyme (CK-MB) is also elevated. Although increased CK-MB measures have been associated with impaired left ventricular function, its use as a marker of cardiac dysfunction related to neurologic injury is not felt to be appropriate as it is a nonspecific marker and seen elevated in a variety of other conditions (Ripoll et al., 2017). Serum levels of B-type natriuretic peptide may also be elevated in patients with acute cerebrovascular disease and are more specific for cardiac dysfunction (Levine, 2007; Ripoll et al., 2017). If these markers are elevated, trending values are generally recommended to ensure values are reducing or normalizing.

The suspected mechanism of cardiac dysfunction following acute cerebrovascular disease is complex and likely involves multiple separate but interrelated processes. Due to focal injury and mechanical factors, critical pathways of the neuroendocrine and autonomic nervous systems are disrupted, leading to an inappropriate influence on the cardiovascular system (Hinson and Sheth, 2012). Excess catecholamine appears to play a pivotal role in the pathophysiology as well (Samuels, 2007; Krishnamoorthy et al., 2016; Chen et al., 2017). Catecholamines increase cardiac energy demand via both direct stimulation of adrenergic receptors of the heart and systemic vasoconstriction, which increases cardiac afterload and workload (Porto et al., 2013; Krishnamoorthy et al., 2016; Chen et al., 2017). When a mismatch of energy demand and supply occurs, a number of linked biochemical reactions are triggered, which ultimately lead to cellular toxicity via increased intracellular calcium and increased reactive oxygen species (Chen et al., 2017). Studies have also suggested brain injury generates a neuroinflammatory response. In the setting of disruption of the blood–brain barrier, proinflammatory and immunologically active substances are released into the systemic circulatory system (Chen et al., 2017). This then can trigger a systemic inflammatory response syndrome, which may also lead to cardiac dysfunction or failure (Krishnamoorthy et al., 2016).

Seizures and status epilepticus

Cardiac arrhythmias are the most common complication seen related to seizures, but EKG changes, cardiac biomarker release, and stress-induced cardiomyopathy have also been described (Scorza et al., 2009; Sevcencu and Struijk, 2010; Duplyakov et al., 2014; Eggleston et al., 2014; van der Lende et al., 2016; Ripoll et al., 2017; Shmuely et al., 2017).

Various cardiac arrhythmias have been described in relation to epilepsy and are suspected to be secondary to excessive cortical stimulation of the autonomic networks, disrupting normal autonomic output (Jansen and Lagae, 2010). Increased catecholamine release, especially in cases of convulsive seizures, seems to also play a role (van der Lende et al., 2016). The seizures themselves may induce physiologic changes such as apnea and hypoxia that further provoke a proarrhythmic state (Sevcencu and Struijk, 2010). In addition, there appears to be a potential genetic contribution, mediated through certain genes expressing abnormal electrolyte channels present in both the brain and heart tissue (Shmuely et al., 2017).

Arrhythmias can precede the clinical and electrographic onset of the seizure, be present during the seizure, or follow a seizure during the postictal period (van der Lende et al., 2016; Shmuely et al., 2017). The most commonly reported rhythm is sinus tachycardia, which has been seen in up to 80% of seizures (Jansen and Lagae, 2010; Eggleston et al., 2014; van der Lende et al., 2016). Ictal tachycardia is most often seen in temporal lobe seizures but can be seen in other types of focal seizures as well as generalized seizures (Sevcencu and Struijk, 2010). The tachycardia is more prolonged in those seizures of temporal lobe onset (Jansen and Lagae, 2010; Eggleston et al., 2014). Heart rates can approach 200 beats per minute (bpm) and typically improve with cessation of seizure activity (Sevcencu and Struijk, 2010). Tachycardia leading to ventricular tachycardia or ventricular fibrillation (VF) has also been reported and is thought possibly a cause of sudden unexpected death in epilepsy (SUDEP) (Jansen and Lagae, 2010).

Bradycardia may also be seen during seizures but is less frequent, occurring in approximately 6% of patients. Similar to ictal tachycardia, ictal bradycardia is also more prevalent in temporal lobe seizures (Duplyakov et al., 2014). Heart rates typically range from 20 to 40 beats per minute (Sevcencu and Struijk, 2010). Bradycardia is important to recognize as it may lead to ictal asystole (Jansen and Lagae, 2010; Sevcencu and Struijk, 2010). Ictal asystole has been reported in 0.177% of patients admitted for prolonged EEG monitoring and 0.318% in patients with refractory focal epilepsy. Most often, it occurs during focal seizures of temporal lobe onset with associated impaired awareness and typically occurs within the first 30 s of seizure activity (van der Lende et al., 2016). Clinically, it resembles vasovagal syncope with a transient loss of consciousness and tone, slowing/pause of the heart rate, and decreased blood pressure (Shmuely et al., 2017). Ictal asystole is assumed to be self-limiting and typically lasts between 20 and 60 s (van der Lende et al., 2016). However, it can result in bodily injury due to unexpected falls. Cardiac pacing has been shown to reduce falls and injury in these cases (Sevcencu and Struijk, 2010; van der Lende et al., 2016; Shmuely et al., 2017).

Postictal arrhythmias are less common than ictal arrhythmias and are typically related to generalized convulsive seizures rather than focal seizures (Duplyakov et al., 2014; van der Lende et al., 2016). Postictal asystole, atrial fibrillation, and ventricular tachycardia/fibrillation have all been described and are often more prolonged in duration than arrhythmias that occur during the ictal period (van der Lende et al., 2016). Importantly, postictal arrhythmias are associated with a higher fatality rate and are thought a potential contributor in cases of SUDEP (Duplyakov et al., 2014; van der Lende et al., 2016; Shmuely et al., 2017). Optimization of seizure control is key in avoiding such devastating complications.

EKG changes may also be seen in relation to seizures (Opherk et al., 2002; Tigaran et al., 2003; Stollberger and Finsterer, 2004; Sevcencu and Struijk, 2010). Excluding tachycardia, abnormal EKG findings were seen in 35% of generalized seizures and 15% of focal seizures and are often associated with more prolonged seizures (Nei et al., 2000; Opherk et al., 2002). Shortened QTc intervals, ST-depression, or T-wave inversion have been described (Opherk et al., 2002). An associated elevation in troponin is not a typical finding of seizures and suggests cardiac injury; therefore, elevated troponins should prompt further evaluation, including consultation with a cardiologist (Tigaran et al., 2003; Stollberger and Finsterer, 2004).

Seizures, especially generalized convulsive seizures, may also induce Takotsubo cardiomyopathy/stress-induced cardiomyopathy through a similar mechanism of excessive catecholamine release as described in cases secondary to acute cerebrovascular disease (Shmuely et al., 2017). Patients with seizure-related Takotsubo cardiomyopathy typically present within the first 72 h following the seizures with symptoms suggestive of heart failure and have a more complicated and serious course than patients who suffer from the cardiomyopathy due to other causes (Stollberger et al., 2011). Similar to other causes of stress-induced cardiomyopathy, consultation with cardiology is recommended and follow-up echocardiogram is typically recommended to ensure resolution of the abnormalities.

In general, seizure-related cardiac dysfunction improves as seizure control also improves. However, many antiepileptic medications have adverse cardiac side effects as well (Sevcencu and Struijk, 2010.) For example, phenytoin can result in bradycardia and even asystole and can raise the threshold for malignant arrhythmias like VF (Feldman and Gidal, 2013; Gulditen et al., 2016; Ishizue et al., 2016). Due to its vasodilatory effects and negative inotropic effects, phenytoin can also cause a lowering in blood pressure. Similarly, lacosamide, particularly at higher doses, may cause certain arrhythmias like atrial fibrillation or lead to heart block and bradycardia via its effect on the sinus node (Chinnasami et al., 2013; Kaufman et al., 2013). The influence of antiepileptic medications on cardiac function, especially when used in higher doses or in the acute setting, should not be overlooked.

Patients with refractory status epilepticus (SE) are at an even higher risk of developing cardiac complications with up to two-thirds of patients with SE having signs of cardiac injury (Hawkes and Hocker, 2018). This is likely due to a number of factors. For example, a similar but more prolonged surge in catecholamine release may occur in patients in SE compared to those with seizures of shorter duration. In addition, direct autonomic effects of prolonged seizure activity and metabolic derangements like electrolyte disturbances or metabolic acidosis increase the risk of cardiac injury. The medications often required to ultimately control the seizures (including general anesthetics like propofol) can also have increased risk of cardiotoxicity, as well (Hocker, 2015; Hawkes and Hocker, 2018).

Similar to other types of brain injury, a variety of signs/symptoms of cardiac injury may occur in SE. For example, elevated cardiac biomarkers are not uncommon and most often due to demand ischemia, especially in patients with premorbid coronary artery disease (Hocker et al., 2013; Hawkes and Hocker, 2018). EKG changes occur in almost 40% of patients and most often include nonspecific ST changes or T-wave inversion. Cardiac arrhythmias occur frequently. Sinus tachycardia is the most commonly reported arrhythmia but atrial fibrillation/flutter, sinus bradycardia, and more dangerous arrhythmias like ventricular tachycardia/fibrillation may also occur (Hocker et al., 2013). In addition, transient systolic dysfunction (i.e., stress-induced or Takotsubo cardiomyopathy) may affect up to 50% of patients with SE as well and is thought secondary to hyperadrenergic stimulation (Hawkes and Hocker, 2018). Notably, the presence of cardiac injury in patients with SE has been associated with poor outcomes (Hawkes and Hocker, 2018).

Traumatic brain injury

Similar cardiac dysfunction is seen in patients with traumatic brain injury compared to those with acute cerebrovascular disease, likely related at least in part to the fact that many patients with TBI also have SAH and/or intracerebral hemorrhage. Cardiac injury occurs in approximately 50% of patients with severe TBI, and EKG change abnormalities, left ventricular dysfunction, and release of cardiac biomarkers have been described (Hasanin et al., 2016). Much of the pathophysiology related to these changes is also similar with dysfunction resulting from a combination of sympathetic hyperactivity, catecholamine excess, and neuro-related inflammation (Lim and Smith, 2007; Prathep et al., 2014; Cheah et al., 2017; El-Menyar et al., 2017; Ripoll et al., 2017; Venkata and Kasal, 2018). However, increased intracranial pressure is thought to play a more significant role in TBI (Prathep et al., 2014).

EKG changes are not uncommon in TBI, occurring in up to 60% of patients with severe TBI. The changes are thought to be reflective of the degree of intracranial hypertension (Lim and Smith, 2007; Hasanin et al., 2016). Various changes have been described including QTc prolongation, ST changes, and T-wave inversion or flattening (El-Menyar et al., 2017; Venkata and Kasal, 2018). Prolonged QTc is the most commonly reported finding, especially in cases of traumatic SAH, and may predispose patients to ventricular arrhythmias (Lim and Smith, 2007; Krishnamoorthy et al., 2014). The EKG changes typically evolve over a course of 2 weeks before resolving (Lim and Smith, 2007). The magnitude of ECG changes seems to also correlate with the severity of the brain injury, with greater ST and QT changes associated with a worsened neurologic outcome (Krishnamoorthy et al., 2014). Patients with TBI may also be predisposed to cardiac arrhythmias in the absence of EKG changes, including malignant tachyarrhythmias like ventricular tachycardia or fibrillation due to autonomic dysregulation. In cases of significantly increased ICP, bradycardia may be seen as part of Cushing's reflex (Lim and Smith, 2007; El-Menyar et al., 2017).

Systolic heart dysfunction has also been described in around 13%–21% of patients with TBI with Takotsubo cardiomyopathy occurring in the minority of these patients (Prathep et al., 2014; Cheah et al., 2017; Venkata and Kasal, 2018). The degree of systolic dysfunction seems to correlate with the severity of neurologic injury (El-Menyar et al., 2017). Cardiac function typically improves over a course of several weeks (El-Menyar et al., 2017). However, abnormal echocardiogram findings within 2 weeks from the initial injury have been associated with increased in-hospital mortality (Prathep et al., 2014; Venkata and Kasal, 2018).

Cardiac biomarkers may also be elevated, typically within the first 2 weeks following injury, before normalizing (Prathep et al., 2014). Troponin has been elevated in up to 30% of patients with TBI and has been associated with increased mortality (El-Menyar et al., 2017). CK-MB has been shown to be elevated in 50%–90% of patients with TBI as well and has also been associated with increased morbidity and mortality (El-Menyar et al., 2017). Both elevated troponin and CK-MB levels have been associated with echocardiogram abnormalities and should be trended, and if persistently elevated, should prompt consultation with a cardiologist (Prathep et al., 2014).

SPINAL CORD

Impairment of the autonomic nervous system, including that of the cardiovascular system, is an unfortunate complication of spinal cord injury (SCI) and other causes of myelopathy (Grigorean et al., 2009). Because the parasympathetic input stems from neurons within the brainstem and is relayed to the heart primarily via the vagus nerve, the parasympathetic innervation of the heart remains relatively intact in cases of SCI (Grigorean et al., 2009; Popa et al., 2010; Hagen et al., 2012). The preganglionic neurons of the sympathetic nervous system are located in the intermediolateral cell column within the gray matter of the spinal cord from T1-L2, and thus, SCI largely results in sympathetic dysfunction (Hagen et al., 2012; Hagen, 2015; Phillips and Krassioukov, 2015). The degree of cardiac dysfunction is a direct consequence of the level of spinal cord injury and severity of the damage. Lesions occurring at T6 and above result in the highest degree of cardiac dysfunction. Cardiac complications may occur in the immediate period following the SCI or months to years following the injury (Grigorean et al., 2009; Hagen et al., 2012; Hagen, 2015). Given the scope of this chapter, only complications seen in the acute phase of spinal cord injury will be discussed.

Within the first few minutes following an acute injury to the spinal cord, a massive sympathetic surge occurs due to release of norepinephrine from the adrenal glands. Clinically, patients are severely hypertensive with reflexive bradycardia or tachycardia. These changes typically last for 3–4 min. (Popa et al., 2010; Phillips and Krassioukov, 2015). Following this brief phase, in what is more classically known as the "acute phase" of spinal cord injury, there is an abrupt disruption of sympathetic output to the body, leading to an unopposed parasympathetic systemic response (Popa et al., 2010). This imbalance is further amplified by sensitivity of the vagus nerve (Grigorean et al., 2009; Hagen et al., 2012). Clinically, this results in bradycardia and hypotension, classically known as spinal or neurogenic shock (Grigorean et al., 2009; Popa et al., 2010). In patients with high cervical injuries, 75% will have bradycardia; in those with thoracic injury, bradycardia is less severe. Nearly 100% of patients with high cervical cord injuries will present with hypotension with mean systolic blood pressures typically around 90 mmHg in the supine position; approximately 50% will require vasopressor support (Popa et al., 2010). Notably, it is the bradycardia that discriminates neurogenic shock from hypovolemic shock, which is typically associated with tachycardia. Spinal shock may last days to months with the average duration ranging from 4 to 12 weeks (Popa et al., 2010). Even if the bradycardia resolves, patients often experience low blood pressures and are prone to orthostatic hypotension. Preventative interventions are an important component of treatment of orthostatic hypotension and include ensuring appropriate fluid balance, use of compression bandages/stockings, upright positioning, and avoiding large meals or significant heat stress. Medications such as diuretics should also be avoided. Despite these conservative measures, pharmacologic therapy with midodrine may be required (Phillips and Krassioukov, 2015). Although most often seen in more chronic phases of spinal cord injury, autonomic dysreflexia may also occur in the acute phase as well (Phillips and Krassioukov, 2015). Autonomic dysreflexia occurs in patients with lesions above T6 and is characterized by an acute and unregulated surge in systolic blood pressure of at least 20 mmHg with or without an associated decrease in heart rate in response to afferent visceral or somatic stimulation (such as full bowel or bladder) from below the level of the injury (Grigorean et al., 2009; Hagen et al., 2012; Phillips and Krassioukov, 2015). In addition to hypertension and bradycardia, patients may appear flushed or have extreme headache; if left untreated, significant complications like intracerebral hemorrhage, seizures, cardiac arrhythmias, or even death can occur. Triggers such as distended bladder or bowel and other uncomfortable or painful stimuli should be avoided as much as possible and if present, should be removed as soon as possible to allow for resolution of the episode of autonomic dysfunction. If the episode does not resolve with removal of the trigger, short-term pharmacologic intervention may be necessary. Medications such as nifedipine, carvedilol, and nitropaste have been used (Popa et al., 2010; Phillips and Krassioukov, 2015).

PERIPHERAL NERVE

Cardiovascular dysfunction may be seen in a variety of acute neuropathies with autonomic involvement including acute inflammatory demyelinating polyradiculoneuropathy (AIDP), immune-mediated autonomic neuropathies, or neuronopathies associated with underlying malignancy (paraneoplastic syndromes) or other systemic autoimmune conditions (i.e., Celiac disease, systemic lupus erythematosus, or Sjogren's syndrome) as well as acute autonomic and sensory neuropathy and acute autonomic, sensory and motor neuropathy. (Freeman, 2007; Koike et al., 2013).

Acute inflammatory demyelinating polyradiculoneuropathy

Autonomic dysfunction has been described in up to two-thirds of patients with acute demyelinating polyradiculoneuropathy (Burakgazi and AlMahameed, 2016; Anandan et al., 2017; Zaeem et al., 2019). Cardiac-specific dysfunction is thought due to a variety of potential mechanisms including a direct autoimmune process affecting the sodium channels of the sinoatrial and atrioventricular nodes; particular vulnerability of the vagus nerve to demyelination; overall heightened sympathetic activity with increased catecholamine release; increased catecholamine sensitivity in denervated organs; and impairment of the baroreceptor reflex (Asahina et al., 2002; Burakgazi and AlMahameed, 2016; Zaeem et al., 2019). Cardiac dysfunction can present with a variety of clinical signs including arrhythmias, blood pressure variability, left ventricular systolic dysfunction, and electrocardiographic changes (Asahina et al., 2002; Mukerji et al., 2009; Burakgazi and AlMahameed, 2016).

Sustained sinus tachycardia is the most common cardiac finding of AIDP, occurring in up to 50% of patients, with rates >125 bpm occurring in approximately 25% (Asahina et al., 2002; Burakgazi and AlMahameed, 2016). Atrial and ventricular arrhythmias can also be seen in AIDP, but the most life-threatening arrhythmia is bradycardia, which has been reported in 6%–34% of patients (Asahina et al., 2002; Burakgazi and AlMahameed, 2016; Anandan et al., 2017). Patients with severe, debilitating AIDP, including those requiring mechanical ventilation and those with fluctuant blood pressures with systolic blood pressures varying >85 mmHg, appear to be at highest risk of developing symptomatic bradycardia and heart block (Mukerji et al., 2009; Burakgazi and AlMahameed, 2016). Close monitoring is necessary and temporary or even permanent cardiac pacing may be required (Mukerji et al., 2009).

Variations in blood pressure are common in AIDP. Due to increased sympathetic input, hypertension with increased systolic blood pressures of >25 mg Hg is seen in virtually all patients with AIDP, especially those with respiratory failure (Asahina et al., 2002; Burakgazi and AlMahameed, 2016). Hypotension may also be seen but occurs less commonly (Mukerji et al., 2009; Anandan et al., 2017). Extreme fluctuations portend a poor prognosis (Burakgazi and AlMahameed, 2016). BP fluctuations and variability are usually transient in AIDP (Mukerji et al., 2009). Close monitoring is necessary and supportive and symptomatic treatment may be needed. Judicious use of IV antihypertensive medications may be required in cases of severe hypertension, whereas IV fluid boluses and low-dose vasopressor support may be required in cases of extreme hypotension. In either case, monitoring treatment response is critical given the degree of autonomic sensitivity present in these patients (Mukerji et al., 2009; Burakgazi and AlMahameed, 2016). EKG changes are also not uncommon in patients with AIDP. Changes including QT prolongation, AV block, and ischemic changes such as ST elevation or depression have been described (Burakgazi and AlMahameed, 2016).

Direct myocardial damage may also occur in patients with AIDP and may be related to a direct autoimmune process on the myocardial cells themselves and/or related to increased catecholamine levels and sensitivity (Burakgazi and AlMahameed, 2016). Clinically, patients can be asymptomatic or experience symptoms suggestive of heart failure. Myocarditis may also occur. The frequency of myocardial dysfunction in patients with AIDP is unknown as many patients do not routinely undergo echocardiography. Care is often supportive and most patients recover completely over a period of several weeks (Mukerji et al., 2009; Burakgazi and AlMahameed, 2016).

NEUROMUSCULAR JUNCTION

Myasthenic crisis

Although uncommon, stress-induced cardiomyopathy has been reported in several cases of myasthenic crisis (Bijulal et al., 2009; Peric et al., 2011; Finsterer and Stollberger, 2016; Benjamin et al., 2018; Douglas et al., 2018), it is thought to be provoked by increased sympathetic activity and catecholamine excess related to the stress of the myasthenic crisis itself (Douglas et al., 2018). However, the question of a direct link between cardiomyopathy and myasthenia has also been raised given that myasthenia gravis and myocardial pathology are well known to cooccur, especially in the setting of thymoma (Douglas et al., 2018). Although EKG changes and arrhythmias have been described in patients with myasthenia gravis (Douglas et al., 2018), less is known about these findings in the context of an acute crisis; further work is needed in this area.

CONCLUSION

In summary, cardiac dysfunction may be a consequence of many types of acute neurologic disease and can manifest in a variety of ways clinically including with EKG abnormalities, cardiac biomarker release, impaired ventricular systolic dysfunction, and changes in blood pressure. It is important to recognize these signs and manage them appropriate as they have been associated with increased morbidity and mortality.

References

Al-Qudah ZA, Yacoub HA, Souayah N (2015). Disorders of the autonomic nervous system after hemispheric cerebrovascular disorders: an update. J Vasc Interv Neurol 8: 43–52.

Anandan C, Khuder SA, Koffman BM (2017). Prevalence of autonomic dysfunction in hospitalized patients with Guillain-Barre syndrome. Muscle Nerve 56: 331–333.

Asahina M, Kuwabara S, Suzuki A et al. (2002). Autonomic function in demyelinating and axonal subtypes of Guillain-Barre syndrome. Acta Neurol Scand 105: 44–50.

Benjamin RN, Aaron S, Sivadasan A et al. (2018). The Spectrum of autonomic dysfunction in Myasthenic crisis. Ann Indian Acad Neurol 21: 42–48.

Bijulal S, Harikrishnan S, Namboodiri N et al. (2009). Takotsubo cardiomyopathy in a patient with myasthenia gravis crisis: a rare clinical association. BMJ Case Rep 2009.

Burakgazi AZ, AlMahameed S (2016). Cardiac involvement in peripheral neuropathies. J Clin Neuromuscul Dis 17: 120–128.

Cheah CF, Kofler M, Schiefecker AJ et al. (2017). Takotsubo cardiomyopathy in traumatic brain injury. Neurocrit Care 26: 284–291.

Chen Z, Venkat P, Seyfried D et al. (2017). Brain-heart interaction: cardiac complications after stroke. Circ Res 121: 451–468.

Chinnasami S, Rathore C, Duncan JS (2013). Sinus node dysfunction: an adverse effect of lacosamide. Epilepsia 54: e90–e93.

Douglas TM, Wengrofsky P, Haseeb S et al. (2018). Takotsubo cardiomyopathy mimicking myocardial infarction in a man with Myasthenic crisis: a case report and literature review. Am J Med Case Rep 6: 184–188.

Duplyakov D, Golovina G, Lyukshina N et al. (2014). Syncope, seizure-induced bradycardia and asystole: two cases and review of clinical and pathophysiological features. Seizure 23: 506–511.

Eggleston KS, Olin BD, Fisher RS (2014). Ictal tachycardia: the head-heart connection. Seizure 23: 496–505.

El-Menyar A, Goyal A, Latifi R et al. (2017). Brain-heart interactions in traumatic brain injury. Cardiol Rev 25: 279–288.

Feldman AE, Gidal BE (2013). QTc prolongation by antiepileptic drugs and the risk of torsade de pointes in patients with epilepsy. Epilepsy Behav 26: 421–426.

Fernandez-Ortega JF, Prieto-Palomino MA, Garcia-Caballero M et al. (2012). Paroxysmal sympathetic hyperactivity after traumatic brain injury: clinical and prognostic implications. J Neurotrauma 29: 1364–1370.

Fernandez-Ortega JF, Baguley IJ, Gates TA et al. (2017). Catecholamines and paroxysmal sympathetic hyperactivity after traumatic brain injury. J Neurotrauma 34: 109–114.

Finsterer J, Stollberger C (2016). Stress from myasthenic crisis triggers Takotsubo (broken heart) syndrome. Int J Cardiol 203: 616–617.

Fodstad H, Kelly PJ, Buchfelder M (2006). History of the Cushing reflex. Neurosurgery 59: 1132–1137. discussion 1137.

Freeman R (2007). Autonomic peripheral neuropathy. Neurol Clin 25: 277.

Gordan R, Gwathmey JK, Xie LH (2015). Autonomic and endocrine control of cardiovascular function. World J Cardiol 7: 204–214.

Grigorean VT, Sandu AM, Popescu M et al. (2009). Cardiac dysfunctions following spinal cord injury. J Med Life 2: 133–145.

Guldiken B, Remi J, Noachtar S (2016). Cardiovascular adverse effects of phenytoin. J Neurol 263: 861–870.

Hagen EM (2015). Acute complications of spinal cord injuries. World J Orthop 6: 17–23.

Hagen EM, Rekand T, Gronning M et al. (2012). Cardiovascular complications of spinal cord injury. Tidsskr Nor Laegeforen 132: 1115–1120.

Hasanin A, Kamal A, Amin S et al. (2016). Incidence and outcome of cardiac injury in patients with severe head trauma. Scand J Trauma Resusc Emerg Med 24: 58.

Hawkes MA, Hocker SE (2018). Systemic complications following status epilepticus. Curr Neurol Neurosci Rep 18: 7.

Hinson HE, Sheth KN (2012). Manifestations of the hyperadrenergic state after acute brain injury. Curr Opin Crit Care 18: 139–145.

Hocker S (2015). Systemic complications of status epilepticus–an update. Epilepsy Behav 49: 83–87.

Hocker S, Prasad A, Rabinstein AA (2013). Cardiac injury in refractory status epilepticus. Epilepsia 54: 518–522.

Ishizue N, Niwano S, Saito M et al. (2016). Polytherapy with sodium channel-blocking antiepileptic drugs is associated with arrhythmogenic ST-T abnormality in patients with epilepsy. Seizure 40: 81–87.

Jansen K, Lagae L (2010). Cardiac changes in epilepsy. Seizure 19: 455–460.

Kaufman KR, Velez AE, Wong S et al. (2013). Low-dose lacosamide-induced atrial fibrillation: case analysis with literature review. Epilepsy Behav Case Rep 1: 22–25.

Kerro A, Woods T, Chang JJ (2017). Neurogenic stunned myocardium in subarachnoid hemorrhage. J Crit Care 38: 27–34.

Koike H, Watanabe H, Sobue G (2013). The spectrum of immune-mediated autonomic neuropathies: insights from the clinicopathological features. J Neurol Neurosurg Psychiatry 84: 98–106.

Krishnamoorthy V, Prathep S, Sharma D et al. (2014). Association between electrocardiographic findings and cardiac dysfunction in adult isolated traumatic brain injury. Indian J Crit Care Med 18: 570–574.

Krishnamoorthy V, Mackensen GB, Gibbons EF et al. (2016). Cardiac dysfunction after neurologic injury what do we know and where are we going? Chest 149: 1325–1331.

Levine RL (2007). Neurocardiology. Resuscitation 73: 186–188.

Lim HB, Smith M (2007). Systemic complications after head injury: a clinical review. Anaesthesia 62: 474–482.

Manea MM, Comsa M, Minca A et al. (2015). Brain-heart axis–review article. J Med Life 8: 266–271.

Meyfroidt G, Baguley IJ, Menon DK (2017). Paroxysmal sympathetic hyperactivity: the storm after acute brain injury. Lancet Neurol 16: 721–729.

Mukerji S, Aloka F, Farooq MU et al. (2009). Cardiovascular complications of the Guillain-Barre syndrome. Am J Cardiol 104: 1452–1455.

Murthy SB, Shah S, Rao CP et al. (2015). Neurogenic stunned myocardium following acute subarachnoid hemorrhage: pathophysiology and practical considerations. J Intensive Care Med 30: 318–325.

Nei M, Ho RT, Sperling MR (2000). EKG abnormalities during partial seizures in refractory epilepsy. Epilepsia 41: 542–548.

Opherk C, Coromilas J, Hirsch LJ (2002). Heart rate and EKG changes in 102 seizures: analysis of influencing factors. Epilepsy Res 52: 117–127.

Peric S, Rakocevic-Stojanovic V, Nisic T et al. (2011). Cardiac autonomic control in patients with myasthenia gravis and thymoma. J Neurol Sci 307: 30–33.

Phillips AA, Krassioukov AV (2015). Contemporary cardiovascular concerns after spinal cord injury: mechanisms, Maladaptations, and management. J Neurotrauma 32: 1927–1942.

Popa C, Popa F, Grigorean VT et al. (2010). Vascular dysfunctions following spinal cord injury. J Med Life 3: 275–285.

Porto I, Della Bona R, Leo A et al. (2013). Stress cardiomyopathy (tako-tsubo) triggered by nervous system diseases: a systematic review of the reported cases. Int J Cardiol 167: 2441–2448.

Prathep S, Sharma D, Hallman M et al. (2014). Preliminary report on cardiac dysfunction after isolated traumatic brain injury. Crit Care Med 42: 142–147.

Rabinstein AA (2007). Paroxysmal sympathetic hyperactivity in the neurological intensive care unit. Neurol Res 29: 680–682.

Rincon F, Dhamoon M, Moon Y et al. (2008). Stroke location and association with fatal cardiac outcomes: northern Manhattan study (NOMAS). Stroke 39: 2425–2431.

Ripoll JG, Blackshear JL, Diaz-Gomez JL (2017). Acute cardiac complications in critical brain disease. Neurol Clin 35: 761–783.

Samuels MA (2007). The brain-heart connection. Circulation 116: 77–84.

Scheitz JF, Nolte CH, Doehner W et al. (2018). Stroke-heart syndrome: clinical presentation and underlying mechanisms. Lancet Neurol 17: 1109–1120.

Scorza FA, Arida RM, Cysneiros RM et al. (2009). The brain-heart connection: implications for understanding sudden unexpected death in epilepsy. Cardiol J 16: 394–399.

Sevcencu C, Struijk JJ (2010). Autonomic alterations and cardiac changes in epilepsy. Epilepsia 51: 725–737.

Shmuely S, van der Lende M, Lamberts RJ et al. (2017). The heart of epilepsy: current views and future concepts. Seizure 44: 176–183.

Stollberger C, Finsterer J (2004). Cardiac troponin levels following monitored epileptic seizures. Neurology 62: 1453.

Stollberger C, Wegner C, Finsterer J (2011). Seizure-associated Takotsubo cardiomyopathy. Epilepsia 52: e160–e167.

Tahsili-Fahadan P, Geocadin RG (2017). Heart-brain Axis: effects of neurologic injury on cardiovascular function. Circ Res 120: 559–572.

Tigaran S, Molgaard H, McClelland R et al. (2003). Evidence of cardiac ischemia during seizures in drug refractory epilepsy patients. Neurology 60: 492–495.

van der Lende M, Surges R, Sander JW et al. (2016). Cardiac arrhythmias during or after epileptic seizures. J Neurol Neurosurg Psychiatry 87: 69–74.

Venkata C, Kasal J (2018). Cardiac dysfunction in adult patients with traumatic brain injury: a prospective cohort study. Clin Med Res 16: 57–65.

Wan WH, Ang BT, Wang E (2008). The Cushing response: a case for a review of its role as a physiological reflex. J Clin Neurosci 15: 223–228.

Zaeem Z, Siddiqi ZA, Zochodne DW (2019). Autonomic involvement in Guillain-Barre syndrome: an update. Clin Auton Res 29: 289–299.

Handbook of Clinical Neurology, Vol. 177 (3rd series)
Heart and Neurologic Disease
J. Biller, Editor
https://doi.org/10.1016/B978-0-12-819814-8.00014-7

Chapter 17

Neurologic complications of genetic channelopathies

WAYNE H. FRANKLIN[1]* AND MATTHEW LAUBHAM[2]

[1]*Department of Pediatrics, Loyola University Chicago, Stritch School of Medicine, Maywood, IL, United States*
[2]*Department of Medicine, Ohio State University, Columbus, OH, United States*

Abstract

This chapter describes what a channelopathy is and how mutations in the genes result in different types of clinical abnormalities. It provides a description of common types of cardiac channelopathies with examples of how there are some areas of overlap with sensory-neuromuscular channelopathies. We describe the cardiac channelopathies of Jervell and Lange-Nielson syndrome, Andersen-Tawil syndrome, Timothy syndrome, catecholaminergic polymorphic ventricular tachycardia, Brugada syndrome, and sinoatrial node dysfunction and deafness. We also discuss sudden unexpected death in epilepsy and how it could relate to some cardiac channelopathies.

Channelopathies are abnormalities in the genetic expression of the proteins of ion channels in a cell membrane. The protein abnormalities can be encoded in a specific gene that codes for the protein in the ion channel or it can encode in any of the processes that allow the proteins to be made. This can occur from the sarcoplasmic reticulum, to transcription of the RNA to the cDNA to any other regulatory protein in the cell, which allows the proteins to form (Kim, 2014). These can occur in any cell in the body. Two of the most common types of channelopathies are the long QT syndrome (LQTS) and cystic fibrosis (CF). These occur at 1:2500 for LQTS and 1:2500 for CF in the white population. Most of the LQTS are inherited in an autosomal dominant pattern (Wallace et al., 2019), whereas the CF patients are inherited in an autosomal recessive pattern.

LQTS occurs in heart muscle and causes abnormal repolarization with subsequent torsade de pointes (specific type of polymorphic ventricular tachycardia) causing syncope or sudden cardiac death. The LQTS have abnormalities in several ion channel genes, the most common of which are KCNQ1 (LQTS1), KCNH2 (LQTS2), and SCN5A (LQTS3). However, there are more than 17 different LQTS gene abnormalities to date. LQTS has an abnormal measurement of the corrected QT, which is corrected for the heart rate. The QT is a measure from the beginning of the QRS on Lead II of the ECG to the end of the T-wave. As per Bazett's formula, it is then divided by the square root of the previous RR interval. Usually, the QTc is less than 470ms in females and less than 450ms in males.

CF causes severe lung disease, exocrine, and endocrine abnormalities of the pancreas, inspissated meconium and infertility in males. It has an abnormality in the cystic fibrosis transmembrane conductance regulator (CFTR), which decreases conductance of the chloride ion. There are over 1700 abnormalities in the CFTR gene described, but ΔF508 is the most common. The CFTR gene is expressed in many epithelial cells of organs: lungs, pancreas, intestines, and vas deferens (Farrell et al., 2020).

In this chapter, we review channelopathies in the central and peripheral nervous system and examine the connections to cardiac channelopathies. We can conceptualize a Venn diagram (Fig. 17.1), where there is a set of cardiac channelopathies and another set of neuromuscular channelopathies. There is a subset of both of these, which have both cardiac and neuromuscular

*Correspondence to: Wayne H. Franklin, M.D., M.P.H., M.M.M., Loyola University Medical Center, 2160 South 1st Avenue, Maguire 3312, Maywood, IL 60153, United States. Tel: +1-708-327-9102, Fax: +1-708-327-9107, E-mail: wayne.franklin@lumc.edu

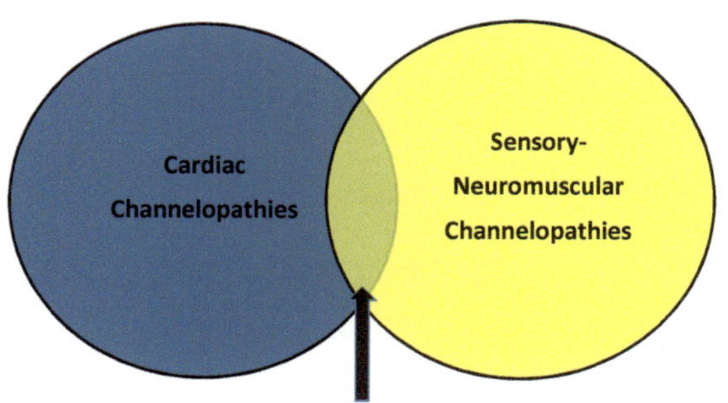

Fig. 17.1. Venn diagram. *Blue*: all cardiac channelopathies. *Yellow*: all sensory-neuromuscular channelopathies.

channelopathies. Both the heart and the nervous system have excitable tissue. It is possible that some of these channelopathies would only be expressed in the heart or in the central or peripheral nervous system. However, it is also likely that with further research, we will learn that many of these channelopathies present both in the heart and in the central and peripheral nervous systems (Hartmann et al., 1999). In addition, the sudden unexplained death in epilepsy (SUDEP) is a phenomenon that needs further clarification. We look into these later.

LONG QT SYNDROME AND DEAFNESS

One of the first types of LQTS was described by Jervell and Lange-Nielson as having both LQTS, seizures, sudden cardiac death, and deafness (Jervell and Lange-Nielson, 1957). This type of LQTS is described as autosomal recessive; however, with formal genetic testing, we now know that most of the parents of these patients actually have LQTS themselves. Although many do not have the phenotype, some do. The patients usually are compound heterozygotes of the genes KCNQ1 and/or KCNE1. About 10% have a de novo gene mutation of one of these genes. These patients have a high risk of sudden cardiac death if their QTc is >550 ms, syncope before age 3 years or male gender and patients with two defects in the KCNQ1 gene have a higher risk of SCD than in the KCNE1 gene (Schwartz et al., 2006).

The cause of deafness is due to dysfunction of I_{Ks} channels in the inner ear. These channels act as K^+ charge carriers for sensory transduction and generation of endocardial potential in the endolymph. This is the same defect causing the LQTS in the heart. Prior to the era of implantable cardioverter defibrillators, the median age of death in Jervell and Lange-Nielson patients was 2.5 years (Faridi et al., 2019).

ANDERSEN-TAWIL SYNDROME (ATS)

This is a rare disorder (11,000,000) characterized by LQTS type 7, periodic paralysis and skeletal abnormalities. ATS has a channelopathy of KCNJ2, which encodes a protein called Kir2.1 (Matthews et al., 2017). It is also possible that the KCNJ5 can have an abnormality as well (Krych et al., 2017). The skeletal abnormalities include a hypoplastic mandible, low-set ears, clinodactyly, and syndactyly of the toes or fingers. The periodic paralysis can occur after exercise and last for hours or days. This paralysis is usually associated with hypokalemia; however it can also have hyperkalemia and normal potassium (Venance et al., 2006; Rajakulendran et al., 2010). One patient was described with concomitant epilepsy and arrhythmias (VT and VF), muscle weakness with a KCNJ2 mutation and skeletal findings consistent with Andersen-Tawil syndrome. Eventually she had a cardioverter-defibrillator implanted (Fryer et al., 2015).

TIMOTHY SYNDROME

Timothy syndrome is also a very rare disorder and has LQTS type 8 and syndactyly. The cardiac manifestations include a prolonged QTc and second-degree atrioventricular block. Most patients have autism spectrum disorder and many have seizures. They have characteristic extracardiac abnormalities: syndactyly, low-set ears, low nasal bridge, small misplaced teeth, immunodeficiency, intermittent hypoglycemia, hypothermia, and myopia (Bidaud and Lory, 2011). The channelopathy with Timothy syndrome is manifest in the L-type calcium channels (Cav 1.2 channels expressed in the CACNA1C gene). Timothy syndrome has a gain of function of the Cav1.2 channel (Liao and Soong, 2010).

CATECHOLAMINERGIC POLYMORPHIC VENTRICULAR TACHYCARDIA (CPVT) AND SEIZURES

CPVT is an abnormality where the baseline ECG is normal; however, when they start experiencing an adrenaline surge as with exercise, they develop a specific type of bidirectional ventricular tachycardia. These patients usually have a mutation of their ryanodine receptor or an absence of calcequestrin 2 (Postma et al., 2002; Lahat et al., 2003). A patient with a ryanodine receptor abnormality, catecholaminergic polymorphic ventricular tachycardia and developmental delay presented with an episode of loss of consciousness. He had a 24 h EEG/ECG and had abnormal spikes during the EEG associated with sinus rhythm. This group was postulating that this patient could be the person representing the mouse knockout model for an abnormal ryanodine receptor model with CPVT and epilepsy (Nagrani et al., 2011). Another patient who was 32 years old presented with generalized seizures. She was diagnosed with a defect in her ryanodine receptor gene. Her brother was diagnosed at age 20 after he survived a cardiac arrest with the same gene abnormality and was diagnosed with CPVT. The patient, however, had several exercise tests without any abnormalities (Yap and Smyth, 2019).

BRUGADA SYNDROME AND SEIZURES

Brugada syndrome was described in 1992 as a right bundle branch block pattern, persistent ST segment elevation, and sudden cardiac death (Brugada and Brugada, 1992). This was noted first in Southeast Asian patients who had sudden unexplained nocturnal death syndrome (SUNDS) (Zheng et al., 2019). Although SUNDS has been around for over 100 years, there are many hypotheses as to what causes it. It is linked most closely to Brugada syndrome; however, there is not total agreement with this. The Brugada syndrome can affect both children and adults. The ECG is sometimes subtle in that one needs to look at V1 and V2 for the right bundle branch block pattern. The LQTS3 has a defect in the SCN5A gene with gain of function. Brugada syndrome has a loss of function of the same gene. Procainamide blocks sodium channels and brings out the coved type complex in patients with Brugada syndrome.

A report of a family where the 5-year-old child had drop attacks and was diagnosed with epilepsy is presented (Parisi et al., 2013). His EEG showed epileptiform activity and he was treated with several different antiepileptic medications until a combination finally stopped the seizures. His father, paternal uncle, and paternal grandmother were all diagnosed with Brugada syndrome. However, his father and uncle when they were children had episodes of drop attacks and had EEGs consistent with epilepsy. The three adults with Brugada syndrome all had episodes of aborted SCD and had cardioverter-defibrillators implanted. The child developed the typical ECG of Brugada syndrome and was having a cardioverter-defibrillator implanted. It is possible that the channelopathy can have different effects at different ages.

SINOATRIAL NODE DYSFUNCTION AND DEAFNESS

Sinus node dysfunction, atrial fibrillation, and epilepsy

One large Pakistani family was noted to have bradycardia and deafness (sinoatrial node dysfunction and deafness). They had a loss of function in the CACNA1D (Cav 1.3) gene. This was a splice variant in Cav 1.3, which is expressed in inner ear hair cells and sinoatrial nodal cells (Liaqat et al., 2019). This report had many consanguineous families in a province of Pakistan.

Other families have sinus node dysfunction that includes bradycardia and atrial fibrillation along with seizures. The hyperpolarization-activated cyclic nucleotide-gated (HCN) channels are expressed in many organs but specifically the heart and the brain. The current in these channels of the heart are expressed as I_f. The current in these channels in the brain are expressed as I_h. The currents are involved in automaticity of the electrical impulses in the heart and in the brain (Rivolta et al., 2020). Since they are involved in the automaticity of the heart that is why these patients have sinus node dysfunction. The automaticity of the neurons potentially can cause seizures.

Sudden unexpected death in epilepsy

SUDEP is the leading cause of death in epilepsy (18%). A wide range of the incidence of SUDEP is from 0.58 to 9.0 per 1000 person-years (Nashef et al., 2012). The universal classification includes:

1. Individual with epilepsy
2. Sudden
3. Unexpected death
4. Occurring in benign circumstances
5. Nontraumatic
6. Nondrowning
7. Witnessed or unwitnessed
8. With or without evidence of a seizure
9. Excluding status epilepticus
10. Postmortem examination does not reveal a cause of death.

There are potentially some cardiac issues as well with molecular autopsy of these patients having 11% of postmortem abnormalities in the sodium and potassium ion channel subunits. Many of these patients have not had complete records and there are often no ECGs for these patients as well. A taskforce of neurologists and cardiologists are involved to try to define the etiology of these deaths (Chahal et al., 2020).

CONCLUSION

This chapter reviewed what channelopathies are and defined cardiac and sensori-neuromuscular channelopathies. Some of each are independent; however, many overlap. We are just at the beginning of defining many future channelopathies, which will have overlap in these organs.

References

Bidaud I, Lory P (2011). Hallmarks of the channelopathies associated with L-type calcium channels: a focus on the Timothy mutations in Cav1.2 channels. Biochimie 93: 2080e2086.

Brugada P, Brugada J (1992). Right bundle branch block, persistent ST segment elevation and sudden cardiac death: a distinct clinical and electrophysiologic syndrome. JACC 20 (6): 1391–1396.

Chahal CAA, Salloum MN, Alahdab F et al. (2020). Systematic review of the genetics of sudden unexpected death in epilepsy: potential overlap with sudden cardiac death and arrhythmia-related genes. J Am Heart Assoc 9: e012264.

Faridi F, Tona R, Brofferio A et al. (2019). Mutational and phenotypic spectra of *KCNE1* deficiency in Jervell and Lange-Nielsen syndrome and Romano-Ward syndrome. Hum Mutat 40 (2): 162–176.

Farrell PM, Rock MJ, Baker MW (2020). The impact of the CFTR gene discovery on cystic fibrosis diagnosis, counseling, and preventive therapy. Genes (Basel) 11: 401–415.

Fryer MD, Kaye G, Tomlinson S (2015). Recurrent syncope in the Andersen Tawil syndrome Cardiac or neurological? Indian Pacing Electrophysiol J 15: 8e161.

Hartmann HA, Colom LV, Sutherland ML et al. (1999). Selective localization of cardiac SCN5A sodium channels in limbic regions of rat brain. Nat Neurosci 2 (7): 593–595.

Jervell A, Lange-Nielson F (1957). Congenital deaf-mutism, functional heart disease with prolongation of the QT interval and sudden death. Am Heart J 54: 59–68.

Kim JB (2014). Channelopathies. Korean J Pediatr 57 (1): 1–18.

Krych M, Biernacka EK, Ponińska J et al. (2017). Andersen-Tawil Syndrome: clinical presentation and predictors of symptomatic arrhythmias—possible role of polymorphisms K897T in *KCNH2* and H558R in *SCN5A* gene. J Cardiol 70 (5): 504–510.

Lahat H, Pras E, Eldar M (2003). RYR2 and CASQ2 mutations in patients suffering from catecholaminergic polymorphic ventricular tachycardia. Circulation 107 (3): e29.

Liao P, Soong TW (2010). CaV1.2 channelopathies: from arrhythmias to autism, bipolar disorder, and immunodeficiency. Eur J Physiol 460: 353–359.

Liaqat K, Schrauwen I, Raza SI et al. (2019). Identification of CACNA1D variants associated with sinoatrial node dysfunction and deafness in additional Pakistani families reveals a clinical significance. J Hum Genet 64 (2): 153–160.

Matthews E, Silwal A, Sud R et al. (2017). Skeletal muscle channelopathies: rare disorders with common pediatric symptoms. J Pediatr 188: 181–185.

Nagrani T, Siyamwala M, Vahid G et al. (2011). Ryanodine calcium channel. A novel channelopathy for seizures. Neurologist 17: 91–94.

Nashef L, So EL, Ryvlin P et al. (2012). Unifying the definitions of sudden unexpected death in epilepsy. Epilepsia 53 (2): 227–233.

Parisi P, Olivab A, Vidal MC et al. (2013). Coexistence of epilepsy and Brugada syndrome in a family with SCN5A mutation. Epilepsy Res 105: 415–418.

Postma AV, Denjoy I, Hoorntje TM et al. (2002). Absence of calsequestrin 2 causes severe forms of catecholaminergic polymorphic ventricular tachycardia. Circ Res 91: e21–e26.

Rajakulendran S, Tan SV, Hanna MG (2010). Muscle weakness, palpitations and a small chin: the Andersen–Tawil syndrome. Pract Neurol 10: 227–231.

Rivolta I, Binda A, Masi A et al. (2020). Cardiac and neuronal HCN channelopathies. Eur J Physiol 472: 931–951.

Schwartz PJ, Spazzolini C, Crotti L et al. (2006). The Jervell and Lange-Nielsen syndrome: natural history, molecular basis, and clinical outcome. Circulation 113: 783–790.

Venance SL, Cannon SC, Fialho D et al. (2006). The primary periodic paralyses: diagnosis, pathogenesis and treatment. Brain 129: 8–17.

Wallace E, Howard L, Liu M et al. (2019). Long QT syndrome: genetics and future perspective. Pediatr Cardiol 40: 1419–1430.

Yap SM, Smyth S (2019). Ryanodine receptor 2 (RYR2) mutation: a potentially novel neurocardiac calcium channelopathy manifesting as primary generalised epilepsy. Seizure 67: 11–14.

Zheng J, Sheng D, Su T et al. (2019). Sudden unexplained nocturnal death syndrome: the hundred years' enigma. J Am Heart Assoc e007837: 7.

Handbook of Clinical Neurology, Vol. 177 (3rd series)
Heart and Neurologic Disease
J. Biller, Editor
https://doi.org/10.1016/B978-0-12-819814-8.00025-1

Chapter 18

Neurological complications of syncope and sudden cardiac arrest

VIJAYAKUMAR JAVALKAR, ABDALLAH AMIREH, AND ROGER E. KELLEY*

Department of Neurology, Ochsner/Louisiana State University Health Sciences Center, Shreveport, LA, United States

Abstract

Syncope is very common and usually comes with enough warning for the person to assume a safer position rather than fall in a potentially dangerous way. Syncope may be associated with pregnancy, for example, but we rarely encounter significant injury related to the potential for an associated fall. In the elderly, however, there are often comorbid factors such as delayed reaction time and other aspects of cognitive impairment, along with gait instability, that can affect the defensive reflexes to the point that brain injury, including subdural or epidural hematoma, is not uncommonly encountered. Sudden syncope without warning can also have both neurological and general physical implications in terms of driving safety, safety operating potentially dangerous equipment or exposure to heights as well as the potential impact for drowning or near-drowning while swimming or taking a bath. Sudden death, from whatever the mechanism, implies cerebral hypoperfusion with the potential consequences of hypoxic–ischemic brain injury.

MECHANISM OF SYNCOPE

Syncope represents a sudden transient loss of consciousness that is accompanied by a loss of postural tone and often leads to a fall. For convenience, it can be divided into pre-syncopal and syncopal stages. It relates to a global but reversible reduction in cerebral perfusion. Syncope vs pre-syncope reflects the severity of the precipitating mechanism. For example, lightheadedness with a near faint, without actual loss of consciousness, would fall into the pre-syncope category. This could, theoretically, lead to stumbling with fall and subsequent injury if the person cannot adequately assume a corrective position. This is most commonly orthostatic in nature and it is not uncommon to see this in patients of advanced age on antihypertensive medication. Cardiac arrhythmias affecting cardiac output can be associated with similar symptoms and the degree of arrhythmia, be it either transient ventricular tachycardia, sick sinus syndrome, or sinoatrial block, has a direct impact on the neurological sequelae.

In a retrospective study of 839 clinic patients presenting with syncope (Al-Busaidi and Jardine, 2020), 42.8% had vasovagal syncope with drug-related postural syncope cited for 26.6%. Only 3.1% were initially identified as cardiac-related with an additional 2.1% later determined to be of cardiac origin. Vasovagal syncope, also known as neurocardiogenic syncope, is believed to be mediated through the Bezold-Jarisch reflex. This represents a paradoxical aberration to normal reflexive response to orthostasis. The overall pathophysiology is encapsulated as follows (Cheshire, 2017): (1) venous pooling; (2) reduction of cardiac filling; (3) hypercontractility of the left ventricle; (4) activation of ventricular mechanoreceptors; (5) projection of the ventricular mechanoreceptors, in addition to limbic structures, to the brainstem; (6) resultant sudden withdrawal of sympathetic vascular tone, peripheral vasodilatation resulting in hypotension accompanied by cerebral hypoperfusion. The triggering factor is typically an overreaction to a stimulus that activates the autonomic nervous system (ANS) such as a significant emotional reaction, intense

*Correspondence to: Roger E. Kelley, MD, Department of Neurology, Ochsner/LSU Health Sciences Center-Shreveport, 1501 Kings Highway, Shreveport, LA 71130, United States. Tel: +1-318-675-7760, Fax: +1-318-675-6382, E-mail: rkelly@lsuhsc.edu

pain, or prolonged standing. This can generate a neural reflex, through the ANS, which can promote an effect on the heart rate and/or reduce vascular tone, that is, a vasodepressor effect.

Syncope related to autonomic dysfunction is increasingly recognized in neurodegenerative diseases such as Parkinson's disease and multisystem atrophy. It is most commonly encountered in reflex-mediated syndromes such as situational syncope but can also be reflective of primary and secondary ANS failure (Adkisson and Benditt, 2015). Naturally, the coexistence of immobility with neurodegenerative disease augments the potential consequences of the falling out. In the cerebrovascular realm, cardiovascular autonomic dysfunction can be related to brainstem infarction such as lateral medullary infarction (Huynh et al., 2018). In such a circumstance, the resultant cerebral hypoperfusion may augment the vascular insult.

Potential explanations for a syncopal-type spell can include vertebrobasilar distribution transient ischemic attack, partial seizure resulting in postural collapse or vertebrobasilar migraine. In the latter, one typically obtains a history of vascular-type headache following the syncope. In addition, various brain structural processes can be associated with syncope such as colloid cyst of the third ventricle. Syncope is not uncommon and is reported to account for 1%–3% of emergency room visits (Casini-Raggi et al., 2002) with roughly 4% attributable to neurological issues (Stone et al., 2010).

Illustrative case

A 26-year-old woman is driving her car near an academic hospital. She suddenly loses control of the car and hits a tree. She is immediately brought to the emergency room and is found to have a normal physical and neurological exam. The EKG is normal as is routine blood work. She reports suddenly feeling lightheaded and is believed to have passed out for roughly 30 s. She is admitted for observation and neurology is consulted. The consulting neurologist confirms a normal neurological exam, identifies this as a vasovagal syncopal spell, and clears her for discharge. The intern instead orders a CT brain scan. The neurologist comes by for follow-up and disputes the need for the CT brain scan and once again clears her for discharge. Several days later she is presented to the neurologist at weekly Neurological Conference by the intern, with a smug look on his face, along with the CT brain scan results. Much to the chagrin of the neurologist, the CT brain scan reveals a prominent colloid cyst of the third ventricle associated with significant obstructive hydrocephalus. She is referred to neurosurgery for a ventriculoperitoneal shunt. This case illustrates the potential for a cerebral mass lesion, such as a colloid cyst of the third ventricle, to have a secondary effect on the heart which can include syncope as well as sudden death. It has been theorized that pressure on the adjacent hypothalamus, with a resultant effect on autonomic regulation, is the mechanism (Turillazzi et al., 2012).

CARDIAC ARRHYTHMIA WITH CEREBRAL HYPOPERFUSION

A transient cardiac arrhythmia can lead to sudden cerebral hypoperfusion with loss of consciousness. The development of symptoms such as lightheadedness preceding syncope occurs when the mean arterial pressure drops below 60 mmHg. Resultant cerebral hypoperfusion, lasting more than 7 s correlates with actual syncope (Weiling et al., 2009). This can lead to a simple collapse with regaining of consciousness once the transient arrhythmia remits. In a vasovagal pre-syncope study of physiological measures (Aebi et al., 2019), the initial changes included decrease in blood pressure and peaking of the heart rate followed by the actual vasovagal reaction of further reduction in blood pressure, heart rate, and cerebral perfusion.

Coma, on the other hand, is related to 5–10 min, or longer, of severe hypoxic–ischemic central nervous system (CNS) insult associated with significant disruption of cerebral autoregulation (Folino, 2007). The mechanisms of brain injury triggered by cardiac arrest and resuscitation are complex and include excitotoxicity, disrupted calcium homeostasis, free radical formation, pathological protease cascades, and activation of cell-death signaling pathways (Neumar et al., 2008). Certain regions of the brain, such as the hippocampus CA1 sector, cerebral cortex, thalamus, and cerebellum are particularly susceptible to hypoxic injury (Romiro and Kumar, 2015). One does not typically expect to see CNS injury with syncope unless there is trauma related to a fall, but syncope can certainly be a harbinger of a more serious outcome.

Illustrative case

A 79-year-old gentleman is referred with the complaint "I feel like I am going out." He is a somewhat nonspecific historian in terms of this complaint, but his family reports actual loss of consciousness for up to a minute on several occasions in recent months. His physical examination, neurological examination, and non-contrast CT brain scan, as well as carotid/vertebral duplex ultrasound study, are all considered fine for his age. During an EEG, he reported a return of the sensation and was found, on the EKG component of the study, to have sinus bradycardia in the '30s. He is found to have sick sinus syndrome with no further spells reported after a cardiac pacemaker is implanted.

Syncope vs seizure

Syncope can be difficult to distinguish from a seizure in certain instances. It has been estimated that seizures have up to a 30% misdiagnosis rate with syncope being the most common explanation for a mistaken diagnosis (Scheepers et al., 1998). In a study of recurrent seizure-like activity, with the use of an implantable loop recorder, up to 20%–30% of patients were found to have occult cardiovascular explanations (Simpson et al., 2000). In a study by Schott et al. (1977), a cardiac cause for seizure-like activity was observed in 20% of patients. This overlap between syncope and seizures can have serious implications for the potential relationship with sudden cardiac death (SCD). In the clinic study of Al-Busaidi and Jardine (2020), the 2.5-year mortality rate for this 5-year study was 5.7% and this included three with sudden unexplained cardiac deaths.

SUDDEN CARDIAC DEATH WITH POSSIBLE NEUROLOGICAL INTERACTION

SCD is defined as death from a sudden and pulseless condition termed as sudden cardiac arrest and can be related to neurological mediated stress on the heart through sympathetic overstimulation (Rabinstein, 2014). There are a number of CNS disorders that can be associated with both syncope and SCD and these can include: epilepsy, subarachnoid hemorrhage, traumatic brain injury, and ischemic stroke (Finisterer and Wahbi, 2014).

Naturally, the consequences of sudden cardiac arrest, assuming resuscitation, are more severe than uncomplicated syncope. Other than global hypoxic–ischemic ischemia, one can see a more discrete infarct pattern. Watershed, also known as borderzone, infarction is recognized as an infarct pattern in areas of overlap of cerebral circulatory territories such as the middle cerebral artery-anterior cerebral artery borderzone and middle cerebral artery-posterior cerebral artery borderzone. Low output cardiac failure or actual cardiac arrest can be associated with such a cerebral infarct pattern (Finisterer and Stollberger, 2016).

From a practical, and precautionary standpoint, certain medications can promote SCD, for example antipsychotic medications (Zhu et al., 2019). This has resulted in concern over such an occurrence when such commonly used agents are prescribed in an effort to promote a manageable situation for the caregiver taking care of an elderly patient with agitation. SCD is now well recognized as a potential complication of epilepsy and is termed sudden unexplained death in epilepsy (SUDEP) (Shorvon and Tomson, 2011). This is most commonly associated with intractable epilepsy. SUDEP is presumably related to cardiac impairment precipitated by the epileptic pathogenesis, which can be associated with changes in heart rate, arrhythmias, asystole, and perhaps other cardiac-related factors (Dlough et al., 2016). There appears to be an overlap of conduction abnormalities within the heart and epileptogenesis (Scorza et al., 2009). For example, there is a reported relationship between seizure-like episodes and EEG abnormalities in patients with long QT syndrome (Gonzalez et al., 2018).

The relationship of genetic mediated cardiac channelopathies and sudden death (Fernandez-Falgueras et al., 2017) may extend to genetic epilepsy disorders such as Dravet Syndrome which has a relatively high incidence of SUDEP (Frasier et al., 2018). Efforts have been made to predict patients most susceptible to SUDEP based upon electrocardiographic features. Chyou et al. (2016) reported the identification of an abnormal ventricular conduction pattern as more common in SUDEP patients compared to controls. Respiratory dysfunction has also been implicated in SUDEP (Nashef et al., 1996), which has led to speculation that epilepsy-related disruption of the functional connectivity of certain brain structures associated with central autonomic control of cardiorespiratory function, termed "the neuro-cardio-respiratory connection," is the culprit (Manolis et al., 2019).

The implications of the "neuro-cardio-respiratory connection" are also an important factor in the sleep disorder realm where there can be a "perfect storm" of contributing factors (Silvani, 2019).

The well-established association of obstructive sleep apnea (OSA) with SCD may be related to such factors as intermittent nocturnal hypoxia, reactive oxygen species, cardiomyocyte metabolic disturbances, myocardial electric heterogeneity, and intrathoracic pressure (Brodovskaya et al., 2018). According to a large longitudinal study reported by Gami et al. (2013), significant OSA was an independent risk factor for SCD with nocturnal hypoxemia of particular concern.

SUMMARY

In summary, syncope is not necessarily a benign presentation. Red flags should include advancing age, risk factors for cardiac disease and stroke, sleep disorder with features of OSA, the potential for epilepsy to be misdiagnosed as syncope, associated factors such as significant headache, as well as atypical presentation. Younger patients with a specific triggering factor for syncope should be distinguished from subjects at higher risk for a more serious mechanism with the potential for SCD.

References

Adkisson WO, Benditt DG (2015). Syncope due to autonomic dysfunction. Diagnosis and management. Med Clin N Am 99: 691–710.

Aebi MR, Bourdillon N, Mezianer HB et al. (2019). Cardiovascular and cerebrovascular responses during a vasovagal reaction without syncope. Front Neurosci 13: 1–8.

Al-Busaidi IS, Jardine DL (2020). Different types of syncope presenting to clinic: do we miss cardiac syncope? Heart Lung Circ 29: 1129–1138.

Brodovskaya TO, Grishina IF, Peretolchina TF et al. (2018). Clues to the pathophysiology of sudden cardiac death in obstructive sleep apnea. Cardiology 140: 247–253.

Casini-Raggi V, Bandinelli G, Lagi A (2002). Vasovagal syncope in emergency room patients: analysis of a metropolitan area registry. Neuroepidemiology 21: 287–291.

Cheshire WP (2017). Syncope. Continuum 23: 335–358 Review article.

Chyou JY, Friedman D, Cerrone M et al. (2016). Electrocardiographic features of sudden unexpected death in epilepsy. Epilepsia 57: 135–139.

Dlough BJ, Gehlbach BK, Richerson GB (2016). Sudden unexpected death in epilepsy: basic mechanisms and clinical implications for prevention. J Neurol Neurosurg Psychiatry 87: 402–413.

Fernandez-Falgueras A, Sarquella-Brugada G, Brugada J et al. (2017). Cardiac channelopathies and sudden death: recent clinical and genetic advances. Biology (Basel) 6: 3390–3397.

Finisterer J, Stollberger C (2016). Neurological complications of cardiac disease (heart-brain disorders). Minerva Med 107: 14–25.

Finisterer J, Wahbi K (2014). CNS-disease affecting the heart: brain–heart disorders. J Neurol Sci 345: 8–14.

Folino AF (2007). Cerebral autoregulation and syncope. Prog Cardiovasc Dis 50: 49–80.

Frasier CR, Zhang H, Offord J et al. (2018). Channelopathy as a SUDEP biomarker in Dravet syndrome patient-derived cardiac myocytes. Stem Cell Reports 11: 626–634.

Gami AS, Olson EJ, Shen WK et al. (2013). Obstructive sleep apnea and the risk of sudden cardiac death: a longitudinal study of 10,701 adults. J Am Coll Cardiol 62: 610–616.

Gonzalez A, Aurien D, Larsson PG et al. (2018). Seizure-like episodes and EEG abnormalities in patients with long QT syndrome. Seizure 61: 214–220.

Huynh TR, Decker B, Fries TJ et al. (2018). Lateral medullary infarction with cardiovascular autonomic dysfunction: an unusual presentation with review of the literature. Clin Auton Res 28: 569–576.

Manolis TA, Manolis AA, Melita H et al. (2019). Sudden unexpected death in epilepsy: the neuro-cardio-respiratory connection. Seizure 64: 65–73.

Nashef L, Walker F, Allen P et al. (1996). Apnea and bradycardia during epileptic seizures: relation to sudden death in epilepsy. J Neurol Neurosurg Psychiatry 60: 3239–3245.

Neumar RW, Nolan JP, Adrie C et al. (2008). Post-cardiac arrest syndrome: epidemiology, pathophysiology, treatment, and prognostication. A consensus statement from the International Liaison Committee on Resuscitation (American Heart Association). Circulation 118: 2452–2483.

Rabinstein AA (2014). Sudden cardiac death. Handb Clin Neurol 119: 19–24.

Romiro JI, Kumar A (2015). Updates on management of anoxic brain injury after cardiac arrest. Mo Med 112: 136–141.

Scheepers B, Clough P, Pickles C (1998). The misdiagnosis of epilepsy: findings of a population study. Seizure 7: 403–406.

Schott GD, McLeod AA, Jewitt DE (1977). Cardiac arrhythmias that masquerade as epilepsy. Br Med J 1: 1454–1457.

Scorza FA, Arida RM, Cysneiros RM et al. (2009). The brain–heart connection: implications in understanding sudden unexpected death in epilepsy. Cardio J 16: 394–399.

Shorvon S, Tomson T (2011). Sudden unexplained death in epilepsy. Lancet 378: 2028–2038.

Silvani A (2019). Sleep disorders, nocturnal blood pressure, and cardiovascular risk: a translational perspective. Auton Neurosci 218: 31–42.

Simpson CS, Barlow MA, Krahn AD et al. (2000). Recurrent seizure diagnosed by the insertable loop recorder. Interv Card Electrophysiol 4: 475–479.

Stone J, Carson A, Duncan R et al. (2010). Who is referred to neurology clinics?—The diagnoses made in 3781 new patients. Clin Neurol Neurosurg 112: 747–751.

Turillazzi E, Bello S, Neri M et al. (2012). Colloid cyst of the third ventricle, hypothalamus, and heart: a dangerous link for sudden death. Diagn Pathol 7: 144–154.

Weiling W, Thijs RD, van Dijk N et al. (2009). Symptoms and signs of syncope: a review of the link between physiology and clinical clues. Brain 132: 2630–2642.

Zhu J, Hou W, Xu Y et al. (2019). Antipsychotic drugs and sudden cardiac death: a literature review of the challenges in the prediction, management and future steps. Psychiatry Res 281: 1–7.

Handbook of Clinical Neurology, Vol. 177 (3rd series)
Heart and Neurologic Disease
J. Biller, Editor
https://doi.org/10.1016/B978-0-12-819814-8.00029-9

Chapter 19

Neurologic complications of cardiac arrest

RICK GILL, MICHAEL TEITCHER, AND SEAN RULAND*

Department of Neurology, Loyola University Chicago, Chicago, Stritch School of Medicine, Maywood, IL, United States

Abstract

Cardiac arrest is a catastrophic event with high morbidity and mortality. Despite advances over time in cardiac arrest management and postresuscitation care, the neurologic consequences of cardiac arrest are frequently devastating to patients and their families. Targeted temperature management is an intervention aimed at limiting postanoxic injury and improving neurologic outcomes following cardiac arrest. Recovery of neurologic function governs long-term outcome after cardiac arrest and prognosticating on the potential for recovery is a heavy burden for physicians. An early and accurate estimate of the potential for recovery can establish realistic expectations and avoid futile care in those destined for a poor outcome. This chapter reviews the epidemiology, pathophysiology, therapeutic interventions, prognostication, and neurologic sequelae of cardiac arrest.

INTRODUCTION

Epidemiology of cardiac arrest

Cardiac arrest is a loss of mechanical cardiac activity with absent clinical signs of systemic circulation. If return of spontaneous circulation (ROSC) is not promptly achieved, sudden cardiac death inevitably ensues. Sudden cardiac death is defined as death without an obvious noncardiac cause that occurs within 1 h of symptom onset or within 24 h of last seen in usual health (Herzog et al., 2017). Cardiac arrest accounts for approximately 15% of all deaths and 50% of all cardiac deaths worldwide (Albert and Stevenson, 2018). In the United States, sudden cardiac death contributed to nearly 380,000 deaths in 2017 (Virani et al., 2020). Owing to differences in underlying pathophysiology and systems of care, the epidemiology can be understood through analysis of two categories: in-hospital cardiac arrest (IHCA) and out-of-hospital cardiac arrest (OHCA).

Out-of-hospital cardiac arrest

In the United States, the annual incidence of OHCA among people of any age treated by emergency medical services (EMS) is 74 per 100,000 people with considerable variability between states. The most common locations in descending order were private residence, public settings, and nursing facilities. About 50% of cases are unwitnessed, 38% of cases are witnessed by a layperson, and only 13% are witnessed by an EMS provider. The initial rhythm was shockable, either ventricular fibrillation (VF) or ventricular tachycardia (VT), in 19% of EMS-treated cases (Virani et al., 2020). In 2018, 28% of EMS-treated patients survived to hospital admission and only 8% survived to discharge with a neurologically favorable outcome defined by a cerebral performance category (CPC) score of 1 or 2 (see Table 19.1) (CARES 2018 Annual Report, 2018).

The global understanding of cardiac arrest as a cause of mortality and morbidity is complicated by a lack of

*Correspondence to: Sean Ruland, D.O., Department of Neurology, Loyola University Chicago Stritch School of Medicine, 2160 S. First Ave., Maguire Center, Room 2700, Maywood, IL 60153, United States. Tel: +1-708-216-6831, Fax: +1-708-216-5617, E-mail: sruland@lumc.edu

Table 19.1

Cerebral performance categories and modified Rankin scale

	Cerebral Performance Categories	Modified Rankin Scale
0		No symptoms at all
1	Good cerebral performance: conscious, alert, able to work, might have mild neurologic or psychologic deficit	No significant disability despite symptoms; able to carry out all usual duties and activities
2	Moderate cerebral disability: conscious, sufficient cerebral function for independent activities of daily life. Able to work in sheltered environment	Slight disability; unable to carry out all previous activities, but able to look after own affairs without assistance
3	Severe cerebral disability: conscious, dependent on others for daily support because of impaired brain function. Ranges from ambulatory state to severe dementia or paralysis	Moderate disability; requiring some help, but able to walk without assistance
4	Coma or vegetative state: any degree of coma without the presence of all brain death criteria. Unawareness, even if appears awake (vegetative state) without interaction with environment; may have spontaneous eye opening and sleep/awake cycles. Cerebral unresponsiveness	Moderately severe disability; unable to walk without assistance and unable to attend to own bodily needs without assistance
5	Brain death: apnea, brainstem areflexia, EEG silence, etc.	Severe disability; bedridden, incontinent and requiring constant nursing care and attention
6		Dead

standard terminology, incomplete documentation on cause of death, and variable reporting methodology. In a systematic review, the estimated incidence of EMS-assessed OHCA per 100,000 population varied by country ranging from 53 in Asia to 113 in Australia (Berdowski et al., 2010). A prospective analysis involving 10,682 cases of OHCA in 27 European countries reported ROSC in 29% (range 9%–50%). Survival to either hospital discharge or 30 days was reported in 10% (range 1%–31%) with substantial regional variability (Gräsner et al., 2016).

In-hospital cardiac arrest

The incidence of IHCA in the United States for 2018 was 10 per 1000 hospital admissions and 2 per 1000 inpatient days (Virani et al., 2020). Of these, 54% occurred in either an intensive care unit, operating room, or emergency department. The remainder occurred in noncritical care areas. The initial rhythm was shockable VF or VT in about 15%. Survival to hospital discharge was 26% with 82% of survivors having a good neurologic outcome (Virani et al., 2020).

Pediatric cardiac arrest

Cardiac arrest is a significant cause of morbidity and mortality in children. In the United States, estimates of pediatric OHCA range from just over 7000 (Virani et al., 2020) to 15,200 cases annually (Holmberg et al., 2019). An overwhelming majority of pediatric OHCA occurs at home. Survival varies by age ranging from 7% for children less than 1 year old, 16% for children 1–12 years old, and 19% for children 13–18 years of age (CARES 2018 Annual Report, 2018).

In 2018, pediatric IHCA in the United States occurred in 13 per 1000 admissions (Virani et al., 2020). In contrast with adults, the initial rhythm was shockable VF or VT in only 9%. Survival to hospital discharge for pediatric IHCA was 41% (Virani et al., 2020).

PATHOPHYSIOLOGY OF BRAIN INJURY IN CARDIAC ARREST

Neurologic injury is a common cause of morbidity and mortality after cardiac arrest. It is the reported cause of death in 68% after OHCA and 23% after IHCA (Laver et al., 2004). In survivors, the extent and severity of brain injury often determine neurologic outcome (Neumar et al., 2008), and outcomes are often reported as neurologic functional status.

The brain receives approximately 15%–20% of cardiac output (Williams and Leggett, 1989), albeit with sex variation, and a gradual decline beginning in the third decade of 1% per decade of life (Xing et al., 2017). Although whole body ischemia leads to multisystem organ dysfunction, the brain is especially vulnerable to ischemia given its limited energy stores and obligate dependence on aerobic glucose metabolism. Cardiac arrest leads to whole brain anoxia and cessation of cerebral perfusion, triggering a cascade of pathologic mechanisms.

Mechanisms of cellular injury

Upon cardiac arrest, circulatory collapse rapidly ensues, with cessation of cerebral circulation. Loss of consciousness occurs within 10 s, loss of electroencephalographic (EEG) activity by 30 s (Pana et al., 2016), and cerebral tissue oxygen tension drops to zero within 2 min (Imberti et al., 2003). Since the brain lacks oxygen stores, adenosine triphosphate is quickly depleted. The cessation of aerobic metabolism leads to cessation of energy-dependent ion channel function necessary to maintain osmotic gradients. Failure of energy-dependent calcium efflux pumps causes accumulation of intracellular calcium, and the accumulation of interstitial potassium ions causes membrane depolarization and rapid intracellular influx of sodium, chloride, and water, leading to cytotoxic edema. Impaired cellular membrane function leads to extracellular leakage of intracellular potassium, hydrogen ions, and glutamate (Greer, 2006). Glutamate activation of alpha amino-3-hydroxyl-5-methyl-4-isoxasolepropionic acid-mediated sodium channels allows further intracellular influx of sodium and worsens edema (Reis et al., 2017). Excess extracellular glutamate also activates *N*-methyl-D-aspartate (NMDA) receptor-mediated calcium ion channels and further raises intracellular calcium concentrations. The net effect, termed glutamate excitotoxicity, results in high intracellular calcium concentrations (Choi, 1985). Intracellular calcium activates catabolic enzymes such as proteases, lipases, and endonucleases, a cascade that leads to cellular breakdown via necrosis and activation of apoptotic pathways (Busl and Greer, 2010). High intracellular calcium concentration also leads to mitochondrial dysfunction and impairs production of adenosine triphosphate (Sekhon et al., 2017). Mitochondrial damage releases cytochrome C that activates apoptotic pathways and accelerates free radical production worsening cellular damage (Reis et al., 2017). Histologically, early changes include nuclear swelling with loss of its basophilic appearance, followed by the appearance of red neurons with pyknotic nuclei about 8–12 h after an ischemic insult (Adams et al., 1966).

Regional metabolic requirements lead to selective vulnerability among various neuronal populations. The CA1 pyramidal neurons of the hippocampus are more likely than other neuronal populations to undergo delayed cell death with prolonged periods of cardiac arrest or persistent postresuscitation hypotension (Petito et al., 1987). This contributes to memory impairment in survivors. Other vulnerable subpopulations include cerebellar Purkinje cells, thalamic reticular neurons, medium-sized striatal neurons, and neocortical layers three, five, and six (Busl and Greer, 2010). The selective vulnerability of the neocortical layers leads to laminar necrosis (Greer, 2006).

Secondary injury

Once circulation is restored, there is risk of reperfusion injury. Animal models suggest increased blood–brain barrier permeability, which worsens brain edema (Pluta et al., 1994). Transient hyperemia contributes to edema, worsens cerebral blood flow (CBF) by increasing intracranial pressure (ICP), and exacerbates ischemia. Reperfusion promotes the production of reactive oxygen species, superoxide anion, and hydroxyl radicals, which overwhelm the endogenous scavenging mechanisms and promote apoptosis (Broughton et al., 2009).

Dysfunction of cerebral autoregulation

Cerebral autoregulation is the process by which the vascular beds match CBF with the metabolic demands of the brain despite fluctuating systemic arterial blood pressure (Lassen, 1959; Paulson et al., 1990). Normal subjects maintain a constant CBF within an average mean arterial pressure (MAP) range of 50–150 mmHg. However, there is substantial heterogeneity in various physiologic and disease states. For example, sympathetic activation in both acute and chronic arterial hypertension leads to an upward shift of both the lower and upper limits (Strandgaard et al., 1973; Markus, 2004). Mechanisms of autoregulation include arteriole vasoconstriction, endothelial-derived vasodilatory factors including nitric oxide, and neurovascular coupling (Paulson et al., 1990; Sekhon and Griesdale, 2017).

During cardiac arrest, MAP drops to zero and cerebral autoregulatory mechanisms fail. Once perfusion is restored, the degree that cerebral autoregulation remains intact varies. Some studies have demonstrated completely absent autoregulation and a linear relationship between MAP and CBF (Nishizawa and Kudoh, 1996). Other studies have suggested zones of preserved autoregulation among various regions and ranges of autoregulation that are narrowed and shifted upward (Sundgreen et al., 2001). The upward shift hypothesis is further supported by evidence that higher MAP in the immediate postresuscitation period is associated with improved neurologic outcomes (Kilgannon et al., 2014; Laurikkala et al., 2016). Given diversity in zones of autoregulation among patients, there is interest in determining individualized blood pressure targets that may optimize cerebral perfusion (Sekhon and Griesdale, 2017).

No-reflow

Hemodynamic collapse in the microvasculature leads to vascular injury and microvascular thrombosis. Accumulation of intracellular sodium causes cytotoxic edema. A corollary is that intravascular fluid egress raises blood viscosity, which in turn promotes fibrin-mediated clot formation. Cytotoxic edema causes perivascular cellular swelling and increased vascular resistance via reduced capillary lumen size by mechanical compression (Ames et al., 1968). Impaired tissue perfusion after blood flow restoration has been termed the "no-reflow" phenomenon.

Inflammatory considerations

Systemic ischemia results in an inflammatory response that contributes to secondary neurologic injury. In the first few hours after resuscitation, cytokine levels including interleukin-6, interleukin-8, and tumor necrosis factor rise in correlation with blood lactate (Adrie et al., 2002). This is a similar immunologic profile to sepsis and contributes to systemic vasodilation, persistent hypotension, and cerebral hypoperfusion after ROSC (Adrie et al., 2002). Increased blood–brain barrier permeability occurs early. In a mouse model, blood–brain barrier permeability was seen within 2 min of ischemia and again 6–24 h postarrest (Pluta et al., 1994). Other mouse models have demonstrated brain infiltration of proinflammatory CD4+ T-lymphocytes and neutrophils within this time frame (Deng et al., 2014). The inflammatory milieu activates microglia contributing to neuronal destruction, a process that begins early and that can last for days (Deng et al., 2014; Xiang et al., 2016).

INTERVENTIONS

Treatment of cardiac arrest and the postcardiac arrest syndrome

Resuscitation has evolved since the advent of standardized CPR in the 1950s. Refinements in the approach to chest compressions and ventilation and the parallel development of intensive care units and critical care medicine have led to cardiac arrest becoming an increasingly survivable condition. Despite this progress, a minority of patients still survive to hospital discharge and withdrawal of life-sustaining treatment (WLST) is often the primary cause of death (Geocadin et al., 2008). The AHA update on advanced cardiovascular life support provides current management algorithms for pulseless VF/VT, PEA, and asystole (Panchal et al., 2019).

Care following ROSC focuses on four areas: (1) determining and treating the cause of the arrest, (2) minimizing brain injury, (3) optimizing cardiopulmonary function, and (4) and managing organ system dysfunction (Neumar et al., 2008). Immediate hemodynamic support is the priority using intravenous fluids, vasopressors, and inotropes as indicated. Postarrest shock is a major cause of death behind brain injury (Lemiale et al., 2013).

The postcardiac arrest phases include immediate care focusing on establishing a secure airway, treating circulatory collapse, and investigations to determine the cause of cardiac arrest. The immediate and early phases (6–12 h) are the most opportune time for interventions to improve neurologic outcomes. Goal-directed high-quality critical care including targeted temperature management (TTM) is critical during these phases. The intermediate phase from 6 to 72 h focuses on limiting the consequences of enduring injury pathways and optimizing organ support; care is often most resource intense during this phase. During the recovery phase (72 h through hospital discharge) neurologic prognostication becomes more reliable and the rehabilitation stage follows hospital discharge (Neumar et al., 2008).

Critical care principles

A patent and secure airway ensures sufficient ventilation and oxygenation following cardiac arrest. Normocarbia is associated with increased hospital survival and favorable neurologic outcome compared to patients with hypercarbia. Despite dysfunctional autoregulation, a response to arterial carbon dioxide tension ($PaCO_2$) persists and a target $PaCO_2$ 35–45 mmHg is reasonable (Williams et al., 2016). Hypocarbia related to hyperventilation leads to cerebral vasoconstriction and ischemia (Buunk et al., 1997). Supplemental oxygenation with a fraction of inspired oxygen (FiO_2) of 100% is appropriate during CPR with rapid titration down after ROSC to avoid hypoxemia throughout the postarrest phases. Hyperoxia after resuscitation is harmful to postischemic neural tissue. Supplemental 100% oxygen in the first hour after ROSC produced worse neurologic outcomes in a preclinical model (Balan et al., 2006). In the first 24–48 h, CBF and the cerebral metabolic demand decrease and targeting an oxygen saturation of 94%–96% is reasonable (Neumar et al., 2008).

A MAP target of greater than 65 mmHg (often used in shock) may be inadequate to maintain cerebral perfusion when autoregulation is impaired. Moreover, prolonged cardiac arrest beyond 15 min leads to the no-reflow phenomenon necessitating a higher MAP. A MAP target of 80–100 mmHg may maintain cerebral perfusion more effectively (Nishizawa and Kudoh, 1996); however, if extensive myocardial injury is present, a MAP 65–100 mmHg may be reasonable (Neumar et al., 2008). If adequate MAP is maintained, positron emission tomography (PET) studies suggest that regional cerebral perfusion will match metabolic activity (Edgren et al., 2003). Good functional recovery was positively associated with MAP during the first 2 h after resuscitation but not with hypertensive reperfusion (MAP > 100 mmHg) within the first minutes of ROSC (Müllner et al., 1996). Macroscopic reperfusion may be hyperemic despite microcirculatory failure, which can worsen cerebral edema and exacerbate reperfusion injury. Fixed microcirculation no-reflow due to presumed thrombosis has been shown to respond to thrombolytic therapy in preclinical models (Fischer et al., 1996).

Coronary artery disease is the cause of cardiac arrest in 80% of cases (Chugh et al., 2008). Emergent coronary revascularization improves outcomes in survivors and should not be delayed because of neurologic status, nor should cardiac catheterization delay TTM initiation. Combining mild induced hypothermia with percutaneous intervention has been shown to be safe and potentially improves neurologic outcomes in survivors (Knafelj et al., 2007). However, thrombolysis using tenecteplase did not improve survival after OHCA (Neumar et al., 2008).

Sedation facilitates TTM and synchronous ventilation while reducing cerebral metabolic rate of oxygen ($CMRO_2$) in comatose survivors. Sedation and neuromuscular blockade help control shivering which left untreated increases cerebral metabolic demand. A systematic approach using a sedation scale such as the Richmond Agitation-Sedation Scale and a shivering scale (e.g., Bedside Shivering Assessment Scale) should be used. Propofol, fentanyl, and midazolam are commonly used medications, but no study has evaluated the optimal regimen. Propofol and midazolam have antiseizure effects, while neuromuscular blockade masks clinical seizures necessitating EEG. Hemodynamic and pharmacokinetic considerations influence the choice of sedation and medications that facilitate a neurologic exam should be prioritized. Shivering management begins with nonpharmacologic strategies such as active cutaneous counter-warming. Antipyretics such as acetaminophen and nonsteroidal antiinflammatory drugs are reasonable first-line medications, but they have limited effectiveness in brain-injured patients. Magnesium reduces smooth muscle tone, and the resultant vasodilation reduces shivering. Opiates such as meperidine effectively reduce shivering and are often combined with anesthetic sedative infusions. Dexmedetomidine, a presynaptic alpha-2-agonist, reduces shivering and when combined with buspirone (a serotonin agonist/antagonist) the effectiveness is increased. Buspirone in combination with low-dose meperidine has similar effects to higher doses of meperidine alone, which is desirable given the epileptogenic effects of meperidine (Jain et al., 2018). Standardized protocols for sedation holidays after the first 24 h or at the completion of rewarming are beneficial.

Fever, hyperglycemia, seizures, and elevated ICP negatively influence neurologic outcome. Core temperature >39°C within the first 72 h following OHCA is associated with a higher risk of brain death. For every degree above 37°C, there is a 2.3-fold increase in the odds of an unfavorable outcome. Maximal temperature during hospitalization of >37.8°C for survivors of OHCA is associated with decreased survival to discharge (Takasu et al., 2001; Zeiner et al., 2001). Maintaining core temperature <37.5°C with a multimodal approach using surface cooling and medications is a reasonable strategy.

Hyperglycemia is common after cardiac arrest. There is a strong nonlinear association between serum glucose levels 12 h after ROSC and neurologic outcomes. Those achieving normoglycemia are more likely to survive with a favorable neurologic outcome (Longstreth et al., 1986; Losert et al., 2008). However, IHCA survival odds were shown to be relatively insensitive to glycemic control in diabetics with decreased survival only associated with severe hyperglycemia (>240 mg/dL); in nondiabetics, survival odds were sensitive to hypoglycemia (<70 mg/dL) (Beiser et al., 2009). The routine administration of dextrose during resuscitation is harmful (Peng et al., 2015).

Severe hypoxic ischemic brain injury leads to diffuse cerebral edema and the risk of brain herniation. There is significant variability among patients who develop cerebral edema with influence from temporal factors related to TTM and rewarming, duration of cardiac arrest, and cerebrovascular dynamics. Noninvasive measurement using optic nerve sheath diameter and transcranial Doppler ultrasonography have demonstrated agreement with invasively monitored ICP without the associated risk (Cardim et al., 2019). However, evidence that standard interventions to treat cerebral edema affect outcome is lacking.

Seizures occur in 5%–15% of patients with ROSC and 10%–40% of those remaining comatose (Neumar et al., 2008). Increased $CMRO_2$ from seizures can lead to secondary injury. While often treatable, seizures and particularly status epilepticus may impair neurologic outcome. Nonconvulsive status epilepticus (NCSE) is common. An observational study showed that most NCSE occurred within the first 12h (Rittenberger et al., 2012). Anesthetic sedation used for TTM and TTM itself may be protective. Patients with non-refractory status epilepticus (non-RSE) can have good neurologic outcomes when effectively treated. Benzodiazepines such as midazolam or lorazepam and anesthetic infusions such as propofol can suppress seizures. Antiepileptic medications such as phenytoin, valproic acid, or levetiracetam are reasonable second-line treatments (Neumar et al., 2008). Myoclonus (brief sudden involuntary jerking movements) is often driven by an injured motor cortex or subcortical structures (Caviness and Brown, 2004) and occurs in 30%–40% of patients (Wijdicks et al., 1994). In cortical myoclonus, treatment targets an impaired inhibitory process with GABAergic medications. Postanoxic myoclonus responds to clonazepam, sodium valproate, and levetiracetam. Primidone and phenobarbital may also be effective but are used infrequently due to sedating effects. Sodium valproate is effective in treating subcortical myoclonus and lamotrigine may act as an adjunct or be used as monotherapy (Caviness and Brown, 2004).

The majority of mortality occurs in the first 24h making early and aggressive protocol-based, goal-directed intensive care vital. A standardized protocol encompassing oxygenation, hemodynamic support, early coronary reperfusion if needed, TTM, seizure management, and glycemic control more than doubled survival to hospital discharge with good neurologic outcome compared to historical controls (56% vs 26%) (Sunde et al., 2007).

Targeted temperature management

TTM in comatose survivors of cardiac arrest is the single most important development in postresuscitation management in the past 2 decades and is recommended across international guidelines. However, challenges remain in determining its optimal implementation (Taccone et al., 2020).

Two landmark trials in 2002 heralded the modern TTM era. The Bernard trial randomized 77 comatose patients who achieved ROSC after OHCA due to VF to therapeutic hypothermia (33°C within 2 h of ROSC for 12h) or normothermia (37°C). Good outcome, defined as discharge to home or a rehabilitation facility, was achieved in 49% of the hypothermia group compared to 26% in the control group (odds ratio 5.25, 95% CI: 1.47–18.76) (Bernard et al., 2002). The Hypothermia After Cardiac Arrest Study Group randomized a population with OHCA due to VF or pulseless VT to cooling (32–34°C for 24h) or normothermia. Fifty-five percent of patients undergoing therapeutic hypothermia achieved a Pittsburgh Cerebral Performance Category of 1 (good recovery) or 2 (moderate disability) compared with 39% in the control group (odds ratio 1.40, 95% CI: 1.08–1.81) (Hypothermia after Cardiac Arrest Study Group, 2002).

In 2010, AHA guidelines for postcardiac arrest care gave a grade IB recommendation for therapeutic hypothermia in comatose survivors of OHCA due to VF and VT. The TTM Trial (Nielsen et al., 2013) randomized unconscious survivors of OHCA, regardless of initial cardiac rhythm, to TTM with a target temperature of 33°C or 36°C for an intervention period of 36h. The more liberal target temperature of 36°C was not inferior for neurologic functional outcome or mortality at 180 days compared to 33°C. This trial addressed two major concerns of the earlier trials. First, aggressive fever control was employed in all patients during the first 72h. Second, a broader population such as those with nonshockable cardiac rhythms (pulseless electrical activity and asystole) was enrolled (Frydland et al., 2015).

Cardiac arrest due to a nonshockable rhythm is often caused by noncardiac etiologies and carries a worse prognosis. The 2019 HYPERION trial demonstrated an increased likelihood of a favorable 90-day neurologic outcome with TTM (33°C for 24h) after OHCA and IHCA with a nonshockable rhythm compared to normothermia 10.2% vs 5.7%). There was no difference in mortality or adverse events between the groups. Approximately three quarters of these patients had OHCA and two-thirds were presumed to have a noncardiac cause (Lascarrou et al., 2019). Registry analyses, albeit with substantial methodological limitations, suggest that TTM after IHCA may be harmful. The majority of IHCA have an initial nonshockable rhythm (PEA or asystole) for which evidence supporting the neuroprotective benefit of TTM remains limited. Additionally, IHCA typically have faster response time perhaps limiting the degree of anoxic brain injury, which may limit the theoretical benefit of TTM. Therefore, optimal temperature management strategy after IHCA warrants further investigation (Chan et al., 2016).

The 2019 AHA guidelines recommend TTM at 32–36°C for all comatose adult survivors of cardiac arrest without contraindications (Class 1 recommendation). However, the most effective approach remains uncertain (Taccone et al., 2020). Cooling should be initiated immediately after achieving ROSC in appropriate patients. For every hour of delay in initiating TTM, there is a 20% increase in the relative risk of death (Mooney et al., 2011). Prehospital induction of cooling with rapidly cooled IV fluid infusions helps achieve TTM sooner once admitted, but it has been associated with the development of pulmonary edema and evidence of benefit is lacking. Similarly, prolonged cooling (>24h) is not beneficial and increases ICU length of stay (Callaway et al., 2015). TTM should not be delayed while awaiting neuroimaging when the etiology of the cardiac arrest is known or reasonably suspected to be nonneurologic. Although fever control alone may be sufficient to improve patient outcomes, antipyretics are largely insufficient in maintaining normothermia and mechanical cooling was used in many normothermia control groups.

Neuroprotective pharmacology and alternative strategies

Several neuroprotective strategies have shown promise in the preclinical setting, often using short-term outcomes, yet none thus far has improved neurologic outcomes in humans after cardiac arrest. Mechanisms typically target injury pathways by attempting to modulate neuronal cell death, alter oxygen free radicals or improve cerebral hemodynamics. As described previously, excess activation of *N*-methyl-D-aspartate leads to glutamate excitotoxicity. However, NMDA receptor antagonism has shown deleterious effects in preclinical models. Xenon and argon gas also modulate these pathways with promising results in animal models. Lamotrigine inhibits glutamate release with a positive effect on neurohistopathological outcomes in preclinical studies. Ischemic postconditioning (brief-controlled ischemia and reperfusion of a limb) may offer global neuroprotective effects. Efforts to enhance the mechanical efficiency of CPR using assistive devices improve coronary and cerebral perfusion. Early hypertonic saline may treat cerebral edema and decrease ICP thereby improving cerebral perfusion in preclinical studies, but this was not associated with improved functional outcomes. Hyperbaric oxygen has shown promise when implemented during or immediately after CPR but remains investigational (Huang et al., 2014; Mangus et al., 2014).

PROGNOSTICATION OF NEUROLOGIC OUTCOME

Introduction

Coma is a state of absent arousal with eyes closed and no response to internal or external stimuli beyond reflex. It is the most frequent immediate disorder of consciousness resulting from hypoxic brain injury. Neurologic function governs long-term outcome after ROSC, and persistent coma is widely accepted as a poor neurologic outcome. Accurately prognosticating neurologic recovery remains a heavy burden for physicians. Systematic clinical assessments of arousal and brainstem function are the foundation of prognostication. The ascending reticular activating system including the cerebral cortex, the thalamus and hypothalamus, and critical regions of the brainstem mediate arousal. Diagnostic tests such as laboratory, neurophysiology, and imaging studies can help support the physical exam findings in a multimodal approach. However, diagnosing irreversible failure of the brain arousal system can be challenging.

Surrogate decision makers must have reliable predictions to establish expectations and make sound decisions. Most early deaths after ROSC occur from WLST (Elmer et al., 2016b). However, delayed awakening, >72 h after ROSC or rewarming, is increasingly observed in the TTM era (Zanyk-McLean et al., 2017). An erroneous prediction of a poor outcome may lead to premature WLST. Up to 20% of patients who undergo WLST might have actually had a chance for neurologic recovery. Conversely, an overly optimistic forecast in a patient with objective predictors of a poor outcome can lead to futile treatment and ongoing emotional strain for families and caregivers. Therefore, early and accurate prognostication is crucial (Muhlhofer and Szaflarski, 2018).

Current measures of neurologic outcome

Survival is an objective and simple outcome in the immediate postarrest period for interventions performed during or immediately after cardiac arrest. However, this offers little insight into longer-term quality of life for survivors (Becker et al., 2011).

Biomarkers of organ dysfunction and impaired systemic physiology may predict longer-term outcome. Validated tools assessing disability across several domains are common endpoints in patient-centered studies. These domains include body structures, functional ability, impairments, limitations in activities of daily living, and societal participation.

The cerebral performance categories (CPC) and the modified Rankin Scale (mRS) are frequently used post-resuscitation outcome measures (Becker et al., 2011; Geocadin et al., 2019). A CPC score of three or greater or a mRS of four or greater have often been used for discriminating unfavorable neurologic outcome from favorable in studies after cardiac arrest (see Table 19.1). Although the CPC assesses multiple domains, it lacks agreement with patient-reported quality of life measures and does not discriminate between mild and moderate postanoxic brain impairments. Therefore, the International Liaison Committee on Resuscitation now recommends using the mRS (Geocadin et al., 2019).

The goal of a prognostic test is to predict a defined outcome accurately. In the case of cardiac arrest, a prognostic test acts as an indirect measure of injury severity (Geocadin et al., 2019). A result that predicts a poor outcome is a true positive if the patient ultimately does not regain meaningful interaction with his environment. Prognostic tests that consist of continuous or ordinal variables often benefit from a defined cutoff to dichotomize results and more readily report sensitivity and specificity (Geocadin et al., 2019).

Several confounders and biases may affect prognostic tests. Broadly speaking, the quality of evidence supporting nearly all tests is weak. Many studies have not reported cause of death (e.g., brain death, cardiac death, etc.), extent of support at the time of death, and occurrence of WLST. A survivor with good neurologic recovery also remains at risk for death from systemic disease (e.g., circulatory collapse, sepsis, or renal failure). A prognostic test that accurately predicts a good neurologic recovery in this population may be hampered in its perceived utility by deaths from systemic causes. Careful appraisal of the dosage, pharmacokinetics and pharmacodynamics of any sedating medications and neuromuscular blockade is critical, particularly in patients treated with TTM (Geocadin et al., 2019; Rossetti et al., 2016). *Post hoc* analysis to determine the threshold cutoff values of continuous variables such as serum biomarker levels or quantified neuroimaging findings require prospective validation.

Finally, the self-fulfilling prophecy, a prediction that becomes true by virtue of predicting it, biases many neurologic prognostication studies. Treatment teams are often aware of test results and may use them when determining care goals. Ideally, patients should not have WLST during the systematic evaluation of a prognostic test and care providers should be blinded to the test results (Geocadin et al., 2019). This remains both impractical and, in many cases, impossible as the test may be a fundamental component of the routine neurologic examination or a test that identifies treatable conditions such as EEG identification of seizures.

Balancing the risk of providing futile care by delaying prognostication vs inappropriate WLST due to premature estimation of a poor prognosis presents a clinical challenge. Most patients who awaken from coma do so within 3.5 days. Delaying prognostication at least 72 h from ROSC can ensure this cohort of patients receives reasonable opportunity to improve. Of the remaining patients, most destined to awaken will do so by the seventh day. Up to a third of patients (Geocadin et al., 2019), particularly those undergoing TTM, awaken beyond 72 h (Zanyk-McLean et al., 2017).

The neurologic examination

The neurologic examination remains an important prognostication tool. Levy et al. correlated the findings of serial neurologic examinations over 14 days with outcome in patients with hypoxic ischemic encephalopathy. Those with poor functional recovery had absent pupillary light reflexes (PLRs) and a motor response to pain in the extremities not better than extensor posturing (Levy et al., 1985). Only 71% of these patients had suffered cardiac arrest, and the definition of coma used was more liberal than the classic definition of coma with eyes closed and no purposeful motor responses. Additionally, the advent of TTM and advances in postresuscitation critical care have rendered the results largely relevant only in a historical context. Yet the neurologic exam remains important to assess damage and dysfunction at the corresponding levels of the neuroaxis. Excluding major confounders such as medication effects and severe metabolic abnormalities is a crucial first step of a thorough clinical examination.

A bilaterally absent PLR beyond 72 h following ROSC correlates strongly with a poor neurologic outcome regardless of TTM. With a false-positive rate (FPR) approaching zero (0.5%, 95% CI: 0%–8%) across several studies, this exam finding has remained a reliable predictor of poor outcome but lacks sensitivity. Automated infrared pupillometry removes subjective assessment and quantifies not only reactivity but also incorporates pupil size, constriction and dilation velocity, and latency into an algorithm to calculate the Neurological Pupil index (NPi). In a prospective study of 456 patients, the FPR on NPi using automated pupillometry was 0% compared to 6% for standard manual assessment by a neurologist. Automated pupillometry also had better sensitivity than the standard manual assessment (Oddo et al., 2018). It maintains high specificity from day 1 through 3 after ROSC, whereas manual assessment has poor specificity on day 1 and 2 with a 10% FPR. Absent corneal reflexes at 72 h also predict a poor outcome but with a higher FPR (5%, 95% CI: 0%–25%) than automated pupillometry. Moreover, this reflex is prone to confounding from sedative medications and neuromuscular blockade. From the Levy criteria to the advent of TTM, absent or extensor motor responses at 72 h were considered reliable predictors of poor outcome. Similar to the corneal reflex, motor responses may be blunted by medications and as many as a quarter of patients receiving TTM with a poor motor exam achieve a good neurologic outcome. Therefore, motor responses can no longer be considered a reliable predictor of outcome (Hawkes and Rabinstein, 2019).

Postanoxic multifocal myoclonus is common following cardiac arrest and can be subdivided as: myoclonic status epilepticus (MSE), subcortical myoclonus, and Lance-Adams syndrome (Elmer et al., 2016a). MSE is typically multifocal or generalized jerking of the body appearing most often within 72 h of CA, which may be spontaneous or stimulus induced and may originate from subcortical structures. In contrast, Lance-Adams syndrome occurs later than MSE, often months or years after CA, may also be generalized or multifocal and is typically induced by intention or stimulus. The clinical appearance of myoclonus will vary with cortical and subcortical injury, with subcortical myoclonus typically involving proximal musculature (Freund and Kaplan, 2017). The 2006 American Academy of Neurology Practice Parameter on prognostication after cardiac arrest identified MSE as being highly predictive of a poor outcome (FPR 0%, 95% CI: 0%–8.8%). MSE is continuous and generalized, persisting for >30 min within 48 h after ROSC. The myoclonic activity is time locked with generalized epileptiform (spike, polyspike or sharp) activity on EEG. If MSE is present during TTM, it carries a grave prognosis (FPR 0%, 95% CI: 0%–3%). Clinical myoclonus should not be interpreted in isolation. EEG patterns associated with multifocal myoclonus include a suppression-burst background with high-spike–wave discharges in lockstep with myoclonic jerks and subcortical myoclonus (twitching without EEG correlation) that carry a poor prognosis. In contrast, patient with a continuous background with narrow, vertex spike–wave discharges in lockstep with myoclonic jerks is associated with better survival, and among survivors a generally favorable outcome (Elmer et al., 2016a).

Neurophysiology: Electroencephalography and somatosensory evoked potentials

EEG is widely available and remains among the most commonly used prognostication tools next to the clinical examination (Muhlhofer and Szaflarski, 2018). Hypoxic–ischemic brain injury is a dynamic process and EEG provides real-time assessment to guide both management and prognostication. Despite extensive supporting literature, there are clinical limitations due to lack of standardized EEG terminology (Hirsch et al., 2013). The AHA and American Clinical Neurophysiology Society endorse the use of continuous or frequent intermittent EEG in comatose patients within 12 h of ROSC to diagnose clinical and subclinical seizures, guide seizure treatment, and assess cortical activity for prognostication. TTM and medications such as midazolam and propofol attenuate background activity and reactivity, suppress epileptiform discharges, and increase the duration of suppression in a burst-suppressed patient. High doses of anesthetic sedation (Hawkes and Rabinstein, 2019), shivering, and electrical interference can make interpretation challenging.

A low-voltage EEG < 20 μV predicts an unfavorable outcome (Sandroni et al., 2013). When all EEG activity is <10 μV throughout the recording, this is considered suppression. Burst suppression consists of suppression for greater than half of the recording (see Fig. 19.1). Suppression periods comprising 10%–49% of the recording constitute a discontinuous background. Burst suppression with identical bursts, isoelectric, or low voltage (<20 μV) observed within 24 h predicted poor outcome at 6 months (FPR 0%, 95% CI: 0%–2%). Increasing burst suppression ratio (the ratio of time in which signal remains <10 μV) may be a useful objective measure associated with poor outcome (Sandroni et al., 2013).

Discriminating benign from malignant-generalized discharges during a burst-suppressed EEG is important. Pharmacologically induced burst suppression in other patient populations often reveals sharp components and epileptiform activity. Hypothermia may also induce periodic epileptiform activity at low temperatures. Discharges in which the morphology of the first 500 ms of each burst appear identical may help distinguish malignant patterns that carry a poor outcome (FPR 0%). These malignant discharges emerge early and may be transient (Muhlhofer and Szaflarski, 2018).

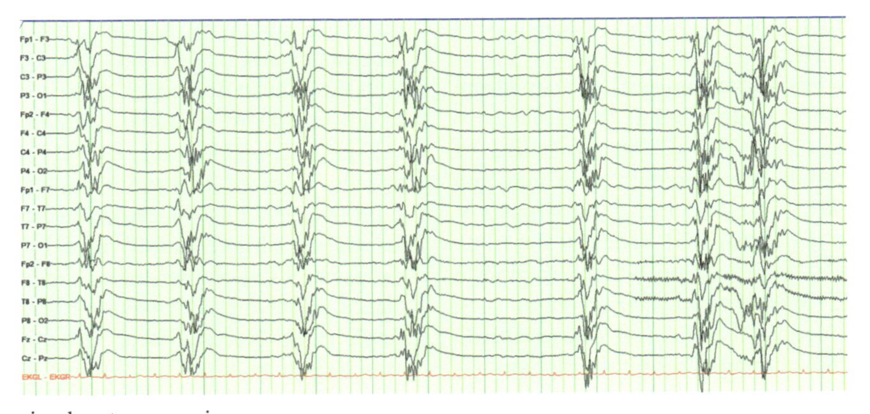

Fig. 19.1. EEG showing burst suppression.

Epileptiform discharges, seizures, and status epilepticus during or after TTM are less predictive of outcome, and seizures should be treated aggressively. A study of RSE after cardiac arrest found that over 50% of patients whose seizures were treated aggressively were alive at 6 months and the majority of survivors had a good neurologic outcome (Hawkes and Rabinstein, 2019). At 72 h after ROSC, the absence of EEG reactivity to external stimuli and persistent burst suppression after rewarming from TTM predict a poor outcome (FPR 0%, 95% CI: 0%–3%). Similarly, persistent intractable status with absent EEG reactivity predicts a poor outcome. Generalized periodic discharges (GPDs) often fall within the ictal-interictal continuum when they occur at a frequency of 1.5–2.5 Hz and are defined as seizure when they occur at >2.5–3 Hz. GPDs are associated with a poor outcome but most often when they accompany a discontinuous or suppressed background with absent variability. GPDs predict seizure activity and are associated with NCSE. It is reasonable to treat GPDs with antiepileptic medications. Many EEG patterns previously thought to predict poor outcome still result in favorable survival. Background continuity and reactivity influence prognosis most significantly. For example, status epilepticus following cardiac arrest was previously considered a strong indicator of poor prognosis. However, when it evolves from a continuous background, it has a better prognosis than when it evolves from burst suppression (Muhlhofer and Szaflarski, 2018). Alpha/theta coma and spindle coma in isolation do not have a strong correlation with outcome.

Background reactivity is an alteration in voltage to auditory or tactile stimuli. In a cohort of 357 patients, the presence or absence of background reactivity was the best discriminator between good and poor outcome (Rossetti et al., 2016). Moreover, background variability (Geocadin et al., 2006), rhythmic delta activity (Hawkes and Rabinstein, 2019), and continuous background, particularly when >20 μV, predict good outcome (Oh et al., 2015).

The bispectral index is an automated tool to measure anesthesia depth, on a scale from 0 (isoelectric) to 100 (normal, awake). A value ≤22 in the setting of hypoxic–ischemic brain injury predicts poor outcome (FPR 0%, 95% CI: 0%–6%) (Seder et al., 2010). Automated quantitative EEG (qEEG) reduces the time needed for analysis but sacrifices the dynamic evolution of traditional EEG. More complex approaches such as amplitude-integrated EEG and entropy-based qEEG are promising prognostic modalities but further validation is required.

Somatosensory evoked potentials (SSEP) are a measure of the short latency (20 ms) cortical response at the postcentral gyrus to a peripherally applied electrical stimulus to a nerve in the forearm. Medications and metabolic derangements affect SSEPs minimally. Their prognostic value is not influenced by TTM (Sandroni et al., 2018), yet electrical interference can confound their interpretation. A meta-analysis of nine studies including 492 patients found bilateral absence of the N20 cortical potential predicted a poor neurologic outcome (FPR of 0.7%, 95% CI: 0.1%–4.7%) (Kamps et al., 2013). Notably, many patients who had preserved N20 responses still had unfavorable outcome. The timing of the study following TTM (normothermia, at 72 h after ROSC or 72 h after normothermia) did not have an impact (Sandroni et al., 2018). The true prognostic value of SSEP remains unknown as test results may influence WLST within a trial creating a self-fulfilling prophecy; improved clinical trial design presents an ethical challenge in this regard.

A cutoff threshold N20 amplitude (≤0.62 μV), as opposed to a dichotomized outcome (absent/present), has improved sensitivity from 30% to 57% (95% CI: 48%–65%) without sacrificing specificity. The middle latency SSEP N70 response has shown promise in predicting a good outcome in those with N20 present (Madl et al., 2000). However, equivocal readings and technically insufficient recordings limited this study and validation is required. Brainstem auditory evoked potentials add little to prognostication, but progression of auditory discrimination from TTM to normothermia may predict good outcome (Tzovara et al., 2016).

Serum biomarkers and cerebral neurochemistry

Serum biomarkers are an objective measure of the severity of brain injury and are not susceptible to the same confounders as neurophysiologic tests and the clinical examination. However, diversity of detection methods, laboratory reagents, and equipment lead to inconsistencies.

Neuron-specific enolase (NSE) is a well-established serum marker of neuronal damage. A threshold value >33 μg/L 24–72 h after ROSC accurately predicted patients destined to have a poor neurologic outcome (FPR 0%, 95% CI: 0%–3%) in an early study. Later studies suggest a higher cutoff value, both with and without the use of TTM. Some studies suggest up trending NSE values in the first 72 h have a higher specificity than a single value (Sandroni et al., 2018). Guidelines endorse sampling at multiple time points to assess trends and reduce false positives. A 2020 meta-analysis of 4806 patients in 42 trials reported a pooled sensitivity of 56% (95% CI: 47%–65%) and specificity of 99%

(95% CI: 98%–100%) regardless of TTM use. No differences were reported in subgroup analyses of OHCA, TTM, and early vs late assessment (Wang et al., 2020). Several studies used cutoff values that were determined post hoc to maximize specificity as opposed to the fixed levels of early trials. However, this aligns with international guidance, which suggests individual hospitals determine internal normal and cutoff values against a reference standard (Nolan et al., 2015). Of note, NSE is also elevated in patients with neuroectodermal tumors and small-cell lung cancer (Rossetti et al., 2016).

S-100B is a protein released not only by injured glial cells but also adipose tissue and muscle. It has a shorter half-life than NSE (2 h vs 24h) and increases quickly after hypoxic ischemic brain injury (Wang et al., 2020). A 2020 meta-analysis found a pooled sensitivity of 62% (95% CI: 46%–78%) and specificity of 97% (95% CI: 92%–100%). Early S-100B level combined with late NSE measurement may be complementary for prognostication and worthy of further evaluation (Wang et al., 2020).

Several other candidate biomarkers have shown promise. Tau protein is a marker of axonal injury and was shown in the TTM trial to have a high specificity for predicting poor outcome (Sandroni et al., 2018). MicroRNAs (miRNAs) have shown potential as a specific marker of cellular damage with the added potential to identify the cell types injured. Compared to other biomarkers, miRNAs have remarkable stability in serum (Hawkes and Rabinstein, 2019).

Regional invasive cerebral oxygenation measurement, near infrared spectroscopy, cerebral oxygen delivery and extraction, lactate and pyruvate measurement via cerebral micodialysis and jugular bulb oximetry have potential but remain investigational (Buunk et al., 1997; Hifumi et al., 2016; Wallin et al., 2018).

Neuroimaging

Computed tomography (CT) and magnetic resonance imaging (MRI) are useful neuroimaging modalities that can aid prognostication after cardiac arrest. Early brain CT is sporadically obtained to assess for intracranial hemorrhage as an infrequent cause of arrest. However, this should not delay TTM initiation. Early loss of gray-white differentiation on CT within 2 h predicts a poor prognosis (FPR 0%, 95% CI: 0%–12%) (Inamasu et al., 2011). Cytotoxic cerebral edema appears radiographically as decreased gray matter density and sulcal effacement. Qualitatively, this is referred to as the "loss of boundary" sign. The quantitative gray-to-white ratio (GWR) in the basal ganglia, centrum semiovale, and high convexity is detectable within 1 h. Sulcal effacement and "pseudo subarachnoid hemorrhage" due to compressed sulci from cerebral edema are additional qualitative findings (Keijzer et al., 2018). Restricted diffusion on brain MRI, quantified with apparent diffusion coefficient (ADC), reliably identifies cytotoxic edema early but attenuates by day 7–10 (see Fig. 19.2). Changes appears on T1- and T2-weighted, and T2-fluid-attenuated inversion recovery sequences day 1–2 remaining visible for days to weeks (Keijzer et al., 2018).

A 2020 meta-analysis including 44 studies analyzed the prognostic value of brain CT and MRI. Among the CT studies, GWR was most frequently used. A few studies used alternative measures such as the modified Alberta Stroke Program Early CT score, Hounsfield unit differences, or optic nerve sheath diameter to complement GWR. The analysis included 2926 patients from 24 studies FPR for predicting a poor neurologic outcome using any CT criteria was 2% (95% CI: 1%–4%). Across all subgroups, which included evaluation of deep brain structures, superficial brain structures, both deep and

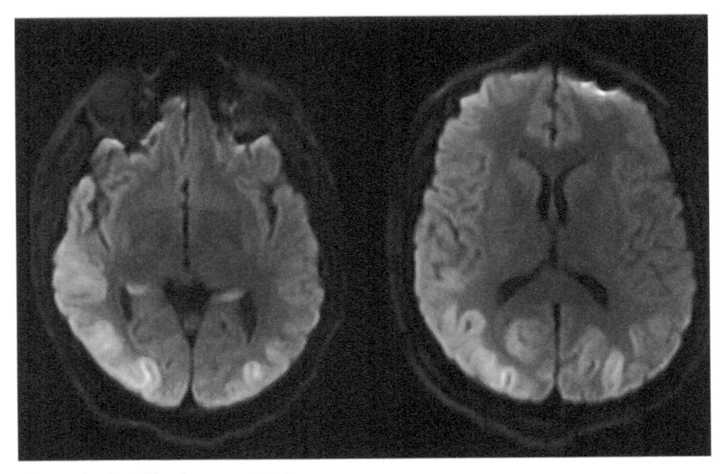

Fig. 19.2. MRI with bilateral cortical diffusion restriction.

superficial structures, use of TTM, CT performed before or after 24h, the FPR ranged from 2% to 6% (95% CI: 1%–13%). Moreover, sensitivity was only 18%–64% (Lopez Soto et al., 2020). Only a small minority of studies used a prespecified cutoff for GWR. Most performed a post hoc analysis to define the cutoff value with the best performance. An earlier analysis found GWR values <1.10 within 24h to have high specificity but highly variable sensitivity. Elapsed time, the generation of scanner, region of interest location, selection method (manual or automated), and reconstruction algorithm can also influence GWR (Keijzer et al., 2018). Finally, quantitative GWR remains infrequent in clinical practice and radiological reports typically provide only a qualitative assessment.

Of the 21 MRI studies, 16 assessed the utility of DWI with similar proportions assessing superficial vs deep structures and 11 studies assessing both. Four studies combined DWI and FLAIR, and all of those assessed deep and superficial structures in combination. ADC quantification was used in five trials, and 650×10^{-6} mm^2/s on day 2–5 was found to be an upper limit threshold for predicting poor prognosis. The timing of MRI examination was much more heterogeneous than in the CT studies. The FPR for all MRI subgroups (5%–15%) was higher than CT, but the sensitivity for predicting a poor outcome was much better (65%–83%). Only the combination of DWI and FLAIR was similar to GWR. When using a prespecified ADC cutoff the FPR was 10% (95% CI: 0.04–0.20). A study with detailed spaciotemporal descriptions of brain injury found cerebral cortex, putamen, and cerebellar injuries within the first 2–3 days predicted poor outcome (Keijzer et al., 2018).

Previous guidelines and meta-analyses cautioned against the use of imaging for prognostication. Furthermore, inclusion of CT in a multimodal model for prognostication did not improve accuracy (Cristia et al., 2014). However, newer studies using different thresholds and time points and newer techniques suggest that it may be necessary to reconsider the utility of neuroimaging. MRI may be useful in patients who remain in persistent coma for several days. CT perfusion imaging has shown promise in a pilot study. Diffusion tensor imaging estimates structural connectivity and has shown potential in patients remaining comatose beyond 7 days. Functional MRI and PET may have a potential role in the future (Velly et al., 2018).

Multimodal approach

It is reasonable to begin a systematic approach to prognostication in patients who remains comatose >72h after cardiac arrest once confounders are considered. EEG monitoring and management of seizures should begin within 12–24h. Serial neurologic examinations with observations such as impaired PLR, corneal reflex or motor responses, early myoclonus, and ancillary neurophysiology studies such as EEG or the SSEP N20 response after rewarming should be used in combination. This may be supplemented with biomarkers and neuroimaging but is not mandatory (Hawkes and Rabinstein, 2019; Rossetti et al., 2016; Sandroni et al., 2018). However, when tests are discordant, prolonged monitoring and treatment is warranted if it aligns with patient-centered directives. Among patients destined to awaken from coma, 10%–30% do so between 48h and 12 days (Paul et al., 2016).

Illness severity scores predict patient outcomes using data from multiple organ systems. The Pittsburgh Cardiac Arrest Category (PCAC) aims to predict survival and functional recovery by including readily available data early after ROSC (Coppler et al., 2015). PCAC may be helpful in providing a baseline prognostic estimate early after cardiac arrest using neurologic, cardiac, and respiratory scoring parameters. Patients are stratified into one of four categories: awake (80% survival, 60% good outcome), coma and mild cardiopulmonary dysfunction (60%, 40%), coma and moderate to severe cardiopulmonary dysfunction (40%, 20%), and coma with at least one absent brainstem reflex (10%, 5%) (Coppler et al., 2015). The cardiac arrest hospital prognosis score includes arrest characteristics such as location of arrest, bystander CPR, initial cardiac rhythm, no-flow time, and organ dysfunction in an elderly population to predict prognosis and select patients who may benefit from early coronary angioplasty (Sauneuf et al., 2020). In addition to novel tests for prognostication, refinement of neurologic illness severity scores represent an opportunity for enhanced prognostication in the future.

Long-term neurologic complications of cardiac arrest

Less than 7% of cardiac arrest survivors return to their previous level of function (Geocadin et al., 2008). Irreversible brain injury is common, and the spectrum of postarrest syndromes correlates with those areas of the brain most susceptible to anoxic injury. These include the cortex, arousal centers, thalamus, and cerebellum. Survivors who immediately awaken may still suffer from acute confusional states and long-term cognitive impairments (Hrishi et al., 2019). Retrograde and anterograde amnestic syndromes are frequently observed (Bass, 1985).

Many patients remain in a state of impaired awareness and interaction, with recovery of crude cycling of the arousal state heralded by periods of eye-opening, known as a vegetative state. When this state remains for at least 30 days, the patient is in a persistent vegetative state.

There are no clear criteria defining when a persistent vegetative state becomes permanent. Those who regain fragments of meaningful interaction and awareness of self or environment are in a minimally conscious state (Posner et al., 2007). The unresponsive wakefulness of the persistent vegetative state carries a particularly poor long-term prognosis for meaningful recovery and should be differentiated from the minimally conscious state (De Salvo et al., 2012).

Global ischemic injury to vulnerable vascular territories such as the middle and anterior cerebral artery border zone leads to bi-brachial weakness (man in the barrel). Bilateral posterior border zone infarctions at the occipitoparietal region cause Bálint syndrome consisting of simultanagnosia, oculomotor apraxia, and optic ataxia. Bilateral occipital cortex infarction causes cortical blindness. Patients with Anton syndrome have denial of the visual impairment and often have visual confabulation.

Among cardiac arrest survivors discharged from hospital, 30%–50% suffer from cognitive and neuropsychiatric impairment (Moulaert et al., 2009). Although long-term cognitive impairment is common in the critically ill, patients with cardiac arrest have particular difficulty with attention and processing speed (Lilja et al., 2015), executive function, memory, and learning (Tiainen et al., 2007). Patients treated with TTM have more than 50% better odds of being cognitively intact or with only mild impairment compared to those who do not undergo TTM (Tiainen et al., 2007). An elevated protein S100 level correlates with working memory impairments and can potentially provide prognostic guidance in that domain (Grubb et al., 2007). Recovery of memory and visuospatial deficits typically occur within the first 3 months, whereas executive function recovery takes up to 10 months or longer to plateau (Lim et al., 2004). Anxiety and posttraumatic stress disorder are also common long-term sequelae (Gamper et al., 2004).

Seizures immediately following cardiac arrest occur in 10%–40% and may impact prognosis. However, among 57,437 discharged survivors of cardiac arrest who never had a seizure in the hospital, the risk of developing late seizures was 1% annually (Morris et al., 2018). Management of late seizures and epilepsy should follow established treatment guidelines. Simple or complex partial seizures may be confused with postanoxic movement disorders or acute confusional states and EEG can be helpful in distinguishing these entities.

Basal ganglia injury manifests as adventitial movement or a parkinsonian syndrome. Chronic postanoxic myoclonus is the Lance-Adams syndrome. It develops days to months after the initial injury and is often triggered by auditory or tactile stimuli. In contrast with acute myoclonus, it carries a more favorable outcome and often spontaneously resolves over months to years. Many patients are eventually able to discontinue treatments such as clonazepam, sodium valproate or levetiracetam (Werhahn et al., 1997).

Cardiac arrest and postresuscitation systemic hypotension may lead to ischemic myelopathy, often in the form of central spinal cord infarction with bilateral spinothalamic sensory deficits and sparing of the posterior columns (Yadav et al., 2018). The incidence of ischemic myelopathy is unknown since spinal cord imaging is not routinely obtained. A neuropathologic study found evidence of ischemic myelopathy in 45% of cases (Duggal and Lach, 2002). Based on anatomical studies, susceptible regions include vascular border zones such as the mid-thoracic region from T4 to T8. However, the same neuropathological study found that the most commonly involved site was the lumbosacral cord (95%). Mid-thoracic level injury occurred in only 8% of cases. Increased lumbosacral vulnerability to global ischemia is thought to be due to higher metabolic demand in this region. In a preclinical model, induced hypothermia was shown to attenuate neuronal loss in the lumbosacral cord (Ahn et al., 2020). However, the effect of TTM on ischemic myelopathy in humans remains unknown.

CONCLUSION

Cardiac arrest is a devastating event with high morbidity and mortality. Significant advances over the years in the approach to resuscitation and postresuscitation care have improved survival and the advent of TTM was a breakthrough in improving neurologic outcomes. Refining the optimal approach to TTM continues to be an active area of research. As more patients continue to survive to a favorable neurologic outcome, improving prognostication to inform decision making of families and caregivers is critical. Future research will continue to elucidate the most appropriate use of neurodiagnostic tools for prognostication.

References

Adams JH, Brierley JB, Connor RC et al. (1966). The effects of systemic hypotension upon the human brain. Clinical and neuropathological observations in 11 cases. Brain J Neurol 89: 235–268. https://doi.org/10.1093/brain/89.2.235.

Adrie C, Adib-Conquy M, Laurent I et al. (2002). Successful cardiopulmonary resuscitation after cardiac arrest as a "Sepsis-like" syndrome. Circulation 106: 562–568. https://doi.org/10.1161/01.CIR.0000023891.80661.AD.

Ahn JH, Lee T-K, Kim B et al. (2020). Therapeutic hypothermia improves hind limb motor outcome and attenuates oxidative stress and neuronal damage in the lumbar spinal cord following cardiac arrest. Antioxidants (Basel) 9: 38. https://doi.org/10.3390/antiox9010038.

Albert CM, Stevenson WG (2018). Cardiovascular collapse, cardiac arrest, and sudden cardiac death. In: JL Jameson, AS Fauci, DL Kasper, SL Hauser, DL Longo, J Loscalzo (Eds.), Harrison's principles of internal medicine. McGraw-Hill Education, New York, NY.

Ames A, Wright RL, Kowada M et al. (1968). Cerebral ischemia. II. The no-reflow phenomenon. Am J Pathol 52: 437–453.

Balan IS, Fiskum G, Hazelton J et al. (2006). Oximetry-guided reoxygenation improves neurological outcome after experimental cardiac arrest. Stroke 37: 3008–3013. https://doi.org/10.1161/01.STR.0000248455.73785.b1.

Bass E (1985). Cardiopulmonary arrest. Pathophysiology and neurologic complications. Ann Intern Med 103: 920–927. https://doi.org/10.7326/0003-4819-103-6-920.

Becker LB, Aufderheide TP, Geocadin RG et al. (2011). Primary outcomes for resuscitation science studies: a consensus statement from the American Heart Association. Circulation 124: 2158–2177. https://doi.org/10.1161/CIR.0b013e3182340239.

Beiser DG, Carr GE, Edelson DP et al. (2009). Derangements in blood glucose following initial resuscitation from in-hospital cardiac arrest: a report from the national registry of cardiopulmonary resuscitation. Resuscitation 80: 624–630. https://doi.org/10.1016/j.resuscitation.2009.02.011.

Berdowski J, Berg RA, Tijssen JGP et al. (2010). Global incidences of out-of-hospital cardiac arrest and survival rates: systematic review of 67 prospective studies. Resuscitation 81: 1479–1487. https://doi.org/10.1016/j.resuscitation.2010.08.006.

Bernard SA, Gray TW, Buist MD et al. (2002). Treatment of comatose survivors of out-of-hospital cardiac arrest with induced hypothermia. N Engl J Med 346: 557–563. https://doi.org/10.1056/NEJMoa003289.

Broughton BRS, Reutens DC, Sobey CG (2009). Apoptotic mechanisms after cerebral ischemia. Stroke 40: e331–e339. https://doi.org/10.1161/STROKEAHA.108.531632.

Busl KM, Greer DM (2010). Hypoxic-ischemic brain injury: pathophysiology, neuropathology and mechanisms. Neuro Rehabilitation 26: 5–13. https://doi.org/10.3233/NRE-2010-0531.

Buunk G, van der Hoeven JG, Meinders AE (1997). Cerebrovascular reactivity in comatose patients resuscitated from a cardiac arrest. Stroke 28: 1569–1573.

Callaway C, Donnino M, Fink E et al. (2015). Part 8: post–cardiac arrest care: 2015 American Heart Association guidelines update for cardiopulmonary resuscitation and emergency cardiovascular care. Circulation 132: S465–S482. https://doi.org/10.1161/CIR.0000000000000262.

Cardim D, Griesdale DE, Ainslie PN et al. (2019). A comparison of non-invasive versus invasive measures of intracranial pressure in hypoxic ischaemic brain injury after cardiac arrest. Resuscitation 137: 221–228. https://doi.org/10.1016/j.resuscitation.2019.01.002.

CARES 2018 Annual Report (2018). [WWW Document], URL. https://mycares.net/sitepages/uploads/2019/2018_flipbook/index.html?page=4(accessed 2.15.20).

Caviness JN, Brown P (2004). Myoclonus: current concepts and recent advances. Lancet Neurol 3: 598–607. https://doi.org/10.1016/S1474-4422(04)00880-4.

Chan PS, Berg RA, Tang Y et al. (2016). Association between therapeutic hypothermia and survival after in-hospital cardiac arrest. JAMA 316: 1375–1382. https://doi.org/10.1001/jama.2016.14380.

Choi DW (1985). Glutamate neurotoxicity in cortical cell culture is calcium dependent. Neurosci Lett 58: 293–297. https://doi.org/10.1016/0304-3940(85)90069-2.

Chugh SS, Reinier K, Teodorescu C et al. (2008). Epidemiology of sudden cardiac death: clinical and research implications. Prog Cardiovasc Dis 51: 213–228. https://doi.org/10.1016/j.pcad.2008.06.003.

Coppler PJ, Elmer J, Calderon L et al. (2015). Validation of the Pittsburgh Cardiac Arrest Category illness severity score. Resuscitation 89: 86–92. https://doi.org/10.1016/j.resuscitation.2015.01.020.

Cristia C, Ho M-L, Levy S et al. (2014). The association between a quantitative computed tomography (CT) measurement of cerebral edema and outcomes in post-cardiac arrest—a validation study. Resuscitation 85: 1348–1353.

De Salvo S, Bramanti P, Marino S (2012). Clinical differentiation and outcome evaluation in vegetative and minimally conscious state patients: the neurophysiological approach. Funct Neurol 27: 155–162.

Deng G, Carter J, Traystman RJ et al. (2014). Pro-inflammatory T-lymphocytes rapidly infiltrate into the brain and contribute to neuronal injury following cardiac arrest and cardiopulmonary resuscitation. J Neuroimmu nol 274: 132–140. https://doi.org/10.1016/j.jneuroim.2014.07.009.

Duggal N, Lach B (2002). Selective vulnerability of the lumbosacral spinal cord after cardiac arrest and hypotension. Stroke 33: 116–121. https://doi.org/10.1161/hs0102.101923.

Edgren E, Enblad P, Grenvik Å et al. (2003). Cerebral blood flow and metabolism after cardiopulmonary resuscitation. A pathophysiologic and prognostic positron emission tomography pilot study. Resuscitation 57: 161–170. https://doi.org/10.1016/S0300-9572(03)00004-2.

Elmer J, Rittenberger JC, Faro J et al. (2016a). Clinically distinct electroencephalographic phenotypes of early myoclonus after cardiac arrest. Ann Neurol 80: 175–184. https://doi.org/10.1002/ana.24697.

Elmer J, Torres C, Aufderheide TP et al. (2016b). Association of early withdrawal of life-sustaining therapy for perceived neurological prognosis with mortality after cardiac arrest. Resuscitation 102: 127–135.

Fischer M, Böttiger BW, Hossmann K-A et al. (1996). Thrombolysis using plasminogen activator and heparin reduces cerebral no-reflow after resuscitation from cardiac arrest: an experimental study in the cat. Intensive Care Med 22: 1214–1223.

Freund B, Kaplan PW (2017). Post-hypoxic myoclonus: differentiating benign and malignant etiologies in diagnosis and prognosis. Clin Neurophysiol Pract 2: 98–102. https://doi.org/10.1016/j.cnp.2017.03.003.

Frydland M, Kjaergaard J, Erlinge D et al. (2015). Target temperature management of 33°C and 36°C in patients with out-of-hospital cardiac arrest with initial non-shockable rhythm—a TTM sub-study. Resuscitation 89: 142–148.

Gamper G, Willeit M, Sterz F et al. (2004). Life after death: posttraumatic stress disorder in survivors of cardiac arrest—prevalence, associated factors, and the influence of sedation and analgesia. Crit Care Med 32: 378–383.

Geocadin RG, Buitrago MM, Torbey MT et al. (2006). Neurologic prognosis and withdrawal of life support after resuscitation from cardiac arrest. Neurology 67: 105–108.

Geocadin RG, Koenig MA, Jia X et al. (2008). Management of brain injury after resuscitation from cardiac arrest. Neurol Clin 26: 487–506. ix. https://doi.org/10.1016/j.ncl.2008.03.015.

Geocadin R, Callaway C, Fink E et al. (2019). Standards for studies of neurological prognostication in comatose survivors of cardiac arrest: a scientific statement from the American Heart Association. Circulation 140: e517–e542. https://doi.org/10.1161/CIR.0000000000000702.

Gräsner J-T, Lefering R, Koster RW et al. (2016). EuReCa ONE-27 Nations, ONE Europe, ONE Registry: a prospective one month analysis of out-of-hospital cardiac arrest outcomes in 27 countries in Europe. Resuscitation 105: 188–195. https://doi.org/10.1016/j.resuscitation.2016.06.004.

Greer D (2006). Mechanisms of injury in hypoxic-ischemic encephalopathy: implications to therapy. Semin Neurol 26: 373–379. https://doi.org/10.1055/s-2006-948317.

Grubb NR, Simpson C, Sherwood RA et al. (2007). Prediction of cognitive dysfunction after resuscitation from out-of-hospital cardiac arrest using serum neuron-specific enolase and protein S-100. Heart 93: 1268–1273.

Hawkes MA, Rabinstein AA (2019). Neurological prognostication after cardiac arrest in the era of target temperature management. Curr Neurol Neurosci Rep 19: 1–8. https://doi.org/10.1007/s11910-019-0922-2.

Herzog E, Herzog L, Aziz E (2017). Cardiopulmonary and cardiocerebral resuscitation. In: V Fuster, RA Harrington, J Narula, ZJ Eapen (Eds.), Hurst's the heart. McGraw-Hill Education, New York, NY.

Hifumi T, Kawakita K, Yoda T et al. (2016). Association of brain metabolites with blood lactate and glucose levels with respect to neurological outcomes after out-of-hospital cardiac arrest: a preliminary microdialysis study. Resuscitation 110: 26–31. https://doi.org/10.1016/j.resuscitation.2016.10.013.

Hirsch LJ, LaRoche SM, Gaspard N et al. (2013). American clinical neurophysiology society's standardized critical care EEG terminology: 2012 version. J Clin Neurophysiol 30: 1–27.

Holmberg MJ, Ross CE, Fitzmaurice GM et al. (2019). Annual incidence of adult and pediatric in-hospital cardiac arrest in the United States. Circ Cardiovasc Qual Outcomes 12: e005580.

Hrishi AP, Prathapadas U, Lionel KR et al. (2019). Anoxic brain injury: the abominable malady. J Neuroanaesth Crit Care 6: 96. https://doi.org/10.1055/s-0039-1688406.

Huang L, Applegate PM, Gatling JW et al. (2014). A systematic review of neuroprotective strategies after cardiac arrest: from bench to bedside (part II-comprehensive protection). Med Gas Res 4: 10. https://doi.org/10.1186/2045-9912-4-10.

Hypothermia after Cardiac Arrest Study Group (2002). Mild therapeutic hypothermia to improve the neurologic outcome after cardiac arrest. N Engl J Med 346: 549–556. https://doi.org/10.1056/NEJMoa012689.

Imberti R, Bellinzona G, Riccardi F et al. (2003). Cerebral perfusion pressure and cerebral tissue oxygen tension in a patient during cardiopulmonary resuscitation. Intensive Care Med 29: 1016–1019. https://doi.org/10.1007/s00134-003-1719-x.

Inamasu J, Miyatake S, Nakatsukasa M et al. (2011). Loss of gray–white matter discrimination as an early CT sign of brain ischemia/hypoxia in victims of asphyxial cardiac arrest. Emerg Radiol 18: 295–298.

Jain A, Gray M, Slisz S et al. (2018). Shivering treatments for targeted temperature management: a review. J Neurosci Nurs J Am Assoc Neurosci Nurses 50: 63–67. https://doi.org/10.1097/JNN.0000000000000340.

Kamps MJA, Horn J, Oddo M et al. (2013). Prognostication of neurologic outcome in cardiac arrest patients after mild therapeutic hypothermia: a meta-analysis of the current literature. Intensive Care Med 39: 1671–1682. https://doi.org/10.1007/s00134-013-3004-y.

Keijzer HM, Hoedemaekers CWE, Meijer FJA et al. (2018). Brain imaging in comatose survivors of cardiac arrest: pathophysiological correlates and prognostic properties. Resuscitation 133: 124–136. https://doi.org/10.1016/j.resuscitation.2018.09.012.

Kilgannon JH, Roberts BW, Jones AE et al. (2014). Arterial blood pressure and neurologic outcome after resuscitation from cardiac arrest. Crit Care Med 42: 2083–2091. https://doi.org/10.1097/CCM.0000000000000406.

Knafelj R, Radsel P, Ploj T et al. (2007). Primary percutaneous coronary intervention and mild induced hypothermia in comatose survivors of ventricular fibrillation with ST-elevation acute myocardial infarction. Resuscitation 74: 227–234. https://doi.org/10.1016/j.resuscitation.2007.01.016.

Lascarrou J-B, Merdji H, Gouge AL et al. (2019). Targeted temperature management for cardiac arrest with nonshockable rhythm. N Engl J Med 381: 2327–2337. https://doi.org/10.1056/NEJMoa1906661.

Lassen NA (1959). Cerebral blood flow and oxygen consumption in man. Physiol Rev 39: 183–238. https://doi.org/10.1152/physrev.1959.39.2.183.

Laurikkala J, Wilkman E, Pettilä V et al. (2016). Mean arterial pressure and vasopressor load after out-of-hospital cardiac arrest: associations with one-year neurologic outcome. Resuscitation 105: 116–122. https://doi.org/10.1016/j.resuscitation.2016.05.026.

Laver S, Farrow C, Turner D et al. (2004). Mode of death after admission to an intensive care unit following cardiac arrest. Intensive Care Med 30: 2126–2128. https://doi.org/10.1007/s00134-004-2425-z.

Lemiale V, Dumas F, Mongardon N et al. (2013). Intensive care unit mortality after cardiac arrest: the relative contribution of shock and brain injury in a large cohort. Intensive Care Med 39: 1972–1980. https://doi.org/10.1007/s00134-013-3043-4.

Levy DE, Caronna JJ, Singer BH et al. (1985). Predicting outcome from hypoxic-Ischemic coma. JAMA 253: 1420–1426.

Lilja G, Nielsen N, Friberg H et al. (2015). Cognitive function in survivors of out-of-hospital cardiac arrest after target temperature management at 33°C versus 36°C. Circulation 131: 1340–1349. https://doi.org/10.1161/CIRCULATIONAHA.114.014414.

Lim C, Alexander MP, LaFleche G et al. (2004). The neurological and cognitive sequelae of cardiac arrest. Neurology 63: 1774–1778. https://doi.org/10.1212/01.wnl.0000144189.83077.8e.

Longstreth WT, Diehr P, Cobb LA et al. (1986). Neurologic outcome and blood glucose levels during out-of-hospital cardiopulmonary resuscitation. Neurology 36: 1186.

Lopez Soto C, Dragoi L, Heyn CC et al. (2020). Imaging for neuroprognostication after cardiac arrest: systematic review and meta-analysis. Neurocrit Care 32: 206–216. https://doi.org/10.1007/s12028-019-00842-0.

Losert H, Sterz F, Roine RO et al. (2008). Strict normoglycaemic blood glucose levels in the therapeutic management of patients within 12h after cardiac arrest might not be necessary. Resuscitation 76: 214–220.

Madl C, Kramer L, Domanovits H et al. (2000). Improved outcome prediction in unconscious cardiac arrest survivors with sensory evoked potentials compared with clinical assessment. Crit Care Med 28: 721–726. https://doi.org/10.1097/00003246-200003000-00020.

Mangus DB, Huang L, Applegate PM et al. (2014). A systematic review of neuroprotective strategies after cardiac arrest: from bench to bedside (part I–protection via specific pathways). Med Gas Res 4: 9. https://doi.org/10.1186/2045-9912-4-9.

Markus HS (2004). Cerebral perfusion and stroke. J Neurol Neurosurg Psychiatry 75: 353–361. https://doi.org/10.1136/jnnp.2003.025825.

Mooney MR, Unger BT, Boland LL et al. (2011). Therapeutic hypothermia after out-of-hospital cardiac arrest: evaluation of a regional system to increase access to cooling. Circulation 124: 206–214. https://doi.org/10.1161/CIRCULATIONAHA.110.986257.

Morris NA, May TL, Motta M et al. (2018). Long-term risk of seizures among cardiac arrest survivors. Resuscitation 129: 94–96. https://doi.org/10.1016/j.resuscitation.2018.06.019.

Moulaert VR, Verbunt JA, van Heugten CM et al. (2009). Cognitive impairments in survivors of out-of-hospital cardiac arrest: a systematic review. Resuscitation 80: 297–305.

Muhlhofer W, Szaflarski J (2018). Prognostic value of EEG in patients after cardiac arrest—an updated review. Curr Neurol Neurosci Rep 18: 1–15. https://doi.org/10.1007/s11910-018-0826-6.

Müllner M, Sterz F, Binder M et al. (1996). Arterial blood pressure after human cardiac arrest and neurological recovery. Stroke 27: 59–62. https://doi.org/10.1161/01.STR.27.1.59.

Neumar RW, Nolan JP, Adrie C et al. (2008). Post-cardiac arrest syndrome: epidemiology, pathophysiology, treatment, and prognostication a consensus statement from the International Liaison Committee on Resuscitation (American Heart Association, Australian and New Zealand Council on Resuscitation, European Resuscitation Council, Heart and Stroke Foundation of Canada, Inter American Heart Foundation, Resuscitation Council of Asia, and the Resuscitation Council of Southern Africa); the American Heart Association Emergency Cardiovascular C-are Committee; the Council on Cardiovascular Surgery and Anesthesia; the Council on Cardiopulmonary, Perioperativ-e, and Critical Care; the Council on Clinical Cardiology; and the Stroke Council. Circulation 118: 2452–2483. https://doi.org/10.1161/CIRCULATIONAHA.108.190652.

Nielsen N, Wetterslev J, Cronberg T et al. (2013). Targeted temperature management at 33°C versus 36°C after cardiac arrest. N Engl J Med 369: 2197–2206. https://doi.org/10.1056/NEJMoa1310519.

Nishizawa H, Kudoh I (1996). Cerebral autoregulation is impaired in patients resuscitated after cardiac arrest. Acta Anaesthesiol Scand 40: 1149–1153. https://doi.org/10.1111/j.1399-6576.1996.tb05579.x.

Nolan JP, Soar J, Cariou A et al. (2015). European Resuscitation Council and European Society of Intensive Care Medicine Guidelines for Post-Resuscitation Care 2015. Resuscitation 95: 202–222. https://doi.org/10.1016/j.resuscitation.2015.07.018.

Oddo M, Sandroni C, Citerio G et al. (2018). Quantitative versus standard pupillary light reflex for early prognostication in comatose cardiac arrest patients: an international prospective multicenter double-blinded study. Intensive Care Med 44: 2102–2111. https://doi.org/10.1007/s00134-018-5448-6.

Oh SH, Park KN, Shon Y-M et al. (2015). Continuous amplitude-integrated electroencephalographic monitoring is a useful prognostic tool for hypothermia-treated cardiac arrest patients. Circulation 132: 1094–1103.

Pana R, Hornby L, Shemie SD et al. (2016). Time to loss of brain function and activity during circulatory arrest. J Crit Care 34: 77–83. https://doi.org/10.1016/j.jcrc.2016.04.001.

Panchal A, Berg K, Hirsch K et al. (2019). 2019 American Heart Association focused update on advanced cardiovascular life support: use of advanced airways, vasopressors, and extracorporeal cardiopulmonary resuscitation during cardiac arrest: an update to the American Heart Association Guidelines for cardiopulmonary resuscitation and emergency cardiovascular care. Circulation 140: e881–e894. https://doi.org/10.1161/CIR.0000000000000732.

Paul M, Bougouin W, Geri G et al. (2016). Delayed awakening after cardiac arrest: prevalence and risk factors in the Parisian registry. Intensive Care Med 42: 1128–1136. https://doi.org/10.1007/s00134-016-4349-9.

Paulson OB, Strandgaard S, Edvinsson L (1990). Cerebral autoregulation. Cerebrovasc Brain Metab Rev 2: 161–192.

Peng TJ, Andersen LW, Saindon BZ et al. (2015). The administration of dextrose during in-hospital cardiac arrest is associated with increased mortality and neurologic morbidity. Crit Care Lond Engl 19: 160. https://doi.org/10.1186/s13054-015-0867-z.

Petito CK, Feldmann E, Pulsinelli WA et al. (1987). Delayed hippocampal damage in humans following cardiorespiratory arrest. Neurology 37: 1281–1286. https://doi.org/10.1212/wnl.37.8.1281.

Pluta R, Lossinsky AS, Wiśniewski HM et al. (1994). Early blood-brain barrier changes in the rat following transient complete cerebral ischemia induced by cardiac arrest. Brain Res 633: 41–52. https://doi.org/10.1016/0006-8993(94)91520-2.

Posner JB, Saper CB, Schiff N et al. (2007). Plum and Posner's diagnosis of stupor and coma, Plum and Posner's diagnosis of stupor and coma, Oxford University Press.

Reis C, Akyol O, Araujo C et al. (2017). Pathophysiology and the monitoring methods for cardiac arrest associated brain injury. Int J Mol Sci 18: 129. https://doi.org/10.3390/ijms18010129.

Rittenberger J, Rittenberger J, Popescu A et al. (2012). Frequency and timing of nonconvulsive status epilepticus in comatose post-cardiac arrest subjects treated with hypothermia. Neurocrit Care 16: 114–122. https://doi.org/10.1007/s12028-011-9565-0.

Rossetti AO, Rabinstein AA, Oddo M (2016). Neurological prognostication of outcome in patients in coma after cardiac arrest. Lancet Neurol 15: 597–609. https://doi.org/10.1016/S1474-4422(16)00015-6.

Sandroni C, Cavallaro F, Callaway CW et al. (2013). Predictors of poor neurological outcome in adult comatose survivors of cardiac arrest: a systematic review and meta-analysis. Part 1: patients not treated with therapeutic hypothermia. Resuscitation 84: 1310–1323.

Sandroni C, D'Arrigo S, Nolan JP (2018). Prognostication after cardiac arrest. Crit Care Lond Engl 22: 150. https://doi.org/10.1186/s13054-018-2060-7.

Sauneuf B, Dupeyrat J, Souloy X et al. (2020). The CAHP (cardiac arrest hospital prognosis) score: a tool for risk stratification after out-of-hospital cardiac arrest in elderly patients. Resuscitation 148: 200–206. https://doi.org/10.1016/j.resuscitation.2020.01.011.

Seder DB, Fraser GL, Robbins T et al. (2010). The bispectral index and suppression ratio are very early predictors of neurological outcome during therapeutic hypothermia after cardiac arrest. Intensive Care Med 36: 281–288.

Sekhon MS, Griesdale DE (2017). Individualized perfusion targets in hypoxic ischemic brain injury after cardiac arrest. Crit Care 21: 259. https://doi.org/10.1186/s13054-017-1832-9.

Sekhon MS, Ainslie PN, Griesdale DE (2017). Clinical pathophysiology of hypoxic ischemic brain injury after cardiac arrest: a "two-hit" model. Crit Care 21: 90. https://doi.org/10.1186/s13054-017-1670-9.

Strandgaard S, Olesen J, Skinhoj E et al. (1973). Autoregulation of brain circulation in severe arterial hypertension. Br Med J1: 507–510. https://doi.org/10.1136/bmj.1.5852.507.

Sunde K, Pytte M, Jacobsen D et al. (2007). Implementation of a standardised treatment protocol for post resuscitation care after out-of-hospital cardiac arrest. Resuscitation 73: 29–39.

Sundgreen C, Larsen FS, Herzog TM et al. (2001). Autoregulation of cerebral blood flow in patients resuscitated from cardiac arrest. Stroke 32: 128–132. https://doi.org/10.1161/01.str.32.1.128.

Taccone FS, Picetti E, Vincent J-L (2020). High quality targeted temperature management (TTM) after cardiac arrest. Crit Care 24: 1–6. https://doi.org/10.1186/s13054-019-2721-1.

Takasu A, Saitoh D, Kaneko N et al. (2001). Hyperthermia: is it an ominous sign after cardiac arrest? Resuscitation 49: 273–277.

Tiainen M, Poutiainen E, Kovala T et al. (2007). Cognitive and neurophysiological outcome of cardiac arrest survivors treated with therapeutic hypothermia. Stroke 38: 2303–2308. https://doi.org/10.1161/STROKEAHA.107.483867.

Tzovara A, Rossetti AO, Juan E et al. (2016). Prediction of awakening from hypothermic postanoxic coma based on auditory discrimination. Ann Neurol 79: 748–757.

Velly L, Perlbarg V, Boulier T et al. (2018). Use of brain diffusion tensor imaging for the prediction of long-term neurological outcomes in patients after cardiac arrest: a multicentre, international, prospective, observational, cohort study. Lancet Neurol 17: 317–326.

Virani SS, Alonso A, Benjamin EJ et al. (2020). Heart disease and stroke statistics-2020 update: a report from the American Heart Association. Circulation 141: e139–e596. https://doi.org/10.1161/CIR.0000000000000757.

Wallin E, Larsson I-M, Nordmark-Grass J et al. (2018). Characteristics of jugular bulb oxygen saturation in patients after cardiac arrest: a prospective study. Acta Anaesthesiol Scand 62: 1237–1245.

Wang C-H, Chang W-T, Su K-I et al. (2020). Neuroprognostic accuracy of blood biomarkers for post-cardiac arrest patients: a systematic review and meta-analysis. Resuscitation 148: 108–117. https://doi.org/10.1016/j.resuscitation.2020.01.006.

Werhahn KJ, Brown P, Thompson PD et al. (1997). The clinical features and prognosis of chronic posthypoxic myoclonus. Mov Disord 12: 216–220.

Wijdicks EF, Parisi JE, Sharbrough FW (1994). Prognostic value of myoclonus status in comatose survivors of cardiac arrest. Ann Neurol Off J Am Neurol Assoc Child Neurol Soc 35: 239–243.

Williams LR, Leggett RW (1989). Reference values for resting blood flow to organs of man. Clin Phys Physiol Meas Off J Hosp Phys Assoc Dtsch Ges Med Phys Eur Fed Organ Med Phys 10: 187–217. https://doi.org/10.1088/0143-0815/10/3/001.

Williams TA, Finn J, Tohira H et al. (2016). The association between arterial carbon dioxide tension and outcomes after cardiac arrest: a systematic review and meta-analysis. Circulation 134.

Xiang Y, Zhao H, Wang J et al. (2016). Inflammatory mechanisms involved in brain injury following cardiac arrest and cardiopulmonary resuscitation. Biomed Rep 5: 11–17. https://doi.org/10.3892/br.2016.677.

Xing C-Y, Tarumi T, Liu J et al. (2017). Distribution of cardiac output to the brain across the adult lifespan. J Cereb Blood Flow Metab 37: 2848–2856. https://doi.org/10.1177/0271678X16676826.

Yadav N, Pendharkar H, Kulkarni GB (2018). Spinal cord infarction: clinical and radiological features. J Stroke Cerebrovasc Dis 27: 2810–2821. https://doi.org/10.1016/j.jstrokecerebrovasdis.2018.06.008.

Zanyk-McLean K, Sawyer KN, Paternoster R et al. (2017). Time to awakening is often delayed in patients who receive targeted temperature management after cardiac arrest. Ther Hypothermia Temp Manag 7: 95–100.

Zeiner A, Holzer M, Sterz F et al. (2001). Hyperthermia after cardiac arrest is associated with an unfavorable neurologic outcome. Arch Intern Med 161: 2007–2012.

Handbook of Clinical Neurology, Vol. 177 (3rd series)
Heart and Neurologic Disease
J. Biller, Editor
https://doi.org/10.1016/B978-0-12-819814-8.00019-6

Chapter 20

Neurologic complications of implantable devices

JORGE G. ORTIZ GARCIA[1], SANDEEP NATHAN[2], AND JAMES R. BRORSON[3]*

[1]*Department of Neurology, University of Oklahoma Health Sciences Center, Oklahoma City, OK, United States*
[2]*Section of Cardiology, Department of Internal Medicine, University of Chicago, Chicago, IL, United States*
[3]*Department of Neurology, University of Chicago, Chicago, IL, United States*

Abstract

Technologies for repairing cardiac structures or sustaining cardiac function with implantable devices have helped patients with an ever-expanding array of cardiac conditions. Patients are surviving and thriving with cardiac conditions that would formerly have been disabling or fatal. With the implantation of devices in the heart, however, comes the inevitable risk of neurological complications. This chapter focuses on devices implanted in the chambers or valves of the heart itself, including prosthetic heart valves, closure devices for patent foramen ovale, atrial appendage occluder devices, short-term implantable circulatory assist devices, and long-term ventricular assist devices, but excluding coronary artery stents or extracardiac devices. Further, it considers the procedural and postprocedural risks of the devices, leaving the discussion of clinical effectiveness of the devices to other chapters of this book.

HEART VALVE REPLACEMENT

As heart valve replacement has evolved over the past 4 decades, mechanical heart valves have been supplanted by biological heart valves in many patient subgroups, and invasive sternotomy approaches by percutaneous valve procedures. However, perioperative stroke has remained a prominent complication. The mechanisms involved in perioperative stroke in heart valve interventions include embolic events in about 60% of the cases, with thrombotic, hypoperfusion, hemorrhagic, and lacunar mechanisms accounting for smaller numbers (Selim, 2007). The reported incidence of stroke with surgical valve replacement ranges from about 5% to 10%, with higher rates when valve surgery is combined with coronary artery bypass graft surgery or when double or triple valve surgery is performed (Selim, 2007).

Risk factors associated with an increased incidence of cerebrovascular complications in the preoperative period include age greater than 70, female gender, history of arterial hypertension, diabetes mellitus, renal failure, cigarette smoking, peripheral artery disease, cardiac disease, history of previous strokes or transient ischemic attacks (TIAs), concomitant carotid artery stenosis, discontinuation of antithrombotic therapy, and atherosclerosis of the ascending aorta. Procedure-related factors also influence the risk of cerebrovascular complications, including the type of surgical procedure, type of anesthesia, duration of surgery, duration of cardiopulmonary bypass, aortic cross-clamp time, and manipulation of large vessels like aortic arch and carotid arteries with atherosclerotic lesions.

With the recent US Food and Drug Administration (FDA) approval of transcatheter aortic valve replacement across the spectrum of patient risk, this percutaneous approach, often utilizing embolic protection devices, is increasingly favored over conventional (open) surgical aortic valve replacement. These technical advances have been associated with declining rates of embolic complications, with a procedural 30-day stroke or TIA rate of 3.3% reported (Jones et al., 2015). Stroke in transcatheter aortic valve replacement is primarily of embolic origin

*Correspondence to: James R. Brorson, M.D., 5841 S Maryland Avenue, MC 2030, Chicago, IL 60637, United States. Tel: +1-773-795-4705, Fax: +1-773-834-5435, E-mail: jbrorson@neurology.bsd.uchicago.edu

and occurs peri-procedurally due to catheter and wire manipulations and as a consequence of advancing bulky devices through an atherosclerotic aortic arch, root, and calcified aortic valve.

With regard to late cerebrovascular complications of heart valve replacement, rates vary significantly depending on the type of prosthetic heart valve implanted. There are three broad categories of mechanical cardiac valves: caged ball, tilting-disk valves, and bileaflet valves. Caged ball valves include Starr-Edwards (American Edwards Laboratories, Santa Ana, CA) and Smeloff-Cutter valves (Cutter Laboratories, Berkeley, CA). Tilting disc valves include the Bjork-Shiley (Pfizer, Inc., New York City, NY), Medtronic-Hall (Medtronic, Inc., Minneapolis, MN), Lillehei-Kaster Omnicarbon (Medical, Inc., Minneapolis, MN), Omniscience (Omniscience, Minneapolis, MN). Bileaflet aortic valves include St. Jude (St. Jude Medical, St. Paul, MN) and Edwards Duromedics (Baxter Healthcare Corp., Santa Ana, CA). It should be noted that many of the aforementioned valves are no longer being implanted as they have been recalled, voluntarily withdrawn from the market, or replaced by newer designs. Tilting disc valves and bileaflet valves have historically shown a far lower incidence of major embolism than caged ball valves which, as mentioned, which have virtually disappeared from clinical usage. Bioprosthetic valves have still lower risks of thromboembolism.

Following the perioperative period, the incidence rates of heart valve thrombosis and major or total embolism without anticoagulation for mechanical heart valves have been estimated to be 1.8 per 100 patient-years (valve thrombosis), 4.0 per 100 patient-years (major embolism), and 8.6 per 100 patient-years (total embolism). With anticoagulation, the incidence of thrombotic complications has been estimated to be 0.2 per 100 patient-years (valve thrombosis), 1.0 per 100 patient-years (major embolism), and 1.8 per 100 patient-years (total embolism) (Cannegieter et al., 1994).

The incidence of bleeding with anticoagulation for mechanical heart valves has been estimated to be 0.5 per 100 patient-years (cerebral bleeding), 1.4 per 100 patient-years (major bleeding), and 1.9 per 100 patient-years (total bleeding) (Cannegieter et al., 1994). Recently, the On-X prosthetic aortic valve (CryoLife, Inc., Kennesaw, GA), a device constructed mainly of thrombo-resistant pyrolytic carbon, became the only FDA- and CE-approved prosthetic heart valve for use with lower intensity of anticoagulation (INR 1.5–2), thus potentially combining the favorable long-term performance profile of a prosthetic valve with reduced bleeding risk (Puskas et al., 2014).

Bioprosthetic valves have a lower risk when they are compared to mechanical valves, but a risk for thrombotic complications is present particularly during the first 3 months after surgery (Tay et al., 2011). Thromboembolic stroke rates after valve repair with a bioprosthetic device have been reported to range from 0.2% to 3.3% per year (Al-Atassi et al., 2012). Stroke rates are as low as 0.2–0.7% per year for patients with normal sinus rhythm. Investigators have theorized that the early thrombogenicity of bioprosthetic valves is attributable to the lack of complete reendothelialization of the suture zone in the early postoperative period (Mérie et al., 2012). For this reason, some degree of antithrombotic therapy is indicated at least in the short-term following bioprosthetic valve replacement.

IMPLANTED DEVICES FOR SECONDARY STROKE PREVENTION

Patent foramen ovale (PFO) and atrial septal defect (ASD) closure devices

Transfemoral device closure of PFO is frequently recommended for secondary stroke prevention following a cryptogenic stroke in young persons. Early case series found low procedural complication rates with PFO device closure (Hong et al., 2003). However, rare procedural complications have been recognized, including device displacement and embolization, cardiac puncture, and device-related thromboembolism. Upon examination of 50 patients with transesophageal echocardiography (TEE) 1 month following placement of a PFO closure device, 5 instances of device thrombus were found, in one occasion requiring open surgical device removal (Anzai et al., 2004). These thrombi were all found in patients with the CardioSEAL device (NMT Medical, Inc., Boston, MA), and none with the Amplatzer septal occluder device (Abbott Structural, Abbott Park, IL/manufactured by AGA Medical Corp., Plymouth, MN).

More definitive safety data has come subsequently from prospective randomized clinical trials testing the clinical effectiveness of devices for stroke prevention. In the CLOSURE I trial, utilizing the STARFlex device, PFO closure was successful in 89% of 402 patients, with 3.2% suffering a major vascular procedural complication (Furlan et al., 2012). Atrial fibrillation occurred in 5.7% of these patients, usually within 30 days of device implantation. Three strokes and two TIAs occurred within 30 days of the procedure, in some cases associated with atrial fibrillation or device-associated thrombus on TEE.

The Amplatzer PFO Occluder device, which is currently most often used in the United States, was similarly tested in prospective randomized trials, the RESPECT and the PC trials (Carroll et al., 2013; Meier et al., 2013). In the RESPECT trial, an overall rate of procedure-related events of 2.4% was found in 499 patients assigned to device closure. In the PC trial, procedural complications occurred in 1.5% of 204 subjects, with no events of device-associated thrombus. Atrial fibrillation did not occur at a significantly elevated rate in either trial with this device, though a small number of transient atrial fibrillation events were observed in the peri-procedural period. Based on the evidence of device safety and effectiveness, the Amplatzer PFO Occluder was approved for use by the US FDA in 2016. A variety of additional nitinol-based solutions are also available in the Amplatzer family of devices for percutaneous treatment of atrial and ventricular septal defects as well as other vascular shunts, each carrying variable but relatively low risks of thrombosis and thromboembolism.

The Gore CARDIOFORM Septal Occluder (W.L. Gore & Associates, Flagstaff, AZ), which replaced the Gore HELEX Septal Occluder (HSO), is another FDA-approved device for closure of secundum ASDs and PFOs, posting an excellent long-term safety profile (Javois et al., 2014).

Left atrial appendage closure devices

Left atrial appendage (LAA) closure is an alternative minimal invasive procedure for stroke prevention in patients with atrial fibrillation who are not candidates for anticoagulation. The LAA closure reduces the risk of thromboembolic complications without the need for long-term anticoagulation. Some of the devices that have been or are currently utilized for LAA closure include the WATCHMAN device (Boston Scientific Corp, Marlborough, MA), Amplatzer Amulet (Investigational, Abbott Laboratories, Abbott Park, IL), and WaveCrest (Investigational, Biosense-Webster, Irvine, CA).

At the time of writing, the WATCHMAN device remains the only FDA-approved implantable percutaneous LAA occlusion device. In the PROTECT AF trial, it has shown noninferiority compared to long-term warfarin therapy for stroke prevention (Holmes et al., 2009), with significant reductions in the rate of hemorrhagic strokes and mortality rate over 3.8 years of follow-up (Reddy et al., 2014). Primary safety events occurred in the WATCHMAN trial in 27 of 463 patients (5.8%) on the day of the procedure (Holmes et al., 2009). Events included pericardial effusion, major bleeding, ischemic stroke, and device embolization. Data from an inpatient nation-wide database survey suggested a somewhat higher rate of complications outside of clinical trial settings, with an overall composite rate of mortality or adverse event totaling 24.3% (Badheka et al., 2015). The percutaneous devices for LAA closure were approved by the US Food and Drug Administration (FDA) in 2015. Patients usually continue on warfarin for 1–2 months after implantation, followed by clopidogrel plus aspirin for 6 months, followed by long-term monotherapy with aspirin.

Rates of stroke and TIA range between 3% and 5% over up to 2 years of follow-up. Ischemic events most of the time are secondary to the thromboembolic or cardioembolic phenomenon. Risk factors accounting for device-associated thrombosis include spontaneous contrast in the LAA cavity, atrial enlargement, high $CHADS_2$ or CHA_2DS_2-VASc score, and poor antithrombotic treatment adherence. LAA morphology has been suggested to be associated with brain lesions and the incidence of clinical stroke or TIAs (Lupercio et al., 2016). Hemorrhagic events are associated with supratherapeutic INR or antithrombotic therapy. Bacterial endocarditis complicated by septic emboli or mycotic aneurysm has also been described in these patients.

IMPLANTABLE MECHANICAL CIRCULATORY SUPPORT DEVICES

Short-term support

Implanted circulatory support devices aimed at augmenting cardiac output as a bridge to cardiac transplantation or as a destination therapy have become increasingly common in advanced cardiac conditions. The mechanisms of cerebrovascular complications in patients with mechanical circulatory support devices can be grouped into three processes: embolic, ischemic, and hemorrhagic mechanisms.

Embolism is the most common cause of cerebrovascular events. The embolism is due to wire manipulation, cross-clamping, new onset of paroxysmal or nonparoxysmal atrial fibrillation, large artery-to-artery embolism in the setting of carotid atherosclerosis, air emboli from cardiopulmonary bypass, calcified debris from large vessels, or prosthetic device deployment. The ischemic mechanisms include hemodynamic instability during the procedure, prolonged cardiopulmonary bypass, and balloon occlusions and vessel manipulation followed by endothelial injury. Hemorrhagic events are due to severe hypertension in the peri-procedural period, anticoagulation or antithrombotic therapies, or hyperperfusion syndrome after decannulations or revascularization of large vessels (Selim, 2007).

In this review, we will focus our discussion on three short-term mechanical circulatory support devices that are frequently used by cardiothoracic surgeons: Intra-aortic balloon pump (IABP), the Impella System, and extracorporeal membrane oxygenation (ECMO).

Intra-aortic balloon pump (IABP)

The use of IABP has been an invaluable asset in the treatment of cardiogenic shock and intractable angina for the last 5 decades. The IABP has a flexible catheter with one lumen that allows for either distal aspiration/flushing or pressure monitoring and a second that permits the periodic delivery and removal of helium gas to a closed polypropylene balloon. A mobile console provides the helium transfer system and control of the inflation/deflation cycle. The catheter is most often inserted via the common femoral artery and advanced under fluoroscopic guidance in the proximal descending aorta. Then, the pumping is initiated, inflating in diastole and collapsing before the next systole, and controlled by the console (Estep et al., 2013).

IABP is often used to manage patients with advanced heart failure (HF) refractory to medical therapy and to preoperatively support patients at high risk of cardiovascular complications by improving myocardial perfusion and lowering cardiac afterload. The IABP has also been used in patients with a high-grade subarachnoid hemorrhage for the management of severe diffuse vasospasm with neurogenic stunned myocardium (Taccone et al., 2009).

The incidence of cerebrovascular complications associated with IABP, whether ischemic or hemorrhagic events, ranges from 1% to 3%. In a randomized trial of the use of IABP in high-risk patients undergoing cardiac surgery, neither an improvement in cerebral hemodynamics nor an increase in the risk of neurological complications was found (Caldas et al., 2019). Other neurological complications associated with IABP include TIAs, silent infarctions (covert stroke), seizures, encephalopathy, postoperative neurocognitive decline (defined as a Montreal Cognitive Assessment (MoCA) lower than 26), and delirium (Parissis et al., 2011). Uncommon complications reported include air embolism, spinal cord infarction, infections, arterial dissections causing embolism, and peripheral neuropathies (Cruz-Flores et al., 2005; Harris et al., 2016).

Impella system

Impella (Abiomed, Inc., Danvers, MA) encompasses a family of technologies that employ a percutaneously inserted, micro-axial pump to augment circulation in a linear (nonpulsatile) fashion. The Impella 2.5, Impella CP, and Impella 5.0 systems assist the left ventricular function by pumping blood from the left ventricle to the aorta. The inflow is placed retrogradely across the aortic valve into the left ventricle. A pump (impeller pump located within a catheter) draws blood out of the left ventricle and ejects it into the ascending aorta. The Impella system is indicated in cardiogenic shock with or without mechanical defects, high risk percutaneous coronary interventions, management of advanced HF refractory to medical management as bridging therapy, ventricular arrhythmias associated with ischemia, and cardiac allograft failure or posttransplant right ventricular failure (Rihal et al., 2015). The main complications are related to the grade of hemolysis and thrombocytopenia caused by the device. Cerebrovascular complications, including ischemic and hemorrhagic strokes, range from 2% to 6.5%, with the occurrence of acute neurological events between 5 and 8 days after Impella implantation (Subramaniam et al., 2019). Cerebrovascular complications are related to inappropriate anticoagulation or thrombocytopenia. Other neurologic complications such as infections, neurocognitive decline, air embolism, and arterial dissections have been reported.

Extracorporeal membrane oxygenation (ECMO)

ECMO is essentially, short-term cardiopulmonary bypass to support patients in severe cardiac and/or respiratory failure and to immediately regain blood pressure and circulation in the setting of cardiac arrest. The patient is anticoagulated with intravenous unfractionated heparin drip and then the cannulas are inserted either percutaneously or via open surgical exposure. For veno-venous (VV) ECMO, venous cannulas are most often placed in the common femoral vein for drainage and right internal jugular vein for infusion. The tip of the femoral cannula should advance to the junction of the inferior vena cava and right atrium, and the tip of the internal jugular cannula should be in the junction of the superior vena cava and right atrium. For veno-arterial (VA) ECMO, the usual configuration involves a venous cannula placed in the inferior vena cava or right atrium for drainage of venous blood and an arterial cannula advanced retrograde through the iliofemoral arterial system to the ostium of the common iliac artery for infusion of oxygenated blood. After the cannulation, the patient is initiated on ECMO and blood flow is titrated until the patient achieves adequate respiratory and hemodynamic parameters like O_2 %Sat > 90% in VA ECMO, PaO_2 > 75% in VV ECMO, venous oxyhemoglobin saturation 25% lower than the arterial saturation, adequate tissue perfusion and normal lactate levels (Merkle et al., 2019).

The vast majority of the information on neurologic complications in ECMO patients comes from the Extracorporeal Life Support Organization (ELSO) Registry, and from case reports and case series from single-center observations. The ELSO registry collects data on ischemic and hemorrhagic cerebrovascular complications, seizures, imaging findings including computed tomography or magnetic resonance imaging, neurosurgical interventions, and brain death on ECMO (Nasr and Rabinstein, 2015).

Neurologic complications after ECMO implantation include ischemic and hemorrhagic strokes, hypoxic encephalopathy, delirium, hearing loss, cerebral edema, seizures, brain death, critical illness neuro-myopathy, and peripheral neuropathies. Cerebral infarctions range between 5% and 26% of the ECMO complications (Lorusso et al., 2016). The etiology is usually multifactorial due to embolic phenomenon, arterial hypotension, large vessel occlusion, thromboembolism, or septic embolism. Ischemic lesions can be associated with electrographic seizures and increased mortality. The risk of having a large watershed cerebral infarction is ca. 5%–10% when the right common carotid artery is cannulated (Martucci et al., 2016). Hemorrhagic stroke also occurs, carrying a high mortality rate. Risk factors associated with an increased incidence of hemorrhagic events are longer duration of ECMO, high activated clotting times, bleeding at other anatomical sites, antithrombotic therapy prior to admission, supratherapeutic INR, and thrombocytopenia. Cerebral edema has been described as a consequence of brain lesions (Fletcher Sandersjöö et al., 2017; Sutter et al., 2018).

Seizures in ECMO patients can present as nonconvulsive status epilepticus. Around 18% of ECMO patients can have electrographic seizures and 61% of those patients have electrographic status epilepticus. Given that patients on ECMO are a high-risk population, seizures should be treated aggressively (Okochi et al., 2018). Sensorineural hearing loss has been reported in 5%–20% of cases, especially in the pediatric population (Fligor et al., 2005). Intravenous sedation, paralytics, hemodynamic instability, prolonged immobilization contribute to critical illness neuropathy/myopathy in ECMO patients. Catastrophic brain injuries can lead to brain death in patients supported on ECMO (Xie et al., 2017).

Long-term support—left ventricular assist devices (LVADs)

Left ventricular assist devices (LVADs) have emerged as the standard of care for bridging or destination therapy in patients with advanced HF refractory to optimal medical therapy. The implantation of LVADs has increased, and with that the incidence of ischemic or hemorrhagic strokes as well. The postoperative cerebrovascular complications are an important cause of morbidity, mortality, and decreased quality of life. At present, there is a lack of data about the risk factors associated with the incidence of cerebrovascular complications in this particular population.

LVAD patients have a particularly significant risk for stroke. Although the etiology of an LVAD-related stroke is commonly embolic in nature, other factors including peri-device clot formation, acquired Von-Willebrand disease, and concomitant infections may increase the risk of ischemic or hemorrhagic strokes (Wille et al., 2014). Despite the significant incidence of stroke in this group of patients, little data exist on the utilization of pharmacological thrombolysis and mechanical thrombectomy; as a matter of fact, this population is often excluded from clinical trials.

Historical context

Orthotopic heart transplantation is regarded as the definitive therapy for patients with advanced HF, but the implementation of this treatment is limited by the lack of sufficient organ donors across the country. Therefore, the advent of mechanical circulatory support therapy, including the ventricular assist devices (VADs), has improved morbidity and mortality rates of this population. Over the last 60 years, innovations associated with device development have resulted in improved VADs that have contributed to an acceptable quality of life for patients with advanced HF. In 1966, the first paracorporeal VAD was utilized in a patient with postcardiopulmonary bypass cardiogenic shock (Frazier, 1999). In 1991, the National Institutes of Health (NIH) approved the first study of implantable continuous-flow VADs. Nowadays, several different LVAD pump modalities exist for application as destination therapy or bridging therapy prior to heart transplantation. After the Organ Transplant Law was revised in 2010, the annual number of patients undergoing heart transplantation markedly increased. Furthermore, medical expense reimbursement for implantation-type LVADs was solidified in 2011, making it possible for patients to wait for transplantation at home. With the improvement in the VADs' performance, a multi-occupational examination and support system for patients was arranged, and the annual number of patients who newly apply for transplantation increased. As a result, the interval until transplantation increased to approximately 2.5 years (Christie et al., 2012). Concomitantly, the incidence of neurological complications in a patient with LVAD implantation has increased considerably in the last few years.

LVAD implantation is currently indicated in patients with severe symptoms despite optimal medical therapy and one or more of the following: left ventricular EF $<25\%$, and a VO_2 peak <12 mL/kg/min (if measured), three or more hospital admissions within the previous 12 months without evident precipitating factors, dependence on inotropic support, progressive hepatic and/or renal failure secondary to hypoperfusion and to increased left ventricular filling pressure (PWP $\geq$ 20 mmHg, systolic pressure 80–90 mmHg and cardiac index ≤ 2 L/min/m^2), or initial right ventricular dysfunction (Ammirati et al., 2014).

Epidemiology of cerebrovascular complications in VAD patients

Four types of LVADs have been implanted, including the Heartmate I with pulsatile flow, the Heartmate II with continuous axial flow, the Heartmate III (Abbott, Abbott Park, IL) with the continuous centrifugal flow, and the HeartWare HVAD (Heartware/Medtronic, Framingham, MA) with continuous centrifugal flow.

With the introduction of continuous-flow devices, the incidence of stroke in patients with LVADs decreased substantially compared to that with pulsatile-flow devices. However, patients remain at a high risk of stroke. Of the neurologic events associated with LVADs, ischemic stroke is most prevalent, but the hemorrhagic stroke has a higher mortality rate. The incidence of ischemic strokes has been reported at between 10% and 30% per year (Wille et al., 2014). Ischemic strokes are distributed as follows: 58% in the right hemisphere, 28% in the left hemisphere, 6.5% in both hemispheres, and 6.5% in the vertebrobasilar system. The predilection for right hemispheric stroke has been explained by the anatomic alignment directing thrombotic material toward the brachiocephalic trunk (Kato et al., 2012a, b). Risk factors for ischemic stroke in patients with LVADs include diabetes, history of the previous stroke, aortic cross-clamping with cardioplegic arrest during the LVAD implantation procedure, and systemic infection (Tsukui et al., 2007).

The use of pulsatile-flow LVADs and continuous-flow LVADs showed the rate of the hemorrhagic stroke to be around 11% per year in the continuous-flow group and 8% per year in the pulsatile-flow group (Slaughter et al., 2009). Risk factors associated with stroke include heparin-induced thrombocytopenia, use of IABP, female gender, primary cardiac diagnosis, high INR, and age younger than 50 years of age (Kirklin et al., 2014) (see Fig. 20.1).

Differences in methodologies for diagnosing and categorizing stroke, VAD types, and the duration of follow-up may account for some of this variability in the incidences. The risk of stroke over time and the impact on survival have not been fully explored in the literature.

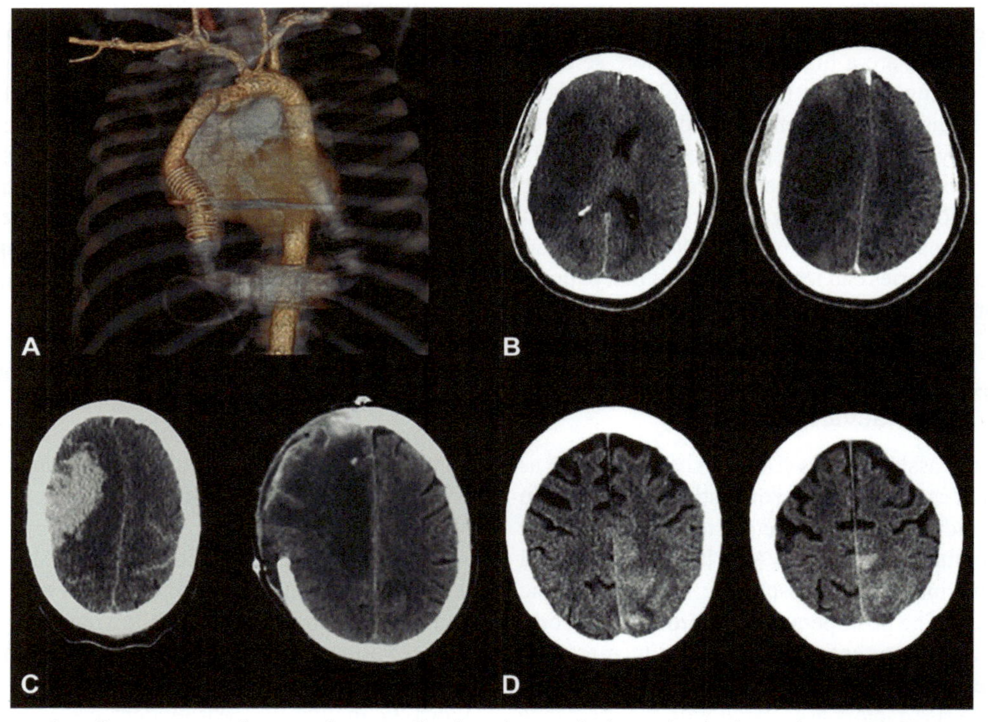

Fig. 20.1. Examples of common cerebrovascular complications in ventricular assist device patients. (A) Computerized tomography (CT) chest shows a three-dimensional reconstruction of a Heartmate type II left ventricular assist device (LVAD). (B) Noncontrast CT head demonstrates a large right hemispheric infarction involving more than 50% territory of the right middle cerebral artery with cerebral edema and left midline shift, in a patient with a Heartmate type II LVAD. (C) Noncontrast CT head showing a lobar right frontal intracerebral hemorrhage with signs of cerebral edema and left midline shift followed by a repeated noncontrast CT head after a right decompressive hemicraniectomy in a patient with a Heartmate type II LVAD. (D) Noncontrast CT head shows a parasagittal cortical nonaneurysmal subarachnoid hemorrhage in a patient with a Heartware LVAD.

Etiology of cerebrovascular complications in VAD patients

The cerebrovascular complications in VAD patients may be grouped into three categories: pump-related complications, patient-related complications, and management-related complications.

Of pump-related complications, VAD thrombosis is the most concerning complication in these patients. Several mechanisms have been proposed, such as foreign body response-related thrombus formation at the titanium junction of an inflow cannula, thrombus formation related to blood flow stagnation between the cardiac apex and a blood-removing cannula, and thrombus formation related to left ventricular collapse associated with a marked increase in the number of pump rotations. Other mechanisms include heat generated by the pump rotor and outflow graft kink or obstruction. A phenomenon called "*suction event*" has been described whereby the walls of the ventricles collapse and contact the inflow cannula causing obstruction and hemodynamic compromise of the patients (Kirklin et al., 2014). Patients on VAD support are at an increased risk of thromboembolic events because of a complex interaction between the nonbiologic surfaces and blood with the activation of the coagulation cascade. The pump becomes a nidus for embolic strokes.

Individual patient-related conditions can render thrombosis and cardioembolism more likely. Cardiac arrhythmias are the most common, especially atrial fibrillation. Infections are also encountered, including bacteremia, driveline infections, and pump pocket infections. These entities can be treated with systemic antibiotics. The cannula and pump are susceptible to biofilm formation and acts as a source for chronic infection and bacteremia, requiring device removal/replacement or prompt transplantation (Thyagarajan et al., 2016).

Management-related complications primarily relate to bleeding related to antithrombotic treatments, patients with implanted VADs require long-term anticoagulation in addition to antiplatelet therapy to prevent thrombosis inside the device. Most postimplantation protocols suggest postoperative unfractionated heparin drip, with subsequent transition to oral vitamin K antagonist (warfarin), accompanied by antiplatelet therapy (aspirin). For cases in which patients cannot tolerate aspirin, the recommendation is to continue on clopidogrel or dipyridamole. The use of anticoagulation predisposes these patients to intracranial hemorrhage, as well as to other sequelae related to bleeding. For patients with second-generation VAD devices, recommendations are to take warfarin with an INR goal of 2–2.5 plus aspirin 81 mg/day. Patients with third-generation VAD devices are recommended to take warfarin with an INR goal of 2.5–3.0 plus aspirin 81–325 mg/day (Hilal et al., 2019). On the other hand, VAD patients can develop acquired von Willebrand syndrome, which seems to be related to the shear stresses the device exerts and to impaired hemostasis of the vascular endothelium (Geisen et al., 2018).

Whether adjustable VAD parameters like *Pump flow* (estimate of cardiac output 4–6 L/min), *Pump speed* (speed at which the VAD rotors are spinning), *Pulse index* (measurement of flow through the pump), and *Power* (amount of wattage the device needs to maintain speed and flow) influence the incidence of ischemic or hemorrhagic cerebrovascular events has not been well established, and this needs further investigation in the near future (Salerno et al., 2014).

Risk factors identified for increasing incidence of events in the perioperative setting include the early thrombus formation, prolonged mechanical circulatory support, and suboptimal antithrombotic regimens. Systemic infections increase the risk of stroke by nearly twofold. Pre-existing cerebrovascular disease, dyslipidemia, diabetes mellitus, and cigarette smoking may play a role as well. Risks increasing intracranial bleeding include cardioembolic hemispheric infarctions (>50% MCA territory) with hemorrhagic transformations, elevated blood pressure (MAP > 90 mmHg), lobar hemorrhages associated with mycotic aneurysm or septic arthritis, and supratherapeutic INR levels (Kato et al., 2012a,b).

Treatment of cerebrovascular complications in VAD patients

There are no guidelines for the treatment of ischemic or hemorrhagic cerebrovascular complications in VAD patients. Decisions are based on case series, case reports, expert opinions, or retrospective data review. The communication and multidisciplinary approach between teams including emergency physicians, neurologists, neurosurgeons, interventionalists, cardiologists, and cardiothoracic surgeons are essential in the decision-making process. Current treatment guidelines for acute ischemic stroke presenting within 4.5 h of symptom onset without any contraindications include intravenous recombinant tissue plasminogen activator (IV rt-PA) at a dose of 0.9 mg/kg over the course of 1 h, but studies supporting this recommendation did not include patients with VADs. The use of IV rt-PA could be contemplated in selected VAD patients if there are no contraindications such as active anticoagulation with elevated INR. In the definitive trials of thrombolysis for acute stroke, anticoagulation was not allowed until 24 h after administration of IV t-PA due to concern for synergistic blood-thinning effect leading to the hemorrhagic conversion of ischemic stroke. Resumption of early anticoagulation in VAD patients after an ischemic stroke must consider the balance between risk and benefits. While we may expect that VAD-related clots are composed of fibrin and denatured proteins that might not respond adequately to the effects of thrombolytic therapy, no study has formally evaluated the efficacy of rt-PA in this population (Starling et al., 2014).

VAD patients have increased risk of hemorrhagic transformation after systemic thrombolysis with intravenous (IV) rt-PA, especially in the setting of underlying infection, concomitant use of anticoagulation with antiplatelet therapy, and acquired von Willebrand's factor deficiency (Kato et al., 2012a,b). If hemorrhagic transformation postsystemic thrombolytic therapy occurs, treatment should include cryoprecipitate. Consideration of antifibrinolytic agents (ϵ-aminocaproic acid or tranexamic acid) is fraught, risking device-related thromboembolic complications.

Mechanical thrombectomy is now established as an adjuvant treatment to IV rt-PA or as a stand-alone treatment in anterior circulation ischemic strokes (Goyal et al., 2016). However, none of the multiple successful trials included VAD patients. There is no inherent reason why the benefits of endovascular thrombectomy would not also occur in patients whose embolic stroke originated as a complication of a VAD device (Kadono et al., 2019).

Considering that VAD patients are at elevated risk of thromboembolic complications presenting with large vessel occlusions, institutions treating significant numbers of these patients would do well to consider an institutional protocol for this special population for management of ischemic strokes. Since patients treated with VAD are commonly anticoagulated, IV rt-PA is contraindicated in most of the cases. Catheter cerebral angiogram followed by mechanical thrombectomy seems to be the most reasonable procedure of choice for urgent management. Some challenges could be encountered during the treatment of VAD patients, including the risk of hemorrhagic transformation similar to that of systemic thrombolysis, the fact there is not arterial pulsation to guide femoral puncture, requiring echo-guided puncture for the intra-arterial access, difficulties with pulse oximetry when pump flow is high, and finally lack of familiarity of anesthesia providers with this type of patient. In addition, there is controversy regarding which type of anesthesia is most efficacious for neuro-interventional procedures, although the literature favors conscious sedation (Simonsen et al., 2018). The optimal blood pressure for VAD patients in the setting of acute ischemia or hemorrhage remains undetermined. Increasing mean arterial pressure (MAP) to above 80 mmHg may potentially maintain perfusion in the ischemic brain tissue (penumbra) until the flow is restored, but paradoxically may decrease flow to the device, and it has been associated with stroke and pump thrombosis (Castagna et al., 2017). In the lack of prospective data or randomized clinical trials, the use of mechanical thrombectomy in this population is supported in the existing retrospective reports with data extrapolated from other sources.

Management of intracranial hemorrhage is also a challenge in VAD patients. Treatment decisions for acute hemorrhagic stroke patients are based on whether the patient has had a primary hemorrhagic stroke (intracerebral hemorrhage or subarachnoid hemorrhage) or hemorrhagic transformation of an ischemic stroke. Other factors include the antithrombotic agents used, the coagulation profile, and the platelet count. The initial approach should be focused on the prevention of hematoma expansion, addressing the reduction of MAP carefully (MAP < 90 mmHg), and reversing coagulopathy. Platelet transfusion and desmopressin could be considered if the hemorrhage is greater than 30 mL or if they have other associated factors for hematoma expansion like evidence of "spot sign" in the CT angiogram of intracranial vasculature. The reversal of anticoagulation should start with discontinuation of warfarin, followed by the administration of vitamin K 10 mg IV over 30 min and fresh frozen plasma or prothrombin complex concentrate to achieve an INR lower than 1.5. In cases of treatment with unfractionated heparin, protamine sulfate should be administered. Recombinant Factor VII remains experimental in coagulopathy-related intracerebral hemorrhage because of the high risk of thrombotic complications. Data suggest that withholding aspirin for at least 7 days and warfarin for 10–14 days is sufficient to reduce the risk of hemorrhage expansion or rebleeding while minimizing the risk of thromboembolic events and pump failure (Wilson et al., 2013).

Stroke is a leading cause of disability in VAD patients, and stroke can impose significant morbidity that could be a relative contraindication for cardiac transplantation. Knowledge about stroke in VAD patients is still limited. There is a strong need for further investigation of risk factors, natural history, complication, and acute management of cerebrovascular complications in patients with VADs.

The aggressive management of these patients can improve their quality of life. An overly conservative and nihilistic approach to this patient population is not commensurate with the aggressive nature of the treatment of VAD placement. Prudent extrapolation of principles of effective stroke interventions derived from the general population will best serve the VAD population until more evidence specific to these patients becomes available.

References

Al-Atassi T, Lam K, Forgie M et al. (2012). Cerebral microembolization after bioprosthetic aortic valve replacement: comparison of warfarin plus aspirin versus aspirin only. Circulation 126: S239–S244.

Ammirati E, Oliva F, Cannata A et al. (2014). Current indications for heart transplantation and left ventricular assist device: a practical point of view. Eur J Intern Med 25: 422–429.

Anzai H, Child J, Natterson B et al. (2004). Incidence of thrombus formation on the CardioSEAL and the Amplatzer interatrial closure devices. Am J Cardiol 93: 426–431.

Badheka AO, Chothani A, Mehta K et al. (2015). Utilization and adverse outcomes of percutaneous left atrial appendage closure for stroke prevention in atrial fibrillation in the United States. Circ Arrhythm Electrophysiol 8: 42–48.

Caldas JR, Panerai RB, Bor-Seng-Shu E et al. (2019). Intra-aortic balloon pump does not influence cerebral hemodynamics and neurological outcomes in high-risk cardiac patients undergoing cardiac surgery: an analysis of the IABCS trial. Ann Intensive Care 9: 130.

Cannegieter SC, Rosendaal FR, Briët E (1994). Thromboembolic and bleeding complications in patients with mechanical heart valve prostheses. Circulation 89: 635–641.

Carroll JD, Saver JL, Thaler DE et al. (2013). Closure of patent foramen ovale versus medical therapy after cryptogenic stroke. N Engl J Med 368: 1092–1100.

Castagna F, Stöhr EJ, Pinsino A et al. (2017). The unique blood pressures and pulsatility of LVAD patients: current challenges and future opportunities. Curr Hypertens Rep 19: 85.

Christie JD, Edwards LB, Kucheryavaya AY et al. (2012). The registry of the international society for heart and lung transplantation: 29th adult lung and heart-lung transplant report—2012. J Heart Lung Transplant 31: 1073–1086.

Cruz-Flores S, Diamond AL, Leira EC (2005). Cerebral air embolism secondary to intra-aortic balloon pump rupture. Neurocrit Care 2: 49–50.

Estep JD, Cordero-Reyes AM, Bhimaraj A et al. (2013). Percutaneous placement of an intra-aortic balloon pump in the left axillary/subclavian position provides safe, ambulatory long-term support as bridge to heart transplantation. JACC Heart Fail 1: 382–388.

Fletcher Sandersjöö A, Bartek Jr J, Thelin AP et al. (2017). Predictors of intracranial hemorrhage in adult patients on extracorporeal membrane oxygenation: an observational cohort study. J Intensive Care 5: 1–10.

Fligor BJ, Neault MW, Mullen CH et al. (2005). Factors associated with sensorineural hearing loss among survivors of extracorporeal membrane oxygenation therapy. Pediatrics 115: 1519–1528.

Frazier OH (1999). Left ventricular assist device as a bridge to partial left ventriculectomy. Eur J Cadiothorac Surg 15: S20–S25.

Furlan AJ, Reisman M, Massaro J et al. (2012). Critique of closure or medical therapy for cryptogenic stroke with patent foramen ovale: the hole truth? N Engl J Med 366: 991–999.

Geisen U, Brehm K, Trummer G et al. (2018). Platelet secretion defects and acquired von Willebrand syndrome in patients with ventricular assist devices. J Am Heart Assoc 7: 1–10.

Goyal M, Menon BK, van Zwam WH et al. (2016). Endovascular thrombectomy after large-vessel ischaemic stroke: a meta-analysis of individual patient data from five randomised trials. Lancet 387: 1723–1731.

Harris M, Karamasis GV, Chotai S et al. (2016). Spinal cord infarction post cardiac arrest in STEMI: a potential complication of intra-aortic balloon pump use. Acute Card Care 18: 18–21.

Hilal T, Mudd J, DeLoughery TG (2019). Hemostatic complications associated with ventricular assist devices. Res Pract Thromb Haemost 3: 589–598.

Holmes DR, Reddy VY, Turi ZG et al. (2009). Percutaneous closure of the left atrial appendage versus warfarin therapy for prevention of stroke in patients with atrial fibrillation: a randomised non-inferiority trial. Lancet 374: 534–542.

Hong TE, Thaler D, Brorson J et al. (2003). Transcatheter closure of patent foramen ovale associated with paradoxical embolism using the Amplatzer PFO Occluder: initial and intermediate-term results of the U.S. Multicenter Clinical Trial. Catheter Cardiovasc Interv 60: 524–528.

Javois AJ, Rome JJ, Jones TK et al. (2014). Results of the U.S. Food and Drug Administration continued access clinical trial of the GORE HELIX septal occluder for secundum atrial septal defect. J Am Coll Cardiol Intv 7: 905–912.

Jones BM, Tuzcu EM, Krishnaswamy A et al. (2015). Neurologic events after transcatheter aortic valve replacement. Intervent Cardiol Clin 4: 83–93.

Kadono Y, Nakamura H, Saito S et al. (2019). Endovascular treatment for large vessel occlusion stroke in patients with ventricular assist devices. J Neurointervent Surg 11: 1205–1209.

Kato TS, Ota T, Schulze PC et al. (2012a). Asymmetric pattern of cerebrovascular lesions in patients after left ventricular assist device implantation. Stroke 43: 872–874.

Kato TS, Schulze PC, Yang J et al. (2012b). Pre-operative and post-operative risk factors associated with neurologic complications in patients with advanced heart failure supported by a left ventricular assist device. J Heart Lung Transplant 31: 1–8.

Kirklin JK, Naftel DC, Kormos RL et al. (2014). Interagency Registry for Mechanically Assisted Circulatory Support (INTERMACS) analysis of pump thrombosis in the HeartMate II left ventricular assist device. J Heart Lung Transplant 33: 12–22.

Lorusso R, Barili F, Mauro MD et al. (2016). In-hospital neurologic complications in adult patients undergoing venoarterial extracorporeal membrane oxygenation: results from the extracorporeal life support organization registry. Crit Care Med 44: e964–e972.

Lupercio F, Carlos Ruiz J, Briceno DF et al. (2016). Left atrial appendage morphology assessment for risk stratification of embolic stroke in patients with atrial fibrillation: a meta-analysis. Heart Rhythm 13: 1402–1409.

Martucci G, Lo Re V, Arcadipane A (2016). Neurological injuries and extracorporeal membrane oxygenation: the challenge of the new ECMO era. Neurol Sci 37: 1133–1136.

Meier B, Kalesan B, Mattle HP et al. (2013). Percutaneous closure of patent foramen ovale in cryptogenic embolism. N Engl J Med 368: 1083–1091.

Mérie C, Køber L, Olsen PS et al. (2012). Association of warfarin therapy duration after bioprosthetic aortic valve replacement with risk of mortality, thromboembolic complications, and bleeding. JAMA 308: 2118–2125.

Merkle J, Azizov F, Fatullayev J et al. (2019). Monitoring of adult patient on venoarterial extracorporeal membrane oxygenation in intensive care medicine. J Thorac Dis 11: S946–S956.

Nasr DM, Rabinstein AA (2015). Neurologic complications of extracorporeal membrane oxygenation. J Clin Neurol 11: 383–389.

Okochi S, Shakoor A, Barton S et al. (2018). Prevalence of seizures in pediatric extracorporeal membrane oxygenation patients as measured by continuous electroencephalography. Pediatr Crit Care Med 19: 1162–1167.

Parissis H, Soo A, Al-Alao B (2011). Intra aortic balloon pump: literature review of risk factors related to complications of the intraaortic balloon pump. J Cardiothorac Surg 6: 147.

Puskas J, Gerdisch M, Nichols D et al. (2014). Reduced anticoagulation after mechanical aortic valve replacement: interim results from the prospective randomized on-X valve anticoagulation clinical trial randomized Food and Drug Administration investigational device exemption trial. J Thorac Cardiovasc Surg 147: 1202–1211.

Reddy VY, Sievert H, Halperin J et al. (2014). Percutaneous left atrial appendage closure vs warfarin for atrial fibrillation a randomized clinical trial. JAMA 312: 1988–1998.

Rihal CS, Naidu SS, Givertz MM et al. (2015). 2015 SCAI/ACC/HFSA/STS clinical expert consensus statement on the use of percutaneous mechanical circulatory support devices in cardiovascular care: endorsed by the American Heart Assocation, the Cardiological Society of India, and Sociedad Latino America de Cardiologia Intervencion; affirmation of value by the Canadian Association of Interventional Cardiology-Association Canadienne de Cardiologie d'intervention. J Am Coll Cardiol 65: e7–e26.

Salerno CT, Sundareswaran KS, Schleeter TP et al. (2014). Early elevations in pump power with the HeartMate II left ventricular assist device do not predict late adverse events. J Heart Lung Transplant 33: 809–815.

Selim M (2007). Perioperative stroke. N Engl J Med 356: 706–713.

Simonsen CZ, Yoo AJ, Sørensen LH et al. (2018). Effect of general anesthesia and conscious sedation during endovascular therapy on infarct growth and clinical outcomes in acute ischemic stroke a randomized clinical trial. JAMA Neurol 75: 470–477.

Slaughter MS, Rogers JG, Milano CA et al. (2009). Advanced heart failure treated with continuous-flow left ventricular assist device. N Engl J Med 361: 2241–2251.

Starling RC, Moazami M, Silvestry SC et al. (2014). Unexpected abrupt increase in left ventricular assist device thrombosis. N Engl J Med 370: 33–40.

Subramaniam AV, Barsness GW, Vallabhajosyula S et al. (2019). Complications of temporary percutaneous mechanical circulatory support for cardiogenic shock: an appraisal of contemporary literature. Cardiol Ther 8: 211–228.

Sutter R, Tisljar K, Marsch S (2018). Acute neurologic complications during extracorporeal membrane oxygenation: a systematic review. Crit Care Med 46: 1506–1513.

Taccone FS, Lubicz B, Piagnerelli M et al. (2009). Cardiogenic shock with stunned myocardium during triple-H therapy treated with intra-aortic balloon pump counterpulsation. Neurocrit Care 10: 76–82.

Tay ELW, Gurvitch R, Wijesinghe N et al. (2011). A high-risk period for cerebrovascular events exists after transcatheter aortic valve implantation. J Am Coll Cardiol Intv 4: 1290–1297.

Thyagarajan B, Kumar MP, Sikachi RR et al. (2016). Endocarditis in left ventricular assist device. Intractable Rare Dis Res 5: 177–184.

Tsukui H, Abla A, Teuteberg JJ et al. (2007). Cerebrovascular accidents in patients with a ventricular assist device. J Thorac Cardiovasc Surg 134: 114–123.

Wille JZ, Demmer RT, Takayama H et al. (2014). Cerebrovascular disease in the era of left ventricular assist devices with continuous flow: risk factors, diagnosis, and treatment. J Heart Lung Transplant 33: 878–887.

Wilson TJ, Stetler Jr. WR, Al-Holou WN et al. (2013). Management of intracranial hemorrhage in patients with left ventricular assist devices. J Neurosurg 118: 1063–1068.

Xie A, Lo P, Yan TD et al. (2017). Neurologic complications of extracorporeal membrane oxygenation: a review. J Cardiothorac Vasc Anesth 31: 1836–1846.

Handbook of Clinical Neurology, Vol. 177 (3rd series)
Heart and Neurologic Disease
J. Biller, Editor
https://doi.org/10.1016/B978-0-12-819814-8.00028-7

Chapter 21

Neurologic complications of diseases of the aorta

STEPHEN W. ENGLISH[1]* AND JAMES P. KLAAS[2]

[1]*Department of Neurology, Mayo Clinic, Jacksonville, FL, United States*
[2]*Department of Neurology, Mayo Clinic, Rochester, MN, United States*

Abstract

Neurologic complications of diseases of the aorta are common, as the brain and spinal cord function is highly dependent on the aorta and its branches for blood supply. Any disease impacting the aorta may have significant impact on the ability to deliver oxygenated blood to the central nervous system, resulting in ischemia—and if prolonged—cerebral and spinal infarct. The breadth of pathology affecting the aorta is diverse and neurologic complications can vary dramatically based on the location, severity, and underlying etiology. This chapter outlines the major pathology of the aorta while highlighting the associated neurologic complications. This chapter covers the entire spectrum of neurologic complications associated with aortic disease by beginning with a detailed overview of the spinal cord vascular anatomy followed by a discussion of the most common aortic pathologies affecting the nervous system, including aortic aneurysm, aortic dissection, aortic atherosclerosis, inflammatory and infectious aortopathies, congenital abnormalities, and aortic surgery.

INTRODUCTION

The aorta is perhaps the most important blood vessel in the human body, serving as the major conduit of oxygenated blood from the heart to the rest of the body. With each heartbeat, pulsatile blood is expelled from the left ventricle through the aortic valve into the ascending aorta, immediately branching into major arteries to supply the heart, head, and upper limbs. From there, the aorta descends through the thorax into the abdominal cavity immediately adjacent to the vertebral column. As such, the aorta is essential for vascular supply to the brain and the spinal cord and diseases of the aorta can result in devastating complications to these structures. The breadth of pathology affecting the aorta is diverse and neurologic complications can vary dramatically based on the location, severity, and underlying etiology. For instance, aortic rupture is universally catastrophic and typically fatal, while aortic atheromatous disease may only result in a mild stroke without significant residual deficits. Recent advances have improved our understanding of aortic pathology, led to new endovascular surgical techniques with fewer complications, and the discovery of neuroprotective strategies to reduce neurologic injury and neurologic sequelae associated with aortic disease. This chapter will cover the entire spectrum of neurologic complications associated with aortic disease by beginning with a detailed overview of the spinal cord vascular anatomy followed by a discussion of the most common aortic pathologies affecting the nervous system, including aortic aneurysm, aortic dissection, aortic atherosclerosis, inflammatory and infectious aortopathies, congenital abnormalities, and aortic surgery.

SPINAL CORD VASCULAR SUPPLY

A basic understanding of the spinal cord anatomy and vascular supply helps provide insights into how aortic

*Correspondence to: Stephen W. English, M.D., M.B.A., Department of Neurology, Mayo Clinic, Jacksonville, FL, United States. Tel: +1-904-953-6869, Fax: +1-904-953-0757, E-mail: stephen.w.english@gmail.com

pathology can contribute to neurologic disease. While normal anatomic variants exist, the aortic arch conventionally gives off three major branches: the brachiocephalic (or innominate) artery, the left common carotid artery, and the left subclavian artery. The brachiocephalic artery branches into the right common carotid artery and the right subclavian, and the vertebral arteries originate from the subclavian arteries.

The spinal cord relies on the majority of its vascular supply from three longitudinally oriented vessels, one anterior spinal artery and two posterior spinal arteries, which course along the surface of the cord. The anterior spinal artery is primarily responsible for perfusion of the anterior two-thirds of the spinal cord, including the ventral gray matter, the corticospinal tracts, and the spinothalamic tracts, whereas the posterior spinal arteries supply the remaining third, largely comprising the dorsal columns. The anterior spinal artery arises from branches of the vertebral arteries near the craniocervical junction, projecting caudally along the ventral surface of the spinal cord to supply. Notably, the caliber of the anterior spinal artery varies along its course and can taper and even become discontinuous in the upper and middle thoracic regions. Therefore, adequate spinal cord perfusion is often dependent upon reinforcement from radially oriented vessels arising from the great arteries of the neck, thorax, and abdomen (Fig. 21.1A). Although bilateral radicular arteries are present at each segmental level, only a few provide meaningful vascular contributions to the spinal cord while the rest terminate within the nerve root, dura, or pial plexus; arteries that perfuse the spinal cord are termed radiculomedullary arteries. Human anatomical studies demonstrate wide variability in radiculomedullary arteries, with anywhere from 2 to

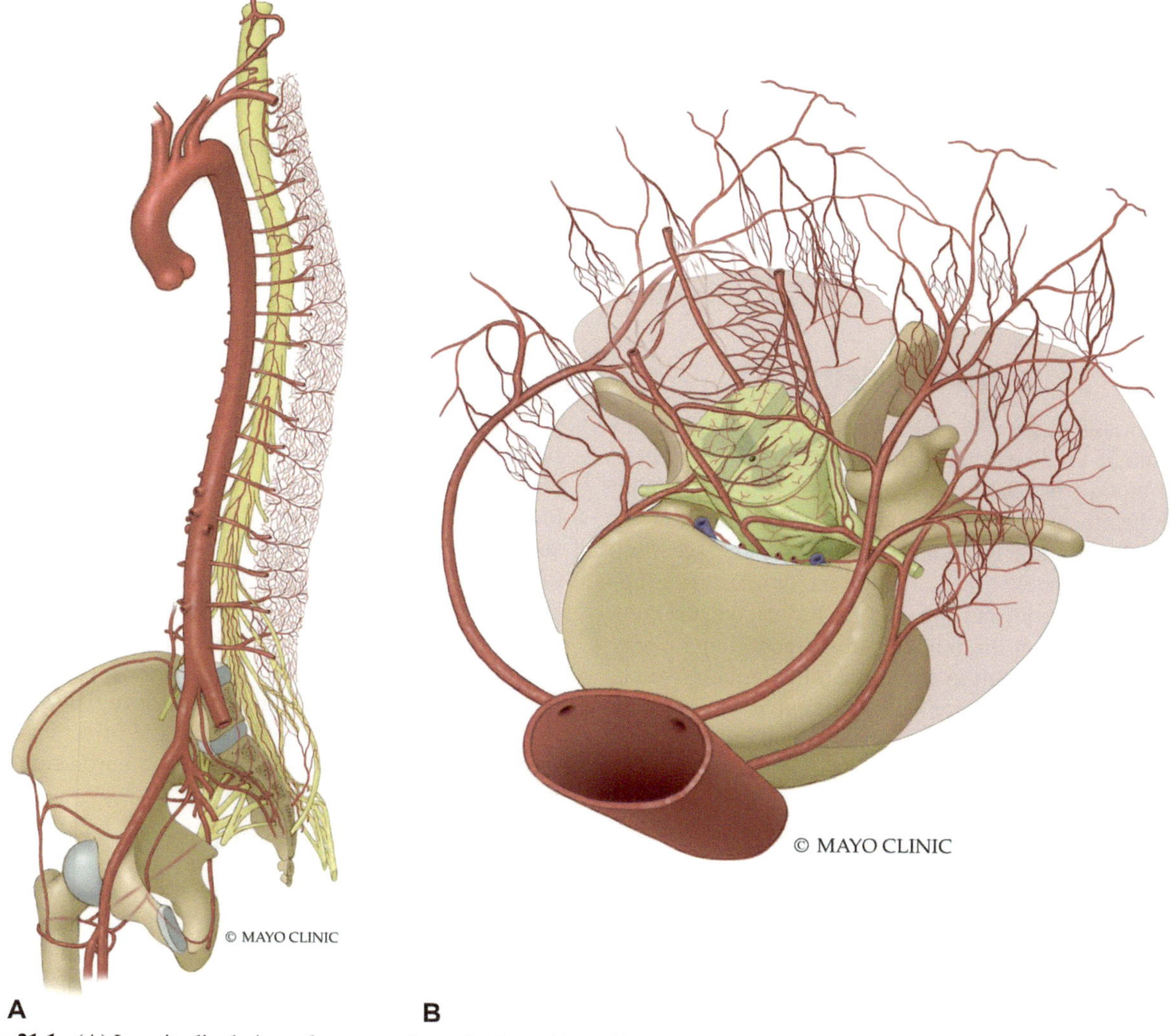

Fig. 21.1. (A) Longitudinal view of segmental arteries branching off aorta to supplement perfusion to the spinal cord and (B) rich anastomotic network of paraspinal muscles and paravertebral tissue, creating a longitudinally connected, flexible system.

17 supplying the spinal cord (Martirosyan et al., 2011; Gao et al., 2013). Contribution from radiculomedullary arteries is most sparse in the thoracic region, where are often only 1–2 main radiculomedullary arteries. The largest of these thoracic vessels is the artery of Adamkiewicz and can found most often between T9 and T12 but reported anywhere between T5 and L3 (Koshino et al., 1999; Colman et al., 2015). Some authors have conferred that the paucity of collaterals in the thoracic cord creates an area of relative vulnerability within the middle thoracic cord (Dommisse, 1974), but evidence from larger studies does not support this notion. Even in aortic surgeries where a clamp is placed proximal to the mid-thoracic feeders, the majority of ischemic events occur in the more metabolically active lumbosacral region rather than thoracic watershed regions (Goodin, 2008).

Posteriorly, paired posterior spinal arteries originate at the upper cervical cord via branches of the posterior inferior cerebellar arteries or the proximal vertebral arteries and course along the posterolateral surface. The posterior spinal arteries similarly rely on vascular reinforcement through posterior radiculomedullary arteries. However, the posterior spinal arteries are typically much smaller in caliber and often become discontinuous during their course; they are better envisioned as an arterial plexus than a linear system (Gillilan, 1958; Colman et al., 2015).

While there is tremendous variability in the extramedullary spinal cord vascular supply, the intramedullary vasculature is comparatively consistent. Major branches off the anterior spinal artery include the sulcal arteries, which pass posteriorly within the anterior longitudinal fissure to supply the deep gray matter and anterior white, and the peripheral arteries that course circumferentially along the anterior surface and perfuse the white matter tracts anteriorly and laterally. The number of sulcal arteries is congruent with the metabolic demand of a given spinal level with anywhere from 2 to 12 vessels per centimeter. The posterior spinal arteries—or posterior pial arterial plexus—penetrate the spinal cord to supply the posterior columns. Despite the rich anastomotic connections within of the peripheral arteries and the posteriorpial arterial plexus, there is often little connection between these two vascular networks (Goodin, 2008; Martirosyan et al., 2011; Colman et al., 2015).

Recent anatomic and physiologic studies demonstrate the importance of a robust network of collateral vessels from the paraspinal muscles and paravertebral tissue (Fig. 21.1B). The spinal arteries are interwoven with these collateral networks throughout their course, and these arteries can experience either anterograde or retrograde flow depending on the functional demand and location of the reinforcing arteries. It is this highly flexible and longitudinally connected system—with redundant autoregulatory mechanisms—that contributes to the spinal cord's relative resistance to ischemia. To highlight this point, permanent disruption of essentially all thoracolumbar intersegmental arteries during endovascular thoracoabdominal aneurysm repair rarely results in spinal cord infarction. This observation strongly supports the concept of abundant collateralization of the spinal cord in the vast majority of patients (Etz et al., 2011; Martirosyan et al., 2011; Colman et al., 2015).

Despite abundant collateralization of spinal vasculature, compromise of the artery of Adamkiewicz can lead to severe neurologic sequelae. Anatomic studies have found wide variability of its course, as approximately one-quarter of patients possess two distinct arteries, and half of these have a bilateral origin (Koshino et al., 1999). Despite its variable location, the artery of Adamkiewicz will typically form a classic "hairpin" turn as it anastomoses with the anterior spinal artery to form a more robust descending vessel (Nijenhuis et al., 2007; Colman et al., 2015). In patients undergoing thoracic aortic aneurysm repair, preoperative identification of the artery of Adamkiewicz can help to reduce the risk of perioperative spinal cord infarction. However, indirect imaging methods such as computerized tomography or magnetic resonance angiography (MRA) can be challenging to interpret because the anterior radiculomedullary vein also follows a similar "hairpin turn" course (Jaspers et al., 2007). Multiple studies have compared imaging techniques with wide variability in the reported detection rates, but recent advances in MRA capabilities utilizing 3-T time-resolved sequences have improved spatial resolution leading to appropriate identification of the artery of Adamkiewicz in 84%–100% of cases (Nijenhuis et al., 2007; Melissano et al., 2009; Mordasini et al., 2012; Takagi et al., 2015).

NEUROLOGIC COMPLICATIONS OF AORTIC DISEASE

Aortic aneurysm

The term aneurysm is derived from the Greek word "aneurusma," which means dilation, as it generally applies to arteries that have weakened and dilated beyond their normal size, resulting in increased risk of hemorrhage from arterial rupture. The aorta can be subdivided based on location between the thoracic compartment, comprised of the ascending aorta, aortic arch, and descending aorta to the level of the diaphragm, and the abdominal compartment below the level of the diaphragm; aneurysms can form at any point along its course. Aortic diameter differs based on location, and the aorta is considered aneurysmal if it experiences enlargement greater than 50% of its expected diameter. In the thoracic portion of the aorta, the cutoff of 4.5 cm is typically used, while the abdominal aorta is consider aneurysmal when it reaches 3 cm in diameter (Johnston et al., 1991; Liddington and Heather, 1992). In addition to size, thoracic and abdominal aortic aneurysms differ in other important respects. Thoracic aortic aneurysms carry an estimated incidence of 5–10 per 100,000 person-years (Kuzmik et al., 2012), typically arise in the sixth to seventh decade (mean age 65) of life, are more prevalent in men (1.7:1 male-to-female ratio) (Bickerstaff et al., 1982), and are more often associated with genetic disorders such as Marfan syndrome, Turner syndrome, or bicuspid aortic valve (Isselbacher, 2005). Additionally, syphilis is an infectious cause of thoracic aortic aneurysms (Roberts et al., 2009). Abdominal aortic aneurysms are much more common with an estimated prevalence of 1.3%–8.9% in men and 1.0%–2.2% in women (Singh et al., 2001; Sakalihasan et al., 2005) and tend to present later in life (mean age 75) (Pleumeekers et al., 1995). Smoking and hypertension are the greatest risk factors for both thoracic and abdominal aortic aneurysms (MacSweeney et al., 1994; Singh et al., 2001; Sakalihasan et al., 2005), and roughly 25% of patients with a thoracic aorta aneurysm will also have an abdominal aortic aneurysm (Bickerstaff et al., 1982). Abdominal aortic aneurysms are best characterized by location and relation to the renal arteries (Fig. 21.2).

The aortic wall is comprised of three distinct layers—the intima, media, and adventitia. The media layer is responsible for the structure and elasticity of the aortic wall; while several infectious and inflammatory conditions can lead to aneurysm formation (discussed later in this chapter), the most common cause is degenerative in nature, primarily affecting the structure and composition of the medial layer (Sakalihasan et al., 2005; Ruddy et al., 2008). The composition of the media layer can vary throughout its course based on the relative composition of elastin, collagen, and smooth muscle cells, and perhaps the most important histological feature of aneurysmal tissue is the fragmentation of elastic fibers and the decreased concentration of elastin within the media (MacSweeney et al., 1992; Dobrin and Mrkvicka, 1994; Sakalihasan et al., 2005; Ruddy et al., 2008). There are distinct pathophysiologic differences driving aneurysmal formation in the thoracic and abdominal aorta, but this is beyond the scope of this chapter. Ultimately, degradation and remodeling of the media through fragmentation of elastic fibers, decreased elastin concentration, and imbalance in collagen synthesis and degradation appear to be major determinants of aneurysmal formation and rupture (Sakalihasan et al., 1993, 2005; Holmes et al., 1995; Ruddy et al., 2008). Aortic aneurysms are typically discovered incidentally with noninvasive imaging for another indication or via routine screening in high-risk patients, but once they are found, they require careful interval evaluation to assess progression (Lederle et al., 2000; Lindholt et al., 2000). Surgical options include either open or endovascular techniques and are generally considered when the patients have

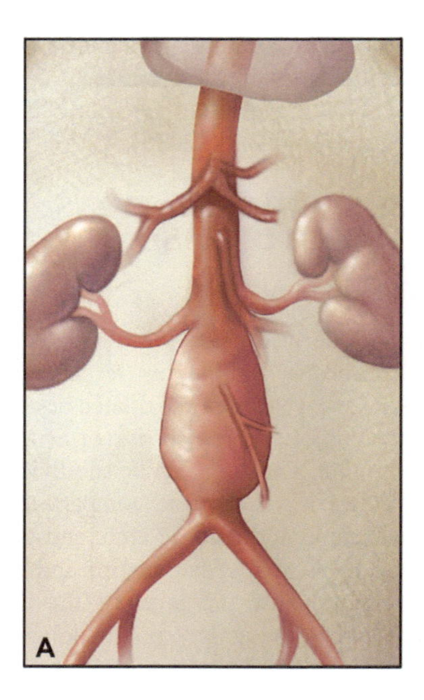

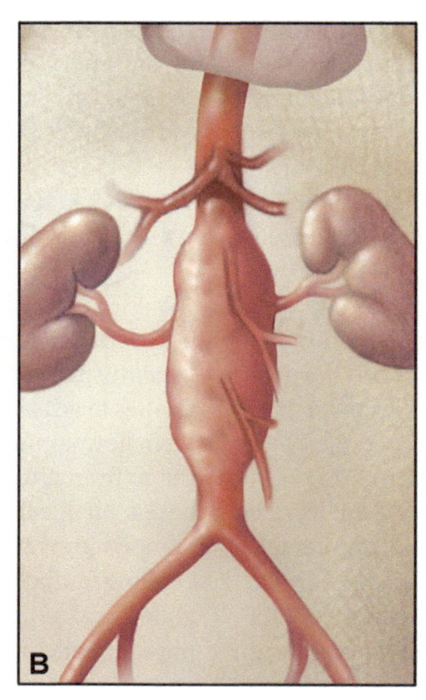

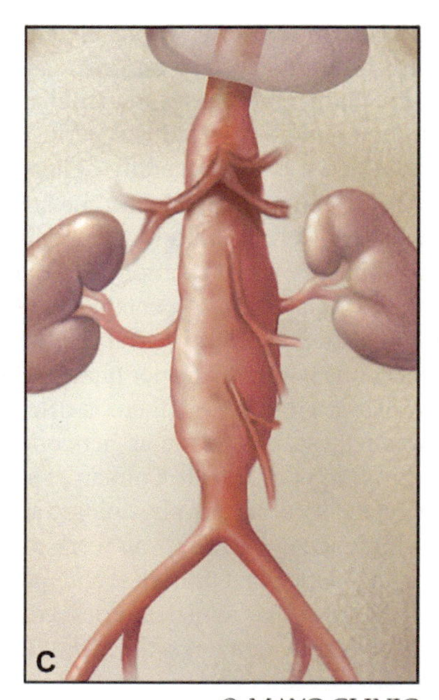

Fig. 21.2. Characterization of aortic aneurysms based on location include (A) infrarenal or juxtarenal, below or adjacent to the renal areas; (B) suprarenal, extending beyond the renal arteries but contained within the abdomen; and (C) thoroacoabdominal, extending into the chest cavity.

rapid progression, symptoms attributable to the aneurysm, or when asymptomatic aneurysms reach a certain size (Isselbacher, 2005; Sakalihasan et al., 2005).

Many of the systemic diseases and risk factors associated with aortic aneurysm (e.g., smoking, hypertension, atherosclerosis, diabetes mellitus) are commonly associated with neurologic disease; however, some neurologic complications have been reported as the direct cause of aortic aneurysms. William Osler was the first to report the association between neurologic symptoms and aortic aneurysms in 1905, when he described a patient with radiating leg pain in the setting of an abdominal aortic aneurysm (Osler, 1905). Over the next century, there have been numerous reports of other neurologic complications, but it is still considered quite rare. Almost invariably, neurologic complications of aortic aneurysms are related to direct compression, as was the case with Dr. Osler's original description. The precise neurologic symptoms are attributable to the site of compression. In the thoracic region, the most common neurologic symptom is hoarseness due to compression of the left recurrent laryngeal nerve, which descends as a branch of the vagus nerve and loops around the aortic arch (Stoob et al., 2004; Gulel et al., 2007; Matteucci et al., 2012). Abdominal aortic aneurysms have been reported to cause lumbosacral plexopathy (Lainez et al., 1989; Ashleigh and Marcuson, 1993; Lacasa et al., 1994; Brett and Hodgetts, 1997), lumbosacral radiculopathy (Wilberger Jr., 1983), and cauda equina syndrome (Nogues et al., 1987; Jauslin et al., 1991). These are documented in the literature in case reports of patients presenting with foot drop, sciatica, meralgia paresthetica, quadriceps weakness, urinary incontinence or retention, constipation, lower extremity sensory loss, and even paraplegia. Mycotic aneurysms, which often form and grow rapidly due to direct infiltration of the causative agent into vessels walls, have been reported to erode nearby vertebral bodies and cause neurologic symptoms through indirect compression from contiguous extension and perhaps more likely to be responsible for cases of cauda equina or conus medullaris syndrome (Boonen et al., 1995; Nadkarni et al., 2009). Perhaps the most severe complications arise in the setting of aortic rupture, which is often fatal, but survivors are at risk of significant spinal cord infarction or cerebral hypoxic–ischemic injury due to systemic hypoperfusion (Kamano et al., 2005).

Aortic dissection

Aortic dissection is a medical and surgical emergency associated with significant morbidity and mortality. It occurs due to a tear in the intimal layer that results in progressive separation of the intima and media from the adventitia, creating a false lumen due to extension of the intimal flap. Aortic dissections can be further classified based on the location and the timing of presentation. The most well-known classification system is the Stanford system, which was created to help guide management. Stanford Type A dissections involve the ascending aorta and arch—regardless of the site of intima tear—while Stanford Type B dissections involve the descending aorta and often arise at the origin of the left subclavian artery (Hershberger and Cho, 2014; Nienaber and Clough, 2015) (Fig. 21.3). Previously, aortic dissections were classified by timing of presentation by acute and chronic with the cutoff being 2 weeks; however, recently a new classification system has further divided into four time domains with a goal of improving prognostication: hyperacute (<24 h), acute (2–7 days), subacute (9–30 days), and chronic (>30 days) (Booher et al., 2013).

Aortic dissection has a reported incidence of approximately three cases per 100,000 per year (Meszaros et al., 2000; Clouse et al., 2004). Mean age at presentation is 63 and is almost twice as common in men than women (Mehta et al., 2002; Nienaber et al., 2004). Risk factors for aortic dissection are diverse and include hypertension, smoking, blunt trauma, vasoactive drugs (e.g., cocaine and amphetamines), pregnancy and delivery, and iatrogenic injury from catheterization or cardiac surgery (Larson and Edwards, 1984; Cregler, 1992; Howard et al., 2013). Hypertension is the most important modifiable risk factor, present in up to 75% of all cases (Baguet et al., 2012). Additionally, up to 20% of cases are associated with a genetic disorder, which typically present at a younger age and include connective tissue disorders such as Marfan's syndrome, Ehlers–Danlos syndrome, Loeys–Dietz syndrome, and Turner's syndrome (Attias et al., 2009; Regalado et al., 2011; Goldfinger et al., 2014). Other hereditary conditions, such as bicuspid aortic valve and aortic coarctation (discussed later), have also been associated with aortic dissection (Roberts and Roberts, 1991; Goldfinger et al., 2014; Nienaber and Clough, 2015). The pathophysiology of aortic dissection differs based on the genetic composition and risk factors of each patient; however, most cases appear to develop without associated vessel dilation, highlighting that other factors must play a role distinct from the medial degeneration or medial necrosis present in aortic aneurysms (Larson and Edwards, 1984; Nienaber and Clough, 2015).

Presenting features of aortic dissection can vary, but acute dissection classically manifests as sudden-onset severe chest pain or back pain without evidence of myocardial infarction. However, up to 15% of cases may present without any symptoms (Howard et al., 2013; Hershberger and Cho, 2014). Further complicating

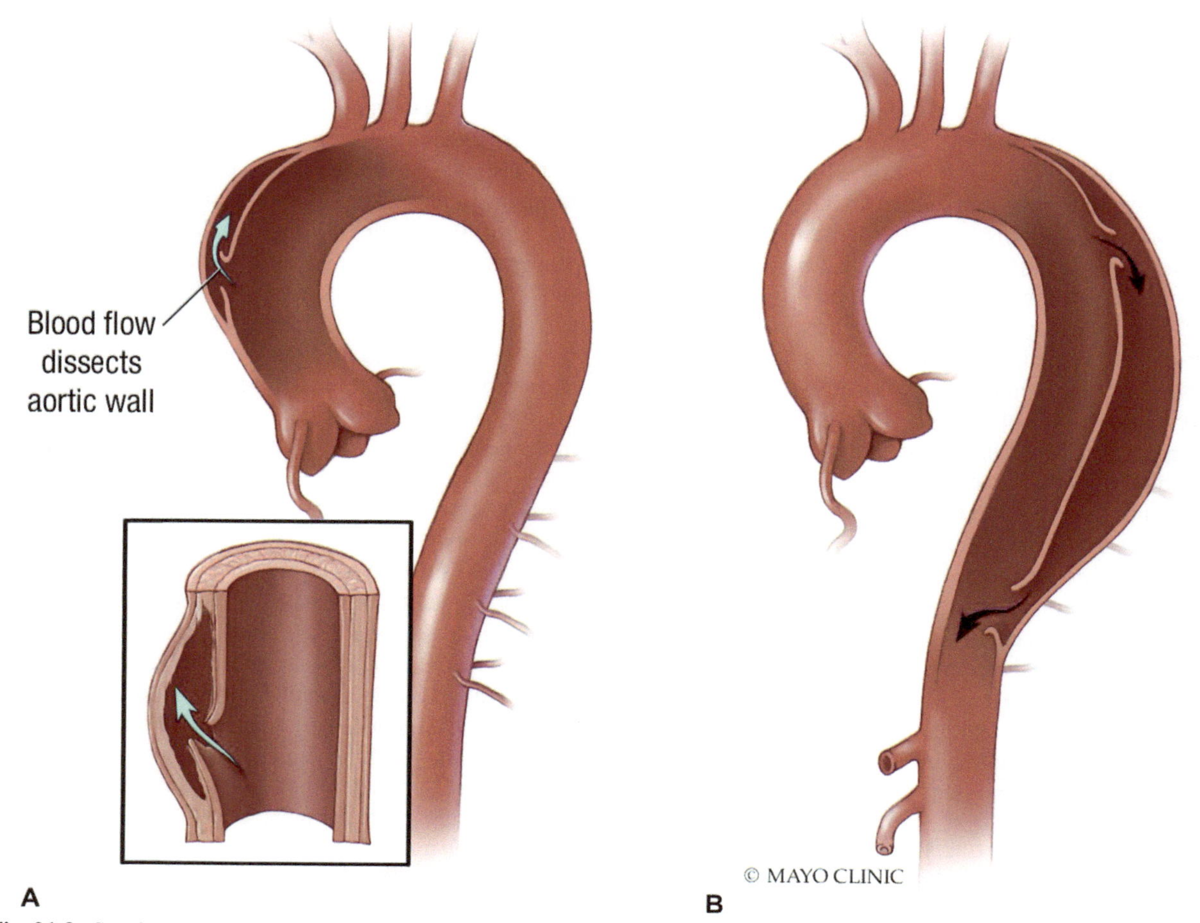

Fig. 21.3. Stanford classification system for aortic dissections based on involvement of the (A) ascending aorta or (B) descending aorta.

acute recognition, between 17% and 40% of patients will present with predominant neurologic symptoms (Guo et al., 2006). Neurologic complications can be broadly classified as either focal or diffuse, are almost invariably related to ischemia, and often differ based on the location of the dissection. Ischemic stroke is the most common presenting neurologic symptom (Blanco et al., 1999; Meszaros et al., 2000; Kazui et al., 2002; Gaul et al., 2007). Acute ischemic stroke in the setting of aortic dissection is typically related to extension of Type A aortic dissection into the cervical vessels, most commonly due to carotid involvement (81.2%) and preferentially affecting the right side (69.2%) (Gaul et al., 2007). However, all vessels may be involved, and it can occur in the absence of involvement of the cervical vessels. Type B dissections can also present with acute ischemic stroke in the setting of retrograde embolism (AlGhamdi et al., 2015). When aortic dissection leads to severe hemodynamic instability, cases of hypoxic ischemic injury from global cerebral hypoperfusion have been described (Blanco et al., 1999; Gaul et al., 2007). Several cases of transient global amnesia have also been reported (Bonnet et al., 2004; Gaul et al., 2007; Mondon et al., 2007).

With the widespread utilization of recombinant tissue plasminogen activator (tPA), there has been increasing effort to screen for aortic dissection with chest X-ray in the acute stroke setting. There are several cases in the literature of morbidity and mortality associated with tPA utilization in the setting of aortic dissection (Fessler and Alberts, 2000), including one case that led to cardiac tamponade and intrapulmonary hemorrhage (Gaul et al., 2007). While these cases are extremely rare in the acute stroke setting, aortic dissection should be considered prior to tPA administration, particularly in the setting of chest pain (Flemming and Brown Jr., 1999). The recommendations regarding the role of chest X-ray in acute stroke evaluation have changed over the years, largely due to the low prevalence of painless aortic dissection and increasing efforts to minimize treatment delays. The most recent guidelines from the American Heart

Association/American Stroke Association state that obtaining a chest X-ray should not delay tPA administration unless there is concern for acute cardiac, aortic, or pulmonary disease (Saber et al., 2016; Powers et al., 2018).

In addition to cerebral ischemia, several other neurologic complications associated with aortic dissection have been described. In Type A dissections, extension into the carotid arteries can cause a Horner's syndrome (Condon and Rose, 1969), and akin to aortic aneurysm, a compressive recurrent laryngeal nerve palsy can result in hoarseness (Khan et al., 1999). When the dissection involves the descending thoracic aorta, spinal cord ischemia can develop due to compromised perfusion from the radiculomedullary branches (Spittell et al., 1993; Zalewski et al., 2019). As a result, patients may present with deficits attributable to spinal cord infarction including acute paraparesis (DeBakey et al., 1982; Spittell et al., 1993), cauda equina or conus medullaris syndrome (Patel et al., 2002), Brown-Sequard syndrome, or a progressive myelopathy (Holloway et al., 1993). In the event, the dissection extends into the iliac arteries beyond the aortic bifurcation, occlusion, or hypoperfusion to the distal region can occur resulting in lower extremity neurologic complications including ischemic neuropathies and lumbosacral plexopathies, which have been reported in 6%–11% of aortic dissections (Alvarez Sabin et al., 1989; Lefebvre et al., 1995; Blanco et al., 1999; Gaul et al., 2007). Distal symptoms tend to occur first and include paresthesia, pain, and pallor, but with prolonged ischemia, the neurologic symptoms will move more proximally to affect the entire lower extremity.

Aortic atherosclerosis

Aortic atherosclerosis may play a role in the pathogenesis of aortic aneurysm and aortic dissection, but it can also independently cause neurologic complications through thromboembolism and atheroemobolism to the brain and spinal cord. There are limited data to provide specific incidence of stroke secondary to aortic atherosclerosis, but it has been shown that the presence of aortic plaques is higher in patients with cryptogenic stroke (Harloff et al., 2010). Recent literature focused on cryptogenic stroke and embolic stroke of undetermined source highlights the importance of low-burden paroxysmal atrial fibrillation and mild-to-moderate carotid atherosclerosis but concludes that the pathogenic role of aortic atherosclerosis is likely underrecognized as a stroke mechanism (Harloff et al., 2010; Katsanos et al., 2014; Saver, 2016). Several studies assessing the role of aortic arch atherosclerosis in ischemic stroke found that mobile plaques, plaques greater than 4 mm, and complex plaques (noncalcified, lipid-rich, ulcerated) are higher risk for embolization (Amarenco et al., 1994; Russo et al., 2009) (Fig. 21.4). One study—The Aortic Arch Related Cerebral Hazard trial—was a prospective, randomized control trial assessing antiplatelet vs anticoagulation for stroke prevention in patients with aortic plaque greater than 4 mm; conclusions were difficult to make as the trial was underpowered, but there did not appear to be a difference in stroke prevention between the two treatment groups (Amarenco et al., 2014).

While aortic arch atherosclerosis is more commonly implicated in neurologic disease, patients with descending aortic atherosclerosis may present with either cerebral or spinal cord infarction. Several studies have highlighted the role of descending aortic atherosclerosis as a potential cause in cryptogenic stroke, as patients may experience retrograde flow in the aortic arch in certain physiologic conditions, including aortic regurgitation, increased aortic stiffness, and decreased heart rate (Harloff et al., 2010; Hashimoto and Ito, 2013). Aortic stiffness with age is of particular importance, as the elastic expansion of aorta during systole (the Windkessel function) results in the diastolic recoil of the aorta and forward propagation of blood; increased aortic stiffness limits elasticity and creates opportunity for retrograde flow in early diastole (Hashimoto and Ito, 2013). Similarly, aortic atherosclerosis may lead to either atheroembolism to the spinal arteries or stenosis/occlusion of the ostia of radicular arteries supplying the spinal cord, and recent insights into spinal infarction highlight these possible mechanisms (Weidauer et al., 2015; Zalewski et al., 2019).

Inflammatory aortitis

Aortitis is broadly used to describe inflammation of the aortic wall but is reserved for primary inflammatory causes. As such, secondary aortic inflammation, most commonly associated with aortic atherosclerosis and a major factor in the development of aortic aneurysm and dissection, is not considered aortitis. There are multiple causes of aortitis that can cause direct neurologic sequelae as part of systemic inflammatory process or indirect neurologic sequelae due to complications associated with aortitis. The most common causes of inflammatory aortitis are giant cell arteritis and Takayasu arteritis, but other causes include granulomatosis with polyangitis, Behçet disease, sarcoidosis, Cogan syndrome, and other rheumatologic diseases including rheumatoid arthritis and systemic lupus erythematosus (Maleszewski, 2015) (Table 21.1).

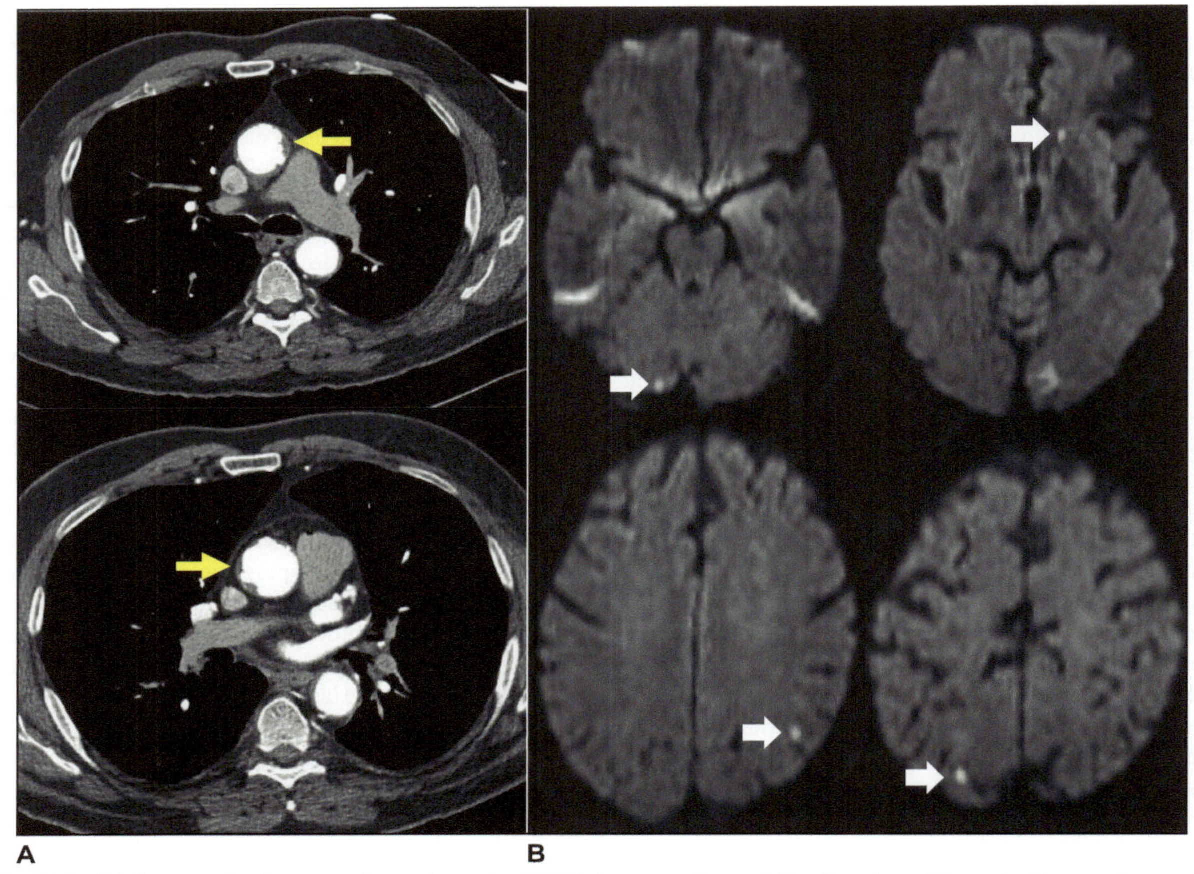

Fig. 21.4. (A) Computerized tomography angiography (CTA) demonstrating multifocal aortic wall irregularities consistent with noncalcified aortic plaques (*yellow arrows*) in a patient with small embolic strokes (*white arrows*) in TA vascular territories. (B) Evaluation included CTA head and neck revealing mild bilateral internal carotid artery atherosclerosis and normal transesophageal echocardiography without evidence of left atrial dilation, patent foramen ovale, or intracardiac thrombus.

Table 21.1

Inflammatory and immune-mediated causes of aortitis

Inflammatory

- Giant cell arteritis
- Takayasu arteritis
- Systemic lupus erythematosus
- Sarcoidosis
- Behçet disease
- Cogan syndrome
- Isolated aortitis

Immune-mediated

- Rheumatoid arthritis
- ANCA-associated vasculitis
  - o Granulomatosis with polyangiitis
  - o Eosinophilic granulomatosis with polyangiitis
  - o Polyarteritis nodosum
  - o Microscopic polyangiitis
- HLA-B27-associated spondyloarthropathies
  - o Ankylosing spondylitis
  - o Reiter syndrome

Giant cell arteritis

Giant cell arteritis affects the aorta, its direct branches, and medium-sized muscular arteries. Aortic involvement is relatively common, ranging from 10% to 40% in most series (Nuenninghoff et al., 2003), but typically occurs years after onset of disease (Nesher et al., 2004; Butler et al., 2010). The median time from initial diagnosis to documented aortic complications can be delayed anywhere from 3 to 8 years (Nuenninghoff et al., 2003; Bongartz and Matteson, 2006). It is considered a disease of the elderly with mean age at diagnosis of 75, twice as common in women, and more frequently seen in patients with Scandinavian descent (Salvarani et al., 2004). Incidence has remained stable over the past 20 years with an average annual incidence of 18.8 cases per 100,000 persons over 50 years of age (Salvarani et al., 2004). It is characterized by a granulomatous inflammatory infiltrate with multinucleated giant cells and has a predisposition for the media (Maleszewski, 2015). Notably, giant cell arteritis has a significantly higher predilection for the thoracic aorta, as patients are 17 times more likely to develop thoracic aortic aneurysms than the general population vs 2.4 times more likely to develop abdominal aortic aneurysms (Evans et al., 1995). Signs and symptoms associated with aortic involvement include aortic aneurysm formation, aortic dissection, aortic rupture, extremity claudication, and cerebrovascular symptoms—particularly when the arch and associated branching vessels become involved. Aortic stenosis and occlusion, however, are quite rare. Population-based studies suggest that the incidence of aortic aneurysm or dissection in giant cell arteritis is 15%–18% (Nuenninghoff et al., 2003; Butler et al., 2010), while involvement of the great vessels of the neck including the innominate, common carotid, and subclavian arteries occurs in 10%–15% of cases (Salvarani et al., 2002; Bongartz and Matteson, 2006).

Neurologic complications generally arise from involvement of the extracranial carotid arteries and posterior ciliary arteries supplying the optic nerve. New-onset headache related to systemic inflammation and temporal artery inflammation is the most common neurologic symptom at presentation. Jaw claudication due to involvement of the external carotid branches occurs in approximately one-third of patients but is a characteristic symptom. Visual symptoms have been reported in one-quarter of patients with biopsy-confirmed giant cell arteritis, and permanent vision loss remains the most feared complication. Vision loss is classically secondary to anterior ischemic optic neuropathy due to intimal hyperplasia and thrombosis of the ciliary arteries supplying the optic nerve (Nesher et al., 2004). In one study, this mechanism accounted for vision loss in 91.7% of cases, while the majority of the remaining cases were due to central retinal artery occlusion (Loddenkemper et al., 2007). Once vision is lost, it is often irreversible. Ischemic stroke from stenosis or occlusion of extracranial cervical vessels can occur in 3%–4% of cases (Nesher et al., 2004; Borg and Dasgupta, 2008). Rarely, involvement of the intracranial vessels has been reported, but this is extremely uncommon and limited to small case series (Alsolaimani et al., 2016).

Detection of giant cell arteritis, particularly in the early stages, can be difficult due to the lack of reliable biomarkers. Elevation of inflammatory markers, specifically erythrocyte sedimentation rate, is suggestive but can be negative in 5%–24% of biopsy-proven cases (Salvarani and Hunder, 2001). Biopsy of the temporal artery should be obtained in all suspected patients, but this can be negative in 50% of cases (Bongartz and Matteson, 2006). Vessel angiography may be supportive, particularly if there is evidence of multiple areas of focal tapering, poststenotic dilation, and enhanced collateralization (Butler et al., 2010). More recently, positron emission tomography has been utilized with high sensitivity to provide evidence of early inflammation of extracranial large vessel vasculitis, even before aneurysmal dilation or stenosis develops (Khan and Dasgupta, 2015). The mainstay of treatment remains corticosteroids and any patient with suspicion of giant cell arteritis should be initiated on treatment early to reduce the risk of vision loss. However, there is mounting evidence to support the use of tocilizumab, an interleukin-6 receptor inhibitor, particularly in severe cases, steroid-refractory cases, or cases with frequently relapses (Dejaco et al., 2017; Stone et al., 2017). However, tocilizumab has large poor central nervous system penetration, and alternative agents should be considered in patients that have aggressive intracranial disease (Cox et al., 2019).

Takayasu arteritis

In contrast to giant cell arteritis, Takayasu arteritis is much less common with an incidence of approximately 2.6 patients per million (Hall et al., 1985). It has a strong predilection for young women of Asian descent, with an average age at diagnosis between 25 and 30 years and an 8:1 female to male ratio (Kerr et al., 1994; Mwipatayi et al., 2005). It has been nicknamed the "pulseless disease" due to its tendency to cause stenosis and obstruction of the aortic arch and branching vessels and thus can have catastrophic complications. Pathologically, patients with Takayasu arteritis have granulomatous inflammation of both the media and adventitia with multinucleated giant cells; early in the disease course it can look quite similar to giant cell arteritis (Maleszewski, 2015). However, with time the diseases clearly diverge, as the aorta is almost universally involved in the disease.

Given the association with systemic inflammation, patients with Takayasu arteritis will often present with vague systemic features such as fever, malaise, anorexia, myalgias, and arthralgias. Nearly all patients with the disease either present or develop complications due to involvement of the large arteries including hypertension from involvement of the suprarenal aorta, renal artery, or baroreceptor dysfunction, pulse deficits, or extremity claudication (Kerr et al., 1994; Mwipatayi et al., 2005). The abdominal artery is the most common site of involvement, followed by the thoracic aorta and the aortic arch. Postinflammatory stenotic lesions are characteristic, but aortic aneurysms have been detected in nearly half of cases (Sueyoshi et al., 2000; Mwipatayi et al., 2005). Rapid aneurysmal expansion and rupture, intramural hematoma, and aortic dissection have been seen in multiple case series, highlighting the severity of the disease (Gornik and Creager, 2008). In addition to the aorta, direct branches including the subclavian, innominate, renal, common carotid, vertebral, and mesenteric arteries may also be implicated in 31%–54% of cases (Matsumura et al., 1991; Sueyoshi et al., 2000; Mwipatayi et al., 2005; Kim et al., 2012). Approximately 40% of patients develop cardiac abnormalities including myocardial infarction, angina pectoris, and acute aortic insufficiency related to involvement of the aortic root and ostia of coronary arteries (Kerr et al., 1994).

Neurologic complications associated with Takayasu arteritis are secondary to cerebral and spinal cord ischemia due to involvement of aortic arch and branching vessels. However, symptoms are more frequently attributed to global hypoperfusion from stenosis and occlusion of the cervical vessels (Fig. 21.5), rather than embolic phenomenon. Patients report symptoms of presyncope/syncope, lightheadedness, dizziness, and headache during the course of the disease much more commonly than stroke or transient ischemic attack. Cerebral ischemia has been observed in 5%–20% of patients, but one study observed that stroke accounted for 9.5% of deaths in patients with Takayasu arteritis (Kerr et al., 1994; Mwipatayi et al., 2005). Visual disturbances have been described secondary to vertebrobasilar insufficiency or due to central retinal hypoperfusion in up to 30% of cases (Kerr et al., 1994).

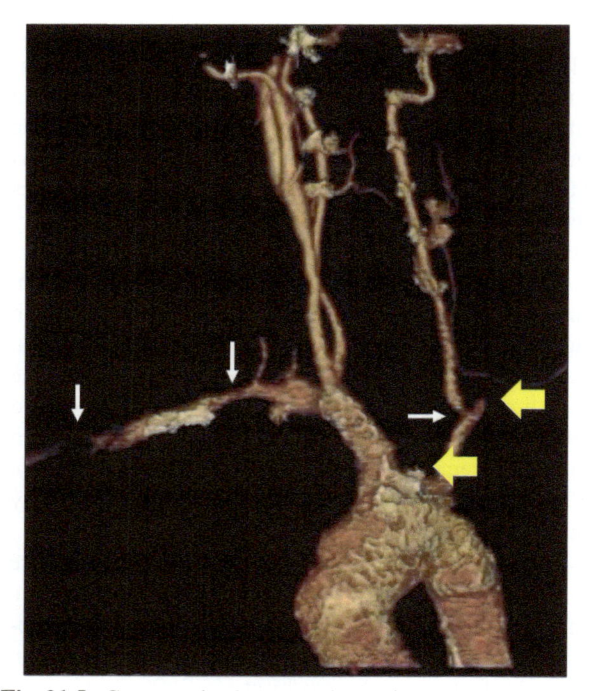

Fig. 21.5. Computerized topography angiography reconstruction demonstrating postinflammatory stenosis of the right subclavian artery and left vertebral artery origin (*white arrows*) and occlusive disease of the left common carotid artery origin and left subclavian arteries (*yellow block arrows*) in a patient with Takayasu arteritis.

Specific diagnostic criteria for Takayasu arteritis have been established by the American College of Rheumatology focusing on patient age, blood pressure, and pulse variation differences (suggestive arterial stenosis or occlusion), the presence of aortic or subclavian bruit, and imaging findings (Arend et al., 1990), but imaging advances including noninvasive angiography, positron emission tomography, and contrast-enhanced ultrasonography have improved recognition (Seyahi, 2017). Treatment of Takayasu arteritis is similar to giant cell arteritis, but it appears to be more responsive to steroids, with remission rates in 40%–60% of cases (Kerr et al., 1994). Steroid-sparing agents including cyclophosphamide, azathioprine, methotrexate, and mycophenolate mofetil have all been used successfully (Shetty et al., 1998; Daina et al., 1999; Ozen et al., 2007), but there is mounting evidence that biologic agents including rituximab and tocilizumab may be beneficial in refractory cases (Seyahi, 2017).

Other rheumatologic diseases

Clinicians should have a low threshold to consider systemic inflammatory disorders or systemic vasculitis in patients with features of aortitis. An in-depth discussion of all causes including granulomatosis with polyangiitis, Behçet disease, sarcoidosis, Cogan syndrome, IgG4-related disease, rheumatoid arthritis, and systemic lupus erythematosus is beyond the scope of this chapter, but each of these has been reported to affect the aorta (Gornik and Creager, 2008; Maleszewski, 2015). Patients may present with acute aortic insufficiency, arterial occlusive disease, aortic dissection, or aortic aneurysm, and these etiologies should be strongly considered in the absence of traditional risk factors or evidence of other systemic manifestations. Neurologic complications can be associated with the underlying inflammatory condition or as a result of aortic involvement. Diagnosis can often be reached through careful review of systemic involvement (e.g., oral or genital ulcers in Behçet disease, malar rash in systemic lupus erythematosus, polyarthritis in rheumatoid arthritis), presence of serum biomarkers, and noninvasive imaging. However, some cases may require a targeted biopsy for tissue diagnosis and requires a high index of suspicion to pursue more invasive testing.

Infectious aortitis

Most cases of aortitis occur in the setting of systemic inflammatory disease, but consideration of an infectious cause should not be overlooked as treatment is vastly different. The most commonly associated organisms include Staphylococcal and Salmonella species, Streptococcus pneumonia, tertiary syphilis, and tuberculosis

(Foote et al., 2005; Maleszewski, 2015). Pathogenesis differs based on etiology, and broad-spectrum antibiotics are typically initiated immediately and tailored once microbiology results are available. Regardless of underlying pathogen, complications of aortic involvement include aneurysm, dissection, and rupture. Neurologic complications of these syndromes have been discussed early. However, prompt recognition of infectious aortitis is perhaps even more critical due to rapid aneurysmal expansion and rupture risk associated with mycotic aneurysms (Reddy et al., 1991; Foote et al., 2005).

Bacterial aortitis

Aortic involvement in bacterial infection is quite uncommon, but complications can be devastating. Bacterial aortitis may arise from multiple mechanisms including direct implantation on intimal surface from endocarditis or blood stream infection, direct extension from an extravascular site, aortic trauma, or embolization to the vasa vasorum (Foote et al., 2005; Maleszewski, 2015). As discussed above, early recognition is critical but diagnosis can be challenging. Antibiotics should be started as soon as a diagnosis is suspected, immediately after blood cultures are obtained. Mortality remains high and surgical management should be strongly considered, particularly in Gram-negative infective aortitis to reduce risk of rupture (Reddy et al., 1991; Foote et al., 2005). Neurologic complications may be related to direct spread of the infection (e.g., meningitis, menigoencephalitis, pyogenic abscess), embolic phenomenon leading to ischemic stroke or hemorrhage from mycotic aneurysm, or secondary to global cerebral hypoperfusion in the setting of septic shock or aortic rupture.

Syphilitic aortitis

Prior to the discovery of penicillin, tertiary syphilis was the most common cause of thoracic aortic aneurysm and cardiovascular complications were the leading cause of death. Although exceedingly rare in the antibiotic era, it remains an important diagnostic consideration given the effectiveness of treatment with intravenous antibiotics. Following a variable latent period often ranging from 10 to 30 years, a characteristic hallmark of tertiary syphilis is the development of an obliterative endarteritis due to spirochete invasion of the vasa vasorum. This leads to death and necrosis of the elastic fibers and connective tissue of the aortic media, weakness of the aortic wall, and subsequent aneurysmal degeneration and dilation (Roberts et al., 2009). Due to these cardiovascular complications, tertiary syphilis is often fatal if not recognized and treated emergently.

Similar to other inflammatory aortopathies, neurologic symptoms may arise as a consequence of aortic involvement or as a direct result of tertiary syphilis. Neurosyphilis may present at any stage of the infection. In the earliest stages, spirochetes invade the CSF and can result in a transient meningitis in the first year of infection but may occur at any time. Syphilitic meningitis may cause headache, nuchal rigidity, depressed level of consciousness, hydrocephalus, and cerebrovascular complications due to arteritis of the cerebral or leptomeningeal vessels (Ghanem et al., 2008). Meningovascular syphilis can occur early in conjunction with syphilitic meningitis or in secondary syphilis often 4–7 years following infection. Patients present with acute or chronic ischemia and neurologic deficits referable to the area of ischemia (Hershberger and Cho, 2014). Ocular involvement is common and usually manifests as a posterior uveitis or panuveitis but can present as optic neuropathy, interstitial keratitis, anterior uveitis, or retinal vasculitis (Moradi et al., 2015). Late neurosyphilis can present with progressive cognitive difficult and dementia, pupillary abnormalities (Argyll-Robertson pupils), or most commonly as tabes dorsalis with progressive sensory ataxia due to the predilection for the posterior columns.

Aortic coarctation

Coarctation of the aorta refers to the congenital narrowing of the aorta and although it can occur at any point along its course, it is characteristically found at the level of the ligamentum arteriosum. The ligamentum arteriosum is the embryologic remnant of the ductus arteriosus, which serves to shunt blood away from the lungs during fetal development and typically closes near the time of birth. This is of pathologic significance, as it has been hypothesized that the oxygen-sensitive smooth muscle cells of the ductus arteriosus can become incorporated into the aortic wall, leading to the congential development of aortic stenosis—or coarctation—at birth. Aortic coarctiation can be classified based on anatomic relation to the ligamentum arteriosum as either preductal, postductal or ductal; most commonly, it occurs in the immediate postductal aorta (Fig. 21.6).

Coarctation of the aorta has been reported in 19/1000 births and is regarded as the most commonly missed congenital heart disease. Children are often asymptomatic aside from the presence of decreased femoral pulses and the early development of hypertension, making it difficult to diagnosis in isolation. It affects males two to five times more commonly than females and can be associated with multiple other congenital anomalies. Approximately 50% of patients with aortic coarctation have associated bicuspid aortic valve, but it also highly associated with Turner syndrome (35% of patients) and can be seen in patients with ventricular septal defects (Liberthson et al., 1979; Dijkema et al., 2017).

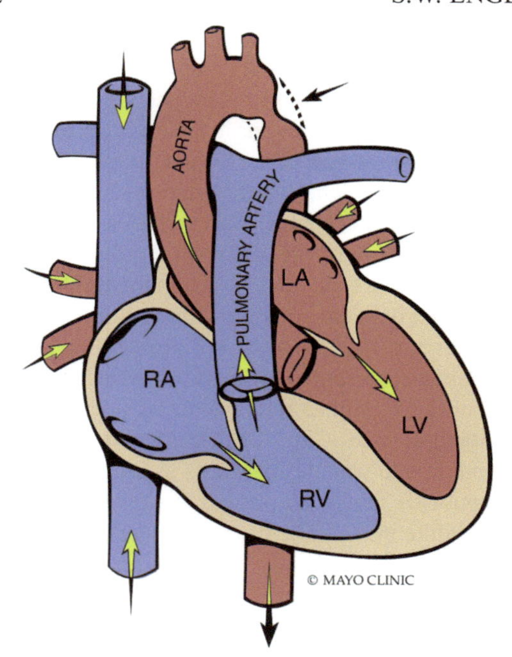

Fig. 21.6. Illustration of postductal aortic stenosis most commonly seen in patients with aortic coarctation (*black arrow*).

Aortic coarctation is associated with early mortality, but additional genetic or congenital defects play a significant role in outcomes and detection rates.

Adults with aortic coarctation can be diagnosed as part of routine health checks prior to symptom onset, often secondary to either the discovery of a cardiac murmur or unexplained hypertension. Vague complaints of headache, cold extremities, abdominal pain, and lower extremity fatigue with exercise may be noted (Dijkema et al., 2017). However, lower extremity claudication is typically only seen in patients with concomitant abdominal aortic coarctation. Patients may present with progressive functional decline secondary to progressive heart failure, left ventricular hypertrophy, or in rare cases, left ventricular dilation (Liberthson et al., 1979). Death in patients who go unrecognized or do not undergo surgical or endovascular repair is typically attributed to progressive heart failure, but coronary artery disease, aortic rupture or dissection, concomitant aortic valve disease, infective endocarditis, and intracranial hemorrhage have all been implicated (Webb et al., 2019).

Neurologic symptoms are routinely associated with either chronic systemic hypertension or alterations of aortic blood flow. Approximately 10% of patients with aortic coarctation will harbor an intracranial aneurysm, most often in the Circle of Willis (Connolly et al., 2003). Both aneurysmal hemorrhage and hypertensive intracerebral hemorrhage can occur and are often fatal. Alterations in blood flow can result in neurologic compression, steal, or arterial collateralization. In the setting of arterial steal phenomenon, patients with reduced perfusion of the descending thoracic aorta can develop collateral flow from the anterior spinal artery, resulting in retrograde flow through the radiculomedullary arteries and ultimately shunting blood away from the spinal cord. There are numerous cases in the literature of patients developing spinal cord ischemia due to this mechanism (Kendall and Andrew, 1972; Gill et al., 2011). Alternatively—and perhaps more commonly—patients can develop aneurysmal dilation of the spinal and radiculomedullary arteries due to progressive collateralization. Aneurysmal dilation of these vessels have been implicated in cases of compressive myelopathy, compressive radiculopathy, hematomyelia, and hematoma (Iwata et al., 1997; Tsutsumi et al., 1998; Sharma and Kumar, 2010; Tsai et al., 2016).

Treatment of aortic coarctation is indicated in symptomatic patients or in those with large concentration gradients (>30 mmHg). Options include both open surgical repair or endovascular angioplasty and stenting. Generally, surgery is preferred in younger patients, particularly those less than 1 year of age. Endovascular approaches are an option in older children or adults without calcified disease and are routinely the treatment approach of choice in recurrent stenosis (Webb et al., 2019). Regardless of treatment, the presence of aortic coarctation impacts expected life span and 75% of patients will have hypertension if treated after 30 years of age. In those treated prior to 14 years of age, 91% were alive at 20 years, but in those treated after 14 years of age, only 79% were alive 20 years later (Cohen et al., 1989).

NEUROLOGIC COMPLICATIONS OF AORTIC SURGERY

Advances in endovascular techniques have rapidly changed the surgical approach to aortic disease, but both open surgical and endovascular approaches remain common methods of aortic intervention. Despite the intricate collateralization of the spinal cord, spinal cord infarct following aortic intervention remains one of the most feared complications of either intervention technique. Mechanisms may relate to direct occlusion of the segmental blood supply or to perioperative hypotension and spinal hypoperfusion. Perioperative spinal cord infarction is of great importance due to the significant impact on patient morbidity and mortality, necessitating careful screening and monitoring in the preoperative and perioperative setting to reduce the risk of this dreaded complication.

Risk factors for the development of spinal cord ischemia can be divided into two groups: patient factors and operative factors. Patient factors can include older age, baseline peripheral arterial disease, chronic kidney disease, proximal aneurysm location, and collateral circulation (Svensson et al., 1993; Cambria et al., 2002; Zalewski et al., 2018). Older age, poor spinal collateralization, and preexisting peripheral arterial disease all contribute to vulnerability in the setting of hypoperfusion. Proximal aneurysms, particularly those of the aortic arch and descending aorta, contribute to surgical factors discussed later, but they can be a source of embolization from high atherosclerotic burden in the perioperative setting. Surgical factors associated with spinal cord ischemia include total aortic cross-clamp time, extent of aortic repair, aortic rupture, and perioperative hypotension (Svensson et al., 1993; Cambria et al., 2002; Piazza et al., 2019). In endovascular cases, the main risk for spinal cord ischemia relates to obstruction or compromise of a critical segmental blood vessel; despite remaining a relatively rare complication, some centers place emphasis on preoperative mapping of identify the Artery of Adamkiewicz to avoid this potential complication (Mordasini et al., 2012). Conversely, in open surgical repair special attention is paid to the cross-clamp time to avoid prolonged hypoperfusion, with prolonged times associated with higher risk of spinal cord ischemia.

Neurologic complications are closely linked to the factors discussed above and vary based on the nature of disease and extent of repair. Thoracoabdominal aortic aneurysm repair carries the highest risk for spinal cord ischemia due to the length of implicated aorta. In the early years of open surgical interventions, the incidence of spinal cord ischemia was as high as 31% in cases extending from the proximal descending aorta to the iliac bifurcation, while only 4% in cases limited to the abdominal aorta (Svensson et al., 1993). More recently, endovascular techniques help to obviate the need for aortic clamping and large surgical incisions, leading to lower complication rates. Endovascular approaches have evolved and can now be considered in complex thoracoabdominal aneurysms due to stents that are customized to the patient's anatomy with fenestrations and branch grafts to allow adequate perfusion to the kidneys, liver, spleen, and gastrointestinal tract (Fig. 21.7). Furthermore, various perioperative monitoring techniques and interventions have been developed to reduce the associated neurologic morbidity from spinal cord ischemia. These developments have led to a dramatic reduction in rates of paraplegia associated with thoracoabdominal aortic repair, with rates as low as 3.8%–10% in more recent case series (Coselli et al., 2007; Greenberg et al., 2010; Conrad et al., 2011).

Neuroprotective strategies leading to this dramatic shift focus on interventions to allow for early detection of ischemic injury and improvement in spinal perfusion. The most commonly implemented strategies include use of vasoactive medications to increase mean arterial pressure (and improve spinal perfusion) and cerebrospinal

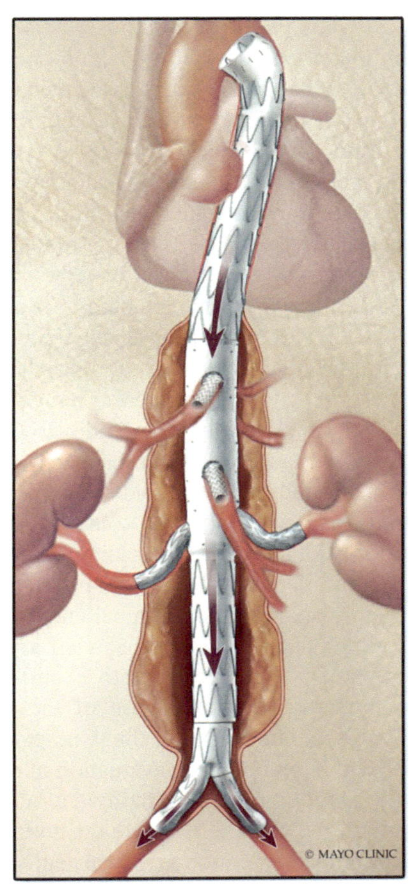

Fig. 21.7. Illustration of endovascular stent graft for thoracoabdominal aortic aneurysm demonstrating the use of fenestrations and branch grafts to preserve blood flow to the organs of the abdominal compartment.

fluid drainage via lumbar drain to reduce spinal vascular resistance. One study found that postoperative use of vasopressors and lumbar drain in patients with paraparesis led to significant neurologic improvement in 21% of patients (Zalewski et al., 2018). Additional techniques include intraoperative monitoring of somatosensory evoked potentials and motor evoked potentials with reimplantation of intercostal arteries when changes occur, iliofemoral conduits, and hypothermia (Augoustides et al., 2014; Sugiura et al., 2017). It is important to note that delayed paraplegia can develop in a small subset of patients, thought to be related to transient hypotension in the setting of relative intramedullary penumbra, with cases reported up to 10 months following surgery (Cho et al., 2008; Zalewski et al., 2018).

Although spinal cord ischemia is a major concern in patients undergoing aortic intervention, cerebral complications including seizure, cerebral edema, and stroke can occur. Seizures and cerebral edema are felt to be related to postoperative hyperperfusion syndrome in patients undergoing bypass grafting for occlusive disease of the aortic arch. Increased blood flow to the brain following revascularization has been clearly associated with these complications, an etiology known as cerebral hyperperfusion syndrome (Fields et al., 2006; Kim et al., 2012). Tight blood pressure control and consideration of staged interventions are important in cases of comorbid aortic and carotid occlusive disease. Ischemic stroke may result from periprocedural embolic event or cerebral hypoperfusion in the setting of aortic clamping or perioperative hypotension. The risk of stroke is closely tied to the burden or aortic arch atherosclerosis in aortic arch repair, the location of clamping in open surgical cases, and the location of the proximal endograft in endovascular cases. Closer proximity to aortic arch is associated with a higher stroke risk. One study revealed that stroke risk was 2% when clamp was placed adjacent to the left subclavian artery, but only 0.4% when the clamp was placed at the level of the diaphragm (Schmittling et al., 2000). Similarly, deployment of the proximal end of the aortic endograft proximal to the left subclavian artery has a stroke correlation with perioperative stroke, likely related to intraarterial manipulation of the arch with catheters, wires, balloons, and grafts (Cho et al., 2006). Due to these risks, preoperative evaluation of the carotid arteries with Doppler ultrasonography and noninvasive angiography of the chest to evaluate the aortic arch are employed prior to intervention to reduce neurologic morbidity of these procedures. Additionally, a rare syndrome resembling progressive supranuclear palsy has been described. The course is biphasic, marked by early vertical supranuclear gaze palsy and mild gait unsteadiness followed by slowly progressive supranuclear gaze dysfunction, mixed unsteady gait disorder, dysarthria, and sometimes dysphagia in the weeks to months following surgery (Mokri et al., 2004). While the precise mechanism remains unclear, this syndrome is believed to be a consequence of hypoxemic perioperative stress in patients who may share common predisposing vulnerability.

CONCLUSION

Aortic disease can manifest in numerous ways and has been associated with underlying congenital, genetic, vascular, traumatic, inflammatory, or infectious processes. As discussed above, aortic diseases can lead to an array of neurologic complications either directly, indirectly, or secondary to surgical intervention. Although these neurologic complications only appear in a small subset of patients with aortic disease, clinicians should be aware of these potential neurologic manifestations, particularly given the high morbidity and mortality associated with these neurologic events. There have been significant advancements in diagnostic tools, medical treatment, and surgical techniques in the last few decades to help the cause. Knowledge of these, coupled with adequate screening, prevention, and early recognition of potential neurologic complications of aortic disease is vital to improve patient outcomes.

References

AlGhamdi A, Alqahtani S, Ricketti M et al. (2015). Early acute ischaemic stroke in two patients with acute type B aortic dissection: an unusual complication. BMJ Case Rep 2015. https://doi.org/10.1136/bcr-2015-210021.

Alsolaimani RS, Bhavsar SV, Khalidi NA et al. (2016). Severe intracranial involvement in giant cell arteritis: 5 cases and literature review. J Rheumatol 43: 648–656. https://doi.org/10.3899/jrheum.150143.

Alvarez Sabin J, Vazquez J, Sala A et al. (1989). Neurologic manifestations of dissecting aneurysms of the aorta. Med Clin (Barc) 92: 447–449.

Amarenco P, Cohen A, Tzourio C et al. (1994). Atherosclerotic disease of the aortic arch and the risk of ischemic stroke. N Engl J Med 331: 1474–1479. https://doi.org/10.1056/nejm199412013312202.

Amarenco P, Davis S, Jones EF et al. (2014). Clopidogrel plus aspirin versus warfarin in patients with stroke and aortic arch plaques. Stroke 45: 1248–1257. https://doi.org/10.1161/strokeaha.113.004251.

Arend WP, Michel BA, Bloch DA et al. (1990). The American College of Rheumatology 1990 criteria for the classification of Takayasu arteritis. Arthritis Rheum 33: 1129–1134. https://doi.org/10.1002/art.1780330811.

Ashleigh RJ, Marcuson RW (1993). False aortic aneurysm presenting as sciatic nerve root pain. Eur J Vasc Surg 7: 214–216.

Attias D, Stheneur C, Roy C et al. (2009). Comparison of clinical presentations and outcomes between patients with TGFBR2 and FBN1 mutations in Marfan syndrome and related disorders. Circulation 120: 2541–2549. https://doi.org/10.1161/circulationaha.109.887042.

Augoustides JG, Stone ME, Drenger B (2014). Novel approaches to spinal cord protection during thoracoabdominal aortic interventions. Curr Opin Anaesthesiol 27: 98–105. https://doi.org/10.1097/aco.0000000000000033.

Baguet JP, Chavanon O, Sessa C et al. (2012). European Society of Hypertension scientific newsletter: hypertension and aortic diseases. J Hypertens 30: 440–443. https://doi.org/10.1097/HJH.0b013e32834f867a.

Bickerstaff LK, Pairolero PC, Hollier LH et al. (1982). Thoracic aortic aneurysms: a population-based study. Surgery 92: 1103–1108.

Blanco M, Diez-Tejedor E, Larrea JL et al. (1999). Neurologic complications of type I aortic dissection. Acta Neurol Scand 99: 232–235. https://doi.org/10.1111/j.1600-0404.1999.tb07352.x.

Bongartz T, Matteson EL (2006). Large-vessel involvement in giant cell arteritis. Curr Opin Rheumatol 18: 10–17.

Bonnet P, Niclot P, Chaussin F et al. (2004). A puzzling case of transient global amnesia. Lancet 364: 554. https://doi.org/10.1016/s0140-6736(04)16817-7.

Booher AM, Isselbacher EM, Nienaber CA et al. (2013). The IRAD classification system for characterizing survival after aortic dissection. Am J Med 126: 730.e19–24. https://doi.org/10.1016/j.amjmed.2013.01.020.

Boonen A, Ghesquiere B, Westhovens R et al. (1995). Vertebral fracture induced by chronic contained rupture of aortic aneurysm. Ann Rheum Dis 54: 437–438. https://doi.org/10.1136/ard.54.5.437-b.

Borg FA, Dasgupta B (2008). Giant cell arteritis: not just a stroke in the eye. Br J Hosp Med (Lond) 69: 608–609. https://doi.org/10.12968/hmed.2008.69.11.31635.

Brett A, Hodgetts T (1997). Abdominal aortic aneurysm presenting as meralgia paraesthetica. J Accid Emerg Med 14: 49–51. https://doi.org/10.1136/emj.14.1.49.

Butler N, Mundy J, Shah P (2010). Aortic complications of giant cell arteritis: a diagnostic and management dilemma. J Card Surg 25: 572–581. https://doi.org/10.1111/j.1540-8191.2010.01100.x.

Cambria RP, Clouse WD, Davison JK et al. (2002). Thoracoabdominal aneurysm repair: results with 337 operations performed over a 15-year interval. Ann Surg 236: 471–479. discussion 479. https://doi.org/10.1097/00000658-200210000-00010.

Cho JS, Haider SE, Makaroun MS (2006). US multicenter trials of endoprostheses for the endovascular treatment of descending thoracic aneurysms. J Vasc Surg 43: 12a–19a. https://doi.org/10.1016/j.jvs.2005.10.056.

Cho JS, Rhee RY, Makaroun MS (2008). Delayed paraplegia 10 months after endovascular repair of thoracic aortic aneurysm. J Vasc Surg 47: 625–628. https://doi.org/10.1016/j.jvs.2007.09.042.

Clouse WD, Hallett Jr JW, Schaff HV et al. (2004). Acute aortic dissection: population-based incidence compared with degenerative aortic aneurysm rupture. Mayo Clin Proc 79: 176–180. https://doi.org/10.4065/79.2.176.

Cohen M, Fuster V, Steele PM et al. (1989). Coarctation of the aorta. Long-term follow-up and prediction of outcome after surgical correction. Circulation 80: 840–845. https://doi.org/10.1161/01.cir.80.4.840.

Colman MW, Hornicek FJ, Schwab JH (2015). Spinal cord blood supply and its surgical implications. J Am Acad Orthop Surg 23: 581–591. https://doi.org/10.5435/jaaos-d-14-00219.

Condon JR, Rose FC (1969). The neurological manifestations of dissecting aneurysm of the aorta. Postgrad Med J 45: 419–422. https://doi.org/10.1136/pgmj.45.525.419.

Connolly HM, Huston J, Brown RD et al. (2003). Intracranial aneurysms in patients with coarctation of the aorta: a prospective magnetic resonance angiographic study of 100 patients. Mayo Clin Proc 78: 1491–1499. https://doi.org/10.4065/78.12.1491. http://www.sciencedirect.com/science/article/pii/S0025619611627456.

Conrad MF, Ergul EA, Patel VI et al. (2011). Evolution of operative strategies in open thoracoabdominal aneurysm repair. J Vasc Surg 53: 1195–1201.e1. https://doi.org/10.1016/j.jvs.2010.11.055.

Coselli JS, Bozinovski J, LeMaire SA (2007). Open surgical repair of 2286 thoracoabdominal aortic aneurysms. Ann Thorac Surg 83: S862–S864. discussion S890–2. https://doi.org/10.1016/j.athoracsur.2006.10.088.

Cox B, Fulgham J, Klaas J (2019). Recurrent stroke in giant cell arteritis despite immunotherapy (P4.3-028). Neurology 92: P4.3-028.

Cregler LL (1992). Aortic dissection and cocaine use. Am Heart J 124: 1665. https://doi.org/10.1016/0002-8703(92)90109-9.

Daina E, Schieppati A, Remuzzi G (1999). Mycophenolate mofetil for the treatment of Takayasu arteritis: report of three cases. Ann Intern Med 130: 422–426. https://doi.org/10.7326/0003-4819-130-5-199903020-00013.

DeBakey ME, McCollum CH, Crawford ES et al. (1982). Dissection and dissecting aneurysms of the aorta: twenty-year follow-up of five hundred twenty-seven patients treated surgically. Surgery 92: 1118–1134.

Dejaco C, Brouwer E, Mason JC et al. (2017). Giant cell arteritis and polymyalgia rheumatica: current challenges and opportunities. Nat Rev Rheumatol 13: 578–592. https://doi.org/10.1038/nrrheum.2017.142.

Dijkema EJ, Leiner T, Grotenhuis HB (2017). Diagnosis, imaging and clinical management of aortic coarctation. Heart 103: 1148–1155. https://doi.org/10.1136/heartjnl-2017-311173. https://heart.bmj.com/content/heartjnl/103/15/1148.full.pdf.

Dobrin PB, Mrkvicka R (1994). Failure of elastin or collagen as possible critical connective tissue alterations underlying aneurysmal dilatation. Cardiovasc Surg 2: 484–488.

Dommisse GF (1974). The blood supply of the spinal cord. A critical vascular zone in spinal surgery. J Bone Joint Surg Br 56: 225–235.

Etz CD, Kari FA, Mueller CS et al. (2011). The collateral network concept: a reassessment of the anatomy of spinal cord perfusion. J Thorac Cardiovasc Surg 141: 1020–1028. https://doi.org/10.1016/j.jtcvs.2010.06.023.

Evans JM, O'Fallon WM, Hunder GG (1995). Increased incidence of aortic aneurysm and dissection in giant cell (temporal) arteritis. A population-based study. Ann Intern Med 122: 502–507. https://doi.org/10.7326/0003-4819-122-7-199504010-00004.

Fessler AJ, Alberts MJ (2000). Stroke treatment with tissue plasminogen activator in the setting of aortic dissection. Neurology 54: 1010. https://doi.org/10.1212/wnl.54.4.1010.

Fields CE, Bower TC, Cooper LT et al. (2006). Takayasu's arteritis: operative results and influence of disease activity. J Vasc Surg 43: 64–71. https://doi.org/10.1016/j.jvs.2005.10.010.

Flemming KD, Brown Jr. RD (1999). Acute cerebral infarction caused by aortic dissection: caution in the thrombolytic era. Stroke 30: 477–478. https://doi.org/10.1161/01.str.30.2.477.

Foote EA, Postier RG, Greenfield RA et al. (2005). Infectious aortitis. Curr Treat Options Cardiovasc Med 7: 89–97.

Gao L, Wang L, Su B et al. (2013). The vascular supply to the spinal cord and its relationship to anterior spine surgical approaches. Spine J 13: 966–973. https://doi.org/10.1016/j.spinee.2013.03.017.

Gaul C, Dietrich W, Friedrich I et al. (2007). Neurological symptoms in type A aortic dissections. Stroke 38: 292–297. https://doi.org/10.1161/01.STR.0000254594.33408.b1.

Ghanem KG, Moore RD, Rompalo AM et al. (2008). Neurosyphilis in a clinical cohort of HIV-1-infected patients. AIDS 22: 1145–1151. https://doi.org/10.1097/QAD.0b013e32830184df.

Gill M, Pathak HC, Singh P et al. (2011). A case of aortic coarctation presenting with quadriparesis due to dilated tortuous anterior spinal artery. Neurol India 59: 317–318. https://doi.org/10.4103/0028-3886.79174.

Gillilan LA (1958). The arterial blood supply of the human spinal cord. J Comp Neurol 110: 75–103.

Goldfinger JZ, Halperin JL, Marin ML et al. (2014). Thoracic aortic aneurysm and dissection. J Am Coll Cardiol 64: 1725–1739. https://doi.org/10.1016/j.jacc.2014.08.025.

Goodin DS (2008). Chapter 2—Neurological complications of aortic disease and surgery. In: MJ Aminoff (Ed.), Neurology and general medicine, 4th ed. Churchill Livingstone, Philadelphia, pp. 23–44.

Gornik HL, Creager MA (2008). Aortitis. Circulation 117: 3039–3051. https://doi.org/10.1161/circulationaha.107.760686.

Greenberg R, Eagleton M, Mastracci T (2010). Branched endografts for thoracoabdominal aneurysms. J Thorac Cardiovasc Surg 140: S171–S178. https://doi.org/10.1016/j.jtcvs.2010.07.061.

Gulel O, Elmali M, Demir S et al. (2007). Ortner's syndrome associated with aortic arch aneurysm. Clin Res Cardiol 96: 49–50. https://doi.org/10.1007/s00392-006-0454-z.

Guo DC, Papke CL, He R et al. (2006). Pathogenesis of thoracic and abdominal aortic aneurysms. Ann N Y Acad Sci 1085: 339–352. https://doi.org/10.1196/annals.1383.013.

Hall S, Barr W, Lie JT et al. (1985). Takayasu arteritis. A study of 32 North American patients. Medicine (Baltimore) 64: 89–99.

Harloff A, Simon J, Brendecke S et al. (2010). Complex plaques in the proximal descending aorta: an underestimated embolic source of stroke. Stroke 41: 1145–1150. https://doi.org/10.1161/strokeaha.109.577775.

Hashimoto J, Ito S (2013). Aortic stiffness determines diastolic blood flow reversal in the descending thoracic aorta: potential implication for retrograde embolic stroke in hypertension. Hypertension 62: 542–549. https://doi.org/10.1161/hypertensionaha.113.01318.

Hershberger R, Cho JS (2014). Neurologic complications of aortic diseases and aortic surgery. Handb Clin Neurol 119: 223–238. https://doi.org/10.1016/b978-0-7020-4086-3.00016-3.

Holloway SF, Fayad PB, Kalb RG et al. (1993). Painless aortic dissection presenting as a progressive myelopathy. J Neurol Sci 120: 141–144. https://doi.org/10.1016/0022-510x(93)90265-z.

Holmes DR, Liao S, Parks WC et al. (1995). Medial neovascularization in abdominal aortic aneurysms: a histopathologic marker of aneurysmal degeneration with pathophysiologic implications. J Vasc Surg 21: 761–771. discussion 771–772.

Howard DP, Banerjee A, Fairhead JF et al. (2013). Population-based study of incidence and outcome of acute aortic dissection and premorbid risk factor control: 10-year results from the Oxford Vascular Study. Circulation 127: 2031–2037. https://doi.org/10.1161/circulationaha.112.000483.

Isselbacher EM (2005). Thoracic and abdominal aortic aneurysms. Circulation 111: 816–828. https://doi.org/10.1161/01.Cir.0000154569.08857.7a.

Iwata A, Takahashi Y, Ohgi K et al. (1997). A case of spinal hemorrhage associated with abdominal aortic coarctation. Rinsho Shinkeigaku 37: 413–416.

Jaspers K, Nijenhuis RJ, Backes WH (2007). Differentiation of spinal cord arteries and veins by time-resolved MR angiography. J Magn Reson Imaging 26: 31–40. https://doi.org/10.1002/jmri.20940.

Jauslin PA, Muller AF, Myers P et al. (1991). Cauda equina syndrome associated with an aorto-caval fistula. Eur J Vasc Surg 5: 471–473.

Johnston KW, Rutherford RB, Tilson MD et al. (1991). Suggested standards+ for reporting on arterial aneurysms. Subcommittee on Reporting Standards for Arterial Aneurysms, Ad Hoc Committee on Reporting Standards, Society for Vascular Surgery and North American Chapter, International Society for Cardiovascular Surgery. J Vasc Surg 13: 452–458.

Kamano S, Yonezawa I, Arai Y et al. (2005). Acute abdominal aortic aneurysm rupture presenting as transient paralysis of the lower legs: a case report. J Emerg Med 29: 53–55. https://doi.org/10.1016/j.jemermed.2005.01.012.

Katsanos AH, Giannopoulos S, Kosmidou M et al. (2014). Complex atheromatous plaques in the descending aorta and the risk of stroke: a systematic review and meta-analysis. Stroke 45: 1764–1770. https://doi.org/10.1161/strokeaha.114.005190.

Kazui T, Washiyama N, Bashar AH et al. (2002). Surgical outcome of acute type A aortic dissection: analysis of risk factors. Ann Thorac Surg 74: 75–81. discussion 81–82. https://doi.org/10.1016/s0003-4975(02)03603-2.

Kendall BE, Andrew J (1972). Neurogenic intermittent claudication associated with aortic steal from the anterior spinal artery complicating coarctation of the aorta. J Neurosurg 37: 89–94. https://doi.org/10.3171/jns.1972.37.1.0089.

Kerr GS, Hallahan CW, Giordano J et al. (1994). Takayasu arteritis. Ann Intern Med 120: 919–929. https://doi.org/10.7326/0003-4819-120-11-199406010-00004.

Khan A, Dasgupta B (2015). Imaging in giant cell arteritis. Curr Rheumatol Rep 17: 52. https://doi.org/10.1007/s11926-015-0527-y.

Khan IA, Wattanasauwan N, Ansari AW (1999). Painless aortic dissection presenting as hoarseness of voice: cardiovocal syndrome: Ortner's syndrome. Am J Emerg Med 17: 361–363.

Kim YW, Kim DI, Park YJ et al. (2012). Surgical bypass vs endovascular treatment for patients with supra-aortic arterial occlusive disease due to Takayasu arteritis. J Vasc Surg 55: 693–700. https://doi.org/10.1016/j.jvs.2011.09.051.

Koshino T, Murakami G, Morishita K et al. (1999). Does the Adamkiewicz artery originate from the larger segmental arteries? J Thorac Cardiovasc Surg 117: 898–905. https://doi.org/10.1016/s0022-5223(99)70369-7.

Kuzmik GA, Sang AX, Elefteriades JA (2012). Natural history of thoracic aortic aneurysms. J Vasc Surg 56: 565–571. https://doi.org/10.1016/j.jvs.2012.04.053.

Lacasa J, Ruiz F, de Escalante B et al. (1994). Lumbosacral plexopathy from aortic aneurysm. An Med Intern 11: 105–106.

Lainez JM, Yaya R, Lluch V et al. (1989). Lumbosacral plexopathy caused by aneurysms of the abdominal aorta. Med Clin (Barc) 92: 462–464.

Larson EW, Edwards WD (1984). Risk factors for aortic dissection: a necropsy study of 161 cases. Am J Cardiol 53: 849–855. https://doi.org/10.1016/0002-9149(84)90418-1.

Lederle FA, Johnson GR, Wilson SE et al. (2000). The aneurysm detection and management study screening program: validation cohort and final results. Aneurysm detection and management veterans affairs cooperative study investigators. Arch Intern Med 160: 1425–1430. https://doi.org/10.1001/archinte.160.10.1425.

Lefebvre V, Leduc JJ, Choteau PH (1995). Painless ischaemic lumbosacral plexopathy and aortic dissection. J Neurol Neurosurg Psychiatry 58: 641. https://doi.org/10.1136/jnnp.58.5.641.

Liberthson RR, Glenn Pennington D, Jacobs ML et al. (1979). Coarctation of the aorta: review of 234 patients and clarification of management problems. Am J Cardiol 43: 835–840. https://doi.org/10.1016/0002-9149(79)90086-9. http://www.sciencedirect.com/science/article/pii/0002914979900869.

Liddington MI, Heather BP (1992). The relationship between aortic diameter and body habitus. Eur J Vasc Surg 6: 89–92.

Lindholt JS, Vammen S, Juul S et al. (2000). Optimal interval screening and surveillance of abdominal aortic aneurysms. Eur J Vasc Endovasc Surg 20: 369–373. https://doi.org/10.1053/ejvs.2000.1191.

Loddenkemper T, Sharma P, Katzan I et al. (2007). Risk factors for early visual deterioration in temporal arteritis. J Neurol Neurosurg Psychiatry 78: 1255–1259. https://doi.org/10.1136/jnnp.2006.113787.

MacSweeney ST, Young G, Greenhalgh RM et al. (1992). Mechanical properties of the aneurysmal aorta. Br J Surg 79: 1281–1284. https://doi.org/10.1002/bjs.1800791211.

MacSweeney ST, Ellis M, Worrell PC et al. (1994). Smoking and growth rate of small abdominal aortic aneurysms. Lancet 344: 651–652. https://doi.org/10.1016/s0140-6736(94)92087-7.

Maleszewski JJ (2015). Inflammatory ascending aortic disease: perspectives from pathology. J Thorac Cardiovasc Surg 149: S176–S183. https://doi.org/10.1016/j.jtcvs.2014.07.046.

Martirosyan NL, Feuerstein JS, Theodore N et al. (2011). Blood supply and vascular reactivity of the spinal cord under normal and pathological conditions. J Neurosurg Spine 15: 238–251. https://doi.org/10.3171/2011.4.spine10543.

Matsumura K, Hirano T, Takeda K et al. (1991). Incidence of aneurysms in Takayasu's arteritis. Angiology 42: 308–315. https://doi.org/10.1177/000331979104200408.

Matteucci ML, Rescigno G, Capestro F et al. (2012). Aortic arch patch aortoplasty for Ortner's syndrome in the age of endovascular stented grafts. Tex Heart Inst J 39: 401–404.

Mehta RH, O'Gara PT, Bossone E et al. (2002). Acute type A aortic dissection in the elderly: clinical characteristics, management, and outcomes in the current era. J Am Coll Cardiol 40: 685–692. https://doi.org/10.1016/s0735-1097(02)02005-3.

Melissano G, Bertoglio L, Civelli V et al. (2009). Demonstration of the Adamkiewicz artery by multidetector computed tomography angiography analysed with the open-source software OsiriX. Eur J Vasc Endovasc Surg 37: 395–400. https://doi.org/10.1016/j.ejvs.2008.12.022.

Meszaros I, Morocz J, Szlavi J et al. (2000). Epidemiology and clinicopathology of aortic dissection. Chest 117: 1271–1278. https://doi.org/10.1378/chest.117.5.1271.

Mokri B, Eric Ahlskog J, Fulgham JR et al. (2004). Syndrome resembling PSP after surgical repair of ascending aorta dissection or aneurysm. Neurology 62: 971–973. https://doi.org/10.1212/01.Wnl.0000115170.40838.9b. https://n.neurology.org/content/neurology/62/6/971.full.pdf.

Mondon K, Blechet C, Gochard A et al. (2007). Transient global amnesia caused by painless aortic dissection. Emerg Med J 24: 63–64. https://doi.org/10.1136/emj.2006.040881.

Moradi A, Salek S, Daniel E et al. (2015). Clinical features and incidence rates of ocular complications in patients with ocular syphilis. Am J Ophthalmol 159: 334–343.e1. https://doi.org/10.1016/j.ajo.2014.10.030.

Mordasini P, El-Koussy M, Schmidli J et al. (2012). Preoperative mapping of arterial spinal supply using 3.0-T MR angiography with an intravasal contrast medium and high-spatial-resolution steady-state. Eur J Radiol 81: 979–984. https://doi.org/10.1016/j.ejrad.2011.02.025.

Mwipatayi BP, Jeffery PC, Beningfield SJ et al. (2005). Takayasu arteritis: clinical features and management: report of 272 cases. ANZ J Surg 75: 110–117. https://doi.org/10.1111/j.1445-2197.2005.03312.x.

Nadkarni NA, Yousef SR, Jagiasi KA et al. (2009). Aortic aneurysm presenting as conus-cauda syndrome. Neurol India 57: 519–520. https://doi.org/10.4103/0028-3886.55580.

Nesher G, Berkun Y, Mates M et al. (2004). Risk factors for cranial ischemic complications in giant cell arteritis. Medicine (Baltimore) 83: 114–122. https://doi.org/10.1097/01.md.0000119761.27564.c9.

Nienaber CA, Clough RE (2015). Management of acute aortic dissection. Lancet 385: 800–811. https://doi.org/10.1016/s0140-6736(14)61005-9.

Nienaber CA, Fattori R, Mehta RH et al. (2004). Gender-related differences in acute aortic dissection. Circulation 109: 3014–3021. https://doi.org/10.1161/01.Cir.0000130644.78677.2c.

Nijenhuis RJ, Jacobs MJ, Jaspers K et al. (2007). Comparison of magnetic resonance with computed tomography angiography for preoperative localization of the Adamkiewicz artery in thoracoabdominal aortic aneurysm patients.

J Vasc Surg 45: 677–685. https://doi.org/10.1016/j.jvs.2006.11.046.

Nogues M, Starkstein S, Berthier M et al. (1987). Cauda equina claudication and abdominal aorta aneurysm. Medicina (B Aires) 47: 331–332.

Nuenninghoff DM, Hunder GG, Christianson TJ et al. (2003). Mortality of large-artery complication (aortic aneurysm, aortic dissection, and/or large-artery stenosis) in patients with giant cell arteritis: a population-based study over 50 years. Arthritis Rheum 48: 3532–3537. https://doi.org/10.1002/art.11480.

Osler W (1905). Aneurysm of the abdominal aorta. Lancet 2: 1089.

Ozen S, Duzova A, Bakkaloglu A et al. (2007). Takayasu arteritis in children: preliminary experience with cyclophosphamide induction and corticosteroids followed by methotrexate. J Pediatr 150: 72–76. https://doi.org/10.1016/j.jpeds.2006.10.059.

Patel NM, Noel CR, Weiner BK (2002). Aortic dissection presenting as an acute cauda equina syndrome: a case report. J Bone Joint Surg Am 84: 1430–1432. https://doi.org/10.2106/00004623-200208000-00019.

Piazza M, Squizzato F, Milan L et al. (2019). Incidence and predictors of neurological complications following thoracic endovascular aneurysm repair in the global registry for endovascular aortic treatment. Eur J Vasc Endovasc Surg 58 (4): 512–519. https://doi.org/10.1016/j.ejvs.2019.05.011.

Pleumeekers HJ, Hoes AW, van der Does E et al. (1995). Aneurysms of the abdominal aorta in older adults. The Rotterdam study. Am J Epidemiol 142: 1291–1299. https://doi.org/10.1093/oxfordjournals.aje.a117596.

Powers WJ, Rabinstein AA, Ackerson T et al. (2018). 2018 guidelines for the early management of patients with acute ischemic stroke: a guideline for healthcare professionals from the American Heart Association/American Stroke Association. Stroke 49: e46–e110. https://doi.org/10.1161/str.0000000000000158.

Reddy DJ, Shepard AD, Evans JR et al. (1991). Management of infected aortoiliac aneurysms. Arch Surg 126: 873–878. discussion 878–879. https://doi.org/10.1001/archsurg.1991.01410310083012.

Regalado ES, Guo DC, Villamizar C et al. (2011). Exome sequencing identifies SMAD3 mutations as a cause of familial thoracic aortic aneurysm and dissection with intracranial and other arterial aneurysms. Circ Res 109: 680–686. https://doi.org/10.1161/circresaha.111.248161.

Roberts CS, Roberts WC (1991). Dissection of the aorta associated with congenital malformation of the aortic valve. J Am Coll Cardiol 17: 712–716. https://doi.org/10.1016/s0735-1097(10)80188-3.

Roberts WC, Ko JM, Vowels TJ (2009). Natural history of syphilitic aortitis. Am J Cardiol 104: 1578–1587. https://doi.org/10.1016/j.amjcard.2009.07.031.

Ruddy JM, Jones JA, Spinale FG et al. (2008). Regional heterogeneity within the aorta: relevance to aneurysm disease. J Thorac Cardiovasc Surg 136: 1123–1130. https://doi.org/10.1016/j.jtcvs.2008.06.027.

Russo C, Jin Z, Rundek T et al. (2009). Atherosclerotic disease of the proximal aorta and the risk of vascular events in a population-based cohort: the Aortic Plaques and Risk of Ischemic Stroke (APRIS) study. Stroke 40: 2313–2318. https://doi.org/10.1161/strokeaha.109.548313.

Saber H, Silver B, Santillan A et al. (2016). Role of emergent chest radiography in evaluation of hyperacute stroke. Neurology 87: 782–785. https://doi.org/10.1212/wnl.0000000000002964.

Sakalihasan N, Heyeres A, Nusgens BV et al. (1993). Modifications of the extracellular matrix of aneurysmal abdominal aortas as a function of their size. Eur J Vasc Surg 7: 633–637.

Sakalihasan N, Limet R, Defawe OD (2005). Abdominal aortic aneurysm. Lancet 365: 1577–1589. https://doi.org/10.1016/s0140-6736(05)66459-8.

Salvarani C, Hunder GG (2001). Giant cell arteritis with low erythrocyte sedimentation rate: frequency of occurence in a population-based study. Arthritis Rheum 45: 140–145. https://doi.org/10.1002/1529-0131(200104)45:2<140::Aid-anr166>3.0.Co;2-2.

Salvarani C, Cantini F, Boiardi L et al. (2002). Polymyalgia rheumatica and giant-cell arteritis. N Engl J Med 347: 261–271. https://doi.org/10.1056/NEJMra011913.

Salvarani C, Crowson CS, O'Fallon WM et al. (2004). Reappraisal of the epidemiology of giant cell arteritis in Olmsted County, Minnesota, over a fifty-year period. Arthritis Rheum 51: 264–268. https://doi.org/10.1002/art.20227.

Saver JL (2016). Cryptogenic stroke. N Engl J Med 375: e26. https://doi.org/10.1056/NEJMc1609156.

Schmittling ZC, LeMaire SA, Köksoy C et al. (2000). Risk factors associated with stroke during thoracoabdominal aortic aneurysm repair. Ann Thorac Surg 70: 1792. https://doi.org/10.1016/S0003-4975(00)02098-1. https://doi.org/10.1016/S0003-4975(00)02098-1.

Seyahi E (2017). Takayasu arteritis: an update. Curr Opin Rheumatol 29: 51–56. https://doi.org/10.1097/bor.0000000000000343.

Sharma S, Kumar S (2010). Hematomyelia due to anterior spinal artery aneurysm in a patient with coarctation of aorta. Neurol India 58: 675–676. https://doi.org/10.4103/0028-3886.68697.

Shetty AK, Stopa AR, Gedalia A (1998). Low-dose methotrexate as a steroid-sparing agent in a child with Takayasu's arteritis. Clin Exp Rheumatol 16: 335–336.

Singh K, Bonaa KH, Jacobsen BK et al. (2001). Prevalence of and risk factors for abdominal aortic aneurysms in a population-based study: the Tromso study. Am J Epidemiol 154: 236–244. https://doi.org/10.1093/aje/154.3.236.

Spittell PC, Spittell Jr JA, Joyce JW et al. (1993). Clinical features and differential diagnosis of aortic dissection: experience with 236 cases (1980 through 1990). Mayo Clin Proc 68: 642–651. https://doi.org/10.1016/s0025-6196(12)60599-0.

Stone JH, Tuckwell K, Dimonaco S et al. (2017). Trial of tocilizumab in giant-cell arteritis. N Engl J Med 377: 317–328. https://doi.org/10.1056/NEJMoa1613849.

Stoob K, Alkadhi H, Lachat M et al. (2004). Resolution of hoarseness after endovascular repair of thoracic aortic aneurysm: a case of Ortner's syndrome. Ann Otol Rhinol Laryngol 113: 43–45. https://doi.org/10.1177/000348940411300109.

Sueyoshi E, Sakamoto I, Hayashi K (2000). Aortic aneurysms in patients with Takayasu's arteritis: CT evaluation. Am J Roentgenol 175: 1727–1733. https://doi.org/10.2214/ajr.175.6.1751727.

Sugiura J, Oshima H, Abe T et al. (2017). The efficacy and risk of cerebrospinal fluid drainage for thoracoabdominal aortic aneurysm repair: a retrospective observational comparison between drainage and non-drainage. Interact Cardiovasc Thorac Surg 24: 609–614. https://doi.org/10.1093/icvts/ivw436.

Svensson LG, Crawford ES, Hess KR et al. (1993). Experience with 1509 patients undergoing thoracoabdominal aortic operations. J Vasc Surg 17: 357–368. discussion 368–370.

Takagi H, Ota H, Natsuaki Y et al. (2015). Identifying the Adamkiewicz artery using 3-T time-resolved magnetic resonance angiography: its role in addition to multidetector computed tomography angiography. Jpn J Radiol 33: 749–756. https://doi.org/10.1007/s11604-015-0490-6.

Tsai YD, Hsu CW, Hsu CC et al. (2016). Paraplegia caused by aortic coarctation complicated with spinal epidural hemorrhage. Am J Emerg Med 34: 680.e1–2. https://doi.org/10.1016/j.ajem.2015.06.057.

Tsutsumi K, Nagata K, Terashi H et al. (1998). A case of aortic coarctation presenting with Brown-Sequard syndrome due to radicular artery aneurysm. Rinsho Shinkeigaku 38: 625–630.

Webb G, Smallhorn J, Therrian J et al. (2019). Congenital heart disease in the adult and pediatric patient. In: DL Mann, DP Zipes, P Libby, RO Bonow, fouding editor and online editor Eugene Braunwald (Eds.), Braunwald's heart disease: a textbook of cardiovascular medicine, 11 ed. Elsevier/Saunders, Philadelphia, PA.

Weidauer S, Nichtweiss M, Hattingen E et al. (2015). Spinal cord ischemia: aetiology, clinical syndromes and imaging features. Neuroradiology 57: 241–257. https://doi.org/10.1007/s00234-014-1464-6.

Wilberger Jr JE (1983). Lumbosacral radiculopathy secondary to abdominal aortic aneurysms. Report of three cases. J Neurosurg 58: 965–967. https://doi.org/10.3171/jns.1983.58.6.0965.

Zalewski NL, Rabinstein AA, Krecke KN et al. (2018). Spinal cord infarction: clinical and imaging insights from the periprocedural setting. J Neurol Sci 388: 162–167. https://doi.org/10.1016/j.jns.2018.03.029.

Zalewski NL, Rabinstein AA, Krecke KN et al. (2019). Characteristics of spontaneous spinal cord infarction and proposed diagnostic criteria. JAMA Neurol 76: 56–63. https://doi.org/10.1001/jamaneurol.2018.2734.

Handbook of Clinical Neurology, Vol. 177 (3rd series)
Heart and Neurologic Disease
J. Biller, Editor
https://doi.org/10.1016/B978-0-12-819814-8.00033-0

Chapter 22

Cerebral aneurysms and cervical artery dissection: Neurological complications and genetic associations

ANDREW M. SOUTHERLAND*, ILANA E. GREEN, AND BRADFORD B. WORRALL

Departments of Neurology and Public Health Sciences, University of Virginia, Charlottesville, VA, United States

Abstract

Dissections and aneurysms are two of the more common nonatherosclerotic arteriopathies of the cerebrovascular system and a significant contributor to neurovascular complications, particularly in the young. Specifically, ruptured intracranial aneurysms (IA) account for nearly 500,000 cases of subarachnoid hemorrhage annually with a 30-day mortality approaching 40% and survivors suffering often permanent neurologic deficits and disability. Unruptured IAs require dedicated assessment of risk and often warrant serial radiologic monitoring. Cervical artery dissection, affecting the carotid and vertebral arteries, accounts for nearly 20% of strokes in young and middle-aged adults. While approximately 70% of cervical artery dissection (CeAD) cases present with stroke or TIA, additional neurologic complications include severe headache and neck pain, oculosympathetic defect (i.e., partial Horner's syndrome), acute vestibular syndrome, and rarely lower cranial nerve palsies. Both aneurysms and dissections of the cerebrovascular system may occur frequently in patients with syndromic connective tissue disorders; however, the majority of cases are spontaneously occurring or mildly heritable with both polygenic and environmental associations. Fibromuscular dysplasia, in particular, is commonly associated with both risk of CeAD and IA formation. Further research is needed to better understand the pathophysiology of both IA and CeAD to better understand risk, improve treatments, and prevent devastating neurologic complications.

INTRODUCTION

Among the nonatherosclerotic vasculopathies, intracranial aneurysms (IA) and cervical artery dissection (CeAD) are among the more common in the cerebrovascular system. Although numerous monogenic connective tissue disorders have a heightened risk for IA and CeAD, the vast majority occur in patients without a genetic syndrome. Moreover, the association with several related conditions, such as fibromuscular dysplasia, and the identification of single nucleotide polymorphisms (SNPs) associated with IA and CeAD suggest a polygenic association in many cases. Familial associations further support an underlying genetic association in the pathogenesis of IA in particular. In this chapter, we will review the genetic associations and neurologic complications of IA and CeAD.

NEUROLOGIC COMPLICATIONS OF INTRACRANIAL ANEURYSMS

Saccular intracranial aneurysms, involving all three layers of the vessel wall, are the most prevalent cerebral large-vessel, nonatherosclerotic arteriopathy (Southerland et al., 2013). These aneurysms, also known as berry aneurysms, commonly occur at arterial bifurcations and occur most frequently in the anterior circulation.

In the general population, the estimated prevalence of unruptured intracranial (UIA) aneurysm without comorbidity is 3.2%, with a mean age of 50 years at diagnosis, and an approximately equal distribution in women and men. In patients with autosomal dominant polycystic kidney disease or in those with a positive family history of ruptured intracranial aneurysm, the prevalence of unruptured,

*Correspondence to: Andrew M. Southerland, MD, MSc, FAHA, Department of Neurology, University of Virginia, Charlottesville VA 22908, United States. Tel: +1-434-243-9988, Fax: +1-434-982-1726, E-mail: as5ef@hscmail.mcc.virginia.edu

asymptomatic intracranial aneurysm approaches 7% (Vlak et al., 2011). Nonmodifiable risk factors include age, sex, ethnicity and genomics potentially influence growth and rupture of aneurysm (Muehlschlegel, 2018). Most intracranial aneurysms are asymptomatic until they rupture. Current AHA/ASA guidelines suggest screening for intracranial aneurysms in any patient with two or more first degree relatives with a history of aneurysmal subarachnoid hemorrhage (SAH) (Thompson et al., 2015).

Cerebrovascular complications of IA

The most serious complication of IA is rupture leading to SAH. The 30-day mortality rate for SAH is 35% and survivors often suffer permanent cognitive and mental health disabilities (Muehlschlegel, 2018). The 1- and 5-year risk of rupture of UIAs are approximately 1.5% and 3.5%, respectively (Greving et al., 2014). The average age of patients at time of aneurysmal rupture is 53 years. Modifiable risk factors for SAH include hypertension, smoking, heavy alcohol use, and sympathomimetic drug use (e.g., cocaine) (Muehlschlegel, 2018). At least 80% of patients presenting with acute SAH who are able to give a history describe an extremely sudden "worst headache of my life" or "thunderclap headache" which immediately reaches maximum intensity (Bassi et al., 1991; Connolly et al., 2012). In 10%–43% of patients, a sentinel headache or bleed precedes the definitive aneurysm rupture, which most often occurs 2–8 weeks prior (Polmear, 2003; Connolly et al., 2012). Other common signs and symptoms of rupture attributable to increased intracranial pressure and meningeal irritation include nausea or vomiting, loss of consciousness, and nuchal rigidity.

In SAH, noncontrast computed tomography (CT) scan commonly reveals hyperdense material in the subarachnoid space especially around the circle of Willis or Sylvian fissure. The sensitivity of CT in detecting SAH within the first 3 days approaches 100% (Dupont et al., 2010). Traditionally, if CT is negative and clinical suspicion for SAH remains high, or after 5–7 days of suspected SAH, lumbar puncture revealing xanthochromia can confirm the diagnosis. More recently, with advances in magnetic resonance imaging (MRI) and the development of new imaging sequences, use of MRI may obviate the need for lumbar puncture in patients with negative CT findings (Inoue et al., 2013; Verma et al., 2013; Hodel et al., 2015).

Cerebrovascular imaging to identify location of suspected ruptured aneurysm is important for interventional planning and most commonly performed via computed tomography arteriography (CTA). Failure to identify an aneurysm on noninvasive imaging generally warrants traditional digital subtraction cerebral angiography given the superior spatial contrast resolution. Many protocols also include repeat imaging after an interval of time given the potential for the acute blood to obscure small aneurysms.

Management per International Subarachnoid Aneurysm Trial comparing endovascular coiling to surgical clipping in the acute setting showed significantly higher odds of disability-free survival at 1 year, lower risk of long-term epilepsy, and overall better neurologic outcomes at 10 years for endovascular-treated patients (Molyneux et al., 2005). Therefore, surgical clipping is generally reserved for situations where the vascular anatomy is not amenable to coiling or as a rescue treatment when coiling procedures are unsuccessful. More recent advances in flow diversion devices have provided another treatment option in complicated cases, such as wide neck or fusiform aneurysms (Briganti et al., 2015).

After initial SAH, patients are at heightened risk of morbidity and mortality from rebleeding, systemic inflammatory response syndrome (SIRS), and cerebral vasospasm. Patients who have not received interventional therapy have a 4%–13% risk of rebleeding within the first 24h after initial rupture, while patients who received intervention have a 0.9%–2.9% risk (Connolly et al., 2012). In the acute period after SAH, 75% of patients suffer from SIRS related to elevated to elevated levels of inflammatory cytokines. SIRS may also be associated with post-SAH seizures (Claassen et al., 2014).

One of the more concerning neurologic sequelae of SAH is delayed cerebral ischemia (DCI) that results from a complex response to the hemorrhage including inflammation, microthrombosis, cerebral vasospasm, microcirculatory dysfunction, and cortical spreading depolarization. DCI occurs in 20%–30% of hospitalized patients with SAH (Rouanet and Silva, 2019; Hurth et al., 2020). Cerebral vasospasm manifests in 20%–50% of patients with SAH (Bauer and Rasmussen, 2014). DCI and cerebral vasospasm are interrelated but distinct complications of SAH usually concur, but not always (Woitzik et al., 2012). Stroke secondary to cerebral vasospasm accounts for approximately 50% of early deaths in patients who survived initial SAH. Standard of care for prevention of post-SAH vasospasm is oral calcium channel blockers, commonly nimodipine.[3] A large randomized controlled trial performed in 1989 demonstrated a 34% reduction in ischemic stroke and a 40% reduction of poor outcome in patients treated with nimodipine compared to placebo (Pickard et al., 1989). Management of cerebral vasospasm remains somewhat variable but traditionally involved HHH therapy (i.e., hypertension, hypervolemia, and hemodilution) or acute treatment with endovascular therapy such as intraarterial vasodilator administration (e.g., verapamil) or balloon angioplasty (Bauer and Rasmussen, 2014). Management decisions are often based on provider preference and characteristics of each individual vasospastic presentation. DCI and neurologic outcomes may be influenced by genetic factors (Heinsberg et al., 2020).

Noncerebrovascular complications of IA

UIAs may also present with structural neurologic complications such as headaches, seizures, visual deficits, or cranial neuropathies (Wiebers et al., 2003). These structural neurologic symptoms are often indications of a large or growing UIA, which often warrants intervention. However, management of small, asymptomatic UIA remains controversial due to an absence of high-quality prospective randomized control trial data. The International Study of Unruptured Intracranial Aneurysms suggested that aneurysms <7 mm in the anterior circulation had a 0% chance of rupture at 5 years (Wiebers et al., 2003). However, this data has been countered by findings that most ruptured IAs are <7 mm, although this is still likely a small percentage of all small UIAs.

Management

Regarding intervention, the CURES (Canadian Unruptured Endovascular Versus Surgery) trial addressing this issue is currently ongoing. An early analysis performed for slow accrual showed that new neurologic deficits and hospitalizations beyond 5 days after procedure were more common in the surgical clipping group compared to the endovascular coiling group, but there was no difference in overall morbidity or mortality at 1 year between groups (Darsaut et al., 2017).

At this time, individualized management decisions of UIAs should be based on current understanding of risk stratification variables such as presence of hypertension, sex, age, family history, and aneurysm characteristics. Using the Delphi method, Etminan et al. developed a multidisciplinary consensus to aid treatment decisions based on an unruptured intracranial aneurysm treatment score including the aforementioned risk factors (Etminan et al., 2015). Future research is needed to identify additional biomarkers or characteristics to further risk stratify small UIAs.

NEUROLOGIC COMPLICATIONS OF CERVICAL ARTERY DISSECTION

Cervical artery dissection (CeAD) is a common cause of stroke in young adults, with a mean age in the fourth to fifth decade (Debette and Leys, 2009). The estimated incidence of CeAD in the general population is 2–3 per 100,000, though the detection rate may be increasing with advances in the availability and quality of noninvasive vascular imaging (Giroud et al., 1994). Risk for CeAD is also higher in people with predisposing connective tissue vasculopathies, such as Ehlers-Danlos syndrome and fibromuscular dysplasia (Lee et al., 2006). CeAD is defined by the presence of a mural hematoma in the wall of a cervical (carotid or vertebral) artery resulting from either an intimal tear or direct bleeding within the wall secondary to a ruptured vasa vasorum (Debette and Leys, 2009). This intramural hematoma often results in tapered stenosis of the mid-cervical portion of the carotid artery (Fig. 22.1) or most commonly in the V3 segment of the vertebral artery.

CeAD can be categorized into cases associated with major neck trauma (e.g., motor vehicle accidents, cervical spine trauma), accounting for approximately 4% of cases, to those that are more spontaneous, or associated with lesser amounts of cervical exertion (e.g., coughing, retching, noncontact sports), accounting for the majority of cases (Galyfos et al., 2016). The pathophysiology of spontaneous CeAD remains poorly understood. Theoretically, spontaneous CeAD likely reflects a systemic vasculopathy contributing to underlying weakness of the vessel wall and in the setting of agonistic environmental factors, such as acute infection or minor trauma, results in arterial dissection.

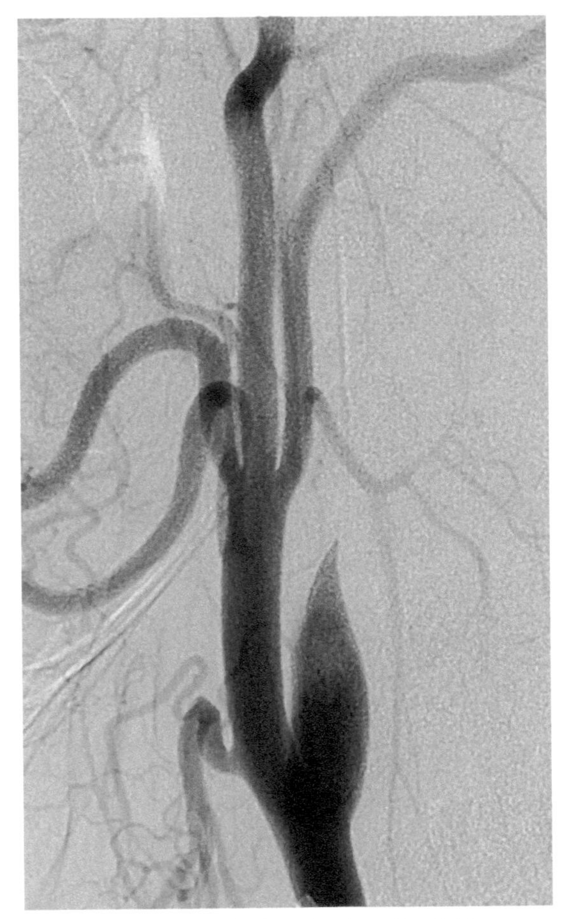

Fig. 22.1. Smooth tapered "flame" or "candle wick" appearance of carotid artery dissection on digital subtraction angiography.

Cerebrovascular complications of CeAD

Approximately 70% of CeAD results in stroke or TIA that may be delayed by days or weeks after arterial dissection. Ischemic stroke results from embolization of intraluminal thrombi that have developed at the site of intimal tear. Rarely, a dissection extending into the intracranial part of the artery can lead to subarachnoid hemorrhage (Debette and Leys, 2009). Acute ischemic stroke suspected to be secondary to CeAD should be managed like any other acute ischemic stroke, considering the appropriateness of tissue plasminogen activator (tPA) and/or mechanical thrombectomy. The CADISS study (Cervical Artery Dissection in Stroke Study), comparing antiplatelet vs anticoagulation after CeAD and stroke recurrence, resulted in a class IIa, level B recommendation equating antiplatelet and anticoagulation for 3–6 months after CeAD (Kennedy et al., 2012; Markus et al., 2015). Endovascular therapy for CeAD such as stenting is generally reserved for patients with recalcitrant stenosis or pseudoaneurysm who fail antithrombotic treatment and have contraindications against anticoagulation (Peng et al., 2017). In most cases, the dissected artery will heal on its own over weeks to months, correlating with an estimated long term risk of recurrence approaching that of the general population, from 1% to 3% (Schievink et al., 1994; Kennedy et al., 2012; Debette, 2014; Markus et al., 2015; Compter et al., 2018).

About one-third of patients with CeAD develop a pseudoaneurysm or dissecting aneurysm (DA), also referred to as a false or pseudoaneurysm (Larsson et al., 2017). A DA is any type of resulting aneurysm that does not include all three vessel wall layers. The presence of a DA can often ignite concern for increased risk of subsequent stroke by serving as a source of embolization or potential for expansion and compressive symptoms; however, this is currently unfounded. Subgroup analysis of patients with DA compared to those without DA in the aforementioned CADISS study showed a benign prognosis in both groups and suggests no need for interventional management in the majority of cases (Larsson et al., 2017).

Noncerebrovascular complications of CeAD

Noncerebrovascular neurologic complications of CeAD include local signs and symptoms referable to the arterial dissection, such as headache/neck pain, pulsatile tinnitus, oculosympathetic defect (i.e., Horner's syndrome, ipsilateral ptosis, and miosis), and less commonly, lower cranial nerve palsies. Both carotid and vertebral dissections often cause significant ipsilateral face or neck pain and a nonspecific headache or can produce refractory migraine or cluster headache (Debette and Leys, 2009). A Horner's syndrome can develop in the setting of a carotid artery dissection due to disruption of ascending third-order oculosympathetic tracts along the internal carotid artery. Rarely, dysarthria and dysphagia can result from ischemia or injury to the glossopharyngeal or vagus nerves.

MONOGENIC SYNDROMES ASSOCIATED WITH ANEURYSMS AND DISSECTION

There are numerous monogenic syndromes that compromise the structural elements of the arterial wall, comprised of the intima, media, and adventitia, leading to arteriopathy and predisposition to aneurysms and dissections in a variety of arterial beds (Fig. 22.2). The most well known of these syndromes include Ehlers-Danlos Syndrome (type IV), Marfan Syndrome, and Adult Polycystic Kidney Disease, with 20+ more monogenic syndromes having been identified (Table 22.1). Sequelae of cervico-cephalic arteriopathy can result in serious neurologic complications including ischemic stroke, intraparenchymal hemorrhage, and subarachnoid hemorrhage (Southerland et al., 2013).

Neuromuscular manifestations of connective tissue disease

In addition to the cerebrovascular complications of connective tissue disease, these monogenic syndromes can be associated with neurological manifestations as well. Similar to the connective tissue instability of the arterial

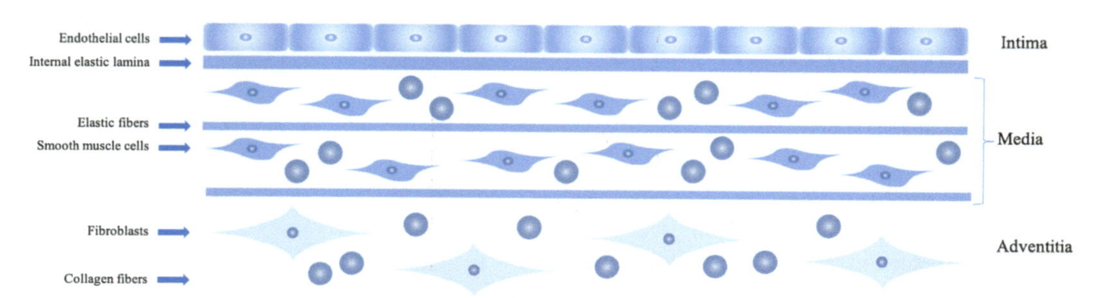

Fig. 22.2. Ultrastructure of arterial wall. Adapted from Southerland AM, Meschia JF, Worrall BB (2013). Shared associations of nonatherosclerotic, large-vessel, cerebrovascular arteriopathies: considering intracranial aneurysms, cervical artery dissection, moyamoya disease, and fibromuscular dysplasia. Curr Opin Neurol 26: 13–28.

Table 22.1

Monogenic syndromes with overlapping cerebrocervical and extra-cerebral arteriopathies

Name (abbreviation; OMIM listing)	Gene/locus	Inheritance	Cerebrocervical arteriopathy	Extracerebral arteriopathy	Other features
Vascular Ehlers-Danlos Type IV (vEDS; 130050)	COL3A1	AD	IA, CeAD	TAA, AAA, aortic dissection	Joint and dermal manifestations, prone to spontaneous rupture of bowel and large arteries
Marfan syndrome (MFS; 154700)	FBN1AD	AD	IA, CeAD	TAA, AAA, aortic dissection	Hallmark skeletal, ocular, and cardiovascular features. Arachnodactyly and subluxation of the lenses
Arterial tortuosity syndrome (ATS; 208050)	SLC2A10	AR	IA, CeAD	TAA, AAA	Generalized tortuosity and elongation of all major arteries, soft skin, joint laxity, severe keratoconus
Adult polycystic kidney disease (PKD1 173900)	PKD1	AD	IA, CeAD	TAA, AAA	Renal cysts, liver cysts
Adult polycystic kidney disease (AKD3; 600666)	Unknown	AD	IA	Unknown	Renal cysts, liver cysts
Loeys-Dietz syndrome type 1A (LDS1A; 608967)	TGFBR1	AD	IA, CeAD	TAA, AAA, aortic dissection, other vessel dissection	Triad of arterial tortuosity and aneurysms, hypertelorism, bifid uvula/cleft palate; pregnancy complications
Loeys-Dietz syndrome type 1B (LDS1B; 610380)	TGFBR2	AD	IA, CeAD	TAA, AAA, aortic dissection, other vessel dissection	Indistinguishable from LDS1A
Loeys-Dietz syndrome type 2A (LDS2A; 610380)	TGFBR1	AD	IA, CeAD	TAA, AAA, aortic dissection, other vessel dissection	Phenotypically similar to vEDS, bifid uvula is usually only craniofacial feature
Loeys-Dietz syndrome type 2B (LDS2B; 610380)	TGFBR2	AD	IA, CeAD	TAA, AAA, aortic dissection, other vessel dissection	Indistinguishable from LDS2A
Loeys-Dietz syndrome type 3 (LDS3; 613795)	SMAD3	AD	IA, CeAD	TAA, AAA, aortic dissection, other vessel dissection	Previously known as aneurysm–osteoarthritis syndrome; congenital heart disease
Loeys-Dietz syndrome type 4 (LDS4; 614816)	TGFB2	AD	IA, CeAD	TAA, AAA	Skeletal manifestations, bicuspid aortic valve, arterial tortuosity, arachnodactyly, scoliosis, club feet and thin skin with easy bruising and striae
Osteogenesis imperfecta type 1 (OI1; 166200)	COL1A1	AD	CeAD, FMD	TAA, AAA	Multiple bone fractures, hearing loss, blue sclera
Alpha-1 antitrypsin deficiency (613490)	SERPINA1	AR	FMD	None	Emphysema, liver disease
Pseudoxanthoma elasticum (PXE; 264800)	ABCC6; polymoprhisms in XYLT1 and XYLT2 modify severity	AR, pseudodominant	IA, CeAD	AAA	Mineralized and fragmented elastic fibers in the skin, vascular walls, and Burch membrane in the eye

Continued

Table 22.1

Continued

Name (abbreviation; OMIM listing)	Gene/locus	Inheritance	Cerebrocervical arteriopathy	Extracerebral arteriopathy	Other features
Microcephalic osteodysplastic primordial dwarfism type II (MOPD2; 210720)	PCNT, pericentrin, 21q22	AR or compound heterozygous	IA, moyamoya	TAA, AAA	Postnatal dwarfism with microcephaly and dysmorphia
Neuro-fibromatosis type 1 (NF1; 162200)	NF1; 17q11.2	AD	IA, moyamoya	Coarctation of thoracic and abdominal aorta, venous and arterial aneurysms	Aortic aneurysms, moyamoya
Grange syndrome (602531)	Unknown	Unclear	IA, moyamoya	TAA, AAA, venous and arterial aneurysms	Stenosis or occlusion renal, abdominal, and cerebral arteries. Cerebral aneurysms, congenital heart defects, brachydactyly, syndactyly, bone fragility, and learning disabilities
Hereditary angiopathy with nephropathy, aneurysms, and muscle cramps (HANAC; 611773)	COL4A1	AD	OA	TAA, AAA, arterial aneurysms	Associated with a small vessel arteriopathy and risk of ICH
Alport syndrome X-linked (ATS; 301050)	COL4A5	AD	CeAD, FMD, moyamoya	TAA, AAA	Progressive glomerulonephropathy, variable sensorineural hearing loss, and variable ocular anomalies
Stenosis, aneurysm, moyamoya, and stroke (SAMS)	SAMHD1	AR or compound heterozygous	IA, moyamoya	Aortic aneurysm	Cerebral vasculopathy and early onset stroke
Homocyst(*e*)inuria (236200)	CBS	AR	IA, moyamoya	Aortic dissection, TAA, AAA	Mydriasis, patent ductus arteriosis, hypotonic bladder, malrotation, and hyperperistalsis of the gastrointestinal tract
Cutis Laxa type 1A (ARCL1A; 219100)	FBLN5	AR	IA, FMD	TAA, AAA, aortic dissection	Phenotypically similar to vEDS, multiple diverticula (esophagus, duodenum, ileum, bladder)

Adapted from Southerland AM, Meschia JF, Worrall BB (2013). Shared associations of nonatherosclerotic, large-vessel, cerebrovascular arteriopathies: considering intracranial aneurysms, cervical artery dissection, moyamoya disease, and fibromuscular dysplasia. Curr Opin Neurol 26: 13–28.

OMIM, Online Mendelian Inheritance in Man database (www.ncbi.nlm.nih.gov/omim). *AAA*, abdominal aortic aneurysm; *AD*, autosomal dominant; *AR*, autosomal recessive; *CeAD*, cervical artery dissection; *FMD*, fibromuscular dysplasia; *IA*, intracranial aneurysm; *ICH*, intracerebral hemorrhage; *TAA*, thoracic aortic aneurysm.

wall that predisposes to aneurysm formation and dissection, the connective tissue makeup of the peripheral nerves may also be affected including the epineurium and perineurium. Thus, patients with connective tissue vasculopathy, such as Ehlers-Danlos syndrome, may also develop sensory disturbances and neuropathic symptoms reflecting underlying peripheral neuropathy (Henderson et al., 2017). Musculoskeletal involvement may also contribute to neuromuscular injury from joint instability, including brachial or lumbar plexopathies related to recurrent shoulder or hip dislocation. Instability at the craniocervical junction warrants close monitoring for myelopathic signs and symptoms as well. On the neurological exam, patients often exhibit such signs of laxity in muscle tone and motor dexterity, hyperreflexia due to tendon exposure and/or spinal stenosis, and early vibratory extinction or loss of pin prick in the distal extremities. Diagnostic workup for signs of myelopathy or peripheral neuropathy may warrant cervical spine imaging to rule out atlanto-axial instability, or electromyography and nerve conduction studies to further assess the peripheral nervous system.

Headache disorders associated with connective tissue vasculopathy

In addition to the thunderclap/ictal headache that is a hallmark sign of IA rupture, and aforementioned neck pain and migraine that often results from CeAD, connective tissue vasculopathy may also be associated with more chronic headache syndromes. These may include cervicogenic headaches associated with occipital neuralgia or craniocervical disease, such as cervical arthropathy, spondylosis, or Chiari malformations, all of which have been characterized in monogenic connective tissue disease. Additionally, episodic or chronic migraine are also common comorbidities of connective tissue disease, likely reflective of direct vascular pathophysiology in the migraine pathway and/or an underlying genetic predisposition as mentioned later in the chapter.

Recently, Ehlers-Danlos syndrome has been purported in association with pseudotumor cerebri or idiopathic intracranial hypertension (IIH). While somewhat controversial, this is thought to be related to a predisposition to cerebral venous sinus stenosis, often in the transverse sinuses, contributing outflow obstruction from the venous sinus system. As a result, venous sinus stenting is considered a treatment option for patients with connective tissue vasculopathy to relieve elevated intracranial pressure and headaches associated with IIH. In patients with connective tissue disease and chronic headaches, one should consider IIH in the differential diagnosis and explore it through proper history (e.g., positional component to headaches) and neurological examination to rule out signs of elevated intracranial pressuring including papilledema (e.g., fundoscopic exam) or cranial nerve VI injury (e.g., horizontal dysconjucacy). Diagnostic evaluation to rule out elevated intracranial pressure may also be warranted through lumbar puncture or in some cases cerebral angiography to assess for pressure gradients associated with venous sinus stenosis. Prospective research is needed to verify these pathological associations and determine the true safety and efficacy of venous sinus stenting in this setting.

POLYGENIC ASSOCIATIONS OF ANEURYSMS AND DISSECTION

Fibromuscular dysplasia

Fibromuscular dysplasia (FMD) is a relatively common noninflammatory, nonatherosclerotic disease of small to medium-sized arteries associated with arterial dissection, aneurysm, and tortuosity, often described as a "string of beads" on imaging (Fig. 22.3) (Gornik et al., 2019a,b). The average age of diagnosis is approximately 52 years,

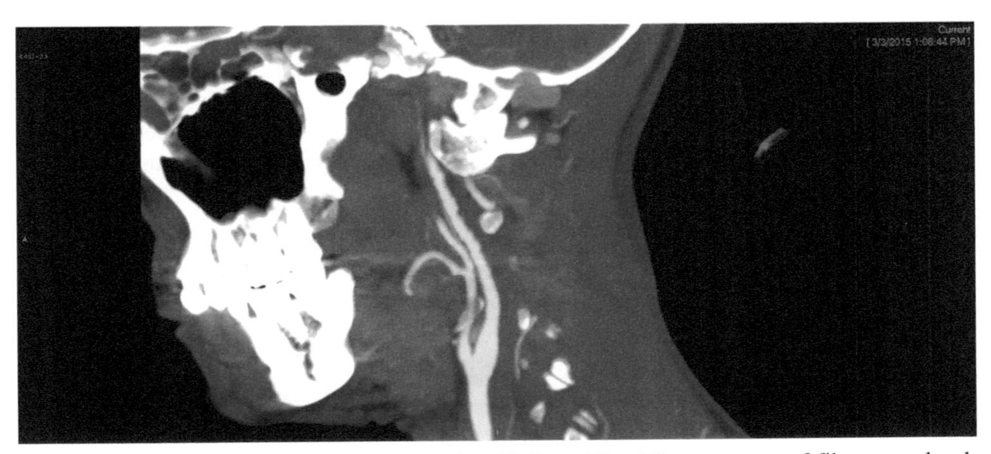

Fig. 22.3. Computed tomography angiography demonstrating "String of Beads" appearance of fibromuscular dysplasia in the internal carotid artery.

and women are more commonly affected than men (Olin et al., 2012). It is difficult to identify the precise prevalence and natural history of FMD because it is often incidentally identified in asymptomatic individuals. The clinical presentation of FMD depends on the affected vascular bed. The two most common sites of FMD are the extracranial carotid and vertebral arteries and the renal arteries (Olin et al., 2012). Diagnosis is made with vascular imaging, most frequently computed tomographic angiography and magnetic resonance angiography (MRA). The prognosis for a patient with FMD varies from patient to patient, generally depending on number and severity of symptoms. Asymptomatic FMD diagnosed incidentally has a favorable natural history. Regardless of symptomatic or radiographic severity at diagnosis, consensus statements suggest a one-time head-to-pelvis screening (CTA or MRA) for aneurysms, dissections, and additional arterial beds affected by FMD (Gornik et al., 2019a,b).

FMD, of any arterial site, is associated with an increased prevalence of intracranial aneurysms, particularly in women (Cloft et al., 1998). Of all women who underwent intracranial imaging in the US FMD registry, 13% had at least one IA, and 4% had multiple IAs. However, the exact prevalence of IA in the setting of FMD is unknown because of sampling bias in registry data. There is some data to suggest that the risk of IA rupture may be elevated at smaller sizes, though this is controversial. Similar to in the general population, cigarette smoking is strongly associated with IA formation and growth in individuals with FMD (Cloft et al., 1998; Kadian-Dodov et al., 2016; Touzé et al., 2019). In observational studies, the coprevalence of CeAD and cerebrocervical FMD ranges from 5% to 15% and is higher in those presenting with multiple CeAD (Southerland et al., 2013; Béjot et al., 2014; Touzé et al., 2019).

Headaches and other neurologic manifestations of FMD

Though cervical artery FMD is often diagnosed incidentally, patients may also present with neurologic symptoms including headaches and pulsatile tinnitus. According to large registries, 70% of FMD patients report headaches (Olin et al., 2012). In a multivariable analysis from the US FMD Registry, FMD patients reporting headaches were more likely to present at a younger age, have a history of CeAD, and report additional symptoms of dizziness and pulsatile tinnitus (Wells et al., 2020). The most predominant headache subtype is migraine, followed to a lesser degree by tension-type headaches. The heightened coprevalence of migraine in patients with FMD remains poorly understood, although likely reflects dysregulation of cerebral blood flow mediated by neurovascular alterations in the vessel wall, and a genetic association between FMD and migraine as noted in the following section.

Management of headache resulting from FMD is largely driven by accurate characterization of headache subtype but may require prophylactic treatment with an increased headache frequency. Vasoconstrictive medications, such as ergots and triptans, should be avoided or used with heightened vigilance given a putative increased risk for stroke or dissection in FMD patients. Vasoconstrictive medications are absolutely contraindicated in FMD patients with history of spontaneous coronary artery dissection (SCAD) (O'connor et al., 2015).

Pulsatile tinnitus or "whooshing" is reported in up to 40% of patients with cervical artery FMD and can be disabling in some instances. Management may be challenging but includes good blood pressure control and potentially sound distraction therapy to aid with sleeping. Other less specific neurologic symptoms of FMD include dizziness or lightheadedness, sensory disturbance or neuropathic pain, blurry vision, and cognitive fogging. These latter symptoms may also reflect fluctuations in blood pressure related to FMD, in which case evaluation and management of secondary hypertension due to renal artery involvement is warranted.

PHACTR1/EDN1, coronary artery dissection

Recently, clinical and genetic association studies have found a link between CeAD, FMD, and spontaneous coronary artery dissection. These entities affect a similar demographic of younger, otherwise healthy adults (Tweet et al., 2012).

Based on the results of genome-wide association studies, CeAD, FMD, SCAD, and even migraine share a common susceptibility locus, an intronic single nucleotide polymorphism (rs9349379) in the *phosphatase and actin regulator 1* (*PHACTR1*) gene. Interestingly, the presence of this risk allele is inversely associated with coronary artery disease and coronary calcification (Table 22.2). PHACTR1 expression was recently found to be different in carotid artery atheroma versus macroscopically normal vessel tissue (Green et al., 2020). There is controversy about whether regulation of PHACTR1 itself plays a role or if it acts through downstream regulation of *endothelin 1* (*EDN1*), a potent vasoconstrictor (Debette et al., 2015; Kiando et al., 2016; Adlam et al., 2019).

CONCLUSION/SUMMARY

IA and CeAD are two of the more common nonatherosclerotic arteriopathies affecting the cerebrovasculature. While the majority of aneurysms and dissections occur

Table 22.2

Related vascular phenotypes associated with PHACTR1 rs9349379

PHACTR1 rs9349379		
Allele	Increased risk	Decreased risk
A	Cervical artery dissection Fibromuscular dysplasia Coronary artery dissection Migraine without aura	Coronary artery disease Coronary calcification
G	Coronary artery disease Coronary calcification	Cervical artery dissection Fibromuscular dysplasia Coronary artery dissection Migraine without aura

in people without a clear genetic syndrome, there are a number of monogenic connective tissues disorders, polygenic associations, and SNPs associated with heightened risk for both entities. Both IA and CeAD may have devastating complications when presenting with SAH and ischemic stroke, respectively. However, these entities are also commonly discovered incidentally or in association with local signs and symptoms, such as headache. Therefore, one must have a thorough understanding of the natural history, risk factors, and clinical manifestations of these vasculopathies to guide proper treatment recommendations and monitoring. Additionally, it is important for clinicians to be aware of the wide range of additional neurologic manifestations associated with these connective tissue vasculopathies, including headache disorders and peripheral neuromuscular disease (Fig. 22.4). Further research is needed to better understand the genetic and environmental associations of IA and CeAD in order to better prevent and treat their cerebrovascular complications.

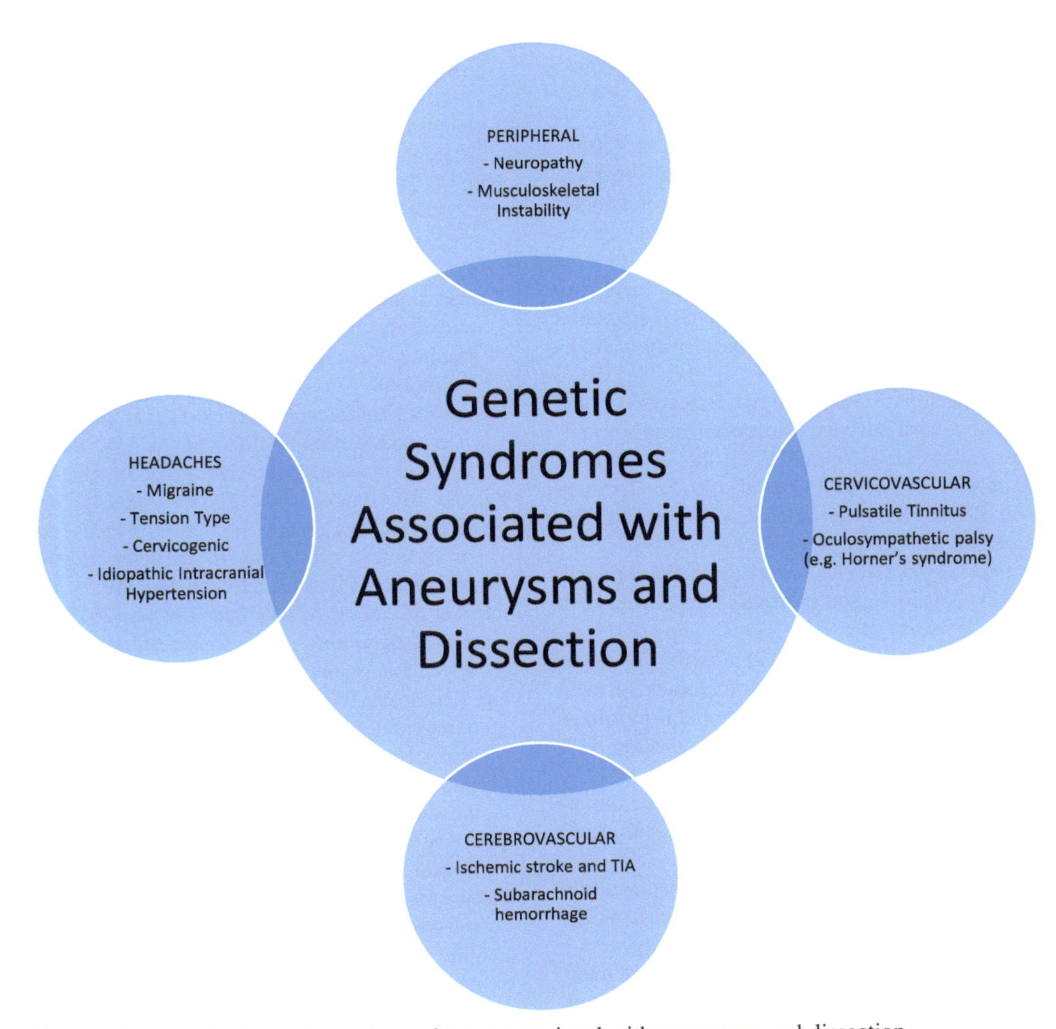

Fig. 22.4. Neurologic complications of genetic syndromes associated with aneurysms and dissection.

References

Adlam D, Olson TM, Combaret N et al. (2019). Association of the PHACTR1/EDN1 genetic locus with spontaneous coronary artery dissection. J Am Coll Cardiol 73: 58–66.

Bassi P, Bandera R, Loiero M et al. (1991). Warning signs in subarachnoid hemorrhage: a cooperative study. Acta Neurol Scand 84: 277–281.

Bauer AM, Rasmussen PA (2014). Treatment of intracranial vasospasm following subarachnoid hemorrhage. Front Neurol 5: 72.

Briganti F, Leone G, Marseglia M et al. (2015). Endovascular treatment of cerebral aneurysms using flow-diverter devices: A systematic review. Neuroradiol J 28: 365–375.

Béjot Y, Aboa-Eboulé C, Debette S et al. (2014). Characteristics and outcomes of patients with multiple cervical artery dissection. Stroke 45: 37–41.

Claassen J, Albers D, Schmidt JM et al. (2014). Nonconvulsive seizures in subarachnoid hemorrhage link inflammation and outcome. Ann Neurol 75: 771–781.

Cloft HJ, Kallmes DF, Kallmes MH et al. (1998). Prevalence of cerebral aneurysms in patients with fibromuscular dysplasia: a reassessment. J Neurosurg 88: 436–440.

Compter A, Schilling S, Vaineau CJ et al. (2018). Determinants and outcome of multiple and early recurrent cervical artery dissections. Neurology 91: e769–e780.

Connolly ES, Rabinstein AA, Carhuapoma JR et al. (2012). Guidelines for the management of aneurysmal subarachnoid hemorrhage: a guideline for healthcare professionals from the American Heart Association/american Stroke Association. Stroke 43: 1711–1737.

Darsaut TE, Findlay JM, Magro E et al. (2017). Surgical clipping or endovascular coiling for unruptured intracranial aneurysms: a pragmatic randomised trial. J Neurol Neurosurg Psychiatry 88: 663–668.

Debette S (2014). Pathophysiology and risk factors of cervical artery dissection: what have we learnt from large hospital-based cohorts? Curr Opin Neurol 27: 20–28.

Debette S, Kamatani Y, Metso TM et al. (2015). Common variation in PHACTR1 is associated with susceptibility to cervical artery dissection. Nat Genet 47: 78–83.

Debette S, Leys D (2009). Cervical-artery dissections: predisposing factors, diagnosis, and outcome. Lancet Neurol 8: 668–678.

Dupont SA, Lanzino G, Wijdicks EF et al. (2010). The use of clinical and routine imaging data to differentiate between aneurysmal and nonaneurysmal subarachnoid hemorrhage prior to angiography. Clinical article. J Neurosurg 113: 790–794.

Etminan n, Brown rD, Beseoglu K et al. (2015). The unruptured intracranial aneurysm treatment score: a multidisciplinary consensus. Neurology 85: 881–889.

Galyfos G, Filis K, Sigala F et al. (2016). Traumatic carotid artery dissection: a different entity without specific guidelines. Vasc Specialist Int 32: 1–5.

Giroud M, Fayolle H, André N et al. (1994). Incidence of internal carotid artery dissection in the community of Dijon. J Neurol Neurosurg Psychiatry 57: 1443.

Gornik HL, Persu A, Adlam D et al. (2019a). First international consensus on the diagnosis and management of fibromuscular dysplasia. Vasc Med 24: 164–189.

Gornik HL, PERSU A, Adlam D et al. (2019b). First international consensus on the diagnosis and management of fibromuscular dysplasia. J Hypertens 37: 229–252.

Green IE, Williams SR, Sale MM et al. (2020). Differential expression of PHACTR1 in atheromatous versus normal carotid artery tissue. J Clin Neurosci 74: 265–267.

Greving JP, Wermer MJ, Brown RD et al. (2014). Development of the PHASES score for prediction of risk of rupture of intracranial aneurysms: a pooled analysis of six prospective cohort studies. Lancet Neurol 13: 59–66.

Heinsberg LW, Arockiaraj AI, Crago EA et al. (2020). Genetic variability and trajectories of DNA methylation may support a role for HAMP in patient outcomes after aneurysmal subarachnoid hemorrhage. Neurocrit Care 32: 550–563.

Henderson FC, Austin C, Benzel E et al. (2017). Neurological and spinal manifestations of the Ehlers-Danlos syndromes. Am J Med Genet C Semin Med Genet 175: 195–211.

Hodel J, Aboukais R, Dutouquet B et al. (2015). Double inversion recovery MR sequence for the detection of subacute subarachnoid hemorrhage. AJNR Am J Neuroradiol 36: 251–258.

Hurth H, Birkenhauer U, Steiner J et al. (2020). Delayed cerebral ischemia in patients with aneurysmal subarachnoid hemorrhage - serum D-dimer and C-reactive protein as early markers. J Stroke Cerebrovasc Dis 29: 104558.

Inoue T, Takada S, Shimizu H et al. (2013). Signal changes on T2*-weighted magnetic resonance imaging from the acute to chronic phases in patients with subarachnoid hemorrhage. Cerebrovasc Dis 36: 421–429.

Kadian-Dodov D, Gornik HL, Gu X et al. (2016). Dissection and aneurysm in patients with fibromuscular dysplasia: findings from the U.S. registry for FMD. J Am Coll Cardiol 68: 176–185.

Kennedy F, Lanfranconi S, Hicks C et al. (2012). Antiplatelets vs anticoagulation for dissection: CADISS nonrandomized arm and meta-analysis. Neurology 79: 686–689.

Kiando SR, Tucker NR, Castro-Vega LJ et al. (2016). PHACTR1 is a genetic susceptibility locus for fibromuscular dysplasia supporting its complex genetic pattern of inheritance. PLoS Genet 12: e1006367.

Larsson SC, King A, Madigan J et al. (2017). Prognosis of carotid dissecting aneurysms: results from Cadiss and a systematic review. Neurology 88: 646–652.

Lee VH, Brown RD, Mandrekar JN et al. (2006). Incidence and outcome of cervical artery dissection: a population-based study. Neurology 67: 1809–1812.

Markus HS, Hayter E, Levi C et al. (2015). Antiplatelet treatment compared with anticoagulation treatment for cervical artery dissection (CADISS): a randomised trial. Lancet Neurol 14: 361–367.

Molyneux AJ, Kerr RS, Yu LM et al. (2005). International subarachnoid aneurysm trial (ISAT) of neurosurgical clipping versus endovascular coiling in 2143 patients with ruptured intracranial aneurysms: a randomised comparison of effects on survival, dependency, seizures, rebleeding, subgroups, and aneurysm occlusion. Lancet 366: 809–817.

Muehlschlegel S (2018). Subarachnoid hemorrhage. Continuum (Minneap Minn) 24: 1623–1657.

O'connor SC, Poria N, Gornik HL (2015). Fibromuscular dysplasia: an update for the headache clinician. Headache 55: 748–755.

Olin JW, Froehlich J, Gu X et al. (2012). The United States registry for fibromuscular dysplasia: results in the first 447 patients. Circulation 125: 3182–3190.

Peng J, Liu Z, Luo C et al. (2017). Treatment of cervical artery dissection: Antithrombotics, thrombolysis, and endovascular therapy. Biomed Res Int 2017: 3072098.

Pickard JD, Murray GD, Illingworth R et al. (1989). Effect of oral nimodipine on cerebral infarction and outcome after subarachnoid haemorrhage: British aneurysm nimodipine trial. BMJ 298: 636–642.

Polmear A (2003). Sentinel headaches in aneurysmal subarachnoid haemorrhage: what is the true incidence? A systematic review. Cephalalgia 23: 935–941.

Rouanet C, Silva GS (2019). Aneurysmal subarachnoid hemorrhage: current concepts and updates. Arq Neuropsiquiatr 77: 806–814.

Schievink WI, Mokri B, O'fallon WM (1994). Recurrent spontaneous cervical-artery dissection. N Engl J Med 330: 393–397.

Southerland AM, Meschia JF, Worrall BB (2013). Shared associations of nonatherosclerotic, large-vessel, cerebrovascular arteriopathies: considering intracranial aneurysms, cervical artery dissection, moyamoya disease and fibromuscular dysplasia. Curr Opin Neurol 26: 13–28.

Thompson BG, Brown RD, Amin-Hanjani S et al. (2015). Guidelines for the Management of Patients with Unruptured Intracranial Aneurysms: A guideline for healthcare professionals from the American Heart Association/American Stroke Association. Stroke 46: 2368–2400.

Touzé E, Southerland AM, Boulanger M et al. (2019). Fibromuscular dysplasia and its neurologic manifestations: A systematic review. JAMA Neurol 76: 217–226.

Tweet MS, Hayes SN, Pitta SR et al. (2012). Clinical features, management, and prognosis of spontaneous coronary artery dissection. Circulation 126: 579–588.

Verma RK, Kottke R, Andereggen L et al. (2013). Detecting subarachnoid hemorrhage: comparison of combined FLAIR/SWI versus CT. Eur J Radiol 82: 1539–1545.

Vlak MH, Algra A, Brandenburg R et al. (2011). Prevalence of unruptured intracranial aneurysms, with emphasis on sex, age, comorbidity, country, and time period: a systematic review and meta-analysis. Lancet Neurol 10: 626–636.

Wells BJ, Modi RD, Gu X et al. (2020). Clinical associations of headaches among patients with fibromuscular dysplasia: a report from the US registry for fibromuscular dysplasia. Vasc Med 25: 348–350.

Wiebers DO, Whisnant JP, Huston J et al. (2003). Unruptured intracranial aneurysms: natural history, clinical outcome, and risks of surgical and endovascular treatment. Lancet 362: 103–110.

Woitzik J, Dreier JP, Hecht N et al. (2012). Delayed cerebral ischemia and spreading depolarization in absence of angiographic vasospasm after subarachnoid hemorrhage. J Cereb Blood Flow Metab 32: 203–212.

Handbook of Clinical Neurology, Vol. 177 (3rd series)
Heart and Neurologic Disease
J. Biller, Editor
https://doi.org/10.1016/B978-0-12-819814-8.00018-4

Chapter 23

Neurological complications of systemic hypertension

DEVIN LOEWENSTEIN[1] AND MARK RABBAT[2]*

[1]*Department of Medicine, Division of Cardiology, Rush University Medical Center, Chicago, IL, United States*

[2]*Department of Medicine, Division of Cardiology, Loyola University Medical Center, Maywood, IL, United States*

Abstract

Systemic hypertension is the most common, most easily diagnosed, and one of the most reversible risk factors for neurologic pathology. Acute severe hypertension above a mean arterial pressure of approximately 150 mmHg exceeds the brain's autoregulatory capacity and results in increased cerebral blood flow leading to hypertensive encephalopathy. Chronic hypertension predisposes to cerebral vasculature atherosclerosis, medial hypertrophy, luminal narrowing, endothelial dysfunction, impaired arterial relaxation, and decreased ability to augment cerebral blood flow at low blood pressures. The pathologic effects of hypertension increase stroke risk by three- to fivefold. With three-fourths of strokes incident events, primary prevention is essential. Multiple studies have demonstrated the benefit of blood pressure lowering in reducing incident and recurrent strokes. Even more, hypertension is a risk factor for cognitive impairment and dementia through multifactorial mechanisms including vascular compromise, cerebral small vessel disease, white matter disease (leukoaraiosis), cerebral microbleeds, cerebral atrophy, amyloid plaque deposition, and neurofibrillary tangles. In patients without hypotension, treatment with antihypertensives slows progression and assuages the degree of cognitive decline. While the choice of antihypertensive did not make a significant difference in most cognitive outcome studies, some large meta-analyses have pointed to angiotensin receptor blockers as the favored agent. Because of the well-documented morbidity and mortality associated with unchecked hypertension, treating and preventing hypertension are universally critical pillars in healthcare.

INTRODUCTION AND BACKGROUND

Systemic arterial hypertension (or hypertension) is an important risk factor for neurologic pathology. In fact, it is the most common, most easily diagnosed, and one of the most reversible risk factors (Blacher et al., 2016). Acute hypertension is associated with hypertensive encephalopathy and eclampsia. Chronic hypertension is associated with ischemic stroke, intracerebral hemorrhage, subarachnoid hemorrhage, atherosclerosis, periventricular white matter disease, and dementia. High blood pressure is the greatest single contributor to the global burden of disease, accounting for two-thirds of all strokes and half of all ischemic heart disease worldwide (Dahlof, 2007; Lim et al., 2012). Hypertension remains the leading risk factor for death globally. Importantly, antihypertensive treatment can substantially reduce the prevalence and severity of these neurologic diseases.

Both systolic and diastolic blood pressures are clinically relevant. In the general population, both roughly follow a normal distribution. In the majority of the world and up until 2017 in the United States, hypertension was defined as a SBP of 140 mmHg or greater or a diastolic blood pressure of 90 mmHg or greater. By this definition, over 1 billion people worldwide are afflicted. With the rising prevalence of obesity and advanced age, by the year 2025, 1.5 billion people are projected to have hypertension (Poulter et al., 2015). In 2017, the American College of Cardiology and American Heart Association (AHA) released a new set of hypertension guidelines

*Correspondence to: Mark Rabbat, Loyola University Medical Center, Building 110, Room 6268, 2160 South First Avenue, Maywood, IL 60153, United States. Tel: +1-708-327-2747, Fax: +1-708-406-1516, E-mail: mrabbat@lumc.edu

lowering the threshold between prehypertension and hypertension to 130 mmHg systolic and 80 mmHg diastolic blood pressure (Whelton et al., 2018). The change means 46% of US adults are identified as having high blood pressure, compared with 32% under the previous definition.

The effect of hypertension on organ dysfunction is well documented. Uniquely, blood flow to the brain is highly regulated to maintain a consistent level of flow for stable brain function. The mechanism by which this is accomplished is termed autoregulation. In autoregulation, the physiologic response to acute elevation in blood pressure is constriction of small arteries in the brain to counteract the increase in flow that would take place as a result of the higher aortic and extracranial arterial blood pressure and thereby protect the more distal, thin-walled vessels from pressure loads they would otherwise be unable to handle. Acute severe hypertension above a mean arterial pressure (MAP) of approximately 150 mmHg exceeds the autoregulatory capacity and results in increased cerebral blood flow. When thin-walled vessels like capillaries are exposed to the increased pressure and flow, endothelial injury may occur resulting in breakdown of the blood–brain barrier (Johansson, 1999). Chronic hypertension leads to remodeling of the cerebral vasculature with medial hypertrophy and luminal narrowing. This results in a brain-protective shift of the autoregulation curve to allow for constancy of flow at a higher window of blood pressures (Heistad and Baumbach, 1992). However, the deleterious consequences of this vascular remodeling are endothelial dysfunction, impaired arterial relaxation, and decreased ability to augment cerebral blood flow at lower blood pressures. Chronic hypertension also predisposes to atherosclerosis, which can affect the entire arterial path from aorta to intracerebral small vessels.

HYPERTENSIVE ENCEPHALOPATHY

Encephalopathy may occur when BP exceeds the upper limit of cerebral autoregulation. In patients without preexisting chronic hypertension, cerebral blood blow is maintained constant over a range of mean arterial pressure from 60 to 150 mmHg. The autoregulatory window is shifted toward higher MAPs in patients with chronic hypertension (Van den Born et al., 2011).

A major risk factor for hypertensive encephalopathy is the development of a hyperadrenergic state. Examples include abrupt antihypertensive discontinuation, use of stimulant medications, pheochromocytoma, tyramine ingestion with monoamine oxidase inhibitors, and lower gastrointestinal irritation in spinal cord injury patients with paraplegia (Pentel, 1984). Other risk factors include acute or chronic kidney injury, aortic coarctation, and renal artery stenosis (Sheth et al., 1996).

Hypertensive encephalopathy is diagnosed when there is reduced level of consciousness, agitation, delirium, seizures, or cortical blindness in the setting of acute severe elevation in blood pressure. It may present initially with headache and later precipitate nausea and vomiting. Focal neurologic signs are uncommon and if present, suggest a stroke rather than encephalopathy. Brain imaging showing areas of cerebral edema confirms the diagnosis of encephalopathy. It is important to perform rapid brain imaging to exclude alternate causes for both the encephalopathy and the hypertension such as stroke or structural lesions, which would significantly alter the course of treatment. Typically, the edema localizes to the posterior brain regions perfused by the posterior circulation. This is because the vertebrobasilar circulation has a less potent autoregulatory mechanism than the carotid circulation. With timely antihypertensive treatment, the edema usually reverses with complete recovery of neurologic deficits seen within 2 weeks. (Van den Born et al., 2011).

Hypertensive encephalopathy is a neurologic emergency that can lead to death if untreated. The initial management should include administration of intravenous antihypertensive therapy to rapidly lower the MAP by 10% over the first hour of treatment and by an additional 15% over the next 12 h to a BP of no less than 160/110 mmHg. Overly-rapid normalization of BP carries the risk of cerebral hypoperfusion. Oral agents should be added and BP can be further reduced over the next 48 h. Infusion of saline is often necessary to counteract the volume loss caused by reflex natriuresis in the face of severe hypertension (Van den Born et al., 2011).

Hypertensive encephalopathy is a recognized cause of posterior reversible encephalopathy syndrome (PRES). Often, there is concomitant use of cyclosporine or tacrolimus with hypertensive response or bevacizumab or bortezomib. Alteration in the permeability of cerebral vascular endothelium cells has been proposed as the pathophysiologic mechanism of PRES. Often, lowering the blood pressure and discontinuing the immunosuppressant reverses the process (Hinchey et al., 1996).

STROKE

Each year in the United States, over 750,000 patients suffer a stroke, and 1 in 7 will die as a direct result. For the survivors, stroke is a leading cause of significant long-term disability. Roughly 78% of strokes are incident events, which makes primary prevention critical (Mozaffarian et al., 2016).

Of all the known modifiable risk factors for stroke, hypertension is likely to be the most important owing to its high prevalence, the three- to fivefold increase in stroke risk it portends, and the fact that it is readily amenable to lifestyle and pharmacologic intervention (Sacco, 1995). Multiple studies have demonstrated the benefit of BP lowering in stroke reduction, with overall reduction more important than any specific means or antihypertensive agent used to achieve that reduction (Chobanian et al., 2003; Goldstein et al., 2011). The benefits of treating hypertension extend similarly to the elderly with isolated elevations in systolic blood pressure (Staessen et al., 1997).

The systolic blood pressure intervention trial (SPRINT) compared treatment to a SBP target under 120 mmHg to a more conservative target under 140 mmHg in older adult patients at elevated risk for cardiovascular events and baseline SBP ranging 130–180 mmHg. A nonsignificant decrease in annual stroke rate was seen in the intensive BP lowering arm (1.5% vs 1.3% annual stroke rate, hazard ratio [HR] 0.89; 95% confidence interval [CI] 0.63–1.25; $P = 0.50$), at the expense of increased treatment related adverse effects (Wright et al., 2015). Brain magnetic resonance imaging was performed on a subset of participants at baseline ($n = 670$) and at 4 years of follow-up (after a median intervention period of 3.40 years; $n = 449$). The treatment arm with the more aggressive SBP target of less than 120 mmHg showed a significantly smaller increase in cerebral white matter lesion volume (0.92 vs 1.45 cm^3), a surrogate for small vessel ischemic disease per study authors (Nasrallah et al., 2019).

In the HOPE-3 trial, people at intermediate risk for CVD were randomized in a two by two factorial design to rosuvastatin, hydrochlorothiazide candesartan, and placebo. With fixed drug doses not titrated to any particular BP target, antihypertensive therapy failed to show a significant stroke reduction (Lonn et al., 2016).

Since roughly one in five stroke survivors will have a second stroke within 4 years, secondary prevention is paramount (Mozaffarian et al., 2016). A meta-analysis of 16 randomized control trials of BP lowering for secondary prevention of stroke involving over 40,000 patients showed that for every 10 mmHg reduction in SBP, there was a relative risk reduction of 33% (95% CI 9%–51%) in recurrent stroke (Arima and Chalmers, 2011).

Based on the 2019 update to the 2018 Guidelines for the Early Management of Acute Ischemic Stroke from the AHA/American Stroke Association, blood pressure should be lowered cautiously in the event of acute ischemic stroke to avoid further ischemic insult to the potentially salvageable peri-infarct brain tissue (known as the ischemic penumbra). If the stroke cannot be treated with thrombolytic therapy or mechanical thrombectomy, blood pressure should not be allowed to exceed 210/110 mmHg and should not be decreased acutely by more than 15% from presentation BP unless necessitated by a comorbid condition. If the patient is eligible for thrombolytic treatment, BP needs to be lowered to under 185/110 mmHg. In patients for whom mechanical thrombectomy is planned and who have not received IV fibrinolytic therapy, it is reasonable to maintain BP below 185/110 mmHg before the procedure. The usefulness of drug-induced hypertension in patients with acute ischemic stroke is not well established (Jauch et al., 2013; Powers et al., 2019).

Stroke survivors are at high risk for recurrence. Reducing SBP by more than 10 mmHg can significantly reduce this risk (Davis and Donnan, 2008). The PROGRESS (Perindopril Protection Against Recurrent Stroke Study) trial demonstrated that in patients who had survived ischemic or hemorrhagic stroke, reducing SBP by 12 mmHg to under 135 mmHg with an angiotensin converting enzyme inhibitor (ACEI)/diuretic combination reduced the relative risk for recurrent ischemic stroke by 35% and recurrent hemorrhagic stroke by 76% compared to placebo. However, lesser reductions in SBP had no significant effect (MacMahon et al., 2001). This was further supported by the PROFESS (Prevention Regimen for Effectively Avoiding Second Strokes) trial in which a 4 mmHg reduction in SBP using angiotensin receptor blocker (ARB) monotherapy vs placebo showed no statistical benefit (Yusuf et al., 2008). Starting or restarting antihypertensive medications has been shown to be associated with improved blood pressure control after discharge in two trials (Robinson et al., 2010; He et al., 2014). Therefore, it is reasonable to start or restart antihypertensive medications in the hospital when the patient remains hypertensive and is neurologically stable (Robinson et al., 2010; He et al., 2014).

Contrary to the blood pressure targets in ischemic stroke, blood pressure in hemorrhagic stroked should be lowered to under 140 mmHg systolic. This target is based on the INTERACT2 (Intensive BP Reduction in Acute Hemorrhage Trial) study, which demonstrated improved functional outcomes without more adverse events in patients for whom SBP was targeted to less than 140 mmHg vs the more conservative goal of less than 180 mmHg (Anderson et al., 2013). As for ischemic stroke, antihypertensive agents of choice include nicardipine, clevidipine, and labetalol, while nitroprusside and hydralazine should be avoided due to risk of paradoxical increase in intracranial pressure (Wang et al., 2015; Powers et al., 2019).

INTRACRANIAL ATHEROSCLEROSIS AND HYPERTROPHIC REMODELING

Hypertension confers a two- to threefold higher risk of intracranial atherosclerotic disease. It promotes formation of atherosclerotic plaque in cerebral arteries and arterioles (Lammie, 2002; Dahlof, 2007). Smoking and diabetes, and to a lesser degree hyperlipidemia, are also important risk factors for intracranial atherosclerosis (Ingall et al., 1991).

In patients with intracranial atherosclerosis, there is a theoretical concern for worsening of cerebral hypoperfusion distal to a stenotic lesion when blood pressure is lowered. However, there is no evidence to suggest management with higher blood pressure targets would be favored. There has been no evidence that treating to standard blood pressure targets in these patients has increased risk of distal infarction (Turan et al., 2007).

Hypertension is also known to induce hypertrophic remodeling of cerebral arteries, which is an adaptive response to reduce vessel wall stress and protect smaller downstream vasculature from pressure overload. As smooth muscle cell hypertrophy, inward encroachment narrows the vessel lumen. Increased collagen content and rigidity of the vessel wall also are involved in vascular stiffening (Baumbach and Heistad, 1988). These adaptive effects to hypertension actually result in the deleteriously wide pulse pressure and predisposition for vascular insufficiency (Waldstein et al., 2004; Benjo et al., 2007; Matthiassen et al., 2007). They also cause a rightward shift in the pressure-flow curve of cerebrovascular autoregulation, increasing the susceptibility of the brain to cerebral ischemia when the blood pressure acutely drops (Baumbach and Heistad, 1988; Immink et al., 2004).

COGNITIVE IMPAIRMENT AND DEMENTIA

Hypertension is a recognized risk factor for cognitive impairment (Launer et al., 1995; Skoog et al., 1996; Iadecola, 2014). The mechanism by which hypertension contributes to cognitive impairment is multifactorial. As discussed earlier, hypertension is associated with increased risk of stroke. Vascular injury to the brain results in interruption of cognitive networks leading to cognitive impairment. Cerebral small vessel disease (cSVD) is a group of pathological processes affecting the smaller arteries, arterioles, venules, and capillaries of the brain. It is a significant contributor to cognitive decline among other disorders (Pantoni, 2010; Huijts et al., 2014). Less clear cut is the association of hypertension with neurodegenerative conditions. In the Honolulu-Asia Aging Study, there was a direct association between characteristic pathologic findings of Alzheimer disease (AD) and chronic systemic hypertension (Petrovitch et al., 2000). Considering the overlap of risk factors accounting for AD and cSVD, clinically differentiating the two can be challenging (Rincon and Wright, 2014), with the resulting estimated proportion of dementia caused by cSVD ranging from 36% to 67% (Chui, 2001). Additional studies have demonstrated that hypertension is a powerful risk factor for AD (Iadecola, 2004; Skoog and Gustafson, 2006; Kelley and Petersen, 2007). Cerebral atrophy, amyloid plaques, and neurofibrillary tangles are also known to be more prevalent in patients with a history of midlife hypertension (Skoog and Gustafson, 2006).

In the presence of hypertension, cerebral microcirculation can be harmed by mechanisms including change in blood flow and direct vascular damage (Fiehl et al., 2006). Hypertension is also associated with cerebral microbleeds (Greenberg et al., 2009), leukoaraiosis or white matter disease (Verhaaren et al., 2013), and increased amyloid beta deposition (Rodrigue et al., 2013), all of which are linked to cognitive impairment. It has also been shown that patients with an exaggerated variability of blood pressure exhibit reduced cerebral blood flow and increased ischemic insults (O'Sullivan et al., 2002; Fernando et al., 2006). In addition, hypertension has also been shown to result in abnormally low densities of arterioles and capillaries, thought secondary to reduced cerebral blood flow (De La Torre, 2012).

While it has been widely accepted that hypertension confers greater risk of cognitive impairment, treating hypertension to target lower blood pressure has historically had disappointing and inconsistent results. Analogous to the above concern that treating to a lower blood pressure target would induce infarction distal to intracranial arterial stenoses, it has been suggested that aggressively lowering blood pressure could worsen cognition. As medical management and risk factor modulation have evolved, this association has been more readily elucidated. In a trial of patients with a history of stroke, antihypertensive therapy with calcium channel blocker or diuretic was associated with lower dementia risk (Tzourio et al., 2003). A meta-analysis of longitudinal studies demonstrated reduced risk of cognitive decline in patients treated with antihypertensives (Chang-Quan et al., 2011). Other studies have shown that controlling blood pressure reverses some of the deleterious effects on the cerebral vasculature and mitigates cognitive dysfunction (Peila et al., 2006).

Thus, treatment of hypertension is accepted as a major pillar in prevention or slowing of cognitive decline and dementia (Dahlof, 2007; Messerli et al., 2007). Interclass differences between antihypertensive agents have been observed in some trials but not others. According to some, ACEIs and ARBs conferred the greatest protection against cognitive decline and dementia (Dahlof, 2007; Messerli et al., 2007), while others have shown ARBs derive greater benefit than ACEIs. The renin-angiotensin-aldosterone system (RAAS) has been implicated in dementia through mechanisms independent of cerebral blood flow and vascular resistance, including tau phosphorylation, amyloid metabolism, and oxidative stress. Furthermore, angiotensin II inhibits the release of acetylcholine, thus potentiating cognitive dysfunction (Kuan et al., 2016). Notably, humans have two distinct types of angiotensin II receptor. ARBs specifically block activity at the harmful type 1 receptor (AT1), while ACEIs block activity at both AT1 and the neurobeneficial AT2 receptors. This is one postulated pathophysiologic mechanism that would suggest a neuroprotective benefit of RAAS agents and specifically ARBs over ACEIs (Kehoe et al., 2009).

Interestingly, in one murine study, valsartan reduced the degree of cerebral amyloid plaque deposition and improved behavioral performance despite not significantly affecting arterial pressure (Wang et al., 2007). In a study of 819,491 hypertensive adults aged 65 or older taking ARBs, lisinopril, or other antihypertensives and without a prior diagnosis of AD or dementia, the use of ARBs was associated with a 19% lower risk of developing AD or dementia vs lisinopril. For other antihypertensives, the risk reduction was 16% and 24%, respectively. Also promising was the significantly lower risk of death and nursing home admission for ARB users compared to those on lisinopril or other antihypertensives (Li et al., 2010). In a 2017 study, Ho and colleagues evaluated whether use of ARBs was associated with improved memory preservation compared with other antihypertensive medications. They followed 1626 dementia-free adults aged 55–91 years from three groups: hypertensive patients on ARBs, hypertensive patients on non-ARBs, and patients without hypertension. Over roughly 3 years of follow-up, hypertensive patients in the ARB group had similar cognitive outcomes to normotensive patients, while hypertensive patients treated with non-ARBs had worse outcomes (Ho and Nation, 2017).

In a meta-analysis of six long-term prospective cohort studies involving 31,090 dementia-free participants, the authors assessed association of different classes of antihypertensives to incident dementia and AD. During follow-up, 3728 and 1741 participants developed dementia and AD, respectively. In adults with elevated blood pressure at baseline, those on any antihypertensive medication regardless of pharmacologic class had a lower risk for developing all-cause dementia (HR 0.88; 95% CI 0.79–0.98) and AD (HR 0.85; 95% CI 0.74–0.99). No significant difference was noted for participants with normal baseline blood pressure (Ding et al., 2018). Another notable study, the SPRINT MIND study, showed that aggressive lowering of SBP to 120 mmHg significantly reduced the risk for mild cognitive impairment compared with targeting 140 mmHg over a median treatment period of 3.3 years and medial follow-up of 5.11 years (14.6 vs 18.3 cases per 1000 person-years; HR 0.81; 95% CI 0.69–0.95). SPRINT MIND followed 9361 cognitively healthy hypertensive older adults with increased cardiovascular risk but without diagnosed diabetes, dementia, or stroke. Additionally, there was a nonsignificant reduction in the primary outcome of probable dementia (7.2 vs 8.6 cases per 1000 person-years; HR 0.83; 95% CI 0.67–1.04) (Williamson et al., 2019). The choice of antihypertensive in this study did not seem to make a significant difference in outcome.

Other prospective cohort studies have helped to define risk of hypertension and treatment by age group. One study evaluated 4761 participants with 24-year follow-up and blood pressure measurements at midlife and late in life. Compared to those who remained normotensive from midlife through late-life, participants with midlife hypertension and late-life hypotension experienced the highest incident risk of dementia (HR 1.62; 95% CI 1.11–2.37). The high-low blood pressure pattern was associated with even higher dementia risk than hypertension in both midlife and late-life (HR 1.49; 95% CI 1.06–2.08) (Walker et al., 2019). Another study examined persons aged 75 years or older from the Integrated Systematic Care for Older Persons database over 1-year follow-up. Those on antihypertensive therapy with SBP 130 mmHg or greater showed less cognitive decline compared to those with SBP less than 130 mmHg, without loss of daily functioning or quality of life. This effect was predominantly seen in participants with complex medical problems (Streit et al., 2019). These studies exemplify the evidence suggesting late-life hypotension, especially in medically complex patients, could accelerate cognitive decline, and should be avoided.

References

Anderson CS, Heeley E, Huang Y et al. (2013). Rapid blood-pressure lowering in patients with acute intracerebral hemorrhage. N Engl J Med 368: 2355–2365.

Arima H, Chalmers J (2011). Progress. Prevention of recurrent stroke. J Clin Hypertens 13: 369–702.

Baumbach GL, Heistad DD (1988). Cerebral circulation in chronic arterial hypertension. Hypertension 89–95.

Benjo A, Thompson RE, Fine D et al. (2007). Pulse pressure is an age-independent predictor of stroke development after cardiac surgery. Hypertension 630–635.

Blacher J, Levy BI, Mourad JJ et al. (2016). From epidemiological transition to modern cardiovascular epidemiology: hypertension in the 21st century. Lancet 338: 530–532.

Chang-Quan H, Hui W, Chao-Min W et al. (2011). The association of antihypertensive medication use with risk of cognitive decline and dementia: a meta-analysis of longitudinal studies. Int J Clin Pract 691–701.

Chobanian AV, Bakris GL, Black HR et al. (2003). The seventh report of the joint National Committee on prevention, detection, evaluation, and treatment of high blood pressure: the JNC 7 report. JAMA 289: 2560–2571.

Chui H (2001). Dementia due to subcortical ischemic vascular disease. Clin Cornerstone 40–51.

Dahlof B (2007). Prevention of stroke in patients with hypertension. Am J Cardiol 17J–24J.

Davis SM, Donnan GA (2008). Clinical practice. Secondary prevention after ischemic stroke and cardiovascular events. N Engl J Med 366: 1914–1922.

De La Torre JC (2012). Cardiovascular risk factors promote brain hypoperfusion leading to cognitive decline and dementia. Cardiovasc Psychiatray Neurol 2012: 367516.

Ding J, Davis-Plourde K, Sedaghat S et al. (2018). Use of blood pressure-lowering drugs and risk of incident dementia and Alzheimer's disease in older people with and without high blood pressure: a meta-analysis of individual participant data from prospective cohort studies. J Alzheimers Dis 1045–1046.

Fernando MS, Simpson JE, Matthews F et al. (2006). MRC Cognitive Function and Ageing Neuropathology Study Group: white matter lesions in an unselected cohort of the elderly: molecular pathology suggests origin from chronic hypoperfusion injury. Stroke 1391–1398.

Fiehl F, Liaudet L, Waeber B et al. (2006). Hypertension: a disease of the microcirculation? Hypertension 1012–1017.

Goldstein LB, Bushnell CD, Adams RJ et al. (2011). Guidelines for the primary prevention of stroke. A guideline for healthcare professionals from the American Heart Association/ American Stroke Association. Stroke 42: 517–584.

Greenberg SM, Verjooij MW, Cordonnier C et al. (2009). Microbleed Study Group: cerebral microbleeds: a guide to detection and interpretation. Lancet Neurol 165–174.

He J, Zhang Y, Xu T et al. (2014). CATIS Investigators. Effects of immediate blood pressure reduction on death and major disability in patients with acute ischemic stroke: the CATIS randomized clinical trial. JAMA 11: 479–489.

Heistad DD, Baumbach GL (1992). Cerebral vascular changes during chronic hypertension: good guys and bad guys. J Hypertens Suppl 10: S71.

Hinchey J, Chaves C, Appignani B et al. (1996). A reversible posterior leukoencephalopathy syndrome. N Engl J Med 334: 494.

Ho JK, Nation DA (2017). Alzheimer's disease neuroimaging initiative: memory is preserved in older adults taking AT1 receptor blockers. Alzheimers Res Ther 9: 33.

Huijts M, Duits A, Staals J et al. (2014). Basal ganglia enlarged perivascular spaces are linked to cognitive function in patients with cerebral small vessel disease. Curr Neurovasc Res 136–141.

Iadecola C (2004). Neurovascular regulation in the normal brain and in Alzheimer's disease. Nat Rev Neurosci 347–360.

Iadecola C (2014). Hypertension and dementia. Hypertension 91–99.

Immink RV, van den Born BJ, van Montfrans GA et al. (2004). Impaired cerebral autoregulation in patients with malignant hypertension. Circulation 2241–2245.

Ingall TJ, Homer D, Baker Jr HL et al. (1991). Predictors of intracranial carotid artery atherosclerosis. Duration of cigarette smoking and hypertension are more powerful than serum lipids. Arch Neurol 48: 687.

Jauch EC, Saver JL, Adams Jr JP et al. (2013). Guidelines for the early management of patients with acute ischemic stroke: a guideline for healthcare professionals from the American Heart Association/American Stroke Association. Stroke 44: 870–947.

Johansson BB (1999). Hypertension mechanisms causing stroke. Clin Exp Pharmacol Physiol 26: 563.

Kehoe PG, Miners S, Love S (2009). Angiotensins in Alzheimer's disease—friend or foe? Trends Neurosci 619–628.

Kelley BJ, Petersen RC (2007). Alzheimer's disease and mild cognitive impairment. Neurol Clin 577–609.

Kuan YC, Huang KW, Yen DJ et al. (2016). Angiotensin-converting enzyme inhibitors and angiotensin II receptor blocker reduced dementia risk in patients with diabetes mellitus and hypertension. Int J Cardiol 462–466.

Lammie GA (2002). Hypertensive cerebral small vessel disease and stroke. Brain Pathol 358–370.

Launer LJ, Masaki K, Petrovitch H et al. (1995). The association between midlife blood pressure levels and late life cognitive function. JAMA 274: 1846.

Li NC, Lee A, Whitmer RA et al. (2010). Use of angiotensin receptor blockers and risk of dementia in a predominantly male population: prospective cohort analysis. BMJ 340–350.

Lim SS, Vos T, Flaxman AD et al. (2012). A comparative risk assessment of burden of disease and injury attributable to 67 risk factors and risk factors clusters in 21 regions, 1990–2010: a systematic analysis of the Global Gurden of Disease Study 2010. Lancet 380: 224.

Lonn EM, Bosch J, Lopez-Jamarillo P et al. (2016). Blood-pressure lowering in intermediate-risk persons without cardiovascular disease. N Engl J Med 374: 2009–2020.

MacMahon S, Neal B, Tzourio C et al. (2001). Randomised trial of a perindopril-based blood-pressure-lowering regimen among 6,105 individuals with previous stroek or transient ischaemic attack. Lancet 358: 1033–1041.

Matthiassen ON, Buus NH, Sihm I et al. (2007). Small artery structure is an independent predictor of cardiovascular events in essential hypertension. J Hypertens 1021–1026.

Messerli FH, Williams B, Ritz E (2007). Essential hypertension. Lancet 591–603.

Mozaffarian D, Benjamin EJ, Go AS et al. (2016). Heart disease and stroke statistics—2016 update. Circulation 133: e38–e60.

Nasrallah IM, Pajewski NM, Auchus AP et al. (2019). Association of Intensive vs standard blood pressure control with cerebral White matter lesions. JAMA 322: 524–534.

O'Sullivan M, Lythgoe DJ, Pereira AC et al. (2002). Patterns of cerebral blood flow reduction in patients with ischemic leukoaraiosis. Neurology 321–326.

Pantoni L (2010). Cerebral small vessel disease: from pathogenesis and clinical characteristics to therapeutic challenges. Lancet Neurol 689–701.

Peila R, White LR, Masaki K et al. (2006). Reducing the risk of dementia. Efficacy of long-term treatment of hypertension. Stroke 1165–1170.

Pentel P (1984). Toxicity of over-the-counter stimulants. JAMA 252: 1898.

Petrovitch H, White LR, Izmirilian G et al. (2000). Midlife blood pressure and neuritic plaques, neurofibrillary tangles, and brain weight at death: the HAAS (Honolulu-Asia Aging Study). Neurobiol Aging 21: 57.

Poulter NR, Prabhakaran D, Calufield M (2015). Hypertension. Lancet 386: 801–812.

Powers WJ, Rabinstein AA, Ackerson T et al. (2019). Guidelines for the early management of patients with acute ischemic stroke: 2019 update to the 2018 guidelines for the early management of acute ischemic stroke: A guideline for healthcare professionals from the American Heart Association/American Stroke Association. Stroke 50: e344–e418. https://doi.org/10.1161/STR.0000000000000211.

Rincon F, Wright CB (2014). Current pathophysiological concepts in cerebral small vessel disease. Front Aging Neurosci 6. 24. https://doi.org/10.3389/fnagi.2014.00024.

Robinson TG, Potter JF, Ford GA et al. (2010). Effects of antihypertensive treatment after acute stroke in the Continue or Stop Post-Stroke antihypertensives Collaborative Study (COSSACS): a prospective, randomised, open, blinded-endpoint trial. Lancet Neurol 9: 767–775.

Rodrigue KM, Rieck JR, Kennedy KM et al. (2013). Risk factors for beta-amyloid deposition in healthy aging: vascular and genetic effects. JAMA Neurol 600–606.

Sacco RL (1995). Risk factors and outcomes for ischemic stroke. Neurology 45: S10.

Sheth RD, Riggs JE, Bodenstenier JB et al. (1996). Parietal occipital edema in hypertensive encephalopathy: a pathogenic mechanism. Eur Neurol 36: 25.

Skoog I, Gustafson D (2006). Update on hypertension and Alzheimer's disease. Neurol Res 605–611.

Skoog I, Lernfelt B, Landahl S et al. (1996). 15-year longitudinal study of blood pressure and dementia. Lancet 347: 1141.

Staessen JA, Gafard R, Thijs L et al. (1997). Randomised double-blind comparison of placebo and active treatment for older patients with isolated systolic hypertension: the Systolic Hypertension in Europe (Syst-Eur) Trial Investigators. Lancet 350: 757.

Streit S, Poortvliet R, den Elzen W et al. (2019). Systolic blood pressure and cognitive decline in older adults with hypertension. Ann Fam Med 100–107.

Turan TN, Cotsonis G, Lynn MJ et al. (2007). Relationship between blood pressure and stroke recurrence in patients with intracranial atherosclerosis. Circulation 115: 2969.

Tzourio C, Anderson C, Chapman N et al. (2003). Effects of blood pressure lowering with perindopril and indapamide on dementia and cognitive decline in patients with cerebrovascular disease. Arch Intern Med 163: 1069.

Van den Born BJ, Beutler JJ, Gaillard CA et al. (2011). Dutch guideline for the management of hypertensive crisis: 2010 revision. Neth J Med 69: 248–255.

Verhaaren BF, Vernooij MW, de Boer R et al. (2013). High blood pressure and cerebral white matter lesion progression in the general population. Hypertension 1354–1359.

Waldstein SR, Rice SC, Thayer JF et al. (2004). Pulse pressure and pulse wave velocity are related to cognitive decline in the Baltimore Longitudinal Study of Aging. Hypertension 625–630.

Walker KA, Sharrett AR, Wu A et al. (2019). Association of Midlife to late-life blood PRessure patterns with incident dementia. JAMA 535–545.

Wang J, Ho L, Chen L et al. (2007). Valsartan lowers brain beta-amyloid protein levels and improves spatial learning in a mouse model of Alzheimers disease. J Clin Invest 3393–3402.

Wang X, Arima H, Heeley E et al. (2015). Magnitude of blood pressure reduction and clinical outcomes in acute intracerebral hemorrhage: intensive blood pressure reduction in acute cerebral hemorrhage trial study. Hypertension 65: 1026–1032.

Whelton PK, Carey RM, Aronow WS et al. (2018). 2017 ACC/AHA/AAPA/ABC/ACPM/AGS/APhA/ASH/ASPC/NMA/PCNA Guideline for the prevention, detection, evaluation, and management of high blood pressure in Adults. J Am Coll Cardiol 71: e127–e248.

Williamson JD, Pajewski NM, Auchus AP et al. (2019). Effect of intensive vs standard blood pressure control on probable dementia. JAMA 553–561.

Wright Jr JT, Williamson JD, Whelton PK et al. (2015). A randomized trial of intensive versus standard blood pressure control. N Engl J Med 373: 2103–2116.

Yusuf S, Diener HC, Sacco RL et al. (2008). Telmisartan to prevent recurrent stroke and cardiovascular events. N Engl J Med 359: 1235–1237.

Handbook of Clinical Neurology, Vol. 177 (3rd series)
Heart and Neurologic Disease
J. Biller, Editor
https://doi.org/10.1016/B978-0-12-819814-8.00017-2

Chapter 24

Neurologic complications of venous thromboembolism

MICHAEL J. SCHNECK*

Department of Neurology, Loyola University Chicago, Stritch School of Medicine, Maywood, IL, United States

Abstract

Patients with neurologic disease are at high risk of venous thromboembolism (VTE). The converse risk of neurological complications in concert or following peripheral VTE is rarely considered. The major neurologic complication following pulmonary embolism or peripheral VTE is intracranial hemorrhage that occurs following anticoagulation therapy for VTE. Ischemic stroke may occur concomitantly with VTE. VT occurrence may also be a marker for a future increased risk of stroke. Peripheral neuropathy may occur or neuropathic pain may occur as a result of thrombophlebitis following deep venous thrombosis. Other sequelae of VTE are somewhat theoretical including the discovery of central nervous system malignancies after incident VTE.

INTRODUCTION

Reviews of the association between venous thromboembolism (VTE) and neurologic disease typically focus on superficial venous thrombosis (SVT), deep venous thrombosis (DVT), and/or pulmonary embolism (PE) as a sequelae or complication of neurologic disease. For example, a 2014 review in the *Handbook of Clinical Neurology* by this author discusses the relationship of VTE in various neurologic diseases (Schneck, 2014). Other than the complication of intracranial hemorrhage in the context of anticoagulation treatment post VTE, there is not an extensive discussion of other neurologic sequelae of VTE, however. This particular chapter attempts to explore some of those issues. [Note that the special case of cerebral venous sinus thrombosis is not considered this chapter]. Central nervous system (CNS) complications of VTE are mainly related to cerebrovascular disease. The diagnosis of CNS malignancy following VTE is also considered. Finally, peripheral nervous system complications are mainly related to neuropathic pain or nerve entrapment.

RISK OF ISCHEMIC CEREBROVASCULAR DISEASE AFTER VTE

Patients with unprovoked DVT (as opposed to provoked DVT related to trauma or immobility) are at increased risk of symptomatic arterial cardiovascular events including myocardial infarction (MI), ischemic stroke, peripheral artery disease, and sudden death (Sørensen et al., 2007). In a case-comparison study of a cohort of 138 patients with unprovoked DVT, compared with 123 patients with secondary, or provoked, DVT, the hazard ratio (HR) for arterial atherosclerotic events, such as stroke or MI, for the unprovoked group was 2.89 (95% CI 1.06–7.88; $P = 0.038$), compared with those with a provoked DVT (Bilora et al., 2017).

In a report from the Danish National Registry, 25,199 patients with DVT, 16,925 patients with PE, and 134,566 population controls without known cardiovascular disease from a 2-year population based cohort study were studied for the risk of subsequent stroke or MI. The relative risk (RR) of MI for patients with DVT was 1.6 (95% CI 1.35–1.91), and the RR for stroke was 2.19

*Correspondence to: Michael J. Schneck, M.D., Department of Neurology, Loyola University Medical Center, Maguire Building Suite 2700, 2160 South First Avenue, Maywood, IL 60153, United States. Tel: +1-708-216-2662, Fax: +1-708-216-5617, E-mail: mschneck@lumc.edu

(95% CI 1.85–2.60) in the first year after DVT. For patients with PE, the RR was 2.60 (95% CI 2.14–3.14) for MI and 2.93 (95% CI 2.34–3.66) for stroke with a 20 year RR increase of 20%–40% for arterial cardiovascular events following VTE. In this study, the RRs were similar for provoked or unprovoked DVT or PE (Sørensen et al., 2007). SVT was also associated with an increased risk of subsequent arterial cardiovascular events (Cannegieter et al., 2015). In that study, from the Danish National Registry of Hospitalized Patients, 10,973 patients were identified with a first-time diagnosis of SVT between 1980 and 2012. The authors reported an HR for stroke of 1.3 (95% CI 1.2–1.4). There was a similar 1.2 HR for acute MI and 1.3 HR for sudden death with the highest risk of cardiovascular events in the period shortly after SVT.

In the Registro Informatizado de Enfermedad TromboEmbólica registry of 23,370 patients (12,397 initially presenting with PE and 10,973 with DVT), the mortality for recurrent PE was lower the mortality for arterial ischemic events in patients with incident VTE (Madridano et al., 2015). There were 59 patients during the mean follow-up of 9.3 months who subsequently developed PE, whereas 162 had arterial ischemic events; there were 86 ischemic strokes and 53 MI in this cohort. Of the patients who had a recurrent vascular event, 3 died of hemorrhagic complications, 53 died of ischemic events, and only 29 died of PE.

Simultaneous PE and acute ischemic stroke is an uncommon event (see Fig. 24.1). There are a number of case reports and small case series describing these occurrences and their treatment (Allport et al., 2008; Barros-Gomes et al., 2018; Lak et al., 2020; Lapostolle et al., 2003; Lio et al., 2020; Naidoo and Hift, 2011; Pavesi et al., 2008; Vindis et al., 2018; Zhang et al., 2017). Patients with concomitant PE and acute ischemic stroke can be successfully treated with intravenous tissue plasminogen activator (tPA) and/or mechanical thrombectomy (Vindis et al., 2018). The presence of a patent foramen ovale (PFO) may also put VTE patients at subsequent risk for ischemic strokes (LeMoigne et al., 2019). Of 78 patients with acute PE, at baseline 50%

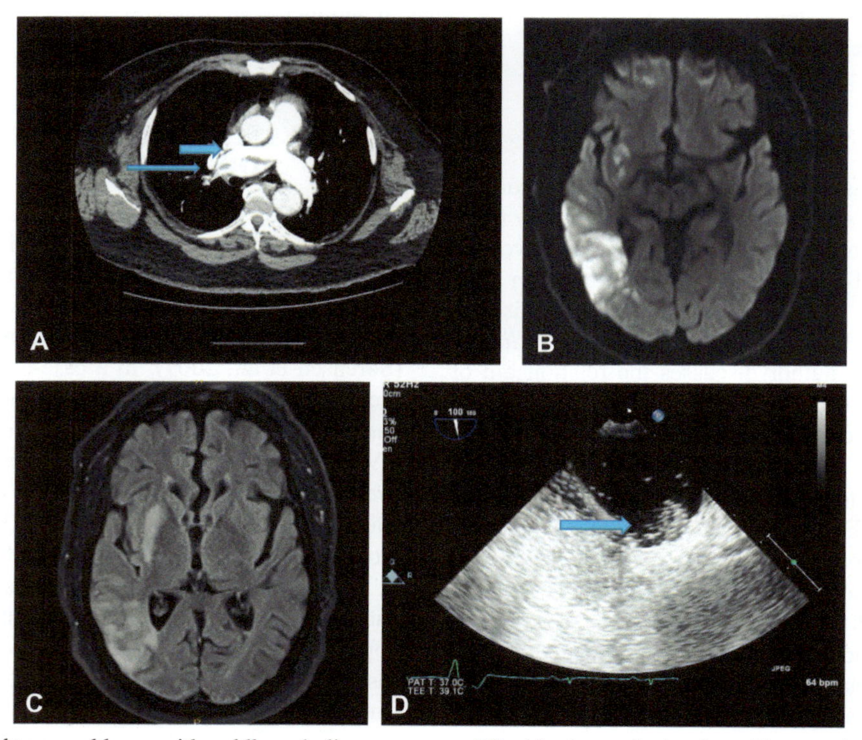

Fig. 24.1. Fifty-eight-year-old man with saddle embolism as seen on CT with a large clot burden with extension into the lobar and all bilateral segmental arteries (A). The patient also had right hemispheric cortical stroke (diffusion-weighted MR (B) and FLAIR MR (C) images below) and was found to have a patent foramen ovale (PFO) as shown on the TEE below (D). The patient was successfully treated with anticoagulation for the pulmonary embolism. At a later date, he underwent closure of the PFO: (A) CT perfusion study demonstrating saddle pulmonary embolism with cut off of lobar pulmonary arteries as well; (B) diffusion-weighted image (DWI) of the right hemispheric embolic infarct involving the right parietal cortical region; (C) fluid-attenuated inversion recovery (FLAIR) image of right hemispheric embolic infarct involving the regions of the right external capsule and right parietal cortex; and (D) Transesophageal echo showing agitated saline "bubbles" passing from the right to left atrium via a patent foramen ovale.

had ischemic brain lesions on a screening magnetic resonance imaging (MRI). Thirty-one of these patients had a PFO. On the baseline assessments, there was no relevant difference in the number of patients with ischemic brain lesions, with or without PFO. At 12 months, 58 of these patients had a follow-up MRI; on the follow-up MRIs, there were 9/58 patients with new ischemic brain lesions on MRI. In this follow-up cohort, however, the rate of new ischemic brain lesions was significantly higher ($P = 0.008$) in those with concomitant PFO (7/9; 78%) vs those without PFO (2/9; 22%) By contrast, a comparison of 34 patients with PFO and PE compared with 120 patients without PFO and PE did not report a statistically significant difference between the two groups (Roy et al., 2020). In this study, 35.3% of the PFO plus PE group had a stroke and 39.2% of the PFO without PE group had a stroke ($P = 0.682$). The PFO plus PE group has a higher mean body mass index, lower left ventricular ejection fraction, lower red blood cell count, and higher frequency of congestive heart failure; however, suggesting these patients had greater medical comorbidities. In patients with symptomatic PE, those with PFO also had a subsequent higher risk of ischemic stroke: 21.4% of patients (9/42) had a recent ischemic stroke, confirmed by MRI in the PFO group vs 5.5% (15/273) in the non-PFO group supporting the hypothesis of paradoxical embolism as a result of VTE in many patients with PFO-associated ischemic stroke (LeMoigne et al., 2019; Schmidt, 2019).

Among patients with an acute PE and PFO, vigilance for new neurologic deficits should be increased, with a low threshold for brain imaging (Zietz et al., 2020). There is an increased association of DVT and stroke in patients with PFO (Cramer et al., 2003). This association reflects the evidence that venous thrombosis is common in adults, and venous thrombi generate numerous migrating emboli, so that right-to-left shunting can expose patients to substantial ischemic stroke risk. Silent venous thrombus formation is common. Estimates are that peripheral venous thrombi may occur at an annual rate of 4%, with a possible 10% occurrence rate during prolonged air travel (Kiernan et al., 2009; Saver et al., 2018; Zietz et al., 2020).

Given that cryptogenic stroke, in the context of PFO or other right to left cardiac shunts, may occur in association with VTE, many clinicians will screen patients with embolic stroke of uncertain etiology (ESUS) for DVT (Saver et al., 2018). When clinically suspected, the threshold for VTE testing should be low (Zietz et al., 2020). The absence of DVT on noninvasive testing in ESUS patients, done in the first few days poststroke, does not, however, exclude the possibility of embolism arising from a venous source causing a stroke via a PFO or atrial septal defect. This is because a small clot that has embolized might not be identifiable on noninvasive testing. Other considerations would include prolonged immobility due to extended travel, illness, or surgical procedures, laboratory data suggestive of a hypercoagulable state, history of prior VTE and the presence of anatomic sources that might contribute to VTE, including the May–Thurner syndrome (Kiernan et al., 2009; Zietz et al., 2020),

The May–Thurner syndrome, resulting in ilio-femoral DVT, occurs as a result of anatomical variant in which the right common iliac artery overlies and compresses the left common iliac vein against the lumbar spine. Peters et al. report that this anatomic variant may be present in over 20% of the population but is rarely considered as a potential source of VTE (Peters, 2012). In a retrospective study of 470 patients who had PFO closure for cryptogenic stroke, 30 patients (6.3%) had MR venography that was suggestive of May–Thurner syndrome (Kiernan et al., 2009). Patients were predominantly female (80%); 54.1% of these women were on hormonal therapies. Twelve (40%) of the patients also had an abnormal laboratory hypercoagulable profile. Only two patients had a prior history of a chronic venous stasis disorder.

One other consideration regarding concomitant ischemic stroke and PE is the impact of the COVID-19 pandemic. Patients with COVID-19 disease due to the recently identified SARS-CoV-2 virus are at higher risk of thromboembolic events (Gill et al., 2020). DVT or PE should be considered in all COVID-19 stroke patients even in the absence of a right-to-left cardiac shunt, as the SARs-CoV-2 virus predisposes patients to both venous and arterial thrombosis.

INTRACRANIAL HEMORRHAGE FOLLOWING VTE

Pharmacologic options for VTE prophylaxis include unfractionated heparin (UFH), low-molecular-weight heparins (LMWH), vitamin K antagonists (VKA), such as warfarin, and direct oral anticoauglants (DOAC) specifically including antifactor Xa inhibitors and direct thrombin inhibitors (Milling and Ziebell, 2020). DOAC are indicated for the prevention of embolic events in patients with nonvalvular atrial fibrillation. These agents are also beneficial in the treatment of VTE patients with a lesser risk of intracranial and systemic hemorrhage as compared with warfarin or LMWH (Lamsam et al., 2018). The main problems with DOAC are the cost of the agents and the difficulty of reversing their anticoagulant effects in the context of life-threatening bleeding (Cotton, 2011). The current American College of Chest Physician (ACCP) Guidelines recommend DOAC (Grade 2B recommendation) as long-term anticoagulant therapy for patients with VTE and no malignancy as opposed to VKA or LMWH. For VTE in cancer patients, the ACCP guidelines recommended LMWH over DOAC or VKA (Kearon et al., 2016). This latter recommendation for LMWH above DOAC has been called into question though DOAC were still associated with a greater bleeding risk as compared with LMWH in cancer patients with VTE (Li et al., 2016). Overall, DOAC have become the first-line treatment for prevention of recurrent events in DVT or PE patients because of a lower risk of major bleeding complications, and the greater ease of use (Di Nisio et al., 2016). A pooled meta-analysis suggested that the risk of bleeding was lowest with rivaroxaban and apixaban, as compared with LMWH, UFH, and/or VKA (Castellucci et al., 2014). This analysis did not report whether the specific risk of intracranial hemorrhage was also reduced. Table 24.1 lists the major anticoagulant drug classes used in the prophylaxis or treatment of VTE, and the reversal agents for each of these drug classes (Dhakal et al., 2017).

The risk–benefit trade-off of anticoagulant VTE prophylaxis vs major bleeding is complex. For every 1000 patients who receive heparin prophylaxis, 91 VTE may be prevented, 35 of which are proximal DVT or PE and 9 to 18 of which are symptomatic DVT or PE. This benefit is associated with an increase of 7 ICH and 28 more minor bleeds per 1000 patients (Hamilton et al., 2011). A meta-analysis reported that the recurrent 3-month incident VTE case fatality rate was 11.3% (95% CI 8–15.2) whereas the case fatality rate for major bleeding was an equally high 11.3% (95% CI 7.5–15.9) (Carrier et al., 2010). A meta-analysis of neurosurgical patients noted that DVT incidence was greater for cranial procedures (7.7%) as compared spinal procedures (1/5%) (Epstein, 2005). UFH or LMWH appeared to reduce the incidence of PE, as opposed to DVT but with an increased risk of both minor and major hemorrhages. The rate was 2%–4% in a cranial series, 3.4% minor and 3.4% major hemorrhages in a combined cranial/spinal series, and a 0.7% incidence of major/minor hemorrhages in a spinal series (Khan et al., 2018). In our own experience of 1638 patients, the use of subcutaneous UFH for VTE prophylaxis was associated with a reduction in the DVT rate from 16% to 9% without any increased risk in serious hemorrhagic complications. PE was uncommon in our population as only 0.8% of patients had an identified PE. There was no evidence of surgical site or intracranial hemorrhage associated with VTE prophylaxis in any of our patients (Khaldi et al., 2011).

Table 24.1

Reversal agents for anticoagulant associated intracranial hemorrhage

Drug	Reversal agents
Vitamin K antagonists (e.g., warfarin)	• Vitamin K • 4-Factor prothrombin complex concentrate (PCC) • Fresh frozen plasma (FFP) • May consider factor VIIa
Factor Xa inhibitors (e.g., apixaban, rivaroxaban)	• Andexanet alfa • 4-Factor PCC (especially if andexanet alfa unavailable)
Direct thrombin inhibitors (e.g., Dabigatran)	• Idarucizumab • If bleeding persists: consider 4-factor PCC • May also consider hemodialysis and activated charcoal
Heparins (e.g., unfractionated heparin, enoxaparin, dalteparin)	• Protamine sulfate
Tissue plasminogen activators (e.g., alteplase)	• Pooled cryoprecipitate • Platelet transfusion • Tranexamic acid

The HAS-BLED score, originally designed for prediction of hemorrhagic risk following anticoagulant therapy in patients with atrial fibrillation, can also be used for assessment of bleeding risk in VTE patients (Brown et al., 2018). In a large retrospective administrative cohort database, a one-point increase in the HAS-BLED score was associated with a 20%–30% increased bleeding risk. Patients with cancer comorbidities were at particularly high increased bleeding risks.

Thrombolytic use in VTE patients is associated with lower all-cause mortality (odds ratio, OR 0.53; 95% CI 0.32–0.88) compared with anticoagulants but with greater risks of major bleeding (OR 2.73; 95% CI 1.91–3.91) and ICH (OR 4.63; 95% CI 1.78–12.04). Thrombolysis was also associated with a lower risk of recurrent PE (OR 0.40; 95% CI 0.22–0.74). In intermediate-risk PE trials, thrombolysis was associated with lower mortality (OR 0.48; 95% CI 0.25–0.92), but again with increased major bleeding events (OR 3.19; 95% CI 2.07–4.92), thrombolysis was associated with a higher ICH rate (OR 4.63; 95% CI 1.78–12.04) (Chatterjee et al., 2014). In a 10-year Nationwide Inpatient Sample (1998–2008) of the United States Agency for Healthcare Research and Quality administrative database cohort, 2,237,600 patients were discharged with a diagnosis of PE (Stein et al., 2012). For those who received thrombolytic therapy for PE, ICH prevalence was 430 of 49,500 (0.9%). Prevalence of ICH in these PE patients increased with agent and was less common in patients with a primary PE diagnosis (250/39,300 [0.6%]) as opposed to patients where PE was a secondary diagnosis (180/10,300 [1.7%], $P < 0.0001$). Younger patients (age 65 or less) or patients without kidney disease also had a lower risk of ICH following PE thrombolysis.

VTE AS THE INITIAL MANIFESTATION OF BRAIN TUMORS

Trousseau syndrome is the development of VTE as an initial manifestation of an underlying malignancy (Metharom et al., 2019). First described in 1865 by Armand Trousseau, this syndrome is most typically seen in gastrointestinal cancers. Ironically, Trousseau himself, subsequent to his description of this phenomenon, developed an upper extremity DVT as the initial manifestation of gastric cancer (Metharom et al., 2019).

The presumed basis for the Trousseau syndrome is the development of a hypercoagulable state associated with malignancy. Patients with systemic or brain malignancies are at higher risk for both arterial and venous thrombosis (Hultcrantz et al., 2018). The mechanism for cancer associated thromboembolism is presumably the release of various tissue factors (especially growth factors and tumor necrosis factors), other cytokines and phospholipids by the malignancy, particularly mucinous tumors, leading to thrombus generation. Cancer-associated VTE is associated with a five times increased risk as compared to VTE in the general population (Caio et al., 2019). Hematologic malignancies, lung cancers, and GI malignancies are most typically associated with this increased risk of hypercoagulability. Despite the increased risk, imaging for occult malignancy is not currently recommended (Khorana et al., 2016; Van Es et al., 2017). Furthermore, primary VTE prophylaxis for ambulatory cancer patients is also not recommended except perhaps in hospital.

Brain tumors are also associated with a high hypercoagulable risk. The risk of VTE is particularly high for patients with CNS malignancies, particularly for those patients with high-grade astrocytomas. In particular, risk factors for VTE among patients with malignant glioma include patient age, tumor size, and ABO blood group; the HR for thrombosis was 2.7 for blood group A and 9.4 for blood group AB (Streiff et al., 2004). Higher plasma levels of coagulation-related molecules, specifically D-dimer, homocysteine, lp (a), VEGF, tPA, and PAI-1, are present in patients with astrocytoma that might contribute to an increased VTE risk (Sciacca et al., 2004). Yet, there seems to be a paucity of identifiable literature wherein a brain cancer was discovered in r which the initial manifestation of that cancer was a hypercoagulable state manifesting as VTE. Thus, there is no clearly described phenomenon of a CNS variant of Trousseau syndrome.

PERIPHERAL NERVOUS SYSTEM COMPLICATIONS OF VTE

Entrapment neuropathies may occur following retroperitoneal hematoma resulting from anticoagulation as part of VTE treatment. Femoral neuropathy, as a result of retroperitoneal hemorrhage, most commonly occurs following pelvic and lower extremity trauma but can occur as a complication of antithrombotic therapies. While the retroperitoneum is an uncommon site for nontraumatic anticoagulation-related hemorrhage, these hemorrhages can occur in upwards of 6% of patients on anticoagulants including those patients treated for VTE (Alberty-Ryöppy et al., 1985; Ho et al., 2003; Macauley et al., 2017). Diabetic neuropathic bladder has also reported as a rare complication of ilio-femoral venous compression with clinical features of acute DVT (Jennings et al., 1988). There are scattered reports of entrapment neuropathies due to venous thrombosis or hemorrhagic complications of venous thrombosis involving the median and ulnar nerves (Nkele, 1986; Grossman and Becker, 1996). Peroneal nerve palsy has also occurred as a rare complication of intermittent compression devices for VTE prophylaxis (McGrory and Burke, 2000). In general the phenomenon of entrapment neuropathy secondary to SVT or DVT appears to be relatively uncommon, however.

In addition to direct nerve injury, or compression from hematoma or venous thrombosis, neuropathic pain may also occur as a result of thrombophlebitis or post-thrombotic syndrome resulting from superficial or DVT (Moustafa et al., 2018). This may occur as a result of inflammation around the nerve, a particular report described involvement of the internal saphenous nerve (Langeron, 1992). Peripheral nerve demyelination may also contribute to the painful sensations of postvenous thrombosis thrombophlebitis. In addition to anticoagulation, other treatments include elastic compression stocking, exercise, and symptomatic pain control primarily with analgesics and antiinflammatory agents (Cesarone et al., 2007).

CONCLUSIONS

With the exception of intracranial hemorrhage in the context of anticoagulation for treatment of DVT or PE, neurologic sequelae of VTE are rare. While the harm of anticoagulated ICH is a concern, the risk–benefit ratio favors both acute thrombolysis and prophylaxis against incident or recurrent VTE. The other main neurologic problems following a VTE event are concomitant acute ischemic stroke or increased future risk of ischemic stroke. That risk is increased by the presence of a hypercoagulable state or right-to-left cardiac shunt. Other neurologic complications related to VTE are rare.

References

Alberty-Ryöppy A, Juntunen J, Salmi T (1985). Femoral neuropathy following anticoagulant therapy for "economy class syndrome" in a young woman. A case report. Acta Chir Scand 151: 643–645.

Allport LE et al. (2008). Thrombolysis for concomitant acute stroke and pulmonary embolism. J Clin Neurosci 15: 917–920 PMID: 18474426.

Barros-Gomes S, El Sabbagh A, Eleid MF et al. (2018). Concomitant acute stroke, pulmonary and myocardial infarction due to in-transient thrombus across a patent foramen ovale. Echo Res Pract 5: I9–I10.

Bilora F, Ceresa M, Milan M et al. (2017). The impact of deep vein thrombosis on the risk of subsequent cardiovascular events: a 14-year follow-up study. Int Angiol 36: 156–159.

Brown JD, Goodin AJ, Lip GYH et al. (2018). Risk stratification for bleeding complications in patients with venous thromboembolism: application of the HAS_BLED core during the first six months of anticoagulant treatment. J Am Heart Assoc 7: e007901.

Caio JF, Luciana TK, Morinaga J (2019). Cancer-associated thrombosis: the when, how and why. Eur Respir Rev 28: 180119. https://doi.org/10.1183/16000617.0119-2018.

Cannegieter SC, Horváth-Puhóe E, Schmidt M (2015). Risk of venous and arterial thrombotic events in patients diagnosed with superficial vein thrombosis: a nationwide cohort study. Blood 125: 229–235.

Carrier M, Le Gal G, Wells PS et al. (2010). Systematic review: case-fatality rates of recurrent venous thromboembolism and major bleeding events among patients treated for venous thromboembolism. Ann Intern Med 152: 578–589.

Castellucci LA, Cameron C, Le Gal G et al. (2014). Clinical and safety outcomes associated with treatment of acute venous thromboembolism: a systematic review and meta-analysis. JAMA 312: 1122–1135.

Cesarone MR, Belcaaro G, Augus G et al. (2007). Management of superficial vein thrombosis and thrombophlebitis: status and expert opinion document. Angiology 58: 7S–14S.

Chatterjee S, Chakraborty A, Weinberg I et al. (2014). Thrombolysis for pulmonary embolism and risk of all-cause mortality, major bleeding, and intracranial hemorrhage. A meta-analysis. JAMA 311: 2414–2421.

Cramer SC, Maki JH, Waitches GM et al. (2003). Paradoxical emboli from calf and pelvic veins in cryptogenic stroke. J Neuroimaging 13: 218–223, PMID: 12889167.

Dhakal P, Rayamajhi S, Verma V et al. (2017). Reversal of anticoagulation and management of bleeding in patients on anticoagulants. Clin Appl Thromb Hemost 23: 410–415.

Di Nisio M, van Es N, Buller HR (2016). Deep vein thrombosis and pulmonary embolism. Lancet 388: 3060–3073.

Epstein NE (2005). Review of the risks and benefits of differing prophylaxis regimens for the treatment of deep venous thrombosis and pulmonary embolism in neurosurgery. Surg Neurol 64: 295–301.

Gill I, Chan S, Fitzpatrick D (2020). COVID-19-associated pulmonary and cerebral thromboembolic disease. Radiol Case Rep 15: 1242–1249.

Grossman JA, Becker GA (1996). Ulnar neuropathy caused by a thrombosed ulnar vein. Case report and literature review. Ann Chir Main Memb Super 15: 244–247.

Hamilton MG, Yee WH, Hull RD et al. (2011). Venous thromboembolism prophylaxis in patients undergoing cranial neurosurgery: a systematic review and meta-analysis. Neurosurgery 68: 571–581.

Ho KJ, Gawley SD, Young MR (2003). Psoas haematoma and femoral neuropathy associated with enoxaparin therapy. MR Int J Clin Pract 57: 553–554.

Hultcrantz M, Bjorkhom M, Dickman PW et al. (2018). Risk for arterial and venous thrombosis in patients with myeloproliferative neoplasms: a population-based cohort study. Ann Intern Med 168: 317–325.

Jennings AM, Walker M, Ward JD (1988). Diabetic neuropathic bladder associated with clinical features of iliofemoral venous thrombosis. Diabet Med 5: 391–392.

Kearon C, Aki EA, Ornelas J et al. (2016). Antithrombotic therapy for VTE disease: CHEST guideline and expert panel report. Chest 19: 315–352.

Khaldi A, Helo N, Schneck MJ (2011). Origatano TC venous thromboembolism: deep venous thrombosis and pulmonary embolism in a neurosurgical population. J Neurosurg 114: 40–46.

Khan NR, Patel PG, Sharpe JP et al. (2018). Chemical venous thromboembolism prophylaxis in neurosurgical patients: an updated systematic review and meta-analysis. J Neurosurg 129: 906–915.

Khorana AA, Carrier M, Garcia DA et al. (2016). Guidance for the prevention and treatment of cancer-associated venous thromboembolism. J Thromb Thrombolysis 41: 81–91.

Kiernan TH, Yan BP, Cubeddu RJ et al. (2009). May-Thurner syndrome in patients with cryptogenic stroke and patent foramen ovale: an important clinical association. Stroke 40: 1502–1504.

Lak HM, Ahmed T, Nair R et al. (2020). Simultaneous multifocal paradoxical embolism in an elderly patient with patent foramen ovale: a case report. Cureus 12: e6992.

Lamsam L, Sussman ES, Iyer AK et al. (2018). Intracranial hemorrhage in deep vein thrombosis/pulmonary embolus patients without atrial fibrillation direct oral anticoagulants versus warfarin. Stroke 49: 1866–1871.

Langeron P (1992). Painful manifestations of the sequelae of phlebitis. Phlebologie 45: 51–58.

Lapostolle F, Borron SW, Surget V (2003). Stroke associated with pulmonary embolism after air travel. Neurology 60: 1983–1985.

LeMoigne E, Timsit S, Ben Salem D et al. (2019). Patent foramen ovale and ischemic stroke in patients with pulmonary embolism: a prospective cohort study. Ann Intern Med 170: 756–763. https://doi.org/10.7326/M18-3485.

Li A, Garcia DA, Lyman GH et al. (2016). Direct oral anticoagulant (DOAC) versus low-molecular-weight heparin (LMWH) for treatment of cancer associated thrombosis (CAT): a systematic review and meta-analysis. Thrombosis Research 173: 158–163.

Lio KU, Jiménez D, Moores L et al. (2020). Clinical conundrum: concomitant high-risk pulmonary embolism and acute ischemic stroke. Emerg Radiol 27: 433–439.

Macauley P, Soni P, Akkad I et al. (2017). Bilateral femoral neuropathy following psoas muscle hematomas caused by enoxaparin therapy. Am J Case Rep 18: 937–940.

Madridano O, del Toro J, Lorenzo A et al. (2015). RIETE investigators subsequent arterial ischemic events in patients receiving anticoagulant therapy for venous thromboembolism. J Vasc Surg Venous Lymphat Disord 3: 135–141.

McGrory BJ, Burke DW (2000). Peroneal nerve palsy following intermittent sequential pneumatic compression. Orthopedics 23: 1103–1105.

Metharom P, Falasca M, Berndt MC (2019). The history of armand trousseau and cancer-associated thrombosis. Cancers (Basel) 11: 158.

Milling JRTJ, Ziebell CM (2020). A review of oral anticoagulants, old and new, in major bleeding and the need for urgent surgery. Trends Cardiovasc Med 30: 86–90.

Moustafa A, Alim HM, Chowdhury MA et al. (2018). Postthrombotic syndrome: long-term sequela of deep venous thrombosis. Am J Med Sci 356: 152–158.

Naidoo P, Hift R (2011). Massive pulmonary thromboembolism and stroke. Case Rep 2011: 398571. https://doi.org/10.1155/2011/398571.

Nkele CJ (1986). Acute carpal tunnel syndrome resulting from haemorrhage into the carpal tunnel in a patient on warfarin. Hand Surg Br 11: 455–456.

Pavesi PC, Pedone C, Crisci M et al. (2008). Concomitant submassive pulmonary embolism and paradoxical embolic stroke after a long flight: which is the optimal treatment? J Cardiovasc Med 9: 1070–1073.

Roy S, Le H, Ayobamidele B et al. (2020). Risk of stroke in patients with patent foramen ovale who had pulmonary embolism. J Clin Med Res 12: 190–199.

Saver J, Mattle JL, Thaler D (2018). Patent foramen ovale closure versus medical therapy for cryptogenic ischemic stroke. A topical review. Stroke 49: 1541–1548.

Schmidt MR (2019). Patent foramen ovale: a villain in pulmonary embolism? Ann Intern Med. 170: 805–806. https://doi.org/10.7326/M19-1089.

Schneck MJ (2014). Chapter 20—Venous thromboembolism in neurologic disease (Chapter 20). In: J Biller, JM Ferro (Eds.), Neurologic aspects of systemic disease part I. Handbook of clinical neurology. vol. 119. Elsevier. 289–304.

Sciacca FL, Ciusani E, Silvani A et al. (2004). Genetic and plasma markers of venous thromboembolism in patients with high grade glioma. Clin Cancer Res 10: 1312–1317. https://doi.org/10.1158/1078-0432.ccr-03-0198.

Sørensen HT, Horvath-Puho E, Pedersen L et al. (2007). Venous thromboembolism and subsequent hospitalization due to acute arterial cardiovascular events: a 20-year cohort study. Lancet 370: 1773–1779. https://doi.org/10.1016/S0140-6736(07)61745-0.

Stein PD, Matta F, Steinberger DS et al. (2012). Intracerebral hemorrhage with thrombolytic therapy for acute pulmonary embolism. Am J Med 125: 50–56.

Streiff MB, Segal J, Grossman SA et al. (2004). ABO blood group is a potent risk factor for venous thromboembolism in patients with malignant gliomas. Cancer 100: 1717–1723. https://doi.org/10.1002/cncr.20150.

Van Es N, Legal G, Otten H-M et al. (2017). Screening for cancer in patients with unprovoked venous thromboembolism: protocol for a systematic review and individual patient data meta-analysis. BMJ Open 7: e015562.

Vindis D, Hutyra M, Sanak D et al. (2018). Patent foramen ovale and the risk of cerebral infarcts in acute pulmonary embolism—a prospective observational study. J Stroke Cerebrovasc Dis 27: 357–364.

Zhang HY, Zhang Y, Cao YJ et al. (2017). Acute inferior ST-segment elevation myocardial infarction and previous cryptogenic stroke caused by a paradoxical embolism with a concomitant pulmonary embolism. J Geriatr Cardiol 14: 421–424.

Zietz A, Sutter R, De Marchis GM (2020). Deep vein thrombosis and pulmonary embolism among patients with a cryptogenic stroke linked to patent foramen ovale—a review of the literature. Front Neurol 11: 336. https://doi.org/10.3389/fneur.2020.00336.

Handbook of Clinical Neurology, Vol. 177 (3rd series)
Heart and Neurologic Disease
J. Biller, Editor
https://doi.org/10.1016/B978-0-12-819814-8.00031-7

Chapter 25

Neurologic complications of cardiac disease in athletes

SARKIS MORALES-VIDAL[1]*, ROBERT LICHTENBERG[2], AND CHRISTINE WOODS[1]

[1]*Department of Neurology, Loyola University Chicago, Stritch School of Medicine, Maywood, IL, United States*
[2]*Heart Care Centers of Illinois, Mokena, IL, United States*

Abstract

Athletic participation at all levels of proficiency is an encouraged activity. Physicians evaluating athletes are tasked with assessing the benefits and risks of participating in vigorous physical activity and should engage in shared decision making with the athlete. Identifying the neurologic sequelae is an essential part of the assessment that is often not covered. This chapter will review the association of a wide range of cardiac disorders that can be related to or associated with subsequent neurologic sequelae, along with a brief overview of recommendations for management. Prevalent neurological complications of cardiac disease in athletes include stroke and seizures. There are also certain channelopathies that result in concurrent cardiac dysrhythmias and epilepsy. In addition, physiologic cardiac rhythm changes and the athlete's heart are discussed in the context of the differential diagnoses of subsequent cardiac and neurologic disease. The primary objective of this chapter is to prepare the physician for accurate recognition of cardiac disease in athletes that could result in neurologic complications if not diagnosed and managed early on.

INTRODUCTION

Engaging in athletic events is common by individuals who are considered athletes and those who are not. Although there is no standard definition of an athlete, the term originates from the Greek root "athlos" meaning achievement. An athlete is an individual who competes in one or more sports involving physical strength, speed, or endurance (Araujo, 2016). Athletes may be professionals or amateurs. They may have known cardiovascular abnormalities or clinically silent cardiovascular disease. It has been estimated that in the United States (US) more than 35 million youths between 5 and 18 years of age participate in organized sports including 8 million competing in high school varsity sports (Lawless et al., 2014). While participation in sports is considered safe in general, certain medical disorders are associated with an increased risk of neurological complications. Multiple guidelines have been developed to address the risks and benefits of sport participation and to define the safety of athletic participation (Maron et al., 2015). Neurologic manifestations of cardiac disease in athletes may have a direct impact to the nervous system (e.g., stroke) or may be associated with a related neurological disorder (e.g., channelopathies manifesting with cardiac arrhythmias or seizures).

Sudden cardiac arrest is the most recognizable and feared complication of cardiac disease among athletes (Meagan et al., 2016; Landry et al., 2017). The causes of sudden cardiac death/events among athletes vary based on studied population. Among patients older than 35 years of age, this is primarily the result of atherosclerotic coronary artery disease. A much different group of disorders is observed among those younger than 35 years of age. A major theme for most of these disorders is a predilection for life-threatening ventricular arrhythmias (England et al., 2015; Garritano and Willmarth, 2015; Maron et al., 2015; Trahan and Simone, 2015; Rigatelli and Zuin, 2019). Data from the US registry, US military, Italian registry, and the National Collegiate Athletic Association noted that the

*Correspondence to: Sarkis Morales-Vidal, M.D., Associate Professor, Department of Neurology, Loyola University Medical Center, Maywood, IL, United States. Tel: +1-312-330-0963, Fax: +1-708-216-5617, E-mail: sarkis.moralesvidal@lumc.edu

most prominent causes of sudden cardiac death among athletes are hypertrophic cardiomyopathy (HCM), sudden arrhythmic death syndrome, arrhythmogenic right ventricular dysplasia (ARVD), coronary artery anomalies, ischemic heart disease, aortic dissection, channelopathies, and myocarditis among others (Corrado et al., 2003; D'Silva and Papadakis, 2015).

ATHLETE'S HEART

The athlete's cardiovascular system undergoes physiology as well as structural changes as an adaptive mechanism to intense physical activity. The so-called "athlete heart syndrome" (AHS) describes the adaptive physiologic changes commonly observed among athletes seeking clearance for sports participation (Rich and Havens, 2004). The return of structural and physiologic states to pertaining to the following deconditioning is another diagnostic feature of the AHS. Cardiac changes, known as athlete's heart (AH), are the most prominent among all the observed adaptive mechanisms. Conversely, incorrect diagnosis could result in increased mortality in athletes with underlying cardiac disease or disqualify athletes without cardiac disease from participating in sporting activities.

Thus, it is critical to differentiate the athlete's heart from pathologic changes that may warrant restriction from sports participation. The AH results in increased left ventricular (LV) diastolic capacity and increased in LV wall thickness. This physiologic hypertrophy observed in AH is more prominent in certain sports, including distance running, cycling, rowing, or canoeing. One of the most important cardiac diseases to differentiate from the AH is HCM (Corrado et al., 2003; D'Silva and Papadakis, 2015). Changes that support diagnosis of AH include left ventricular wall thickness (LVWT) <12 mm; a homogeneous pattern of LVWT; LV cavity >55 mm; and normal LV filling and a predictable reduction in LVWT of 2–5 mm within 3 months of deconditioning. Changes that support HCM include LVWT >16 mm (on average 20 mm; however, it could be >50 mm); a heterogeneous LVWT pattern; LV cavity <45 mm; left atrial enlargement; abnormal LV filling; family history of HCM; and inverted early peak/atrial contraction (E/A) ratio on transmittal Doppler ultrasound.

The European and American Societies/Associations recommend that affected athletes should have comprehensive evaluation in order to determine their eligibility to engage in competitive sports (Maron et al., 2015; Pelliccia et al., 2019). As an example, athletes affected by dilated cardiomyopathy (DCM) or LV noncompaction cardiomyopathy are considered at significant risk of sudden cardiac death during exercise.

CARDIAC RHYTHM CHANGES AMONG ATHLETES

Intense physical activity may lead to serious electrophysiological changes in athletes (Walker et al., 2010). Electrocardiogram (EKG) changes could be seen in up to 40% of participating athletes. The most commonly observed EKG changes are those indicating LV hypertrophy such as increased voltage and left axis deviation. Other changes include sinus bradycardia, inverted T-waves, first degree atrioventricular block, and QRS prolongation. Identification of true pathology based on these EKG changes in an asymptomatic athlete is relatively scant. For instance, inverted T-waves, particularly in precordial leads V1–V3, are very difficult to interpret among asymptomatic athletes. Nonetheless, diffusely distributed and deeply inverted T-waves in athletes should prompt further investigation of an underlying cardiomyopathy.

There is a growing awareness for exercise-related cardiac arrhythmias. Exercise induced increased cardiac load/output, electrolyte imbalances, autonomic changes, and catecholamine release may trigger an arrhythmia. In such cases, the athlete should be considered for additional evaluation with electrophysiologic studies and to potentially limit activities.

The relationship of AH and atrial fibrillation (AF) is complex. AF and other cardiac arrhythmias may mimic a primary underlying neurologic disorder (Pfammatter et al., 1995). Numerous studies have shown that endurance athletes are more likely to develop AF than nonathletes. The type, intensity, and amount of sport activities appear to influence the risk of developing AF (Guasch et al., 2018). High-intensity endurance training is associated with an increased risk of AF. Lifetime hours of participation also correlate with the risk of developing AF. Endurance athletes are 5.3 times more likely to develop AF (Abdulla and Nielsen, 2009). In comparison, the rate of arterial hypertension confers a 1.8-fold increase AF risk (Ogunsua et al., 2015).

Participation in endurance sports has increased quite dramatically over the past decades, raising the possibility of an increase rate in sport-associated AF cases (Opondo et al., 2018). Women appear relatively protected, although this is probably a reflection of lower number of women in studies (Stergiou and Duncan, 2018). AF risk may be more influenced by duration as compared with sport intensity. Among a cohort study of 50,000 skiers participating in a 90 km endurance race over a 10-year period, faster completion time (a surrogate of intensity) plus the number of previous races completed (a surrogate of duration of sports activity) portended an incremental increase in AF occurrence (Andersen et al., 2013).

The mechanisms by which endurance sports promote the development of AF are unclear. Proposed mechanisms include structural left atrium remodeling, elevated left atrial pressure, cardiac inflammation, myocardial fibrosis, increased vagal tone, and genetic predisposition (Guasch et al., 2018). Left atrial remodeling is of interest since left atrial enlargement is associated with the development and progression of AF among nonathletes. Athletes have been shown to have an increased incidence of left atrial enlargement (Król et al., 2016). Left atrial enlargement in former endurance athletes may well persist for many years following cessation of endurance training. The consequences of endurance sport training may not become apparent for years after cessation of training. Moreover, it is not clear whether the structural remodeling in athletes is permanent or how much of the remodeling is reversible following cessation of endurance exercise (Mihl et al., 2008). Endurance sports may result in permanent structural remodeling potentially leading to a lifelong risk of developing AF.

Management of AF in athletes involves counseling and consideration for pharmacologic or other type of intervention (Raju and Kalman, 2018). The role of exercise in promoting AF should be discussed, acknowledging that there is no hard evidence to guide decision making on this point. It is reasonable to recommend reducing physical activity for athletes practicing endurance sports in order to decrease the frequency and burden of AF. However, there is limited evidence to support this strategy. Athletes with AF should consider moderating, but not stopping, endurance exercise training (Perez-Quilis et al., 2017). For symptomatic athletes, a trial of medical therapy is usually required. The most effective antiarrhythmic is the combination of flecainide with a beta blocker. Amiodarone side effects generally preclude its use. Careful consideration of the athlete's risk of stroke and risks from anticoagulation should be made. Risk score stratification (e.g., CHA2DS2-VASc) is helpful. In general, athletes tend to be healthy and the risks of stroke are low. The type of sport practiced introduces obvious individualized management considerations (e.g., bleeding risk in a competitive mountain biker) regarding oral anticoagulation therapy.

NEUROLOGICAL MANIFESTATION OF CARDIAC DISEASE AMONG ATHLETES

Stroke

Regular physical activity has been shown to decrease the stroke risk (Wendel-Vos et al., 2004). The benefits of regular physical activity include lower long-term mean arterial blood pressure (Jennings et al., 1986), improved lipid profile, decreased in the inflammatory response (Ford, 2002), and better glycemic control (Praet et al., 2008).

Athletes often have physiological changes following physical deconditioning. The physiologic responses during vigorous exercise among deconditioned athletes may have negative health consequences. Bouts of physical activity result in an increased in sympathetic nervous system activity with increased heart rate and systolic blood pressure (SBP). This leads to increased shear stress and changes in the endogenous thrombotic–fibrinolytic balance (Lee et al., 2003). These bouts of rigorous physical activity following deconditioning may trigger ischemic strokes (Mostofsky et al., 2011) and subarachnoid hemorrhage (Anderson et al., 2003). Nonetheless, stroke may also occur following vigorous exercise among nondeconditioned athletes (Thompson et al., 2007).

Cardiac diseases that increase the risk of sudden cardiac death in athletes could potentially increase the risk of ischemic stroke in this population. Most notable are HCM, DCM, noncompaction cardiomyopathy, and cardiac arrythmias.

Ischemic stroke could be the initial presentation of an underlying woven coronary artery anomaly (Fujita et al., 2017; Akcay and Soylu, 2018). In addition, coronary artery aneurysms may coexist with brain aneurysms (Yang et al., 2012). Noncyanotic congenital heart disease may remain asymptomatic for long periods of time and encompass a wide variety of conditions including ventricular or atrial septal defects and aortic stenosis among others. Congenital coronary artery anomaly (CCAA) is an abnormal penetration of a coronary artery into the ventricular wall leading to fibrous scars and replacement fibrosis. CCAA is the cause of ~12% of sudden deaths among athletes.

Anabolic androgenic steroids as performance enhancing drug could also be the underlying culprit of a stroke among athletes (Frankle et al., 1988) as a result of prothrombotic mechanisms including decrease in high-density lipoprotein levels and increase in low-density lipoprotein/cholesterol levels.

Cardiovascular abnormalities are among the most common cause of mortality in patients with Marfan syndrome (von Kodolitsch et al., 2015). The most common neurological manifestation in these patients results from ischemic strokes due to embolic phenomena from underlying dissecting aortic aneurysm or valvular heart disease-related arrhythmias (Zalzstein et al., 2003). The reported frequency of ischemic stroke ranges from 3.5% to 20% (Schievink et al., 1997). Patients with neurovascular events are more likely to have underlying AF (22.2% vs 3.2%) and prosthetic heart valves. In addition, patients with Marfan's syndrome have an increased risk of brain aneurysms but not an increased risk of SAH (Schievink et al., 1997).

Loeys–Dietz syndrome (LDS) is another important disorder to recognize among athletes (Thijssen et al., 2019). LDS is an autosomal dominant disease associated with cerebral and aortic aneurysms and dissections (Zenteno et al., 2014.). Two types of LDS have been recognized. Type 1 consists of arterial tortuosity, hypertelorism, and bifid palate or uvula, and type 2 manifests with cutaneous abnormalities but without craniofacial anomalies. Fig. 25.1 shows cerebral vasculature tortuosity commonly seen in patients with LDS.

Coarctation of the aorta may be asymptomatic or may result in high blood pressure in the upper limbs and low blood pressure in the lower limbs. The severity of the coarctation is determined by arm/leg pressure gradient, exercise testing, echocardiogram, and MRI studies. Aortic coarctation is often associated with bicuspid aortic valve. There is also a recognized association with brain aneurysms (Singh et al., 2010), as well as an increased risk of ischemic and hemorrhagic stroke (Pickard et al., 2018).

Most patients, except for those with mild aortic coarctation, will undergo intervention (surgical repair or percutaneous balloon angioplasty and stenting) at some point. AHA and American College of Cardiology guidelines provided detailed recommendations for patients with coartcation of the aorta practicing sports (Van Hare et al., 2015) are listed below.

1. Athletes with untreated coarctation of the aorta without significant dilatation of the ascending aorta (z-score $\leq$ 3.0; a score of 3.0 equals 3 standard deviations from the mean for patient size), a normal exercise test and resting SBP gradient <20 mmHg, and a peak SBP < 95th percentile of predicted with exercise can participate in all competitive sports (Class I; Level C). Athletes with untreated coarctation of the aorta with a SBP gradient >20 mm Hg or exercise-induced hypertension (a peak SBP > 95th percentile of predicted with exercise) or with significant ascending aortic dilation (z-score > 3.0) may be considered for participation only in low-intensity class IA sports (*Class IIb; Level of Evidence C*).
2. Athletes with aortic coarctation of the aorta who are >3 months past surgical repair or stent placement with <20 mm Hg blood pressure gradient at rest, as well as a normal exercise test with no significant dilation of the ascending aorta (z-score < 3.0), no aneurysm at the site of coarctation intervention, and no significant concomitant aortic valve disease may be considered for participation in competitive sports but with the exception of high-intensity static exercise (classes IIIA, IIIB, and IIIC), as well as sports that pose a danger of bodily collision (*Class IIb; Level of Evidence C*). Athletes with evidence of significant aortic dilation or aneurysm formation (not yet at a size to need surgical repair) may be considered for participation only in low-intensity (classes IA and IB) sports (*Class IIb; Level of Evidence C*).
3. Athletes with coarctation and without significant ascending aortic dilation (z-score $\leq$ 3.0) with a normal exercise test and a resting SBP gradient <20 mmHg between the upper and lower limbs and a peak SBP not exceeding the 95th percentile of predicted with exercise can participate in all competitive sports (*Class I; Level of Evidence C*).

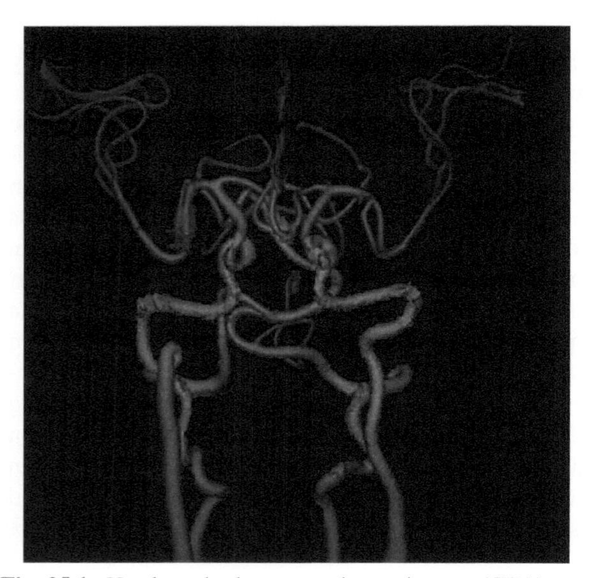

Fig. 25.1. Head cerebral tomography angiogram (CTA).

Seizures

Seizures are seldom the initial clinical manifestation of an underlying cardiac disease among athletes. However, conditions should be considered as well. Exercise-induced syncope and generalized seizures could be the initial presentation of an underlying woven CAA (Fujita et al., 2017; Akcay and Soylu, 2018). Seizure-like episodes may also be the initial manifestation of ARVD (Kimber et al., 2018; Parikh et al., 2019), a cardiac disorder where an association with Charcot–Marie tooth type 2B1 has been reported (Liang et al., 2016).

Certain channelopathies result in epilepsy and cardiac disease and should be considered among athletes with cardiac arrhythmias, seizures, or neuropathies (Imbrici et al., 2016). Potassium, sodium, and calcium channelopathies with a long/short QT syndrome or Brugada syndrome have been associated with epilepsy syndromes, ataxic syndromes, and polyneuropathies. Brugada syndrome is characterized by unique ECG changes in the ST-Segment in the right precordial lead V1. The clinical presentation is usually cardiac arrest, syncope, or palpitations; however, an association between epilepsy and Brugada syndrome has been reported (Fauchier et al., 2000). A specific mutation in the KCNQ1 ion channel has been coexpressed in the heart and brain and implicated in both seizures and arrhythmias (Goldman et al., 2009). Potassium channelopathies with long/short QT syndrome include KCNQ1, KCNH2, KCNE1, and KCNJ2. Those associated with Brugada syndrome include KCNE3 and HCN4. Sodium channelopathies associated with long/short QT syndrome include SCN9A, SCN10A, and SCN11A; those associated with Brugada syndrome include SCN5A, SCN1B, and SCN3B. CACNA1C is associated with both long/short QT syndrome and Brugada syndrome.

SUMMARY

Athletes have a different cardiac anatomy and physiology compared to nonathletes. It is crucial to differentiate the AH from underlying cardiac disease. Timely and accurate recognition of cardiac disease may reduce neurological complications among athletes. AF associated with endurance sports have likely increased in prevalence. Ischemic stroke, exercise-induced syncope, and generalized seizures could be the initial presentation of an underlying congenital CAA. Marfan syndrome, LDS, and coarctation of the aorta are among the aortic disorders to be considered in athletes. Finally, sodium, potassium, and calcium channelopathies may result in an array of neurological and cardiac disorders.

REFERENCES

Abdulla J, Nielsen JR (2009). Is the risk of atrial fibrillation higher in athletes than in the general population? A systematic review and meta-analysis. Europace 11: 1156–1159. https://doi.org/10.1093/europace/eup197. Epub 2009 Jul 24.

Akcay M, Soylu K (2018). An unusual etiology of ischemic stroke: woven coronary artery anomaly. J Saudi Heart Assoc 30: 316–318. https://doi.org/10.1016/j.jsha.2018.05.001. Epub 2018 May 24.

Andersen K, Farahmand B, Ahlbom A et al. (2013). Risk of arrhythmias in 52 755 long-distance cross-country skiers: a cohort study. Eur Heart J. 34: 3624–3631.

Anderson C, Ni Mhurchu C, Scott D et al. (2003). Triggers of subarachnoid hemorrhage: role of physical exertion, smoking, and alcohol in the Australasian Cooperative Research on Subarachnoid Hemorrhage Study (ACROSS). Stroke 34: 1771–1776.

Araujo CGS (2016). The terms "Athlete" and "Exercisers". Expert analysis. https://www.acc.org/latest-in cardiology/articles/2016/06/27/07/06/the-terms-athlete-and-exercisers. Jun 27.

Corrado D, Basso C, Rizzoli G et al. (2003). Does sports activity enhance the risk of sudden death in adolescents and young adults? J Am Coll Cardiol 42: 1959–1963.

D'Silva A, Papadakis M (2015). Sudden cardiac death in athletes. Eur Cardiol 10: 48–53.

England H, Hoffman C, Hodgman T et al. (2015). Effectiveness of automated external defibrillators in high schools in greater Boston. Am J Cardiol 95: 1484–1486.

Fauchier L, Babuty D, Cosnay P (2000). Epilepsy, Brugada syndrome and the risk of sudden unexpected death. J Neurol 247: 643–644.

Ford ES (2002). Does exercise reduce inflammation? Physical activity and C-reactive protein among U.S. adults. Epidemiology 13: 561–568.

Frankle MA, Eichberg R, Zachariah SB (1988). Anabolic androgenic steroids and a stroke in an athlete: case report. Arch Phys Med Rehabil 69: 632–633.

Fujita S, Sato A, Nagata Y et al. (2017). Congenital left main coronary artery atresia presenting as syncope and generalized seizure during exercise in a 13-year-old boy. J Cardiol Cases 16: 126–130.

Garritano NF, Willmarth SM (2015). Student athletes, sudden cardiac death, and lifesaving legislation: a review of the literature. J Pediatr Health Care 29: 233–242.

Goldman AM, Glasscock E, Yoo J et al. (2009). Arrhythmia in heart and brain: KCNQ1 mutations link epilepsy and sudden unexplained death. Sci Transl Med 1: 2–6.

Guasch E, Mont L, Sitges M (2018). Mechanisms of atrial fibrillation in athletes: what we know and what we do not know. Neth Heart J 26: 133–145.

Imbrici P, Liantonio A, Camerino GM et al. (2016). Therapeutic approaches to genetic ion channellopathies and perspective in drug discovery. Front Pharmacol 7: 121.

Jennings G, Nelson L, Nestel P et al. (1986). The effects of changes in physical activity on major cardiovascular risk factors, hemodynamics, sympathetic function, and glucose utilization in man: a controlled study of four levels of activity. Circulation 73: 30–40.

Kimber JR et al. (2018). Syncopal episodes of arrhythmogenic right ventricular cardiomyopathy in a patient with preexisting seizure disorder. Cureus 10: 6.e2760. https://doi.org/10.7759/cureus.2760.

Król W, Jędrzejewska I, Konopka M et al. (2016). Left atrial enlargement in young high-level endurance athletes—another sign of athlete's heart? J Hum Kinet 53: 81–90.

Landry CH, Allan KS, Connelly KA et al. (2017). Rescu investigators. Sudden cardiac arrest during participation in competitive sports. N Engl J Med 377: 1943–1953. https://doi.org/10.1056/NEJMoa1615710.

Lawless CE, Asplund C, Asif IM et al. (2014). Protecting the heart of the American athlete: proceedings of the American College of Cardiology Sports and Exercise Cardiology Think Tank. J Am Coll Cardiol 60: 2146–2171.

Lee CD, Folsom AR, Blair SN (2003). Physical activity and stroke risk: a meta-analysis. Stroke 34: 2475–2481.

Liang JJ, Grogan M, Ackerman MJ et al. (2016). LMNA-mediated arrhythmogenic right ventricular cardiomyopathy and Charcot-Marie-tooth type 2B1: a patient-discovered unifying diagnosis. J Cardiovasc Electrophysiol 27: 868–871.

Maron BJ, Udelson JE, Bonow RO et al. (2015). American Heart Association Electrocardiography and Arrhythmias Committee of Council on Clinical Cardiology, Council on Cardiovascular Disease in Young, Council on Cardiovascular and Stroke Nursing, Council on Functional Genomics and Translational Biology, and American College of

Cardiology. Eligibility and Disqualification Recommendations for Competitive Athletes With Cardiovascular Abnormalities: Task Force 3: Hypertrophic Cardiomyopathy, Arrhythmogenic Right Ventricular Cardiomyopathy and Other Cardiomyopathies, and Myocarditis: A Scientific Statement From the American Heart Association and American College of Cardiology. Circulation 132: e273–e280.

Meagan MW, Hutter AM, Weiner RB (2016). Sudden cardiac death in athletes. J Methodist Debakey Cardiovasc J 12: 76–80.

Mihl C, Dassen WR, Kuipers H (2008). Cardiac remodeling: concentric versus eccentric hypertrophy in strength and endurance athletes. Neth Heart J 16: 129–133.

Mostofsky E, Laier E, Levitan EB et al. (2011). Physical activity and onset of acute ischemic stroke: the stroke onset study. Am J Epidemiol 173: 330–336. https://doi.org/10.1093/aje/kwq369.

Ogunsua AA, Shaikh AY, Ahmed M et al. (2015). Atrial fibrillation and hypertension: mechanistic, epidemiologic, and treatment parallels. Methodist Debakey Cardiovasc J 11: 228–234.

Opondo MA, Aiad N, Cain MA et al. (2018). Does high-intensity endurance training increase the risk of atrial fibrillation? A longitudinal study of left atrial structure and function. Circ Arrhythm Electrophysiol 11: e005598.

Parikh JM, Ganeshwala G, Mathew N et al. (2019). Young stroke: an unusual presentation of arrhythmogenic right ventricular dysplasia. Neurol India 67: 1528–1531. https://doi.org/10.4103/0028-3886.273639.

Pelliccia A, Solberg EE, Papadakis M et al. (2019). Recommendations for participation in competitive and leisure time sport in athletes with cardiomyopathies, myocarditis, and pericarditis: position statement of the Sport Cardiology Section of the European Association of Preventive Cardiology (EAPC). Eur Heart J 40: 19–33.

Perez-Quilis C, Lippi G, Cervellin G et al. (2017). Exercising recommendations for paroxysmal AF in young and middle-aged athletes (PAFIYAMA) syndrome. Ann Transl Med 5: 24.

Pfammatter JP, Donati F, Dürig P et al. (1995). Cardiac arrhythmias mimicking primary neurological disorders: a difficult diagnostic situation. Acta Paediatr 84: 569–572.

Pickard SS, Gauvreau K, Gurvitz M et al. (2018). Stroke in adults with coarctation of the aorta: a national population-based study. J Am Heart Assoc 7: e009072.

Praet SF, van Rooij ES, Wijtvliet A et al. (2008). Brisk walking compared with an individualized medical fitness program for patients with type 2 diabetes: a randomized controlled trial. Diabetologia 51: 736–746.

Raju H, Kalman JM (2018). Management of atrial fibrillation in the athlete. Heart Lung Circ 27: 1086–1092.

Rich BS, Havens SA (2004). The athletic heart syndrome. Curr Sports Med Rep 3: 84–88.

Rigatelli G, Zuin M (2019). Aborted sudden death in an athlete. J Am College Cardiol Cardiovasc Interv 12: e173–e174.

Schievink WI, Parisi JE, Piepgras DG et al. (1997). Intracranial aneurysms in Marfan's syndrome: an autopsy study. Neurosurgery 41: 866–870.

Singh PK, Marzo A, Staicu C et al. (2010). The effects of aortic coarctation on cerebral hemodynamics and its importance in the etiopathogenesis of intracranial aneurysms. J Vasc Interv Neurol 3: 17–30.

Stergiou D, Duncan E (2018). Atrial fibrillation (AF) in endurance athletes: a complicated affair. Curr Treat Options Cardiovasc Med 20: 98.

Thijssen CGE, Bons LR, Gökalp AL et al. (2019). Exercise and sports participation in patients with thoracic aortic disease: a review. Expert Rev Cardiovasc Ther 17: 251–266.

Thompson PD, Franklin BA, Balady GJ et al. (2007). Exercise and acute cardiovascular events placing the risks into perspective: a scientific statement from the American Heart Association Council on Nutrition, Physical Activity, and Metabolism and the Council on Clinical Cardiology. Circulation 115: 2358–2368.

Trahan M, Simone S (2015). Aborted sudden cardiac death in a 14 year old athlete: the anomalous coronary artery. J Pediatr Health Care 28: 366–371.

Van Hare GF, Ackerman MJ, Evangelista JA et al. (2015). American Heart Association Electrocardiography and Arrhythmias Committee of Council on Clinical Cardiology, Council on Cardiovascular Disease in Young, Council on Cardiovascular and Stroke Nursing, Council on Functional Genomics and Translational Biology, and American College of Cardiology. Eligibility and Disqualification Recommendations for Competitive Athletes With Cardiovascular Abnormalities: Task Force 4: Congenital Heart Disease: A Scientific Statement From the American Heart Association and American College of Cardiology. Circulation 132: e281–e291.

von Kodolitsch Y, De Backer J, Schüler H et al. (2015). Perspectives on the revised Ghent criteria for the diagnosis of Marfan syndrome. Appl Clin Genet 8: 137–155.

Walker J, Calkins H, Nazarian S (2010). Evaluation of cardiac arrhythmia among athletes. Am J Med 123: 1075–1081.a.

Wendel-Vos GC, Schuit AJ, Feskens EJ et al. (2004). Physical activity and stroke. A meta-analysis of observational data. Int J Epidemiol 33: 787–798.

Yang EH, Kapoor N, Gheissari A et al. (2012). Coronary and intracerebral arterial aneurysms in a young adult with acute coronary syndrome. Tex Heart Inst J 39: 380–383.

Zalzstein E, Hamilton R, Zucker N et al. (2003). Aortic dissection in children and young adults: diagnosis, patients at risk, and outcomes. Cardiol Young 13: 341–344.

Zenteno M, Lee A, Alvis-Miranda HR et al. (2014). Cerebral aneurysm in a patient with Loeys-Dietz syndrome. J Neurosci Rural Pract 5: 198–199.

Handbook of Clinical Neurology, Vol. 177 (3rd series)
Heart and Neurologic Disease
J. Biller, Editor
https://doi.org/10.1016/B978-0-12-819814-8.00016-0

Chapter 26

Cerebrovascular manifestations of tumors of the heart

HAROLD P. ADAMS Jr*

Division of Cerebrovascular Diseases, Department of Neurology, Carver College of Medicine, University of Iowa Hospitals and Clinics, University of Iowa, Iowa City, IA, United States

Abstract

Primary tumors of the heart, most commonly myxoma, are an uncommon cause of ischemic stroke and intracranial aneurysms. The tumors may occur in any age group but are most frequently detected in middle-aged persons with an atypical or cryptogenic stroke. While some patients will have a history of cardiac or constitutional symptoms, in many cases ischemic stroke will be the initial manifestation of the cardiac mass. Myxomas are the most common cardiac tumors, and valvular fibroelastoma is also a potential cardiac cause of stroke. Among patients with stroke, the most common location for a myxoma is the left atrium. Elevations of inflammatory markers provide clue for a myxoma. Cardiac imaging is the most definitive diagnostic study. Treatment centers on surgical removal of the cardiac mass may be curative.

INTRODUCTION

Although tumors of the heart are an uncommon cause of neurologic symptoms including stroke, these lesions need to be considered in the etiologic differential diagnosis of patients with otherwise unexplained cerebrovascular events. Overall, neurologic manifestations occur in approximately 15%–45% of patients with cardiac tumors (Rosario et al., 2019). While intracranial hemorrhage, seizures, or encephalopathy may occur, ischemic stroke is the most common neurologic complication.

While the pathology of most of these tumors is benign, their effects are in many ways clinically malignant. Table 26.1 lists a classification of cardiac tumors. Approximately 80% of primary cardiac tumors are myxomas (Baikoussis et al., 2015; Rosario et al., 2019). Malignant tumors of the heart do occur but many are secondary to tumors arising elsewhere in the body (Nomoto et al., 2017; Rosario et al., 2019). Cardiac tumors may be diagnosed in men and women of all ages including children (Mariano et al., 2009). Because of the low prevalence of cardiac tumors, most of the information about their diagnosis, prognosis, and treatment is based on case reports, small clinical series, or reviews of the literature. Information about management is limited, for example, there are no clinical trials. This chapter is organized describing general findings of cardiac tumors in addition to more specific features of different tumor types.

GENERAL CLINICAL PRESENTATIONS OF CARDIAC TUMORS

The classic triad of cardiac tumors consists of symptoms of cardiac obstruction, constitutional symptoms, and evidence of embolization. In fact, the classic triad is rarely found (Roeltgen and Kidwell, 2014; Rosario et al., 2019). The features vary depending on the location of the tumor, the tumor type, and pathological findings. Tumors arising on the right side of the heart most commonly cause valvular dysfunction, pulmonary emboli, or metastatic disease in the lungs (Rosario et al., 2019). A majority of tumors arise on the left side of the heart. These lesions produce valvular dysfunction, obstruction of cardiac outflow, and embolization to the brain or other organs. The emboli may be pieces of tumor or tumor intermixed with thrombus. In addition, metastatic tumors

*Correspondence to: Harold P. Adams Jr, M.D., Department of Neurology, Carver College of Medicine, University of Iowa, 200 Hawkins Drive, Iowa City, IA 52242, United States. Tel: +1-319-356-4110, Fax: +1-319-356-4505, E-mail: harold-adams@uiowa.edu

Table 26.1

Classification of tumors of the heart

Benign primary cardiac tumors
Myxoma
Fibroelastoma
Fibroma
Rhabdomyoma
Teratoma
Lipoma
Paraganglioma
Malignant primary cardiac tumors
Angiosarcoma
Fibrosarcoma
Rhabdomyosarcoma
Liposarcoma
Metastatic cardiac tumors
Carcinoma
Leukemia
Sarcomas

may develop secondary to embolization of tumor tissue. Tumors that are solid in consistency are more likely to present with cardiac complaints, while papillary tumors, which are friable, are associated with embolism.

Cardiac dysfunction is most commonly associated with tumors arising on the septum and those tumors that are generally larger in size (Pinede et al., 2001). Patients develop evidence of valvular dysfunction, in particular mitral regurgitation, and congestive heart failure. Shortness of breath is often a chief complaint (Barnes et al., 2014; Nehaj et al., 2018). Patients also have decreased energy and limited exercise tolerance. Syncope has been reported in up to 50% of patients (Rosario et al., 2019). A dramatic event may be sudden death due to obstruction of outflow. This scenario is most common with large tumors near the mitral orifice. Most patients do not have any specific abnormalities on cardiac examination. Some patients may have diastolic or systolic murmurs. A common murmur is that associated with mitral regurgitation. A classic sign of a large cardiac tumor found on auscultation is a precordial plop. However, this change is rarely heard (Pinede et al., 2001).

A broad range of nonspecific symptoms may occur among patients with cardiac tumors, in particular myxomas (Smith et al., 2012; Roeltgen and Kidwell, 2014). Approximately, one-half of the patients have fatigue, weight loss, pain, low grade fever, or skin eruptions that may be transient and that may mimic autoimmune diseases such as systemic lupus erythematosus. These constitutional phenomena are thought to be associated with the production of the cytokines IL-6, VEGF, and bFGF (Smith et al., 2012). A cardiac myxoma may produce findings that suggest a systemic vasculitis (Moreno-Arino et al., 2016). Approximately 10% of patients have cutaneous changes suggestive of vasculitis including Raynaud syndrome, livedo reticularis, splinter hemorrhages, or malar eruptions. A large number of hematologic abnormalities are also found. These include leukocytosis, thrombocytosis, or anemia. Patients may have elevations in erythrocyte sedimentation rate (ESR) or C-reactive protein. Increased antiphospholipid antibodies or decreased factor VII levels also may be found (Smith et al., 2012). Because of the variety of findings, including some response to treatment with steroids, that suggest a vasculitis, the delay in diagnosis may be months in duration (Moreno-Arino et al., 2016).

Embolic events occur in approximately 50% of patients and may be the initial manifestation of the tumor (Elbardissi et al., 2009; Barnes et al., 2014). Besides acute ischemic stroke or transient ischemic attack (TIA), emboli may also affect the kidney, heart, limbs, and viscera (Wang et al., 2018; Wu et al., 2018). Embolization is linked with tumor location in the left atrium or the aortic valve (Elbardissi et al., 2009). The likelihood for embolization is highest with myxoma or fibroelastoma. While the size of the tumor does not predict embolization, increased mobility or a friable surface are recognized features that are related to the development of emboli (Rosario et al., 2019). Because the emboli may contain neoplastic cells, secondary formation of aneurysms or tumors may develop. These complications, which often cause serious neurological events including intracranial hemorrhage, may develop weeks or months after an ischemic stroke.

Evaluation

The evaluation of a patient with a neurologic event thought to be secondary to a cardiac tumor includes assessment of the brain, the cerebral vasculature, the heart, and the blood. The hematologic tests include measurements of ESR, C-reactive protein, and other inflammatory markers. Findings include evidence of the syndrome of disseminated intravascular coagulation: thrombocytopenia, elevated prothrombin time, elevated activated partial thromboplastin time, reduced fibrinogen, and elevated fibrin split products. Brain imaging, either computed tomography (CT) or magnetic resonance imaging (MRI) is performed for to find evidence of hemorrhagic or ischemic stroke. Given the source of the emboli, imaging studies often show abnormalities in all vascular territories. While most embolic events to the brain involve small pieces of tumor or clot, evaluation of the larger extra- or intracranial vessels is appropriate when assessing a patient with a major stroke. Vascular imaging of the brain is also done to screen for tumor-related aneurysms. Although larger aneurysms may be seen by computed tomographic angiography (CTA) or magnetic resonance angiography (MRA), conventional angiography remains the best choice to screen for cerebral aneurysms complicating a cardiac tumor.

Prior to the development of modern imaging of the heart, cardiac tumors were rarely diagnosed prior to autopsy. With the development of echocardiography (transthoracic/TTE or transesophageal/TEE) and the application of CT or MRI to examine the structure of the heart, the ability of physicians to diagnose intracardiac tumors has become facile. Echocardiography now is the first line imaging study (Wen et al., 2018). Nomoto et al. (2017) reported that echocardiography had a diagnostic accuracy of 80%. Because TEE better visualizes the left atrium, which is the most common site for a cardiac tumor, the yield of TEE is superior to TTE. The findings are influenced by the type of tumor and the mass needs to be differentiated from an intracardiac thrombus. The type of tumor also impacts the echocardiographic findings. Many tumors will appear heterogenetic or amorphous. Rarely, calcification can be found. Although CT or MRI may be more sensitive than echocardiography, the role of cardiac imaging is usually restricted to cases in which the echocardiographic findings are inconclusive. Advantages of cardiac imaging are increased ability to differentiate tumor from clot and detection of tumor extending outside the heart (Haji and Nasis, 2017).

Myxoma

Myxomas are the most common primary cardiac tumors (approximately 80%) and much of the previous discussion is influenced by this tumor type. Myxomas arise from mesenchymal cells of the endocardial surface. The tumor consists of a mixture of a mucopolysaccharide cytoplasm and polygonal cells. In addition, the tumor may have cysts, areas of hemorrhage, and calcifications. While myxomas may be found in persons of any age from childhood to the elderly, the mean age for discovery is among persons aged 40–60 (Al-Mateen et al., 2003; Wen et al., 2018). There appears to be a higher prevalence in women. In most cases, a single tumor is identified although multiple intracardiac masses may be identified. The most common site is the left atrium; in one series of 99 patients, 92 had left atrial lesions (Boyacioglu et al., 2017). Tumors most commonly arise from the interatrial septum in the left atrium (He Dk et al., 2015). Cases of a tumor attached by a pedicle to the interatrial septum have been reported (Rosario et al., 2019). Other sites include the mitral valve, left ventricle, and the right side of the heart (Kong et al., 2015).

Carney syndrome is found in approximately 3% of patients with myxoma, and approximately 30% of patients with Carney syndrome will have a myxoma discovered (Rosario et al., 2019). This autosomal dominant mutation of the PRKARIA (protein kinase, cAMP-dependent, regulatory, type I, alpha) gene and protein are involved in the regulation of cell metabolism. Two subvarieties of Carney syndrome have been described (Lee et al., 2018). The major clinical criteria for diagnosis include spotty skin pigmentation, acromegaly, thyroid cancer in childhood, breast myxomatosis, melanotic Schwannoma, blue nevi, and cardiac myxoma (Domanski et al., 2016; Lee et al., 2018). Because of the many noncardiac features, the syndrome is usually diagnosed in childhood. Children with Carney syndrome are at risk for embolic stroke (Domanski et al., 2016).

Two major subtypes of myxoma are identified by pathology. Approximately 65% of tumors are solid and may have hemorrhage within the mass (Swartz et al., 2006). The risk of embolization from this type of tumor is relatively low, and they are more likely to cause outflow obstruction (Barnes et al., 2014). The remaining tumors are papillary and friable; these are associated with embolization of fragments of the tumor or secondary thrombi (Perez Andreu et al., 2013; Boyacioglu et al., 2017; Wen et al., 2018). The tumors have a wide range in size from less than 1 to 15 cm. In the series of Perez Andreu et al. (2013), the mean size among patients with neurologic symptoms was 4.1 cm. The weights of myxomas removed at the time of surgery range from 15 to 180 g. Boyacioglu et al. (2017) found that the risk of embolization was higher among patients with smaller-sized tumors. A Chinese study reported that an atypical location or an irregular surface was associated with an increased risk for embolization (He Dk et al., 2015). Rarely, a myxoma may become infected, which could lead to secondary embolization.

The natural history of atrial myxomas is unpredictable. Many are discovered incidentally during imaging of the heart. Most patients with myxomas do not have all the features of the triad of cardiac problems, a nonspecific inflammatory illness, and embolization (Table 26.2). Cardiac symptoms include dyspnea, chest pain, and sudden death. Arrhythmias are rare. Sudden death is ascribed to a tumor, which is large and often on a pedicle, obstructing the mitral orifice (Swartz et al., 2006). This situation is usually seen with solid tumors. The cardiac examination usually is normal (Knepper et al., 1988). Auscultatory abnormalities may mimic those of mitral valve disease. The previously described tumor plop is a rare finding. The electrocardiogram may be normal or show changes consistent with left atrial hypertrophy.

Embolization to the brain is the most common neurologic presentation (Knepper et al., 1988; Perez Andreu

Table 26.2

Clinical features of cardiac myxoma

Cardiac
Cardiac arrhythmias and syncope
Congestive heart failure
Reduced stamina
A plop sound on auscultation
Sudden death
Systemic
Fever, fatigue, malaise
Skin eruptions
Elevated inflammatory markers
Embolism
Ischemic stroke/transient ischemic attack
Aneurysm
Metastatic disease
Encephalopathy

et al., 2013; Rosario et al., 2019). Overall, stroke will occur in approximately 25% of cases and it may be the initial clinical feature (Knepper et al., 1988; Brinjikji et al., 2015). While ischemic stroke may happen in men or women of any age, the likelihood of an event is greater in young adults (Rosario et al., 2019). As a result, cardiac myxoma should be considered as a potential etiology of stroke in a young adult who does not have an obvious explanation or traditional risk factors. Because the tumors often are quite small, patients may not have any cardiac or constitutional complaints.

Most of the ischemic strokes are in the territory of the middle cerebral artery; as a result, the most frequent findings are aphasia, neglect, hemiparesis, unilateral sensory loss, gaze preference, and visual field deficits (Perez Andreu et al., 2013; Liao et al., 2015; Wen et al., 2018). The most common scenario is multiple areas of brain injury due to obstruction of several branch cortical vessels. Occlusion of major arteries from large pieces of tumor may result in a multilobar infarction. Because multiple emboli may impact many areas of the brain simultaneously, some patients may have findings suggestive of an acute delirium or psychosis (Ekinci and Donnan, 2004; Jain et al., 2014). Less common features are syncope, seizures, and headache (Ekinci and Donnan, 2004; Rosario et al., 2019). Simultaneously, emboli may produce symptoms in the limbs, viscera, or heart.

The incidence of intracranial aneurysms complicating myxomas is unknown and may be underestimated (Rosario et al., 2019). In one series of 47 patients with myxoma, seven had aneurysms detected (Brinjikji et al., 2015). The pathogenesis of the aneurysms involves perivascular damage and invasion by tumor cells, ischemic damage to the vessel walls, and altered flow dynamics (Herbst et al., 2005). The pathologic changes lead to vascular dilation and aneurysm formation. The aneurysms are usually fusiform in shape and located on distal vessels (Liao et al., 2015). Because of the association of multiple distal aneurysms and myxoma, the tumor should be considered as a potential etiology of the intracranial lesions even if other evidence for myxoma is not apparent (Liao et al., 2015; Zheng et al., 2015). The neoplastic aneurysms are a potential delayed complication among patients who have had stroke, although cases of myxomatous aneurysms in patients with no history of an ischemic stroke have been reported (Ashalatha et al., 2005; Roeltgen and Kidwell, 2014; Takenouchi et al., 2014). The interval from the stroke until discovery of the aneurysm may be several months and may be recognized even after successful cardiac surgery (Roeltgen and Kidwell, 2014).

The approach to screening of patients for the presence of aneurysms is uncertain. CT may show hyperdense lesion in distal cortical arteries. The changes may be secondary to tumor in the arterial wall (Rosario et al., 2019). Although screening with MRA or CTA may be performed, the small size and distal locations of the aneurysms mean that angiography is the best method for detecting and monitoring the evolution of the lesions. Because of the uncertain prognosis with outcomes ranging from spontaneous resolution of the lesion to rupture, management is not established. Options include conservative observation, surgery, endovascular occlusion, chemotherapy, and radiotherapy (Zheng et al., 2015).

Besides developing into an aneurysm, a myxomatous embolus may also evolve into a metastatic tumor. In one series of 47 patients, 2 had brain tumors detected during long-term follow-up (Brinjikji et al., 2015). The most common scenario is discovery of a hemorrhagic mass by brain imaging (Rosario et al., 2019). The tumors may be multiple in location and often do not appear until several months after the original diagnosis. The potential for metastatic myxoma is another reason to do follow-up imaging of the brain (Rosario et al., 2019). Patients with metastatic myxoma usually are treated with chemotherapy or radiotherapy.

The emergency treatment of patients with acute ischemic stroke secondary to myxoma is similar to that prescribed to other patients. In fact, the presence of the myxoma as the cause of the stroke likely will not be known at the time of the administration of the acute interventions. Experience with intravenous thrombolysis is limited. Most of the literature involves reports from single cases. Alteplase has been given to some patients with improvement noted (Acampa et al., 2014). On the other hand, the composition of the embolus may portend a lack of success to alteplase. The agent should not be expected to help lyse an embolus consisting primarily of tumor tissue. The risk of bleeding with intravenous thrombolysis probably is similar to that associated with administration of the agents in patients with acute ischemic stroke secondary to other sources of embolism. Presumably, the risk of bleeding would be related to the presence of a myxomatous aneurysm. However, this is a delayed complication of myxoma and likely will not be a major concern at the time of the original diagnosis of ischemic stroke. Pharmacological thrombolysis has been combined with endovascular interventions. Chung et al. (2016) reported successful removal of a myxomatous embolus in a patient with acute ischemic stroke. Other early management includes prevention and treatment of medical or neurological complications, general supportive care, and rehabilitation.

Surgical resection of the tumor is the fundamental treatment of patients with myxoma. Because there is a risk of recurrent embolization, surgery is recommended as soon as possible (Knepper et al., 1988). The goal is to remove the tumor in its entirety and in one piece. However, papillary tumors may need to be taken out piecemeal. Although robotic endoscopic resection is an option, most patients have open heart surgery with cardiopulmonary bypass (Moss et al., 2016). While this is a major operation, the surgery has relatively low morbidity and mortality (Labauge et al., 1993). The 30-day mortality is approximately 3% (Baikoussis et al., 2015; Zheng et al., 2015). A history of embolization does not affect the outcomes following surgery (Boyacioglu et al., 2017). The long-term prognosis of patients having successful surgery is excellent (Knepper et al., 1988). Local recurrence of myxoma is uncommon, but it may result from an incomplete resection or the presence of multiple tumors, which may occur in some genetic disorders (Rosario et al., 2019). There is no clear evidence for the use of oral anticoagulants on a long-term basis and medical management should not be considered as an alternative to surgical treatment of the myxoma.

Fibroelastoma

The prevalence of fibroelastoma is approximately 0.02%–0.45% in the general population (Kumar et al., 2016). As such, it is the most common valvular tumor of the heart. The tumors are more common on the left side of the heart and are usually detected in persons in their 60s (Tamin et al., 2015). They occur in both men and women and are usually diagnosed by echocardiography (Gowda et al., 2003). Usually, fibroelastoma is an incidental finding, but up to 50% of patients it will have neurological symptoms (Tamin et al., 2015; Carino et al., 2017). Stroke may be the initial symptom of the tumor. The small size of the fibroelastoma means that cardiac symptoms are uncommon.

Pathologically, a fibroelastoma is benign and the size ranges from 2 to 70 mm (Gowda et al., 2003; Val-Bernal et al., 2013). A solitary tumor located on the valvular surface is found in most patients (Val-Bernal et al., 2013; Carino et al., 2017). The tumors are located primarily on the valve surface and are most commonly affecting the aortic and mitral valves (Val-Bernal et al., 2013; Carino et al., 2017). In many patients, concurrent valvular disease of other origins may be present.

Most of the symptoms are secondary to embolization (Gegouskov et al., 2008; Val-Bernal et al., 2013). These embolic events are secondary to shedding of pieces of tumor. Tumor mobility is linked to an increased risk of embolic phenomena (Gowda et al., 2003). Besides stroke or TIA, embolization to the heart (coronary arteries), sudden death, syncope, or peripheral embolization have been reported (Gowda et al., 2003). Heart failure may develop secondary to the associated valvular dysfunction. The prognosis in patients with fibroelastoma relates to the mobility of the tumor, and patients may have recurrent embolic events without surgery (Tamin et al., 2015).

Because of the relatively rare nature of fibroelastoma, no controlled trials are available to guide plans for management in the setting of acute ischemic stroke or for long-term treatment to prevent recurrent embolization. Surgical resection of the mass on the valve is recommended and it appears to be relatively safe (Gegouskov et al., 2008; Tamin et al., 2015; Carino et al., 2017). Some patients have received long-term anticoagulant therapy when surgery is not done. The utility of this strategy has not been demonstrated. Either open heart surgery or one of the new endovascular procedures probably is the best treatment option.

Rhabdomyoma

Rhabdomyoma is the most common primary tumor of the heart diagnosed in childhood (Tzani et al., 2017). It may be detected by echocardiography in a fetus, neonate, or young child (Degueldre et al., 2010; Yinon et al., 2010; Moreno-Arino et al., 2016). There is a strong link between cardiac rhabdomyoma and tuberous sclerosis (Quek et al., 1998; Lee et al., 2018). This autosomal dominant disease with mutations in tumor suppression genes produces cardiac tumors in 50% of cases (Lee et al., 2018). The cardiac tumor may be the first detected abnormality of tuberous sclerosis (Lee et al., 2018).

Most cases are asymptomatic (Quek et al., 1998). Some children may have cardiac arrhythmias, such as supraventricular tachycardia, particularly in the neonatal period (Degueldre et al., 2010). Other symptoms include syncope, TIA, or stroke. Outflow obstruction may cause sudden death (Mariano et al., 2009). The prognosis for a child with a rhabdomyoma generally is poor (Tzani et al., 2017).

In part, the poor prognosis is connected with the high likelihood of neurodevelopmental abnormalities associated with tuberous sclerosis (Yinon et al., 2010). Increased fetal and neonatal mortality is correlated with the earlier diagnosis of the tumor and its size (Yinon et al., 2010). Treatment is surgical resection of the mass. Prenatal counseling about tuberous sclerosis is also recommended.

OTHER PRIMARY TUMORS

Cardiac fibroma is one of the features of the Gorlin syndrome; 3%–5% of people with Gorlin syndrome have cardiac fibromas, and a similar ratio is found among patients with cardiac fibromas (Lee et al., 2018). This autosomal dominant mutation in the tumor suppressor gene is associated with nevoid basal cell carcinoma, tumors of the jaw, medulloblastoma, and calcification of the falx cerebri. The presence of the cardiac fibroma is a minor criterion for diagnosis of Gorlin syndrome. Fibromas, which are the second most common primary cardiac tumor, are most commonly found on the intraventricular septum or wall (Lee et al., 2018). Calcified lesions in both the ventricular wall and within the ventricle may be detected by echocardiography.

A paraganglioma is a very rare primary cardiac tumor that may occur in isolation or with similar tumors elsewhere in the body (Khan et al., 2013). The left atrium is the most common site, but they can arise in any location. Local invasion of the heart or pericardium is common (Khan et al., 2013). The tumor may be associated with a number of genetic diseases including neurofibromatosis or von Hippel–Lindau disease. The tumors occur in both men and women, and the mean age of diagnosis is approximately 40 (Khan et al., 2013). Chest pain, poorly controlled hypertension, dyspnea, palpitations, diaphoresis, and headaches are the most common findings. Stroke has been described as a complication (Hayek et al., 2007). The tumor is usually discovered by cardiac imaging. Because these tumors are producing catecholamines, elevated levels of these transmitters are found in approximately 75% of patients (Khan et al., 2013). Surgery is the preferred approach to management.

Although very rare, malignant tumors of the heart may be the source of emboli or metastases (Sun et al., 2017; Taguchi, 2018). Stroke is a potential complication and may be the initial symptom (Caballero, 2008). Most of these tumors are sarcomas that arise from any tissue in the heart (Neragi-Miandoab et al., 2007; Caballero, 2008). A myosarcoma is the malignant version of a myxoma. Other tumors include liposarcoma and rhabdomyosarcoma (Mitomi et al., 2011; Serdaroglu et al., 2011). Clinical features often relate to obstruction of blood flow or interference with valve function. Local invasion of the tumor may cause secondary arrhythmias. Overall, the prognosis among patients with primary malignancies is very poor (Neragi-Miandoab et al., 2007; Taguchi, 2018). Delays in diagnosis, extensive local spread, and high rates of metastatic lesions are reasons for the poor prognosis. In many cases, surgical resection may not be feasible (Neragi-Miandoab et al., 2007). The risk of surgical complications is high (Dias et al., 2014). Chemotherapy and radiotherapy have been prescribed (Neragi-Miandoab et al., 2007).

METASTATIC HEART TUMORS

Metastatic lesions in the heart may be visualized by cardiac imaging among patients with a wide range of cancers. The most common sources are sarcomas, melanoma, gastrointestinal carcinomas, carcinoma of breast, and lung cancers (Pun et al., 2016). In many patients, multiple cardiac masses are detected with the most common site being the right ventricle, which reflects the hematogenous spread of the cancer cells. Most patients will have evidence of metastases in other organs (Pun et al., 2016). In addition, direct invasion of the heart and pericardium from lung cancers may happen. Embolization producing stroke has been reported (Dimitrovic et al., 2016). Stroke has been the presenting symptom of these advanced cancers. The prognosis is poorer than among people with stage IV cancer without cardiac involvement (Pun et al., 2016). Treatment is focused on the underlying malignancy.

CONCLUSIONS

Primary or metastatic tumors of the heart are an uncommon cause of cerebrovascular events. Most physicians will go through their entire career without ever diagnosing or treating a patient with a stroke secondary to a cardiac tumor. Still, the frequency of cardiac tumors, in particular myxomas, is sufficiently high that they should be considered in the etiologic differential diagnosis among patients with atypical stroke. Although the classical triad of cardiac symptoms, vague constitutional complaints, and embolization are associated with myxomas, most patients will not have all these findings. An ischemic stroke may be the initial or only symptom of a cardiac tumor. The emboli, which may be pieces of tumor or associated clot, often are disseminated in multiple vascular territories. No specific ischemic stroke syndrome is associated with cardiac tumors. A cardiac tumor is also a potential explanation among patients with cryptogenic stroke. Elevations in ESR, C-reactive protein, and D-dimer provide clues that a myxoma may be underlying reason for the stroke. The use of cardiac imaging, particularly echocardiography, has greatly facilitated detection of these lesions. TEE is part of the assessment of patients with cryptogenic stroke because it is able to visualize the left atrium, which is the site of most myxomas that produce neurological symptoms. TEE is also the best way to detect a fibroelastoma on the mitral or aortic valves. Because the diagnosis of an underlying cardiac tumor is unlikely at the time of acute treatment of an ischemic stroke with reperfusion interventions, patients likely will be treated with some degree of success depending on the composition of the embolus. Endovascular extraction could be successful because it may be able to remove a piece of the tumor. The risk of bleeding appears not to be great.

Surgical removal of the tumor is the cornerstone of management. Although endovascular procedures have been done, open heart surgery remains the preferred choice. The morbidity and mortality associated with the surgery is relatively low. The role of anticoagulants or antiplatelet agents in prevention of recurrent stroke is limited and should not be considered as an alternative to surgery.

Physicians should also remain alert to delayed cerebrovascular complications including the development of aneurysms, which may cause hemorrhage. These complications may occur among patients who do not have a history of embolization and may arise after surgical treatment of the tumor.

References

Acampa M, Guideri F, Tassi R et al. (2014). Thrombolytic treatment of cardiac myxoma-induced ischemic stroke: a review. Curr Drug Saf 9: 83–88.

Al-Mateen M, Hood M, Trippel D et al. (2003). Cerebral embolism from atrial myxoma in pediatric patients. Pediatrics 112: 162–167.

Ashalatha R, Moosa A, Gupta AK et al. (2005). Cerebral aneurysms in atrial myxoma: a delayed, rare manifestation. Neurol India 53: 216–218.

Baikoussis NG, Papakonstantinou NA, Dedeilias P et al. (2015). Cardiac tumors: a retrospective multicenter institutional study. J BUON 20: 1115–1123.

Barnes H, Conaglen P, Russell P et al. (2014). Clinicopathological and surgical experience with primary cardiac tumors. Asian Cardiovasc Thorac Ann 22: 1054–1058.

Boyacioglu K, Kalender M, Donmez AA et al. (2017). Outcomes follwoing embolization in patients with cardiac myxoma. J Cardiac Surg 32: 621–626.

Brinjikji W, Morris JM, Brown RD et al. (2015). Neuroimaging findings in cardiac myxoma patients: a single-center case series of 47 patients. Cerebrovasc Dis 40: 35–44.

Caballero PE (2008). Left atrial sarcoma presenting as cerebral infarction. Neurologist 14: 131–133.

Carino D, Nicolini F, Molardi A et al. (2017). Unusual locations for cardiac papillary fibroelastomas. J Heart Valve Dis 26: 226–230.

Chung YS, Lee WJ, Hong J et al. (2016). Mechanical thrombectomy in cardiac myxoma stroke: a case report and review of the literature. Acta Neurochir 158: 1083–1088.

Degueldre SC, Chockalingam P, Mivelaz Y et al. (2010). Considerations for prenatal counselling of patients with cardiac rhabdomyomas based on their cardiac and neurologic outcomes. Cardiol Young 20: 18–24.

Dias RR, Fernandes F, Ramires FJ et al. (2014). Mortality and embolic potential of cardiac tumors. Arg Bras Cardiol 103: 13–18.

Dimitrovic A, Breitenfeld T, Supanc V et al. (2016). Stroke caused by lung cancer invading the left atrium. J Stroke Cerebrovasc Dis 25: 66–68.

Domanski O, Dubois R, Jegou B (2016). Ischemic stroke due to a cardiac myxoma. Pediatr Neurol 65: 94–95.

Ekinci EI, Donnan GA (2004). Neurological manifestations of cardiac myxoma: a review of the literature and report of cases. Intern Med J 34: 243–249.

Elbardissi AW, Dearani JA, Daly RC et al. (2009). Embolic potential of cardiac tumors and outcome after resection: a case-control study. Stroke 40: 156–162.

Gegouskov V, Kadner A, Engelberger L et al. (2008). Papillary fibroelastoma of the heart. Heart Surg Forum 11: 333–339.

Gowda RM, Khan IA, Nair CK et al. (2003). Cardiac papillary fibroelastoma: a comprehensive analysis of 725 cases. Am Heart J 146: 404–410.

Haji K, Nasis A (2017). Radiological characteristics of artrial myxoma in cardiac computed tomography. J Cardiovasc Tomogr 11: 234–236.

Hayek ER, Hughes MM, Speakman ED et al. (2007). Cardiac paraganglioma presenting with acute myocardial infarction and stroke. Ann Thorac Surg 83: 1882–1884.

He Dk ZY, Liang Y, Sx Y et al. (2015). Risk factors for embolism in cardiac myxoma: a retrospective analysis. Med Sci Monit 21: 1146–1154.

Herbst M, Wattjes MP, Urbach H et al. (2005). Cerebral embolism from left atrial myxoma leading to cerebral and retinal aneurysms: a case report. Am J Neuroradiol 26: 666–669.

Jain RS, Nagpal K, Jain R et al. (2014). Acute psychosis presenting as a sole manifestation of left atrial myxoma: a new paradigm. Am J Emerg Med 32: 3–5.

Khan MF, Datta S, Chisti MM et al. (2013). Cardiac paraganglioma: clinical presentation, diagnostic approach and factors affecting short and long-term outcomes. Int J Cardiol 166: 315–320.

Knepper LE, Biller J, Adams Jr HP et al. (1988). Neurologic manifestations of atrial myxoma. A 12-year experience and review. Stroke 19: 1435–1440.

Kong Y, Li H, Wang J et al. (2015). Left ventricular myxoma leading to stroke: a rare case report. Medicine (Baltimore) 94: 1913.

Kumar V, Soni P, Hashmi A et al. (2016). Aortic valve fibroelastoma: a rare cause of stroke. BMJ Case Rep. https://doi.org/10.1136/bcr-2016-217631.

Labauge P, Messner-Pellenc P, Blard JM et al. (1993). Neurologic complications of cardiac myxomas. Presse Med 22: 1317–1321.

Lee E, Mahani MG, Lu JC et al. (2018). Primary cardiac tumors associated with genetic syndromes: a comprehensive review. Pediatr Radiol 48: 156–164.

Liao WH, Ramkalawan D, Liu JL et al. (2015). The imaging features of neurologic complications of left atrial myxomas. Eur J Radiol 84: 933–939.

Mariano A, Pita A, Leon R et al. (2009). Primary cardiac tumors in children: a 16-year experience. Rev Port Cardiol 28: 279–288.

Mitomi M, Kimura K, Iguchi Y et al. (2011). A case of stroke due to tumor emboli associated with metastatic cardiac liposarcoma. Intern Med 50: 1489–1491.

Moreno-Arino M, Ortiz-Santamaria V, Deudero Infante A et al. (2016). A classic mimicker of systemic vasculitis. Rheumatol Clin 12: 103–106.

Moss E, Halkos ME, Miller JS et al. (2016). Comparison of endoscopic robotic versus sternotomy approach for the resection of left atrial tumors. Innovations (Phila) 11: 274–277.

Nehaj F, Sokol J, Mokan M et al. (2018). Outcomes of patients with newly diagnosed cardiac myxoma: a retrospective multicentric study. Biomed Res Int. https://doi.org/10.1155/2018/8320793.

Neragi-Miandoab S, Kim J, Vlahakes GJ (2007). Malignant tumours of the heart: a review of tumour type, diagnosis, and therapy. Clin Oncol (R Coll Radiol) 19: 748–756.

Nomoto N, Tani T, Kim K et al. (2017). Primiary and metastic cardiac tumors: echocardiographic diagnosis, treatment and prognosis in a 15 year single center study. J Cardiothorac Surg 12: 103.

Perez Andreu J, Parrilla G, Arribas JM et al. (2013). Neurological manifestations of cardiac myxoma: experience in a referral hospital. Neurologia 28: 529–534.

Pinede L, Duhaut P, Loire R (2001). Clinical presentation of left atrial cardiac myxoma. a series of 112 consecutive cases. Medicine (Baltimore) 80: 159–172.

Pun SC, Plodkowski A, Matasar MJ et al. (2016). Pattern and prognostic implications of cardiac metastases among patients with advanced systemic cancer assessed with cardiac magnetic resonance imaging. J Am Heart Assoc 5. https://doi.org/10.1161/JAHA.116.003368.

Quek SC, Yip W, Quek ST et al. (1998). Cardiac manifestations in tuberous sclerosis: a 10-year review. J Paediatr Child Health 34: 283–287.

Roeltgen D, Kidwell CS (2014). Neurologic complications of cardiac tumors. Clin Neurol 119: 209–222.

Rosario M, Fonseca AC, Sotero FD et al. (2019). Neurological complications of cardiac tumors. Curr Neurol Neurosci Rep 19: 15.

Serdaroglu G, Yilmaz S, Ulger Z et al. (2011). A rare cause of recurrent stroke in childhood: left atrial rhabdomyosarcoma. Acta Paediatr 100: 189–191.

Smith M, Chaudhry MA, Lozano P et al. (2012). Cardiac myxoma induced paraneoplastic syndromes: a review of the literature. Eur J Intern Med 23: 669–673.

Sun YP, Wang X, Gao YS et al. (2017). Primary cardiac sarcoma complicated with cerebral infarction and brain metastasis: a case report and literature review. Cancer Biomark 21: 247–250.

Swartz MF, Lutz CJ, Chandan VS et al. (2006). Atrial myxomas: pathologic types, tumor location, and presenting symptoms. J Cardiothorac Surg 21: 435–440.

Taguchi S (2018). Comprehensive review of the epidemiology and treatments for malignant adult cardiac tumors. Gen Thorac Cardiovasc Surg 66: 257–262.

Takenouchi T, Sasaki A, Takahashi T (2014). Multiple cerebral aneurysms after myxomatous stroke. Arch Dis Child 99: 849.

Tamin SS, Maleszewski JJ, Scott CG et al. (2015). Prognostic and bioepidemiologic implications of papillary fibroelastomas. J Am Coll Cardiol 65: 2420–2429.

Tzani A, Doulamis IP, Mylonas KS et al. (2017). Cardiac tumors in pediatric patients: a systematic review. World J Pediatr Congenit Heart Surg 8: 624–632.

Val-Bernal JF, Mayorga M, Garijo MF et al. (2013). Cardiac papillary fibroelastoma: retrospective clinicopathologic study of 17 tumors with resection at a single institution and literature review. Pathol Res Pract 209: 208–214.

Wang Q, Yang F, Zhu F et al. (2018). A case report of left atrial myxoma-induced acute myocardial infarction and successive stroke. Medicine (Baltimore) 97: 13451.

Wen XY, Chen YM, Yu LL et al. (2018). Neurological manifestations of atrial myxoma: a retrospective analysis. Oncol Lett 16: 4635–4639.

Wu Y, Fu XM, Liao XB et al. (2018). Stroke and peripheral embolisms in a pediatric patient with giant atrial myxoma: case report and review of current literature. Medicine (Baltimore) 97: 11653.

Yinon Y, Chitayat D, Blaser S et al. (2010). Fetal cardiac tumors: a single-center experience of 40 cases. Prenat Diagn 30: 941–949.

Zheng J, Li S, Cao Y et al. (2015). Multiple cerebral myxomatous aneurysms: what is the optimal treatment? J Stroke Cerebrovasc Dis 24: 232–238.

Handbook of Clinical Neurology, Vol. 177 (3rd series)
Heart and Neurologic Disease
J. Biller, Editor
https://doi.org/10.1016/B978-0-12-819814-8.00032-9

Chapter 27

Stroke in pregnancy

AMANDA OPASKAR, REYANNA MASSAQUOI, AND CATHY SILA*

Department of Neurology, University Hospitals Cleveland Medical Center, Cleveland, OH, United States

Abstract

Stroke in pregnancy is rare and has a wide range of etiologies and implications on stroke management that differ from nonpregnant individuals. The highest risk of stroke is during the third trimester and puerperium period, where hypertensive disorders of pregnancy occur; however, stroke can occur at any point during pregnancy. In this chapter, we will provide an overview of the epidemiology of stroke in pregnancy and then review the specific etiologies of ischemic and hemorrhagic stroke as they relate to pregnant women. Finally, we discuss the process of acute stroke evaluation in pregnancy and the management of women after stroke with regard to long-term risk factors, medications, and implications in future pregnancies. Throughout the chapter, we highlight relevant guidelines from the American Heart Association and American Stroke Association and key literature on stroke in pregnancy.

EPIDEMIOLOGY

Stroke is defined as a vascular event caused by an acute focal injury to the central nervous system (CNS), and is the second most common cause of death, and the third leading cause of disability worldwide (Sacco et al., 2013; Johnson et al., 2016). Stroke is further subdivided into classifications of ischemic stroke accounting for about 80% of all strokes and hemorrhagic stroke. An estimated one in 10 strokes occur in young adults, with an increase in incidence up to 40% within the past few decades attributed to rising rates of obesity and hypertension (Ekker et al., 2018).

Pregnancy confers an increased risk of stroke, with an estimated incidence of 34 per 100,000 pregnant woman-years, with the highest rates occurring in the puerperium (relative risk 8.7 ischemic, 28.3 hemorrhagic) compared to during pregnancy (relative risk 0.7 ischemic, 2.5 hemorrhagic) (Yoshida et al., 2017; Karjalainen et al., 2019). Pregnancy associated stroke is the most common cause of serious long-term disability following pregnancy, has impacts on subsequent pregnancies, and increases the lifetime risk of stroke (Kaplovitch and Anand, 2018; Wu et al., 2020). Longitudinal studies of patients with a history of preeclampsia have shown an 80% increased risk of stroke later in life (Kaplovitch and Anand, 2018).

Risk factors for arterial ischemic stroke in pregnancy include thrombophilias, structural heart disease, paradoxical embolism, as well as the spectrum of hypertensive disorders (Bushnell et al., 2014; Yoshida et al., 2017; Kaplovitch and Anand, 2018; Karjalainen et al., 2019). Hemorrhagic stroke is associated with higher rate of morbidity and mortality, accounting for 5%–12% of all maternal deaths during pregnancy and puerperium (Porras et al., 2017). In nonpregnant cohorts, hemorrhagic stroke rates differ based on geographic and ethnic groups with higher rates in Asian cohorts. This has also been demonstrated in pregnant populations with retrospective cohort studies showing the proportion of hemorrhagic stroke compared to combined stroke of 73.5% in a Japanese cohort, compared to 29%–45% in the United States, 38% in Canada, and 52% in France (Yoshida et al., 2017).

To address the importance of stroke in women, the American Heart Association and American Stroke

*Correspondence to: Cathy Sila, M.D., Gilbert W Humphrey Professor and Chair, Department of Neurology, University Hospitals Cleveland Medical Center, Cleveland, OH, United States. Tel: +1-216-844-5505; Fax: +1-216-844-7443, E-mail: cathy.sila@uhhospitals.org

Association published guidelines in 2014 for the prevention of stroke in women, with specific sections addressing pregnancy-associated stroke (Bushnell et al., 2014).

ETIOLOGY

Ischemic stroke

Ischemic stroke involves an arterial or venous occlusion blocking blood flow to an area of the brain or spinal cord resulting in ischemia and then infarction. Clinical features include focal neurological deficits that follow specific patterns based on the vascular territory. Symptoms include impairments in language, vision, swallowing, strength, sensation, and coordination.

The highest incidence of ischemic stroke in pregnancy occurs in the puerperium followed by the third trimester. Risk factors include older age and those with pregnancy-related hypertensive disorders (Cauldwell et al., 2018).

Cardioembolic ischemic stroke

Hemodynamic changes and hypercoagulable state

Pregnancy is a state of high metabolic demand and involves changes within the cardiovascular system to support the developing fetus. Cardiovascular changes include an increase of up to 50% of the cardiac output as well as increases in blood volume, stroke volume, and heart rate with a decrease in blood pressure (Ouzounian and Elkayam, 2012; Grear and Bushnell, 2013).

Hematologic changes primarily occur in the third trimester and early puerperium. These changes result in a hypercoaguable state due to increases in levels of procoagulant factors (VII, VIII, IX, X, XII, and von Willebrand factor), activated protein C, fibrinogen, and decreased levels of factor XIII and protein S (Grear and Bushnell, 2013; Katz and Beilin, 2015). Additionally, pregnancy is associated with the production of antiphospholipid antibodies, a group of immunoglobulins associated with vascular thrombocytosis, thrombocytopenia, and fetal loss (Bushnell, 2016). Antiphosopholipid antibodies, which include lupus anticoagulant, anticardiolipin, and antibeta 2 glycoprotein, cause platelet activation, prothrombin activation, and interference with the APC pathway resulting in both small- and large-vessel thrombosis (Bushnell, 2016).

Interpreting the presence of antiphospholipid antibodies can be challenging as these antibodies can occur in up to 12%–50% of normal patients and can be induced in the setting of stress and viral infections. Therefore, these antibodies must be positive on repeated testing 6–12 weeks apart to confirm diagnosis, with higher risk of thrombosis occurring in patients with a prior history of thromboembolism and high IgG anticardiolipin titers typically >40 GPL units (Bushnell, 2016).

Mechanical heart valves

Mechanical heart valves have a high risk of mechanical valve thrombosis and subsequent thromboembolic events. Pregnant women with mechanical heart valves are at high risk for complications due to the hemodynamic changes of pregnancy and need for anticoagulation, which carries risks of hemorrhage, teratogenicity, and fetal toxicity. A European cohort of pregnant women with mechanical heart valves found a 1.4% risk of maternal mortality and 42% risk of serious adverse events during pregnancy. Additionally, vitamin K antagonist use in the first trimester was associated miscarriage (28.6%) and late fetal death (7.1%) (Van Hagen et al., 2015).

Risk of thromboembolism is higher in mitral valve than aortic valve replacement, due to increase in left atrial size and atrial fibrillation, as well as with multiple prosthetic valves and caged-ball valves. The optimal treatment is vitamin K antagonists as these medications have the lowest risk of valve thrombosis. The goal INR is 2.5–3.5 for mitral valves and 2.0–3.0 for aortic valves. Higher anticoagulation targets of INR 3–4.5 were studied in the AREVA trial and were shown to be equivalent to standard INR targets (2.0–3.0) with increased risk of harm from hemorrhagic complications (Acar et al., 1996; Chesebro et al., 1996; Furie and Khan, 2016). However, due to the risk of fetal toxicity, in low-risk patients, vitamin K antagonists should be transitioned to heparin-based therapy, typically low-molecular-weight heparin, during the first trimester. Then, vitamin K antagonists can be resumed during the second trimester and continued until 36 weeks before being transitioned back to low-molecular-weight therapy for delivery (Van Hagen et al., 2015). Anticoagulation transitions require close monitoring as they are the highest risk period for stroke (Cauldwell et al., 2018).

Patients with mechanical heart valves who have an ischemic stroke while on therapeutic anticoagulation should be investigated with a transesophageal echocardiogram to evaluate for thrombi or infective endocarditis. In certain cases, adding an antiplatelet agent (low-dose aspirin) to systemic anticoagulation may reduce recurrent stroke if there is concurrent cerebrovascular disease or valve-related thrombi (Furie and Khan, 2016).

Cardiomyopathy

Peripartum cardiomyopathy (PPCM, also called pregnancy-associated cardiomyopathy) is a rare complication of pregnancy that occurs in the third trimester or puerperium, where patients with no history of underlying heart disease develop new heart failure with reduced ejection fraction (EF), defined as left ventricular ejection fraction (LVEF) <45%. The estimated incidence is 18–333 cases per 100,000 live births but carries a high mortality up to 20% (Block and Biller, 2014). The etiology of PPCM remains unclear although may be related to preeclampsia (Block and Biller, 2014; Arany and Elkayam, 2016). Risk factors include age greater than 30, African descent, multifetal gestation, history of preeclampsia or postpartum hypertension, and cocaine abuse (Arany and Elkayam, 2016). Additionally, the increase in cardiac demands of pregnancy can unmask preexisting cardiomyopathies leading to clot formation and subsequent embolization (Arany and Elkayam, 2016).

In cohorts of nonpregnant patient's, cardiomyopathy has a 1.3%–3.5% risk of stroke per year with up to a 58% increase in thromboembolic events for 10% every decrease in EF (Greer et al., 2016). Risk factors for stroke in cardiomyopathy include high concordant incidence of atrial fibrillation, as well as decreased apical flow. Several trials have evaluated the optimal primary stroke prevention for heart failure including the Survival and Ventricular Enlargement Study, Warfarin and Antiplatelet Therapy in Chronic Heart Failure (WATCH), and the Warfarin versus Aspirin in Reduced Ejection Fraction (WARCEF) trials. The primary endpoint of these trials showed no significant difference between antiplatelet therapy and anticoagulation, but both WATCH and WARCEF suggested a benefit for warfarin. The WARCEF trial evaluated patients with an EF of less than 35%; warfarin use resulted in a statistically significant reduction in ischemic stroke but also had an increase in major hemorrhage. Current practice supports consideration of warfarin treatment for patients with cardiomyopathy with reduced EF of less than 35% (Greer et al., 2016). Once started, anticoagulation should be continued until at least 2 months postpartum to reduce the risk of thromboembolism (Arany and Elkayam, 2016).

Arrhythmia

Arrhythmia is the most common cardiac complication of pregnancy. This is thought to be due to the hypervolemic state in pregnancy increasing heart size, myocardial stretch, and subsequent ion channel activation (Knotts and Garan, 2014). Identification, workup, and treatment of cardiac arrhythmias are important in pregnancy as certain arrhythmias increase the risk of ischemic stroke. Atrial fibrillation is the most common mechanism of cardioembolic stroke, and the risk can be substantially decreased with appropriate treatment with anticoagulation if identified (Kernan et al., 2014).

Cryptogenic ischemic stroke

Paradoxical embolism with pelvic venous thrombosis

Paradoxical embolism is a rare cause of stroke but an important consideration in stroke in young patients. A persistent connection between the right and left chambers via a patent foramen ovale (PFO) allow venous thrombi to cross into the arterial system causing an ischemic stroke. The risk of paradoxical embolism score was developed to help predict the presence of a PFO in patients with cryptogenic stroke (Kent et al., 2013). However, a PFO is common in the general population occurring in 20%–25% of all patients, which requires careful evaluation to identify the causal relationship of a PFO in each patient after an ischemic stroke (Greer et al., 2016).

A PFO is determined routinely on a transthoracic echocardiogram (TTE) with the addition of a bubble study. The bubble study is performed using agitated saline administered through an IV with the ultrasound probe on the heart to detect microbubbles passing from the right to left side of the heart. High-risk features of a PFO include PFO associated with an atrial septal aneurysm (15.2% risk of stroke in 4 years compared to 2.3% with PFO alone) and larger PFO defects with higher degree of shunting (measured by number of microbubbles) (Greer et al., 2016). Transesophageal echocardiogram is more sensitive for PFO; however, this is an invasive study that should be reserved for patients with high clinical suspicion with negative TTE or when TTE is technically difficult.

All patients with a PFO should have noninvasive vascular imaging to rule out venous thrombosis with lower extremity Doppler ultrasound and MR or computed tomography (CT) venogram of the pelvis. In pregnancy, the pelvic venous plexus becomes compressed, which leads to increased stasis and subsequent thrombus formation. Studies have shown that there is an increased incidence of pelvic vein thrombosis in patients with PFO and cryptogenic stroke (Osgood et al., 2015).

Treatment of PFO involves appropriate anticoagulation if there is any identifiable venous thrombosis. A PFO can also be closed by a transcatheter closure device. In 2017, the REDUCE clinical trial: PFO closure or antiplatelet therapy for cryptogenic stroke showed that PFO closure with subsequent antiplatelet therapy had a reduced risk of stroke compared to antiplatelet therapy alone. It is important to note that as PFO closure is an invasive procedure and is associated with higher risks of periprocedural device complications and subsequent atrial fibrillation (Sondergaard et al., 2017).

Amniotic fluid embolism

Amniotic fluid embolism (AFE) is a very rare complication of pregnancy which is characterized by the entry of amniotic fluid, fetal cells, hair, and other debris into the maternal pulmonary circulation leading to cardiovascular collapse. The increased intrathoracic pressure during labors can lead to a right to left shunt through a PFO (Kaur et al., 2016). It can also cause a proinflammatory response. AFE commonly occurs during delivery or immediately postpartum. There has been a linkage between AFE and maternal strokes. In a cohort of 600,000 deliveries in Australia, there were 20 cases of AFE reported, of which four women had cerebral infarction felt to be due to the hypercoagulable state caused by AFE. There have also been case reports of embolic strokes during pregnancy due to the presence of both AFE and PFO (Kaur et al., 2016).

Ischemic stroke due to carotid and vertebral artery dissection

Spontaneous cervical artery dissections account for 10%–25% of ischemic stroke in young patients. It is a rare cause of stroke in pregnancy (Adams Jr. et al., 1995). In an epidemiologic study of 2470 pregnancy-related strokes, the overall relative risk of cerebral infarction during the 6 weeks after pregnancy was estimated to be 5.4 (95% CI 2.9–10.0); notably there was only one case of carotid dissection (Kittner et al., 1996).

Clinical features include ipsilateral headache with pain localized to the frontal, retroorbital, or anterior cervical regions with carotid dissection and posterior cervical pain mimicking meningismus with vertebral dissection. The pain typically precedes cerebral ischemic events by days to weeks. Associated features include throbbing, nausea, vomiting, and photophobia and ipsilateral Horner's which can mimic migraine or cluster headache, respectively (Schievink, 2001).

Data regarding optimal management is limited, but most patients are treated with antiplatelet therapy or oral anticoagulants (Schievink, 2001; Lyrer and Engleter, 2010). In 2015, the randomized controlled trial called CADISS was published, which compared antiplatelet treatment with anticoagulation for cervical artery dissection. The trial enrolled 250 patients with extracranial dissection, 118 carotid, and 132 vertebral and showed no significant difference in efficacy of antiplatelet vs anticoagulant medication. Thus, in most cases, antiplatelet treatment with aspirin is sufficient unless there is another indication for anticoagulation such as an associated intraluminal thrombus (CADISS Trial Investigators et al., 2015).

The risk of recurrent ischemic events is largely confined to patients with ischemic symptoms at presentation and is uncommon in patients who present with headache or Horner's syndrome in isolation. Most recurrent ischemic events occur within the first month, overall 10%–15% for TIAs and 1%–2% for ischemic stroke (Schievink, 2001). Antithrombotic therapies are typically continued until healing, or recanalization has occurred. Overall, approximately 60% of vessels recanalize (Nedeltchev et al., 2009).

In patients with a documented cervical artery dissection, the risk of recurrent dissection is low, from 0.3% per year in epidemiologic studies to 1% per year in case series with the highest risk in the first month after the dissection. Although underlying connective tissue disorders such as fibromuscular dysplasia or Ehlers-Danlos syndrome or nonspecific vascular abnormalities such as ectasia and arterial redundancy have accounted for the majority of recurrences in some studies, other series have described recurrences in otherwise normal, albeit somewhat younger, individuals (Schievink et al., 1994; Bassetti et al., 1996).

Our approach to patients with cervical artery dissection who wish to become pregnant, based on the available data, is to delay conception until the major risk of recurrent ischemic events has passed and recanalization or pseudoaneurysm healing has had a chance to occur, a minimum of 1–3 months but better if 6–12 months, so that antithrombotic therapy can be discontinued during pregnancy. Patients with symptomatic stenoses or the uncommon expanding pseudoaneurysms failing medical therapy should be considered for endovascular repair and conception delayed until such time as antithrombotic therapies can be more safely discontinued.

All patients should consult with their obstetrician prior to conception for good prenatal care, particularly monitoring and control of blood pressure. For the majority of patients with no evidence of an underlying arteriopathy or family history of dissection would advise routine obstetrical care. However, for those with evidence of an underlying arteriopathy or family history of dissection where the risk of recurrent dissection may be greater, we recommend a more prudent approach with consideration to a cesarean section to achieve a controlled delivery that avoids excessive straining. High-risk patients with Ehlers-Danlos IV require special counseling prior to conception as their pregnancy and delivery can be particularly dangerous. Other high-risk patients would include patients with active symptoms, persistent severe or hemodynamically vulnerable stenosis, or intracranial/intradural extensions of their dissections (Wiebers and Mokri, 1985; De Bray et al., 2000; Kernan et al., 2014).

Migraine headache

Migraine headaches are common affecting one-fifth of the United States population (Kurth et al., 2016). The exact mechanism is unknown but has been linked to the cerebrovascular system.

Migraine is defined by the International Headache Society, ICHD-3 classification, as a unilateral headache lasting for 4–72 h with associated photophobia, phonophobia and/or nausea, and vomiting. Migraine headache with aura refers to a primarily positive phenomenon that can occur proceeding or during a headache and consists of either visual changes, speech changes, numbness, or weakness lasting from 5 to 60 min (Olsen et al., 2018).

Migraine headache with aura confers an increased risk of stroke. A large prospective health study of female nurses followed for more than 20 years demonstrated a 50% increased risk of major cardiovascular disease including myocardial infarction, stroke, and coronary artery procedures including an increased risk of cardiovascular disease mortality (Kurth et al., 2016). In the Women's Health Study, migraine with aura was associated with a risk of hemorrhagic stroke, with a stronger association in the subset of women with fatal hemorrhagic stroke and in women <55 years of age. Another cohort of migraine patients again demonstrated a large association with hemorrhagic stroke (OR 9.1; 95% CI 3.0–27.8); however, in the pregnant population, the risk of vascular disease was closely associated with a concomitant diagnosis of preeclampsia/eclampsia (Bushnell et al., 2014).

It is important to recognize migraine headache with aura and distinguish this from other secondary headaches due to cerebrovascular disorders such as ischemic or hemorrhagic stroke. Migraine with aura will classically cause positive phenomenon with either bright lights in the vision or paresthesias in the arms as opposed to negative phenomenon that occur in stroke such as loss of vision or numbness (Chou, 2018).

In a pregnant woman, any new focal neurological deficit or change to headache pattern should be investigated. In an emergent setting, an initial STAT CT brain without contrast can be done to assess for headache. Additional imaging should be obtained based on clinical suspicion but includes MRI brain without contrast and determination of need for vascular imaging such as MRA or MR venography (MRV).

Intracerebral hemorrhage

Intracerebral hemorrhage is the rupture of a blood vessel resulting in a collection of blood in the brain or surrounding structures. Clinically, patients with intracerebral hemorrhage present with seizures, headaches, or focal neurological deficits. The etiology of hemorrhagic stroke in pregnancy includes hypertensive disorders of pregnancy (preeclampsia/eclampsia and pregnancy-induced hypertension), vascular malformations such as arteriovenous malformations (AVMs) or aneurysm rupture, or venous infarction with hemorrhagic transformation from cerebral venous thrombosis (CVT). These etiologies are addressed below.

Preeclampsia/eclampsia

Preeclampsia and eclampsia occur primarily during the third trimester and puerperium period and have neurovascular and systemic complications. There is a higher risk of stroke due to preclampsia in the puerperium period often because women are not monitored as closely as they were during the third trimester.

Pathophysiology of preeclampsia is related to poor placental perfusion leading to multiorgan endotheliopathy resulting in loss of central cerebral vascular autoregulation causing cerebrovascular complications including posterior reversible encephalopathy syndromes, hemorrhagic, and ischemic strokes (Bushnell et al., 2014; Kaplovitch and Anand, 2018).

Preeclampsia is defined as new onset hypertension ≥140/90 and proteinuria (≥300 mg of protein in a 24-h urine specimen) and can be classified as early onset (before 37 weeks gestation) or late onset (after 37 weeks gestation). Eclampsia is defined as an unprovoked seizure in the setting of preeclampsia (Block and Biller, 2014; Bushnell et al., 2014). This is in contrast to pregnancy-induced (gestational) hypertension, which is diagnosed as hypertension without proteinuria or other signs and symptoms of preeclampsia (Bushnell et al., 2014).

In women with preeclampsia, risk of stroke has been associated with comorbid conditions including chronic hypertension, coagulopathy, prothrombotic conditions such as infection or inherited coagulopathy, and underlying vascular malformations (Miller et al., 2017). However, systemic complications of preeclampsia include renal failure, pulmonary edema, disseminated intravascular coagulation, and HELLP (hemolysis, elevated liver enzymes, or low platelets), among others (Block and Biller, 2014; Bushnell et al., 2014).

The primary management of preeclampsia includes magnesium sulfate and blood pressure control. Magnesium sulfate has been shown in several clinical trials to be beneficial for prevention and treatment seizures and preventing complications such as stroke in preeclampsia. Aggressive blood pressure control is paramount as stroke can occur at moderately elevated blood pressure controls in preeclampsia (Bushnell et al., 2014). The first line treatment of severe hypertension is labetalol and any blood pressure medications that cross the placenta should be avoided including angiotensin-converting enzyme inhibitors and angiotensin receptor blockers (Bushnell et al., 2014; Kaplovitch and Anand, 2018).

In patients at high risk for preeclampsia who have chronic hypertension or pregnancy-related hypertension, low-dose aspirin taken from the 12th week until delivery and calcium repletion 1000 mg daily have been shown to be beneficial in preventing preeclampsia (Bushnell et al., 2014; Kaplovitch and Anand, 2018). Vitamin D3 supplementation has limited data in regards to its effectiveness and is not currently recommended by the American Stroke Association (Bushnell et al., 2014). Antihypertensive medications such as methyldopa, labetalol, and nifedipine are preferred during pregnancy and antihypertensive medications contraindicated during pregnancy include atenolol, direct renin inhibitors, and ARBs.

Reversible cerebral vasoconstriction syndrome

Reversible cerebral vasoconstriction syndrome (RCVS) is a rare cause of stroke in pregnancy and the puerperium. RCVS is an acute angiitis of the CNS affecting the large and medium-sized artery (Bougousslavsky et al., 1989; Calado et al., 2004). It follows a normal pregnancy and is not associated with features of preeclampsia. RCVS has been associated with use of sympathomimetic medications such as selective serotonin reuptake inhibitors (SSRIs), ergotamine, triptans, and crack/cocaine use. Clinical features include a severe "thunderclap" headache, vomiting, seizures, and focal neurological deficits (Call et al., 1988).

Radiologic imaging with MRA/CTA or angiogram shows multivessel, often bilaterally, narrowing and dilation of the cerebral vessels (Call et al., 1988). Much confusion is incited by the typical radiology report indicating evidence of "vasculitis," which has led to unnecessary and expensive tests for autoimmune disorders and even brain biopsies. The reversible nature of the condition or temporary improvement with intraarterial injections of vasodilators supports the vasoconstrictive etiology (Hajj-Ali et al., 2002). RCVS can be complicated by ischemic or hemorrhagic stroke and the prognosis varies, with many having a good outcome with supportive care.

Management includes treatment of the vasoconstriction with the calcium channel blocker verapamil and magnesium supplementation. Medications that can promote vasospasm should be discontinued including SSRIs, triptans, ergotamines, pseudoephredrine, selective norepinephrine reuptake inhibitors, dopamine agonists, as well as illicit drugs—cocaine and amphmetamines. Additionally, symptomatic treatment is indicated with blood pressure control, pain control (may require short-term oral narcotic medications), and seizure management (Nowak et al., 2003).

Outcomes are generally good and the majority of patients return to their neurological baseline although some may continue to have chronic headaches (John et al., 2016).

Vascular malformations

Intracranial vascular malformations include cavernous malformations, AVMs, saccular aneurysms, and more rarely developmental venous anomalies and capillary telangiectasia.

Cavernous malformations are a collection of dilated capillary vessels that cause small focal hemorrhages due to hemosiderin deposition around the malformation. They are relatively common, occur in sporadic or in familial forms, and have an overall good prognosis depending on location. Clinical presentations include focal neurological deficits and seizures (Lanzino and Spetzler, 2007). The risk of hemorrhage is low, about 1%–4% per year. Management is primarily conservative with antiepileptic medications for seizures and routine monitoring. Surgical removal is reserved for specific cases with recurrent hemorrhage or refractory seizures (Lanzino and Spetzler, 2007).

Brain AVMs are a formed from a direct connection of artery to vein with no intervening capillary bed (Derdeyn et al., 2017). Brain AVMs have a relatively high risk of intracranial hemorrhage of about 2%–4% per year if untreated, with higher rates once an AVM has ruptured (Derdeyn et al., 2017; Ruigrok, 2020).

Several retrospective studies have evaluated the rates of AVM rupture in pregnancy. Liu et al. reviewed a Chinese cohort of 774 female patients aged 18–40 years, 452 pregnancies where 393 had ICH and 381 had no ICH. They found no significant increased risk of ICH during pregnancy or puerperium period (Liu et al., 2019). However, a North American Cohort had a 5.7% rate of AVM rupture during pregnancy and puerperium vs 1.3% risk during the nonpregnant period. Additionally, those with a ruptured AVM had a rebleed rate during pregnancy of 25% (Porras et al., 2017).

Aneurysms are dilations of the surface an artery that occur at major branching points in the Circle of Willis or in the anterior circulation. Unruptured aneurysms are relatively common and have been found in at least 3% of the population (Ruigrok, 2020). However, most are small and are at low risk for complications as the risk of aneurysm rupture is related to aneurysm size and location (posterior circulation, posterior communicating, and anterior communicating arteries) as well as patient factors including younger age (<50 years), hypertension, cigarette smoking, personal or family history of SAH, and country of origin (Finnish or Japanese heritage) (Ruigrok, 2020).

Ruptured intracranial aneurysms result in subarachnoid hemorrhage and are a neurological emergency with a high mortality rate with up to 20% of patients dying prior to presenting to the hospital. Clinically, subarachnoid hemorrhage presents as a severe headache described as the "worst headache of my life" associated with meningismus, altered level of consciousness, focal neurological deficits, and seizures. The severity of subarachnoid hemorrhage is graded clinically using the Hunt Hess Scale (1–5 level of consciousness, severity of headache, and focal neurological deficits) and radiologically using the modified Fisher scale (amount and location of hemorrhage ± intraventricular hemorrhage) (Ruigrok, 2020).

AVMs and aneurysms are reliably diagnosed with CT or MR angiography but digital subtraction angiography (DSA) remains the gold standard. DSA delivers a range of 0.17–2.8 mGy well below the threshold of 50 mGy accepted maximum for fetal exposure and may be considered during pregnancy if needed. Two cohorts, Liu et al. and Porras et al. have reviewed AVMs in pregnant patients and have shown no radiation-induced fetal malformations (Porras et al., 2017; Liu et al., 2019).

Cerebral venous thrombosis

Cerebral venous thrombosis (CVT) is a venous thrombosis that forms in one of the cortical veins or cerebral venous sinuses. CVT accounts for 0.5%–1% of all strokes, with an increased odds of 30% to 13-fold higher (ORs, 1.3–13) in pregnancy and the puerperium, with one US population-based study showing an incidence of CVT in 11.6 per 100,000 deliveries (Bushnell et al., 2014; Ferro and Canhao, 2016).

The primary presenting symptom is a headache related to increased intracranial pressure. A CVT can also lead to venous infarction and hemorrhage, which presents as seizures and focal neurological deficits. Risk factors for CVT are similar to systemic venous thromboembolism (VTE) and include prothrombotic conditions of which pregnancy and the puerperium period are risk factors. The greatest risk for CVT in pregnancy occurs in the third trimester and the first 4 weeks postpartum. There is also a higher risk with cesarean delivery, which has been shown after controlling for other risk factors (Bushnell et al., 2014).

In a review of CVT in pregnancy, Cantu et al. showed that pregnant patients with CVT present with more acute rather than subacute headache and had a more favorable outcome, although reasons for this remain unclear. In their cohort, 80% achieved a good outcome and had a lower mortality rate compared to nonpregnancy-related CVT (Cantu and Barinagarrementeria, 1993).

CVT can be diagnosed utilizing either MRI, CT, or catheter-based angiogram. The initial imaging is usually a CT head without contrast as this can be performed rapidly and can exclude other cerebral disorders. CT findings are nonspecific and can be normal in up to 30% of patients with CVT (Ferro and Canhao, 2016). MR venography is the preferred method for diagnosis of CVT as it is noninvasive, does not involve radiation, and allows for imaging of the brain parenchyma. MRV can be done without contrast in a 2D time-of-flight sequence, which will demonstrate absence of flow in the thrombosed vessel (Ferro and Canhao, 2016). The diagnostic yield is increased with the use of gadolinium contrast; however, gadolinium contrast is avoided in pregnancy due to concern for fetal toxicity. CT venogram can also be done which is traditionally obtained by doing a CT angiogram with delayed imaging acquisition to the venous phase. The benefits of a CT venogram are the increased availability of CT scanners, fast image acquisition time, lack of contraindications due to ferromagnetic devices, and low cost. However, this requires both radiation and use of iodinated contrast, which is typically avoided in pregnant patients. Anatomic variations including hypoplasia or duplication of the sinus or hypoplasia/aplasia of the transverse sinuses can lend to diagnostic uncertainty. Catheter-based angiograms are the gold standard and can help clarify venous sinus anatomy and evaluate for other vascular abnormalities such as aneurysms or dural AV Fistulas. This invasive study is reserved for diagnostic uncertainty or treatment purposes (Ferro and Canhao, 2016).

Treatment of acute CVT is based on preventing propagation of the thrombus, which can cause progressive infarction, hemorrhage, and cerebral edema with the goal to recanalize the occluded sinus or vein (Ferro and Canhao, 2016). Treatment of acute CVT is anticoagulation with a heparin-based therapy either IV unfractionated (UFH) or low-molecular-weight heparin (LMWH) and then transition to oral anticoagulation. The choice of anticoagulation is based on size of clot and stability of the patient. UFH is preferred in patients with higher risk of bleeding as this can be rapidly stopped and reversed in the setting of any hemorrhagic complications. However, UFH has many challenges achieving consistent systemic anticoagulation without supra- or subtherapeutic levels requiring close monitoring of aPTT levels. AHA/ASA guidelines allow for treating acute CVT in pregnancy with full dose LMWH rather than UFH unless high-risk cases or LMWH is contraindicated (i.e., renal failure) (Bushnell et al., 2014). Catheter-based thrombolysis is reserved for severe cases, where there is poor prognosis, deep vein involvement, or clot propagation despite anticoagulation treatment (Ferro and Canhao, 2016).

The duration of anticoagulation follows traditional anticoagulation for venous thromboembolism (VTE) and is typically 3 months for a provoked VTE where the trigger is eliminated, and longer, typically 6–12 months, for an unprovoked VTE. Imaging is repeated at 3 month intervals until resolution of the thrombus occurs (Kearon et al., 2012; Bushnell et al., 2014). AHA/ASA Guidelines recommend continuing LMWH in full anticoagulant doses throughout the pregnancy and LMWH or VKA with target INR 2–3 for ≥6 weeks postpartum (minimum duration of therapy 6 months) (Bushnell et al., 2014).

Recurrence rates are low and typically occur within the first year and are more common in severe coagulopathies such as antiphospholipid antibody syndrome or other thrombophilias. Prospective studies have shown higher rates of a recurrent systemic VTE rather than recurrent CVT (Bushnell et al., 2014).

In regards to future pregnancies, the risk of complications is low, but it is reasonable to initiate prophylaxis with LMWH during future pregnancies and the postpartum period (Bushnell et al., 2014).

MANAGEMENT

Acute stroke management

The initial evaluation for an acute stroke includes an initial assessment of the airway, breathing, and circulation as hemorrhagic strokes and large ischemic strokes can cause brain compression and compromise the airway. Then, a standardized neurological assessment such as the NIH Stroke Scale should be performed to help categorize the severity of the symptoms along with initial vital signs including blood pressure and a point of care blood glucose. This should be obtained followed by an emergent noncontrast CT of the head, which is used to evaluate for hemorrhage or early ischemic changes (Powers et al., 2018).

Patients with concern for ischemic stroke should be evaluated for candidacy for IV thrombolytic (IV tPA) and endovascular therapy. There is a short time window for eligibility for this therapy, due to the rapid progression of ischemia as well as risk of hemorrhage. IV thrombolytic therapy should be administered within 4.5 h and mechanical thrombectomy within 24 h with earlier therapy resulting in improved outcomes.

Thrombolytic therapy is a pregnancy category C drug and has been traditionally avoided in pregnancy due to concern for maternal and fetal complications. Previous studies of thrombolysis with streptokinase in pregnant women have demonstrated an increased rate of maternal and fetal death and maternal hemorrhage (Johnson et al., 2015). However, rtPA has been used in several case reports for acute myocardial infarction, pulmonary embolism, DVT, stroke, and other life-threatening indications with low rates of complications primarily placental hematoma and uterine bleeding and one report of spontaneous abortion leading to intrauterine fetal death (Johnson et al., 2015). Another study reviewed 11 women who received tPA for stroke while pregnant, and 9 during the first trimester. Five patients received IV tPA and 6 patients had intraarterial tPA. Of those who received IV tPA, there was only one intrauterine hematoma and no symptomatic intracranial hemorrhage (Aleu et al., 2007).

With this information, the AHA/ASA Acute Stroke management guidelines have adapted the following statement:

> *"IV alteplase administration may be considered in pregnancy when the anticipated benefits of treating moderate or severe stroke outweigh the anticipated increased risk of uterine bleeding. Class IIb; LOE C-LD. The safety and efficacy of IV alteplase in early postpartum period (<14 d after delivery) have not been well established. Class IIb; LOE C-LD." (Powers et al., 2018)*[1]

For ischemic and hemorrhagic stroke evaluation, the next step is to obtain emergent noninvasive intracranial vascular imaging (CTA or MRA time of flight). This can identify large vessel occlusions for endovascular therapy in ischemic strokes or identification of vascular malformations in hemorrhagic strokes (Powers et al., 2018).

Neurosurgical management of AVM/aneurysm

National guidelines recommend an emergent evaluation for intracranial hemorrhages due to the risk of rapid hematoma expansion and early deterioration (Hemphill 3rd et al., 2015). The initial evaluation process for intracranial hemorrhage is similar to ischemic strokes. This includes a baseline neurological assessment with a severity score, rapid neuroimaging with CT or MRI, and then noninvasive vascular imaging (CTA/CTV, MRA/MRV) to identify structural lesions (vascular malformations or tumors) (Hemphill 3rd et al., 2015).

Early and rapid blood pressure control can help reduce hematoma expansion and improve mortality. Guidelines encourage treatment of SBP to a goal of <140 mmHg for ICH patients presenting with SBP between 150 and 220 mmHg and without contraindication to acute BP treatment (Hemphill 3rd et al., 2015).

Institutional protocols should be followed for reversal of anticoagulation therapy. Reversal of antiplatelet therapy is controversial, and the guidelines note that there is uncertain evidence regarding the use of platelet transfusion (Hemphill 3rd et al., 2015).

Patients should be monitored closely for complications and are at high risk for venous thromboembolism and aspiration pneumonia, among others. Intermittent pneumatic compression should be used on initial hospital admission for prevention of venous thromboembolism and LMWH or UFH can be started within 1–4 days of documented cessation of bleeding (Hemphill 3rd et al., 2015).

Treatment options for AVMs

The goal of AVM treatment is elimination of the nidus and arteriovenous shunt to reduce/eliminate risk of recurrent hemorrhage (Derdeyn et al., 2017).

AVMs are traditionally classified using the Spetzler Martin Grading Scale, which estimates the risk of open neurosurgery by evaluating the AVM size, pattern of venous drainage, and eloquence of brain location. Grade I AVMs are small, superficial, and located in noneloquent brain. Grade IV or V is large, deep, and situated in neurologically critical areas. Grade VI AVMs are considered not operable (Spetzler and Martin, 1986).

Treatment options for AVMs include observation, craniotomy and resection, stereotactic radiosurgery, endovascular embolization, or a combination of the previously mentioned termed multimodal therapy (Derdeyn et al., 2017).

---

[1]Reprinted with permission, Circulation. 2018;49:e46–e99. ©2018 American Heart Association, Inc.

Grades V and VI are generally managed conservatively because of high predicted morbidity and mortality. Emergency AVM resection has been recommended for patients with signs of brain herniation and lesions graded I–IV. For patients presenting late in pregnancy with lower risk AVMs, surgery can usually be deferred until after delivery (Porras et al., 2017).

Treatment options for aneurysms

The management of aneurysms focuses on securing the aneurysm and preventing and treating complications. Aneurysms are secured by either endovascular coiling or surgical clipping determined based on features of the aneurysm as well as patient factors including age and other comorbidities. Ruptured aneurysms result in subarachnoid hemorrhage and have multiple complications including acute hydrocephalus, vasospasm and delayed cerebral ischemia, seizures, and cerebral salt wasting that require management in specialized neurological intensive care units (Suarez, 2015).

Secondary stroke prevention

Ischemic stroke workup is directed toward identifying the underlying etiology. Intra- and extracranial vascular imaging should be obtained utilizing CT angiography, MR angiography, or combination of carotid ultrasound and transcranial Dopplers. A transthoracic echocardiogram (TTE) is performed to evaluate for cardioembolic sources and should include an agitated saline injection, aka bubble study, to evaluate for intracardiac shunt (e.g., PFO). If a PFO is present, then a venous thromboembolic workup should be pursued including evaluation for lower extremity deep vein thrombosis with Doppler ultrasound and consideration of pelvic venous imaging with CT or MR venogram (Grear and Bushnell, 2013).

Laboratory for vascular risk factor should be assessed with a fasting lipid panel, hemoglobin A1C, and thyroid function screen. Additional screening for inherited and acquired thrombophilia should be pursued for cryptogenic ischemic stroke, which includes evaluation for antiphospholipid antibodies—lupus anticoagulant, anticardiolipin, and antibeta2 glycoprotein, as well as for genetic thrombophilia including factor V deficiency, antithrombin deficiency, factor VIII excess, hyperhomocysteinemia, protein C and S deficiency, and prothrombin 20,210 mutation (Grear and Bushnell, 2013).

Secondary stroke prevention in women with a history of stroke in pregnancy includes vascular risk factor modification as identified during the workup and antithrombotic medications. As previously stated, there is a higher incidence of stroke in women with history of preeclampsia and pregnancy-related hypertension. Therefore, it is important to counsel patients on long-term blood pressure management. The optimal blood pressure target is unclear, but the current ASA/AHA stroke prevention guidelines recommend BP control to <140/90 mmHg unless there is a lacunar stroke where it is reasonable to target a systolic pressure <130 mmHg (Kernan et al., 2014; Caso et al., 2017).

In regards to antithrombotic medications, treatment is determined based on the underlying stroke etiology. Single antiplatelet agent such as aspirin 81 mg daily is the most common used therapy. Dual antiplatelet therapy is reserved for specific situations in high-risk patients and for short periods of time due to long-term bleeding risks. Anticoagulation therapy is reserved for patients with embolic stroke, thrombus, or other systemic thrombophilias (Kernan et al., 2014).

A history of stroke has implications on future pregnancies including optimal preventative treatment, screening, delivery, and postpartum monitoring. Recent consensus guidelines in Canada provided specific recommendations for future pregnancy in women with a history of stroke and can be referenced in the paper by Caso et al. (2017). This consensus statement recommended that women be categorized as low or high risk to determine need for antiplatelet vs anticoagulant medications, respectively (Caso et al., 2017).

Aspirin use has been shown to be safe during the second and third trimesters of pregnancy for the mother and fetus. There is no data on the use of other antiplatelet therapy (Caso et al., 2017).

Low-molecular-weight heparin is the preferred method of anticoagulation in pregnancy based on favorable side-effect profile, and low risk to fetus as heparin does not cross the placenta. For patients with high-risk conditions that require anticoagulation outside of pregnancy, LMWH or adjusted-dose UFH should be continued throughout pregnancy until the 13th week, when oral anticoagulation with VKA is initiated until close to delivery, followed by resumption of LMWH or UHF after delivery (Kernan et al., 2014).

Warfarin crosses the placenta and has risk for fetal teratogenicity and therefore is avoided in pregnancy. The new oral anticoagulant medications have no data regards to safety in pregnancy and therefore should be avoided (Caso et al., 2017).

For delivery, the focus should remain on obstetric indications rather than on history of stroke although antithrombotic and anticoagulant medications may increase the bleeding risk (Caso et al., 2017). Deliveries are classified as minor general surgery and can follow similar guidelines for resuming anticoagulant therapy (Caso et al., 2017).

There is variable data regarding safety of antiplatelet and anticoagulant medications in breastfeeding. Low-dose aspirin appears to be safe; however, high dose aspirin can be excreted in breast milk and lead to severe side effects such as bleeding or Reye's syndrome. Warfarin is not known to be excreted in breast milk. Plavix and the new oral anticoagulant medications have not been studied (Caso et al., 2017).

Additional, SSRIs should be avoided in any patient with a history of RCVS due to risk of ischemic and hemorrhagic stroke (Caso et al., 2017).

References

Acar J, Lung B, Boissel JP et al. (1996). Multicenter randomized comparison of low-dose versus standard-dose anticoagulation in patients with mechanical prosthetic heart valves. Circulaton 94: 2107–2112.

Adams Jr HP, Kappelle J, Biller J et al. (1995). Ischemic stroke in young adults: experience in 329 patients enrolled in the Iowa registry of stroke in young adults. Arch Neurol 52: 491–495.

Aleu A, Mellado P, Lichy C et al. (2007). Hemorrhagic complications after off-label thrombolysis for ischemic stroke. Stroke 38: 417–422.

Arany Z, Elkayam U (2016). Peripartum cardiomyopathy. Circulation 133: 1397–1409.

Bassetti C, Carruzzo A, Sturzenegger M et al. (1996). Recurrence of cervical artery dissection. A prospective study of 81 patients. Stroke 27: 1804–1807.

Block HS, Biller J (2014). Neurology of pregnancy. Handb Clin Neurol 121: 1595–1622.

Bougousslavsky J, Despland PA, Regli F et al. (1989). Postpartum cerebral angiopathy; reversible vasoconstriction assessed by TCD. Eur Neurol 29: 102–105.

Bushnell C (2016). Chapter 42: Hematologic disorders and stroke. In: JC Grotta, GW Albers, SE Kasner, EH Lo, AD Mendelow, RL Sacco, LKS Wang (Eds.), Stroke: Pathophysiology, diagnosis, and management, sixth edn, Elsevier, 680–694.

Bushnell C, McCullough LD, Awad IA et al. (2014). Guidelines for the prevention of stroke in women: a statement for healthcare professionals from the American Heart Association/ American Stroke Association. Stroke 45: 1545–1588.

CADISS Trial Investigators, Markus HS, Hayler E et al. (2015). Antiplatelet treatment compared with anticoagulation treatment for cervical artery dissection (CADISS): a randomised trial. Lancet Neurol 14: 361–367.

Calado S, Vale-Santos J, Lima C et al. (2004). Postpartum cerebral angiopathy: vasospasm, vasculitis, or both? Cerebrovasc Dis 18: 340–341.

Call G, Fleming MC, Sealfon S et al. (1988). Reversible cerebral segmental vasoconstriction. Stroke 19: 1159–1170.

Cantu C, Barinagarrementeria F (1993). Cerebral venous thrombosis associated with pregnancy and puerperium: review of 67 cases. Stroke 24: 1880–1884.

Caso V, Falomi A, Bushnell CD et al. (2017). Pregnancy, hormonal treatments for infertility, contraception, and menopause in women after ischemic stroke. A consenus document. Stroke 48: 501–506.

Cauldwell M, Rudd A, Nelson-Piercy C (2018). Management of stroke and pregnancy. Eur Stroke J 3: 227–236.

Chesebro JH, Valentin F et al. (1996). Optimal antithrombotic therapy for mechanical prosthetic heart valves. Circulation 94: 2055–2056.

Chou D (2018). Secondary headache syndromes. Continuum (Minneap Minn) 24: 1179–1191.

De Bray JM, Guillon B, Bouilliat J et al. (2000). Cervical artery dissections in the puerperium; pathogenic hypotheses concerning seven observations. Cerebrovasc Dis 10: 158–159.

Derdeyn CP, Zipfel GJ, Albuquerque FC et al. (2017). Management of brain arteriovenous malformations. A scientific statement for healthcare professionals from the American Heart Association/American Stroke Association. Stroke 48: e200–e224.

Ekker MS, Boot EM, Singhal AB et al. (2018). Epidemiology, aetiology, and management of ischaemic stroke in young adults. Lancet Neurol 17: 790–801.

Ferro JM, Canhao P (2016). Chapter 45: Cerebral venous thrombosis. In: JC Grotta, GW Albers, SE Kasner, EH Lo, AD Mendelow, RL Sacco, LKS Wang (Eds.), Stroke: Pathophysiology, diagnosis, and management, sixth edn, Elsevier, 716–719.

Furie K, Khan M (2016). Chapter 62: Secondary prevention of cardioembolic stroke. In: JC Grotta, GW Albers, SE Kasner, EH Lo, AD Mendelow, RL Sacco, LKS Wang (Eds.), Stroke: Pathophysiology, diagnosis, and management, sixth edn, Elsevier, 1014–1029.

Grear KE, Bushnell CD (2013). Stroke and pregnancy: clinical presentation, evaluation, treatment and epidemiology. Clin Obstet Gynecol 56: 350–359.

Greer DM, Homma S, Furie K (2016). Chapter 32: Cardiac diseases. In: JC Grotta, GW Albers, SE Kasner, EH Lo, AD Mendelow, RL Sacco, LKS Wang (Eds.), Stroke: Pathophysiology, diagnosis, and management, sixth edn, Elsevier, 563–575.

Hajj-Ali RA, Furlan A, Abou-Chebel A et al. (2002). Benign angiopathy of the central nervous system: cohort of 16 patients with long-term. Arthitis Rheum Followup 47: 662–669.

Hemphill 3rd JC, Greenbert SM, Anderson CS et al. (2015). Guidelines for the management of spontaneous intracerebral hemorrhage. A guideline for healthcare professionals from the American Heart Association/American Stroke Association. Stroke 46: 2032–2060.

John S, Singhal AB, Calabrese L et al. (2016). Long-term outcomes after reversible cerebral vasoconstriction syndrome. Cephalgia 36: 387.

Johnson DM, Kramer DC, Cohen E et al. (2015). Thrombolytic therapy for acute stroke in late pregnancy with intra-arterial recombinant tissue plasminogen activator. Stroke 36: e53–e55.

Johnson W, Onuma O, Owolabi M et al. (2016). Stroke: a global response is needed. Bull World Health Organ 94: 634-634A. https://doi.org/10.2471/BLT.16.181636.

Kaplovitch E, Anand SS (2018). Stroke in women: recognizing opportunities for prevention and treatment. Stroke 49: 515–517.

Karjalainen L, Tikkanen M, Rantanen K et al. (2019). Pregnancy-associated stroke—a systematic review of subsequent pregnancies and maternal health. BMC Pregnancy Childbirth 19: 187.

Katz D, Beilin Y (2015). Disorders of coagulation in pregnancy. Br J Anaesth 115: ii75–88.

Kaur K, Bhardwaj M, Kumar P et al. (2016). Amniotic fluid embolism. J Anesthesiol Clin Pharmacol 32: 153–159.

Kearon C, Akl E, Comerota AJ et al. (2012). Antithrombotic therapy for VTE disease. Antithrombotic therapy and prevention of thrombosis, 9th ed: American College of Chest Physicians Evidenced-Based Clinical Practice Guidelines. Chest 141: e419S–e496S.

Kent DM, Ruthazer R, Weimar C et al. (2013). An index to identify stroke-related vs incidental patent foramen ovale in cryptogenic stroke. Neurology 81: 619–625.

Kernan WN, Ovbiagele B, Black HR (2014). Guidelines for prevention of stroke in patients with ischemic stroke or TIA. American Heart Assn/American Stroke Assn Guideline. Stroke 45: 2160–2236.

Kittner SJ, Stern BJ, Feeser BR et al. (1996). Pregnancy and the risk of stroke. N Engl J Med 335: 768–774.

Knotts RJ, Garan H (2014). Cardiac arrhythmias in pregnancy. Semin Perinatol 385: 285–288.

Kurth T, Winter AC, Eliassen AH et al. (2016). Migraine and risk of cardiovascular disease in women: prospective cohort study. BMJ 353: i2610.

Lanzino G, Spetzler R (2007). Cavernous malformations of the brain and spinal cord, first edn Thieme.

Liu S, Chan W-S, Ray JG et al. (2019). stroke and cerebrovascular disease in pregnancy: incidence, temporal trends, and risk factors. Stroke 50: 13–20.

Lyrer P, Engleter S (2010). Antithrombotic drugs for carotid artery dissection. Cochrane Database Syst Rev 1: CD000255.

Miller EC, Gatollari HJ, Too G et al. (2017). Risk factors for pregnancy-associated stroke in women with preeclampsia. Stroke 48: 1752–1759.

Nedeltchev K, Bickel S, Arnold M et al. (2009). Recanalization of spontaneous carotid artery dissection. Stroke 40: 499–504.

Nowak DA, Rodiek SO, Henneken S et al. (2003). Reversible segmental cerebral vasoconstriction (Call-Fleming syndrome): are calcium channel inhibitors a potential treatment option? Cephalgia 23: 218.

Olsen J et al. (2018). The international classification of headache disorders, 3rd edition. Cephalgia 38: 1–211.

Osgood M, Budman E, Carandang R et al. (2015). Prevalence of pelvic vein pathology in patients with cryptogenic stroke and patent foramen ovale undergoing MRV pelvis. Cerebrovasc Dis 39: 216–223.

Ouzounian JG, Elkayam U (2012). Physiologic changes during normal pregnancy and delivery. Cardiol Clin 30: 317–329.

Porras JL, Yang W, Philadelphia E et al. (2017). Hemorrhage risk of brain arteriovenous malformations during pregnancy and puerperium in a north American cohort. Stroke 48: 1507–1513.

Powers WJ, Rabinstein AA, Ackerson T et al. (2018). 2018 guidelines for the early management of patients with acute ischemic stroke. A guideline for healthcare professionals from the American Heart Association/American Stroke Association. Stroke 49: e46–e49.

Ruigrok YM (2020). Management of unruptured cerebral aneurysms and arteriovenous malformations. Continuum (Minneap Minn) 26: 478–498.

Sacco RL, Kasner SE, Broderick JP et al. (2013). An updated definition of stroke for the 21st century. A statement for healthcare professionals from the American Heart Association/American Stroke Association. Stroke 44: 2064–2089.

Schievink WI (2001). Spontaneous dissection of the carotid and vertebral arteries. N Engl J Med 344: 898–906.

Schievink WI, Mokri B, O'Fallon WM (1994). Recurrent spontaneous cervical-artery dissection. N Engl J Med 330: 393–397.

Sondergaard L, Kasner S, Rhodes J et al. (2017). Patent foramen ovale closure or antiplatelet therapy for cryptogenic stroke. NEJM 377: 1033–1042.

Spetzler RF, Martin NA (1986). A proposed grading system for arteriovenous malformations. J Neurosurg 65: 476–483.

Suarez JI (2015). Diagnosis and management of subarachnoid hemorrhage. Continuum (Minneap Minn) 21: 1263–1287.

Van Hagen IM, Roos-Hesselink JW, Ruys TPE et al. (2015). Pregnancy in women with a mechanical heart valve. Data of the European Society of Cardiology Registry of pregnancy and cardiac disease (ROPAC). Circulation 132: 132–142.

Wiebers DO, Mokri B (1985). Internal carotid artery dissection after childbirth. Stroke 16: 956–959.

Wu P, Jordan KP, Chew-Graham CA et al. (2020). Temporal trends in pregnancy-associated stroke and its outcomes among women with hypertensive disorders of pregnancy. JAHA 9: e016182.

Yoshida K, Takahashi JC, Takenobu Y et al. (2017). Strokes associated with pregnancy and puerperium: a nationwide study by the Japan stroke society. Stroke 48: 276–282.

Handbook of Clinical Neurology, Vol. 177 (3rd series)
Heart and Neurologic Disease
J. Biller, Editor
https://doi.org/10.1016/B978-0-12-819814-8.00015-9

Chapter 28

Hemodynamics in acute stroke: Cerebral and cardiac complications

POURIA MOSHAYEDI AND DAVID S. LIEBESKIND*

Department of Neurology, Comprehensive Stroke Center, University of California Los Angeles, Los Angeles, CA, United States

Abstract

Hemodynamics is the study of blood flow, where parameters have been defined to quantify blood flow and the relationship with systemic circulatory changes. Understanding these perfusion parameters, the relationship between different blood flow variables and the implications for ischemic injury are outlined in the ensuing discussion. This chapter focuses on the hemodynamic changes that occur in ischemic stroke, and their contribution to ischemic stroke pathophysiology. We discuss the interaction between cardiovascular response and hemodynamic changes in stroke. Studying hemodynamic changes has a key role in stroke prevention, therapeutic implications and prognostic importance in acute ischemic stroke: preexisting hemodynamic and autoregulatory impairments predict the occurrence of stroke. Hemodynamic failure predisposes to the formation of thromboemboli and accelerates infarction due to impairing compensatory mechanisms. In ischemic stroke involving occlusion of a large vessel, persistent collateral circulation leads to preservation of ischemic penumbra and therefore justifying endovascular thrombectomy. Following thrombectomy, impaired autoregulation may lead to reperfusion injury and hemorrhage.

INTRODUCTION

Hemodynamics is the study of blood flow, where parameters have been defined to quantify blood flow and the relationship with systemic circulatory changes. Understanding these perfusion parameters, the relationship between different blood flow variables and the implications for ischemic injury are outlined in the ensuing discussion. This chapter focuses on the hemodynamic changes that occur in ischemic stroke, and their contribution to ischemic stroke pathophysiology. We discuss the interaction between cardiovascular response and hemodynamic changes in stroke. Studying hemodynamic changes has a key role in stroke prevention, therapeutic implications and prognostic importance in acute ischemic stroke: preexisting hemodynamic and autoregulatory impairments predict the occurrence of stroke (Silvestrini et al., 1996). Hemodynamic failure predisposes to the formation of thromboemboli (Caplan and Hennerici, 1998) and accelerates infarction due to impairing compensatory mechanisms. In ischemic stroke involving occlusion of a large vessel, persistent collateral circulation leads to preservation of ischemic penumbra and therefore justifying endovascular thrombectomy (EVT; Piedade et al., 2019). Following thrombectomy, impaired autoregulation may lead to reperfusion injury and hemorrhage (Vitt et al., 2019). We have summarized the fundamental hemodynamic or perfusion parameters in Table 28.1, along with their definition, modifying factors and measurement techniques.

HEMODYNAMIC CHANGES IN ACUTE ISCHEMIC STROKE

In order for blood to provide oxygen and nutrient to the brain tissue a constant "flow" or supply is necessary. Therefore, cerebral blood flow (CBF) may be the most relevant hemodynamic parameter to understand ischemic

*Correspondence to: David S. Liebeskind, Neuroscience Research Building, 635 Charles E Young Drive South, Suite 225 Los Angeles, CA, 90095-7334, United States. Tel: +1-310-963-5539, E-mail: dliebeskind@mednet.ucla.edu

Table 28.1

Parameters of systemic and cerebral hemodynamics

Parameters	Definition	Varying modifiers	Means of measurement
Systemic blood pressure	Pressure exerted by blood on arterial walls in the body	Intravascular volume, vascular tone, vascular compliance and cardiac stroke volume	Commonly by exerting counter-pressure on limb arteries by sphygmomanometer, also by intravascular catheters
Mean arterial pressure	Average blood pressure in a cardiac cycle	Systemic blood pressure during cardiac systole and diastole	Summation of systolic blood pressure and twice diastolic blood pressure divided by 3
Systemic blood flow	Amount of total body blood stream moving from high towards low pressure points	Systemic perfusion pressure and systemic vascular resistance	Doppler ultrasound probes, electromagnetic flowmeters or radioactive tracers
Systemic vascular resistance	Resistance against systemic blood flow	Arterial elasticity, arteriolar diameter, blood viscosity and tissue pressure	Mathematical division of systemic perfusion pressure by systemic blood flow
Cerebral perfusion pressure	Pressure gradient driving cerebral blood perfusion	Mean arterial pressure and intracranial pressure	Difference between the mean arterial and intracranial pressures
Cerebral blood flow	Volume of blood moving through the mass of brain tissue in a given time	Cerebral perfusion pressure and cerebral vascular resistance	Radioactive tracers, CT perfusion, MR perfusion and transcranial Doppler sonography
Cerebral vascular resistance	Resistance against cerebral blood flow	Small pial and precapillary arteries diameter, blood viscosity and intracranial pressure	Mathematical division of cerebral perfusion pressure by cerebral blood flow
Cerebral blood volume	Blood volume within the mass of brain tissue	Diameter and capacitance of cerebral blood vessels	Radioactive tracers, contrasted CT or MRI
Mean transit time	Average time blood stays within a given volume of microcirculation	Cerebral perfusion pressure and diameter of cerebral arteries and arterioles	Mathematical division of cerebral blood volume by cerebral blood flow
Cerebral oxygen extraction fraction/cerebral metabolic rate of oxygen	Ratio of oxygen that the brain tissue takes up from oxygenated blood	Neuronal activity, mean transit time, blood oxygenation level	Radioactive tracers, blood oxygenation level-dependent sequence of MRI, and near-infrared spectroscopy
Cerebral autoregulation indices	Cerebral vessels capability to maintain constant cerebral blood flow over a range of cerebral perfusion pressure levels	Degree of hypoperfusion, vasomotor reactivity and degree of collateral circulation	Correlation between spontaneous fluctuations of cerebral perfusion pressure and cerebral blood flow

stroke pathophysiology (Amukotuwa et al., 2019). Blood flows across a blood pressure gradient. The difference between mean arterial pressure (MAP) and intracranial pressure (ICP), known as cerebral perfusion pressure (CPP), provides the required pressure gradient for cerebral blood to flow in the healthy brain. In the setting of acute ischemic stroke; however, blood flow changes or perfusion abnormalities undergo several alterations.

Hemodynamics of ischemic stroke: An overview

Ischemic stroke arterial occlusion or high-grade stenosis decreases CPP and therefore diminishes downstream blood flow. A drop in the arterial pressure causes a compensatory increase in the systemic blood pressure (Willmot et al., 2004). As the resultant pressure gradient develops across collateral anastomoses bridging normal arteries with the downstream territory of an occluded vessel, collateral flow is recruited in order to compensate for the drop in the blood flow (Piedade et al., 2019). If collateral circulation is not recruited or is insufficient, CBF continues to fall. As a compensatory measure, cerebral autoregulation dilates arteries in order to achieve two goals (Rapela and Green, 1964; Derdeyn et al., 2002): (1) to preserve blood flow by increasing the luminal cross-section and augmenting intravascular blood volume; and (2) to increase mean transit time (MTT), the average time blood stays within a given volume of capillary circulation, in order to allow the brain tissue increasing the oxygen extraction fraction (OEF) and maintain the cerebral metabolic rate of oxygen ($CMRO_2$).

If CPP continues to drop, autoregulatory mechanisms are exhausted and blood flow passively falls. OEF initially surges but cannot sustain brain tissue demands. The subsequent cessation of oxygen and nutrients impairs the process of oxidative phosphorylation and energy production in the brain tissue, which in turn leads to a cascade of biochemical changes and eventually, cell death: membrane ionic and water imbalance (detected by diffusion-weighted imaging or DWI), membrane depolarization, excitatory neurotransmitter release, influx of calcium, generation of oxygen free radicals, and ultimately disintegration of cellular membranes (Wechsler, 2011). Hemodynamics of infarcted brain tissue is notable for a severe reduction of CBF, as well as vascular collapse and marked reduction of blood volume in the infarcted tissue. In addition to a reduction in cerebral blood volume (CBV) on computed tomography (CT) perfusion, studies have shown that a decrease in relative CBF below 32% can reliably delineate infarcted brain tissue (Amukotuwa et al., 2019).

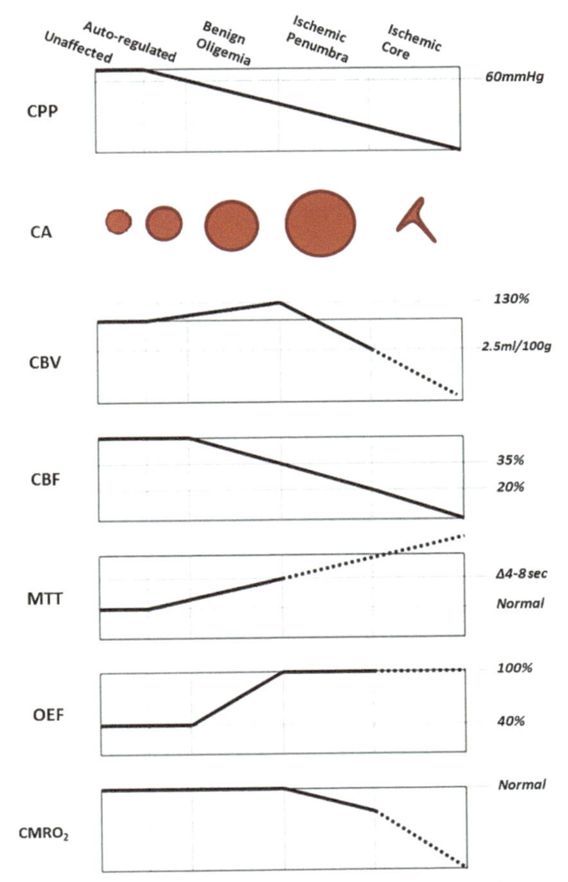

Fig. 28.1. Schematic representation of hemodynamic stages of the brain tissue across ischemic process as cerebral perfusion (CPP) progressively diminishes. Hemodynamic stages have been labelled across the x-axis. Y axis represents changes in hemodynamic parameters. Thick dotted curves indicate variable values. Changes in cerebral autoregulation (CA) has shown as changes in vessel diameter. Other variables include *CBF*, cerebral blood flow; *CBV*, cerebral blood volume; *$CMRO_2$*, cerebral metabolic rate of oxygen; *MTT*, mean transit time; *OEF*, oxygen extraction fraction. Reproduced from Lee, D.H., Kang, D.W., Ahn, J.S., et al., 2005. Imaging of the ischemic penumbra in acute stroke. Korean J Radiol 6, 64–74 with some modifications.

Ischemic spectrum: From benign oligemia to infarction

After any blood flow impairment, the brain tissue transitions across a spectrum of pathophysiologic, hemodynamic states. Fig. 28.1 summarizes stages of the ischemic process. Following initial decrease in CPP ("Auto-regulated" in Fig. 28.1), compensatory mechanisms (cardiovascular response, collateral perfusion and cerebral autoregulation) are recruited to avoid a drop in CPP. Stage 2, or "benign oligemia" in Fig. 28.1, is marked by declining CPP beyond the capacity of compensatory mechanisms to preserve CBF and therefore blood flow decreases (i.e., oligemia). However, compensatory mechanisms are not yet exhausted and by increasing OEF, the oxygen metabolic rate is relatively preserved (i.e., benign). As CPP continues to drop, cardiovascular response and collateral perfusion are exhausted; however, cerebral autoregulation slows down CBF decline by vasodilation and increasing CBV. MTT is further prolonged to maximize OEF, but eventually fails and oxygen metabolic rate ($CMRO_2$) starts to diminish, therefore marking the beginning of ischemia. Stage 3 is called "ischemic penumbra" (Fig. 28.1), as the degree of ischemia is not severe enough to cause irreversible damage. However, ischemic penumbra is dysfunctional and causes neurological impairments. In fact, the initial volume of ischemic penumbra is better correlated with neurological deficits than the initial infarcted tissue in acute stroke (Barber et al., 1998). Ischemia progresses and the brain tissue eventually suffers irreversible damage (stage 4, or "ischemic infarct" in Fig. 28.1).

In addition to the temporal progression of brain tissue across the ischemia spectrum, it is important to understand a spatial pattern of ischemic changes: the territory of an affected artery is impacted differently based on metabolic demands and availability of collateral perfusion. For instance, in middle cerebral artery (MCA) occlusion, the basal ganglia are frequently infarcted at a more rapid pace than surrounding areas (Fig. 28.2), given the lack of collateral supply to penetrating lenticulostriate arteries. Ischemic penumbra and benign oligemia are located adjacent to the infarcted tissue, and if brain hemodynamics continues to deteriorate those regions also become inevitably infarcted. To improve the clinical outcome of ischemic stroke, it is therefore important to understand compensatory mechanisms involved in the preservation of oligemic or ischemic tissue to augment and prevent their failure.

In the next three sections, we will elaborate on three hemodynamic processes involved in acute ischemic stroke to preserve cerebral hemodynamics and therefore brain tissue viability: cardiovascular response, collateral circulation and cerebral autoregulation.

Blood pressure and cardiac response in ischemic stroke

Acute ischemic stroke is a sudden occlusive disease of blood flow. Therefore, the cardiovascular system compensates to maintain blood flow by increasing systemic blood pressure. Hypertension is found in 52% (Willmot et al., 2004) to 75% (Bath et al., 2003) of patients presenting with acute ischemic stroke. It may be initially challenging to discriminate a higher prevalence of premorbid hypertension in acute stroke patients from an acute hypertensive response. Crude comparisons find acute stroke patients with hypertension to be more prevalent than patients with premorbid hypertension (Qureshi, 2008). But the data on spontaneous decrease in blood pressure in the following 10–60 hours after stroke (Wallace and Levy, 1981; Aslanyan et al., 2003) are more convincing and suggest that hypertension in a majority of acute stroke patients is an acute compensatory response. In addition, the commonly noted blood pressure drop following successful thrombectomy further supports hypertension partly being a reaction to acute ischemic stroke (Mattle et al., 2005).

The hypertensive response is thought to be triggered directly by brain ischemic damage: insular cortex has an activating role for cardiovascular responses (Meyer et al., 2004) through nucleus tractus solitarius and the ventrolateral medulla (Nason and Mason, 2004), with cingulate cortex, amygdala, and hypothalamus modifying the response. Prefrontal cortex, on the other hand, has an inhibitory role in cardiovascular response (Hilz et al., 2006). It is interesting to know that the sympathetic drive is lateralized to the right hemisphere damage, while parasympathetic tone is lateralized to the left-sided stroke (Oppenheimer, 1993). In addition to the brain tissue damage, muscle paralysis (Ichiyama et al., 2004) and changes in nitric oxide release (Resstel and Correa, 2006) are

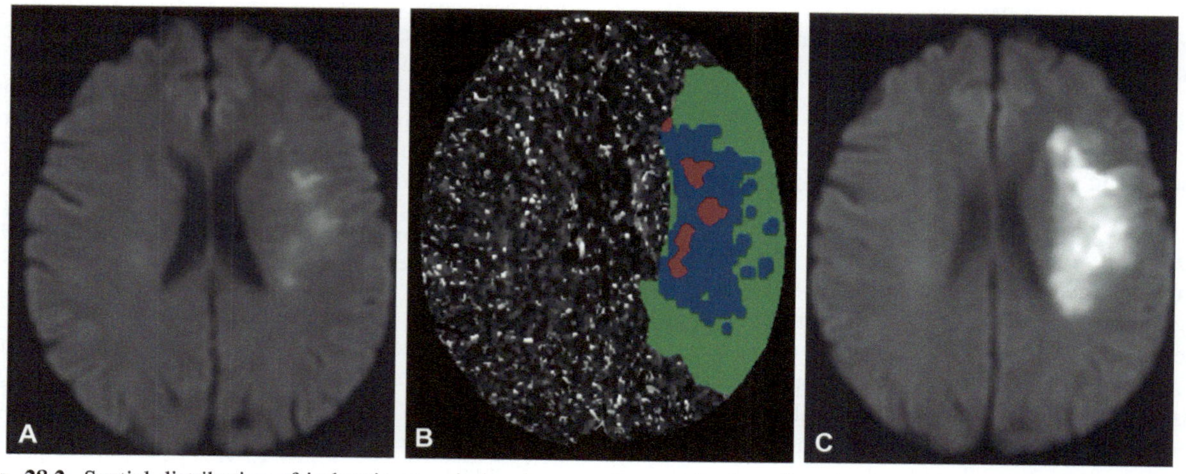

Fig. 28.2. Spatial distribution of ischemic core, ischemic penumbra and benign oligemic tissue in a patient with left middle cerebral artery occlusion who was not revascularized. (A) The baseline axial diffusion-weighted image (DWI) shows ischemic core. CT-mean transit time (MTT) map (B) has been segmented for MTT longer than 12 s and further post-processed for benign oligemia (*green overlay*) and at-risk ischemic penumbra (*blue overlay*), with baseline ischemic core overlaid in *red*. Follow up DWI sequence 44 h after the baseline CTP (C) shows infarction of ischemic penumbra. Reproduced from Kamalian, S., Kamalian, S., Konstas, A.A., et al., 2012. CT perfusion mean transit time maps optimally distinguish benign oligemia from true "at-risk" ischemic penumbra, but thresholds vary by postprocessing technique. AJNR Am J Neuroradiol 33, 545-549 with modifications.

thought to drive cardiovascular response to acute ischemic stroke. In addition to cardiac sympathetic activity, sympathoadrenal axis activation and release of renin with subsequent vasoconstriction of arterioles further contribute to augmentation of blood pressure (Barron et al., 1994).

Cardiovascular response to acute stroke can also be maladaptive. Increased sympathetic tone and catecholamine release, or increased parasympathetic drive can induce a variety of arrhythmias, some predisposing stroke patients to sudden cardiac death: atrial fibrillation, atrial flutter, sinus tachycardia, atrial and ventricular premature complexes, ventricular tachycardia, bradyarrhythmia, second- and third-degree blocks, prolonged QT syndrome and Brugada syndrome (Taggart et al., 2011; Kallmunzer et al., 2012). Some of these arrhythmias, namely long QT and subsequent ventricular polymorphic tachycardia, have been particularly associated with sudden death after stroke (Kumar et al., 2012).

Acute stroke also induces transient cardiac systolic dysfunction, known as neurogenic stunned myocardium, marked by hypokinetic ventricular motion in the absence of any coronary perfusion deficits (Agewall et al., 2011; Biso et al., 2017). It is thought to be caused by coronary vasospasm secondary to catecholamine release (Wang et al., 1997), increased cardiac demand and resultant ischemic (Nguyen and Zaroff, 2009), or direct toxicity of catecholamines on myocardium (Samuels, 2007). In addition to systolic dysfunction, patients with acute stroke have a concomitant diastolic dysfunction with increased left ventricular end-diastolic pressure that may lead to endothelia injury and subsequent hypercoagulability (Manea et al., 2015). However, a causal relation to acute stroke is unclear.

Troponin elevation in acute stroke

There is evidence of cardiac troponin (cTn) enzyme elevation in as many as 53% patients with acute stroke (Faiz et al., 2014), suggesting some degree of myocardial injury. However, only 6% of patients meet the criteria for acute myocardial infarction (MI). It is important to note that elevated cTn, especially using the contemporary high sensitivity assays, does not necessarily signify coronary artery disease. It is helpful to know while majority of troponin is bound to myocardial cytoskeleton, about 8% is soluble in the cytoplasm (Bleier et al., 1998). The cytosolic pool could be released into the serum by changes in cellular wall permeability secondary to myocardial stretch or mild ischemia (White, 2011) without cardiomyocyte necrosis.

cTn elevation is correlated with strokes involving insular cortex (Jensen et al., 2007; Scheitz et al., 2012), suggesting sympathetic over-activation through increasing cardiac demand (with or without a fixed coronary stenosis), coronary vasospasm or direct catecholamine toxicity is the likely mechanism for elevated cTn. In addition to insular stroke, cardiac enzyme elevation is associated with age, diabetes, hypercholesterolemia, chronic kidney disease, premorbid coronary artery disease and congestive heart failure (Hasirci et al., 2013; Scheitz et al., 2015).

There is no standard recommendation on interpreting elevation cTn in acute stroke patients or any guideline for further coronary work up. Absence of chest pain and dyspnea, lack of electrocardiographic evidence of MI and stable or decreasing levels of cTn are suggesting against MI. It is important to consider other noncoronary causes of cTn elevation, such as sepsis or pulmonary emboli (Scheitz et al., 2015). Elevated cTn has also prognostic values and predicts a poor clinical outcome (Sandhu et al., 2008).

Blood pressure management in acute stroke

Retrospective analysis of several large randomized trials has revealed a U-shaped relationship between blood pressure and clinical outcome. Patients in the IST trial with systolic blood pressure (SBP) 140–179 mmHg had better clinical outcome compared with lower or higher SBPs (Leonardi-Bee et al., 2002). SBP <120 or SBP >200 was particularly associated with worse clinical outcome in the IST cohort. Another study from Mayo Clinic has suggested SBP range of 156–220 is best correlated with an improved clinical outcome, with twofold increase in the mortality in hypotensive patients (Stead et al., 2005). Castillo and colleagues have revealed acute stroke patients presenting with SBP around 180 mmHg have the most optimum clinical outcome, and a sudden decrease (>20 mmHg) drop in SBP has a strong correlation with larger ischemic core sizes (Castillo et al., 2004). These observational information, although pointing towards, do not provide evidence for BP lowering to intermediate values in acute stroke patients. CHHIPS was a randomized double-blinded trial on acute stroke patients, including both ischemic and hemorrhagic (Potter et al., 2009). In patients with hypertension (SBP >160 mmHg) within 24 h after stroke onset, a decrease in SBP (21; IQR[17–25] was associated with no change in mortality or clinical deterioration at 2 weeks; however, 3-month mortality decreased by more than 50% in intervention arm. In CATIS, a randomized single-blinded trial on acute ischemic stroke patients recruited within 48 h of stroke onset revealed that a mean 9.1–9.3 mmHg decrease in SBP did not lead to any changes in death or major disability in 14 days or 3 months (He et al., 2014).

We review blood pressure management in three different scenarios of acute ischemic stroke:

- Patients not candidate for thrombolysis or EVT: Current AHA guidelines advises permissive hypertension to 220/120 (Powers et al., 2018). In patients with contraindications for permissive hypertension, such as those with acute coronary disease, heart failure or eclampsia, no more than 15% initial blood pressure decrease is recommended to avoid hemodynamic disturbances leading to ischemic core growth.
- Patients receiving thrombolysis: Current guidelines recommend blood pressure should be controlled below 185/110 in patient receiving thrombolysis. Those BP values were mainly extrapolated from the NINDS alteplase trial (The National Institute of Neurological and Stroke rt, 1995), but also supported by retrospective analyses of cohorts that revealed patients with higher BP values have a higher risk of intracranial hemorrhage (The National Institute of Neurological and Stroke rt, 1995; Lopez-Yunez et al., 2001; Lansberg et al., 2007). For instance, patients treated with alteplase with SBP >170 mmHg had twofold increase in intracranial hemorrhage risk compared to those with SBP between 141 and 150 mmHg (Ahmed et al., 2009). However, caution should be practiced with lowering blood pressure in this patient's population, as an abrupt (short-term decrease in SBP >30 mmHg) or significant (overall SBP >60 mmHg) BP decrease was associated with a twofold increase in mortality (Silver et al., 2008).
- Patients undergoing EVT: Baseline BP in patients undergoing EVT has the same U-shaped correlation with clinical outcome irrespective of recanalization status: Baseline SBP below 110 mmHg or higher than 180 mmHg was associated with higher mortality rates (Maier et al., 2017). Beyond those observation studies, informed guidelines for BP control before EVT are missing. It is generally recommended to maintain BP below 185/110 mmHg, but caution must be practiced in lowering BP prior to revascularization as collateral circulation and cerebral autoregulation rely on systemic blood pressure to sustain viability of ischemic penumbra. In fact, an intraprocedural decrease in MAP was highly associated with permanent neurological impairment (John et al., 2016).
- Blood pressure control following thrombectomy has to strike the balance between hypoperfusion and reperfusion injury. In cases of poor recanalization, penumbra may sustain on persistent higher BP. In revascularized patients, on the other hand, rate of reperfusion injury is particularly high in hypertensive patients (Zaidat et al., 2013). In an interesting observation study, among patients with successful recanalization who had intracranial hemorrhage, mean SBP was lower compared to nonrevascularized patients with intracranial (170 vs 196 mmHg), suggesting that revascularization expectedly decreases the BP threshold of reperfusion injury (Mistry et al., 2017). Randomized controlled trials in post-EVT BP management are lacking, but in a prospective registry of acute stroke patients, a moderate (BP <160/90 mmHg) or intensive (BP <140/90 mmHg) post-EVT BP control was associated with a far better 3-month mortality rate (6.5%) compared with patients with higher BP (28.7%) (Goyal et al., 2017). Based on more recent EVT trials, an SBP limit of 140 mmHg for patients with good recanalization appears safe (Albers et al., 2018; Nogueira et al., 2018), while in other patients BP goals have to be determined based on the degree of reperfusion, size of infarct and other comorbidities.

Collateral circulation in ischemic stroke

Collateral circulation, as a built-in anastomosed artery-to-artery or arteriole-to-arteriole network, exists in different organs, such as the coronary arteries and peripheral circulation. Collateral vessels can potentially provide the tissue with blood flow while the artery normally supplying the blood is occluded. Therefore, collateral circulation protects the tissue from ischemia in acute ischemic stroke. Cerebral collaterals were first described by Sir Thomas Willis in the 17th century: "The cephalic arteries, whether they be carotids or vertebrals, communicate one with the other reciprocally in various ways…This we have demonstrated by injecting dark substances in only one branch and observing that the whole brain becomes colored" (Willis, 1664).

Collateral perfusion leads to augmentation of CBF in the area affected by arterial occlusion. In addition, collateral circulation facilities clearance of fragmented thrombi into more distal vessels (Caplan and Hennerici, 1998; Wang et al., 2001). The later mechanism may explain how better collaterals are associated with a more successful revascularization following EVT (Liebeskind et al., 2014; Sheth et al., 2016). Collateral circulation also influences the delivery of thrombolytic therapy, leading to an improved early recanalization in large vessel occlusion (LVO) patients with enhanced collaterals (Seners et al., 2019). Collaterals also determine the response to perfusion-enhancing strategies: although the overall results of aortic occlusion to improve stroke outcome (SENTIS trial) were not striking, a subset of patients with an intact circle of Willis showed promising results (Schellinger et al., 2013). In the late subacute to chronic phase after brain infarction, collateral circulation has been implemented in clearing infarcted tissue debris (Manoonkitiwongsa et al., 2001).

Collateral anatomy

Cerebral collaterals are divided into three anatomic categories:

a. Circle of Willis: It is a circle of interconnected arteries at the base of brain. It provides cross-connections across anterior circulation arteries (including anterior cerebral arteries or ACAs, MCAs and internal carotid arteries or ICAs), with vertebrobasilar arteries. In the circle of Willis, medium-size feed-arteries are macroscopically connected providing a high flow collateral blood supply. However, capacity of the circle of Willis depends on size and continuity of its component arteries, where there are variations in the population: A complete circle of Willis is only present in 25% (Zhou et al., 2016) to 36% (Hartkamp et al., 1999) of individuals, and it is associated with less severe disability following stroke. A complete circle of Willis is considered type I (Fig. 28.3A). Other variations include a complete anterior half but an incomplete posterior half (Type II; Fig. 28.3B), an incomplete anterior half and a complete posterior half (Type III; Fig. 28.3C) and an incomplete anterior and posterior halves (Type IV; Fig. 28.3D). Reported prevalence of other types of circle of Willis are 57%, 3% and 15% for types II, III and VI, respectively (Zhou et al., 2016).

Recent studies have shown that age, diabetes and hypertension are correlated with an incomplete circle of Willis (Eaton et al., 2020) especially reducing posterior communicating artery diameter (Faber et al., 2019), suggesting impaired collaterals as an explanation for more severe stroke outcomes in older patients with comorbidities. In addition, women have a slightly higher rate of complete circle of Willis than men (43.8% vs 31.2%) across all age groups (Zaninovich et al., 2017).

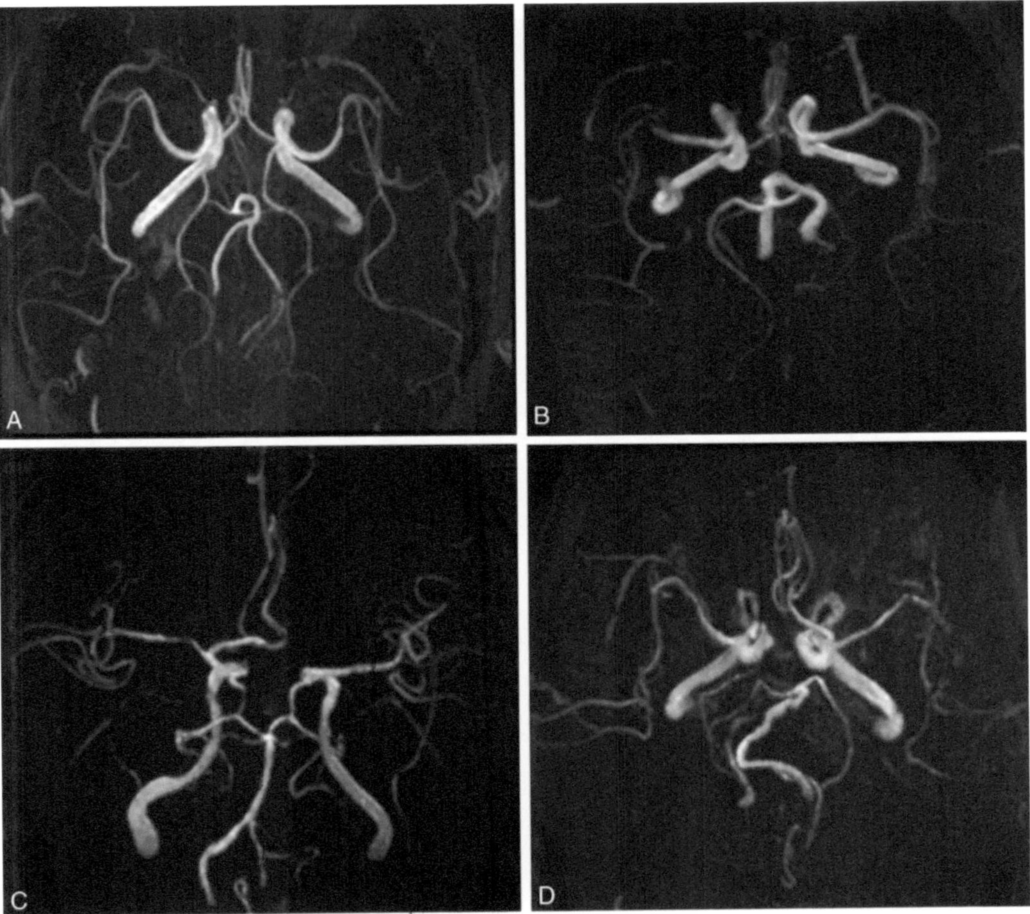

Fig. 28.3. Types of the circle of Willis. (A) A complete circle of Willis or type I; (B) a complete anterior half but an incomplete posterior half or type II; (C) an incomplete anterior half but a complete posterior half or type III; and (D) an incomplete anterior and posterior halves or type IV. Reproduced from Zhou, H., Sun, J., Ji, X., et al., 2016. Correlation between the integrity of the circle of willis and the severity of initial noncardiac cerebral infarction and clinical prognosis. Medicine (Baltimore) 95, e2892 with permission.

Different components of the circle of Willis have different weights in stroke prevention. An anterior nonfunctional collateral pathway is more commonly associated with anterior circulation stroke while no such correlation was found for the posterior segments of the circle of Willis (Hoksbergen et al., 2003). In patients with ICA occlusion and stroke, odds ratio of a dysfunctional anterior or posterior components of the circle of Willis is 7.33 and 3.0, respectively, suggesting a more substantial role for anterior segments of the circle of Willis to prevent stroke.

b. Pial (or leptomeningeal) collaterals: Those are arteriole-to-arteriole anastomoses cross-connecting arterioles at distal branches (i.e., "crowns") of large cerebral vessels. Pial collaterals bridge the following pairs of intracranial arteries (Fig. 28.4): ACA-MCA, MCA-PCA, PCA-SCA (superior cerebellar artery) and pairs of long circumferential cerebellar arteries (SCA, anterior inferior cerebellar artery or AICA and posterior inferior cerebellar artery or PICA), but among those, collaterals between ACA and MCA are higher in numbers and size (Liebeskind, 2003). In contrast to the circle of Willis, pial collaterals are considered "microvascular" anastomoses connecting arterioles 50–250 μm in diameter, and therefore provide a slow retrograde filling of the recipient artery. Such retrograde filling is visible on the cerebral angiogram, as well as delayed perfusion on multiphase CT angiogram (Menon et al., 2015) and perfusion scans (Ip and Liebeskind, 2014).

The extent of pial collaterals is predictive of infarct size and clinical outcome (Ringelstein et al., 1992; Kucinski et al., 2003; Christoforidis et al., 2005; Angermaier et al., 2011; Seeta Ramaiah et al., 2014; van den Wijngaard et al., 2015). As expected from neurovascular anatomy, the rescuing role of pial collaterals is more pronounced in intracranial ICA or main branch MCA occlusions (Menon et al., 2011). Following MCA occlusion, poor pial collaterals are more strongly associated with a higher burden of cortical infarctions compared with deep structures infarctions (Verma et al., 2015). In MCA or PCA occlusions basal ganglia or thalamus, respectively, are often infarcted due to a poor collateral perfusion in penetrating arteries while pial collaterals support viability of cortices. Recruitment of pial collaterals is influenced by the circle of Willis: a complete circle of Willis in ICA occlusion is associated with better pial collaterals (Millesi et al., 2019). In addition, pial collateral

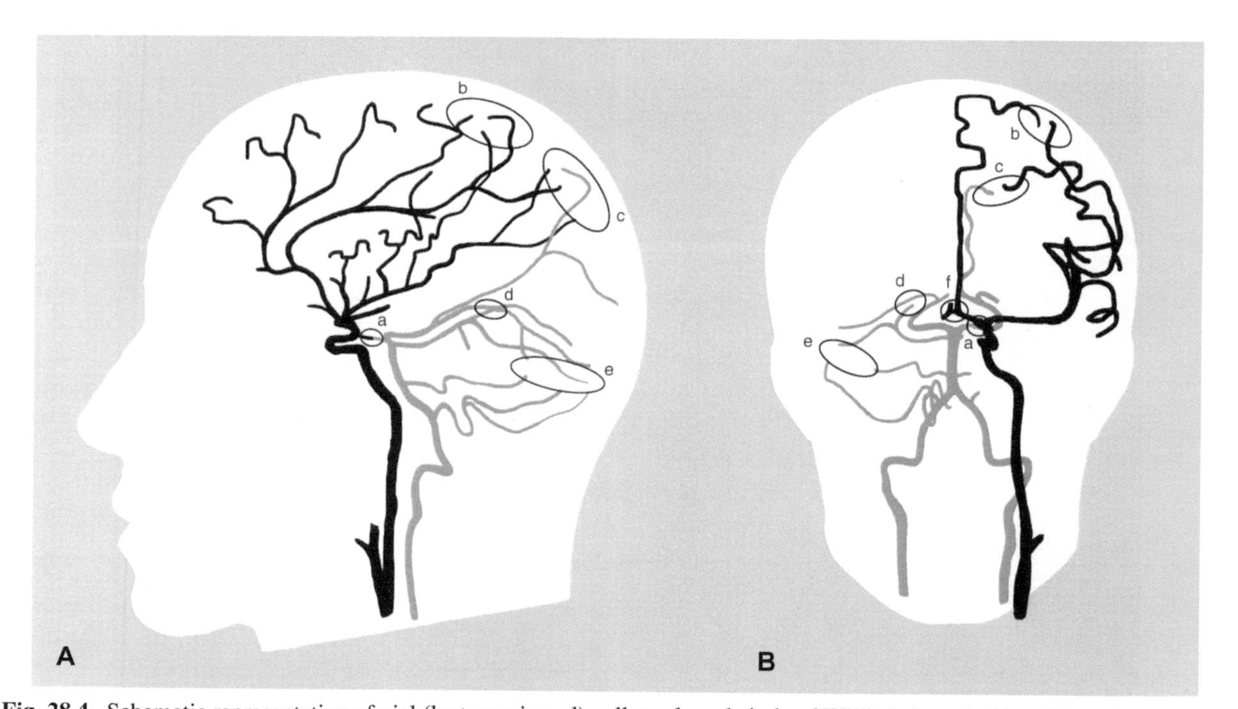

Fig. 28.4. Schematic representation of pial (leptomeningeal) collaterals and circle of Willis in lateral (A) and frontal (B) views. Posterior communicating artery (a) and anterior communicating artery (f) are shown. Anastomoses between the anterior and middle cerebral arteries (b), between the posterior and middle cerebral arteries (c), tectal plexus between posterior cerebral and superior cerebellar arteries (d), and anastomoses between distal cerebellar arteries (e) are also marked. Adapted from Shuaib, A., Butcher, K., Mohammad, A.A., et al., 2011. Collateral blood vessels in acute ischaemic stroke: a potential therapeutic target. Lancet Neurol 10, 909–921.

recruitment in anterior circulation occlusion is associated with the potency of anterior communicating artery and absence of ipsilateral posterior communicating artery (Millesi et al., 2019).

Pial collaterals decline with age, and comorbidities such as metabolic syndrome, dyslipidemia and hyperuricemia (Menon et al., 2013; Nannoni et al., 2019). There is an association between poor pial collaterals and a lower peripheral blood concentrations of certain long noncoding RNAs in patients with acute stroke, suggesting a potential genetic role for pial collaterals regulation (Wu et al., 2019). Pial collaterals are more plastic and augment in response to chronic ischemia (Schaper, 2009; van Royen et al., 2009), likely due to their smaller sizes and less complex structure allowing more vasoplasticity.

c. External–internal carotid arteries collaterals: Those are anastomoses between distal branches of extracranial carotid arteries and those of ICAs. The most significant extracranial–intracranial anastomoses include (Geibprasert et al., 2009):
   - Petrosal ICA: Receiving collaterals from internal maxillary artery branches (Vidian artery).
   - Meningohypophyseal trunk of ICA: Receiving collaterals from ascending pharyngeal artery branches (superior pharyngeal artery, hypoglossal branches and jugular branches).
   - Ophthalmic artery: Receiving collaterals from internal maxillary artery branches (middle meningeal artery, anterior deep temporal artery), facial artery branches and superficial temporal artery branches (frontal branch).
   - Inferolateral trunk of ICA: Receiving collaterals from internal maxillary artery branches (middle meningeal artery, accessory meningeal artery and the artery of foramen rotundum).
   - Vertebral arteries: Receiving collaterals from ascending pharyngeal artery branches (Odontoid arch), muscular branches of the occipital artery, and ascending and deep cervical artery branches.

A schematic figure summarizes highlights of extracranial–intracranial collaterals (Fig. 28.5). Valuable review articles have also been referenced (Liebeskind, 2003; Geibprasert et al., 2009) for more in-depth reading.

Extracranial–intracranial collateral blood flow is mainly recruited in internal carotid occlusions. In patients with ICA occlusion presence of collateral blood flow through ipsilateral external carotid artery is associated with less neurologic symptoms (Countee and Vijayanathan, 1979; Macchi et al., 2002; van Laar et al., 2008) or evidence of tissue ischemia (Yamauchi et al., 2004). In patients with ICA occlusion, a concomitant ECA stenosis is associated with neurologic symptoms (Dalainas et al., 2012), and ECA revascularization leads to clinical improvement (Countee and Vijayanathan, 1979; Xu et al., 2010).

The microvascular nature of extracranial–intracranial anastomoses enables them to be plastic and gradually grow in parallel to progression of atherosclerotic ICA stenosis. Among different anatomic locations of extracranial–intracranial anastomosis (see above), retrograde ophthalmic artery flow in ICA stenosis or occlusion is the most encountered in clinical settings. Ophthalmic artery retrograde flow correlates with the degree of ICA stenosis (Park et al., 2019), suggesting collateralization via ophthalmic artery is a compensatory mechanism. Treating ICA stenosis with stenting (Ishii et al., 2016) or endarterectomy (Wang et al., 2016) leads to a reversal of retrograde flow in ophthalmic artery marking de-recruitment of collateral pathway after ICA revascularization. Retrograde flow patterns in ophthalmic artery have been associated with ophthalmic ischemic syndrome (Wang et al., 2017) and neovascularization of the iris, later known as *rubeosis iridis* (Oller et al., 2012).

Collaterals: Recruitment, persistence and collapse

The existing collateral anastomoses in healthy tissue encounters blood flow in opposite directions from two feeding arteries. A "to-and-fro" flow across anastomoses has been speculated to explain how thrombosis is prevented at baseline (Trzeciakowski and Chilian, 2008; Toriumi et al., 2009; Chalothorn and Faber, 2010). Such bidirectional flow can occur by a slight difference in the feeding arteries' distances from the heart leading to an asynchrony in blood flow cycle between two anastomosed arterial territories.

Recruitment of collateral circulation is triggered by development of blood pressure gradients across an arterial anastomosis through drop in the "receiving" artery's blood pressure downstream of an arterial stenosis or occlusion (Meyer and Denny-Brown, 1957). Some of the preexisting arterial anastomoses are patent and immediately recruited following blood pressure gradient. Some literature has named those "primary" collaterals (Liebeskind, 2003) and their rapid recruitment has been demonstrated following 15 to 30 s of MCA balloon occlusion (Qureshi et al., 2008). "Secondary" collaterals, in contrast, are anatomically present but recruited with a delay as their development involves metabolic and neuro-hormonal mechanisms (Liebeskind, 2003).

Persistence of collateral flow, ischemia and infarction trigger "tertiary" collateral formation in the ensuing days to weeks that involves formation of new arteries (arteriogenesis) or microvasculature (angiogenesis) to support tissue viability in chronic ischemia, such as progressive

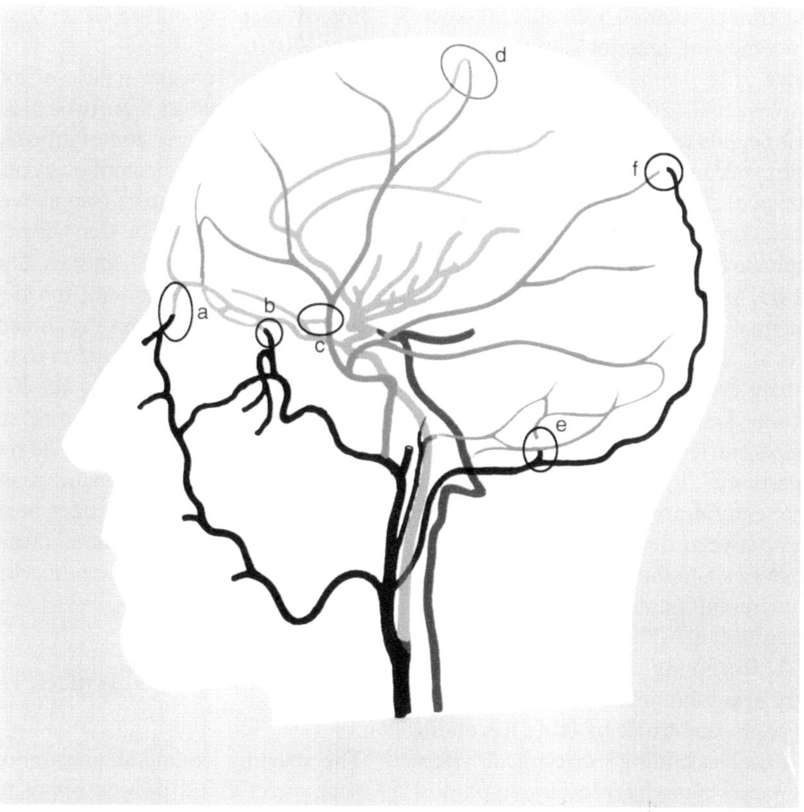

Fig. 28.5. Schematic representation of collateral anastomoses between extracranial and intracranial arterial circulation. (a) facial artery anastomosis with ophthalmic artery, (b) maxillary artery branches anastomosis with internal carotid artery and ophthalmic artery branches, (c) middle meningeal artery anastomosis with internal carotid artery branches, (d) dural anterior anastomosis of middle meningeal artery, (e) occipital artery anastomoses with vertebral artery and (f) occipital artery anastomosis through parietal foramen. Reproduced from Shuaib, A., Butcher, K., Mohammad, A.A., et al. 2011. Collateral blood vessels in acute ischaemic stroke: a potential therapeutic target. Lancet Neurol 10, 909–921.

focal atherosclerotic disease (Iwasawa et al., 2016). Robust collaterals in patients with sudden occlusion of the main branch MCA who also have concomitant ipsilateral ICA severe stenosis (Pienimaki et al., 2020) support the notion of the gradual development of tertiary collaterals in parallel to gradually narrowing proximal ICA stenosis. In general, collaterals are more robust in atherosclerotic causes of stroke compared with cardioembolic etiologies (Rebello et al., 2017; Guglielmi et al., 2019) indicating a gradual increase in collateral capacity in atherosclerotic disease.

Initial blood circulation across existing collateral anastomoses triggers a complex molecular and cellular response to promote further recruitment and persistence of collateral circulation. Blood shear stress activates mechanosensory cascades in vascular endothelial cells that lead to an enhanced transcription of proliferative and regenerative factors, such as brain-derived neurotrophic factor (Nakahashi et al., 2000; Prigent-Tessier et al., 2013) or endothelial transforming growth factor beta-1 (Ohno et al., 1995). The subsequent proliferation and development of endothelial and smooth muscle cells contribute to arteriogenesis and development of new collaterals. Fluid shear stress also activates endothelial nitric oxide synthase that leads to a subsequent early collateral recruitment and persistence (Jung et al., 2012). In addition, hypoxia triggers hypoxia-inducible factors, which in turn promotes secretion of growth factors such as vascular endothelial growth factor leading to proliferation and development of endothelial cells, smooth muscle cells and pericytes into formation of new vasculature (Iwasawa et al., 2016). The process of collateral recruitment is therefore driven by cerebral hemodynamic compromise and hypoxemia, and it leads to the initial expansion of collateral perfusion following ischemic stroke (Toyoda et al., 1994).

Animals and clinical studies have shown significant variability in the extent of collateral recruitment. Animal studies have identified genes, such as *Candq1* (Wang et al., 2010) and *Dce1* (Sealock et al., 2014), affecting collateralization; however, there is a 50-fold difference in collateral perfusion within inbred mice (Wang et al., 2012) suggesting other factors yet to be discovered. Clinical studies have shown age, male gender and premorbidities such as hypertension and diabetes to negatively affect collateral recruitment after ischemic stroke (Wang et al., 2012; Wiegers et al., 2020). Relation of collateral perfusion and baseline small vessel disease, known as leukoaraiosis, is controversial as some studies have shown association (Mark et al., 2020) while others have not (Sanossian et al., 2011). Baseline statin use has been correlated with enhanced collateral circulation following acute MCA occlusion (Lee et al., 2014).

Collateral circulation preserves ischemic penumbra following acute LVO strokes. Therefore the fate of penumbra tissue in nonrevascularized patients has been used as a proxy for the persistence of collateral circulation. Two late thrombectomy trials, DAWN and DEFUSE 3, have shown penumbra tissue is eventually infarcted if cerebral perfusion is not restored by mechanical thrombectomy (Albers et al., 2018; Nogueira et al., 2018). Therefore collateral circulation, mostly pial anastomoses, are generally thought to be nonsustainable in patients with acute stroke. In contrast, collateral perfusion through the circle of Willis is more robust and may be preserved for 6–12 months after ICA occlusion (Rutgers et al., 2000).

It is important to note while collateral perfusion plays a crucial role in preservation of ischemic penumbra, other compensatory mechanisms such as cerebral autoregulation and systematic blood pressure are also involved. Therefore, one should be cautious in concluding collateral failure from ischemic penumbra demise. To purely study collateral circulation, animal studies have provided us with more focused information on the persistence of collaterals. Studying leptomeningeal collaterals in rats following MCA occlusion revealed a very dynamic collateral response (Wang et al., 2012). Three patterns of collateral perfusion were identified within 180 minutes following MCA occlusion: Some collateral channels were persistent, some collapsed, and some provided intermittent perfusion. These findings highlight the complexity and our limited knowledge of collateral perfusion maintenance.

Cerebral autoregulation

Cerebral autoregulation is the cerebral vessels' rapid response to maintain a steady cerebral blood perfusion across a range of CPP changes, and to accommodate CBF to changes in the brain metabolic demands (Jordan and Powers, 2012). Brain tissue encompassing 2% of body weight but as a highly metabolic organ receives ~20% of cardiac output (Clarke and Sokoloff, 1989). Such a discrepancy between tissue size and volume of blood perfusion signifies the importance of a constant and efficient regulation of CBF to avoid ischemia in a highly metabolic organ, or blood–brain barrier (BBB) breakdown due to hyper-perfusion. Cerebral autoregulation is one of the compensatory mechanisms in acute ischemic stroke to preserve CBF and the brain tissue viability. In addition, cerebral autoregulation affects outcome of revascularization therapies since an impaired cerebral autoregulation causes reperfusion injury manifested as brain edema or hemorrhage (Xiong et al., 2017). Therefore studying cerebral autoregulation could improve outcome in ischemic stroke following thrombolytic or EVT administration.

Cerebral autoregulation is exerted through changes in vessel diameter to control cerebral vascular resistance (CVR) and therefore maintain a steady CBF across a range of CPPs ("CA" in Fig. 28.1). To understand the anatomy of cerebral autoregulation it is important to know pial arteries and parenchymal penetrating arteries (Fig. 28.6) exert the most of, and contribute almost equally to, the CVR (Faraci and Heistad, 1990; Liu et al., 2013). Cerebral autoregulation is exerted through a complex interplay between the following mechanisms:

a. Nitric oxide released by endothelial cells: It is thought to partially contribute to the vasodilatory response to hypertension. Treating patients with L-NMMA, an inhibitor of nitric oxide synthase, led to an increased vascular tone and caused ~17% decrease in autoregulatory capacity (White et al., 2000).
b. Neurogenic factors: Autonomic neurovascular control plays a major role in cerebral autoregulation, and pharmacologic blocking experiments estimate

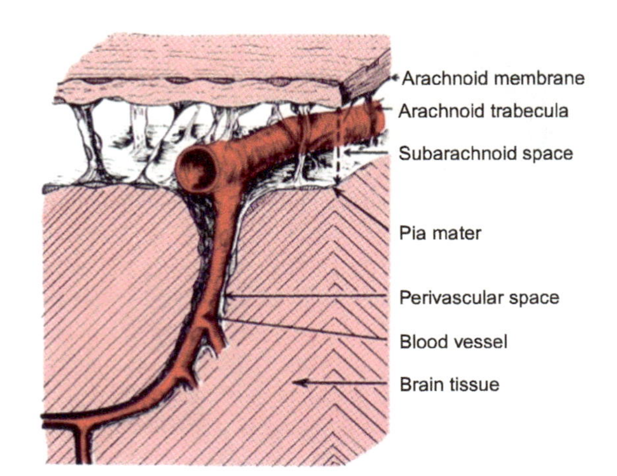

Fig. 28.6. Schematic representation of leptomeningeal arteries giving rise to parenchymal arteries and arterioles. Reproduced from Ranson, S.W., Clark, S.L., 1961. The anatomy of the nervous system. W.B Saunders Co., Philadelphia.

neurogenic factors contribute to ~62% of cerebral autoregulation capacity (Hamner and Tan, 2014). Adrenergic inhibition by prazosin (Ogoh et al., 2008) or use of anticholinergics (Hamner et al., 2012) impair cerebral autoregulation, suggesting both sympathetic and cholinergic mechanisms are involved (Hamel, 2006).

Pial arteries are "extrinsically" innervated by the peripheral autonomic system involving superior cervical ganglion, sphenopalatine ganglion (SPG), otic ganglion, and trigeminal ganglion (Iadecola, 2004; Hamel, 2006). Parenchymal arteries and arterioles, in contrast, receive "intrinsic" neural control from subcortical neurons located within CNS nuclei such as locus coeruleus, basal forebrain raphe nucleus, or local cortical interneurons (Iadecola, 2004; Hamel, 2006). A close feedback system between local neurons and the parenchymal vasculature forms the neurovascular unit facilitating hemodynamic responses to changes in local metabolic demands (Rosengarten et al., 2001).

Such segmental discrepancies in autonomic innervation lead to a heterogeneous segmental vascular response to neural mediators. Norepinephrine (in the presence of alpha-adrenoreceptors) and serotonin are shown to cause vasoconstriction in pial arteries, but due to an abundance of beta-adrenoreceptors in parenchymal vessels (Lincoln, 1995) norepinephrine may cause vasodilation (Cipolla et al., 2004). In addition to a segmental heterogeneity, there are regional differences in autonomic innervation: sympathetic nerves have a denser presence on the anterior circulation compared with vertebrobasilar arteries and their branches (Edvinsson et al., 1976), which has led to a differential response to hypertension: Cerebral autoregulatory response to hypertension is generally shown to be more effective in the posterior circulation through mounting a vasoconstrictive response and moderating CBF (Faraci et al., 1987).

c. Myogenic factor: Intrinsic responses by the smooth muscle cell layer of arteries and arterioles play a crucial role in cerebral autoregulation. They involve stretch-sensitive ion channels (Bayliss, 1902) activating Ca^{2+} influx and changes in vessel tone in response to alterations in blood pressure (Jackson, 2000; Hill et al., 2006). Myogenic response is thought to be more instrumental in mediating responses to large changes in blood pressure, while neurogenic mechanisms exert control over smaller changes (Hamner and Tan, 2014). Parenchymal arteries have a higher basal vascular tone compared with pial arteries (Cipolla et al., 2004). A higher basal tone makes parenchymal arteries less reactive to changes in blood pressure and it is thought to act as a protective mechanism for brain microvasculature against rapid and drastic blood pressure changes.

Cerebral autoregulation in ischemic stroke

Cerebral autoregulation does not act in isolation, but it functions in a complex interplay with cardiovascular response and collateral recruitment in acute ischemic stroke to improve cerebral hemodynamics and avoid irreversible brain damage. For instance, studies have shown collateral circulation through the circle of Willis is associated with improved cerebral autoregulation in intracranial artery stenosis (Guo et al., 2018), suggesting collateral blood flow may improve autoregulatory response. In addition, revascularization of ICA stenosis improves cerebral autoregulation (Tang et al., 2008; Mense et al., 2010; Semenyutin et al., 2017).

As cardiovascular and collateral responses fail in acute ischemic stroke and CPP continues to drop, cerebral autoregulation mounts a vasodilatory response to preserve CBF, and prolong MTT to enhance OEF in order to maintain $CMRO_2$ (Fig. 28.1). Therefore cerebral autoregulation is thought to be the final hemodynamic defense line before brain infarction. Hence, some degrees of vasomotor dysregulation are expected in stroke. Studies have shown cerebral autoregulation is impaired in acute ischemic stroke for 1–2 weeks (Dawson et al., 2003; Intharakham et al., 2019). Other studies, allowing for differences in measurement techniques, estimate vasomotor dysregulation may persist for 3 months (Kwan et al., 2004) to 6 months (Hu et al., 2008) after stroke. Baseline demographic and premorbid status of acute stroke patients may modify cerebral vasoregulation. Ipsilateral impaired autoregulation 1 to 3 days after acute stroke is independently predicted by older age, nonadministration of tPA, large artery atherosclerosis stroke subtype and a higher serum uric acid level (Ma et al., 2018). Etiology of ischemic stroke also affects the pattern of autoregulatory failure: cerebral autoregulation is diminished ipsilaterally in LVO strokes, while impaired autoregulation is bilateral in small vessel stroke (Immink et al., 2005; Guo et al., 2014). To explain a global impairment in small vessel stroke some mention the context of microangiopathy in which small vessel strokes take place (Purkayastha et al., 2014).

Cerebral autoregulation has also been studied in individuals without stroke. Younger women (26 ± 4 years) may have better cerebral autoregulation compared with men (Favre and Serrador, 2019). Several preexisting conditions increasing stroke risk are also associated with diminished cerebral autoregulation response. Those include chronic hypertension (Paulson et al., 1990), diabetes (Bentsen et al., 1975; Mankovsky et al., 2003), hypercholesterolemia (Meyer et al., 1987), metabolic syndrome (Giannopoulos et al., 2010), atrial fibrillation (Junejo et al., 2019) and obstructive sleep apnea (Urbano et al., 2008; Oz et al., 2017), suggesting diminished autoregulatory response may mediate the impact of stroke risk factors. In cross-sectional studies of patients with ICA stenosis an impaired cerebral autoregulation is associated with stroke risk (Ju et al., 2018); however, prospective studies are required to provide evidence for a predictive role of autoregulation.

Studying cerebral autoregulation determines prognosis in acute ischemic stroke. An impaired autoregulation is associated with a larger infarction (Reinhard et al., 2012) and a worse clinical outcome (Aoi et al., 2012; Reinhard et al., 2012; Chi et al., 2018) in acute stroke. Failure in cerebral autoregulation within 6 hours of MCA stroke independently predicts clinical outcome, and is associated with a larger infarct size (Castro et al., 2017b).

Cerebral autoregulation and reperfusion injury

Administration of tPA and mechanical thrombectomy are very effective treatment options for acute stroke and they lead to significant functional improvements in patients. However, blood perfusion in the tissue damaged by ischemia causes a secondary pathophysiology known as reperfusion injury manifested as cerebral edema and hemorrhagic conversion (Savitz et al., 2017). Ischemia-reperfusion damage is believed to clinically antagonize the beneficial effect of recanalization. In addition, it leads to clinical deterioration in a subset of patients. For instance, the clinical worsening after tPA administration is generally associated with brain hemorrhage into the area of infarction that occurs in 4%–7% of patients (Savitz et al., 2017). Among the patients receiving mechanical thrombectomy 6%–7% experience symptomatic intracranial hemorrhage (Albers et al., 2018; Nogueira et al., 2018). Different nonpharmacologic (remote conditioning and hypothermia) and pharmacologic (edaravone and citicoline) methods have been suggested but they were clinically impractical or ineffective (Sun et al., 2018). This includes the URICO-ICTUS trial studying uric acid administration in stroke patients treated with tPA (Chamorro et al., 2014), which did not improve clinical outcomes.

The BBB damaged by ischemia plays a role in reperfusion injury allowing penetration of red blood cells and serum proteins into the infarcted brain tissue. This causes cerebral edema and hemorrhagic conversion of the infarction, both appreciable on serial brain scans (Warach and Latour, 2004). Edema further contributes to ischemic damage by increasing the oxygen and nutrient diffusion distance, as well as physically compressing micro-vessels due to a rise in tissue pressure (Kalogeris et al., 2016). Reperfusion-induced damage to the pericytes also impairs microcirculation leading to "reflow arrest" phenomenon (Ames et al., 1968). Ischemia-reperfusion further damages BBB and ischemic tissue by releasing reactive oxygen species, and it also causes secondary damage through the release of excitotoxic neurotransmitters, programmed cell death, intracellular Ca^{2+} accumulation and lipolysis (Sun et al., 2018), followed by triggering of innate and adaptive immunity in the brain (Eltzschig and Eckle, 2011).

Impaired cerebral autoregulation in stroke is also thought to be an adjunctive cause for reperfusion injury and hemorrhagic transformation. Stroke core is irreversibly affected by ischemia and therefore in the common hemorrhagic transformation of infarct core it is difficult to dissect out the roles of damaged BBB from impaired cerebral autoregulation. But the role of impaired cerebral autoregulation may be more significant in cases of hemorrhagic transformation in ischemic penumbra after reperfusion (Fig. 28.7). Disturbances in cerebral autoregulation in fact correlate with edema and hemorrhagic conversion, imaging manifestation of reperfusion injury in acute ischemic stroke (Dohmen et al., 2007; Castro et al., 2017a).

IMPLICATIONS OF HEMODYNAMICS FOR SELECTING PATIENTS FOR REVASCULARIZATION PROCEDURES

Assessing hemodynamic state in acute ischemic stroke can be utilized to identify patients who may benefit from revascularization therapies. Thrombolysis and EVT aim to restore the blood supply into the ischemic tissue and salvage the ischemic penumbra (Darby et al., 1999; Rocha and Jovin, 2017). It is therefore important to choose patients with salvageable penumbra tissue. The time from stroke onset was initially used to screen patients for EVT; however, it assumes a uniform pace of pathophysiologic clock for development of infarct across patients. However, a recent study on acute ischemic stroke patients due to LVO showed a very diverse range of infarcted tissue volumes 24 h after stroke onset, suggesting variability in pathophysiologic processes driving brain tissue death (Rocha et al., 2019). Therefore the current time-based criterion to include or exclude patients for acute revascularization therapies lacks the precision expected for such high potency therapies. In addition, as a result of a strict universal time-based criterion the majority of patients are deprived of treatment due to missing conventional therapeutic time windows (Adeoye et al., 2011; Jadhav et al., 2018).

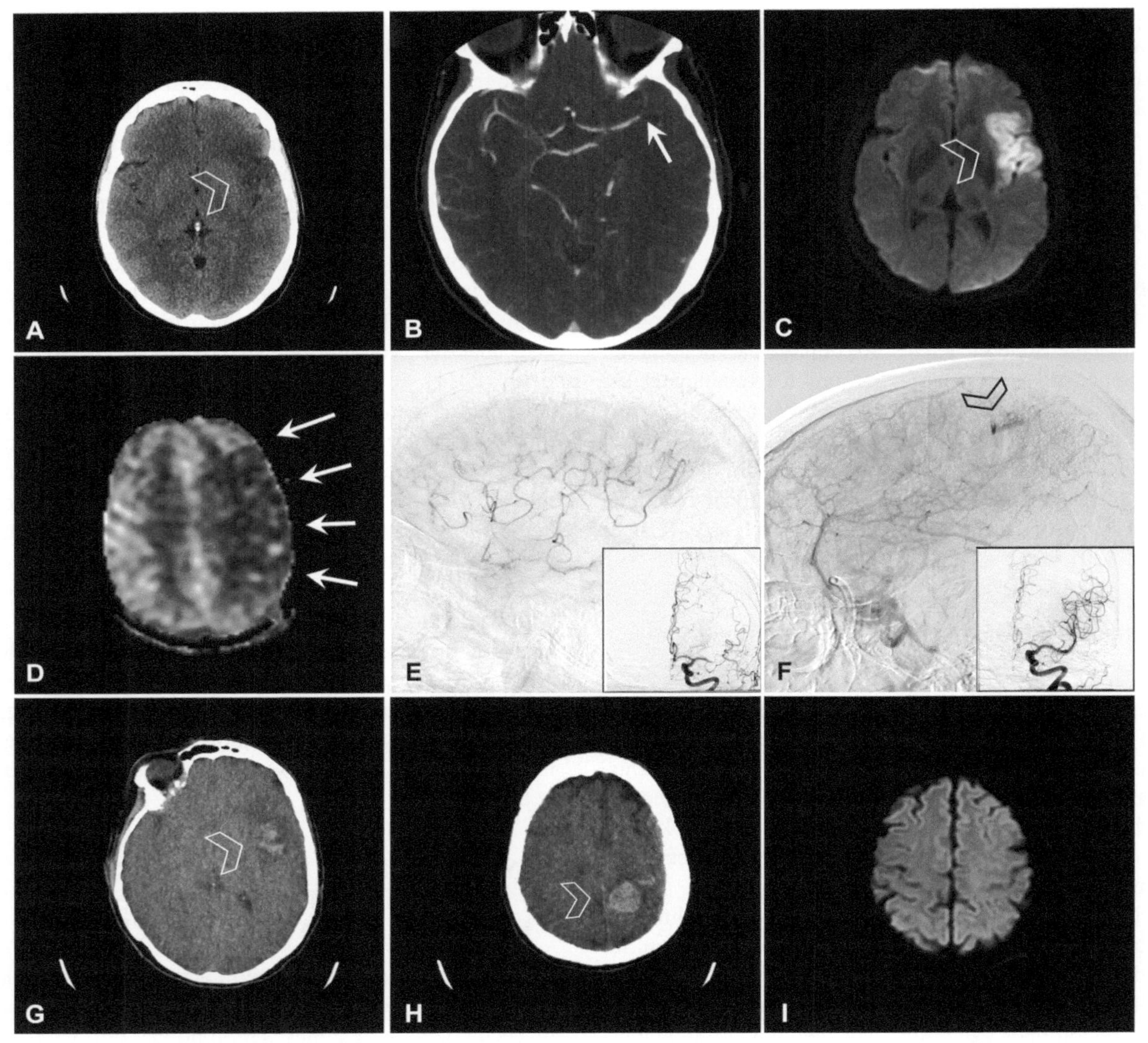

Fig. 28.7. Reperfusion hemorrhage in the ischemic penumbra. (A) Initial head CT shows hypodensity evolving in the left frontal and insular lobes (*arrowhead*). (B) CT angiogram demonstrates an abrupt cutoff in the distal left middle cerebral artery (MCA), first segment (*arrow*). (C) Brain MRI reveals diffusion restriction indicating acute infarction in the left frontal and insular lobes (*arrowhead*). (D) Arterial spin labeling shows reduced blood perfusion in the left MCA territory (*arrows*). Late phase sagittal plane cerebral angiograms of the left anterior circulation before (E) and after (F) thrombectomy (Insets show early phase anteroposterior plane angiograms). Note the contrast extravasation in the left parietal region (*arrowhead* in F). Repeat head CT 2.5h after reperfusion reveals hemorrhage into the infarct core (*arrowhead* in G) and the left parietal cortex (*arrowhead* in H), the latter corresponds to the area of contrast extravasation in cerebral angiogram (*arrowhead* in F). (I) Brain MRI sequence corresponding to the left parietal region shows no stroke, which in addition to the perfusion scan (D) indicates the area of hemorrhage is located within penumbral tissue.

To substitute the universal time-based selection criteria with a patient-specific approach studies have focused on the volume of penumbra. The baseline ischemic penumbra volume within 24h has a stronger correlation with acute neurological deficits compared with the baseline infarcted core volume (Barber et al., 1998), indicating hypoperfusion is the dominant pathophysiology for acute stroke clinical presentation. The baseline penumbra volume in acute stroke also correlates with clinical outcome and the final infarct volume at 84 days (Barber et al., 1998), suggesting the initial volume of ischemic penumbra in acute stroke correlates with both acute clinical presentation and ultimate infarct burden and disability. These further validate replacing the concept of time with the ischemic penumbra tissue in choosing patients for revascularization therapies.

Based on the availability of imaging modalities there are few approaches in estimating the extent of ischemic penumbra in acute stroke: hypoperfused tissue volume can be directly measured by CT or MR perfusion after subtracting the infarct core. In an alternative approach, patient's clinical exam can be utilized as a surrogate for hypoperfused tissue and a mismatch between clinical exam and infarct core suggests presence of some dysfunctional brain tissue due to hypoperfusion causing clinical deficit, however not yet infarcted and therefore salvageable. This approach is supported by the finding that the volume of penumbra tissue is strongly associated with clinical deficits (Barber et al., 1998). The later approach can be particularly adapted in facilities where perfusion scans are not available. These approaches have been validated in the two landmark trials expanding thrombectomy time window beyond 6 hours (Albers et al., 2018; Nogueira et al., 2018). Clinical diffusion mismatch and perfusion-diffusion mismatch have been shown to be comparable in selecting patients for reperfusion therapies for a good long-term clinical outcome (Bivard et al., 2019).

Other approaches have utilized the site of LVO or extent of collaterals to estimate hypoperfused tissue and therefore select patients for revascularization. Presence of MRA-DWI mismatch profiles in acute stroke patients, defined as ICA or main branch of MCA occlusion and a small infarct core, compared with patients without MRA-DWI mismatch has been correlated with a significant clinical benefit and 90-days outcome following revascularization (Mishra et al., 2014). In this method, more proximal levels of arterial occlusion is a proxy for higher volumes of perfusion deficit (Kidwell et al., 2013). On the other hand, collateral-based triage of patients for EVT was comparable with perfusion-based approach in selecting patients for a favorable clinical outcome (Kim et al., 2019). It is, however, important to note this study was performed as *post-hoc* analysis of a cohort of 93 late-presenting LVO patients already selected for EVT based on clinical-core mismatch or perfusion-core mismatch. Therefore one cannot conclude if collateral grading is an equally acceptable selection method for late-presenting LVO patients, since the patients were already triaged by perfusion criteria. In addition, the agreement between perfusion-based and collateral-based selection methods was poor ($\kappa = 0.41$). Therefore beyond such correlations, MRA-DWI mismatch or collateral status have not been validated in randomized trials to choose patients for revascularization and our data on their utility is limited.

ENHANCING HEMODYNAMICS IN ACUTE STROKE

Revascularization using thrombolytics or EVT is the centerpiece of acute ischemic stroke treatment. However, methods have been investigated to improve cerebral hemodynamics, enhance and preserve collateral perfusion, and prevent expansion of infarction. Such therapeutics can be implemented as a bridge to revascularization therapy or as the sole option in cases when EVT or thrombolytics are not indicated.

Blood pressure augmentation

Augmenting blood pressure is the most accessible treatment modality in neuro-intensive care of stroke patients. Blood pressure can be simply raised by intravascular volume infusion or pharmacologic vasopressors. Blood pressure enhancement in rats after MCA occlusion enhances collateral recruitment, which in turn improves blood flow and oxygenation leading to smaller infarction sizes (Shin et al., 2008). Pharmacologic enhancement of blood pressure has been shown to be relatively safe (Koenig et al., 2006), but few number of small randomized trials with a heterogeneous population of enrolled patients (Meier et al., 1991; Saxena et al., 1998; Hillis et al., 2003) limits conclusive assessment of risks and benefits (Mistri et al., 2006). There is currently no standardized recommendation to use blood pressure augmentation for acute stroke care (Powers et al., 2019), but it can be employed on a case-by-case basis.

Plasma expansion and hemodilution

Infusion of molecules retained in the intravascular space, such as dextran, hydroxyethyl starch or albumin, can theoretically improve blood perfusion by increasing intravascular volume, as well as decreasing blood viscosity and subsequently diminishing vascular resistance. Despite the initial data on improvement of clinical outcome in acute stroke patients receiving dextran (Strand et al., 1984), no further clinical trials have validated this approach. Two controlled clinical studies on volume expansion and hemodilution by hydroxyethyl starch showed no clinical benefit in acute stroke patients (Rudolf and Group, 2002; Woessner et al., 2003). Administration of albumin in acute stroke has been the subject of ALIAS study, which was terminated early due to concerns over side effects and futility (Ginsberg et al., 2013). Subsequent analysis did not reveal any clinical benefit, but it showed sixfold increase in pulmonary edema in albumin-treated patients (Martin et al., 2016). Therefore, none of the strategies to expand intravascular volume has formally entered the acute care of ischemic stroke.

External counterpulsation

External counterpulsation or ECP involves three pairs of pneumonic cuffs around calves, lower thighs, and upper thighs triggering a distal-to-proximal rhythmic inflation during cardiac diastole and releasing pressure during cardiac systole, therefore increasing the venous return and augmenting cerebral perfusion. ECP increases systemic blood pressure in healthy control subjects and stroke patients due to LVO (Lin et al., 2014). In addition, ECP enhances CBF in stroke patients but not matched controls (Lin et al., 2014). This may indicate impaired cerebral autoregulation in acute stroke. CUFFS was a randomized clinical trial studied clinical outcome in acute patients with occluded MCA who underwent ECP (Guluma et al., 2015). Administration of ECP was associated with an improved early neurologic exam, but no changes in long term clinical indices.

Partial aortic occlusion

Impeding perfusion through descending aorta increases blood pressure in aortic arch and therefore augments cerebral hemodynamics. SANTIS trial has studied partial aortic occlusion by NeuroFlo device in acute stroke patients with small to medium size cortical infarctions (Shuaib et al., 2011). Although partial aortic occlusion improved cerebral hemodynamics (Liebeskind, 2008), it failed to improved long-term disability (Shuaib et al., 2011). Subgroup analysis, however, revealed application of NeuroFlo in patients older than 70, patients with underlying atrial fibrillation, those presented very early (<6h), or patients with mild to moderate neurological disability (NIHSS < 14) improved functional outcome (Shuaib et al., 2011; Bernardini et al., 2014). In addition, a complete circle of Willis determined if hemodynamics augmentation in SENTIS trial was translated into clinical benefits (Schellinger et al., 2013), indicating increased CPP acts in synergy with collateral circulation.

Near-infrared and electrical stimulation

Transcranial near-infrared laser therapy has been shown to increase CBF and enhance oxidative phosphorylation in experimental stroke models (Lapchak, 2010), which led to three clinical trials (NEST-1, 2 and 3) studying transcranial near-infrared lasers in treating acute stroke patients. NEST-2 (Zivin et al., 2009) did not show any side effects or clinical benefit in reducing long-term disability. However, NEST-1 (Lampl et al., 2007) as well as a combined retrospective analysis of NEST-1 and NEST-2 revealed a significant reduction in disability (Stemer et al., 2010). NEST-3 trials were therefore conducted, but it was terminated after interim analysis indicated futility (Hacke et al., 2014).

The SPG is located posterior to the maxillary sinus and it is the main source of parasympathetic innervation to the vasculature in the anterior cerebral circulation. SPG stimulation has been suggested to improve outcomes in acute ischemic stroke through vasodilation and collateral recruitment, direct neuroprotection, blood-brain-barrier stabilization, and neuroplasticity enhancement (Bahr Hosseini and Saver, 2020). The most recent randomized blinded clinical trial on acute application of SPG stimulation in ischemic stroke, ImpACT-24B, has been recently published and although the initial analysis did not show a significant benefit in the long term clinical outcome, subgroup analysis in patients with cortical involvement has shown a reduction in disability in patients underwent SPG stimulation (Bornstein et al., 2019).

Application of vasodilators

Nitric oxide is among the endogenous mediators of vasodilation and cerebral autoregulation. Experimental studies on the administration of nitric oxide in acute MCA stroke have shown improved collateral recruitment and reduced infarct volume (Terpolilli et al., 2012). Despite studies on clinical application of nitric oxide in hypoxemic respiratory failure and pulmonary hypertension (Kinsella and Abman, 2005), acute sickle cell crisis (Weiner et al., 2003) and acute lung injury (Griffiths and Evans, 2005), we are not aware of any organized clinical study in acute stroke.

Carbon dioxide, another naturally occurring molecule, has been the focus of studies to improve cerebral perfusion in acute stroke. As a vasodilator, carbon dioxide improves collateral recruitment and cerebral perfusion (Grune et al., 2015; Meng and Gelb, 2015). Induced hypercapnia during EVT is currently subject of a randomized clinical trial (SEACOAST-1; NCT03737786).

CONCLUSION

Studying hemodynamic changes in acute ischemic stroke enhances our understanding of stroke pathophysiology and may be used to improve patient care. Individual stroke patients have divergent trajectories in the severity of ischemic stroke evolution and each may progress at a different pace. Supporting cerebral perfusion may slow or decrease ischemic core growth. Extracting individual hemodynamic parameters enhances clinical outcomes by selecting patients most benefiting from revascularization therapies. Meticulous monitoring of hemodynamic parameters of patients after thrombolysis or EVT prevent reperfusion injury and promotes long-term functional outcome.

References

Adeoye O, Hornung R, Khatri P et al. (2011). Recombinant tissue-type plasminogen activator use for ischemic stroke in the United States: a doubling of treatment rates over the course of 5 years. Stroke 42: 1952–1955.

Agewall S, Giannitsis E, Jernberg T et al. (2011). Troponin elevation in coronary vs. non-coronary disease. Eur Heart J 32: 404–411.

Ahmed N, Wahlgren N, Brainin M et al. (2009). Relationship of blood pressure, antihypertensive therapy, and outcome in ischemic stroke treated with intravenous thrombolysis: retrospective analysis from safe implementation of thrombolysis in stroke-international stroke thrombolysis register (SITS-ISTR). Stroke 40: 2442–2449.

Albers GW, Marks MP, Kemp S et al. (2018). Thrombectomy for stroke at 6 to 16 hours with selection by perfusion imaging. N Engl J Med 378: 708–718.

Ames 3rd A, Wright RL, Kowada M et al. (1968). Cerebral ischemia. II. The no-reflow phenomenon. Am J Pathol 52: 437–453.

Amukotuwa S, Straka M, Aksoy D et al. (2019). Cerebral blood flow predicts the infarct core: new insights from contemporaneous diffusion and perfusion imaging. Stroke 50: 2783–2789.

Angermaier A, Langner S, Kirsch M et al. (2011). CT-angiographic collateralization predicts final infarct volume after intra-arterial thrombolysis for acute anterior circulation ischemic stroke. Cerebrovasc Dis 31: 177–184.

Aoi MC, Hu K, Lo MT et al. (2012). Impaired cerebral autoregulation is associated with brain atrophy and worse functional status in chronic ischemic stroke. PLoS One 7: e46794.

Aslanyan S, Fazekas F, Weir CJ et al. (2003). Effect of blood pressure during the acute period of ischemic stroke on stroke outcome: a tertiary analysis of the GAIN International Trial. Stroke 34: 2420–2425.

Bahr Hosseini M, Saver JL (2020). Mechanisms of action of acute and subacute sphenopalatine ganglion stimulation for ischemic stroke. Int J Stroke. https://doi.org/10.1177/1747493020920739.

Barber PA, Darby DG, Desmond PM et al. (1998). Prediction of stroke outcome with echoplanar perfusion- and diffusion-weighted MRI. Neurology 51: 418–426.

Barron SA, Rogovski Z, Hemli J (1994). Autonomic consequences of cerebral hemisphere infarction. Stroke 25: 113–116.

Bath P, Chalmers J, Powers W et al. (2003). International society of hypertension (ISH): statement on the management of blood pressure in acute stroke. J Hypertens 21: 665–672.

Bayliss WM (1902). On the local reactions of the arterial wall to changes of internal pressure. J Physiol 28: 220–231.

Bentsen N, Larsen B, Lassen NA (1975). Chronically impaired autoregulation of cerebral blood flow in long-term diabetics. Stroke 6: 497–502.

Bernardini GL, Schellinger PD, Abou-Chebl A et al. (2014). Efficacy of NeuroFlo device in treatment of patients with atrial fibrillation. J Stroke Cerebrovasc Dis 23: 1457–1461.

Biso S, Wongrakpanich S, Agrawal A et al. (2017). A review of neurogenic stunned myocardium. Cardiovasc Psychiatry Neurol 2017: 5842182.

Bivard A, Huang X, Levi CR et al. (2019). Comparing mismatch strategies for patients being considered for ischemic stroke tenecteplase trials. Int J Stroke 15 (5): 507–515. https://doi.org/10.1177/1747493019884529.

Bleier J, Vorderwinkler KP, Falkensammer J et al. (1998). Different intracellular compartmentations of cardiac troponins and myosin heavy chains: a causal connection to their different early release after myocardial damage. Clin Chem 44: 1912–1918.

Bornstein NM, Saver JL, Diener HC et al. (2019). An injectable implant to stimulate the sphenopalatine ganglion for treatment of acute ischaemic stroke up to 24 h from onset (ImpACT-24B): an international, randomised, double-blind, sham-controlled, pivotal trial. Lancet 394: 219–229.

Caplan LR, Hennerici M (1998). Impaired clearance of emboli (washout) is an important link between hypoperfusion, embolism, and ischemic stroke. Arch Neurol 55: 1475–1482.

Castillo J, Leira R, Garcia MM et al. (2004). Blood pressure decrease during the acute phase of ischemic stroke is associated with brain injury and poor stroke outcome. Stroke 35: 520–526.

Castro P, Azevedo E, Serrador J et al. (2017a). Hemorrhagic transformation and cerebral edema in acute ischemic stroke: Link to cerebral autoregulation. J Neurol Sci 372: 256–261.

Castro P, Serrador JM, Rocha I et al. (2017b). Efficacy of cerebral autoregulation in early ischemic stroke predicts smaller infarcts and better outcome. Front Neurol 8: 113.

Chalothorn D, Faber JE (2010). Formation and maturation of the native cerebral collateral circulation. J Mol Cell Cardiol 49: 251–259.

Chamorro A, Amaro S, Castellanos M et al. (2014). Safety and efficacy of uric acid in patients with acute stroke (URICO-ICTUS): a randomised, double-blind phase 2b/3 trial. Lancet Neurol 13: 453–460.

Chi NF, Hu HH, Wang CY et al. (2018). Dynamic cerebral autoregulation is an independent functional outcome predictor of mild acute ischemic stroke. Stroke 49: 2605–2611.

Christoforidis GA, Mohammad Y, Kehagias D et al. (2005). Angiographic assessment of pial collaterals as a prognostic indicator following intra-arterial thrombolysis for acute ischemic stroke. AJNR Am J Neuroradiol 26: 1789–1797.

Cipolla MJ, Li R, Vitullo L (2004). Perivascular innervation of penetrating brain parenchymal arterioles. J Cardiovasc Pharmacol 44: 1–8.

Clarke DD, Sokoloff L (1989). Circulation and energy metabolism of the brain. In: G Siegel, BV Agrano, RW Albers, PV Molino (Eds.), Basic neurochemistry. Raven Press, New York.

Countee RW, Vijayanathan T (1979). External carotid artery in internal carotid artery occlusion. Angiographic, therapeutic, and prognostic considerations. Stroke 10: 450–460.

Dalainas I, Avgerinos ED, Daskalopoulos ME et al. (2012). The critical role of the external carotid artery in cerebral perfusion of patients with total occlusion of the internal carotid artery. Int Angiol 31: 16–21.

Darby DG, Barber PA, Gerraty RP et al. (1999). Pathophysiological topography of acute ischemia by combined diffusion-weighted and perfusion MRI. Stroke 30: 2043–2052.

Dawson SL, Panerai RB, Potter JF (2003). Serial changes in static and dynamic cerebral autoregulation after acute ischaemic stroke. Cerebrovasc Dis 16: 69–75.

Derdeyn CP, Videen TO, Yundt KD et al. (2002). Variability of cerebral blood volume and oxygen extraction: stages of cerebral haemodynamic impairment revisited. Brain 125: 595–607.

Dohmen C, Bosche B, Graf R et al. (2007). Identification and clinical impact of impaired cerebrovascular autoregulation in patients with malignant middle cerebral artery infarction. Stroke 38: 56–61.

Eaton RG, Shah VS, Dornbos 3rd D et al. (2020). Demographic age-related variation in Circle of Willis completeness assessed by digital subtraction angiography. Brain Circ 6: 31–37.

Edvinsson L, Owman C, Sjoberg NO (1976). Autonomic nerves, mast cells, and amine receptors in human brain vessels. A histochemical and pharmacological study. Brain Res 115: 377–393.

Eltzschig HK, Eckle T (2011). Ischemia and reperfusion—from mechanism to translation. Nat Med 17: 1391–1401.

Faber JE, Zhang H, Rzechorzek W et al. (2019). Genetic and environmental contributions to variation in the posterior communicating collaterals of the circle of Willis. Transl Stroke Res 10: 189–203.

Faiz KW, Thommessen B, Einvik G et al. (2014). Determinants of high sensitivity cardiac troponin T elevation in acute ischemic stroke. BMC Neurol 14: 96.

Faraci FM, Heistad DD (1990). Regulation of large cerebral arteries and cerebral microvascular pressure. Circ Res 66: 8–17.

Faraci FM, Mayhan WG, Heistad DD (1987). Segmental vascular responses to acute hypertension in cerebrum and brain stem. Am J Physiol 252: H738–H742.

Favre ME, Serrador JM (2019). Sex differences in cerebral autoregulation are unaffected by menstrual cycle phase in young, healthy women. Am J Physiol Heart Circ Physiol 316: H920–H933.

Geibprasert S, Pongpech S, Armstrong D et al. (2009). Dangerous extracranial-intracranial anastomoses and supply to the cranial nerves: vessels the neurointerventionalist needs to know. AJNR Am J Neuroradiol 30: 1459–1468.

Giannopoulos S, Boden-Albala B, Choi JH et al. (2010). Metabolic syndrome and cerebral vasomotor reactivity. Eur J Neurol 17: 1457–1462.

Ginsberg MD, Palesch YY, Hill MD et al. (2013). High-dose albumin treatment for acute ischaemic stroke (ALIAS) Part 2: a randomised, double-blind, phase 3, placebo-controlled trial. Lancet Neurol 12: 1049–1058.

Goyal N, Tsivgoulis G, Pandhi A et al. (2017). Blood pressure levels post mechanical thrombectomy and outcomes in large vessel occlusion strokes. Neurology 89: 540–547.

Griffiths MJ, Evans TW (2005). Inhaled nitric oxide therapy in adults. N Engl J Med 353: 2683–2695.

Grune F, Kazmaier S, Stolker RJ et al. (2015). Carbon dioxide induced changes in cerebral blood flow and flow velocity: role of cerebrovascular resistance and effective cerebral perfusion pressure. J Cereb Blood Flow Metab 35: 1470–1477.

Guglielmi V, LeCouffe NE, Zinkstok SM et al. (2019). Collateral circulation and outcome in atherosclerotic versus cardioembolic cerebral large vessel occlusion. Stroke 50: 3360–3368.

Guluma KZ, Liebeskind DS, Raman R et al. (2015). Feasibility and safety of using external counterpulsation to augment cerebral blood flow in acute ischemic stroke-the counterpulsation to upgrade forward flow in stroke (CUFFS) trial. J Stroke Cerebrovasc Dis 24: 2596–2604.

Guo ZN, Liu J, Xing Y et al. (2014). Dynamic cerebral autoregulation is heterogeneous in different subtypes of acute ischemic stroke. PLoS One 9: e93213.

Guo ZN, Sun X, Liu J et al. (2018). The impact of variational primary collaterals on cerebral autoregulation. Front Physiol 9: 759.

Hacke W, Schellinger PD, Albers GW et al. (2014). Transcranial laser therapy in acute stroke treatment: results of neurothera effectiveness and safety trial 3, a phase III clinical end point device trial. Stroke 45: 3187–3193.

Hamel E (2006). Perivascular nerves and the regulation of cerebrovascular tone. J Appl Physiol (1985) 100: 1059–1064.

Hamner JW, Tan CO (2014). Relative contributions of sympathetic, cholinergic, and myogenic mechanisms to cerebral autoregulation. Stroke 45: 1771–1777.

Hamner JW, Tan CO, Tzeng YC et al. (2012). Cholinergic control of the cerebral vasculature in humans. J Physiol 590: 6343–6352.

Hartkamp MJ, van Der Grond J, van Everdingen KJ et al. (1999). Circle of Willis collateral flow investigated by magnetic resonance angiography. Stroke 30: 2671–2678.

Hasirci B, Okay M, Agircan D et al. (2013). Elevated troponin level with negative outcome was found in ischemic stroke. Cardiovasc Psychiatry Neurol 2013: 953672.

He J, Zhang Y, Xu T et al. (2014). Effects of immediate blood pressure reduction on death and major disability in patients with acute ischemic stroke: the CATIS randomized clinical trial. JAMA 311: 479–489.

Hill MA, Davis MJ, Meininger GA et al. (2006). Arteriolar myogenic signalling mechanisms: implications for local vascular function. Clin Hemorheol Microcirc 34: 67–79.

Hillis AE, Ulatowski JA, Barker PB et al. (2003). A pilot randomized trial of induced blood pressure elevation: effects on function and focal perfusion in acute and subacute stroke. Cerebrovasc Dis 16: 236–246.

Hilz MJ, Devinsky O, Szczepanska H et al. (2006). Right ventromedial prefrontal lesions result in paradoxical cardiovascular activation with emotional stimuli. Brain 129: 3343–3355.

Hoksbergen AW, Legemate DA, Csiba L et al. (2003). Absent collateral function of the circle of Willis as risk factor for ischemic stroke. Cerebrovasc Dis 16: 191–198.

Hu K, Peng CK, Czosnyka M et al. (2008). Nonlinear assessment of cerebral autoregulation from spontaneous blood pressure and cerebral blood flow fluctuations. Cardiovasc Eng 8: 60–71.

Iadecola C (2004). Neurovascular regulation in the normal brain and in Alzheimer's disease. Nat Rev Neurosci 5: 347–360.

Ichiyama RM, Waldrop TG, Iwamoto GA (2004). Neurons in and near insular cortex are responsive to muscular contraction and have sympathetic and/or cardiac-related discharge. Brain Res 1008: 273–277.

Immink RV, van Montfrans GA, Stam J et al. (2005). Dynamic cerebral autoregulation in acute lacunar and middle cerebral artery territory ischemic stroke. Stroke 36: 2595–2600.

Intharakham K, Beishon L, Panerai RB et al. (2019). Assessment of cerebral autoregulation in stroke: a systematic review and meta-analysis of studies at rest. J Cereb Blood Flow Metab 39: 2105–2116.

Ip HL, Liebeskind DS (2014). The future of ischemic stroke: flow from prehospital neuroprotection to definitive reperfusion. Interv Neurol 2: 105–117.

Ishii M, Hayashi M, Yagi F et al. (2016). Relationship between the direction of ophthalmic artery blood flow and ocular microcirculation before and after carotid artery stenting. J Ophthalmol 2016: 2530914.

Iwasawa E, Ichijo M, Ishibashi S et al. (2016). Acute development of collateral circulation and therapeutic prospects in ischemic stroke. Neural Regen Res 11: 368–371.

Jackson WF (2000). Ion channels and vascular tone. Hypertension 35: 173–178.

Jadhav AP, Desai SM, Kenmuir CL et al. (2018). Eligibility for endovascular trial enrollment in the 6- to 24-hour time window: analysis of a single comprehensive stroke center. Stroke 49: 1015–1017.

Jensen JK, Kristensen SR, Bak S et al. (2007). Frequency and significance of troponin T elevation in acute ischemic stroke. Am J Cardiol 99: 108–112.

John S, Hazaa W, Uchino K et al. (2016). Lower intraprocedural systolic blood pressure predicts good outcome in patients undergoing endovascular therapy for acute ischemic stroke. Interv Neurol 4: 151–157.

Jordan JD, Powers WJ (2012). Cerebral autoregulation and acute ischemic stroke. Am J Hypertens 25: 946–950.

Ju K, Zhong L, Ni X et al. (2018). Cerebral vasomotor reactivity predicts the development of acute stroke in patients with internal carotid artery stenosis. Neurol Neurochir Pol 52: 374–378.

Junejo RT, Braz ID, Lucas SJ et al. (2019). Neurovascular coupling and cerebral autoregulation in atrial fibrillation. J Cereb Blood Flow Metab. 271678X19870770.

Jung B, Obinata H, Galvani S et al. (2012). Flow-regulated endothelial S1P receptor-1 signaling sustains vascular development. Dev Cell 23: 600–610.

Kallmunzer B, Breuer L, Kahl N et al. (2012). Serious cardiac arrhythmias after stroke: incidence, time course, and predictors—a systematic, prospective analysis. Stroke 43: 2892–2897.

Kalogeris T, Baines CP, Krenz M et al. (2016). Ischemia/reperfusion. Compr Physiol 7: 113–170.

Kidwell CS, Jahan R, Gornbein J et al. (2013). A trial of imaging selection and endovascular treatment for ischemic stroke. N Engl J Med 368: 914–923.

Kim B, Jung C, Nam HS et al. (2019). Comparison between perfusion- and collateral-based triage for endovascular thrombectomy in a late time window. Stroke 50: 3465–3470.

Kinsella JP, Abman SH (2005). Inhaled nitric oxide therapy in children. Paediatr Respir Rev 6: 190–198.

Koenig MA, Geocadin RG, de Grouchy M et al. (2006). Safety of induced hypertension therapy in patients with acute ischemic stroke. Neurocrit Care 4: 3–7.

Kucinski T, Koch C, Eckert B et al. (2003). Collateral circulation is an independent radiological predictor of outcome after thrombolysis in acute ischaemic stroke. Neuroradiology 45: 11–18.

Kumar AP, Babu E, Subrahmanyam D (2012). Cerebrogenic tachyarrhythmia in acute stroke. J Neurosci Rural Pract 3: 204–206.

Kwan J, Lunt M, Jenkinson D (2004). Assessing dynamic cerebral autoregulation after stroke using a novel technique of combining transcranial Doppler ultrasonography and rhythmic handgrip. Blood Press Monit 9: 3–8.

Lampl Y, Zivin JA, Fisher M et al. (2007). Infrared laser therapy for ischemic stroke: a new treatment strategy: results of the NeuroThera Effectiveness and Safety Trial-1 (NEST-1). Stroke 38: 1843–1849.

Lansberg MG, Albers GW, Wijman CA (2007). Symptomatic intracerebral hemorrhage following thrombolytic therapy for acute ischemic stroke: a review of the risk factors. Cerebrovasc Dis 24: 1–10.

Lapchak PA (2010). Taking a light approach to treating acute ischemic stroke patients: transcranial near-infrared laser therapy translational science. Ann Med 42: 576–586.

Lee MJ, Bang OY, Kim SJ et al. (2014). Role of statin in atrial fibrillation-related stroke: an angiographic study for collateral flow. Cerebrovasc Dis 37: 77–84.

Leonardi-Bee J, Bath PM, Phillips SJ et al. (2002). Blood pressure and clinical outcomes in the International Stroke Trial. Stroke 33: 1315–1320.

Liebeskind DS (2003). Collateral circulation. Stroke 34: 2279–2284.

Liebeskind DS (2008). Aortic occlusion for cerebral ischemia: from theory to practice. Curr Cardiol Rep 10: 31–36.

Liebeskind DS, Jahan R, Nogueira RG et al. (2014). Impact of collaterals on successful revascularization in solitaire FR with the intention for thrombectomy. Stroke 45: 2036–2040.

Lin W, Xiong L, Han J et al. (2014). Increasing pressure of external counterpulsation augments blood pressure but not cerebral blood flow velocity in ischemic stroke. J Clin Neurosci 21: 1148–1152.

Lincoln J (1995). Innervation of cerebral arteries by nerves containing 5-hydroxytryptamine and noradrenaline. Pharmacol Ther 68: 473–501.

Liu J, Zhu YS, Hill C et al. (2013). Cerebral autoregulation of blood velocity and volumetric flow during steady-state changes in arterial pressure. Hypertension 62: 973–979.

Lopez-Yunez AM, Bruno A, Williams LS et al. (2001). Protocol violations in community-based rTPA stroke treatment are associated with symptomatic intracerebral hemorrhage. Stroke 32: 12–16.

Ma H, Guo ZN, Jin H et al. (2018). Preliminary study of dynamic cerebral autoregulation in acute ischemic stroke: Association with clinical factors. Front Neurol 9: 1006.

Macchi C, Molino Lova R, Miniati B et al. (2002). Collateral circulation in internal carotid artery occlusion. A study by duplex scan and magnetic resonance angiography. Minerva Cardioangiol 50: 695–700.

Maier B, Gory B, Taylor G et al. (2017). Mortality and disability according to baseline blood pressure in acute ischemic stroke patients treated by thrombectomy: a collaborative pooled analysis. J Am Heart Assoc 6.

Manea MM, Comsa M, Minca A et al. (2015). Brain-heart axis—review article. J Med Life 8: 266–271.

Mankovsky BN, Piolot R, Mankovsky OL et al. (2003). Impairment of cerebral autoregulation in diabetic patients with cardiovascular autonomic neuropathy and orthostatic hypotension. Diabet Med 20: 119–126.

Manoonkitiwongsa PS, Jackson-Friedman C, McMillan PJ et al. (2001). Angiogenesis after stroke is correlated with increased numbers of macrophages: the clean-up hypothesis. J Cereb Blood Flow Metab 21: 1223–1231.

Mark I, Seyedsaadat SM, Benson JC et al. (2020). Leukoaraiosis and collateral blood flow in stroke patients with anterior circulation large vessel occlusion. J Neurointerv Surg.

Martin RH, Yeatts SD, Hill MD et al. (2016). ALIAS (Albumin in Acute Ischemic Stroke) Trials: analysis of the combined data from parts 1 and 2. Stroke 47: 2355–2359.

Mattle HP, Kappeler L, Arnold M et al. (2005). Blood pressure and vessel recanalization in the first hours after ischemic stroke. Stroke 36: 264–268.

Meier F, Wessel G, Thiele R et al. (1991). Induced hypertension as an approach to treating acute cerebrovascular ischaemia: possibilities and limitations. Exp Pathol 42: 257–263.

Meng L, Gelb AW (2015). Regulation of cerebral autoregulation by carbon dioxide. Anesthesiology 122: 196–205.

Menon BK, Smith EE, Modi J et al. (2011). Regional leptomeningeal score on CT angiography predicts clinical and imaging outcomes in patients with acute anterior circulation occlusions. AJNR Am J Neuroradiol 32: 1640–1645.

Menon BK, Smith EE, Coutts SB et al. (2013). Leptomeningeal collaterals are associated with modifiable metabolic risk factors. Ann Neurol 74: 241–248.

Menon BK, d'Esterre CD, Qazi EM et al. (2015). Multiphase CT angiography: a new tool for the imaging triage of patients with acute ischemic stroke. Radiology 275: 510–520.

Mense L, Reimann M, Rudiger H et al. (2010). Autonomic function and cerebral autoregulation in patients undergoing carotid endarterectomy. Circ J 74: 2139–2145.

Meyer JS, Denny-Brown D (1957). The cerebral collateral circulation. I. Factors influencing collateral blood flow. Neurology 7: 447–458.

Meyer JS, Rogers RL, Mortel KF et al. (1987). Hyperlipidemia is a risk factor for decreased cerebral perfusion and stroke. Arch Neurol 44: 418–422.

Meyer S, Strittmatter M, Fischer C et al. (2004). Lateralization in autonomic dysfunction in ischemic stroke involving the insular cortex. Neuroreport 15: 357–361.

Millesi K, Mutzenbach JS, Killer-Oberpfalzer M et al. (2019). Influence of the circle of Willis on leptomeningeal collateral flow in anterior circulation occlusive stroke: friend or foe? J Neurol Sci 396: 69–75.

Mishra NK, Albers GW, Christensen S et al. (2014). Comparison of magnetic resonance imaging mismatch criteria to select patients for endovascular stroke therapy. Stroke 45: 1369–1374.

Mistri AK, Robinson TG, Potter JF (2006). Pressor therapy in acute ischemic stroke: systematic review. Stroke 37: 1565–1571.

Mistry EA, Mistry AM, Nakawah MO et al. (2017). Systolic blood pressure within 24 hours after thrombectomy for acute ischemic stroke correlates with outcome. J Am Heart Assoc 6: e006167.

Nakahashi T, Fujimura H, Altar CA et al. (2000). Vascular endothelial cells synthesize and secrete brain-derived neurotrophic factor. FEBS Lett 470: 113–117.

Nannoni S, Sirimarco G, Cereda CW et al. (2019). Determining factors of better leptomeningeal collaterals: a study of 857 consecutive acute ischemic stroke patients. J Neurol 266: 582–588.

Nason Jr MW, Mason P (2004). Modulation of sympathetic and somatomotor function by the ventromedial medulla. J Neurophysiol 92: 510–522.

Nguyen H, Zaroff JG (2009). Neurogenic stunned myocardium. Curr Neurol Neurosci Rep 9: 486–491.

Nogueira RG, Jadhav AP, Haussen DC et al. (2018). Thrombectomy 6 to 24 hours after stroke with a mismatch between deficit and infarct. N Engl J Med 378: 11–21.

Ogoh S, Brothers RM, Eubank WL et al. (2008). Autonomic neural control of the cerebral vasculature: acute hypotension. Stroke 39: 1979–1987.

Ohno M, Cooke JP, Dzau VJ et al. (1995). Fluid shear stress induces endothelial transforming growth factor beta-1 transcription and production. Modulation by potassium channel blockade. J Clin Invest 95: 1363–1369.

Oller M, Esteban C, Perez P et al. (2012). Rubeosis iridis as a sign of underlying carotid stenosis. J Vasc Surg 56: 1724–1726.

Oppenheimer S (1993). The anatomy and physiology of cortical mechanisms of cardiac control. Stroke 24: I3–I5.

Oz O, Tasdemir S, Akgun H et al. (2017). Decreased cerebral vasomotor reactivity in patients with obstructive sleep apnea syndrome. Sleep Med 30: 88–92.

Park YK, Lee K, Jung BJ et al. (2019). Relationship between ophthalmic artery flow direction and visual deterioration after carotid angioplasty and stenting. J Neurosurg 1–7.

Paulson OB, Strandgaard S, Edvinsson L (1990). Cerebral autoregulation. Cerebrovasc Brain Metab Rev 2: 161–192.

Piedade GS, Schirmer CM, Goren O et al. (2019). Cerebral collateral circulation: a review in the context of ischemic stroke and mechanical thrombectomy. World Neurosurg 122: 33–42.

Pienimaki JP, Sillanpaa N, Jolma P et al. (2020). Carotid artery stenosis is associated with better intracranial collateral circulation in stroke patients. Cerebrovasc Dis 49: 200–205.

Potter JF, Robinson TG, Ford GA et al. (2009). Controlling hypertension and hypotension immediately post-stroke (CHHIPS): a randomised, placebo-controlled, double-blind pilot trial. Lancet Neurol 8: 48–56.

Powers WJ, Rabinstein AA, Ackerson T et al. (2018). Guidelines for the Early Management of Patients With Acute Ischemic Stroke: A Guideline for Healthcare Professionals From the American Heart Association/American Stroke Association. Stroke 49: e46–e110.

Powers WJ, Rabinstein AA, Ackerson T et al. (2019). Guidelines for the Early Management of Patients With Acute Ischemic Stroke: 2019 Update to the 2018 Guidelines for the Early Management of Acute Ischemic Stroke: A Guideline for Healthcare Professionals From the American Heart Association/American Stroke Association. Stroke 50: e344–e418.

Prigent-Tessier A, Quirie A, Maguin-Gate K et al. (2013). Physical training and hypertension have opposite effects on endothelial brain-derived neurotrophic factor expression. Cardiovasc Res 100: 374–382.

Purkayastha S, Fadar O, Mehregan A et al. (2014). Impaired cerebrovascular hemodynamics are associated with cerebral white matter damage. J Cereb Blood Flow Metab 34: 228–234.

Qureshi AI (2008). Acute hypertensive response in patients with stroke: pathophysiology and management. Circulation 118: 176–187.

Qureshi AI, El-Gengaihi A, Hussein HM et al. (2008). Occurence and variability in acute formation of leptomeningeal collaterals in proximal middle cerebral artery occlusion. J Vasc Interv Neurol 1: 70–72.

Rapela CE, Green HD (1964). Autoregulation of canine cerebral blood flow. Circ Res 15: 205–212.

Rebello LC, Bouslama M, Haussen DC et al. (2017). Stroke etiology and collaterals: atheroembolic strokes have greater collateral recruitment than cardioembolic strokes. Eur J Neurol 24: 762–767.

Reinhard M, Rutsch S, Lambeck J et al. (2012). Dynamic cerebral autoregulation associates with infarct size and outcome after ischemic stroke. Acta Neurol Scand 125: 156–162.

Resstel LB, Correa FM (2006). Medial prefrontal cortex NMDA receptors and nitric oxide modulate the parasympathetic component of the baroreflex. Eur J Neurosci 23: 481–488.

Ringelstein EB, Biniek R, Weiller C et al. (1992). Type and extent of hemispheric brain infarctions and clinical outcome in early and delayed middle cerebral artery recanalization. Neurology 42: 289–298.

Rocha M, Jovin TG (2017). Fast versus slow progressors of infarct growth in large vessel occlusion stroke: clinical and research implications. Stroke 48: 2621–2627.

Rocha M, Desai SM, Jadhav AP et al. (2019). Prevalence and temporal distribution of fast and slow progressors of infarct growth in large vessel occlusion stroke. Stroke 50: 2238–2240.

Rosengarten B, Huwendiek O, Kaps M (2001). Neurovascular coupling and cerebral autoregulation can be described in terms of a control system. Ultrasound Med Biol 27: 189–193.

Rudolf J, Group HESiASS (2002). Hydroxyethyl starch for hypervolemic hemodilution in patients with acute ischemic stroke: a randomized, placebo-controlled phase II safety study. Cerebrovasc Dis 14: 33–41.

Rutgers DR, Klijn CJ, Kappelle LJ et al. (2000). A longitudinal study of collateral flow patterns in the circle of Willis and the ophthalmic artery in patients with a symptomatic internal carotid artery occlusion. Stroke 31: 1913–1920.

Samuels MA (2007). The brain-heart connection. Circulation 116: 77–84.

Sandhu R, Aronow WS, Rajdev A et al. (2008). Relation of cardiac troponin I levels with in-hospital mortality in patients with ischemic stroke, intracerebral hemorrhage, and subarachnoid hemorrhage. Am J Cardiol 102: 632–634.

Sanossian N, Ovbiagele B, Saver JL et al. (2011). Leukoaraiosis and collaterals in acute ischemic stroke. J Neuroimaging 21: 232–235.

Savitz SI, Baron JC, Yenari MA et al. (2017). Reconsidering neuroprotection in the reperfusion era. Stroke 48: 3413–3419.

Saxena R, Wijnhoud AD, Man in 't Veld AJ et al. (1998). Effect of diaspirin cross-linked hemoglobin on endothelin-1 and blood pressure in acute ischemic stroke in man. J Hypertens 16: 1459–1465.

Schaper W (2009). Collateral circulation: past and present. Basic Res Cardiol 104: 5–21.

Scheitz JF, Endres M, Mochmann HC et al. (2012). Frequency, determinants and outcome of elevated troponin in acute ischemic stroke patients. Int J Cardiol 157: 239–242.

Scheitz JF, Nolte CH, Laufs U et al. (2015). Application and interpretation of high-sensitivity cardiac troponin assays in patients with acute ischemic stroke. Stroke 46: 1132–1140.

Schellinger PD, Kohrmann M, Liu S et al. (2013). Favorable vascular profile is an independent predictor of outcome: a post hoc analysis of the safety and efficacy of NeuroFlo Technology in Ischemic Stroke trial. Stroke 44: 1606–1608.

Sealock R, Zhang H, Lucitti JL et al. (2014). Congenic fine-mapping identifies a major causal locus for variation in the native collateral circulation and ischemic injury in brain and lower extremity. Circ Res 114: 660–671.

Seeta Ramaiah S, Churilov L, Mitchell P et al. (2014). The impact of arterial collateralization on outcome after intra-arterial therapy for acute ischemic stroke. AJNR Am J Neuroradiol 35: 667–672.

Semenyutin VB, Asaturyan GA, Nikiforova AA et al. (2017). Predictive value of dynamic cerebral autoregulation assessment in surgical management of patients with high-grade carotid artery stenosis. Front Physiol 8: 872.

Seners P, Roca P, Legrand L et al. (2019). Better collaterals are independently associated with post-thrombolysis recanalization before thrombectomy. Stroke 50: 867–872.

Sheth SA, Sanossian N, Hao Q et al. (2016). Collateral flow as causative of good outcomes in endovascular stroke therapy. J Neurointerv Surg 8: 2–7.

Shin HK, Nishimura M, Jones PB et al. (2008). Mild induced hypertension improves blood flow and oxygen metabolism in transient focal cerebral ischemia. Stroke 39: 1548–1555.

Shuaib A, Bornstein NM, Diener HC et al. (2011). Partial aortic occlusion for cerebral perfusion augmentation: safety and efficacy of NeuroFlo in Acute Ischemic Stroke trial. Stroke 42: 1680–1690.

Silver B, Lu M, Morris DC et al. (2008). Blood pressure declines and less favorable outcomes in the NINDS tPA stroke study. J Neurol Sci 271: 61–67.

Silvestrini M, Troisi E, Matteis M et al. (1996). Transcranial Doppler assessment of cerebrovascular reactivity in symptomatic and asymptomatic severe carotid stenosis. Stroke 27: 1970–1973.

Stead LG, Gilmore RM, Decker WW et al. (2005). Initial emergency department blood pressure as predictor of survival after acute ischemic stroke. Neurology 65: 1179–1183.

Stemer AB, Huisa BN, Zivin JA (2010). The evolution of transcranial laser therapy for acute ischemic stroke, including a pooled analysis of NEST-1 and NEST-2. Curr Cardiol Rep 12: 29–33.

Strand T, Asplund K, Eriksson S et al. (1984). A randomized controlled trial of hemodilution therapy in acute ischemic stroke. Stroke 15: 980–989.

Sun MS, Jin H, Sun X et al. (2018). Free radical damage in ischemia-reperfusion injury: an obstacle in acute ischemic stroke after revascularization therapy. Oxid Med Cell Longev 2018: 3804979.

Taggart P, Boyett MR, Logantha S et al. (2011). Anger, emotion, and arrhythmias: from brain to heart. Front Physiol 2: 67.

Tang SC, Huang YW, Shieh JS et al. (2008). Dynamic cerebral autoregulation in carotid stenosis before and after carotid stenting. J Vasc Surg 48: 88–92.

Terpolilli NA, Kim SW, Thal SC et al. (2012). Inhalation of nitric oxide prevents ischemic brain damage in experimental stroke by selective dilatation of collateral arterioles. Circ Res 110: 727–738.

The National Institute of Neurological Disorders and Stroke rt-PA Stroke Study Group (1995). Tissue plasminogen activator for acute ischemic stroke. N Engl J Med 333: 1581–1587.

Toriumi H, Tatarishvili J, Tomita M et al. (2009). Dually supplied T-junctions in arteriolo-arteriolar anastomosis in mice: key to local hemodynamic homeostasis in normal and ischemic states? Stroke 40: 3378–3383.

Toyoda K, Minematsu K, Yamaguchi T (1994). Long-term changes in cerebral blood flow according to different types of ischemic stroke. J Neurol Sci 121: 222–228.

Trzeciakowski J, Chilian WM (2008). Chaotic behavior of the coronary circulation. Med Biol Eng Comput 46: 433–442.

Urbano F, Roux F, Schindler J et al. (2008). Impaired cerebral autoregulation in obstructive sleep apnea. J Appl Physiol (1985) 105: 1852–1857.

van den Wijngaard IR, Boiten J, Holswilder G et al. (2015). Impact of collateral status evaluated by dynamic computed tomographic angiography on clinical outcome in patients with ischemic stroke. Stroke 46: 3398–3404.

van Laar PJ, van der Grond J, Bremmer JP et al. (2008). Assessment of the contribution of the external carotid artery to brain perfusion in patients with internal carotid artery occlusion. Stroke 39: 3003–3008.

van Royen N, Piek JJ, Schaper W et al. (2009). A critical review of clinical arteriogenesis research. J Am Coll Cardiol 55: 17–25.

Verma RK, Gralla J, Klinger-Gratz PP et al. (2015). Infarction distribution pattern in acute stroke may predict the extent of leptomeningeal collaterals. PLoS One 10: e0137292.

Vitt JR, Trillanes M, Hemphill 3rd JC (2019). Management of blood pressure during and after recanalization therapy for acute ischemic stroke. Front Neurol 10: 138.

Wallace JD, Levy LL (1981). Blood pressure after stroke. JAMA 246: 2177–2180.

Wang TD, Wu CC, Lee YT (1997). Myocardial stunning after cerebral infarction. Int J Cardiol 58: 308–311.

Wang CX, Todd KG, Yang Y et al. (2001). Patency of cerebral microvessels after focal embolic stroke in the rat. J Cereb Blood Flow Metab 21: 413–421.

Wang S, Zhang H, Dai X et al. (2010). Genetic architecture underlying variation in extent and remodeling of the collateral circulation. Circ Res 107: 558–568.

Wang Z, Luo W, Zhou F et al. (2012). Dynamic change of collateral flow varying with distribution of regional blood flow in acute ischemic rat cortex. J Biomed Opt 17: 125001.

Wang J, Wang W, Jin B et al. (2016). Improvement in cerebral and ocular hemodynamics early after carotid endarterectomy in patients of severe carotid artery stenosis with or without contralateral carotid occlusion. Biomed Res Int 2016: 2901028.

Wang H, Wang Y, Li H (2017). Multimodality imaging assessment of ocular ischemic syndrome. J Ophthalmol 2017: 4169135.

Warach S, Latour LL (2004). Evidence of reperfusion injury, exacerbated by thrombolytic therapy, in human focal brain ischemia using a novel imaging marker of early blood-brain barrier disruption. Stroke 35: 2659–2661.

Wechsler LR (2011). Intravenous thrombolytic therapy for acute ischemic stroke. New Engl J Med 364: 2138–2146.

Weiner DL, Hibberd PL, Betit P et al. (2003). Preliminary assessment of inhaled nitric oxide for acute vaso-occlusive crisis in pediatric patients with sickle cell disease. JAMA 289: 1136–1142.

White HD (2011). Pathobiology of troponin elevations: do elevations occur with myocardial ischemia as well as necrosis? J Am Coll Cardiol 57: 2406–2408.

White RP, Vallance P, Markus HS (2000). Effect of inhibition of nitric oxide synthase on dynamic cerebral autoregulation in humans. Clin Sci (Lond) 99: 555–560.

Wiegers EJA, Mulder M, Jansen IGH et al. (2020). Clinical and imaging determinants of collateral status in patients with acute ischemic stroke in MR CLEAN trial and registry. Stroke 51: 1493–1502.

Willis T (1664). Cerebri Anatome: Cui Accessit Nervorum Descriptio Et Usus, Flesher, Martyn and Allestry, London.

Willmot M, Leonardi-Bee J, Bath PM (2004). High blood pressure in acute stroke and subsequent outcome: a systematic review. Hypertension 43: 18–24.

Woessner R, Grauer MT, Dieterich HJ et al. (2003). Influence of a long-term, high-dose volume therapy with 6% hydroxyethyl starch 130/0.4 or crystalloid solution on hemodynamics, rheology and hemostasis in patients with acute ischemic stroke. Results of a randomized, placebo-controlled, double-blind study. Pathophysiol Haemost Thromb 33: 121–126.

Wu Q, Li T, Zhu D et al. (2019). Altered expression of long noncoding RNAs in peripheral blood mononuclear cells in patients with impaired leptomeningeal collaterals after acute anterior large vessel occlusions. Ann Transl Med 7: 523.

Xiong L, Liu X, Shang T et al. (2017). Impaired cerebral autoregulation: measurement and application to stroke. J Neurol Neurosurg Psychiatry 88: 520–531.

Xu DS, Abruzzo TA, Albuquerque FC et al. (2010). External carotid artery stenting to treat patients with symptomatic ipsilateral internal carotid artery occlusion: a multicenter case series. Neurosurgery 67: 314–321.

Yamauchi H, Kudoh T, Sugimoto K et al. (2004). Pattern of collaterals, type of infarcts, and haemodynamic impairment in carotid artery occlusion. J Neurol Neurosurg Psychiatry 75: 1697–1701.

Zaidat OO, Yoo AJ, Khatri P et al. (2013). Recommendations on angiographic revascularization grading standards for acute ischemic stroke: a consensus statement. Stroke 44: 2650–2663.

Zaninovich OA, Ramey WL, Walter CM et al. (2017). Completion of the circle of Willis varies by gender, age, and indication for computed tomography angiography. World Neurosurg 106: 953–963.

Zhou H, Sun J, Ji X et al. (2016). Correlation between the integrity of the circle of willis and the severity of initial noncardiac cerebral infarction and clinical prognosis. Medicine (Baltimore) 95: e2892.

Zivin JA, Albers GW, Bornstein N et al. (2009). Effectiveness and safety of transcranial laser therapy for acute ischemic stroke. Stroke 40: 1359–1364.

Handbook of Clinical Neurology, Vol. 177 (3rd series)
Heart and Neurologic Disease
J. Biller, Editor
https://doi.org/10.1016/B978-0-12-819814-8.00020-2

Chapter 29

Neurological complications of cardiovascular drugs

MICHAEL A. KELLY*

Department of Neurology, Loyola University Chicago, Stritch School of Medicine, Maywood, IL, United States

Abstract

Cardiovascular drugs are used to treat arterial hypertension, hyperlipidemia, arrhythmias, heart failure, and coronary artery disease. They also include antiplatelet and anticoagulant drugs that are essential for prevention of cardiogenic embolism. Most neurologic complications of the cardiovascular drugs are minor or transient and are far outweighed by the anticipated benefits of treatment. Other neurologic complications are more serious and require early recognition and management. Overtreatment of arterial hypertension may cause lightheadedness or fatigue but often responds readily to dose adjustment or an alternative drug. Other drug complications may be more troublesome as in myalgia associated with statins or headache associated with vasodilators. The recognized bleeding risk of the antithrombotics requires careful calculation of risk/benefit ratios for individual patients. Many neurologic complications of cardiovascular drugs are well documented in clinical trials with known frequency and severity, but others are rare and recognized only in isolated case reports or small case series. This chapter draws on both sources to report the adverse effects on muscle, nerve, and brain associated with commonly used cardiovascular drugs.

INTRODUCTION

Treatment of cardiovascular disorders employs a wide range of surgical and medical therapies. Although cardiovascular drugs, broadly defined, comprise a large proportion of any pharmacopeia, several specific drugs and classes of drugs have particular usefulness in the management of cardiovascular disease (CVD). Most are well tolerated and have few if any neurologic implications. Others may affect brain, nerve, and muscle. There are many drugs of several classes available to treat arterial hypertension. Atherosclerotic coronary artery disease is responsible for ischemic heart disease in its many forms: the acute coronary syndromes, chronic coronary artery disease, cardiac arrhythmias, and ischemic cardiomyopathy. Medical treatment of ischemic heart disease involves risk factor control, in large part pharmacologic, of arterial hypertension and hypercholesterolemia. Atrial fibrillation is associated with systemic embolism and ischemic stroke and a major part of its management involves antithrombotics, both antiplatelet and anticoagulant drugs. Other sources of cardioembolism include valvular heart disease and ischemic cardiomyopathy. Medical therapies of cardiac arrhythmias affecting cardiac electrophysiology can affect the function of the nervous system in a variety of ways. This review discusses the cardiovascular drugs in common use, particularly those with the potential for neurologic complications.

ANTIHYPERTENSIVE DRUGS

Arterial hypertension is the major treatable risk factor for CVD including coronary artery disease, atrial fibrillation, congestive heart failure, and ischemic and hemorrhagic stroke (Rapsomaniki et al., 2014). Prevention of CVD involves optimization of diet, physical activity, and weight. Many drugs are available for pharmacologic therapy. First-line drugs for arterial hypertension include diuretics, angiotensin converting enzyme inhibitors (ACEIs), angiotensin receptor blockers (ARBs), and dihydropyridine calcium channel blockers (CCBs) (Whelton et al., 2017). Beta-blockers reduce blood

*Correspondence to: Michael A. Kelly, M.D., Department of Neurology, Loyola University Medical Center, 2160 S. First Ave., Maywood, IL 60153, United States. Tel: +1-708-216-5350, Fax: +1-708-216-5617, E-mail: michael.a.kelly@lumc.edu

pressure but may be most useful after myocardial infarction and in patients with heart failure. Vasodilator drugs such as nitroprusside are indicated in hypertensive emergencies.

In general, the antihypertensive drugs are well tolerated with adverse events that, if present, are mild and resolve with cessation of the drug. Neurologic complications are usually few. As antihypertensives, overtreatment or patient sensitivity may result in hypotension with resulting dizziness, lightheadedness, and syncope. Headache is common, most often due to drug-induced vasodilatation. Other neurologic side effects are often particular to the individual drug.

DIURETICS

Thiazide diuretics such as hydrochlorothiazide and thiazide-like chlorthalidone and indapamide act on the Na^{+}/Cl^{-} pump in the distal convoluted tubule of the kidney and increase sodium excretion, thus reducing intravascular volume and blood pressure. They may affect calcium channels causing relaxation of vascular smooth muscle and reduction of peripheral vascular resistance. The thiazide diuretics cause loss of potassium and can cause hypokalemia. Hyponatremia and hypercalcemia are known to occur.

More commonly used in heart failure, loop diuretics such as furosemide inhibit the epithelial sodium channel in the ascending limb of the loop of Henle causing natriuresis and diuresis. Amiloride and triamterene have potassium-sparing actions on the kidneys. Also, potassium sparing are the mineralocorticoid antagonists spironolactone and eplerenone that inhibit aldosterone and reduce blood pressure. Neurologic complications of diuretics arise primarily out of volume depletion, hypokalemia, or metabolic derangements. Dizziness, fatigue, orthostasis, and syncope are recognized adverse events. Headache, confusion, and muscle cramps have been described.

ANGIOTENSIN CONVERTING ENZYME INHIBITORS AND ANGIOTENSIN RECEPTOR BLOCKERS

The renin–angiotensin–aldosterone system plays a critical role in maintenance of blood pressure. Angiotensin II is a potent vasoconstrictor and influences aldosterone production, sympathetic tone, and affects renal salt and water reabsorption. The ACEIs block conversion of angiotensin I to angiotensin II and thereby reduce blood pressure. The ARBs reduce blood pressure by blocking the AT1 receptors on angiotensin II. Both classes of drugs are approved for use in treatment of arterial hypertension and congestive heart failure.

Captopril was approved for use in the United States in 1981. Today, lisinopril is the most commonly prescribed of the ACEIs, followed by enalapril, benazepril, ramipril, and quinapril (DrugStats, 2019). ACEIs are well tolerated, and adverse effects are infrequent. As effective treatments of hypertension, the ACEIs can cause hypotension and the associated symptoms of weakness, fatigue, dizziness, and syncope (Rush and Merrill, 1987). Cough and other respiratory symptoms, possibly related to greater circulating levels of bradykinin, are well recognized occurring in 10% of patients (Matchar et al., 2008; Pinargote et al., 2014). In a large meta-analysis of clinical trials comparing ACEIs and ARBs, the most frequent neurologic complications associated with ACEIs were headache (7.9%) and dizziness (5.4%) (Matchar et al., 2008). Case reports describe fatigue and nausea (McAreavey and Robertson, 1990; Huckell et al., 1993) and myalgia and arthralgia (Peppers, 1995). Disorders of taste occur rarely, most notably with captopril (Unnikrishnan et al., 2004). Visual hallucinations have been described in older patients with preexisting cognitive impairments (Doane and Stuits, 2013). A nonallergenic, bradykinin-related angioedema has been associated with ACEIs, and to a lesser extent, ARBs and fibrinolytic drugs (Inomata, 2012). In a US Veterans Affairs Health Care System study, 0.20% of patients prescribed ACEIs received treatment for angioedema (Miller et al., 2008). An antihypertensive trial found angioedema in 0.68% of patients prescribed enalapril (Kostis et al., 2005). ACEIs increase the risk of orolingual angioedema in patients receiving fibrinolytics for acute ischemic stroke thought due to increased bradykinin production (Engelter et al., 2005).

Losartan is the most commonly prescribed of the ARBs (DrugStats, 2019). Other ARBs approved for use in treatment of arterial hypertension include valsartan, olmesartan, irbesartan, telmisartan, candesartan, and eprosartan. The above-mentioned meta-analysis of clinical trials comparing ACEIs and ARBs reported headache in 6.3% and dizziness in 4.3% of patients taking ARBs, similar to the frequencies seen with ACEIs (Matchar et al., 2008). In a review of double-blind, placebo-controlled, clinical trials, headache was reported in 14.1%, dizziness in 4.1%, and asthenia or fatigue in 3.8% of patients receiving losartan (Goldberg et al., 1995). A trial comparing losartan to atenolol found dizziness in 17% and asthenia/fatigue in 15% of patients receiving losartan, not significantly different from patients receiving atenolol (Dahlöf et al., 2002). Cough is reported in 3%, less frequent than with ACEIs (Matchar et al., 2008). A reported patient repeatedly developed muscle pain, generalized weakness, and CK elevation in association with exposures to telmisartan (Barvaliya et al., 2015). Case reports describe dysgeusia with losartan (Heeringa and van Puijenbroek, 1998) and eprosartan (Castells et al., 2002). Visual hallucinations have been described with losartan (Heeringa and van Puijenbroek, 1998; Doane and Stuits, 2013).

Aliskiren is a direct renin inhibitor that binds to renin and blocks conversion of angiotensin I to angiotensin II. Neurologic adverse events are few, primarily headache (2%–8%), fatigue (2%–4%), and dizziness (1%–5%) (Strasser et al., 2007).

CALCIUM CHANNEL BLOCKERS

The CCBs are approved for use in arterial hypertension, angina, and cardiac arrhythmias. CCBs bind to receptors on L-type voltage-gated calcium channels of vascular smooth muscle cells, cardiac myocytes, and sinoatrial and atrioventricular nodes to reduce depolarization-induced influx of extracellular calcium. The vascular-selective dihydropyridine CCBs include amlodipine, nifedipine, felodipine, isradipine, nicardipine, nimodipine, nisoldipine, and clevidipine. CCBs relax peripheral vascular smooth muscle and thereby reduce vascular resistance and blood pressure. A reflex sympathetic activation may occur resulting in increased heart rate. The nondihydropyridine CCBs verapamil and diltiazem have additional action on cardiac muscle to reduce contractility and on the A–V node to decrease heart rate. Unlike the dihydropridine CCBs, the nondihydropyridine CCBs are used in treatment of ventricular and atrial tachyarrhythmias. The CCBs are inhibitors of the cytochrome P-450 3A4 enzyme (CYP3A4) and may increase levels of drugs such as carbamazepine and phenytoin that are metabolized by it. Conversely, carbamazepine and phenytoin are CYP34A inducers and may lower plasma concentration of drugs such as the CCBs that are likewise metabolized by it.

The CCBs are well tolerated and neurologic complications are few. The most common side effects are headache, flushing, and dizziness thought to be related to peripheral vasodilatation. Clinical trial data finds no detectable differences among CCBs in the frequency of these adverse events (McDonagh et al., 2005). A postmarketing surveillance study found withdrawal of nifedipine for headache in 4.9% of patients (Fallowfield et al., 1993). A prescription event-monitoring study found headache in 7.5% of patients taking nifedipine and 1.6% of patients taking diltiazem (Kubota et al., 1995).

Myoclonus has been described in several case reports after overdosage (Vadlamudi and Wijdicks, 2002) and after weeks (De Medina et al., 1986; Pedro-Botet et al., 1989), months (Hicks and Abraham, 1985), and years of treatment (Swanoski et al., 2011). Seizures have been reported with verapamil after intravenous infusion in a child (Maiteh and Daoud, 2001) and after intraarterial injection for vasospasm in an adult (Westhout and Nwagwu, 2007). Delirium and psychosis have been described with verapamil and diltiazem (Jacobsen et al., 1987; Busche, 1988; Binder et al., 1991). Case reports have associated parkinsonism with verapamil (García-Albea et al., 1993) and amlodipine (Sempere et al., 1995), although a neuroprotective effect of CCBs has been considered by others (Ritz et al., 2010; Lee et al., 2014).

ALPHA 1-ADRENERGIC RECEPTOR BLOCKERS

The α1-adrenergic receptor blockers reduce sympathetic tone and vascular resistance thereby lowering blood pressure. Prazosin, doxazosin, and terazosin are used in treatment of arterial hypertension.

The α1-adrenergic receptor blockers are generally well tolerated with few side effects. Orthostatic hypotension and syncope can occur. The most common neurologic adverse effects are dizziness, headache, asthenia, fatigue, and somnolence occurring in 10%–20% of patients (Grimm Jr, 1989; Carruthers, 1994). In a prostatic hypertrophy study of doxazosin, dizziness and fatigue were more common with doxazosin than placebo (11% vs 7%, and 6% vs 3%, respectively) (MacDonald et al., 2004). Confusion, abnormal behaviors, visual hallucinations, and paranoia have been reported in patients with renal failure taking prazosin (Chin et al., 1986). Prazosin has also been reported to show benefit in treatment of behavioral problems in patients with Alzheimer's disease (Wang et al., 2009).

BETA-ADRENERGIC BLOCKERS

Beta-blockers are a large class of drugs defined by their ability to block β1 adrenergic receptors, thereby reducing vascular resistance and blood pressure, cardiac contractility, and heart rate. Nonspecific β-blockers such as carvedilol, labetalol, andosol, propranolol, and sotalol also block β2 receptors. Carvedilol and labetalol in addition block α1 receptors. Pindolol has intrinsic sympathomimetic activity. Sotalol is a β-blocker used primarily for its antiarrhythmic properties.

Adverse effects of β-blockers are primarily due to their effect on the heart and include congestive heart failure, bradycardia, hypotension and associated dizziness, drowsiness, fatigue, and syncope. A meta-analysis of seven randomized trials found depressive symptom to be common but without significant difference between β-blockers and placebo (20.1% and 20.5%, respectively) (Ko et al., 2002). The same analysis found a small but significant increase in fatigue in patients taking β-blockers over those taking placebo (33.4% vs 30.4%). A Cochrane review found no significant difference in depression or fatigue among patients taking β-blockers or placebo or no treatment (Wiysonge et al., 2012). Delirium has been described in case reports with propranolol, atenolol, and metoprolol (Fisher et al., 2002; Huffman and Stern, 2007) and visual hallucinations with propranolol (Fleminger, 1978) and metoprolol (Goldner, 2012). Falls have been associated with the nonselective β-blockers (Ham et al., 2017). Case reports and analysis of 30 patients drawn from a population-based administrative database suggest an increased risk of hospitalization for myopathy, including rhabdomyolysis, in patients taking β-blockers, especially propranolol (Setoguchi et al., 2010). A network meta-analysis found cutaneous vasoconstriction, in 7.0% of patients receiving β-blockers and 4.6% receiving placebo (Khouri et al., 2016).

VASODILATORS

Hydralazine lowers blood pressure by arteriolar dilatation and reduction in peripheral resistance. It is associated with reflex sympathetic activity. The mechanism of action is not known but may relate to movement of intracellular calcium. It is indicated for the treatment of essential arterial hypertension and heart failure and has been used in treatment of severe hypertension in pregnancy. The most common neurologic adverse effect is headache related to the vasodilatation (Cohn et al., 2011). This may be associated with dizziness, tachycardia, and hypotension. A case report describes a sensorimotor polyneuropathy with profound weakness thought to have been related to hydralazine (Tsujimoto et al., 1981). A drug-induced lupus syndrome can occur and manifests as asthenia, myalgia, and arthralgia (Stratton, 1985).

Minoxidil activates potassium channels resulting in hyperpolarization and relaxation of arteriolar smooth muscle cells. Like other vasodilators it may by associated with reflex sympathetic activity causing flushing, palpitations, and headache (Cohn et al., 2011). Fenoldopam is a selective postsynaptic dopamine agonist (acting on D1-receptors) that decreases peripheral vascular resistance and increases renal blood flow with resulting diuresis and natriuresis and reduction of blood pressure. It is an intravenously administered drug indicated for severe arterial hypertension. It can be associated with restlessness and anxiety (Taylor et al., 1999).

Nitroprusside and oral nitrates release nitric oxide resulting in vascular dilatation. Nitroprusside is an intravenously administered drug that acts on arterioles and venules. It is indicated for the treatment of arterial hypertension and heart failure. Oral nitrates reduce blood pressure but are more commonly used for angina and heart failure. Adverse events are primarily those of vasodilation and hypotension. Headache is common. As part of the development of a migraine model, a trial of nitroprusside in healthy volunteers produced headache in all of five subjects (Guo et al., 2013). Intracranial pressure elevation, presumed due to intracranial vasodilatation, has been documented in series of neurosurgical patients (Turner et al., 1977; Cottrell et al., 1978). With prolonged administration, metabolic conversion of nitroprusside to cyanide and thiocyanate may result in nausea, fatigue, confusion, and psychosis (Harmon and Wohlreich, 1995).

CHOLESTEROL-LOWERING DRUGS

Elevated low-density lipoprotein cholesterol (LDL-C) is a major risk factor for atherosclerosis. Lowering LDL-C reduces the risk of coronary artery disease, myocardial infarction, and ischemic stroke.

Reduction of circulating LDL-C levels can be accomplished by a variety of mechanisms. Reduced dietary intake of saturated fats and cholesterol leads to an increase in LDL-C receptors and a decrease of plasma LDL-C levels. Drugs that impede absorption or accelerate catabolism of cholesterol also increase LDL-C receptor number and decrease plasma LDL-C levels.

STATINS

Statins are competitive inhibitors of 3-hydroxy-3-methylglutaryl-coenzyme A (HMGCoA) reductase which converts HMGCoA to mevalonate, a precursor of sterols, including cholesterol, and is the rate-limiting step in hepatic cholesterol biosynthesis. Statins are the first-line treatment of hypercholesterolemia. Lovastatin, first approved for use in the United States in 1987, was soon followed by approval of pravastatin, simvastatin, and later by fluvastatin, rosuvastatin, atorvastatin, and pitavastatin. All are metabolized by the liver but vary by requirements for transformation to an active form, percentage of protein-binding, metabolism by the cytochrome P450 system, drug interactions, potential side effects, and half-life duration. Statins lower elevated triglyceride levels, increase high-density lipoprotein cholesterol (HDL-C) levels, and decrease LDL-C levels. Based on the percentage of decrease in the plasma LDL-C level, the intensity of statin dosing is described as low (<30%), moderate (30%–50%), or high (>50%). Neurologic adverse events primarily involve muscle. Concern has been raised about possible neurocognitive effects, neuropathy, and the risk of intracerebral hemorrhage.

Myopathy

Muscle-related complaints are common, reported by 25% of patients taking statins in an internet survey (Cohen et al., 2012). Statin-associated muscle symptoms include pain, stiffness, cramps, fatigue, and weakness. A meta-analysis of 26 clinical trials involving statins in which muscle-related symptoms were reported found 12.7% of treated patients and 12.4% of placebo patients reported muscle-related problems (Ganga et al., 2014). The STOMP (Effect of Statin on Skeletal Muscle Function and Performance) trial randomized 420 participants to 6 months of high-dose atorvastatin or placebo. Myalgia was reported in 9.4% of patients receiving atorvastatin and 4.6% of patients receiving placebo (Parker et al., 2013). Statin-associated myopathy with significant CK elevation (>10x ULN) is rare, occurring in about 0.1% of patients (Newman et al., 2019). Biopsy-proven myopathy without CK elevation has been described in a report of four patients (Phillips et al., 2002). The risk of statin-associated rhabdomyolysis with CK elevation >40x ULN is estimated to occur in about 0.01% of patients (Newman et al., 2019). A study of insurance claim data found the myopathy-related hospitalization rate for patients taking monotherapy with atorvastatin, pravastatin, or simvastatin to be 0.44 per 10,000 patient-years (Graham et al., 2004). An immune-mediated necrotizing myopathy both with and without plasma anti-HMGCoA antibodies has been associated with statins. Multiple case reports describe this myositis as profound shoulder and pelvic girdle weakness with markedly elevated CK levels and muscle biopsies demonstrating necrosis with regenerating muscle fibers and minimal inflammation (Nazir et al., 2017). Risk of myopathy is increased with higher doses and conditions that impair catabolism including patients of older age and those with renal and hepatic dysfunction (Holbrook et al., 2011). Cardiovascular drugs that interfere with statin metabolism include fibrates, amiodarone, and warfarin. The etiology of statin-associated muscle disorders is poorly understood. Proposed mechanisms implicate mitochondrial dysfunction, oxidative stress, impaired mevalonate metabolism, and genetic factors (Selva-O'Callaghan et al., 2018).

Behavior and cognition

Statins have been associated in case reports with behavioral changes including irritability, impulsivity, violent behavior, and depression (Golomb et al., 2004). Others have reported cases of cognitive problems including memory loss (Orsi et al., 2001; Galatti et al., 2006; Suraweera et al., 2016). A review of the literature and MedWatch reports found 60 patients reported to have statin-associated memory loss (Wagstaff et al., 2003). In a survey of participants in a statin trial, 75% reported cognitive problems including amnestic episodes, (Evans and Golomb, 2009). Psychological tests of attention and psychomotor speed found minor decrements in patients taking lovastatin and simvastatin (Muldoon et al., 2000, 2004). The US Food and Drug Administration issued a warning in 2012 requiring drug labels to include the potential for memory loss and confusion side effects (FDA Drug Safety Communication, 2012). In contrast, an Italian review of reported adverse drug events with statins found more insomnia but less somnolence, agitation, confusion, and hallucinations in patients receiving statins (Tuccori et al., 2008). The randomized, controlled PROSPER (Pravastatin in Elderly Individual at Risk of Vascular Disease) trial found no significant effect on cognitive function over 3 years of follow-up (Shepherd et al., 2002). A review by the Statin Cognitive Safety Task Force in 2014 concluded that statins are not associated with neurocognitive adverse events (Rojas-Fernandez et al., 2014).

Neuropathy

Various forms of neuropathy have been associated with use of statins. A 1994 case report described a patient with paresthesias of hands and feet temporally related to use of lovastatin and again to pravastatin (Jacobs, 1994). A sensorimotor polyneuropathy has been described in case reports with simvastatin both soon after it began to be taken and 2 years after initiation of the drug (Phan et al., 1995). A sensory and autonomic ganglionopathy has been described (Novak et al., 2015). In the US National Health and Nutrition Examination Survey data, polyneuropathy is more prevalent in participants taking statins than those not taking statins (23.5% vs 13.5%; $P<0.01$) (Tierney et al., 2013). A multivariate logistic regression analysis of these data found a significant association between statin use and polyneuropathy after adjustment for age, race, diabetes, and other factors (adjusted OR 1.3, 95% CI 1.1–1.6; $P=0.04$). A population-based case–control study in Italy suggested an association between statin and use and polyneuropathy (Corrao et al., 2004). Contrary to these reports, a similar population-based case–control study in Denmark found no relationship between statin use, whether recent, long term, or high intensity, and polyneuropathy (Svendsen et al., 2017). An analysis of a US hospital system database found an insignificant increase in polyneuropathy among statin users (OR 1.30, 95% CI 0.3–2.1; $P=0.27$) (Anderson et al., 2005).

Hemorrhagic stroke

The SPARCL (Stroke Prevention with Aggressive Reduction in Cholesterol) trial demonstrated significant reduction in ischemic stroke in patients with a history of stroke, either ischemic or hemorrhagic, taking high-dose atorvastatin (Amarenco et al., 2006). The rate of intracerebral hemorrhage, however, was increased in the treatment group (HR 1.66; 95% CI 1.08–2.55). A meta-analyses of randomized trials revealed a nonsignificant increase in risk of hemorrhagic stroke in five trials of higher vs lower statin doses (RR 1.21, 99% CI 0.76–1.91; $P=0.3$) and in 26 trials of statin vs control or low-dose statin (RR 1.12, 95% CI 0.93–1.35; $P=0.2$) (Baigent et al., 2010). Analyzed data did not include the SPARCL results. A large meta-analysis that included SPARCL found no increase in statin-associated risk of hemorrhage in randomized trials (RR 1.10, 95% CI 0.86–1.41), cohort studies (RR 0.94, 95% CI 0.81–1.10), or case–control studies in which patients had a history of cerebrovascular disease (RR 0.60, 95% CI 0.41–0.88) (Hackam et al., 2011). The risk of hemorrhage was not influenced by the magnitude of cholesterol reduction.

CHOLESTEROL ABSORPTION INHIBITOR, EZETIMIBE

Ezetimibe inhibits intestinal absorption of cholesterol by binding Niemann-Pick C1-like 1 protein, the major intestinal transporter of dietary cholesterol. It is used as an adjunctive treatment of hypercholesterolemia. Because it increases hepatic synthesis of cholesterol, it is commonly combined with a statin.

IMPROVE-IT (Improved Reduction of Outcomes: Vytorin Efficacy International Trial) demonstrated reduced LDL-C levels and improved cardiovascular outcomes from the addition of ezetemibe to simvastatin in patients with a recent acute coronary syndrome (Cannon et al., 2015). Nonsignificant differences were found between simvastatin monotherapy and simvastatin-ezetimibe combination therapy in frequency of myopathy (0.1% vs 0.2%; $P=0.32$), rhabdomyolysis and myopathy (both 0.3%), and myopathy, rhabdomyolysis, or myalgia with CK elevation $\geq$5x ULN (both 0.6%). A trial of the PCSK9 inhibitor evolocumab and ezetimibe in statin-intolerant patients reported adverse events of myalgia, muscle spasms, headache, and fatigue, but withdrawal from the study for muscle-related events occurred more often with placebo (7.6%) than with ezetimibe (6.8%) (Nissen et al., 2016). An analysis of pooled data from four phase 3 clinical trials investigating evolocumab found ezetimibe controls had more muscle-related adverse events than placebo controls (7.8% vs 2.9%), less CK elevation (0.6% vs 0.7%), and more neurocognitive adverse events (0.6% vs 0.0%) (Stroes et al., 2018).

FIBRIC ACID DERIVATIVES

Fibrates activate the nuclear transcription factor peroxisome proliferator-activated receptor-alpha that regulates genes that control lipid metabolism resulting in decreased plasma triglycerides, very-low-density lipoprotein cholesterol and LDL-C and increased plasma HDL-C (Lipid-lowering drugs, 2019). Fibric acid derivatives in use for treatment of hypercholesterolemia are fenofibrate, fenofibric acid, and gemfibrozil. Gemfibrozil is primarily used in familial hypertriglyceridemias.

The fibric acid derivatives are well tolerated. Reported neurologic adverse events associated with fibric acid derivatives include headache, dizziness, fatigue, arthralgias, limb pain, and muscle disorders. Randomized clinical trials involving fenofibrate found no significant differences between treatment and controls in myalgia, muscle cramps, or creatine kinase levels (Keech et al., 2005; Grundy et al., 2006; Arai et al., 2018). Gemfibrozil has been associated with myopathy when used in conjunction with statins (Curtin and Jones, 2007).

PCSK9 INHIBITORS

PCSK9 inhibitors are monoclonal antibodies delivered as subcutaneous injections that inhibit proprotein convertase subtilisin-kexin type 9 increasing LDL-C receptor numbers and clearance of plasma LDL-C. Evolocumab and alirocumab are approved as adjunctive treatment of hypercholesterolemia and for prevention of myocardial infarction and ischemic stroke in patients with established CVD.

The FOURIER (Further Cardiovascular Outcomes Research with PCSK9 Inhibition in Subjects with Elevated Risk) trial was a randomized, placebo-controlled trial of patients with atherosclerotic CVD who were receiving high-intensity statins that compared evolocumab and placebo with median follow up of 2.2 years (Sabatine et al., 2017). There were no significant differences between treatment and placebo groups in in muscle-related events (5.0% vs 4.8%), rhabdomyolysis (0.1% vs 0.1%), or neurocognitive events (1.6% vs 1.5%). An analysis of pooled data from four phase 3 clinical trials comparing two doses of evolocumab (140 mg biweekly and 420 mg monthly) and placebo found muscle-related adverse events more common in the treatment group (3.5% vs 3.8% vs 2.9%, respectively) and neurocognitive effects rare in all groups (0.1%, 0.1%, and 0.0%, respectively) (Stroes et al., 2018).

In an efficacy and safety study of alirocumab, patients at risk for cardiovascular events and receiving maximum therapy with statins were randomized to treatment with alirocumab or placebo. Myalgia was more common in the treatment group (5.4% vs 2.9%, $P=0.006$) (Robinson et al., 2015). Nonsignificant differences were found between treatment and placebo in neurocognitive disorder (confusion, amnesia, or memory disorder) (1.2% and 0.5%, respectively, $P=0.17$). The ODYSSEY OUTCOMES (Evaluation of Cardiovascular Outcomes after an Acute Coronary Syndrome with Alirocumab) trial was a randomized controlled trial of alirocumab vs placebo in patients with acute coronary syndrome within 12 months and taking statins who were followed for a median of 2.8 years (Schwartz et al., 2018). There were no significant differences between the treatment and placebo groups in neurocognitive disorders (1.5% vs 1.8%) or hemorrhagic stroke (<0.01% vs 0.20%).

ANTIARRHYTHMIC DRUGS

Disruption of the exquisite conduction system of the heart results in atrial, nodal, and ventricular arrhythmias. Atrial flutter and atrial fibrillation, in addition to their influence on cardiac function, are well recognized for their embolic stroke potential. The antiarrhythmic drugs (AADs) act by any of several mechanisms, some by multiple mechanisms. The widely used classification scheme is that of Vaughan-Williams: class I (sodium channel blockade), class II (B-adrenergic receptor blockade), class III (potassium channel blockade), and class IV (calcium channel blockade). Minor neurologic complications associated with these drugs are many. Serious complications are rare. In clinical trials and clinical practice, common adverse events associated with these drugs include headache, dizziness, vertigo, fatigue, weakness, anxiety, and depression. Intoxication can cause confusion, depressed consciousness and coma, seizures, and death. Most can initiate or exacerbate arrhythmias (proarrhythmic), worsen heart failure, and cause hypotension. Most undergo hepatic metabolism and can induce liver dysfunction. Many affect the gastrointestinal system and cause nausea, vomiting, and diarrhea. Neurologically, several can cause impairment of vision, others tremor, polyneuropathy, and altered cognition. Undesirable drug–drug interactions are often a result of effects on hepatic metabolism of the AAD or its effect on other drugs.

Amiodarone

Amiodarone, the most commonly prescribed AAD, is used in the treatment of ventricular arrhythmias and for rhythm control in atrial fibrillation. The drug blocks sodium channels and affects potassium and calcium channels prolonging PR, QRS, and QT intervals. It has alpha- and beta-blocking properties. Amiodarone is highly lipophilic, concentrates in many tissues, and is slow to be eliminated.

Adverse events associated with amiodarone are cumulative and dose related and involve lungs, heart, liver, thyroid, and skin. Early amiodarone trials using high doses noted rare patients with numbness, ataxia, and vision symptoms (Cairns et al., 1991). Neurologic side effects were described in 29 of 54 (54%) consecutive patients started on amiodarone for ventricular tachycardia (Charness et al., 1984). Postural tremor was the earliest and most common side effect occurring in 39%. Three patients with prior tremor worsened on amiodarone treatment. Falls and ataxia of gait were noted in 37% and neuropathy with tingling of hands and feet developed in 6%. A series of 102 patients was examined neurologically before and intermittently while receiving amiodarone (Palakurthy et al., 1987). Tremor was noted in 43%, peripheral neuropathy in 10%, and gait ataxia in 7%.

More recent reports describe a lower frequency of neurotoxic effects. A meta-analysis of four trials of low-dose amiodarone (<330 mg daily) vs. placebo found the frequency of neurologic adverse effects of any kind in patients taking amiodarone (4.6%) to be twice that of those taking placebo (1.9%) (OR 2.0, 95% CI 1.085–3.697; $P=0.03$) (Vorperian et al., 1997). An analysis of 707 Olmstead County patients taking amiodarone found attributable neurotoxic effects in 2.8% of patients (Orr and Ahlskog, 2009). Effects included tremor, gait ataxia, peripheral neuropathy, and cognitive impairment.

Amiodarone-associated optic neuropathy is a slowly progressive loss of visual acuity, most commonly bilateral, with disc edema (Macaluso et al., 1999; Purvin et al., 2006; Passman et al., 2012; Wang and Cheng, 2016). A Taiwan population-based cohort study of 6175 patients on amiodarone found 17 cases of amiodarone-associated optic neuropathy, twice the frequency seen in controls, with a male predominance (Cheng et al., 2015). Slow improvement was seen in 58% of a series of 296 patients following cessation of drug (Passman et al., 2012). Pseudotumor cerebri with papilledema and elevated lumbar opening pressure has been associated with amiodarone in several case reports (Fikkers et al., 1986; Van Zandijcke and Dewachter, 1986; Grogan and Narkun, 1987; Kristin and Ulbig, 2001). Vision may also be impaired by amiodarone-associated corneal changes, cataracts, and maculopathy (Wang and Cheng, 2016).

Sensorimotor polyneuropathy has been reported since the introduction of amiodarone (Lustman and Monseu, 1974). Although usually mild, weakness and sensory loss can be profound. Patients may develop an associated sensory gait ataxia (Martinez-Arizala et al., 1983; Palakurthy et al., 1987; Chaubey et al., 2013). Electrophysiological studies demonstrate a variety of abnormalities, both axon loss and demyelination. Nerve biopsies show demyelination and may show lysosomal inclusions in axons, Schwann cells, and perineural tissues (Pellissier et al., 1984; Abarbanel et al., 1987; Pulipaka et al., 2002; Kang et al., 2007). Myopathic proximal weakness has been reported with biopsies demonstrating a noninflammatory vacuolar myopathy (Pulipaka et al., 2002; Flanagan et al., 2012).

Case reports describe parkinsonism (Ishida et al., 2010; Bondon-Guitton et al., 2011), progressive supranuclear palsy (Mattos et al., 2009), depression (Ambrose and Salib, 1999), and delirium (Trohman et al., 1988; Athwal et al., 2003; Foley and Bugg, 2010).

Disopyramide

Disopyramide is a sodium channel blocking drug that decreases cardiac automaticity, increases refractory periods and slows conduction. Anticholinergic properties are largely responsible for its adverse effects (Morady et al., 1982; Teichman, 1985). These include constipation, urinary retention, mouth dryness, and blurred vision. In an amiodarone comparison trial, 41 patients received disopyramide, of which 4 stopped taking the drug due to anticholinergic side effects (Villani et al., 1992). Case reports describe a sensory and a sensorimotor polyneuropathy (Dawkins and Gibson, 1978; Briani et al., 2002). Several reports describe altered mentation and psychosis thought to be related to the drug's anticholinergic properties (Falk et al., 1977; Padfield et al., 1977; Ahmad et al., 1979). Myasthenia gravis can be exacerbated by disopyramide (Hirose et al., 2008).

Dofetilide

Dofetilide is a potassium blocking drug that prolongs action potentials in the atria and ventricles and is used to treat atrial fibrillation and atrial flutter. Noncardiac adverse events are few. In a randomized, placebo-controlled trial of dofetilide in chronic atrial fibrillation or flutter, there were no significant differences in serious AEs between the treatment (2.9%) and placebo groups (2.3%) (Singh et al., 2000). A review noted headache and dizziness to be a common side effect (Lenz and Hilleman, 2000). A trial of high-dose administration (2500 μg daily) in nine patients caused headache, vasodilatation, and syncope in one patient (Allen et al., 2000). Lower doses had no such effect. Case reports describe trigeminal neuralgia (Maluli, 2015) and facial weakness (Zhang and Steckman, 2019).

Dronedarone

Dronedarone is a derivative of amiodarone with many of the same effects on sodium and potassium channels and cardiac function. It is used in the treatment of atrial fibrillation. It is associated with fewer pulmonary, thyroid, and skin adverse effects than amiodarone but is considered less effective (DrugStats, 2019). In a randomized, placebo-controlled trial of dronedarone for atrial fibrillation in 4628 patients, no significant differences in neurologic adverse events were noted between treatment and placebo groups (Hohnloser et al., 2009). No adverse neurologic events were reported in a randomized, placebo-controlled trial of 3236 patients with high-risk permanent atrial fibrillation (Connolly et al., 2011).

Flecainide

Flecainide is a sodium channel blocker that slows conduction and raises depolarization threshold in cardiac tissue. The drug's side effect profile is similar to other sodium blocking AADs but with fewer adverse events overall (Naccarelli et al., 2003). The Flecainide-Quinidine Research Group treated 280 patients with either flecainide or quinidine for chronic premature ventricular contractions and compared the 2 groups (Flecainide-Quinidine Research Group, 1983). Adverse events associated with flecainide were dizziness (30%), blurred vision (28%), nausea (9%), and headache (9%). Two later reviews found similar results (Gentzkow and Sullivan, 1984; Tamargo et al., 2012). In a study of 27 patients in whom flecainide was orally loaded (200 mg), 33% had mild side effects of dizziness, tremor, and headache (Schwartz et al., 1985).

Case reports describe patients with transient psychoses (Bennett, 1997), psychosis with myoclonus (Ghika et al., 1994; Ting et al., 2008), at initiation of oral or IV therapy (Ramhamadany et al., 1986), often with auditory or visual hallucinations (Drerup, 1988; Ghika et al., 1994). Single case reports describe an episode of generalized tonic–clonic seizure (Kennerdy et al., 1989) and dystonia (Linazasoro et al., 1991; Miller and Jankovic, 1992) associated with the drug. A sensorimotor peripheral neuropathy developed in two patients after having taken flecainide for more than 2 years, both improving after the drug was withheld (Palace et al., 1992; Malesker et al., 2005).

Ibutilide

Ibutilide is an intravenously administered potassium channel blocker that prolongs cardiac repolarization and action potential duration resulting in a prolongation of the refractory period and slowing the heart rate while increasing resistance to arrhythmia. Neurologic adverse events are not reported.

Lidocaine

Lidocaine is a sodium channel blocker with antiarrhythmic and anesthetic properties. With toxic blood levels, lightheadedness, dysarthria, blurred vision, muscle twitching, confusion, and seizures can occur (Scott, 1986; Gil-Gouveia and Goadsby, 2009; Rahimi et al., 2018). Conversely, lidocaine has been used in pediatric and adult populations to treat status epilepticus in conjunction with other anticonvulsants (Zeiler et al., 2015). Case reports describe seizures after topical and intramuscular application (Wu et al., 1993; Dorf et al., 2006). Neuropsychiatric side effects include dysphoria, depression with paranoia, and auditory and visual hallucinations (Turner, 1982; Saravay et al., 1987; Gil-Gouveia and Goadsby, 2009).

Mexiletine

Mexiletine, structurally related to lidocaine, blocks sodium channels and is approved for treatment of ventricular arrhythmias. Gastrointestinal side effects are common. Neurologic side effects are largely dose-dependent. In clinical trials of mexiletine for arrhythmias adverse events commonly include dizziness, headache, and tremor (Nygaard et al., 1986). A review of clinical trials of mexiletine for arrhythmias found 10% of patients to have had side effects of tremor, dizziness, or memory loss (Manolis et al., 1990).

Additional understanding of neurologic side effects has been gained from the use of mexiletine for treatment of neuropathic pain and muscle cramps. A review of trials of mexiletine for diabetic neuropathy pain reported mexiletine-associated adverse events of sleep disturbance, headache, shakiness, dizziness, and tiredness (Jarvis and Coukell, 1998). In a trial of mexiletine for treatment of chronic pain of diabetic neuropathy or fibromyalgia, symptoms of dizziness, confusion, and anxiety caused 11 of 61 patients (18%) of patients taking to stop the drug (Romman et al., 2018). In anticipation of another trial for painful diabetic neuropathy, 12 healthy volunteer patients took up to 1350mg of mexiletine daily (Ando et al., 2000). All experienced lightheadedness, muscle twitching, weakness, blurred vision, headache, tremors, poor concentration, or dysphoria. In a cohort of patients with amyotrophic lateral sclerosis, muscle cramps responded to mexiletine although at 900mg daily, 25% of the study subjects experienced dizziness, falls, tremor, or nausea (Weiss et al., 2016). Dizziness, ataxia, or dysarthria was seen in six of nine patients treated with meclizine for torticollis (Ohara et al., 1998). Rare case reports describe transient psychosis (Fernández-Solá et al., 1987; Shailesh et al., 2011).

Procainamide

Procainamide decreases myocardial excitability and conduction velocity and is used in the treatment of ventricular and supraventricular arrhythmias. The major adverse effect is a lupus-like syndrome of arthralgia and myalgia that develops in 15%–20% of patients after long periods of use, often 3 years or more (Vaglio et al., 2018). A myopathy was considered in a patient, having taken procainamide for several years, who slowly developed dysphagia, dysarthria, and facial weakness that resolved on stopping the drug (Agius et al., 1998). A demyelinating polyneuropathy thought related to procainamide was diagnosed in a patient taking procainamide for years who gradually developed a syndrome of weakness and areflexia that resolved with cessation of the drug (Erdem et al., 1998). Sural nerve biopsies on this and a second case demonstrated loss of myelinated fibers and remyelination (Sahenk et al., 1977).

Myopathy may develop acutely. An elderly man developed generalized paralysis 2 weeks after starting procainamide, and a necrotizing myopathy was found on muscle biopsy (Venkayya et al., 1993). After 2 weeks of procainamide, a patient presented with dysphagia and face and limb weakness that resolved on stopping the drug (Niakan et al., 1981). Weakness recurred on a second trial of procainamide. A patient who received intravenous procainamide over days for acute ventricular arrhythmia developed dystonic arm movements, ataxia, elevation of creatinine kinase, and myopathy on muscle biopsy (Lewis et al., 1986). The symptoms resolved on stopping the drug. Procainamide may exacerbate myasthenia gravis (Godley et al., 1990) and induce a transient, myasthenic-like weakness in patients without recognized myasthenia gravis (Drachman and Skom, 1965; Niakan et al., 1981; Miller et al., 1993). Although rare, multiple case reports describe mania, hallucinations, and psychosis on institution of treatment with procainamide (Jelinek et al., 1974; McCrum and Guidry, 1978; Bizjak et al., 1999).

Propafenone

Propafenone is a sodium blocking AAD with modest beta adrenergic actions that prolongs PR and QRS durations and is used in management of supraventricular tachycardia in patients without structural heart disease. A review of propafenone reported neurologic side effects in 20% of patients to include dizziness, paresthesia, ataxia, altered mentation, tremors, and rarely seizures (Shen, 1990). Others have reported disturbance in taste, dizziness, and blurred vision in up to 8% of patients (Funck-Brentano et al., 1990). An investigation of propafenone metabolism in 28 patients found central nervous system side effects of visual blurring, dizziness, and paresthesias in 25% of patients, greatest in poor metabolizers (Siddoway et al., 1987). Drug was held in one patient due to psychosis and another for intolerable lip paresthesias. Propafenone was considered the cause of a painful small fiber polyneuropathy that on sural nerve biopsy demonstrated a mononuclear inflammatory cell infiltrate (Galasso et al., 1995). Case reports describe myoclonus (Chua et al., 1994) and ataxia (Odeh et al., 2000). Others have reported propafenone-related amnesia (Jones et al., 1995), confusion (Ahmad, 1991), mania (Jack, 1985), and psychosis (Pfeffer and Grube, 2001).

Quinidine

Quinidine, first used clinically in the 1920s, still has a role in the treatment of ventricular arrhythmias and atrial fibrillation (Al-Khatib et al., 2018). Quinidine prolongs the Q-T interval and can cause paroxysmal ventricular tachycardia and syncope (quinidine syncope) (Kim and Benowitz, 1990). A clinical trial of quinidine treatment of ventricular arrhythmias reported headache in 14% and dizziness in 11% of participants (Flecainide-Quinidine Research Group, 1983). Trials investigating pseudobulbar affect in patients with amyotrophic lateral sclerosis reported dizziness in 2.7% of patient taking quinidine (Brooks et al., 2004). Cinchonism, named after the tree bark from which quinidine was first derived, is a toxic syndrome of headache, nausea, vomiting, and tinnitus that can progress to hearing loss, vision loss, confusion, and delirium (Huffman and Stern, 2007). Case reports describe transient visual disturbances (Fisher, 1981), blindness (Bacon et al., 1988; Wolf et al., 1992), delirium (Summers et al., 1981; Eisenman and McKegney, 1994) psychosis (Deleu and Schmedding, 1987; Bizjak et al., 1999), and memory loss (Gilbert, 1978; Billig and Buongiorno, 1985). Prolongation of the INR may be seen in patients taking warfarin and quinidine metabolism may be accelerated by antiepileptic drugs such as phenytoin and phenobarbital resulting in decreased serum levels (Grace and Camm, 1998). Because quinidine has anticholinergic effects, it can exacerbate myasthenia gravis (Wittbrodt, 1997). Quinine, a stereoisomer of quinidine, is a flavoring agent in tonic water, a treatment of leg cramps and malaria, and an adulterant in heroin, has a similar side-effect profile (Bateman and Dyson, 1986; Barrocas and Cymet, 2007).

Sotalol

Beta adrenoreceptor blockers prolong atrioventricular node conduction and slow the sinus rate. Rate control in atrial fibrillation can be maintained with bisoprolol, carvedilol, esmolol, metoprolol, nadolol, and propranolol (January et al., 2014). Potential neurologic complications of their use are discussed in the section on antihypertensive drugs. Sotalol is a beta adrenergic antagonist that also acts to block potassium currents. It is used to treat ventricular arrhythmias as well as to prevent recurrence of atrial fibrillation (January et al., 2014). Sotalol is renally excreted and does not rely on the liver for metabolism.

Fatigue, dizziness, and asthenia were the most frequent noncardiac side effects described in a review of sotalol safety in clinical trials (Soyka et al., 1990). A dose effect was noted in a double-blind placebo-controlled trial of sotalol with unspecified nervous system adverse events in none of the patients who received 50 mg of sotalol, in 5% of patients who received 100 or 200 mg, and in 2% of patients who received placebo (Hohnloser et al., 1995). A series of 82 adult patients with congenital heart disease treated with sotalol found 13% to have fatigue or lethargy and 1% to have each of headache, visual disturbance, dizziness, loss of taste, and severe depression (Moore et al., 2019). Case reports describe myalgia (Kieffer et al., 1988), hyperhidrosis (Schmutz et al., 1995), face atrophy after cold exposure (Aho and Haapa, 1982), and muzziness (Lader and Tyrer, 1972). Sotalol may exacerbate myasthenia gravis (Sharifkazemi and Ziya, 2019).

Verapamil and diltiazem

The nondihydropyridine CCBs, verapamil and diltiazem, have antihypertensive and antiarrhythmic activity. Inhibition of calcium-dependent channels reduces nodal automaticity and slows atrioventricular node conduction resulting in a slowed heart rate resistant to catecholamine stimulation. Potential neurologic complications of their use are described in the section of antihypertensive drugs.

ANTIPLATELET DRUGS

Unstable angina, myocardial infarction, and ischemic stroke are thrombotic processes that involve platelet aggregation and fibrin formation. Aspirin effectively impairs platelet aggregation through inhibition of cyclooxygenase. $P2Y_{12}$ receptor inhibitors accomplish the same goal at least as effectively as aspirin by their action on platelet adenosine diphosphate (ADP) receptors. The option of intravenous administration and knowledge of the rapidity of onset of effect directs the choice of drugs in emergent situations. Additional antiplatelet effect can be achieved with glycoprotein IIb/IIIa inhibitors operating through still additional mechanisms. All antiplatelet drugs are associated with bleeding, primarily gastrointestinal but also intracranial.

ASPIRIN

As an antithrombotic, aspirin plays a pivotal role in cardiovascular medicine. It is essential in treatment of acute coronary syndromes, in maintenance of coronary artery stent patency, and in prevention of myocardial infarction and ischemic stroke in patients with atherosclerotic arterial disease.

Aspirin is an irreversible inhibitor of cyclooxygenase, an enzyme critical in the synthesis of thromboxane. Reduction of thromboxane results in inhibition of platelet aggregation and clot formation. The desired therapeutic antithrombotic effect is countered by its associated risk of bleeding. Most aspirin-associated bleeding is gastrointestinal, usually minor but can be life threatening. The major neurologic complication is intracranial bleeding. The risk of intracranial bleeding is influenced by the dose of aspirin and the duration for which it is taken. Patient factors that increase the risk of intracranial bleeding include older age, male sex, arterial hypertension, and cigarette smoking (Whitlock et al., 2016). Risk is also increased by concomitant use of other antithrombotic drugs such as other antiplatelet agents, nonsteroidal antiinflammatory agents, and anticoagulants.

The Antithrombotics Trialists' Collaboration conducted a meta-analysis of randomized controlled trials of aspirin for the prevention of cardiovascular events and found aspirin to be associated with an increase in hemorrhagic stroke risk in six primary prevention trials (0.04%/year in aspirin patients vs 0.03%/year in control patients, HR 1.32; $P=0.05$) and an increase in hemorrhagic stroke risk in 16 secondary prevention trials (0.17%/year vs 0.09%/year; $P=0.07$) (Baigent et al., 2009). The increase in risk of hemorrhagic stroke reached significance when the two trial types were combined (RR 1.39, 95% CI 1.08–1.78; $P=0.01$). Fatal hemorrhagic strokes were more common in the aspirin groups (RR 1.73, 99% CI 0.96–3.13; $P=0.02$).

The US Preventive Services Task Force performed a systematic review of controlled trials, cohort studies, and meta-analyses comparing aspirin with placebo or no treatment for primary prevention of CVD (Whitlock et al., 2016). Regular use of aspirin of 100 mg or less was associated with an increase in the risk of hemorrhagic stroke (OR 1.27, 95% CI 0.96–1.68). Inclusion of studies using higher doses only slightly increased bleeding risk. Excess hemorrhagic stroke events were 0.32 per 1000 person-years of aspirin exposure over baseline bleeding rates. Older age, male sex, and those with CVD risk factors increased risk of intracranial bleeding and hemorrhagic stroke.

A systematic review of 22 randomized, controlled trials of aspirin vs placebo for primary or secondary prevention observed no significant difference in hemorrhagic stroke between aspirin 75–162 and 163–325 mg/day (McQuaid and Laine, 2006). The absolute annual increase for intracranial bleeding attributable to aspirin in this review was 0.03% (95% CI 0.01–0.08). The risk of subdural and extradural hemorrhage may be increased as is the risk of intracerebral and subarachnoid hemorrhage (Huang et al., 2019).

Overall, it appears that aspirin at low and high doses, in both primary and secondary prevention, increases the small risk of intracranial hemorrhage (ICH) by approximately 30%. Aspirin is very well tolerated and beyond dyspepsia and rare bleeding events, causes few side effects. Neurologically, aspirin is associated with Reye's syndrome, now rare (Belay et al., 1999). High daily dosage is well recognized to cause tinnitus and hearing loss (Sheppard et al., 2014). Aspirin toxicity, acute, or chronic (salicylism) can as confusion, agitation, hallucinations, and seizures (Anderson et al., 1976). Cases of aspirin-associated rhabdomyolysis have been reported (Leventhal et al., 1989; Montgomery et al., 1994).

$P2Y_{12}$ RECEPTOR INHIBITORS

The thienopyridine drugs clopidogrel, prasugrel, ticagrelor, and cangrelor impair platelet-mediated thrombosis by inhibition of the $P2Y_{12}$ ADP receptor. $P2Y_{12}$ inhibitors have been shown to be more effective than aspirin in reducing the incidence of myocardial infarction, ischemic stroke, and cardiac death following the acute coronary syndromes of unstable angina and acute myocardial infarction. In patients who undergo percutaneous coronary intervention (PCI), the drugs reduce the incidence of stent thrombosis. Bleeding, however, is also associated with their use, particularly when combined with aspirin. Clinical use of these drugs requires a balancing of the reduced risk of cardiac and brain ischemic events and the increased risk of serious bleeding.

Clopidogrel

Clopidogrel received approval in the United States 1997 for secondary stroke prevention. Current indications include acute coronary syndromes, in conjunction with aspirin, for reduction of acute myocardial infarction and ischemic stroke and to reduce long-term risk of myocardial infarction and ischemic stroke in patients with a history of myocardial infarction, ischemic stroke, or peripheral arterial disease. Clopidogrel is a thienopyridine prodrug biotransformed by the cytochrome P450 system, principally CYP2C19, to an active metabolite that irreversibly blocks the $P2Y_{12}$ component of platelet ADP receptors and prevents activation of the GPIIb/IIIa receptor complex and platelet aggregation. Diminished effect is seen in patients with loss-of-function variants of CYP2C19 and possibly in patients taking drugs such as omeprazole and other proton pump inhibitors (Sherwood et al., 2015). In general, the drug is well tolerated with low but significant bleeding risk.

The CAPRIE (Clopidogrel vs Aspirin in Patients at Risk for Ischemic Events) trial of 1996 assessed the safety and efficacy of clopidogrel (75 mg daily) and aspirin (325 mg daily) in patients with recent MI, recent ischemic stroke, or established peripheral arterial disease (CAPRIE Steering Committee, 1996). Over a follow up of 1.9 years, clopidogrel was more effective than aspirin (RRR 8.7%; $P = 0.04$) in preventing the composite outcome of ischemic stroke, myocardial infarction, or vascular death. The overall incidence of hemorrhagic events of any kind was similar in the clopidogrel and aspirin groups (9.27% vs 9.28%) (Harker et al., 1999). Any gastrointestinal hemorrhage occurred less frequently with clopidogrel than with aspirin (1.99% vs 2.66%; $P < 0.002$). ICH tended to occur less frequently with clopidogrel (0.31% vs 0.42%).

Subsequent clinical trials have investigated clopidogrel and aspirin as dual antiplatelet therapy in patients with acute coronary syndromes (Yusuf et al., 2001), recent ischemic stroke or TIA (Diener et al., 2004; Johnston et al., 2018), CVD or vascular risk factors (Bhatt et al., 2006), atrial fibrillation (ACTIVE Investigators et al., 2009), post-PCI (Watanabe et al., 2019), and other patient populations. Triple therapy using aspirin, clopidogrel, and dipyridamole in patients with ischemic stroke has been investigated (Bath et al., 2018). Across studies, dual antiplatelet therapy with aspirin and clopidogrel reduces ischemic cardiovascular events but increases major bleeding, particularly gastrointestinal. A meta-analysis calculates the hazard ratio for major bleeding of clopidogrel vs aspirin to be 0.85 (95% CI 0.75–1.96) and for clopidogrel plus aspirin vs aspirin to be 1.73 (95% CI 1.38–2.18). The incidence of ICH is similar with either single or dual antiplatelet therapy. Risk of major hemorrhage is influenced by age, blood pressure, ethnicity, and medical factors. Bleeding scores have been developed to assess risk in patients receiving antiplatelet therapy (Hilkens et al., 2017).

Nonbleeding side effects of clopidogrel include dizziness, diarrhea, and rash. One patient taking clopidogrel in the CHARISMA (Clopidogrel for High Atherothrombotic Risk and Ischemic Stabilization, Management, and Avoidance) trial developed thrombotic thrombocytopenic purpura and others have been reported (Bennett et al., 2000; Azarm et al., 2011). Rare thrombocytopenia and neutropenia have been seen in clinical trials at rates similar to aspirin (CAPRIE Steering Committee, 1996; Harker et al., 1999; Yusuf et al., 2001). Multiple case reports have associated clopidogrel with disorders of taste and smell (Golka et al., 2000; Cave et al., 2008; Ksouda et al., 2011). Other case reports have described hallucinations (Founztopoulous et al., 2007; Osuagwu et al., 2016), polyarthritis (Williams and Maloof 3rd, 2014; Ayesha et al., 2019), and drug-induced hypersensitivity reaction with rhabdomyolysis (Ibrahim and Nunley, 2017; Ayesha et al., 2019).

Prasugrel

Prasugrel is an oral thienopyridine prodrug, similar to clopidogrel, which irreversibly blocks $P2Y_{12}$ platelet ADP receptors. Onset of action is more rapid and the antiplatelet effect stronger than clopidogrel. It is approved for use in patients with acute coronary syndrome for whom PCI is planned following coronary angiography. The TRITON-TIMI (Trial to Assess Improvement in Therapeutic Outcomes by Optimizing Platelet Inhibition with Prasugrel–Thrombolysis in Myocardial Infarction) 38 trial demonstrated an increased risk of major bleeding in patients receiving prasugrel over patients receiving clopidogrel (2.4% vs 1.8%, HR 1.32; $P = 0.03$) over 15 months of follow-up (Wiviott et al., 2007). No significant difference was found in intracranial bleeding between patients receiving prasugrel and patients receiving clopidogrel (0.28% vs 0.25%). Post hoc analyses found that patients who were older, had a lower body weight, or had a history of cerebrovascular events had a higher rate of all bleeding and therefore no clinical benefit from prasugrel (10 mg daily) over clopidogrel (75 mg daily). Both prasugrel and clopidogrel increased the risk of intracranial bleeding in patients with a history of ischemic stroke or transient ischemic event. The TRILOGY-ACS (Targeted Platelet Inhibition to Clarify the Optimal Strategy to Medically Manage Acute Coronary Syndromes) trial evaluated prasugrel (10 mg daily) vs clopidogrel (75 mg daily) in patients with unstable angina or myocardial infarction without ST-segment elevation for whom revascularization was not planned. Prasugrel and clopidogrel groups were similar in efficacy and adverse events including major and intracranial bleeding with no age-related increase in bleeding (Roe et al., 2012). Analysis of data from the US Food and Drug Administration Adverse Event Reporting System found more reports of intracranial bleeding for prasugrel and ticagrelor than for clopidogrel (Fahmy et al., 2019). Nonhemorrhagic adverse events are infrequent headache, dizziness, and fatigue.

Ticagrelor

Ticagrelor is an oral $P2Y_{12}$ inhibitor that does not require metabolic activation and binds rapidly and reversibly to the $P2Y_{12}$ receptor. It was approved in the United States in 2011 for use in patients with acute coronary syndrome and for secondary prevention of myocardial infarction. In the PLATO (Platelet Inhibition and Patient Outcomes) trial, patients with acute coronary syndrome were randomized to ticagrelor or clopidogrel and followed for 12 months (Wallentin et al., 2009). ICH occurred more frequently with ticagrelor than with clopidogrel (0.28% vs 0.15%; HR 1.87; $P = 0.06$). In the PEGASUS

TIMI-54 (Prevention of Cardiovascular Events in Patients with Prior Heart Attack Using Ticagrelor Compared to Placebo on a Background of Aspirin–Thrombolysis in Myocardial Infarction 54) trial, patients with a history of myocardial infarction or risk factors were randomized to ticagrelor 60 mg, ticagrelor 90 mg, or placebo and followed for 3 years (Bonaca et al., 2015). Rates of ICH were similar among the groups, 0.61%, 0.56%, and 0.47%, respectively. The ISAR-REACT (Intracoronary Stenting and Antithrombotic Regimen: Rapid Early Action for Coronary Treatment) five trial randomized patients with acute coronary syndromes to ticagrelor or prasugrel and followed them for 1 year (Schüpke et al., 2019). No significant differences were found in Barc 3c hemorrhages (intracranial and intraocular). Bradycardia with symptoms of dizziness or syncope may occur more frequently with ticagrelor than with other $P2Y_{12}$ inhibitors (Low et al., 2018).

Cangrelor

Cangrelor is an intravenous $P2Y_{12}$ inhibitor rapid in onset and short in effect time approved in the United States in 2015 for use in PCI to prevent periprocedural myocardial infarction, and stent thrombosis in patients who have not been treated with another $P2Y_{12}$ inhibitor and are not being given a glycoprotein IIb/IIIa inhibitor. Off label, it has been used as a bridging therapy prior to cardiac surgery. In the CHAMPION (Cangrelor versus Standard Therapy to Achieve Optimal Management of Platelet Inhibition) trials (Bhatt et al., 2009, 2013; Harrington et al., 2009), cangrelor was compared with loading doses of clopidogrel at 48 h. Cangrelor reduced ischemic events of stent thrombosis, myocardial infarction, revascularization, and death. TIMI (thrombolysis in myocardial infarction) major bleeding was seen in 0.1% of each group. ICH was seen in 0.05% of patients receiving cangrelor and 0.02% of patients receiving clopidogrel (HR 2.47, $P=0.52$) (Vaduganathan et al., 2017).

$P2Y_{12}$ inhibitors are important drugs in management of acute coronary syndrome and in long-term prevention of myocardial infarction, ischemic stroke, and cardiovascular death. As potent antiplatelet agents, major bleeding, including intracranial bleeding, is a recognized risk, in most cases outweighed by the antithrombotic benefits. In clinical trials, the incidence of ICH is similar to aspirin. The combination of aspirin and a $P2Y_{12}$ inhibitor has a role in acute coronary syndromes, in patients with coronary stents, possibly in patients with acute ischemic stroke. The accurate assessment of the risks and benefits in these circumstances continues to evolve.

GLYCOPROTEIN IIB/IIIA INHIBITORS

Glycoprotein IIb/IIIa is a platelet transmembrane receptor that when activated causes platelets to aggregate by binding fibrinogen and von Willebrand factor. Three intravenously administered, rapid onset, glycoprotein IIb/IIIa inhibitors are in clinical use as adjunctive antiplatelet agents in selected patients with acute coronary syndromes undergoing PCI. Abciximab is an intravenously administered monoclonal antibody with action against platelet glycoprotein IIb/IIIa receptors. Eptifibatide is a cyclic peptide that blocks the fibrinogen binding site of the receptor. Tirofiban is a nonpeptide, small molecule inhibitor of glycoprotein IIb/IIIa receptors. In acute coronary syndromes, the glycoprotein IIb/IIIa inhibitors are used in conjunction with aspirin and other $P2Y_{12}$ inhibitors. All are associated with bleeding, primarily gastrointestinal.

In a randomized, placebo-controlled, double-blind trial of abciximab in acute ischemic stroke, symptomatic ICH occurred in 5.5% of patients and 0.5% of placebo controls at 5 days after enrollment (Adams Jr et al., 2008). An open-label trial of eptifibatide (Pancioli et al., 2008) and a dose-escalation and safety trial of tirofiban (Siebler et al., 2011) in acute ischemic stroke did not find excess risk of ICH. Glycoprotein IIb/IIIa inhibitors are not recommended for treatment of acute ischemic stroke (Ciccone et al., 2014; Powers et al., 2019). Glycoprotein IIb/IIIa inhibitors have been used following intracranial stent placement for prevention of thromboembolic complications after aneurysm coiling (Brinjikji et al., 2015) and mechanical thrombectomy (Delgado et al., 2019).

ANTICOAGULANT DRUGS

Vitamin K antagonists are recommended for patients with rheumatic heart disease and mechanical heart valves (Whitlock et al., 2012; Nishimura et al., 2017). The direct oral anticoagulants (DOACs) are recommended for prevention of cardiac thromboembolism in selected patients with nonvalvular atrial fibrillation (January et al., 2019). Unfractionated heparin (UFH), low-molecular-weight heparin (LMWH), and fondaparinux are used in the treatment of patients with acute coronary syndromes and during cardiac surgery (Garcia et al., 2012).

VITAMIN K ANTAGONISTS, WARFARIN

Warfarin achieves its anticoagulation effect by competitively inhibiting the activation of vitamin K and thereby preventing synthesis of vitamin K-dependent coagulation factors II, VII, IX, and X and protein C and S. Bleeding is the most important neurologic adverse event. A pooled analysis of early studies of patients with atrial fibrillation treated with warfarin anticoagulation found that the annual rate of ICH was 0.1% in controls and 0.3% in warfarin-treated patients (Atrial Fibrillation Investigators, 1994). A later meta-analysis of 29 randomized trials of antithrombotic drugs for stroke prevention in patients with atrial fibrillation found that warfarin increased the absolute risk of intracranial bleeding by 0.2%/year over aspirin (Hart et al., 2007). Warfarin served as the control in randomized clinical trials of the four available direct thrombin inhibitors for stroke prevention in atrial fibrillation. The annual rate of ICH in the warfarin control group of each of the trials was 0.76%, 0.70%, 0.80%, and 0.85% for dabigatran (Connolly et al., 2009), rivaroxaban (Patel et al., 2011), apixaban (Granger et al., 2011), and edoxaban (Giugliano et al., 2013), respectively.

Within the clinical practice setting of the Kaiser Permanente health care system, ICHs occurred in 0.46% of patients receiving warfarin for atrial fibrillation and 0.23% in patients who not receiving warfarin (Go et al., 2003). No hemorrhagic stroke occurred in a cohort of 408 patients on warfarin for atrial fibrillation followed over 19 months in an anticoagulation clinic (Abdelhafiz and Wheeldon, 2004). The risk of ICH is influenced by the intensity of anticoagulation and patient factors such as older age, arterial hypertension, and Asian race (Shen et al., 2007; Schulman et al., 2008). The addition of aspirin to warfarin may double the risk of ICH (Hart et al., 1999). Scoring systems for the risk of major bleeding have been developed and are being adapted for use in estimation of ICH risk (Lip et al., 2013).

DIRECT ORAL ANTICOAGULANTS

Dabigatran reversibly blocks the active site of thrombin and its conversion of fibrinogen to fibrin and blocks thrombin-associated platelet activations. Apixaban, edoxaban, and rivaroxaban reversibly inhibit factor Xa and its conversion of prothrombin to thrombin resulting in an anticoagulant effect. All the DOACs are indicated for use in prevention of thromboembolism in nonvalvular atrial fibrillation and prevention of deep vein thromboses and pulmonary embolism.

Randomized, controlled trials of the DOACs for prevention of stroke and systemic thromboembolism demonstrate similar efficacy but fewer bleeding complications than warfarin. A meta-analysis of trials of each of the four DOACs totaling 71,683 patients found a reduced rate of ICH (RR 0.48, 95% CI 0.39–0.59; $P<0.0001$) and of hemorrhagic stroke (RR 0.49, 0.38–0.64; $P<0.0001$) in patients receiving DOACs compared to patients receiving warfarin (Ruff et al., 2014).

Observational trials providing a more real-world estimation of bleeding risk are consistent with the clinical trials. Analysis of data from a large US insurance database of patients with AF taking warfarin, apixaban, rivaroxaban, or dabigatran found that risks of stroke or thromboembolism and risk of major bleeding were at least similar if not better than that of warfarin. All DOACs were associated with a lower risk of intracranial bleeding. Annual event rates were 0.29% for apixaban, 0.28% for dabigatran, and 0.44% for rivaroxaban. Annual event rates for hemorrhagic stroke were 0.19% for apixaban, 0.16% for dabigatran, and 0.21% for rivaroxaban (Yao et al., 2016).

Bleeding risks may be higher in patients older than 75 years (Adam et al., 2012). In a systematic review and meta-analysis of 15 studies, mortality and hematoma expansion was similar in DOAC-related and warfarin-related intracerebral hemorrhage (DiRisio et al., 2019). A randomized trial in Japan of rivaroxaban vs rivaroxaban and an antiplatelet drug, either aspirin or a $P2Y_{12}$ inhibitor, in patients with atrial fibrillation and coronary artery disease found excess major bleeding and hemorrhagic stroke in the combination group (Yasuda et al., 2019). The annual incidence of hemorrhagic stroke was 0.18% in the rivaroxaban group and 0.60% in the combination group.

HEPARIN, ENOXAPARIN, AND FONDAPARINUX

UFH is a glycosaminoglycan mixture that has a mean molecular weight of 15,000 and a mean length of 45 saccharide units (Garcia et al., 2012). LMWH, derived from UFH, has a molecular weight of 4000–5000 and a length of about 15 saccharide units. Available forms of LMWHs include bemiparin, dalterparin, danaparoid, enoxaparin, nadroparin, and tinzaparin. Fondaparinux is a synthetic pentasaccharide that has a molecular weight of 1728. Heparin binds to and potentiates antithrombin resulting in inhibition of thrombin and factor Xa. LMWH has less antithrombin activity than UFH and fondaparinux has only antifactor Xa activity. In acute coronary syndromes of unstable angina or myocardial infarction, subcutaneous administration of enoxaparin or fondaparinux is recommended in addition to aspirin and a $P2Y_{12}$ inhibitor (Amsterdam et al., 2014). Selected patients may receive intravenous bivalirudin (a direct thrombin inhibitor) or UFH.

Neurological complications associated with these drugs are primarily those of related to bleeding. The addition of heparin to standard therapy in a randomized trial of 652 patients with acute myocardial infarction found no ICH in the control group and 2 (0.6%) in the heparin group (de Bono et al., 1992). A similar trial investigating fibrinolytic agents found no ICH in patients receiving alteplase alone, whereas ICH occurred in 0.05% of patients receiving streptokinase and subcutaneous heparin and in 0.9% of patients receiving streptokinase and intravenous heparin (GUSTO Angiographic Investigators, 1993). Intravenous heparin was again associated with ICH in the GUSTO (Global Use of Strategies to Open Occluded Coronary Arteries) II trial found in 0.7% of patients treated without thrombolysis and 0.9% of those treated with thrombolysis (Gusto IIa Investigators, 1994). The TIMI trials demonstrated a higher incidence of ICH in patients whose aPTT was maintained at 60–90 s (1.7%) vs patient maintained at 55–85 s (0.9%) (Antman, 1996). In a comparison trial in patients with ST-elevation myocardial infarction, the incidence of ICH with enoxaparin was similar to that of UFH (0.8% vs 0.7%; $P < 0.14$) (Antman et al., 2006). In a trial of fondaparinux vs control (UFH or placebo), the rates of ICH were similar (0.16% for control vs 0.18% for fondaparinux; $P = 0.82$) (Yusuf et al., 2006). Despite benefits overall, administration of heparin, enoxaparin, and fondaparinux increase the risk of ICH in patients with acute myocardial infarction (Sobel, 1994).

Occurrences of spinal epidural hemorrhages have been documented in patients receiving heparin, enoxaparin, and fondaparinux who have had neuraxial anesthesia, spinal puncture, or spine surgery (Horlocker et al., 2010). Heparin is associated with thrombocytopenia. The immune-mediated form is thrombogenic. In a meta-analysis heparin-induced thrombocytopenia (HIT)-related studies, primarily postoperative, the incidence of HIT was 2.6% with UFH and 0.2% with LMWH (Martel et al., 2005). Of patients with HIT, up to half will develop thromboembolic complications, primarily venous (Warkentin and Kelton, 1996). A retrospective series of 225 patients with thromboembolic complications of HIT found most were lower extremity venous thrombosis and pulmonary embolism, but 6.0% had thrombotic stroke and 1.6% had cerebral vein thrombosis (Greinacher et al., 2005).

PARENTERAL THROMBIN INHIBITORS

Bivalirudin is a parenteral synthetic polypeptide that directly inhibits thrombin. It is an anticoagulant used in PCIs and as an alternative to heparin in patients with HIT (Hillegass and Bradford, 2016). Major bleeding may be less than heparin but with an increased risk of thrombotic events (Barria Perez et al., 2016).

Argatroban is a parenteral synthetic compound derived from l-arginine that reversibly binds to thrombin inhibiting fibrin formation and activation of platelets. Like bivalirudin, it is approved for use in PCIs and as an alternative to heparin in patients with HIT. Major and intracranial bleeding risks are similar to bivalirudin (Direct Thrombin Inhibitor Trialists' Collaborative Group, 2002; Sun et al., 2017).

FIBRINOLYTICS

Fibrinolytic therapy is recommended for ST-elevation myocardial infarction when delay to PCI is anticipated (O'Gara et al., 2013). Use of fibrinolytics in ACS is in conjunction with antiplatelet and/or anticoagulant therapies. Fibrinolytic drugs are recombinant forms of tissue-type plasminogen activator and include tenecteplase, reteplase, and alteplase. Fibrinolytics act by converting plasminogen to plasmin inducing a fibrinolytic state. All are approved for treatment of ST-elevation myocardial infarction. Alteplase is also approved for acute ischemic stroke.

The major adverse event associated with fibrinolytic drugs is bleeding. A systematic review of early trials of fibrinolytics in acute myocardial infarction of found nonsignificant differences in the rates of hemorrhagic stroke occurring in 0.94% of patients receiving alteplase, 0.77% receiving reteplase, and 0.93% receiving tenecteplase (Dundar et al., 2003). A pooled analysis of the ASSENT and ASSENT 3 trials that randomized 5917 patients with ST-elevation myocardial infarctions receiving tenecteplase and aspirin to enoxaparin or UFH found ICH to occur in 1.1% of all patients with a trend toward more hemorrhage with enoxaparin (Armstrong et al., 2006). The additional use of glycoprotein IIb/IIIa inhibitors increased the risk several fold (Jinatongthai et al., 2017).

HEART FAILURE DRUGS

Digoxin

Digoxin is a cardiac glycoside that reversibly inhibits Na^+/K^+-ATPase increasing cardiac contractility and decreasing conduction across the atrioventricular node. A reflexive reduction in sympathetic tone and peripheral vascular resistance is seen in patients with heart failure. Associated increased parasympathetic tone can contribute to bradycardia. Digoxin is commonly used for heart failure with reduced ejection fraction and for rate control in atrial fibrillation.

Neurologic complications are well described, particularly with elevated blood levels, and usually mild and reversible. Headache, dizziness, and confusion are most common. Perception of color can be affected; often a yellowing of the visual field (xanthopsia) but other colorations are described (Renard et al., 2015). Patients have reported impaired acuity, a snowy, blizzard affect, light sensitivity, and visual hallucinations (Renard et al., 2015). Chorea has been described as associated with digoxin toxicity (Mulder et al., 1988) as have epileptic convulsions (Kerr et al., 1982). A manufacturing error resulted in intoxication in 179 patients; the majority of whom had neurologic symptoms of vision problems, weakness, dizziness, and psychic complaints (Lely and van Enter, 1970).

Sacubitril/valsartan

Neprilysin acts by decreasing breakdown of vasoactive peptides such as natriuretic peptide with associated improvement in cardiac function and heart failure symptoms. A combination of valsartan, an ARB, and sacubitril, a neprilysin inhibitor, is approved for treatment of heart failure with reduced ejection fraction. The combination drug is generally well tolerated with hypotension and hyperkalemia as the most common serious adverse events (McMurray et al., 2014). Neurologic adverse events include dizziness (6.3%), headache (2.4%), and asthenia (2.1%). No increase in cognition-related events was found on review of published trials (Cannon et al., 2017).

Ivabradine

Ivabradine is drug that reduces sinoatrial node automaticity and heart rate. It is approved for the treatment of heart failure. A trial in patients with heart failure followed patients on ivabradine or placebo for a mean of 2.9 years and found the most common neurologic adverse event to be bradycardia-associated dizziness and fatigue (4.6% vs 1.0%; $P < 0.0001$) (Swedberg et al., 2010). Phosphenes were reported more often by patients receiving ivabradine than placebo (2.8% vs 0.5%; $P < 0.0001$).

ANTIANGINAL DRUGS

Organic nitrates

Nitroglycerin and isosorbide mononitrate and dinitrate form nitric oxide resulting relaxation of vascular smooth muscle, both venous and arterial, and reduction of myocardial oxygen demand and angina. Coronary artery dilatation contributes to a lesser degree.

Hypotension, flushing, and headaches are the common adverse events associated with nitrates. A dose-dependent, generalized headache commonly occurs on infusion or after dermal or oral administration, likely related to nitric oxide induced vasodilatation of cerebral vasculature (Tfelt-Hansen and Tfelt-Hansen, 2009; Bagdy et al., 2010). Within several hours, a second headache may occur, more migrainous in character and more commonly in patients with a history of migraine headaches. Cluster and tension headache may be seen as well (Tfelt-Hansen and Tfelt-Hansen, 2009). Patients with extensive coronary artery disease may experience fewer and less severe headaches than patients with normal or mild coronary artery disease (Erkan et al., 2015). Headaches frequency declines as treatment is continued.

Ranolazine

Ranolazine is approved in the United States and the European Union for treatment of chronic stable angina. Through action on myocyte sodium and potassium ion channels and modulation of fatty acid metabolism, ranolazine reduces cardiac contractility and oxygen consumption (Rayner-Hartley and Sedlak, 2016). Ranolazine is metabolized by the CYP3A4 system and inducers such as carbamazepine and phenytoin may cause reduction in plasma concentrations.

In an open-label safety and tolerability study of 746 patients, receiving ranolazine followed over 2.8 years patients reported neurologic symptoms of dizziness (11.8%), fatigue (7.0%), headache (5.5%), asthenia (4.4%), and vertigo (4.3%) (Koren et al., 2007). A placebo-controlled trial of ranolazine in post-PCI patients found similar adverse neurologic events of dizziness (18.5% vs 9.3%, $P < 0.001$), asthenia (4.2% vs 2.3%, $P = 0.003$), syncope (3.5% vs 1.9%, $P = 0.02$), and vertigo (2.9% vs 1.4%, $P = 0.01$) (Weisz et al., 2016). Case reports describe myalgias, weakness, and CK elevation after administration of ranolazine in patients also receiving statins (Hylton and Ezekiel, 2010; Correa and Landau, 2013; Kassardjian et al., 2014) and CK elevation in a patient not receiving statins (Dein et al., 2018). Elderly patients have experienced ataxia, tremor, and hallucinations after initiation of ranolazine (Southard et al., 2013; Mishra et al., 2014). Generalized myoclonus has been reported (Porhomayon et al., 2013).

FUTURE DIRECTIONS

Available antihypertensive and cholesterol-lowering drugs are effective with good safety profiles. Prevention of coronary and cerebrovascular disease may be more a matter of access to treatment and adherence to a medication regimen than a question of new and better drugs. Surgical procedures and implantable devices may reduce still further the need for pharmacologic arrhythmia control and its associated neurologic complications. Potent antiplatelet and anticoagulant drugs are moderately effective in prevention of thromboembolic events but come with a thus far unavoidable risk of major bleeding. Better tools for assessment of risk and benefit and treatment recommendations tailored to individual patients may be means to minimize hemorrhagic complications.

References

Abarbanel JM, Osiman A, Frisher S et al. (1987). Peripheral neuropathy and cerebellar syndrome associated with amiodarone therapy. Isr J Med Sci 23: 893–895.

Abdelhafiz AH, Wheeldon NM (2004). Results of an open-label, prospective study of anticoagulant therapy for atrial fibrillation in an outpatient anticoagulation clinic. Clin Ther 26: 1470–1478.

ACTIVE Investigators, Connolly SJ, Pogue J et al. (2009). Effect of clopidogrel added to aspirin in patients with atrial fibrillation. N Engl J Med 360: 2066–2078.

Adam SS, McDuffie JR, Ortel TL et al. (2012). Comparative effectiveness of warfarin and new oral anticoagulants for the management of atrial fibrillation and venous thromboembolism: a systematic review. Ann Intern Med 157: 796–807.

Adams Jr HP, Effron MB, Torner J et al. (2008). Emergency administration of abciximab for treatment of patients with acute ischemic stroke: results of an international phase III trial: abciximab in Emergency Treatment of Stroke Trial (AbESTT-II). Stroke 39: 87–99.

Agius MA, Zhu S, Fairclough RH (1998). Antirapsyn antibodies in chronic procainamide-associated myopathy (CPAM). Ann N Y Acad Sci 841: 527–529.

Ahmad S (1991). Metoprolol-induced delirium perpetuated by propafenone. Am Fam Physician 44: 1142.

Ahmad S, Sheikh AI, Meeran MK (1979). Disopyramide-induced acute psychosis. Chest 76: 712.

Aho K, Haapa K (1982). Facial atrophy during sotalol treatment. J Neurol Neurosurg Psychiatry 45: 179.

Al-Khatib SM, Stevenson WG, Ackerman MJ et al. (2018). 2017 AHA/ACC/HRS guideline for management of patients with ventricular arrhythmias and the prevention of sudden cardiac death. Circulation 138: e210–e271.

Allen MJ, Nichols DJ, Oliver SD (2000). The pharmacokinetics and pharmacodynamics of oral dofetilide after twice daily and three times daily dosing. Br J Clin Pharmacol 50: 247–253.

Amarenco P, Bogousslavsky J, Callahan 3rd A et al. (2006). High-dose atorvastatin after stroke or transient ischemic attack [published correction appears in N Engl J Med. 2018; 378:2450]. N Engl J Med 355: 549–559.

Ambrose A, Salib E (1999). Amiodarone-induced depression. Br J Psychiatry 174: 366–367.

Amsterdam EA, Wenger NK, Brindis RG et al. (2014). 2014 AHA/ACC guideline for the management of patients with non-ST-elevation acute coronary syndromes: a report of the American College of Cardiology/American Heart Association Task Force on Practice Guidelines [published correction appears in Circulation. 2014 Dec 23;130(25): e433-4. Dosage error in article text]. Circulation 130: e344–e426.

Anderson RJ, Potts DE, Gabow PA et al. (1976). Unrecognized adult salicylate intoxication. Ann Intern Med 85: 745–748.

Anderson JL, Muhlestein JB, Bair TL et al. (2005). Do statins increase the risk of idiopathic polyneuropathy? Am J Cardiol 95: 1097–1099.

Ando K, Wallace MS, Braun J et al. (2000). Effect of oral mexiletine on capsaicin-induced allodynia and hyperalgesia: a double-blind, placebo-controlled, crossover study. Reg Anesth Pain Med 25: 468–474.

GUSTO Angiographic Investigators (1993). The effects of tissue plasminogen activator, streptokinase, or both on coronary-artery patency, ventricular function, and survival after acute myocardial infarction [published correction appears in N Engl J Med 1994 Feb 17;330(7):516]. N Engl J Med 329: 1615–1622.

Antman EM (1996). Hirudin in acute myocardial infarction. Thrombolysis and thrombin inhibition in myocardial infarction (TIMI) 9B trial. Circulation 94: 911–921.

Antman EM, Morrow DA, McCabe CH et al. (2006). Enoxaparin versus unfractionated heparin with fibrinolysis for ST-elevation myocardial infarction. N Engl J Med 354: 1477–1488.

Arai H, Yamashita S, Yokote K et al. (2018). Efficacy and safety of pemafibrate versus fenofibrate in patients with high triglyceride and low HDL cholesterol levels: a multicenter, placebo-controlled, double-blind, randomized trial. J Atheroscler Thromb 25: 521–538.

Armstrong PW, Chang WC, Wallentin L et al. (2006). Efficacy and safety of unfractionated heparin versus enoxaparin: a pooled analysis of ASSENT-3 and -3 PLUS data [published correction appears in CMAJ. 2006 Jun 20;174(13):1874]. CMAJ 174: 1421–1426.

Athwal H, Murphy Jr G, Chun S (2003). Amiodarone-induced delirium. Am J Geriatr Psychiatry 11: 696–697.

Atrial Fibrillation Investigators (1994). Risk factors for stroke and efficacy of antithrombotic therapy in atrial fibrillation: analysis of pooled data from five randomized controlled trials. Arch Intern Med 154: 1449–1457.

Ayesha B, Varghese J, Stafford H (2019). Clopidogrel-associated migratory inflammatory polyarthritis. Am J Case Rep 20: 489–492.

Azarm T, Sohrabi A, Mohajer H et al. (2011). Thrombotic thrombocytopenic purpura associated with clopidogrel: a case report and review of the literature. J Res Med Sci 16: 353–357.

Bacon P, Spalton DJ, Smith SE (1988). Blindness from quinine toxicity. Br J Ophthalmol 72: 219–224.

Bagdy G, Riba P, Kecskeméti V et al. (2010). Headache-type adverse effects of NO donors: vasodilation and beyond. Br J Pharmacol 160: 20–35.

Baigent C, Blackwell L, Collins R et al. (2009). Antithrombotic trialists' collaboration. Aspirin in the primary and secondary prevention of vascular disease: collaborative meta-analysis of individual participant data from randomised trials. Lancet 373: 1849–1860.

Baigent C, Blackwell L, Emberson J et al. (2010). Efficacy and safety of more intensive lowering of LDL cholesterol: a meta-analysis of data from 170,000 participants in 26 randomised trials. Lancet 76: 1670–1681.

Barria Perez AE, Rao SV, Jolly SJ et al. (2016). Meta-analysis of effects of bivalirudin versus heparin on myocardial ischemic and bleeding outcomes after percutaneous coronary intervention. Am J Cardiol 117: 1256–1266.

Barrocas AM, Cymet T (2007). Cinchonism in a patient taking quinine for leg cramps. Compr Ther 33: 162–163.
Barvaliya MJ, Naik VN, Shah AC et al. (2015). Safety of olmesartan in a patient with telmisartan-induced myotoxicity: a case report. Br J Clin Pharmacol 79: 1034–1036.
Bateman DN, Dyson EH (1986). Quinine toxicity. Adverse Drug React Acute Poisoning Rev 5: 215–233.
Bath PM, Woodhouse LJ, Appleton JP et al. (2018). Antiplatelet therapy with aspirin, clopidogrel, and dipyridamole versus clopidogrel alone or aspirin and dipyridamole in patients with acute cerebral ischaemia (TARDIS): a randomised, open-label, phase 3 superiority trial. Lancet 391: 850–859.
Belay ED, Bresee JS, Holman RC et al. (1999). Reye's syndrome in the United States from 1981 through 1997. N Engl J Med 340: 1377–1382.
Bennett MI (1997). Paranoid psychosis due to flecainide toxicity in malignant neuropathic pain. Pain 70: 93–94.
Bennett CL, Connors JM, Carwile JM et al. (2000). Thrombotic thrombocytopenic purpura associated with clopidogrel. N Engl J Med 342: 1773–1777.
Bhatt DL, Fox KA, Hacke W et al. (2006). Clopidogrel and aspirin versus aspirin alone for the prevention of atherothrombotic events. N Engl J Med 354: 1706–1717.
Bhatt DL, Lincoff AM, Gibson CM et al. (2009). Intravenous platelet blockade with cangrelor during PCI. N Engl J Med 361: 2330–2341.
Bhatt DL, Stone GW, Mahaffey KW et al. (2013). Effect of platelet inhibition with cangrelor during PCI on ischemic events. N Engl J Med 368: 1303–1313.
Billig N, Buongiorno P (1985). Quinidine-induced organic mental disorders. J Am Geriatr Soc 33: 504–506.
Binder EF, Cayabyab L, Ritchie DJ et al. (1991). Diltiazem-induced psychosis and a possible diltiazem-lithium interaction. Arch Intern Med 151: 373–374.
Bizjak ED, Nolan Jr PE, Brody EA et al. (1999). Procainamide-induced psychosis: a case report and review of the literature. Ann Pharmacother 33: 948–951.
Bonaca MP, Bhatt DL, Cohen M et al. (2015). Long-term use of ticagrelor in patients with prior myocardial infarction. N Engl J Med 372: 1791–1800.
Bondon-Guitton E, Perez-Lloret S, Bagheri H et al. (2011). Drug-induced parkinsonism, bradykinesia: a review of 17 years' experience in a regional pharmacovigilance center in France. Mov Disord 26: 2226–2231.
Briani C, Zara G, Negrin P (2002). Disopyramide-induced neuropathy. Neurology 58: 663.
Brinjikji W, Morales-Valero SF, Murad MH et al. (2015). Rescue treatment of thromboembolic complications during endovascular treatment of cerebral aneurysms: a meta-analysis. Am J Neuroradiol 36: 121–125.
Brooks BR, Thisted RA, Appel SH et al. (2004). Treatment of pseudobulbar affect in ALS with dextromethorphan/quinidine: a randomized trial. Neurology 63: 1364–1370.
Busche CJ (1988). Organic psychosis caused by diltiazem. J R Soc Med 81: 296–297.
Cairns JA, Connolly SJ, Gent M et al. (1991). Post-myocardial infarction mortality in patients with ventricular premature depolarizations. Canadian Amiodarone Myocardial Infarction Arrhythmia Trial Pilot Study. Circulation 84: 550–557.
Cannon CP, Blazing MA, Giugliano RP et al. (2015). Ezetimibe added to statin therapy after acute coronary syndromes. N Engl J Med 372: 2387–2397.
Cannon JA, Shen L, Jhund PS et al. (2017). Dementia-related adverse events in PARADIGM-HF and other trials in heart failure with reduced ejection fraction. Eur J Heart Fail 19: 129–137.
CAPRIE Steering Committee (1996). A randomised, blinded, trial of clopidogrel versus aspirin in patients at risk of ischaemic events (CAPRIE). CAPRIE Steering Committee. Lancet 348: 1329–1339.
Carruthers SG (1994). Adverse effects of alpha 1-adrenergic blocking drugs. Drug Saf 11: 12–20.
Castells X, Rodoreda I, Pedrós C et al. (2002). Drug points: dysgeusia and burning mouth syndrome by eprosartan. BMJ 325: 1277.
Cave AJ, Cox DW, Vicaruddin O (2008). Loss of taste with clopidogrel. Can Fam Physician 54: 195–196.
Charness ME, Morady F, Scheinman MM (1984). Frequent neurologic toxicity associated with amiodarone therapy. Neurology 34: 669–671.
Chaubey VK, Chhabra L, Kapila A (2013). Ataxia: a diagnostic perplexity and management dilemma. BMJ Case Rep 2013: bcr2013200575.
Cheng HC, Yeh HJ, Huang N et al. (2015). Amiodarone-associated optic neuropathy: a nationwide study. Ophthalmology 122: 2553–2559.
Chin DK, Ho AK, Tse CY (1986). Neuropsychiatric complications related to use of prazosin in patients with renal failure. Br Med J (Clin Res Ed) 293: 1347.
Chua TP, Farrell T, Lipkin DP (1994). Myoclonus associated with propafenone. BMJ 308: 113.
Ciccone A, Motto C, Abraha I et al. (2014). Glycoprotein IIb-IIIa inhibitors for acute ischaemic stroke. Cochrane Database Syst Rev CD005208.
Cohen JD, Brinton EA, Ito MK et al. (2012). Understanding statin use in America and gaps in patient education (USAGE): an internet-based survey of 10,138 current and former statin users. J Clin Lipidol 6: 208–215.
Cohn JN, McInnes GT, Shepherd AM (2011). Direct-acting vasodilators. J Clin Hypertens (Greenwich) 13: 690–692.
Connolly SJ, Ezekowitz MD, Yusuf S et al. (2009). Dabigatran versus warfarin in patients with atrial fibrillation [published correction appears in N Engl J Med. 2010 Nov 4;363(19):1877]. N Engl J Med 361: 1139–1151.
Connolly SJ, Camm AJ, Halperin JL et al. (2011). Dronedarone in high-risk permanent atrial fibrillation [published correction appears in N Engl J Med 2012; 366:672]. N Engl J Med 365: 2268–2276.
Corrao G, Zambon A, Bertù L et al. (2004). Lipid lowering drugs prescription and the risk of peripheral neuropathy: an exploratory case-control study using automated databases. J Epidemiol Community Health 58: 1047–1051.
Correa D, Landau M (2013). Ranolazine-induced myopathy in a patient on chronic statin therapy. J Clin Neuromuscul Dis 14: 114–116.

Cottrell JE, Patel K, Turndorf H et al. (1978). Intracranial pressure changes induced by sodium nitroprusside in patients with intracranial mass lesions. J Neurosurg 48: 329–331.

Curtin PO, Jones WN (2007). Therapeutic rationale of combining therapy with gemfibrozil and simvastatin. J Am Pharm Assoc (2003) 47: 140–146.

Dahlöf B, Devereux RB, Kjeldsen SE et al. (2002). Cardiovascular morbidity and mortality in the losartan intervention for endpoint reduction in hypertension study (LIFE): a randomised trial against atenolol. Lancet 359: 995–1003.

Dawkins KD, Gibson J (1978). Peripheral neuropathy with disopyramide. Lancet 1: 329.

de Bono DP, Simoons ML, Tijssen J et al. (1992). Effect of early intravenous heparin on coronary patency, infarct size, and bleeding complications after alteplase thrombolysis: results of a randomised double blind European Cooperative Study Group trial. Br Heart J 67: 122–128.

De Medina A, Biasini O, Rivera A et al. (1986). Nifedipine and myoclonic dystonia. Ann Intern Med 104: 125.

Dein E, Manno R, Syed A et al. (2018). Ranolazine-induced elevation of creatinine kinase in the absence of statin usage. Cureus 10: e2832.

Deleu D, Schmedding E (1987). Acute psychosis as idiosyncratic reaction to quinidine: report of two cases. Br Med J (Clin Res Ed) 294: 1001–1002.

Delgado F, Oteros R, Jimenez-Gomez E et al. (2019). Half bolus dose of intravenous abciximab is safe and effective in the setting of acute stroke endovascular treatment. J Neurointerv Surg 11: 147–152.

Diener HC, Bogousslavsky J, Brass LM et al. (2004). Aspirin and clopidogrel compared with clopidogrel alone after recent ischaemic stroke or transient ischaemic attack in high-risk patients (MATCH): randomised, double-blind, placebo-controlled trial. Lancet 364: 331–337.

Direct Thrombin Inhibitor Trialists' Collaborative Group (2002). Direct thrombin inhibitors in acute coronary syndromes: principal results of a meta-analysis based on individual patients' data. Lancet 359: 294–302.

DiRisio AC, Harary M, Muskens IS et al. (2019). Outcomes of intraparenchymal hemorrhage after direct oral anticoagulant or vitamin K antagonist therapy: a systematic review and meta-analysis. J Clin Neurosci 62: 188–194.

Doane J, Stuits B (2013). Visual hallucinations related to angiotensin-converting enzyme inhibitor use: case reports and review. J Clin Hypertens 15: 230–233.

Dorf E, Kuntz AF, Kelsey J et al. (2006). Lidocaine-induced altered mental status and seizure after hematoma block. J Emerg Med 31: 251–253.

Drachman DA, Skom JH (1965). Procainamide—a hazard in myasthenia gravis. Arch Neurol 13: 316–320.

Drerup U (1988). Central nervous system side effects due to anti-arrhythmia therapy. Psychotic depression due to flecainide. Dtsch Med Wochenschr 113: 386–388.

DrugStats (2019). Drugs for atrial fibrillation. Med Lett Drugs Ther 61: 137–144. https://clincalc.com/DrugStats/ Last accessed December 4, 2019.

Dundar Y, Hill R, Dickson R et al. (2003). Comparative efficacy of thrombolytics in acute myocardial infarction: a systematic review. QJM 96: 103–113.

Eisenman DP, McKegney FP (1994). Delirium at therapeutic serum concentrations of digoxin and quinidine. Psychosomatics 35: 91–93.

Engelter ST, Fluri F, Buitrago-Téllez C et al. (2005). Life-threatening orolingual angioedema during thrombolysis in acute ischemic stroke. J Neurol 252: 1167–1170.

Erdem S, Freimer ML, O'Dorisio T et al. (1998). Procainamide-induced chronic inflammatory demyelinating polyradiculoneuropathy. Neurology 50: 824–825.

Erkan H, Kırış G, Korkmaz L et al. (2015). Relationship between nitrate-induced headache and coronary artery lesion complexity. Med Princ Pract 24: 560–564.

Evans MA, Golomb BA (2009). Statin-associated adverse cognitive effects: survey results from 171 patients. Pharmacotherapy 29: 800–811.

Fahmy AI, Mekkawy MA, Abou-Ali A (2019). Evaluation of adverse events involving bleeding associated with oral P2Y12 inhibitors use in the Food and Drug Administration adverse event reporting system. Int J Clin Pharmacol Ther 57: 175–181.

Falk RH, Nisbet PA, Gray TJ (1977). Mental distress in patient on disopyramide. Lancet 1: 858–859.

Fallowfield JM, Blenkinsopp J, Raza A et al. (1993). Post-marketing surveillance of lisinopril in general practice in the UK. Br J Clin Pract 47: 296–304.

FDA Drug Safety Communication (February 28, 2012). Important safety label changes to cholesterol-lowering statin drugs, Food and Drug Administration, Silver Spring, MD. https://www.fda.gov/drugs/drug-safety-and-availability/fda-drug-safety-communication-important-safety-label-changes-cholesterol-lowering-statin-drugs. Date last accessed December 30, 2019.

Fernández-Solá J, Ponz E, Seguí J et al. (1987). Psicosis exógena reversible durante el tratamiento oral con mexiletine [Reversible exogenous psychosis during oral treatment with mexiletine]. Med Clin (Barc) 89: 530.

Fikkers BG, Bogousslavsky J, Regli F et al. (1986). Pseudotumor cerebri with amiodarone. J Neurol Neurosurg Psychiatry 49: 606.

Fisher CM (1981). Visual disturbances associated with quinidine and quinine. Neurology 31: 1569–1571.

Fisher AA, Davis M, Jeffery I (2002). Acute delirium induced by metoprolol. Cardiovasc Drugs Ther 16: 161–165.

Flanagan EP, Harper CM, St Louis EK et al. (2012). Amiodarone-associated neuromyopathy: a report of four cases. Eur J Neurol 19: e50–e51.

Flecainide-Quinidine Research Group (1983). Flecainide versus quinidine for treatment of chronic ventricular arrhythmias. A multicenter clinical trial. Circulation 67: 1117–1123.

Fleminger R (1978). Visual hallucinations and illusions with propranolol. Br Med J 1: 1182.

Foley KT, Bugg KS (2010). Separate episodes of delirium associated with levetiracetam and amiodarone treatment in an elderly woman. Am J Geriatr Pharmacother 8: 170–174.

Founztopoulous E, Mavroudis C, Nadar SK et al. (2007). Case report: an unusual complication of clopidogrel. Int J Cardiol 115: e27–e28.

Funck-Brentano C, Kroemer HK, Lee JT et al. (1990). Propafenone. N Engl J Med 322: 518–525.

Galasso PJ, Stanton MS, Vogel H (1995). Propafenone-induced peripheral neuropathy. Mayo Clin Proc 70: 469–472.

Galatti L, Polimeni G, Salvo F et al. (2006). Short-term memory loss associated with rosuvastatin. Pharmacotherapy 26: 1190–1192.

Ganga HV, Slim HB, Thompson PD (2014). A systematic review of statin-induced muscle problems in clinical trials. Am Heart J 168: 6–15.

Garcia DA, Baglin TP, Weitz JI et al. (2012). Parenteral anticoagulants: antithrombotic therapy and prevention of thrombosis, 9th ed: American College of Chest Physicians Evidence-Based Clinical Practice Guidelines [published correction appears in Chest. 2012 May;141 (5):1369. Dosage error in article text] [published correction appears in Chest. 2013 Aug;144(2):721. Dosage error in article text]. Chest 141: e24S–e43S.

García-Albea E, Jiménez-Jiménez FJ, Ayuso-Peralta L et al. (1993). Parkinsonism unmasked by verapamil. Clin Neuropharmacol 16: 263–265.

Gentzkow GD, Sullivan JY (1984). Extracardiac adverse effects of flecainide. Am J Cardiol 53: 101B–105B.

Ghika J, Goy JJ, Naegeli C et al. (1994). Acute reversible ataxo-myoclonic encephalopathy with flecainide therapy. Schweiz Arch Neurol Psychiatr (1985) 145: 4–6.

Gilbert GJ (1978). Quinidine dementia. Am J Cardiol 41: 791.

Gil-Gouveia R, Goadsby PJ (2009). Neuropsychiatric side-effects of lidocaine: examples from the treatment of headache and a review. Cephalalgia 29: 496–508.

Giugliano RP, Ruff CT, Braunwald E et al. (2013). Edoxaban versus warfarin in patients with atrial fibrillation. N Engl J Med 369: 2093–2104.

Go AS, Hylek EM, Chang Y et al. (2003). Anticoagulation therapy for stroke prevention in atrial fibrillation: how well do randomized trials translate into clinical practice? JAMA 290: 2685–2692.

Godley PJ, Morton TA, Karboski JA et al. (1990). Procainamide-induced myasthenic crisis. Ther Drug Monit 12: 411–414.

Goldberg AI, Dunlay MC, Sweet CS (1995). Safety and tolerability of losartan compared with atenolol, felodipine and angiotensin converting enzyme inhibitors. J Hypertens Suppl 13: S77–S80.

Goldner JA (2012). Metoprolol-induced visual hallucinations: a case series. J Med Case Rep 6: 65.

Golka K, Roth E, Huber J et al. (2000). Reversible ageusia as an effect of clopidogrel treatment. Lancet 355: 465–466.

Golomb BA, Kane T, Dimsdale JE (2004). Severe irritability associated with statin cholesterol-lowering drugs. QJM 97: 229–235.

Grace AA, Camm AJ (1998). Quinidine. N Engl J Med 338: 35–45.

Graham DJ, Staffa JA, Shatin D et al. (2004). Incidence of hospitalized rhabdomyolysis in patients treated with lipid-lowering drugs. JAMA 292: 2585–2590.

Granger CB, Alexander JH, McMurray JJ et al. (2011). Apixaban versus warfarin in patients with atrial fibrillation. N Engl J Med 365: 981–992.

Greinacher A, Farner B, Kroll H et al. (2005). Clinical features of heparin-induced thrombocytopenia including risk factors for thrombosis. A retrospective analysis of 408 patients. Thromb Haemost 94: 132–135.

Grimm Jr RH (1989). Alpha 1-antagonists in the treatment of hypertension. Hypertension 13: I131–I136.

Grogan WA, Narkun DM (1987). Pseudotumor cerebri with amiodarone. J Neurol Neurosurg Psychiatry 50: 651.

Grundy SM, Vega GL, Yuan Z et al. (2006). Effectiveness and tolerability of simvastatin plus fenofibrate for combined hyperlipidemia (the SAFARI trial) [published correction appears in Am J Cardiol. 2006 Aug 1;98(3):427-8]. Am J Cardiol 95: 462–468.

Guo S, Ashina M, Olesen J et al. (2013). The effect of sodium nitroprusside on cerebral hemodynamics and headache in healthy subjects. Cephalalgia 33: 301–307.

Hackam DG, Woodward M, Newby LK et al. (2011). Statins and intracerebral hemorrhage: collaborative systematic review and meta-analysis. Circulation 124: 2233–2242.

Ham AC, van Dijk SC, Swart KMA et al. (2017). Beta-blocker use and fall risk in older individuals: original results from two studies with meta-analysis. Br J Clin Pharmacol 83: 2292–2302.

Harker LA, Boissel JP, Pilgrim AJ et al. (1999). Comparative safety and tolerability of clopidogrel and aspirin: results from CAPRIE. CAPRIE Steering Committee and Investigators Clopidogrel versus aspirin in patients at risk of ischaemic events. Drug Saf 21: 325–335.

Harmon C, Wohlreich MM (1995). Sodium nitroprusside-induced delirium. Psychosomatics 36: 83–85.

Harrington RA, Stone GW, McNulty S et al. (2009). Platelet inhibition with cangrelor in patients undergoing PCI. N Engl J Med 361: 2318–2329.

Hart RG, Benavente O, Pearce LA (1999). Increased risk of intracranial hemorrhage when aspirin is combined with warfarin: a meta-analysis and hypothesis. Cerebrovasc Dis 9: 215–217.

Hart RG, Pearce LA, Aguilar MI (2007). Meta-analysis: antithrombotic therapy to prevent stroke in patients who have nonvalvular atrial fibrillation. Ann Intern Med 146: 857–867.

Heeringa M, van Puijenbroek EP (1998). Reversible dysgeusia attributed to losartan. Ann Intern Med 129: 72.

Hicks CB, Abraham K (1985). Verapamil and myoclonic dystonia. Ann Intern Med 103: 154.

Hilkens NA, Algra A, Diener HC et al. (2017). Predicting major bleeding in patients with noncardioembolic stroke on antiplatelets: S_2TOP-BLEED. Neurology 89: 936–943.

Hillegass WB, Bradford GS (2016). Risk guided use of the direct thrombin inhibitor bivalirudin: insights from recent trials and analyses. J Thorac Dis 8: E1034–E1040.

Hirose K, Yamaguchi H, Oshima Y et al. (2008). Severe respiratory failure and torsades de pointes induced by disopyramide in a patient with myasthenia gravis. Intern Med 47: 1703–1708.

Hohnloser SH, Meinertz T, Stubbs P et al. (1995). Efficacy and safety of d-sotalol, a pure class III antiarrhythmic compound, in patients with symptomatic complex ventricular ectopy. Results of a multicenter, randomized, double-blind, placebo-controlled dose-finding study. The d-Sotalol PVC Study Group. Circulation 92: 1517–1525.

Hohnloser SH, Crijns H, Eickels M et al. (2009). Effect of dronedarone on cardiovascular events in atrial fibrillation. N Engl J Med 360: 668–678.

Holbrook A, Wright M, Sung M et al. (2011). Statin-associated rhabdomyolysis: is there a dose-response relationship? Can J Cardiol 27: 146–151.

Horlocker TT, Wedel DJ, Rowlingson JC et al. (2010). Regional anesthesia in the patient receiving antithrombotic or thrombolytic therapy: American Society of Regional Anesthesia and Pain Medicine Evidence-Based Guidelines (third edition). Reg Anesth Pain Med 35: 64–101.

Huang W-Y, Saver J, Wu YL et al. (2019). Frequency of intracranial hemorrhage with low-dose aspirin in individuals without symptomatic cardiovascular disease: a systematic review and meta-analysis. JAMA Neurol 76: 906–914.

Huckell VF, Bélanger LG, Kazimirski M et al. (1993). Lisinopril in the treatment of hypertension: a Canadian postmarketing surveillance study. Clin Ther 15: 407–422.

Huffman JC, Stern TA (2007). Neuropsychiatric consequences of cardiovascular medications. Dialogues Clin Neurosci 9: 29–45.

Hylton AC, Ezekiel TO (2010). Rhabdomyolysis in a patient receiving ranolazine and simvastatin. Am J Health Syst Pharm 67: 1829–1831.

Ibrahim M, Nunley DL (2017). Two catastrophes in one patient: drug reaction with eosinophilia and systemic symptoms and toxic shock syndrome. Cureus 9: e1359.

GUSTO IIa Investigators (1994). Randomized trial of intravenous heparin versus recombinant hirudin for acute coronary syndromes. The Global Use of Strategies to Open Occluded Coronary Arteries (GUSTO) IIa Investigators. Circulation 90: 1631–1637.

Inomata N (2012). Recent advances in drug-induced angioedema. Allergol Int 61: 545–557.

Ishida S, Sugino M, Hosokawa T et al. (2010). Amiodarone-induced liver cirrhosis and parkinsonism: a case report. Clin Neuropathol 29: 84–88.

Jack RA (1985). A case of mania secondary to propafenone. J Clin Psychiatry 46: 104–105.

Jacobs MB (1994). HMG-CoA reductase inhibitor therapy and peripheral neuropathy. Ann Intern Med 120: 970.

Jacobsen FM, Sack DA, James SP (1987). Delirium induced by verapamil. Am J Psychiatry 144: 248.

January CT, Wann LS, Alpert JS et al. (2014). 2014 AHA/ACC/HRS guideline for the management of patients with atrial fibrillation: a report of the American College of Cardiology/American Heart Association Task Force on practice guidelines and the Heart Rhythm Society [published correction appears in Circulation. 2014 Dec 2;130 (23):e272-4]. Circulation 130: e199–e267.

January CT, Wann LS, Calkins H et al. (2019). 2019 AHA/ACC/HRS Focused Update of the 2014 AHA/ACC/HRS guideline for the management of patients with atrial fibrillation: a report of the American College of Cardiology/American Heart Association Task Force on Clinical Practice Guidelines and the Heart Rhythm Society in Collaboration With the Society of Thoracic Surgeons. Circulation 140: e125–e151.

Jarvis B, Coukell AJ (1998). Mexiletine. A review of its therapeutic use in painful diabetic neuropathy. Drugs 56: 691–707.

Jelinek MV, Lohrbauer L, Lown B (1974). Antiarrhythmic drug therapy for sporadic ventricular ectopic arrhythmias. Circulation 49: 659–666.

Jinatongthai P, Kongwatcharapong J, Foo CY et al. (2017). Comparative efficacy and safety of reperfusion therapy with fibrinolytic agents in patients with ST-segment elevation myocardial infarction: a systematic review and network meta-analysis. Lancet 390: 747–759.

Johnston SC, Easton JD, Farrant M et al. (2018). Clopidogrel and aspirin in acute ischemic stroke and high-risk TIA. N Engl J Med 379: 215–225.

Jones RJ, Brace SR, Vander Tuin EL (1995). Probable propafenone-induced transient global amnesia. Ann Pharmacother 29: 586–590.

Kang HM, Kang YS, Kim SH et al. (2007). Amiodarone-induced hepatitis and polyneuropathy. Korean J Intern Med 22: 225–229.

Kassardjian CD, Tian X, Vladutiu G et al. (2014). Myopathy during treatment with the antianginal drug ranolazine. J Neurol Sci 347: 380–382.

Keech A, Simes RJ, Barter P et al. (2005). Effects of long-term fenofibrate therapy on cardiovascular events in 9795 people with type 2 diabetes mellitus (the FIELD study): randomised controlled trial [published correction appears in Lancet. 2006 Oct 21;368(9545):1420] [published correction appears in Lancet. 2006 Oct 21;368(9545):1415]. Lancet 366: 1849–1861.

Kennerdy A, Thomas P, Sheridan DJ (1989). Generalized seizures as the presentation of flecainide toxicity. Eur Heart J 10: 950–954.

Kerr DJ, Elliott HL, Hillis WS (1982). Epileptiform seizures and electroencephalographic abnormalities as manifestations of digoxin toxicity. Br Med J (Clin Res Ed) 284: 162–163.

Khouri C, Jouve T, Blaise S et al. (2016). Peripheral vasoconstriction induced by β-adrenoceptor blockers: a systematic review and a network meta-analysis. Br J Clin Pharmacol 82: 549–560.

Kieffer I, Dellinger A, Sibille M et al. (1988). Myalgia induced by sotalol. A case. Presse Med 17: 1215.

Kim SY, Benowitz NL (1990). Poisoning due to class IA antiarrhythmic drugs. Quinidine, procainamide and disopyramide. Drug Saf 5: 393–420.

Ko DT, Hebert PR, Coffey CS et al. (2002). Beta-blocker therapy and symptoms of depression, fatigue, and sexual dysfunction. JAMA 288: 351–357.

Koren MJ, Crager MR, Sweeney M (2007). Long-term safety of a novel antianginal agent in patients with severe chronic stable angina: the ranolazine open label experience (ROLE). J Am Coll Cardiol 49: 1027–1034.

Kostis JB, Kim HJ, Rusnak J et al. (2005). Incidence and characteristics of angioedema associated with enalapril. Arch Intern Med 165: 1637–1642.

Kristin N, Ulbig M (2001). Acute papilledema. A 69-year-old patient with acute bilateral papilledema. Ophthalmologe 98: 212–213.

Ksouda K, Affes H, Hammami B et al. (2011). Ageusia as a side effect of clopidogrel treatment. Indian J Pharmacol 43: 350–351.

Kubota K, Pearce GL, Inman WH (1995). Vasodilation-related adverse events in diltiazem and dihydropyridine calcium antagonists studied by prescription-event monitoring. Eur J Clin Pharmacol 48: 1–7.

Lader MH, Tyrer PJ (1972). Central and peripheral effects of propranolol and sotalol in normal human subjects. Br J Pharmacol 45: 557–560.

Lee YC, Lin CH, Wu RM et al. (2014). Antihypertensive agents and risk of Parkinson's disease: a nationwide cohort study. PLoS One 9: e98961.

Lely AH, van Enter CH (1970). Large-scale digitoxin intoxication. Br Med J 3: 737–740.

Lenz TL, Hilleman DE (2000). Dofetilide: a new antiarrhythmic agent approved for conversion and/or maintenance of atrial fibrillation/atrial flutter. Drugs Today (Barc) 36: 759–771.

Leventhal LJ, Kuritsky L, Ginsburg R et al. (1989). Salicylate-induced rhabdomyolysis. Am J Emerg 7: 409–410.

Lewis CA, Boheimer N, Rose P et al. (1986). Myopathy after short term administration of procainamide. Br Med J (Clin Res Ed) 292: 593–594.

Linazasoro G, Martí-Massó JF, Urtasun M et al. (1991). Paroxysmal dystonia induced by exercise secondary to flecainide. Neurologia 6: 344.

Lip GY, Lin HJ, Hsu HC et al. (2013). Comparative assessment of the HAS-BLED score with other published bleeding risk scoring schemes, for intracranial haemorrhage risk in a non-atrial fibrillation population: the Chin-Shan Community Cohort Study. Int J Cardiol 168: 1832–1836.

Low A, Leong K, Sharma A et al. (2018). Ticagrelor-associated ventricular pauses: a case report and literature review. Eur Heart J Case Rep 3: yty156.

Lustman F, Monseu G (1974). Letter: amiodarone and neurological side-effects. Lancet 1: 568.

Macaluso DC, Shults WT, Fraunfelder FT (1999). Features of amiodarone-induced optic neuropathy. Am J Ophthalmol 127: 610–612.

MacDonald R, Wilt TJ, Howe RW (2004). Doxazosin for treating lower urinary tract symptoms compatible with benign prostatic obstruction: a systematic review of efficacy and adverse effects. BJU Int 94: 1263–1270.

Maiteh M, Daoud AS (2001). Myoclonic seizure following intravenous verapamil injection: case report and review of the literature. Ann Trop Paediatr 21: 271–272.

Malesker MA, Sojka SG, Fagan NL (2005). Flecainide-induced neuropathy. Ann Pharmacother 39: 1580.

Maluli HA (2015). Dofetilide induced trigeminal neuralgia. Indian J Pharm 47: 336–337.

Manolis AS, Deering TF, Cameron J et al. (1990). Mexiletine: pharmacology and therapeutic use. Clin Cardiol 13: 349–359.

Martel N, Lee J, Wells PS (2005). Risk for heparin-induced thrombocytopenia with unfractionated and low-molecular-weight heparin thromboprophylaxis: a meta-analysis. Blood 106: 2710–2715.

Martinez-Arizala A, Sobol SM, McCarty GE et al. (1983). Amiodarone neuropathy. Neurology 33: 643–645.

Matchar DB, McCrory DC, Orlando LA et al. (2008). Systematic review: comparative effectiveness of angiotensin-converting enzyme inhibitors and angiotensin II receptor blockers for treating essential hypertension. Ann Intern Med 148: 16–29.

Mattos JP, Nicaretta DH, Rosso AL (2009). Progressive supranuclear palsy-like syndrome induced by amiodarone and flunarizine. Arq Neuropsiquiatr 67: 909–910.

McAreavey D, Robertson JI (1990). Angiotensin converting enzyme inhibitors and moderate hypertension. Drugs 40: 326–345.

McCrum ID, Guidry JR (1978). Procainamide-induced psychosis. JAMA 240: 1265–1266.

McDonagh MS, Eden KB, Peterson K (2005). Drug class review: calcium channel blockers: final report [Internet], Oregon Health & Science University, Portland, OR Available from: https://www.ncbi.nlm.nih.gov/books/NBK10474/.

McMurray JJ, Packer M, Desai AS et al. (2014). Angiotensin-neprilysin inhibition versus enalapril in heart failure. N Engl J Med 371: 993–1004.

McQuaid KR, Laine L (2006). Systematic review and meta-analysis of adverse events of low-dose aspirin and clopidogrel in randomized controlled trials. Am J Med 119: 624–638.

Lipid-lowering drugs Med Lett Drugs Ther 61: 17–24.

Miller LG, Jankovic J (1992). Persistent dystonia possibly induced by flecainide. Mov Disord 7: 62–63.

Miller CD, Oleshansky MA, Gibson KF et al. (1993). Procainamide-induced myasthenia-like weakness and dysphagia. Ther Drug Monit 15: 251–254.

Miller DR, Oliveria SA, Berlowitz DR et al. (2008). Angioedema incidence in US veterans initiating angiotensin-converting enzyme inhibitors. Hypertension 51: 1624–1630.

Mishra A, Pandya HV, Dave N et al. (2014). A rare debilitating neurological adverse effect of ranolazine due to drug interaction with clarithromycin. Indian J Pharm 46: 547–548.

Montgomery H, Porter JC, Bradley RD (1994). Salicylate intoxication causing a severe systemic inflammatory response and rhabdomyolysis. Am J Emerg Med 12: 531–532.

Moore BM, Cordina RL, McGuire MA et al. (2019). Efficacy and adverse effects of sotalol in adults with congenital heart disease. Int J Cardiol 274: 74–79.

Morady F, Scheinman MM, Desai J (1982). Disopyramide. Ann Intern Med 96: 337–343.

Mulder LJ, van der Mast RC, Meerwaldt JD (1988). Generalised chorea due to digoxin toxicity [published correction appears in Br Med J (Clin Res Ed) 1988;297:562]. Br Med J (Clin Res Ed) 296: 1262.

Muldoon MF, Barger SD, Ryan CM et al. (2000). Effects of lovastatin on cognitive function and psychological well-being. Am J Med 108: 538–546.

Muldoon MF, Ryan CM, Sereika SM et al. (2004). Randomized trial of the effects of simvastatin on cognitive functioning in hypercholesterolemic adults. Am J Med 117: 823–829.

Naccarelli GV, Wolbrette DL, Khan M et al. (2003). Old and new antiarrhythmic drugs for converting and maintaining sinus rhythm in atrial fibrillation: comparative efficacy and results of trials. Am J Cardiol 91: 15D–26D.

Nazir S, Lohani S, Tachamo N et al. (2017). Statin-associated autoimmune myopathy: a systematic review of 100 cases. J Clin Rheumatol 23: 149–154.

Newman CB, Preiss D, Tobert JA et al. (2019). Statin safety and associated adverse events: a scientific statement from the American Heart Association [published correction appears in Arterioscler Thromb Vasc Biol. 2019 May;39(5):e158]. Arterioscler Thromb Vasc Biol 39: e38–e81.

Niakan E, Bertorini TE, Acchiardo SR (1981). Procainamide-induced myasthenia-like weakness in a patient with peripheral neuropathy. Arch Neurol 38: 378–379.

Nishimura RA, Otto CM, Bonow RO et al. (2017). 2017 AHA/ACC focused update of the 2014 AHA/ACC guideline for the management of patients with valvular heart disease: a report of the American College of Cardiology/American Heart Association Task Force on Clinical Practice Guidelines. Circulation 135: e1159–e1195.

Nissen SE, Stroes E, Dent-Acosta RE et al. (2016). Efficacy and tolerability of evolocumab vs ezetimibe in patients with muscle-related statin intolerance: the GAUSS-3 randomized clinical trial. JAMA 315: 1580–1590.

Novak P, Pimentel DA, Sundar B et al. (2015). Association of statins with sensory and autonomic ganglionopathy. Front Aging Neurosci 7: 191.

Nygaard TW, Sellers TD, Cook TS et al. (1986). Adverse reactions to antiarrhythmic drugs during therapy for ventricular arrhythmias. JAMA 256: 55–57.

Odeh M, Seligmann H, Oliven A (2000). Propafenone-induced ataxia: report of three cases. Am J Med Sci 320: 151–153.

O'Gara PT, Kushner FG, Ascheim DD et al. (2013). 2013 ACCF/AHA guideline for the management of ST-elevation myocardial infarction: a report of the American College of Cardiology Foundation/American Heart Association Task Force on Practice Guidelines [published correction appears in Circulation. 2013 Dec 24;128(25):e481]. Circulation 127: e362–e425.

Ohara S, Hayashi R, Momoi HN et al. (1998). Mexiletine in the treatment of spasmodic torticollis. Mov Disord 13: 934–940.

Orr CF, Ahlskog JE (2009). Frequency, characteristics, and risk factors for amiodarone neurotoxicity. Arch Neurol 66: 865–869.

Orsi A, Sherman O, Woldeselassie Z (2001). Simvastatin-associated memory loss. Pharmacotherapy 21: 767–769.

Osuagwu FC, Parashar S, Amalraj B et al. (2016). Clopidogrel-induced auditory and visual hallucinations. Prim Care Companion CNS Disord 18. https://doi.org/10.4088/PCC.15l01894.

Padfield PL, Smith DA, Fitzsimons EJ et al. (1977). Disopyramide and acute psychosis. Lancet 1: 1152.

Palace J, Shah R, Clough C (1992). Flecainide induced peripheral neuropathy. BMJ 305: 810.

Palakurthy PR, Iyer V, Meckler RJ (1987). Unusual neurotoxicity associated with amiodarone therapy. Arch Intern Med 147: 881–884.

Pancioli AM, Broderick J, Brott T et al. (2008). The combined approach to lysis utilizing eptifibatide and rt-PA in acute ischemic stroke: the CLEAR stroke trial. Stroke 39: 3268–3276.

Parker BA, Capizzi JA, Grimaldi AS et al. (2013). Effect of statins on skeletal muscle function. Circulation 127: 96–103.

Passman RS, Bennett CL, Purpura JM et al. (2012). Amiodarone-associated optic neuropathy: a critical review. Am J Med 125: 447–453.

Patel MR, Mahaffey KW, Garg J et al. (2011). Rivaroxaban versus warfarin in nonvalvular atrial fibrillation. N Engl J Med 365: 883–891.

Pedro-Botet ML, Bonal J, Caralps A (1989). Nifedipine and myoclonic disorders. Nephron 51: 281.

Pellissier JF, Pouget J, Cros D (1984). Peripheral neuropathy induced by amiodarone chlorhydrate. A clinicopathological study. J Neurol Sci 63: 251–266.

Peppers MP (1995). Myalgia and arthralgia associated with enalapril and ramipril. Am J Health Syst Pharm 52: 203–204.

Pfeffer F, Grube M (2001). An organic psychosis due to a venlafaxine-propafenone interaction. Int J Psychiatry Med 31: 427–432.

Phan T, McLeod JG, Pollard JD et al. (1995). Peripheral neuropathy associated with simvastatin. J Neurol Neurosurg Psychiatry 58: 625–628.

Phillips PS, Haas RH, Bannykh S et al. (2002). Statin-associated myopathy with normal creatine kinase levels. Ann Intern Med 137: 581–585.

Pinargote P, Guillen D, Guarderas JC (2014). ACE inhibitors: upper respiratory symptoms. BMJ Case Rep 2014: bcr2014205462.

Porhomayon J, Zadeii G, Yarahmadi A (2013). A rare neurological complication of ranolazine. Case Rep Neurol Med 2013: 451206.

Powers WJ, Rabinstein AA, Ackerson T et al. (2019). 2018 guidelines for the early management of patients with acute ischemic stroke: 2019 update to the 2018 guidelines for the early management of acute ischemic stroke. A guideline for healthcare professionals from the American Heart Association/American Stroke Association. Stroke 50: e344–e418.

Pulipaka U, Lacomis D, Omalu B (2002). Amiodarone-induced neuromyopathy: three cases and a review of the literature. J Clin Neuromuscul Dis 3: 97–105.

Purvin V, Kawasaki A, Borruat FX (2006). Optic neuropathy in patients using amiodarone. Arch Ophthalmol 124: 696–701.

Rahimi M, Elmi M, Hassanian-Moghaddam H et al. (2018). Acute lidocaine toxicity; a case series. Emerg (Tehran) 6: e38.

Ramhamadany E, Mackenzie S, Ramsdale DR (1986). Dysarthria and visual hallucinations due to flecainide toxicity. Postgrad Med J 62: 61–62.

Rapsomaniki E, Timmis A, George J et al. (2014). Blood pressure and incidence of twelve cardiovascular diseases: lifetime risks, healthy life-years lost, and age-specific associations in 1·25 million people. Lancet 383: 1899–1911.

Rayner-Hartley E, Sedlak T (2016). Ranolazine: a contemporary review. J Am Heart Assoc 5: e003196.

Renard D, Rubli E, Voide N et al. (2015). Spectrum of digoxin-induced ocular toxicity: a case report and literature review. BMC Res Notes 8: 368.

Ritz B, Rhodes SL, Qian L et al. (2010). L-type calcium channel blockers and Parkinson disease in Denmark. Ann Neurol 67: 600–606.

Robinson JG, Farnier M, Krempf M et al. (2015). Efficacy and safety of alirocumab in reducing lipids and cardiovascular events. N Engl J Med 372: 1489–1499.

Roe MT, Armstrong PW, Fox KA et al. (2012). Prasugrel versus clopidogrel for acute coronary syndromes without revascularization. N Engl J Med 367: 1297–1309.

Rojas-Fernandez CH, Goldstein LB, Levey AI et al. (2014). The National Lipid Association's safety task force. An assessment by the Statin Cognitive Safety Task Force: 2014 update. J Clin Lipidol 8: S5–S16.

Romman A, Salama-Hanna J, Dwivedi S (2018). Mexiletine usage in a chronic pain clinic: indications, tolerability, and side effects. Pain Physician 21: e573–e579.

Ruff CT, Giugliano RP, Braunwald E et al. (2014). Comparison of the efficacy and safety of new oral anticoagulants with warfarin in patients with atrial fibrillation: a meta-analysis of randomised trials. Lancet 383: 955–962.

Rush JE, Merrill DD (1987). The safety and tolerability of lisinopril in clinical trials. J Cardiovasc Pharmacol 9: S99–S107.

Sabatine MS, Giugliano RP, Keech AC et al. (2017). Evolocumab and clinical outcomes in patients with cardiovascular disease. N Engl J Med 376: 1713–1722.

Sahenk Z, Mendell JR, Rossio JL et al. (1977). Polyradiculoneuropathy accompanying procainamide-induced lupus erythematosus: evidence for drug-induced enhanced sensitization to peripheral nerve myelin. Ann Neurol 1: 378–384.

Saravay SM, Marke J, Steinberg MD et al. (1987). "Doom anxiety" and delirium in lidocaine toxicity. Am J Psychiatry 144: 159–163.

Schmutz JL, Houet C, Trechot P et al. (1995). Sweating and beta-adrenoceptor antagonists. Dermatology 190: 86.

Schulman S, Beyth RJ, Kearon C et al. (2008). Hemorrhagic complications of anticoagulant and thrombolytic treatment: American College of Chest Physicians Evidence-Based Clinical Practice Guidelines (8th edition). Chest 133: 257S–298S.

Schüpke S, Neumann FJ, Menichelli M et al. (2019). Ticagrelor or prasugrel in patients with acute coronary syndromes. N Engl J Med 381: 1524–1534.

Schwartz PJ, Facchini M, Bonazzi O et al. (1985). Valutazione della flecainide nella terapia delle aritmie ventricolari croniche con il metodo del carico orale acuto [Evaluation of flecainide in the therapy of chronic ventricular arrhythmia using the acute oral load method]. G Ital Cardiol 15: 273–282.

Schwartz GG, Steg PG, Szarek M et al. (2018). Alirocumab and cardiovascular outcomes after acute coronary syndrome. N Engl J Med 379: 2097–2107.

Scott DB (1986). Toxic effects of local anaesthetic agents on the central nervous system. Br J Anaesth 58: 732–735.

Selva-O'Callaghan A, Alvarado-Cardenas M, Pinal-Fernández I et al. (2018). Statin-induced myalgia and myositis: an update on pathogenesis and clinical recommendations. Expert Rev Clin Immunol 14: 215–224.

Sempere AP, Duarte J, Cabezas C et al. (1995). Parkinsonism induced by amlodipine. Mov Disord 10: 115–116.

Setoguchi S, Higgins JM, Mogun H et al. (2010). Propranolol and the risk of hospitalized myopathy: translating chemical genomics findings into population-level hypotheses. Am Heart J 159: 428–433.

Shailesh F, Singla S, Sureddi R et al. (2011). A patient with mexiletine-related psychosis. Int J Gen Med 4: 765–766.

Sharifkazemi M, Ziya F (2019). Sotalol-induced generalized and ocular myasthenia gravis. Clin Case Rep 7: 1831–1832.

Shen EN (1990). Propafenone: a promising new antiarrhythmic agent. Chest 98: 434–441.

Shen AY, Yao JF, Brar SS et al. (2007). Racial/ethnic differences in the risk of intracranial hemorrhage among patients with atrial fibrillation. J Am Coll Cardiol 50: 309–315.

Shepherd J, Blauw GJ, Murphy MB et al. (2002). Pravastatin in elderly individuals at risk of vascular disease (PROSPER): a randomised controlled trial. Lancet 360: 1623–1630.

Sheppard A, Hayes SH, Chen GD et al. (2014). Review of salicylate-induced hearing loss, neurotoxicity, tinnitus and neuropathophysiology. Acta Otorhinolaryngol Ital 34: 79–93.

Sherwood MW, Melloni C, Jones WS et al. (2015). Individual proton pump inhibitors and outcomes in patients with coronary artery disease on dual antiplatelet therapy: a systematic review. J Am Heart Assoc 4: e002245.

Siddoway LA, Thompson KA, McAllister CB et al. (1987). Polymorphism of propafenone metabolism and disposition in man: clinical and pharmacokinetic consequences. Circulation 75: 785–791.

Siebler M, Hennerici MG, Schneider D et al. (2011). Safety of tirofiban in acute ischemic stroke: the SaTIS trial. Stroke 42: 2388–2392.

Singh S, Zoble RG, Yellen L et al. (2000). Efficacy and safety of oral dofetilide in converting to and maintaining sinus rhythm in patients with chronic atrial fibrillation or atrial flutter: the symptomatic atrial fibrillation investigative research on dofetilide (SAFIRE-D) study. Circulation 102: 2385–2390.

Sobel BE (1994). Intracranial bleeding, fibrinolysis, and anticoagulation. Causal connections and clinical implications. Circulation 90: 2147–2152.

Southard RA, Blum RM, Bui AH et al. (2013). Neurologic adverse effects of ranolazine in an elderly patient with renal impairment. Pharmacotherapy 33: e9–e13.

Soyka LF, Wirtz C, Spangenberg RB (1990). Clinical safety profile of sotalol in patients with arrhythmias. Am J Cardiol 65: 74A–81A discussion 82A-83A.

Strasser RH, Puig JG, Farsang C et al. (2007). A comparison of the tolerability of the direct renin inhibitor aliskiren and lisinopril in patients with severe hypertension. J Hum Hypertens 21: 780–787.

Stratton MA (1985). Drug-induced systemic lupus erythematosus. Clin Pharm 4: 657–663.

Stroes E, Robinson JG, Raal FJ et al. (2018). Consistent LDL-C response with evolocumab among patient subgroups in PROFICIO: a pooled analysis of 3146 patients from phase 3 studies. Clin Cardiol 41: 1328–1335.

Summers WK, Allen RE, Pitts Jr FN (1981). Does physostigmine reverse quinidine delirium? West J Med 135: 411–414.

Sun Z, Lan X, Li S et al. (2017). Comparisons of argatroban to lepirudin and bivalirudin in the treatment of heparin-induced thrombocytopenia: a systematic review and meta-analysis. Int J Hematol 106: 476–483.

Suraweera C, de Silva V, Hanwella R (2016). Simvastatin-induced cognitive dysfunction: two case reports. J Med Case Rep 10: 83.

Svendsen TK, Nørregaard Hansen P, García Rodríguez LA et al. (2017). Statins and polyneuropathy revisited: case-control study in Denmark, 1999-2013. Br J Clin Pharmacol 83: 2087–2095.

Swanoski MT, Chen JS, Monson MH (2011). Myoclonus associated with long-term use of diltiazem. Am J Health Syst Pharm 68: 1707–1710.

Swedberg K, Komajda M, Böhm M et al. (2010). Ivabradine and outcomes in chronic heart failure (SHIFT): a randomised placebo-controlled study [published correction appears in Lancet. 2010 Dec 11;376(9757):1988.]. Lancet 376: 875–885.

Tamargo J, Capucci A, Mabo P (2012). Safety of flecainide. Drug Saf 35: 273–289.

Taylor AA, Mangoo-Karim R, Ballard KD et al. (1999). Sustained hemodynamic effects of the selective dopamine-1 agonist, fenoldopam, during 48-hour infusions in hypertensive patients: a dose-tolerability study. J Clin Pharmacol 39: 471–479.

Teichman S (1985). The anticholinergic side effects of disopyramide and controlled-release disopyramide. Angiology 36: 767–771.

Tfelt-Hansen PC, Tfelt-Hansen J (2009). Nitroglycerin headache and nitroglycerin-induced primary headaches from 1846 and onwards: a historical overview and an update. Headache 49: 445–456.

Tierney EF, Thurman DJ, Beckles GL et al. (2013). Association of statin use with peripheral neuropathy in the U.S. population 40 years of age or older. J Diabetes 5: 207–215.

Ting SM, Lee D, Maclean D et al. (2008). Paranoid psychosis and myoclonus: flecainide toxicity in renal failure. Cardiology 111: 83–86.

Trohman RG, Castellanos D, Castellanos A et al. (1988). Amiodarone-induced delirium. Ann Intern Med 108: 68–69.

Tsujimoto G, Horai Y, Ishizaki T et al. (1981). Hydralazine-induced peripheral neuropathy seen in a Japanese slow acetylator patient. Br J Clin Pharmacol 11: 622–625.

Tuccori M, Lapi F, Testi A et al. (2008). Statin-associated psychiatric adverse events: a case/non-case evaluation of an Italian database of spontaneous adverse drug reaction reporting. Drug Saf 31: 1115–1123.

Turner WM (1982). Lidocaine and psychotic reactions. Ann Intern Med 97: 149–150.

Turner JM, Powell D, Gibson RM et al. (1977). Intracranial pressure changes in neurosurgical patients during hypotension induced with sodium nitroprusside or trimetaphan. Br J Anaesth 49: 419–425.

Unnikrishnan D, Murakonda P, Dharmarajan TS (2004). If it is not cough, it must be dysgeusia: differing adverse effects of angiotensin-converting enzyme inhibitors in the same individual. J Am Med Dir Assoc 5: 107–110.

Vadlamudi L, Wijdicks EF (2002). Multifocal myoclonus due to verapamil overdose. Neurology 58: 984.

Vaduganathan M, Harrington RA, Stone GW et al. (2017). Cangrelor with and without glycoprotein iib/iiia inhibitors in patients undergoing percutaneous coronary intervention. J Am Coll Cardiol 69: 176–185.

Vaglio A, Grayson PC, Fenaroli P et al. (2018). Drug-induced lupus: traditional and new concepts. Autoimmun Rev 17: 912–918.

Van Zandijcke M, Dewachter A (1986). Pseudotumor cerebri with amiodarone. J Neurol Neurosurg Psychiatry 49: 1463–1464.

Venkayya RV, Poole RM, Pentz WH (1993). Respiratory failure from procainamide-induced myopathy. Ann Intern Med 119: 345–346.

Villani R, Zoletti F, Veniani M et al. (1992). Confronto fra amiodarone e disopiramide in formulazione retard nella prevenzione delle recidive di fibrillazione atriale sintomatica [A comparison between amiodarone and disopyramide in a delayed-release formulation in the prevention of recurrences of symptomatic atrial fibrillation]. Clin Ter 140: 35–39.

Vorperian VR, Havighurst TC, Miller S et al. (1997). Adverse effects of low dose amiodarone: a meta-analysis. J Am Coll Cardiol 30: 791–798.

Wagstaff LR, Mitton MW, Arvik BM et al. (2003). Statin-associated memory loss: analysis of 60 case reports and review of the literature. Pharmacotherapy 23: 871–880.

Wallentin L, Becker RC, Budaj A et al. (2009). Ticagrelor versus clopidogrel in patients with acute coronary syndromes. N Engl J Med 361: 1045–1057.

Wang AG, Cheng HC (2016). Amiodarone-associated optic neuropathy: clinical review. Neuroophthalmology 41: 55–58.

Wang LY, Shofer JB, Rohde K et al. (2009). Prazosin for the treatment of behavioral symptoms in patients with Alzheimer disease with agitation and aggression. Am J Geriatr Psychiatry 17: 744–751.

Warkentin TE, Kelton JG (1996). A 14-year study of heparin-induced thrombocytopenia. Am J Med 101: 502–507.

Watanabe H, Domei T, Morimoto T et al. (2019). Effect of 1-month dual antiplatelet therapy followed by clopidogrel vs 12-month dual antiplatelet therapy on cardiovascular and bleeding events in patients receiving PCI: the

STOPDAPT-2 randomized clinical trial. JAMA 321: 2414–2427.

Weiss MD, Macklin EA, Simmons Z et al. (2016). A randomized trial of mexiletine in ALS: safety and effects on muscle cramps and progression. Neurology 86: 1474–1481.

Weisz G, Généreux P, Iñiguez A et al. (2016). Ranolazine in patients with incomplete revascularisation after percutaneous coronary intervention (RIVER-PCI): a multicentre, randomised, double-blind, placebo-controlled trial. Lancet 387: 136–145.

Westhout FD, Nwagwu CI (2007). Intra-arterial verapamil-induced seizures: case report and review of the literature. Surg Neurol 67: 483–486.

Whelton PK, Carey RM, Aronow WS et al. (2017). 2017 ACC/AHA/AAPA/ABC/ACPM/AGS/APhA/ASH/ASPC/NMA/PCNA guideline for the prevention, detection, evaluation, and management of high blood pressure in adults: a report of the American College of Cardiology/American Heart Association task force on clinical practice guidelines. Hypertension 71: e13–e115.

Whitlock RP, Sun JC, Fremes SE et al. (2012). Antithrombotic and thrombolytic therapy for valvular disease: antithrombotic therapy and prevention of thrombosis, 9th ed: american college of chest physicians evidence-based clinical practice guidelines. Chest 141: e576S–e600S.

Whitlock EP, Burda BU, Williams SB et al. (2016). Bleeding risks with aspirin use for primary prevention in adults: a systematic review for the U.S. Preventive Services Task Force. Ann Intern Med 164: 826–835.

Williams MF, Maloof 3rd JA (2014). Resolution of clopidogrel-associated polyarthritis after conversion to prasugrel. Am J Health Syst Pharm 71: 1097–1100.

Wittbrodt ET (1997). Drugs and myasthenia gravis. An update. Arch Intern Med 157: 399–408.

Wiviott SD, Braunwald E, McCabe CH et al. (2007). Prasugrel versus clopidogrel in patients with acute coronary syndromes. N Engl J Med 357: 2001–2015.

Wiysonge CS, Bradley HA, Volmink J et al. (2012). Beta-blockers for hypertension. Cochrane Database Syst Rev 11: CD002003.

Wolf LR, Otten EJ, Spadafora MP (1992). Cinchonism: two case reports and review of acute quinine toxicity and treatment. J Emerg Med 10: 295–301.

Wu FL, Razzaghi A, Souney PF (1993). Seizure after lidocaine for bronchoscopy: case report and review of the use of lidocaine in airway anesthesia. Pharmacotherapy 13: 72–78.

Yao X, Abraham NS, Sangaralingham LR et al. (2016). Effectiveness and safety of dabigatran, rivaroxaban, and apixaban versus warfarin in nonvalvular atrial fibrillation. J Am Heart Assoc 5: e003725.

Yasuda S, Kaikita K, Akao M et al. (2019). Antithrombotic therapy for atrial fibrillation with stable coronary disease. N Engl J Med 381: 1103–1113.

Yusuf S, Zhao F, Mehta SR et al. (2001). Effects of clopidogrel in addition to aspirin in patients with acute coronary syndromes without ST-segment elevation [published correction appears in N Engl J Med 2001 Dec 6;345 (23):1716] [published correction appears in N Engl J Med 2001 Nov 15;345(20):1506]. N Engl J Med 345: 494–502.

Yusuf S, Mehta SR, Chrolavicius S et al. (2006). Effects of fondaparinux on mortality and reinfarction in patients with acute ST-segment elevation myocardial infarction: the OASIS-6 randomized trial. JAMA 295: 1519–1530.

Zeiler FA, Zeiler KJ, Kazina CJ et al. (2015). Lidocaine for status epilepticus in adults. Seizure 31: 41–48.

Zhang L, Steckman D (2019). Stroke or side effect? dofetilide associated facial paralysis after direct current cardioversion for atrial fibrillation. BMJ Case Rep 12: e227705.

Handbook of Clinical Neurology, Vol. 177 (3rd series)
Heart and Neurologic Disease
J. Biller, Editor
https://doi.org/10.1016/B978-0-12-819814-8.00021-4

Chapter 30

The role of biomarkers and neuroimaging in ischemic/hemorrhagic risk assessment for cardiovascular/ cerebrovascular disease prevention

ELIF GOKCAL, MITCHELL J. HORN, AND M. EDIP GUROL*

Department of Neurology, Massachusetts General Hospital, Boston, MA, United States

Abstract

Stroke prevention in patients with atrial fibrillation is arguably one of the fastest developing areas in preventive medicine. The increasing use of direct oral anticoagulants and nonpharmacologic methods such as left atrial appendage closure for stroke prevention in these patients has increased clinicians' options for optimal care. Platelet antiaggregants are also commonly used in other ischemic cardiovascular and or cerebrovascular conditions. Long term use of oral anticoagulants for atrial fibrillation is associated with elevated risks of major bleeds including especially brain hemorrhages, which are known to have extremely poor outcomes. Neuroimaging and other biomarkers have been validated to stratify brain hemorrhage risk among older adults. A thorough understanding of these biomarkers is essential for selection of appropriate anticoagulant or left atrial appendage closure for stroke prevention in patients with atrial fibrillation. This article will address advances in the stratification of ischemic and hemorrhagic stroke risk among patients with atrial fibrillation and other conditions.

INTRODUCTION

Stroke is a leading cause of death and functional impairment worldwide. About 795,000 people have a new or recurrent ischemic or hemorrhagic stroke each year in the United States (Mozaffarian et al., 2016). Despite considerable advances in acute reperfusion therapies for selected acute ischemic stroke patients, data show that effective prevention strategies remain the best approach for reducing the burden of stroke related death/disability (Kernan et al., 2014; Meschia et al., 2014).

Cardiac embolism is responsible for approximately one-third of all ischemic strokes (Topcuoglu et al., 2018). Atrial fibrillation (AF)—the prototype cause of cardiac embolism—accounts for a fivefold increase of ischemic stroke risk (Benjamin et al., 2017). The mechanism of ischemic stroke in nonvalvular AF (NVAF) mainly results from embolization from left atrial appendage (LAA) thrombi (Blackshear and Odell, 1996). Based on pooled data from four large contemporary randomized clinical trials (RCTs), embolic stroke is eight times higher than systemic embolism (Bekwelem et al., 2015). Ischemic strokes attributed to AF are typically more severe, have higher recurrence rate and poorer long-term outcome than non-AF-related ischemic strokes (Kimura et al., 2005; Marini et al., 2005). Therefore, both primary and secondary stroke prevention strategies are of the outmost importance to decrease morbidity/mortality in AF patients.

Life-long oral anticoagulant (OAC) use has been the gold standard for stroke prevention in AF patients. However, vitamin K antagonists such as warfarin and the direct oral anticoagulants (DOACs) increase the intracerebral

*Correspondence to: M. Edip Gurol, MD, MSc, Massachusetts General Hospital, High Ischemic Hemorrhagic Risk Stroke Prevention Program, 175 Cambridge Street, Suite 300, Boston, MA, 02114, United States. Tel: +1-617-726-5362, Fax: +1-617-643-5346, E-mail: edip@mail.harvard.edu

hemorrhage (ICH) risk by two- to fivefold even among patient populations at low baseline intracerebral hemorrhage risk (Gurol, 2018). OAC-related ICHs have an estimated mortality rate of ~50%, making OAC-ICH one of the most lethal and disabling condition among common medical emergencies (Fang et al., 2007). Real-world data show that OACs are globally underutilized, mostly due to the fear of ICH (Kakkar et al., 2013; Piazza et al., 2016). Conversely some clinicians underestimate the risk of ICH, focusing only on thromboembolic prevention (Baczek et al., 2012; Gamra et al., 2014). Effective nonpharmacologic stroke prevention measures circumventing the need for lifelong oral anticoagulation exist such as left atrial appendage closure (LAAC) with the US Food and Drug Administration (FDA)-approved WATCHMAN (Boston Scientific, Marlborough, MA, USA) device for NVAF. Therefore, an accurate stratification of ICH risk is essential for appropriate patient selection. The recent publication of a RCT showing similar ischemic event rates between NVAF patients randomized to LAAC or DOACs support the use of LAAC in appropriate high hemorrhage risk populations (Osmancik et al., 2020). Overall, the optimal approach for stroke prevention is based on identification of all conditions associated with higher ischemic and hemorrhagic stroke risk and thorough consideration for both drug-based and nonpharmacologic options.

Moreover, recent years have seen major progress in our understanding of the risks of embolic/hemorrhagic stroke using biomarkers and neuroimaging in particular (Gokcal et al., 2018). In this chapter, we discuss relevant factors associated with embolic as well as hemorrhagic stroke risk in patients with NVAF and review all available options for both ischemic and hemorrhagic stroke prevention. We also briefly discuss issues pertaining to risk markers and conditions requiring platelet antiaggregant use.

STRATIFICATION OF EMBOLIC STROKE RISK IN AF

AF is associated with advanced age, arterial hypertension, diabetes mellitus, and coronary artery disease, which are established risk factors for other stroke subtypes as well. Stroke risk in AF increases in the presence of additional risk factors. Currently, the most commonly used risk-stratification schemes for AF-related embolism are based on the presence of clinical variables. The CHA_2DS_2-VASc [congestive heart failure, hypertension, age $\geq$75 years (doubled), diabetes mellitus, prior stroke or TIA or thromboembolism (doubled), vascular disease, age 65–74 years, sex (female) category] scoring system is the most commonly used risk stratification score. Current guidelines recommend the use of CHA_2DS_2-VASc to differentiate high-risk from "truly low-risk" AF patients; the latter having annual embolic stroke rates of less than 1% (Camm et al., 2012; January et al., 2014a). The current US guidelines recommend long-term OAC use or LAAC with a WATCHMAN in patients with CHA_2DS_2-VASc of 2 or more for men, or 3 or more for women (January et al., 2019). Current guidelines do not recommend OAC in younger AF patients without coexisting cardiovascular risk factors, if the projected risk is lower than 1% per year. The CHA_2DS_2-VASc score is easy to use, but its accuracy in predicting embolic events has been relatively low in validation efforts outside of its initial development/validation datasets (Singer et al., 2013). Furthermore, it is known to overestimate the risk until a score of 3–4 is reached (Singer et al., 2013).

The ATRIA (anticoagulation and risk factors in atrial fibrillation) score is a computerized scoring system, developed using long-term follow-up data from the Kaiser Permanente Northern California (Singer et al., 2013). The clinical variables used are similar to the CHA_2DS_2-VASc, but the ATRIA score includes a measure of renal function by estimation of the glomerular filtration rate, an important biologic variable missing in other risk stratification scores. The calculation of the ATRIA score has important differences from the CHA_2DS_2-VASc, emphasizing older age and past history of ischemic stroke. Advantages of the ATRIA score are better prediction of ischemic stroke risk and enhanced ability for predicting severe strokes.

Since these risk-stratification schemes have shown a modest predictive value for identifying high-risk patients and also a tendency to overestimate embolic risk, recent years have seen considerable amount of work aiming to improve thromboembolic risk prediction focusing on plasma biomarkers—particularly high-sensitivity troponin (hs-T) and N-terminal fragment B-type natriuretic peptide (nt-BNP) (Hijazi et al., 2013, 2014a,b). As a result, a recently developed biomarker-based stroke risk score [ABC (age, biomarkers [hs-T and nt-BNP], and clinical history of stroke/transient ischemic attack (TIA))] has been validated in a variety of clinical trial cohorts (Hijazi et al., 2016; Oldgren et al., 2016). However, recent real-world data reported no improvement in stroke prediction after adding biomarkers to the CHA_2DS_2-VASc score (Rivera-Caravaca et al., 2019). Thus, it is important to balance the pros and cons of different ischemic risk stratification schemes with a focus on improving risk prediction using clinical, laboratory, and neuroimaging data.

Although some clinical factors and biomarkers are associated with an increased risk of embolic stroke, they do not offer any information about the pathophysiologic factors predisposing to blood stasis in the left atrium and particularly the LAA, which is currently recognized as the major source (>95%) of cardiac thrombus formation in these patients. A number of studies are currently investigating the relationship of echocardiographic parameters representing LA function and morphology (LA dilatation, spontaneous echocardiographic contrast, LAA emptying velocity, or LAA morphology) with thromboembolic risk (Providencia et al., 2013; Lupercio et al., 2016; Vinereanu et al., 2017; Leung et al., 2018).

Beyond the clinical variables, brain magnetic resonance imaging (MRI) provides information pertaining to symptomatic/silent lesions that might be relevant to the prediction of future ischemic and/or hemorrhagic strokes. Population-based studies have been using brain MRI for years to further understand mechanisms and progression of brain injury in older adults. Several studies showed an association of AF with MRI-defined brain infarctions, white matter hyperintensities (WMH) or cerebral microbleeds (CMB) known to be related to future stroke risk, cognitive impairment, and death (Vermeer et al., 2007).

MRI-defined brain infarctions have been reported in 28%–90% of patients with AF without prior strokes. Moreover, AF was an independent risk factor for MRI-defined brain infarctions in different populations (Haeusler et al., 2014). However, the exact pathophysiology of MRI-defined brain infarctions in AF is uncertain. Besides AF-related embolism and hypoperfusion proposed mechanisms include concomitant cerebral small vessel disease (cSVD). In a longitudinal study including patients who had brain MRI for routine health checkups, the incidence of symptomatic ischemic strokes was significantly higher among patients with AF with MRI-defined brain infarctions at baseline compared to those without (5.6% per year vs 2.7% per year, hazard ratio 1.787, 95% CI 1.08–2.93, $P = 0.022$) (Cha et al., 2014). However, current guidelines do not recommend OAC/LAAC based on the presence of silent brain infarctions in patients with low clinical risk stratification scores, as the utility of these imaging-defined lesions has not been validated in well-designed RCTs in this population.

STRATIFICATION OF HEMORRHAGIC STROKE RISK IN AF

Patients with AF also have an increased bleeding risk with the long-term use of warfarin or DOACs. The risk ranges between 2.1% and 3.6% per year for major bleeds, and 15%–25% per year for all bleeding types (major, clinically relevant, and minor). Similar to thromboembolic risk, bleeding risk depends on the presence of various risk factors. A number of bleeding stratification risk scores have been developed to assist predicting bleeding risk in patients treated with OACs. The HAS-BLED (hypertension, abnormal renal/liver function [1 point each], stroke, bleeding history, or labile international normalized ratio (INR), elderly [65 years], drugs/alcohol concomitantly [1 point each]) score is one of these bleeding risk stratification schemes (Lane and Lip, 2012). However, since OAC-related hemorrhages are heterogeneous in terms of site, severity, and outcomes, "hemorrhagic risk scores" including HAS-BLED are much less helpful in identifying high-risk patients for a specific type of bleeding. In a recent prospective large multicenter study, the ability of the HAS-BLED score to predict ICH was less than chance (C-index 0.41, 95% CI 0.29–0.5) (Wilson et al., 2018). Therefore, all patients with AF should be evaluated individually for every potential risk factor that might increase the bleeding risk before and after starting OAC. LAAC is considered in NVAF patients at higher than usual bleeding risk. Because bleeding risk is also a dynamic process, its risk can change over time. Thus, all relevant factors that might predispose to hemorrhagic events need to be reviewed in every follow-up visit and nonpharmacologic methods considered, if necessary.

Of all OAC-related hemorrhages, ICH is by far the most feared complication. OAC-ICH is a devastating condition with a high in-hospital mortality (52% for OAC-ICH vs 25.8% for other ICHs) (Rosand et al., 2004). Approximately 70% of OAC-ICHs result from ruptured arteries/arterioles weakened by cSVDs (Steiner et al., 2006). Therefore, ICH survivors or patients with neuroimaging markers of cSVDs that increase ICH risk constitute the most challenging NVAF population in terms of stroke prevention. Optimal risk-stratification requires identifying these high ICH risk markers in patients who are candidates for long-term OAC treatment.

The most common forms of cSVDs among the elderly are hypertensive cSVD (HTN-cSVD) and cerebral amyloid angiopathy (CAA). HTN-cSVD refers to the type of cSVD associated with arterial hypertension as well as other well-known vascular risk factors that predominantly affect small perforating end arteries supplying the deep gray nuclei and adjacent white matter (Charidimou et al., 2016). Pathologically, this type of cSVD is characterized by fibrinoid necrosis, lipohyalinosis, microatheroma or microaneurysm formation in the affected arteries. Vessel wall thickening and restriction of the lumen result in tissue hypoperfusion and ischemia that contributes to WMH, lacunar infarcts, and microinfarcts, while fibrinoid necrosis can cause rupture of microaneurysms, leading to hemorrhagic changes such as deep ICH and CMBs (Wardlaw et al., 2013). CAA is the second most common form of cSVD in older adults, a pathology that shows an increased prevalence rate in parallel with advanced age (Gurol et al., 2016). Pathologically, CAA is associated with β-amyloid deposition within small-to-medium sized arteries and arterioles predominantly located in the cortex and leptomeningeal space leading to smooth muscle degeneration, vessel wall thickening, focal wall fragmentation, microaneurysm formation, and perivascular leakage of blood products in the affected vessels (Wardlaw et al., 2013).

Although both types of cSVDs present similar neuroimaging markers, their distribution and ICH risks differ. CAA may be radiologically diagnosed, using the Boston criteria, in the presence of lobar ICH and strictly lobar CMBs in patients 55 years or older, as long as alternative diagnoses are ruled out (Knudsen et al., 2001) (Fig. 30.1A). Cortical superficial siderosis (cSS), characterized by linear hemosiderin deposits over the cortex or within sulci, has also been validated as a hemorrhagic

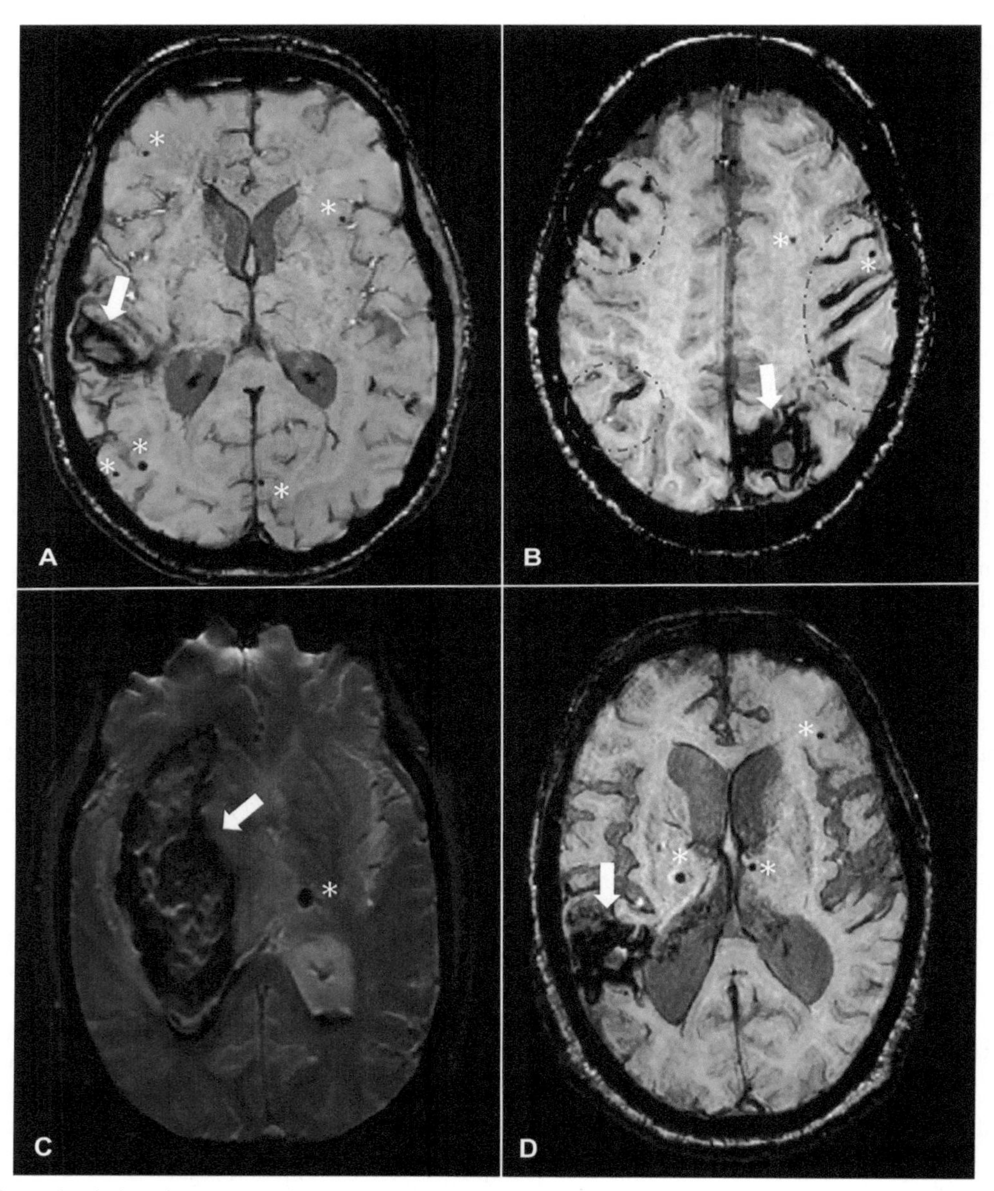

Fig. 30.1. Hemorrhagic imaging markers in different types of cerebral small vessel diseases. (A) Patient with CAA with a lobar ICH (*arrow*, right temporal) and strictly lobar CMBs (*stars*). (B) Patient with CAA with a lobar ICH (*arrow*, left occipital), strictly lobar CMBs (*stars*), and multifocal cSS (*circle/ellipses*). (C) Patient with HTN-cSVD with a deep ICH (*arrow*, right basal ganglia) and a CMB located in deep brain region (*star*, left thalamus). (D) Patient with mixed ICH/CMB with a lobar ICH (*arrow*, right parietotemporal), lobar CMB (*star*, left frontal) as well as deeply located CMBs (*stars*, right and left thalamus). All MRI images are SWI-MRI sequences. The *arrow* represents ICH, the *star* represents CMB, and the *oval shape* represents cSS. *CAA*, cerebral amyloid angiopathy; *CMB*, cerebral microbleeds; *cSS*, cortical superficial siderosis; *HTN-cSVD*, hypertensive cerebral small vessel disease; *ICH*, intracerebral hemorrhage; *SWI-MRI*, susceptibility-weighted imaging-magnetic resonance imaging sequence.

neuroimaging marker of CAA increasing the diagnostic specificity of the Boston criteria (Linn et al., 2010) (Fig. 30.1B). Conversely, the presence of topographically deep ICH and deeply located CMBs supports HTN-cSVD (Pantoni, 2010) (Fig. 30.1C). However, about 20%–57.5% of ICHs in different populations have concomitant hemorrhagic lesions in both lobar and deep regions, known as mixed location ICH/CMB (Pasi et al., 2018; Tsai et al., 2019) (Fig. 30.1D). In addition to the hemorrhagic markers visible on MRI, positron emission tomography (PET) with amyloid tracers can also help differentiate CAA from HTN-cSVD. Florbetapir, an ^{18}F-based PET tracer, is shown to bind vascular amyloid. Florbetapir PET is validated to help diagnose probable CAA among cognitively healthy patients with a sensitivity of 100% and specificity of 89%. (Fig. 30.2) (Gurol et al., 2016). Amyloid PET can help to confirm or rule out CAA in patients with mixed location (lobar and deep)

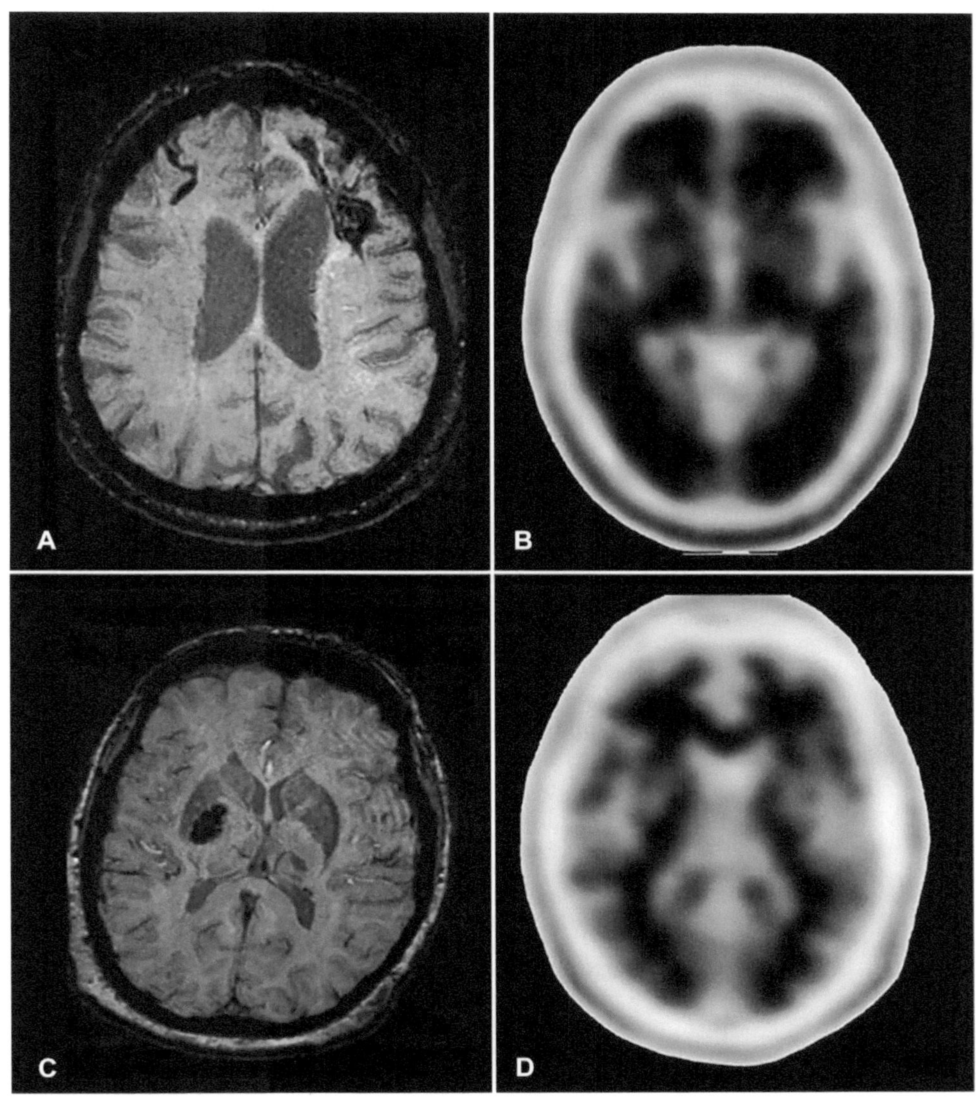

Fig. 30.2. Florbetapir PET for diagnosis of cerebral amyloid angiopathy. (A–B) represent a patient who had a left frontal intracerebral hemorrhage (ICH) and right frontal cortical superficial siderosis on susceptibility-weighted imaging (SWI, Panel A), fulfilling neuroradiological criteria for cerebral amyloid angiopathy (CAA). Florbetapir PET (Panel B) in this CAA patient shows loss of contrast between the cortex and subcortical white matter (positive scan), indicating intense Florbetapir uptake in the cortex because of high vascular amyloid load. The patient in (C–D) had a hypertensive deep right basal ganglia hemorrhage visible on SWI MRI (Panel C). In this patient's Florbetapir PET (Panel D), the contrast between cortex and subcortical structures is preserved (negative scan), confirming low amyloid load in the cortex. Modified from Gurol ME, Becker JA, Fotiadis P, et al. (2016). Florbetapir-PET to diagnose cerebral amyloid angiopathy: A prospective study. Neurology 87: 2043–2049 with permission. Copyright ©2016, Wolters Kluwer Health, Inc.

ICH and/or CMBs or other situations, where CAA is suspected but not diagnosed based on MRI findings alone. Although the FDA approved the use of this tracer for detection of amyloid, the Centers for Medicare & Medicaid Services (CMS) did not approve reimbursement and only few insurance systems are paying for this test. Until CMS approval, the cost is the limiting factor for more widespread use of Florbetapir PET. The risk of future ICH according to imaging markers and SVD types is presented in Table 30.1.

Unfortunately, no RCT has clarified the risk of ICH recurrence with OAC therapy among ICH survivors. The location of ICH is often considered in decision making given the fact that lobar ICH related to CAA demonstrate a much higher recurrence rate than deep ICH due to HTN-cSVD. Current guidelines recommend avoiding warfarin treatment following lobar ICH, whereas resuming OAC therapy might be considered after nonlobar ICH particularly among patients at high risk for thromboembolic stroke and/or low risk of ICH recurrence (Hemphill et al., 2015). In the real world, clinicians tend to avoid OACs or prescribe DOACs instead of resuming warfarin if patients experience a warfarin-related ICH since these newer therapies have lower ICH rates among patients at low baseline risk. However, the safety of these agents in this particular patient population is not known due to the fact that RCTs did not include patients with ICH history. Thus, LAAC with the FDA-approved WATCHMAN device may be a strong alternative among ICH survivors and among patients with high ICH risk based on markers.

Patients with CAA with strictly lobar CMBs in the absence of ICH also show a high rate of first-time ICH upon follow-up (5 per 100 person-years). Warfarin use has been reported to be a predictor of first ICH independently of other conventional risk factors in CAA patients with strictly lobar CMBs without ICH (van Etten et al., 2014). CMBs have been reported in ~25% of patients who had an ischemic stroke or transient ischemic attack (Wilson et al., 2016). The future risk of ICH has been reported to be sixfold higher in patients with ischemic stroke or TIA with CMBs as compared to those without CMBs. The risk increased up to 14-fold when the number of CMBs was 5 or more (Wilson et al., 2016). The presence of CMBs was also more common among patients with warfarin-related ICH as compared to anticoagulated patients without ICH (Lee et al., 2009). A recent multicenter prospective observational cohort study showed an independent association between CMBs and symptomatic ICH risk in patients with AF anticoagulated with warfarin or DOAC after recent ischemic stroke or TIA. The rate of ICH increased as the CMB burden increased (Wilson et al., 2018). Furthermore, the type of OAC used did not significantly affect the hazard of symptomatic ICH associated with CMB presence (hazard ratio interaction term 0.88; 95% CI 0.04–17.13, $P = 0.92$).

Approximately, 60% of CAA patients have cSS that has been consistently reported as an important marker for high ICH risk regardless of history of previous ICH (Roongpiboonsopit et al., 2016; Charidimou et al., 2017b). Furthermore, annual incidence rates of ICH increase up to 26.9% even without any anticoagulation if the CAA patient has multifocal/disseminated cSS (Charidimou et al., 2017a). The cSS multifocality score is a validated tool that can help further stratify the risk of recurrent ICH in patients who survived a CAA-related lobar ICH (Fig. 30.3). Due to high ICH risks, nonanticoagulant strategies such as LAAC should be considered in NVAF patients with cSS (Gurol, 2018).

White matter hyperintensities on FLAIR MRI are another imaging marker of that might be observed in patients with stroke as well as in the general population (Fig. 30.4). In a longitudinal study of community-based stroke-free population, MRI-defined WMH burden was found to be an independent risk factor for spontaneous ICH, even in the absence of anticoagulant therapy

Table 30.1

Neuroimaging markers of SVDs and risk of incident ICH in different scenarios

Imaging marker	Type of SVD	Risk of ICH
General population over 55 years of age		0.08% per year
ICH and CMBs strictly in cortical/lobar regions	Probable CAA	Average = ~10% per year
ICH and CMBs strictly in cortical/lobar regions + cSS	Probable CAA	Up to 26.9% per year for multifocal cSS
>1 CMBs strictly in cortical/lobar regions without ICH	Probable CAA	5% per year
>1 CMBs strictly in cortical/lobar regions + cortical superficial siderosis without ICH	Probable CAA	19% at 5-year follow-up
ICH/CMBs in deep brain regions	HTN-cSVD	Average = ~2% per year
ICH/CMBs in lobar and deep brain regions concomitantly	Mixed ICH/CMB	5.1% per year

CMB, cerebral microbleed; *cSS*, cortical superficial siderosis; *HTN-cSVD*, hypertensive cerebral small vessel disease; *ICH*, intracerebral hemorrhage.

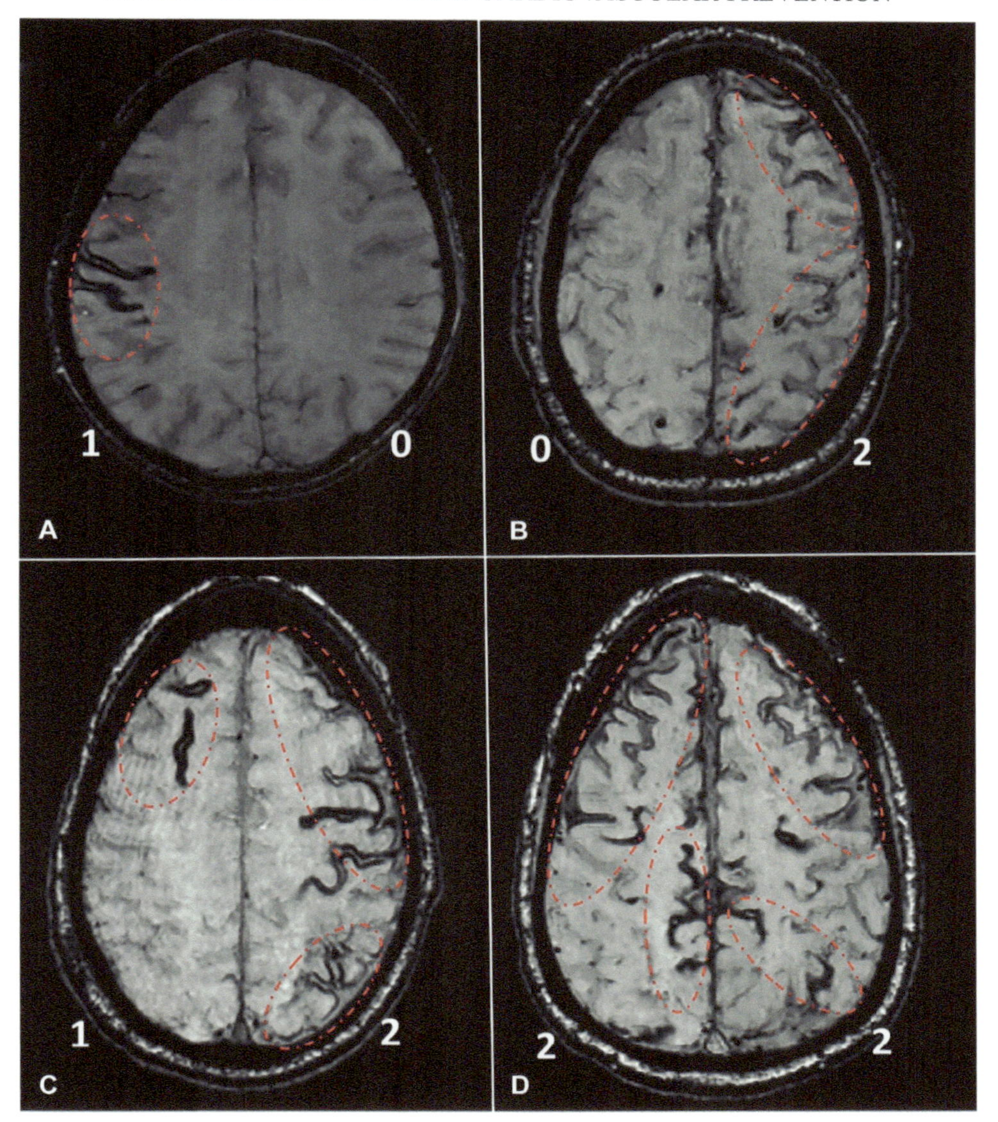

Fig. 30.3. Representative examples of the cortical superficial siderosis (cSS) multifocality score. To assess multifocality of cSS, each hemisphere is scored separately for cSS, as following: 0: none; 1: 1–3 immediately adjacent sulci with cSS; and 2: >3 immediately adjacent sulci (disseminated) or >1 nonadjacent sulci with cSS. The total score is derived by adding the right and left hemisphere scores (range 0–4). Examples are presented on the figure, and the score for each hemisphere (0,1,2) is provided in bottom corners of each patient scan. (A) Patient with a total cSS score of 1 (1 for right and 0 for left hemisphere). (B) Patient with a total cSS score of 2 (0 for right and 2 for left hemisphere). (C) Patient with a total cSS score of 3 (1 for right and 2 for left hemisphere). (D) Patient with a total cSS score of 4 (2 for right and 2 for left hemisphere). All MRI images are SWI-MRI sequences. *SWI-MRI*, susceptibility-weighted imaging-magnetic resonance imaging.

(Folsom et al., 2012). Other studies have shown that the presence and severity of leukoaraiosis was an independent risk factor for warfarin-related and spontaneous ICH, respectively, in patients with ischemic stroke or CAA (Smith et al., 2002, 2004). A recent multicenter observational study prospectively followed 937 patients with cardioembolic strokes who were older than 65 years of age and who were started on OACs (warfarin or DOAC) to identify imaging predictors of future ICH. Patients with moderate/severe WMH and those who had CMBs at baseline MRI had higher risk of ICH during follow-up. The rate of ICH was the highest in patients with both CMBs and moderate/severe WMH (Marti-Fabregas et al., 2019). About two-thirds of the study cohort were started on warfarin, whereas one-third were on DOACs. The type of OAC did not influence the ICH risk. Overall, CMBs, cSS, and moderate-to-severe white matter disease are imaging markers easily identifiable on

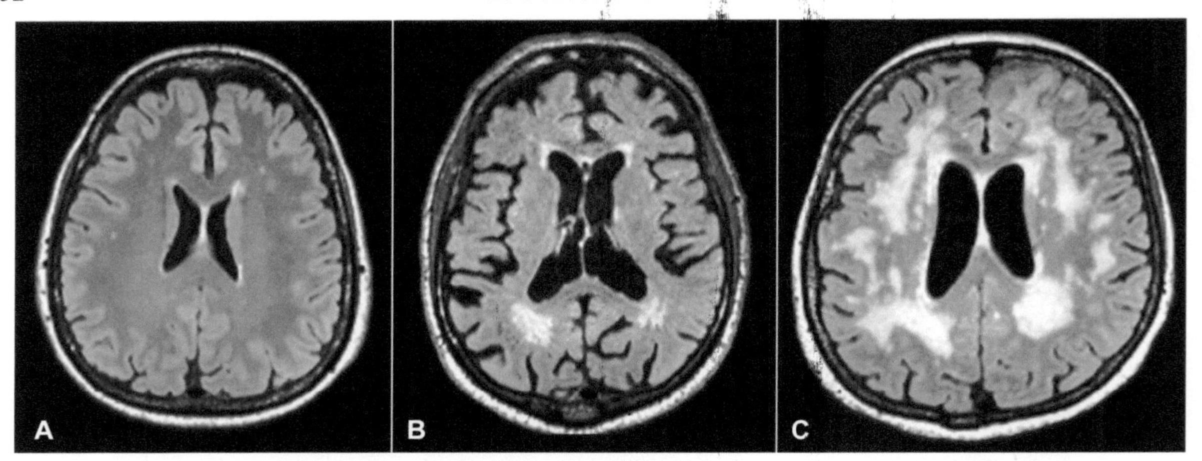

Fig. 30.4. White matter hyperintensities (WMH) on fluid attenuated inversion recovery (FLAIR) MRI sequence. (A) Mild FLAIR white matter hyperintensities (WMH), (B) moderate WMH, and (C) severe WMH.

brain MRI scans that can sensitively predict future ICH risk, thus assisting clinicians select LAAC vs lifelong OAC use in NVAF patients.

PREVENTION OF ISCHEMIC STROKE IN PATIENTS WITH AF AT LOW HEMORRHAGIC RISK

This section summarizes data supporting the use of warfarin or DOACs in NVAF. All RCTs systematically excluded patients with a past history of ICH as well as patients with a high hemorrhage risk. The RCT data summarized below should be viewed with these important caveats in mind. Warfarin has been used to prevent stroke and systemic embolism in patients with AF since the 1950s. A meta-analysis of RCTs from the 1990s showed that adjusted-dose warfarin reduced stroke or systemic embolism by 64% and all-cause mortality by 26% when compared to placebo (Hart et al., 2007). Despite its efficacy, warfarin has several limitations such as narrow therapeutic window, need for frequent blood draws to keep the international normalized ratio within the therapeutic range, several interactions with food and other drugs, and most importantly, an increased risk of hemorrhagic events, particularly ICH. Furthermore, as previously discussed, warfarin-related ICHs have a much higher risk of in-hospital mortality and poorer outcomes when compared to ICHs unrelated to OAC use (Rosand et al., 2004). These factors prompted the search for other easier to use and safer alternatives. The DOACs have been approved by the FDA for stroke prevention in NVAF since 2010. A meta-analysis of four landmark RCTs [RE-LY (dabigatran vs warfarin), ROCKET-AF (rivaroxaban vs warfarin), ARISTOTLE (apixaban vs warfarin), and ENGAGE AF–TIMI 48 (edoxaban vs warfarin)] in NVAF showed a reduction in hemorrhagic stroke risk by 50% with DOACs compared to warfarin (Ruff et al., 2014). As a result, DOACs have been increasingly used for NVAF-related stroke prevention in the United States and worldwide. However, warfarin remains as the only option for patients with mechanical heart valves with a target INR of 2.0–3.0 or 2.5–3.5, based on type and location of the heart valve prosthesis. Likewise, warfarin is the only OAC indicated for patients with antiphospholipid syndrome.

Current guidelines recommend DOACs and warfarin as first-line agents for stroke prevention in patients with NVAF who can tolerate long-term OAC use. Selecting the most appropriate DOAC should also be individualized based on specific features of every medication, as well as patients' characteristics (Gokcal et al., 2018; Topcuoglu et al., 2018). The rates of DOACs discontinuation were high, similar to warfarin in all RCTs, and mostly due to hemorrhagic complications (Topcuoglu et al., 2018).

PREVENTION OF ISCHEMIC AND HEMORRHAGIC STROKES/EVENTS IN PATIENTS WITH AF AT HIGH HEMORRHAGIC RISK

Long-term use of OACs as first-line treatment for embolic prevention is an evidence-based approach in AF patients without high hemorrhage risk at baseline. This recommendation does not apply to patients with high bleeding risk, especially high ICH risk, as these were not included in the previously discussed RCTs, and moreover, because OACs known tendency to cause hemorrhages. Current guidelines recommend individualized shared decision making in choosing appropriate therapy after discussion of absolute and relative risks of stroke and bleeding as well as all alternative management options (January et al., 2014b). An FDA approved alternative that obviates the need for lifelong OAC use in NVAF patients with increased bleeding risk is closure of the LAA. At present, the only FDA approved device used for percutaneous LAAC is the WATCHMAN device (Boston Scientific, Natick, Massachusetts). The implantation of WATCHMAN device is typically performed by cardiac electrophysiologists and interventional cardiologists using transesophageal or intracardiac echocardiography. Local or general anesthesia is chosen based on the patient's characteristics and implanter preference. After the procedure, warfarin with a target INR of 2–3 and aspirin 81 mg/day are used until follow-up transesophageal echocardiography, usually performed 45 days after implantation. When a good LAA seal (no connection or any jet larger than 5 mm in diameter between LAA and left atrium) is shown on echocardiography, warfarin therapy is switched to a combination of clopidogrel 75 mg/day and aspirin 81–325 mg/day for 4.5 months. After 6 months, patients are kept on aspirin only (Gurol, 2018). Although not used in RCTs, alternative regimens include use of a DOAC during the first 6–12 weeks or dual platelet antiaggregant for 3–6 months, switching to aspirin monotherapy afterwards based on single-arm studies and observational data (Reddy et al., 2013; Cohen et al., 2019). The FDA recently updated indications for LAAC with WATCHMAN in a way DOACs can be used for the first 6 weeks after the implant instead of warfarin, so the DOAC use after the procedure is now an FDA-labeled approach (FDA, 2020). The PROTECT AF (Watchman Left Atrial Appendage System for Embolic Protection in Patients with Atrial Fibrillation) study compared the LAAC procedure with warfarin in patients with AF who had a mean $CHADS_2$ of 2.2 for the endpoints of stroke/embolism/death prevention (Holmes et al., 2009). A history of stroke or TIA was present in 19% of patients. In the warfarin arm, the mean time in therapeutic range was 66%, a value higher than the warfarin arms in DOAC studies except for the ENGAGE AF–TIMI 48. During early phases of the trial, rate of procedural complication was relatively high. Despite that, the primary efficacy event rate was 3.0 per 100 patient-years (95% CI 1.9–4.5) in the intervention group vs 4.9 per 100 patient-years (95% CI 2.8–7.1) in the control group (RR: 0.62, 95% CI 0.35–1.25), confirming the noninferiority of LAAC to warfarin stroke prevention, systemic embolism, or cardiovascular/unexplained death. Study results showed similar ischemic stroke rates between the groups, but the ICH risk was significantly decreased with the use of the WATCHMAN device compared to warfarin (0.1 events per 100 patient-years [95% CI 0.0–0.5] vs 1.6 events per 100 patient-years [95% CI 0.6–3.1]). Cardiovascular and unexplained death rates were also significantly lower for the WATCHMAN arm (0.7 events per 100 patient-years [95% CI 0.2–1.5] vs 2.7 events per 100 patient-years [95% CI 1.2–4.4]). Furthermore, upon long-term follow-up (mean duration of 3.8 years), the primary event rate was 2.3 events per 100 patient-years, compared with 3.8 events per 100 patient-years with warfarin. These results fulfill criteria for both noninferiority and superiority, for preventing the combined outcome of stroke, systemic embolism, and cardiovascular death, as well as superiority for cardiovascular and all-cause mortality (Reddy et al., 2014). In the PREVAIL trial (Watchman LAA Closure Device in Patients with Atrial Fibrillation Vs Long Term Warfarin Therapy), there was an improvement in procedural safety, as only 2.2% of subjects experienced an event (Holmes et al., 2014). Based on these results, in March 2015, the FDA approved the use of WATCHMAN device for patients with NVAF at relatively high embolic risk who can use warfarin but who also have a reasonable rationale to avoid long-term anticoagulation.

A recent meta-analysis from two RCTs which included 1114 patients, mean follow-up of 5 years, showed that WATCHMAN significantly reduced hemorrhagic stroke (0.17 per 100 patient-years vs 0.87 per 100 patient-years, hazard ratio 0.20 [95% CI 0.07–0.56], $P=0.002$) while showing similar a trend for higher ischemic stroke or systemic embolism rates (1.6 per 100 patient-years vs 0.95 per 100 patient-years, hazard ratio 1.71 [95% CI 0.94–3.11], $P=0.08$) when compared to warfarin (Reddy et al., 2017a). All-cause stroke or systemic embolism was similar between WATCHMAN and warfarin (hazard ratio 0.96; $P=0.87$). Registry-based data from consecutive procedures in the United States ($n=3822$) and Europe ($n=1021$) showed satisfactory successful deployment rates (95.6% and 98.5%) and decreased procedural complication rates when compared with clinical trials: pericardial effusion/tamponade (1% and 0.3%), device embolization (0.24% and 0.2%), procedural stroke (0.08% and 0.1%), and death within 7 days (0.08% and 0.3%) (Boersma et al., 2016; Reddy et al., 2017b).

The new generation WATCHMAN device (WATCHMAN FLX) approved by the FDA in July 2020 represents a significant improvement over its predecessor as it has more available sizes improving fit and decreasing peridevice leaks, metal and contact point exposure is much lower enhancing the speed and completeness of the seal, and it can be readily recaptured, repositioned, and redeployed. The 7 days major complication rate of WATCHMAN FLX was 0.5%, based on the prospective "Investigational Device Evaluation of the WATCHMAN FLX™ LAA Closure Technology (PINNACLE FLX)" study that resulted into its FDA approval (https://clinicaltrials.gov, 2020). The Amulet IDE study for another endovascular LAAC device, AMPLATZER AMULET

(Abbott Vascular, Abbott Park, IL, USA), has been completed and results are expected within the next 1–2 years. Other endovascular and epicardial LAAC devices being investigated include the Wavecrest device (Coherex Medical, Biosense Webster, Johnson & Johnson, Salt Lake City, UT, USA). All of these devices received "Conformité Européenne" mark approval and are available in Europe. The results of the phase 3 RCTs will further increase our understanding of the merits of LAAC using different devices as the use of implantable devices with different structure/shapes can further improve success rates in patients who have different LAA shapes/sizes.

The Left Atrial Appendage Closure vs. Novel Anticoagulation Agents in Atrial Fibrillation (PRAGUE-17) study compared LAAC (AMPLATZER AMULET or WATCHMAN) to long-term DOAC use head-to-head comparison for the first time (Osmancik et al., 2020). The study showed similar results for the occurrence of the primary composite outcome measures (stroke, TIA, systemic embolism, cardiovascular death, major or nonmajor clinically relevant bleeding, or procedure-/device-related complications) as well as ischemic event rates. After a median 20 months follow-up, the annual rates of the primary outcome were 10.99% with LAAC and 13.42% with DOAC (HR: 0.84; 95% CI 0.53–1.31; $P=0.44$; $P=0.004$ for noninferiority) (Osmancik et al., 2020). Two large-scale RCTs comparing the WATCHMAN FLX to DOACs (CHAMPION AF) and AMPLATZER AMULET to DOACs (CATALYST) are in the process of starting, both aiming to demonstrate noninferiority for ischemic event prevention and superiority of LAAC for major and clinically relevant hemorrhages in NVAF patients.

LAAC is an important method for ischemic and hemorrhagic stroke prevention in NVAF patients who survived any ICH or major/relevant bleed or who have hemorrhage-prone brain conditions such as SVDs that put them at high ICH risk. There are other indications such as cognitive/gait problems, inability to use an OAC regularly, and any bleeding/fall risk that can jeopardize the patient's well-being. Incorporating LAAC into shared decision-making discussions with appropriately selected NVAF patients helps improve the quality of patient care, allowing patients to make informed decisions for their preventive treatment.

RISK STRATIFICATION METHODS AND ANTIPLATELET USE

Platelet antiaggregants constitute the mainstay of arterial ischemic cardiovascular/cerebrovascular event prevention outside of AF, mechanical heart valve replacement, antiphospholipid syndrome, and few other indications that require anticoagulation or other advanced measures. Antiplatelet monotherapy has been used for many decades in hundreds of millions of patients. The indications are diverse ranging from coronary artery disease, ischemic cerebrovascular disease, peripheral vascular disease to other cardiac and vascular conditions. For these important indications, there is typically no alternative to antiplatelets except closure of a patent foramen ovale for secondary prevention after a cryptogenic stroke in patients between 18 and 60 years of age (Topcuoglu et al., 2018). Platelet antiaggregants are associated with a lower hemorrhagic risk when compared to anticoagulants. Long-term use of anticoagulant/antiplatelet combinations as well as dual antiplatelet therapies (DAPT) increase the hemorrhagic risk so limiting the duration of combined antithrombotic use to minimum required timeframes is always the ideal approach. In patients with cSVD-related recent lacunar strokes, the use of DAPT did not reduce recurrent ischemic stroke rate compared to antiplatelet monotherapy with aspirin 325 mg daily (SPS3 Investigators, 2012). Overall, escalating antithrombotic treatment should be avoided if possible in patients with hemorrhage-prone cSVDs, but antiplatelet monotherapy is used when there is a valid indication (Das et al., 2019). The ICH risk with antiplatelets has been reviewed in detail recently (Thon and Gurol, 2016). The risk/benefit ratio of platelet antiaggregant use in patients at high ICH risk has been better studied than anticoagulants in this context, but there is still no extensive RCT-based evidence. The effects of antiplatelet therapy after stroke due to ICH (RESTART) were a randomized open-label trial that excluded all but a very mild increase in the risk of recurrent ICH with antiplatelet therapy for patients on antithrombotic therapy for the prevention of occlusive vascular disease when they developed ICH (Salman et al., 2019). After suffering an ICH, 268 patients with valid indications for antiplatelet were assigned to start and 269 to avoid antiplatelet therapy. Over a median of 2 years follow-up, recurrent ICH rates were 4% in the antiplatelet arm vs 9% in the no-antiplatelet arm (adjusted hazard ratio 0.51 [95% CI 0.25–1.03]; $P=0.060$). Overall major hemorrhagic event rates were also numerically lower in the antiplatelet arm (7% vs 9%, $P=0.27$). When antiplatelet monotherapy is indicated based on current guidelines, such treatment is typically used unless there is a demonstrated excessive bleeding risk in the particular patient. Shared decision-making discussions between treating specialists, the patient and family are always important to determine benefits/risks and understand the patient's preferences. Informing the patients and their families of concurrent risks is the physician's duty, and an open discussion also increases patient satisfaction and therefore long-term compliance.

CONCLUSIONS

Accurate stratification of ICH risk as discussed throughout this article is of paramount importance to advise NVAF patients appropriately for LAAC or lifelong OAC use. Brain MRI imaging provides unprecedented opportunities to detect hemorrhage-prone diseases based on lesions such as microbleeds, superficial siderosis, and white matter disease. These markers can help understand hemorrhage risks even before a potentially fatal/disabling ICH happens. An operational understanding of the ischemic and hemorrhagic risk stratification tools can improve outcomes through selection of optimal preventive measures.

STUDY SUPPORT

This work was made possible by the following NIH grants: Gurol (NS083711, R01NS114526).

REFERENCES

Baczek VL, Chen WT, Kluger J et al. (2012). Predictors of warfarin use in atrial fibrillation in the United States: a systematic review and meta-analysis. BMC Fam Pract 13: 5.

Bekwelem W, Connolly SJ, Halperin JL et al. (2015). Extracranial systemic embolic events in patients with nonvalvular atrial fibrillation: incidence, risk factors, and outcomes. Circulation 132: 796–803.

Benjamin EJ, Blaha MJ, Chiuve SE et al. (2017). Heart disease and stroke statistics-2017 update: a report from the American Heart Association. Circulation 135: e146–e603.

Blackshear JL, Odell JA (1996). Appendage obliteration to reduce stroke in cardiac surgical patients with atrial fibrillation. Ann Thorac Surg 61: 755–759.

Boersma LV, Schmidt B, Betts TR et al. (2016). Implant success and safety of left atrial appendage closure with the WATCHMAN device: peri-procedural outcomes from the EWOLUTION registry. Eur Heart J 37: 2465–2474.

Camm AJ, Lip GY, De Caterina R et al. (2012). 2012 focused update of the ESC guidelines for the management of atrial fibrillation: an update of the 2010 ESC guidelines for the management of atrial fibrillation—developed with the special contribution of the European heart rhythm association. Europace 14: 1385–1413.

Cha MJ, Park HE, Lee MH et al. (2014). Prevalence of and risk factors for silent ischemic stroke in patients with atrial fibrillation as determined by brain magnetic resonance imaging. Am J Cardiol 113: 655–661.

Charidimou A, Boulouis G, Haley K et al. (2016). White matter hyperintensity patterns in cerebral amyloid angiopathy and hypertensive arteriopathy. Neurology 86: 505–511.

Charidimou A, Boulouis G, Roongpiboonsopit D et al. (2017a). Cortical superficial siderosis multifocality in cerebral amyloid angiopathy: a prospective study. Neurology 89: 2128–2135.

Charidimou A, Boulouis G, Xiong L et al. (2017b). Cortical superficial siderosis and first-ever cerebral hemorrhage in cerebral amyloid angiopathy. Neurology 88: 1607–1614.

Cohen JA, Heist EK, Galvin J et al. (2019). A comparison of postprocedural anticoagulation in high-risk patients undergoing WATCHMAN device implantation. Pacing Clin Electrophysiol 42: 1304–1309.

Das AS, Regenhardt RW, Feske SK et al. (2019). Treatment approaches to lacunar stroke. J Stroke Cerebrovasc Dis 28: 2055–2078.

Fang MC, Go AS, Chang Y et al. (2007). Death and disability from warfarin-associated intracranial and extracranial hemorrhages. Am J Med 120: 700–705.

FDA (2020). Available: https://www.fda.gov/medical-devices/recently-approved-devices/watchman-left-atrial-appendage-closure-device-delivery-system-and-watchman-flx-left-atrial-appendageAccessed 08.27. 2020.

Folsom AR, Yatsuya H, Mosley Jr TH et al. (2012). Risk of intraparenchymal hemorrhage with magnetic resonance imaging-defined leukoaraiosis and brain infarcts. Ann Neurol 71: 552–559.

Gamra H, Murin J, Chiang CE et al. (2014). Use of antithrombotics in atrial fibrillation in Africa, Europe, Asia and South America: insights from the International RealiseAF survey. Arch Cardiovasc Dis 107: 77–87.

Gokcal E, Pasi M, Fisher M et al. (2018). Atrial fibrillation for the neurologist: preventing both ischemic and hemorrhagic strokes. Curr Neurol Neurosci Rep 18: 6.

Gurol ME (2018). Nonpharmacological management of atrial fibrillation in patients at high intracranial hemorrhage risk. Stroke 49: 247–254.

Gurol ME, Becker JA, Fotiadis P et al. (2016). Florbetapir-PET to diagnose cerebral amyloid angiopathy: a prospective study. Neurology 87: 2043–2049.

Haeusler KG, Wilson D, Fiebach JB et al. (2014). Brain MRI to personalise atrial fibrillation therapy: current evidence and perspectives. Heart 100: 1408–1413.

Hart RG, Pearce LA, Aguilar MI (2007). Meta-analysis: antithrombotic therapy to prevent stroke in patients who have nonvalvular atrial fibrillation. Ann Intern Med 146: 857–867.

Hemphill 3rd JC, Greenberg SM, Anderson CS et al. (2015). Guidelines for the management of spontaneous intracerebral hemorrhage: a guideline for healthcare professionals from the American Heart Association/American Stroke Association. Stroke 46: 2032–2060.

Hijazi Z, Oldgren J, Siegbahn A et al. (2013). Biomarkers in atrial fibrillation: a clinical review. Eur Heart J 34: 1475–1480.

Hijazi Z, Siegbahn A, Andersson U et al. (2014a). High-sensitivity troponin I for risk assessment in patients with atrial fibrillation: insights from the Apixaban for reduction in stroke and other thromboembolic events in atrial fibrillation (ARISTOTLE) trial. Circulation 129: 625–634.

Hijazi Z, Wallentin L, Siegbahn A et al. (2014b). High-sensitivity troponin T and risk stratification in patients with atrial fibrillation during treatment with apixaban or warfarin. J Am Coll Cardiol 63: 52–61.

Hijazi Z, Lindback J, Alexander JH et al. (2016). The ABC (age, biomarkers, clinical history) stroke risk score: a biomarker-based risk score for predicting stroke in atrial fibrillation. Eur Heart J 37: 1582–1590.

Holmes Jr. DR, Kar S, Price MJ et al. (2014). Prospective randomized evaluation of the watchman left atrial appendage closure device in patients with atrial fibrillation versus long-term warfarin therapy: the PREVAIL trial. J Am Coll Cardiol 64: 1–12.

Holmes DR, Reddy VY, Turi ZG et al. (2009). Percutaneous closure of the left atrial appendage versus warfarin therapy for prevention of stroke in patients with atrial fibrillation: a randomised non-inferiority trial. Lancet 374: 534–542.

https://clinicaltrials.govAvailablehttps://clinicaltrials.gov/ct2/show/NCT02702271Accessed 08.27. 2020.

January CT, Wann LS, Alpert JS et al. (2014a). 2014 AHA/ACC/HRS guideline for the management of patients with atrial fibrillation: a report of the American College of Cardiology/American Heart Association task force on practice guidelines and the Heart Rhythm Society. Circulation 130: e199–e267.

January CT, Wann LS, Alpert JS et al. (2014b). 2014 AHA/ACC/HRS guideline for the management of patients with atrial fibrillation: executive summary: a report of the American College of Cardiology/American Heart Association task force on practice guidelines and the Heart Rhythm Society. Circulation 130: 2071–2104.

January CT, Wann LS, Calkins H et al. (2019). 2019 AHA/ACC/HRS focused update of the 2014 AHA/ACC/HRS guideline for the management of patients with atrial fibrillation: a report of the American College of Cardiology/American Heart Association task force on clinical practice guidelines and the Heart Rhythm Society. J Am Coll Cardiol 74: 104–132.

Kakkar AK, Mueller I, Bassand JP et al. (2013). Risk profiles and antithrombotic treatment of patients newly diagnosed with atrial fibrillation at risk of stroke: perspectives from the international, observational, prospective GARFIELD registry. PLoS One 8: e63479.

Kernan WN, Ovbiagele B, Black HR et al. (2014). Guidelines for the prevention of stroke in patients with stroke and transient ischemic attack: a guideline for healthcare professionals from the American Heart Association/American Stroke Association. Stroke 45: 2160–2236.

Kimura K, Minematsu K, Yamaguchi T et al. (2005). Atrial fibrillation as a predictive factor for severe stroke and early death in 15,831 patients with acute ischaemic stroke. J Neurol Neurosurg Psychiatry 76: 679–683.

Knudsen KA, Rosand J, Karluk D et al. (2001). Clinical diagnosis of cerebral amyloid angiopathy: validation of the Boston criteria. Neurology 56: 537–539.

Lane DA, Lip GY (2012). Use of the CHA(2)DS(2)-VASc and HAS-BLED scores to aid decision making for thromboprophylaxis in nonvalvular atrial fibrillation. Circulation 126: 860–865.

Lee SH, Ryu WS, Roh JK (2009). Cerebral microbleeds are a risk factor for warfarin-related intracerebral hemorrhage. Neurology 72: 171–176.

Leung M, van Rosendael PJ, Abou R et al. (2018). Left atrial function to identify patients with atrial fibrillation at high risk of stroke: new insights from a large registry. Eur Heart J 39: 1416–1425.

Linn J, Halpin A, Demaerel P et al. (2010). Prevalence of superficial siderosis in patients with cerebral amyloid angiopathy. Neurology 74: 1346–1350.

Lupercio F, Carlos Ruiz J, Briceno DF et al. (2016). Left atrial appendage morphology assessment for risk stratification of embolic stroke in patients with atrial fibrillation: a meta-analysis. Heart Rhythm 13: 1402–1409.

Marini C, De Santis F, Sacco S et al. (2005). Contribution of atrial fibrillation to incidence and outcome of ischemic stroke: results from a population-based study. Stroke 36: 1115–1119.

Marti-Fabregas J, Medrano-Martorell S, Merino E et al. (2019). MRI predicts intracranial hemorrhage in patients who receive long-term oral anticoagulation. Neurology 92: e2432–e2443.

Meschia JF, Bushnell C, Boden-Albala B et al. (2014). Guidelines for the primary prevention of stroke: a statement for healthcare professionals from the American Heart Association/American Stroke Association. Stroke 45: 3754–3832.

Mozaffarian D, Benjamin EJ, Go AS et al. (2016). Heart disease and stroke statistics—2016 update: a report from the American Heart Association. Circulation 133: e38–360.

Oldgren J, Hijazi Z, Lindback J et al. (2016). Performance and validation of a novel biomarker-based stroke risk score for atrial fibrillation. Circulation 134: 1697–1707.

Osmancik P, Herman D, Neuzil P et al. (2020). Left atrial appendage closure versus direct oral anticoagulants in high-risk patients with atrial fibrillation. J Am Coll Cardiol 75: 3122–3135.

Pantoni L (2010). Cerebral small vessel disease: from pathogenesis and clinical characteristics to therapeutic challenges. Lancet Neurol 9: 689–701.

Pasi M, Charidimou A, Boulouis G et al. (2018). Mixed-location cerebral hemorrhage/microbleeds: underlying microangiopathy and recurrence risk. Neurology 90: e119–e126.

Piazza G, Karipineni N, Goldberg HS et al. (2016). Underutilization of anticoagulation for stroke prevention in atrial fibrillation. J Am Coll Cardiol 67: 2444–2446.

Providencia R, Trigo J, Paiva L et al. (2013). The role of echocardiography in thromboembolic risk assessment of patients with nonvalvular atrial fibrillation. J Am Soc Echocardiogr 26: 801–812.

Reddy VY, Mobius-Winkler S, Miller MA et al. (2013). Left atrial appendage closure with the watchman device in patients with a contraindication for oral anticoagulation: the ASAP study (ASA Plavix feasibility study with watchman left atrial appendage closure technology). J Am Coll Cardiol 61: 2551–2556.

Reddy VY, Sievert H, Halperin J et al. (2014). Percutaneous left atrial appendage closure vs warfarin for atrial fibrillation: a randomized clinical trial. JAMA 312: 1988–1998.

Reddy VY, Doshi SK, Kar S et al. (2017a). 5-year outcomes after left atrial appendage closure: from the PREVAIL and PROTECT AF trials. J Am Coll Cardiol 70: 2964–2975.

Reddy VY, Gibson DN, Kar S et al. (2017b). Post-approval U.S. experience with left atrial appendage closure for stroke prevention in atrial fibrillation. J Am Coll Cardiol 69: 253–261.

Rivera-Caravaca JM, Marin F, Vilchez JA et al. (2019). Refining stroke and bleeding prediction in atrial fibrillation by adding consecutive biomarkers to clinical risk scores. Stroke 50: 1372–1379.

Roongpiboonsopit D, Charidimou A, William CM et al. (2016). Cortical superficial siderosis predicts early recurrent lobar hemorrhage. Neurology 87: 1863–1870.

Rosand J, Eckman MH, Knudsen KA et al. (2004). The effect of warfarin and intensity of anticoagulation on outcome of intracerebral hemorrhage. Arch Intern Med 164: 880–884.

Ruff CT, Giugliano RP, Braunwald E et al. (2014). Comparison of the efficacy and safety of new oral anticoagulants with warfarin in patients with atrial fibrillation: a meta-analysis of randomised trials. Lancet 383: 955–962.

Salman RA-S, Dennis M, Sandercock P et al. (2019). Effects of antiplatelet therapy after stroke due to intracerebral haemorrhage (RESTART): a randomised, open-label trial. The Lancet 393: 2613–2623.

Singer DE, Chang Y, Borowsky LH et al. (2013). A new risk scheme to predict ischemic stroke and other thromboembolism in atrial fibrillation: the ATRIA study stroke risk score. J Am Heart Assoc 2: e000250.

Smith EE, Rosand J, Knudsen KA et al. (2002). Leukoaraiosis is associated with warfarin-related hemorrhage following ischemic stroke. Neurology 59: 193–197.

Smith EE, Gurol ME, Eng JA et al. (2004). White matter lesions, cognition, and recurrent hemorrhage in lobar intracerebral hemorrhage. Neurology 63: 1606–1612.

SPS3 Investigators (2012). Effects of clopidogrel added to aspirin in patients with recent lacunar stroke. N Engl J Med 367: 817–825.

Steiner T, Rosand J, Diringer M (2006). Intracerebral hemorrhage associated with oral anticoagulant therapy: current practices and unresolved questions. Stroke 37: 256–262.

Thon JM, Gurol ME (2016). Intracranial hemorrhage risk in the era of antithrombotic therapies for ischemic stroke. Curr Treat Options Cardiovasc Med 18: 29.

Topcuoglu MA, Liu L, Kim DE et al. (2018). Updates on prevention of cardioembolic strokes. J Stroke 20: 180–196.

Tsai HH, Pasi M, Tsai LK et al. (2019). Microangiopathy underlying mixed-location intracerebral hemorrhages/ microbleeds: a PiB-PET study. Neurology 92: e774–e781.

van Etten ES, Auriel E, Haley KE et al. (2014). Incidence of symptomatic hemorrhage in patients with lobar microbleeds. Stroke 45: 2280–2285.

Vermeer SE, Longstreth Jr WT, Koudstaal PJ (2007). Silent brain infarcts: a systematic review. Lancet Neurol 6: 611–619.

Vinereanu D, Lopes RD, Mulder H et al. (2017). Echocardiographic risk factors for stroke and outcomes in patients with atrial fibrillation anticoagulated with apixaban or warfarin. Stroke 48: 3266–3273.

Wardlaw JM, Smith EE, Biessels GJ et al. (2013). Neuroimaging standards for research into small vessel disease and its contribution to ageing and neurodegeneration. Lancet Neurol 12: 822–838.

Wilson D, Charidimou A, Ambler G et al. (2016). Recurrent stroke risk and cerebral microbleed burden in ischemic stroke and TIA: a meta-analysis. Neurology 87: 1501–1510.

Wilson D, Ambler G, Shakeshaft C et al. (2018). Cerebral microbleeds and intracranial haemorrhage risk in patients anticoagulated for atrial fibrillation after acute ischaemic stroke or transient ischaemic attack (CROMIS-2): a multicentre observational cohort study. Lancet Neurol 17: 539–547.

Handbook of Clinical Neurology, Vol. 177 (3rd series)
Heart and Neurologic Disease
J. Biller, Editor
https://doi.org/10.1016/B978-0-12-819814-8.00022-6

Chapter 31

Clinical utility of echocardiography in secondary ischemic stroke prevention

WILLIAM J. POWERS*

Department of Neurology, University of North Carolina School of Medicine, Chapel Hill, NC, United States

Abstract

Echocardiography employs ultrasound to evaluate cardiac function, structure and pathology. The clinical value in secondary ischemic stroke prevention depends on identification of associated conditions for which a change in treatment from antiplatelet agents and risk factor intervention leads to improved outcomes. Such therapeutically relevant findings include primarily intracardiac thrombus, valvular heart disease and, in highly selected patients, patent foramen ovale (PFO). Echocardiography in unselected patients with ischemic stroke has a very low yield of therapeutically relevant findings and is not cost-effective. With the exception of PFO, findings on echocardiography that are therapeutically relevant for secondary stroke prevention are almost always associated with history, signs or symptoms of cardiac or systemic disease. Choice of specific echocardiographic modalities should be based on the specific pathology or pathologies that are under consideration for the individual clinical situation. Transthoracic echocardiography (TTE) with agitated saline has comparable accuracy to transesophageal echocardiography (TEE) for PFO detection. For other therapeutically relevant pathologies, with the possible exception of left ventricular thrombus (LVT), TEE is more sensitive than TTE. Professional societies recommend TTE as the initial test but these recommendations do not take cost into account. In contrast, cost-effectiveness studies have determined that the most sensitive echocardiographic modality should be selected as the initial and only test.

INTRODUCTION

Echocardiography (EC) provides a method to evaluate cardiac function, structure and pathology using ultrasound. As with all diagnostic tests, the clinical utility of EC rests whether there is a resultant change in therapy that provides a clinical benefit and the cost-effectiveness of any benefit provided. Despite longstanding and widespread use, data are of medium to low quality to support the clinical benefit and cost effectives of EC used routinely or in defined subgroups of patients with transient ischemic attacks (TIA) or acute ischemic stroke (AIS).

THERAPEUTIC RELEVANCE OF ECHOCARDIOGRAPHIC FINDINGS IN ACUTE CEREBRAL ISCHEMIA

For patients with TIA or AIS, current recommendations based on randomized controlled trials (RCTs) for secondary stroke prevention generally include the use of antiplatelet agents and risk factor interventions for hyperlipidemia, hypertension, tobacco use, and diabetes mellitus. For certain associated conditions, different or additional treatments have been shown to reduce the risk of stroke recurrence, such as substitution of anticoagulation for antiplatelet

*Correspondence to: William J. Powers, M.D., Department of Neurology, University of North Carolina School of Medicine, Room 2133, CB#7025, 170 Manning Drive, Chapel Hill, NC 27514, United States. Tel: +1-919-966-8178, Fax: +1-919-843-4999, E-mail: powersw@neurology.unc.edu

agents for patients with atrial fibrillation and additional carotid endarterectomy or stenting for those with ipsilateral carotid stenosis. The clinical value of EC in secondary ischemic stroke prevention depends on identification of associated conditions for which a change in treatment from antiplatelet agents and risk factor intervention leads to improved outcomes.

Table 31.1 lists the prevalence of echocardiographic findings in patients with TIA or AIS reported in 24 studies from 2002 to 2019 comprising 7188 transthoracic and 5764 transesophageal echocardiograms (Strandberg et al., 2002; Sen et al., 2004; Abreu et al., 2005; Vitebskiy et al., 2005; Chlumsky et al., 2006; Dawn et al., 2006; de Bruijn et al., 2006; Harloff et al., 2006; Douen et al., 2007; Ahmad et al., 2010; Cho et al., 2010; Galougahi et al., 2010; Gupta et al., 2010; Yaghoubi et al., 2011; Kim et al., 2012; Zhang et al., 2012; Secades et al., 2013; Menon et al., 2014; Marino et al., 2016; Pallesen et al., 2016; Miles et al., 2018; Schwartz et al., 2018; Fralick et al., 2019; Wasser et al., 2019). The therapeutic relevance of these findings, based on treatment guidelines issued by professional societies, if they exist, is discussed next.

Cardiac chamber thrombi

Anticoagulant therapy for patients with TIA or AIS who are in sinus rhythm and have left atrial or left ventricular thrombi is recommended by current guidelines (Kernan et al., 2014; Wein et al., 2018). There are no applicable RCTs. Recommendations are based on limited data from observational studies (Level C evidence) (Jacobs et al., 2014).

The presence of thrombi in patients with recent stroke and atrial fibrillation has no therapeutic relevance as they will benefit from oral anticoagulation for secondary prevention regardless (Level A and B evidence) (Kernan et al., 2014; Wein et al., 2018).

Cardiac tumors

Currently, there are no evidence-based guidelines for management of patients with cerebrovascular events in association with cardiac myomas, the most common primary cardiac tumor. A recent study of 13 patients with ischemic stroke or TIA associated with cardiac myxoma concluded that antiplatelet or anticoagulation treatment is not an alternative to surgery and that the time to tumor excision should be kept as short as possible (Stefanou et al., 2018). There are no applicable RCTs. Recommendations are based on limited data (Level C evidence).

Infective endocarditis

Infective endocarditis (IE) requires appropriate antibiotic therapy instituted as quickly as possible based on the results of blood cultures (Nishimura et al., 2014). There are no applicable RCTs. Recommendations are based on extensive observational data (Level B evidence). Neither anticoagulant nor antiplatelet therapy is recommended as adjunctive therapy (Baddour et al., 2015; Habib et al., 2015). Evidence is from a single small RCT and observational studies (Level B evidence).

Nonbacterial thrombotic endocarditis

Anticoagulation with heparin is commonly employed to reduce the risk systemic embolization based on observational studies. Warfarin does not prevent recurrent thromboembolic events (Liu and Frishman, 2016). Although the oral factor Xa inhibitor edoxaban has been shown to be noninferior to low-molecular-weight heparin for preventing recurrent venous thromboembolism or bleeding in patient with cancer, relative efficacy for preventing embolism from nonbacterial thrombotic endocarditis (NBTE) is unknown (Raskob et al., 2018). Surgery may be considered for those with severe valvular dysfunction, large vegetations, or recurrent embolism, despite long-term anticoagulation therapy. Additional research is needed to evaluate the efficacy of corticosteroids in reducing the severity of NBTE, particularly in those with systemic lupus erythematosus (Libman–Sacks endocarditis) (Liu and Frishman, 2016). Recommendations are based on limited data (Level C evidence).

Rheumatic mitral stenosis

For patients with rheumatic mitral stenosis in sinus rhythm who have an embolic event, anticoagulation is recommended in preference to antiplatelet therapy (Kernan et al., 2014; Nishimura et al., 2014; Baumgartner et al., 2017). There are no applicable RCTs. Evidence is from observational studies and expert opinion (Level B and C evidence).

Mitral valve prolapse, mitral annulus calcification, and calcified aortic valve

For patients with TIA or AIS, antiplatelet therapy is recommended for these three conditions (Whitlock et al., 2012; Kernan et al., 2014). There are no applicable RCTs. Recommendations are based on epidemiological studies and expert opinion (Level C evidence).

Patent foramen ovale

Recent guidelines recommend patent foramen ovale (PFO) device closure plus long-term antiplatelet therapy

Table 31.1

Echocardiographic findings in patients with recent cerebral ischemia

	Transthoracic echocardiography		Transesophageal echocardiography	
	Total	Range	Total	Range
Chamber thrombi				
Left atrial appendage	0.10%	0%–0.43% (Strandberg et al., 2002; Douen et al., 2007; Yaghoubi et al., 2011; Secades et al., 2013; Wasser et al., 2019)	5.45%	0.38%–18.54% (Strandberg et al., 2002; Sen et al., 2004; de Bruijn et al., 2006; Harloff et al., 2006; Galougahi et al., 2010; Marino et al., 2016; Schwartz et al., 2018; Wasser et al., 2019)
Left atrial cavity	0.23%	0%–1.08% (Strandberg et al., 2002; de Bruijn et al., 2006; Douen et al., 2007; Yaghoubi et al., 2011; Zhang et al., 2012; Wasser et al., 2019)	0.57%	0%–2.65% (Strandberg et al., 2002; Sen et al., 2004; de Bruijn et al., 2006; Galougahi et al., 2010; Schwartz et al., 2018; Wasser et al., 2019)
Any/all left atrial	0.45%	0%–1.63% (Strandberg et al., 2002; de Bruijn et al., 2006; Douen et al., 2007; Yaghoubi et al., 2011; Zhang et al., 2012; Secades et al., 2013; Menon et al., 2014; Wasser et al., 2019)	2.64%	0.25%–21.19% (Strandberg et al., 2002; Sen et al., 2004; Vitebskiy et al., 2005; Dawn et al., 2006; de Bruijn et al., 2006; Harloff et al., 2006; Cho et al., 2010; Galougahi et al., 2010; Menon et al., 2014; Marino et al., 2016; Pallesen et al., 2016; Schwartz et al., 2018; Wasser et al., 2019)
Left ventricular	1.39%	0%–5.43% (Strandberg et al., 2002; de Bruijn et al., 2006; Douen et al., 2007; Yaghoubi et al., 2011; Zhang et al., 2012; Secades et al., 2013; Menon et al., 2014; Wasser et al., 2019)	0.61%	0%–5.3% (Strandberg et al., 2002; Sen et al., 2004; Vitebskiy et al., 2005; Dawn et al., 2006; de Bruijn et al., 2006; Cho et al., 2010; Galougahi et al., 2010; Menon et al., 2014; Pallesen et al., 2016; Schwartz et al., 2018)
Mass/tumor	0.07%	0%–0.46% (Abreu et al., 2005; Douen et al., 2007; Yaghoubi et al., 2011; Zhang et al., 2012; Secades et al., 2013; Miles et al., 2018; Fralick et al., 2019; Wasser et al., 2019)	0.25%	0%–1.59% (Dawn et al., 2006; Cho et al., 2010; Menon et al., 2014; Pallesen et al., 2016; Schwartz et al., 2018; Wasser et al., 2019)
Valvular disease				
Native valve thrombus	0%	0% (Wasser et al., 2019)	1.41%	0.38%–1.29% (Pallesen et al., 2016; Wasser et al., 2019)
Prosthetic valve thrombus			0.25%	0.25% (Pallesen et al., 2016)
Vegetation/ endocarditis	0.28%	0%–2.61% (Douen et al., 2007; Ahmad et al., 2010; Yaghoubi et al., 2011; Zhang et al., 2012; Secades et al., 2013; Menon et al., 2014; Miles et al., 2018; Fralick et al., 2019; Wasser et al., 2019)	0.87%	0%–8.00% (Dawn et al., 2006; Ahmad et al., 2010; Cho et al., 2010; Galougahi et al., 2010; Menon et al., 2014; Pallesen et al., 2016; Schwartz et al., 2018; Wasser et al., 2019)
NBTE	0.17%	0%–0.54% (Yaghoubi et al., 2011; Zhang et al., 2012)	0.11%	0.11% (Cho et al., 2010)
Mitral stenosis	0.88%	0%–1.26% (Strandberg et al., 2002; Abreu et al., 2005; de Bruijn et al., 2006; Douen et al., 2007; Kim et al., 2012; Zhang et al., 2012; Secades et al., 2013; Fralick et al., 2019)	0.15%	0%–0.91% (Strandberg et al., 2002; de Bruijn et al., 2006; Cho et al., 2010; Schwartz et al., 2018)
Mitral valve prolapse	1.19%	0.98%–4.76% (Strandberg et al., 2002; de Bruijn et al., 2006; Kim et al., 2012; Zhang et al., 2012; Secades et al., 2013)	1.12%	0.16%–4.76% (Strandberg et al., 2002; de Bruijn et al., 2006; Cho et al., 2010)
Mitral annulus calcification	3.00%	0.54%–7.65% (Strandberg et al., 2002; de Bruijn et al., 2006; Kim et al., 2012; Zhang et al., 2012; Secades et al., 2013)	0.64%	0.22%–1.81% (Strandberg et al., 2002; de Bruijn et al., 2006; Cho et al., 2010)
Calcified aortic valve	3.62%	1.36%–6.17% (Strandberg et al., 2002; de Bruijn et al., 2006; Secades et al., 2013)	2.08%	1.36%–3.46% (Strandberg et al., 2002; de Bruijn et al., 2006)

Continued

Table 31.1

Continued

	Transthoracic echocardiography		Transesophageal echocardiography	
	Total	Range	Total	Range
Atrial pathology				
PFO	2.51%	0%–5.19% (Strandberg et al., 2002; de Bruijn et al., 2006; Douen et al., 2007; Ahmad et al., 2010; Kim et al., 2012; Secades et al., 2013; Fralick et al., 2019; Wasser et al., 2019)	12.99%	5.19%–20.10% (Strandberg et al., 2002; Vitebskiy et al., 2005; Chlumsky et al., 2006; de Bruijn et al., 2006; Harloff et al., 2006; Ahmad et al., 2010; Cho et al., 2010; Galougahi et al., 2010; Gupta et al., 2010; Marino et al., 2016; Schwartz et al., 2018; Wasser et al., 2019)
PFO with ASA			3.07%	1.1%–6.45% (Vitebskiy et al., 2005; Gupta et al., 2010; Marino et al., 2016; Wasser et al., 2019)
ASD			0.39%	0.33%–1.00% (Dawn et al., 2006; Cho et al., 2010)
ASA	1.78%	0%–4.54% (Strandberg et al., 2002; de Bruijn et al., 2006; Secades et al., 2013; Wasser et al., 2019)	3.45%	0.49%–9.51% (Strandberg et al., 2002; Vitebskiy et al., 2005; Dawn et al., 2006; de Bruijn et al., 2006; Harloff et al., 2006; Cho et al., 2010; Gupta et al., 2010; Marino et al., 2016; Schwartz et al., 2018; Wasser et al., 2019)
Ventricular pathology				
Reduced LVEF	6.03%	0.98%–11.85% (Strandberg et al., 2002; Abreu et al., 2005; de Bruijn et al., 2006; Douen et al., 2007; Yaghoubi et al., 2011; Zhang et al., 2012; Menon et al., 2014; Miles et al., 2018; Fralick et al., 2019; Wasser et al., 2019)	1.63%	0%–2.16% (Strandberg et al., 2002; de Bruijn et al., 2006; Menon et al., 2014)
Hypo-/akinesia	10.10%	0%–13.3% (Abreu et al., 2005; Kim et al., 2012; Zhang et al., 2012; Miles et al., 2018; Wasser et al., 2019)	1.75%	1.75% (Cho et al., 2010)
Left ventricular aneurysm	0.75%	0%–2.27% (Strandberg et al., 2002; de Bruijn et al., 2006; Zhang et al., 2012; Secades et al., 2013; Wasser et al., 2019)	0.30%	0.23%–0.43% (Strandberg et al., 2002; de Bruijn et al., 2006)
Dilated cardiomyopathy	7.60%	0.33%–19.08% (Abreu et al., 2005; Yaghoubi et al., 2011; Kim et al., 2012)	0.55%	0.55% (Cho et al., 2010)
Severe CHF	5.66%	5.66% (Ahmad et al., 2010)	2.50%	2.50% (Ahmad et al., 2010)
Aortic pathology				
Complex plaque	0.19%	0%–0.99% (Strandberg et al., 2002; de Bruijn et al., 2006; Ahmad et al., 2010; Yaghoubi et al., 2011; Kim et al., 2012; Zhang et al., 2012; Secades et al., 2013; Wasser et al., 2019)	11.30%	2.15%–29.8% (Strandberg et al., 2002; Vitebskiy et al., 2005; Dawn et al., 2006; de Bruijn et al., 2006; Harloff et al., 2006; Ahmad et al., 2010; Cho et al., 2010; Galougahi et al., 2010; Gupta et al., 2010; Marino et al., 2016; Pallesen et al., 2016; Schwartz et al., 2018; Wasser et al., 2019)
Other				
Spontaneous echo contrast	0.25%	0%–1.13% (Strandberg et al., 2002; de Bruijn et al., 2006; Douen et al., 2007; Ahmad et al., 2010; Yaghoubi et al., 2011; Kim et al., 2012; Zhang et al., 2012)	4.55%	0.13%–13.03% (Strandberg et al., 2002; Vitebskiy et al., 2005; Chlumsky et al., 2006; Dawn et al., 2006; de Bruijn et al., 2006; Harloff et al., 2006; Ahmad et al., 2010; Cho et al., 2010; Galougahi et al., 2010; Pallesen et al., 2016)

Abbreviations: *ASA*, atrial septal aneurysm; *ASD*, atrial septal defect; *CHF*, congestive heart failure; *LVEF*, left ventricular ejection fraction; *NBTE*, nonbacterial thrombotic endocarditis; *PFO*, patent foramen ovale.

over antiplatelet therapy alone for carefully-selected patients with a recent ischemic stroke or TIA who are age 18–60 (Wein et al., 2018; Ahmed et al., 2019). The criteria for selection vary between the two guidelines and require careful clinical evaluation and performance of additional tests to exclude other possible associated conditions. These recommendations are based on data from multiple RCTs (Level A evidence). However, the quality of the evidence from these RCTs is low because of many design and execution flaws (Kamel, 2018; McIntyre et al., 2018; Powers, 2018; Powers et al., 2019; Pristipino et al., 2019). The European Stroke Organization Guidelines ESO but not the Canadian Foundation Guidelines also provides recommendations for patients outside the 18–60 age bracket (Ahmed et al., 2019).

In patients between 60 and 65 years, percutaneous closure plus medical therapy instead of antiplatelet therapy alone can be offered (Level B evidence).

Percutaneous closure plus medical therapy can be considered in place of antiplatelet therapy alone also for patients aged <18 and >65 years old on an individual basis (Level C evidence).

Both sets of Guidelines acknowledge that there is insufficient evidence to make a recommendation regarding the comparative effectiveness of PFO closure vs anticoagulant therapy (Wein et al., 2018; Ahmed et al., 2019).

Isolated atrial septal aneurysm

The risk of recurrent stroke in patients with isolated atrial septal aneurysm (ASA) (without PFO) treated with antiplatelet therapy is unmeasurably low (Mas et al., 2001). As a consequence, consideration of alternative therapy does not seem reasonable to this author. Recommendation based on observational study data (Level B evidence).

Left ventricular dysfunction/heart failure

In patients with TIA or AIS in sinus rhythm without a left ventricular thrombus (LVT), there is currently no evidence to suggest advantages of warfarin over antiplatelet drugs in those with dilated cardiomyopathy, reduced left ventricular ejection fraction or heart failure (Kernan et al., 2014; Shantsila and Lip, 2016; Wein et al., 2018). Recommendations are based on evidence from RCTs (Level B evidence). An exception to this may be in the setting of acute anterior ST-elevation myocardial infarction (STEMI) without demonstrable left ventricular mural thrombus; anticoagulation may be considered when anterior apical akinesis or dyskinesis identified by EC (O'Gara et al., 2013; Kernan et al., 2014). There are no applicable RCTs. Recommendations are based on limited data from observational studies (Level C evidence).

Aortic atherosclerotic plaque

For patients with TIA or AIS and aortic arch atheroma, anticoagulant therapy has no demonstrated superiority over antiplatelet therapy (Whitlock et al., 2012; Kernan et al., 2014; Wein et al., 2018). Recommendation is based on data from a single RCT and observation studies (Level B evidence). Surgical removal of aortic arch plaques for the purposes of secondary stroke prevention is not recommended. There are no applicable RCTs. Recommendation is based on limited data from observational studies (Level C evidence).

Spontaneous echo contrast

In patients with mitral stenosis in sinus rhythm, oral anticoagulation should be considered when TEE shows dense spontaneous echocardiographic contrast (Baumgartner et al., 2017). There are no applicable RCTs. Recommendation is based on consensus of opinion of the experts and/or small studies, retrospective studies, registries (Level C evidence). However, patients with mitral stenosis who have had an AIS or TIA qualify for recommendation for anticoagulation so the addition presence not of spontaneous echo contrast (SEC) does not matter.

Prosthetic cardiac valves

Prosthetic valves requires treatment in the setting of valve obstruction or hemodynamic compromise. Recent guidelines provide recommendations for the medical and surgical management of prosthetic valve thrombosis (Baumgartner et al., 2017; Nishimura et al., 2017). There are no applicable RCTs. Recommendations are based on limited data and expert opinion (Level C evidence).

Comment

These evidence-based data on treatment enable classification of cardiac pathologies in to those that are therapeutically relevant and those that are not (Table 31.2). Save for the inclusion of mitral stenosis, this classification is the same as used by Fralick et al. who used the term "clinically actionable" (Fralick et al., 2019). However, even for these pathologies categorized as therapeutically relevant or clinically actionable, the evidence supporting anticoagulation or procedures is of medium to low quality (Levels of evidence B and C or Level A evidence from RCTs with design and execution flaws).

Table 31.2

Echocardiographic findings in patients with recent cerebral ischemia categorized by therapeutic relevance (see Table 31.1 for reference citations)

	Transthoracic echocardiography		Transesophageal echocardiography	
	Total	Range	Total	Range
Therapeutically relevant pathology				
Chamber thrombi				
Left atrial appendage	0.10%	0%–0.43%	5.45%	0.38%–18.54%
Left atrial cavity	0.23%	0%–1.08%	0.57%	0%–2.65%
Any/all left atrial	0.45%	0%–1.63%	2.64%	0.25%–21.19%
Left ventricular	1.39%	0%–5.43%	0.61%	0%–5.3%
Mass/tumor	0.07%	0%–0.46%	0.25%	0%–1.59%
Valvular disease				
Native valve thrombus	0%	0%	1.41%	0.38%–1.29%
Prosthetic valve thrombus			0.25%	0.25%
Vegetation/endocarditis	0.28%	0%–2.61%	0.87%	0%–8.00%
NBTE	0.17%	0%–0.54%	0.11%	0.11%
Mitral stenosis	0.88%	0%–1.26%	0.15%	0%–0.91%
Atrial pathology				
PFO	2.51%	0%–5.19%	12.99%	5.19%–20.10%
PFO with ASA			3.07%	1.1%–6.45%
ASD			0.39%	0.33%–1.00%
Therapeutically irrelevant pathology				
Valvular disease				
Mitral valve prolapse	1.19%	0.98%–4.76%	1.12%	0.16%–4.76%
Mitral annulus calcification	3.00%	0.54%–7.65%	0.64%	0.22%–1.81%
Calcified aortic valve	3.62%	1.36%–6.17%	2.08%	1.36%–3.46%
Atrial pathology				
ASA	1.78%	0%–4.54%	3.45%	0.49%–9.51%
Ventricular pathology				
Reduced LVEF	6.03%	0.98%–11.85%	1.63%	0%–2.16%
Hypo –/akinesia	10.10%	0%–13.3%	1.75%	1.75%
Left ventricular aneurysm	0.75%	0%–2.27%	0.30%	0.23%–0.43%
Dilated cardiomyopathy	7.60%	0.33%–19.08%	0.55%	0.55%
Severe CHF	5.66%	5.66%	2.50%	2.50%
Aortic pathology				
Complex plaque	0.19%	0%–0.99%	11.30%	2.15%–29.8%
Other				
Spontaneous echo contrast	0.25%	0%–1.13%	4.55%	0.13%–13.03%

Abbreviations: *ASA*, atrial septal aneurysm; *ASD*, atrial septal defect; *CHF*, congestive heart failure; *LVEF*, left ventricular ejection fraction; *NBTE*, nonbacterial thrombotic endocarditis; *PFO*, patent foramen ovale; *NBTE*, nonbacterial thrombotic endocarditis; *PFO*, patent foramen ovale; *ASD*, atrial septal defect; *ASA*, atrial septal aneurysm; *LVEF*, left ventricular ejection fraction; *CHF*, congestive heart failure.

TTE VS TTE FOR DETECTION OF THERAPEUTICALLY RELEVANT CARDIAC PATHOLOGIES

Both TTE and TEE are available and extensively used for evaluation of patients with acute cerebral ischemia. TTE is performed noninvasively with an ultrasound transducer placed on the skin of the anterior and lateral chest wall. This transducer emits high-frequency ultrasound waves into the body and then picks up those waves that are reflected back by subcutaneous tissues. These reflected ultrasound waves are converted into moving images of the heart. TTE can be performed in fundamental imaging mode (TTEf), which uses the reflected echoes from the same spectral band of frequencies as that of the emitted ultrasound waves. Most modern TTE is performed using harmonic imaging (TTEh), which employs the reflected waves of the second harmonic of the emitted frequency band to construct images. Image quality varies depending on body habitus, the size of the intercostal spaces, chest deformities, and lung disease such as emphysema. TTE is essentially without risk. TEE is performed with a differently designed ultrasound probe that is swallowed and obtains images of the heart posteriorly through the anterior esophageal wall. Major TEE complications occur in fewer than 1% of procedures. Esophageal perforation occurs in 0.01%–0.03%. Oropharyngeal trauma, dysphagia, and odynophagia occur in 0.1%–13% and are usually rapidly self-limited. Dysphagia after TEE has been associated with aspiration (Holmes et al., 2014; Saric et al., 2016; Stewart and Gilliland, 2018).

In the United States of America charges and costs for the procedures vary widely. However, the rate reimbursed through Medicare by the United States Government in 2019 was the same for both procedures (C8925 and C8929) at $691.75 and slightly more for TEE interpretation ($251.19) that for TTE ($210.47). (http://www.definityimaging.com/pdf/2019_DEFINITY_Medicare_reimbursement_RA7705.pdf Accessed January 3, 2020).

As demonstrated in Table 31.2, studies employing TEE report an equal or greater prevalence of therapeutically important pathologies than those employing TTE, implying greater sensitivity for detection. Similar findings were reported in two series (Table 31.3) in which all subjects had both studies. Particularly notable was the increased prevalence of left atrial thrombi and PFO, but not LVT, detected by TEE.

Holmes et al. performed a systematic review of the diagnostic accuracy of TTEh compared to TEE (Holmes et al., 2014).

Cardiac vegetation		
TTEh	Sensitivity	0.83 (95% CrI 0.62–0.94)
	Specificity	0.96 (95% CrI 0.86–0.99)
Left atrial thrombus (LAT)		
TTEh	Sensitivity	0.79 (95% CrI 0.47–0.94)
	Specificity	1.00 (95% CrI 0.99–1.00)
Left atrial appendage thrombus (LAAT)		
TTEh	Sensitivity	1.00 (95% CrI 0.16–1.00)
	Specificity	1.00 (95% CrI 0.97–1.00)

For detection of LAAT and LAT, three-dimensional TTE with intravenously injected ultrasound provides comparable accuracy to TEE (Karakus et al., 2008; Sallach et al., 2009; Gorani et al., 2013). No studies were identified in the systematic review by Holmes et al. that evaluated the diagnostic accuracy of TTE vs TEE for LVT. TTE is general considered superior to TEE for detection of LVT but the comparison has not been extensively explored (Srichai et al., 2006). Srichai et al. compared contrast-enhanced MRI, TTE and TEE in 361 patients with ischemic heart disease who had surgical and/or pathological confirmation of presence or absence of LVT. LVT was present in 106. In 160 patients with all three imaging modalities performed within 30 days of surgical or pathological confirmation, TEE (sensitivity 40% ± 14% and specificity 96% ± 3.6%) was superior to TTE (23% ± 12% and 96% ± 3.6%, respectively) for thrombus detection, but contrast-enhanced MRI showed the highest sensitivity and specificity (88% ± 9% and 99% ± 2%) (Srichai et al., 2006).

Table 31.3

Direct comparisons of transthoracic echocardiography (TTE) and transesophageal echocardiography (TEE) for detecting therapeutically relevant findings in patients with recent cerebral ischemia

	de Bruijn et al. (2006)		Strandberg et al. (2002)	
Number of subjects	231		441	
	TTE	TEE	TTE	TEE
Chamber ihrombi				
Left atrial appendage	1 (0.5%)	38 (16%)	0	21 (4.8%)
Left atrial cavity	0	1 (0.5%)	2 (0.5%)	2 (0.5%)
Left ventricular	2 (1%)	2 (1%)	2 (0.5%)	2 (0.5%)
Valvular disease				
Mitral stenosis	0	0	1 (0.2%)	1 (0.2%)
Atrial pathology				
PFO	3 (1.5%)	12 (5%)	0	67 (15.2%)

Abbreviation: *PFO*, patent foramen ovale.

Holmes et al. reported comparable accuracy for TTEh and TEE for detection of PFO (Holmes et al., 2014).

TTEh vs TEE	Sensitivity Specificity	0.89 (95% CrI 0.80–0.95) 0.99 (95% CrI 0.97–1.00)

In a separate systematic review and meta-analysis, Mojadidi et al. have also reported comparable sensitivity and specificity of TTEh as compared to TEE for detection of right-to-left intracardiac shunts.

TTEh	Sensitivity Specificity	0.91 (95% CI 0.88–0.93) 0.93 (95% CI 0.91–0.94)

All TTE and TEE were performed with an agent such as agitated saline bubbles that does not pass through the lungs and a maneuver to increase right heart pressure and provoke right-to-left shunting (Mojadidi et al., 2014).

Table 31.4 presents relevant recommendations regarding the use of TTE and TEE for the detection of therapeutically relevant cardiac pathologies from 2010 European Association of Echocardiography Recommendations for Echocardiography Use in the Diagnosis and Management of Cardiac Sources of Embolism, the 2016 American Society of Echocardiography Guidelines for the Use of Echocardiography in the Evaluation of a Cardiac Source of Embolism, and the 2017 Task Force for the Management of Valvular Heart Disease of the European Society of Cardiology (ESC) and the European Association for Cardio-Thoracic Surgery (EACTS) Guidelines for the Management of Valvular Heart Disease (Pepi et al., 2010; Saric et al., 2016; Baumgartner et al., 2017).

Generally, they recommend TTE as the initial or only study, while acknowledging the superior sensitivity of TEE, except for LVT.

CLINICAL CORRELATIONS AND ASSOCIATIONS WITH THERAPEUTICALLY IMPORTANT CARDIAC PATHOLOGIES

Left atrial and left atrial appendage thrombi

In a prospective study of 869 patients with "embolic" stroke or TIA, LAAT was more common in those with atrial fibrillation (39/286, 14%) as compared to those in sinus rhythm (6/583, 1%). Three of the patients in sinus rhythm had mitral stenosis: one had aortic stenosis, one had a dilated cardiomyopathy, and one had coronary heart disease (Omran et al., 2000). In four studies of patients with recent cerebral ischemia that excluded those with AF, left atrial thrombus (LAT) was detected by TEE in 0.5%–8.0% (Yahia et al., 2009; Cho et al., 2010; Menon et al., 2014; Wasser et al., 2019). In a multivariate analysis of LAT occurring in 13/238 patients recent cerebral ischemic events who were in sinus rhythm, LAT formation was independently associated with left ventricular systolic dysfunction (OR 10.6, CI 2.2–51.6, $P < 0.003$), occurring in 5/18 (28%) of those with dysfunction and in 8/225 (4%) of those without (Yahia et al., 2009). Ellis et al. studied 3768 patients referred for TEE to exclude LAAT after embolic events or before cardioversion for atrial fibrillation or atrial flutter. LAAT was present in 199. No LAAT occurred in the 247 patients in sinus rhythm with a structurally normal heart on TTE defined by ejection fraction >50%, absence of left ventricular hypertrophy, absence of right ventricular (RV) hypertrophy or dysfunction, left atrial diameter <4.0 cm, absence of aortic valve stenotic or regurgitant lesions, ≤1+ mitral regurgitation, absence of mitral stenosis, ≤1+ tricuspid regurgitation, absence of tricuspid stenosis, ≤1+ pulmonary regurgitation, absence of pulmonary stenosis, absence of a ventricular septal defect or an atrial septal defect, and absence of a prosthetic valve (Ellis et al., 2006). However, Mugge et al. reported that 7 of 75 LAAT detected by TEE in a cohort of 2000 patients referred for a variety of clinical reasons occurred in patients with sinus rhythm "without morphological abnormalities additional to" LAAT (Mugge et al., 1990).

Left ventricular thrombi

LVT are strongly associated with left ventricular dysfunction. In a study of 62 patients with LVT, all had some underlying left ventricular disease: ischemic cardiomyopathy in 50, dilated cardiomyopathy in 5, stress-induced cardiomyopathy in 3, apical hypertrophic cardiomyopathy in 2, severe aortic stenosis and cardiac arrest in 1 each (Lee et al., 2013). The most common risk factor for LVT is recent myocardial infarction (MI) (Leow et al., 2018). In the thrombolytic era, LVT was observed in 17% of all patients after acute MI and up to 34%–57% in those with anterior MI. Primary percutaneous coronary intervention is associated with a reduction in the incidence of LVT postacute MI to 3% overall and 9% in those with anterior MI (McCarthy et al., 2018). However, LVT occasionally occurs in those without known cardiac disease. In 1833 consecutive patients with cerebral ischemic stroke or TIA in sinus rhythm without a cardiac disease history studied within 7 days after onset, LVT was detected by TEE in 8 (0.4%) (Cho et al., 2010).

Table 31.4

Echocardiographic recommendations from the 2016 American Society of Echocardiography Guideline (ASE) (Saric et al., 2016), the 2010 European Association of Echocardiography (EAE) Recommendations (Pepi et al., 2010), and the 2017 European Society of Cardiology/European Association for Cardio-Thoracic Surgery (ESC/EACTS) Guidelines (Baumgartner et al., 2017)

Suspected left atrial and left atrial appendage thrombus

ASE

"TTE is recommended in patients with suspected LA or LAA thrombus to assess LA size and LV size and function, as well to assess for underlying etiologies of atrial fibrillation and additional risk factors for stroke.

TEE is superior to TTE in assessment of anatomy and function of LAA in a variety of clinical contexts, such as before cardioversion, ablation of atrial arrhythmias, and percutaneous procedures for LAA closure."

EAE

"The presence of thrombi in left atrium (LA) or LV can be detected with TTE, but the most common location for thrombi in patients with AF is the LAA, which cannot be regularly examined by TTE." (text)

Suspected LV thrombus

ASE

"TTE is recommended for the evaluation of patients with underlying cardiac disease known to predispose to LV thrombus formation (such as myocardial infarction or nonischemic cardiomyopathy).

TTE is typically superior to TEE in the assessment of LV apical thrombus."

EAE

"(2) Echocardiography should be used in identifying LV thrombus, and the addition of contrast may increase diagnostic accuracy.

(3) TOE has little to offer in the detection of LV thrombus."

Suspected cardiac tumors

ASE

"Complete TTE is recommended in all patients suspected of having cardiac tumors."

"TEE may be superior to TTE in evaluating cardiac tumors, especially myxomas and PFEs [papillary fibroelastomas]."

EAE (Cardiac myxoma)

"TTE imaging is usually sufficient, although small tumors or those that involve the right heart may require TOE for diagnosis." (text)

Suspected prosthetic valve thrombosis

ASE

"Both TTE and TEE are indicated when suspicion of prosthetic valve thrombosis arises."

EAE

"(1) TTE must be performed in patients with a prosthetic valve and an embolic event.

(2) Owing to its better sensitivity, TOE must be performed in patients with a prosthetic valve and an embolic event, even if TTE is negative.

(3) TOE plays an important role in guiding therapeutic strategy in prosthetic thrombosis, the presence of a large thrombus favoring surgery."

ESC/EACTS

"Transesophageal echocardiography (TOE) should be considered when transthoracic echocardiography (TTE) is of suboptimal quality or when thrombosis, prosthetic valve dysfunction or endocarditis is suspected." (text)

Suspected infective endocarditis

ASE

"TTE is recommended for the following:

Initial evaluation of suspected endocarditis with positive blood culture results or a new murmur."

"TEE is recommended for the following:

To diagnose IE and its complications when clinical suspicion is intermediate or high, regardless of negative results on TTE.

As the first-line modality when complications of IE are suspected, such as abscesses, fistulas, or valve perforation, or when prosthetic valve endocarditis is clinically suspected."

EAE

"(1) TTE must be performed first in suspected IE.

(2) Given to its better sensitivity, TOE must be performed in cases of initially negative TTE with a high level of clinical suspicion, in suspected prosthetic valve endocarditis, and when TTE provides inadequate imaging."

ESC/EACTS

"Transesophageal echocardiography (TOE) should be considered when transthoracic echocardiography (TTE) is of suboptimal quality or when thrombosis, prosthetic valve dysfunction or endocarditis is suspected." (text)

Continued

Table 31.4

Continued

Suspected noninfective endocarditis
ASE
"Transthoracic echocardiographic surveillance in patients with primary antiphospholipid syndrome given the high prevalence of NBTE in these patients.
Transthoracic echocardiographic surveillance in patients with lupus erythematosus with secondary antiphospholipid syndrome."
Mitral stenosis
ESC/EACTS
"TTE usually provides sufficient information for routine management."
"TOE should be performed to exclude LA thrombus before PMC or after an embolic episode." (text)
Suspected paradoxical embolism
ASE
"TTE is recommended for the evaluation of a right-to-left shunt and atrial septal anatomy in a patient who presents with cryptogenic stroke, especially in the setting of elevated right atrial pressure with documented PE or deep venous thrombosis of lower extremities or pelvic veins. If the shunt could not be demonstrated by color Doppler, contrast echocardiography using intravenous injection of agitated saline should be performed at baseline and after provocative maneuvers (such as coughing or Valsalva maneuver).
TEE may be performed if TTE fails to demonstrate a right-to-left shunt."
EAE
"(1) TOE is traditionally the gold standard for the detection of PFO, however in the presence of good image quality, transthoracic echo is sufficient to detect the presence of a PFO. Performance of a valid Valsalve [sic] maneuver or strong cough must be ensured with both methods."

Abbreviations: *AF* atrial fibrillation; *IE*, infective endocarditis; *LA*, left atrium/left atrial; *LAA*, left atrial appendage; *LV*, left ventricle/left ventricular; *NBTE*, nonbacterial thrombotic endocarditis; *PE*, pulmonary embolism; *PFO*, patent foramen ovale; *PMC*, percutaneous mitral commissurotomy; *TEE*, transesophageal echocardiography; *TOE*, transesophageal echocardiography; *TTE*, transthoracic echocardiography.

Any LAT/LAAT or LVT

Sen et al. performed a multivariate analysis to determine the factors associated with cardiac thrombus detected by multiplanar TEE (40 LAAT, 28 LAT, and 8 LVT) in 151 patients who had first symptomatic ischemic stroke or TIA within the previous 7 days. Coronary artery disease [odds ratio (OR) = 3.0, 95% confidence interval (CI) 1.2–7.4); $P = 0.07$] and radiological large stroke (adjusted OR = 2.8, 95% CI 1.2–6.4) were significantly associated with the presence of thrombus (Sen et al., 2004). Fralick et al. analyzed a series of 1272 patients admitted to hospital with stroke or TIA of whom 11 (0.9%) had a cardiac thrombus identified by TTE. In univariate analysis, male sex (OR 9.11, 95% CI 1.74–197.37), heart failure (OR 6.38, 95% CI 1.38–22.60), and coronary artery disease (OR 5.24, 95% CI 1.36–17.60) were associated with thrombus.

Thus, although intracardiac thrombi are strongly associated with cardiac disease, they can rarely occur in its absence.

Mass/tumor

Cardiac tumors are rare and even rarer causes of stroke. Cardiac myxomas are the most common primary cardiac tumor and estimated to cause of 0.5% of ischemic strokes (Saric et al., 2016; Stefanou et al., 2018). Systemic signs (fever, asthenia, weight loss) occur in one-third of cases and cardiac auscultation abnormalities in two-thirds (Pinede et al., 2001). In a systematic literature review of 133 cases of stroke associated with cardiac myxoma, the initial presentation was neurological in 97/109 (90%). The erythrocyte sedimentation rate was increased in 23/25 (median 60 mm/h, range 30–85). C-reactive protein was elevated in 10/16 (median 1.8 mg/L, range 0.1–35. Vascular risk factors (hypertension, smoking, atrial fibrillation, hyperlipidemia, diabetes mellitus, or coronary artery disease) occurred in less than 10% each in this series with a mean age of 42 ± 19 (SD) years (Yuan and Humuruola, 2015). This is in contrast to a series of 13 cases of myxoma with cerebrovascular events with mean age 62 ± 18 (SD) years reported by Stefanou et al. in whom concomitant cardiovascular risk factors were common (hypertension 10, diabetes mellitus 3, atrial fibrillation 2, smoking 2, hyperlipidemia 1 and coronary artery disease 1) (Stefanou et al., 2018).

In Cho et al' s series of 1833 consecutive patients with cerebral ischemic stroke or TIA in sinus rhythm without a cardiac disease history studied within 7 days after onset, an atrial myxoma was detected by TEE in 4 (0.2%) (Cho et al., 2010).

Infective endocarditis

IE classically presents with fever, cardiac signs and symptoms or embolization. In the International Collaboration on Endocarditis prospective cohort study of 2781 patients from 58 sites in 25 countries carried out from

2000 to 2005, admission signs and symptoms were fever >38°C in 96%, elevated C-reactive protein in 62%, elevated erythrocyte sedimentation rate in 61%, new cardiac murmur in 48%, worsening of old cardiac murmur in 20%, hematuria in 26% and embolic events in 17% (Murdoch et al., 2009).

In low-income countries, rheumatic heart disease is the most important risk factor for IE. In higher-income countries, chronic intravenous access, previous IE, recent invasive procedures, degenerative native valve disease, prosthetic valves, intravenous drug use, congenital heart disease, intracardiac electronic implanted devices, hemodialysis, cancer, HIV infection, and postcardiac transplant valvulopathy are implicated in the development of IE (Murdoch et al., 2009; Cahill and Prendergast, 2016; Bin Abdulhak et al., 2018). Clinical presentations that engender a high suspicion for IE include *Staphylococcus aureus* bacteremia, fever in patients with prosthetic heart valves, intracardiac devices, and intravenous drug use (Bin Abdulhak et al., 2018).

NBTE

NBTE is strongly associated with underlying diseases, particularly advanced malignancies and systemic lupus erythematosus (Libman–Sacks endocarditis). Patients with systemic lupus erythematosus and antiphospholipid antibodies (aPL) have a higher prevalence of vegetations, particularly on the mitral valve (Asopa et al., 2007). In two series of patients with acute cerebral ischemic events and cancer, NBTE was detected by EC in 9/51 by TEE and 5/215 by TTE. Intracardiac thrombus was detected in 3/51 and 2/216, respectively (Dutta et al., 2006; Merkler et al., 2015). NBTE may occur with other such as tuberculosis, uremia, infection with the human immunodeficiency virus, and trauma from indwelling pulmonary and central venous catheters (Asopa et al., 2007).

Mitral stenosis

Rheumatic mitral valvular disease remains common in lower income countries (Chandrashekhar et al., 2009). In the United States of America, the prevalence of mitral stenosis detected by EC is estimated as 0.02%–0.2% (Nkomo et al., 2006).

Patent foramen ovale

At autopsy, the overall incidence of PFO is 27.3%, slightly declining with increasing age from 34.3% in the first three decades of life to 20.2% in the 9th and 10th decades (Hagen et al., 1984). Among 974 patients admitted for recent ischemic stroke of all types and referred for transesophageal echocardiography (TEE) in whom the presence or absence of PFO could be determined accurately, PFO was present in 294 patients (30.2%). There was a U-shaped age distribution with the highest prevalences of 38.9% for ages <30 years and 34.8% for ages >79 years and the lowest prevalence of 25.7% at ages 40–49 years (Gupta et al., 2008).

A multicenter cohort study of 1272 patients admitted to hospital with stroke or TIA between 2010 and 2015 identified 66 patients (5.2%) with PFO detected by TTE. In a multivariable logistic regression model assessing the odds of detecting PFO in this series, the presence of any of the following variables decreased the odds of PFO: age older than 60 years (adjusted OR 0.34, 95% CI 0.20–0.57), presence of dyslipidemia (adjusted OR 0.39, 95% CI 0.15–0.84), or history of prior stroke or TIA (adjusted OR 0.31, 95% CI 0.09–0.76 (Fralick et al., 2019).

In a multivariate regression model based on eight studies of patients with stroke not attributed to another known cause (cryptogenic stroke), the prevalence of PFO was determined to be age related with a reduction of 0.72 (0.67–0.77) for each addition decade of life after the third decade. Other factors that reduced the estimated prevalence of PFO were history of hypertension, history of diabetes mellitus, history of previous stroke or TIA, smoking, and no cortical infarct on neuroimaging. Applying this algorithm, the lowest possible estimated prevalence for PFO is 23% (Kent et al., 2013).

Although there are clinical factors that reduce the chance that PFO will be detected by EC, none are sufficient to preclude at least a 20%–25% chance of detection.

COST-EFFECTIVENESS STUDIES

Determining the best strategy for applying EC to secondary stroke prevention requires choosing which patients to study and which modalities to use, based on both costs and outcomes. Such a determination lends itself well to decision-analysis modeling. Three cost-effectiveness analyses using decision-analysis modeling have been published addressing various strategies for employing TTE and TTE in secondary stroke prevention. Due to the availability of necessary data, all three studies restricted the analysis to detection of intracardiac thrombi and the subsequent benefits of anticoagulation over antiplatelet therapy.

McNamara et al. used a Markov model to evaluate the benefits and costs of nine diagnostic strategies involving TTE and TEE, including single and sequential approaches, selective imaging by presence or absence of cardiac history, or no imaging and with anticoagulating all or none. They excluded from analysis patients in atrial fibrillation. LAT was the only indication for anticoagulation. They concluded that TEE alone in only those with cardiac history was most cost-effective at $9000 per quality adjusted life year (QALY). These results were moderately sensitive to efficacy of anticoagulation and incidence of intracranial bleeding during anticoagulation and were mildly sensitive to prevalence of left atrial thrombus, rate of recurrent stroke in patients with thrombus, quality of life after stroke, cost of TEE and specificity of TTE. TTE, alone or in sequence with TEE, was not cost-effective compared with TEE. The base case parameter for thrombus prevalence of 8% was varied from 3% to 17% with 59% of thrombi assumed to be present in patients with a cardiac history (McNamara et al., 1997).

Meenan et al. performed model-based cost-effectiveness analysis for newly diagnosed patients with AIS with parameters based on systematic literature review. The analysis was limited to intracardiac (left atrial or left ventricular) thrombus. They evaluated the same nine strategies as McNamara et al. Only selective TEE in those with "manifest heart disease" (prior MI, congestive heart failure, valvular heart disease, and arrhythmia other than AF) was better than managing patients by standard medical therapy without EC. This strategy had an incremental cost-effectiveness ratio of $137,600/QALY. The base case parameter for thrombus prevalence in patients with cardiac disease of 5% was varied from 1% to 36% and for those without cardiac disease a base case of 0.7% was varied from 0.3% to 3.3%. At a 2% thrombus prevalence—the estimated prevalence among unselected stroke patients from systematic review—no echocardiographic strategy was more cost-effective than no testing (Meenan et al., 2007).

These results differ from McNamara et al. in which selective application of TEE had a much lower cost of $9000 per QALY for the same winning strategy. The reduced costs per QALY can be attributed to different model assumptions by McNamara et al. They assumed thrombus prevalence among unselected patients was assumed to be 8% vs 2% assumed by Meenan et al. TEE was assumed by McNamara et al. in their base case to have perfect accuracy as opposed to sensitivity of 93% and specific of 97% by Meenan et al. A 40% recurrent stroke rate was assumed among patients with thrombus by McNamara et al., substantially higher than Mennan et al.'s estimate of 22%. The cost of TEE was assumed to be $360, substantially lower than $564 estimate by Meenan et al. Finally, TEE complications were not modeled. All these assumptions by McNamara et al. all favor more cost-effective use of TEE to identify patients with thrombus for anticoagulation.

Holmes et al. used a decision-analytic model to estimate cost-effectiveness of different EC strategies in the management of stroke and TIA for patients aged 45, 55, and 65 years. The model is based on all stroke patients being tested but only those with LAT receiving the benefits and harms of treatment. Four strategies were tested: all undergo second harmonic TTE (TTEh), all undergo TEE, all undergo TTEh and those with negative studies undergo TEE, all undergo TTEh and those with positive studies undergo TEE. They did not include strategies of no EC with anticoagulate all or no EC with anticoagulate none as did the previous two analyses. There were no separate analyses for those with and without cardiac disease. Costs were inflation adjusted to 2009–2010 levels. They concluded that TTEh as a single study is the most cost-effective choice for EC and is a cost-effective use of resources compared with TEE when EC of some kind is going to be performed. Cost was £17,541–£22,361/QALY (Holmes et al., 2014). This is equivalent to 2010 $26,986–$34,400 (https://www.poundsterlinglive.com/bank-of-england-spot/historical-spot-exchange-rates/usd/USD-to-GBP-2010 Accessed January 4, 2020).

Holmes et al. used several assumptions different from Meenan et al. that favored the cost-effectiveness of TTEh over TEE. For the base case thrombus prevalence in all, Holmes et al. used 5.45%, similar to the 5% used by Meenan et al. for those with positive cardiac history. Holmes et al. used a sensitivity of 0.7282 for LAT detection by TTEh, whereas Meenan used 0.42 based on older TTEf data. The relative cost of TTEh compared to TEE was 37% in the Holmes analysis as compared to 82% in the Meenan et al. analysis.

These three analyses share two similar conclusions: (1) EC to detect intracardiac thrombus is not cost-effective in the general stroke population, but only if there are cardiac factors that increase the prevalence of thrombus to at least 5%, (2) a single most sensitive test is more cost-effective that a less sensitive test followed by a more sensitive test. This is in contrast to the recommendations in Table 31.4 from professional societies, which often recommend sequential TTE followed by TEE and do not address cost-effectiveness.

SUMMARY AND CONCLUSIONS

EC in unselected patients with stroke has a very low yield of therapeutically relevant findings and is not cost-effective. Except for PFO, the other findings on EC that are therapeutically relevant for secondary stroke prevention are almost always associated with history, signs or symptoms of cardiac or systemic disease. The finding of PFO carries different therapeutic relevance depending on the additional clinical characteristics of the patient.

Except for PFO device closure, data to support treatment of therapeutically relevant EC finding are based on observational studies. The benefits of PFO device closure demonstrated in the RCTs can be presumed only for those patients who meet the eligibility criteria and match the characteristics of those enrolled in those RCTs. Tables 31.5 and 31.6 summarize some of the key eligibility criteria and characteristics of the enrolled subjects from these RCTs. These trials were comprised mostly of patients under the age of 60 with few or no vascular

Table 31.5

Selected eligibility criteria from the PFO device closure randomized controlled trials

	Age (years)	mRS	Vascular exclusion	Cardiac exclusion	Hematologic exclusion	Medical exclusion	Lacune exclusion
CLOSURE (Furlan et al., 2012) (v.5.0)	18–60	≤2	>50% stenosis, dissection	Many other heart lesions	Coagulopathies		
PC Trial (Meier et al., 2013) (v.2)	<60	≤3	"Clinically relevant" atherosclerosis, dissection	Many other heart lesions	Coagulopathies		
RESPECT (Saver et al., 2017) (Rev M)	18–60	≤3	>50% stenosis, dissection	Many other heart lesions	Coagulopathies	Uncontrolled HTN, DM	Lacunar infarct w age ≥50, HTN, DM, or WMHI
REDUCE (Sondergaard et al., 2017) (v.1/15/16)	18–< 60	≤2	>50% stenosis, dissection	Many other heart lesions	Coagulopathies	Uncontrolled HTN, DM	Clinical Lacunar stroke syndrome
CLOSE (Mas et al., 2017) (v.11)	16–60	</=3	≥30 stenosis; occlusion w HTN, DM, HCL, smoking, MI, PVD; arterial dissection	Many other heart lesions	Coagulopathies		Lacunar infarct w DM or HTN or old lacunae or WMHI
DEFENSE-PFO (Lee et al., 2018)[a]	18–80		>50% stenosis, dissection	Many other heart lesions	Coagulopathies		Lacunar infarct or syndrome

[a]https://clinicaltrials.gov/ct2/show/NCT01550588?cond=DEFENSE-PFO&draw=2&rank=1, Accessed January 15, 2020.
Abbreviations: *DM*, diabetes mellitus; *HCL*, hypercholesterolemia; *HTN*, hypertension; *MI*, myocardial infarction; *mRS*, modified Rankin scale; *PVD*, peripheral vascular disease; *WMHI*, white matter hyperintensities; yrs, years.

Table 31.6

Selected patient characteristics from the PFO device closure randomized controlled trials

		Age (yrs)	HTN	DM	Current smoker	HCL	mRS
CLOSURE (Furlan et al., 2012)	Active	46.3 ± 9.6	33.8%		21.5%	47.4%	
	Control	45.7 ± 9.1	28.4%		22.6%	40.9%	
PC Trial (Meier et al., 2013)	Active	44.3 ± 10.2	24.0%	2.5%	25.5%	24.5%	
	Control	44.6 ± 10.1	27.6%	2.9%	22.4%	29.5%	
RESPECT (Saver et al., 2017)	Active	45.7 ± 9.7	32.1%	6.6%	15.0%	39.3%	
	Control	46.2 ± 10.0	31.8%	8.5%	11.4%	40.5%	
REDUCE (Sondergaard et al., 2017)	Active	45.4 ± 9.3	25.4%	4.1%	14.3%		
	Control	44.8 ± 9.6	26.0%	4.5%	11.2%		
CLOSE (Mas et al., 2017)	Active	42.9 ± 10.1	11.3%	1.3%	28.6%	12.6%	0–1 82.8%
	Control	43.8 ± 10.5	10.2%	3.8%	29.4%	15.3%	0–1 80.4%
DEFENSE-PFO (Lee et al., 2018)	Active	49 ± 15	20%	10%	16.7%	30%	0–1 78.3%
	Control	54 ± 12	28.3%	13.3%	26.7%	41.7%	0–1 75%

Abbreviations: *DM*, diabetes mellitus; *HCL*, hypercholesterolemia; *HTN*, hypertension; *mRS*, modified Rankin scale; yrs, years

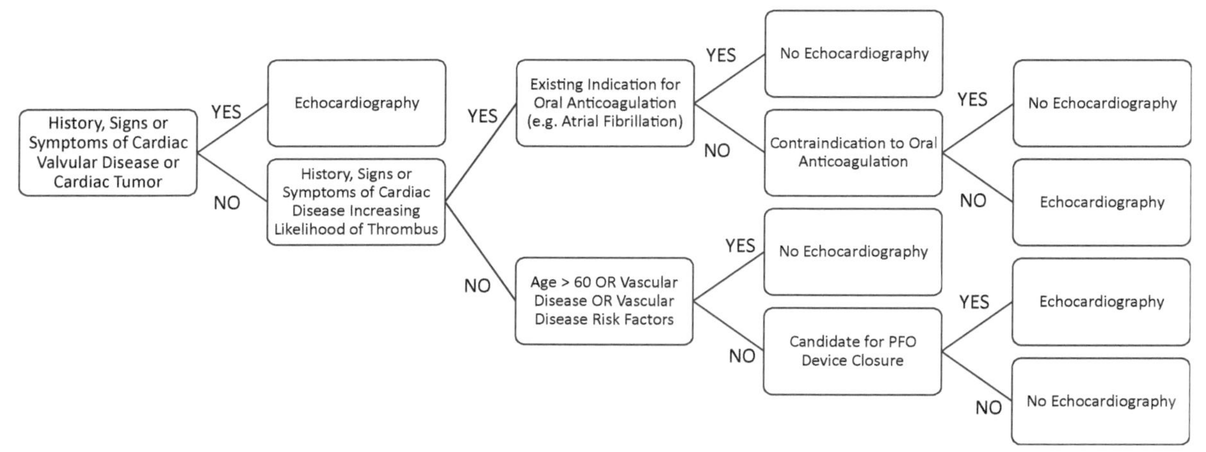

Fig. 31.1. A selective strategy for echocardiography for secondary stroke prevention.

disease risk factors and little or no manifest cerebrovascular disease.

Fig. 31.1 provides a selective strategy to rationally and cost-effectively employ EC for secondary stroke prevention. Choice of the echocardiographic modality should be based on the specific pathology or pathologies that are under consideration for the individual clinical situation, e.g. the use of agitated saline to detect PFO. The most sensitive echocardiographic modality should be selected since the objective is to definitively identify all therapeutically relevant pathology with a single test.

References

Abreu TT, Mateus S, Correia J (2005). Therapy implications of transthoracic echocardiography in acute ischemic stroke patients. Stroke 36: 1565–1566.

Ahmad O, Ahmad KE, Dear KB et al. (2010). Echocardiography in the detection of cardioembolism in a stroke population. J Clin Neurosci 17: 561–565.

Ahmed N, Audebert H, Turc G et al. (2019). Consensus statements and recommendations from the ESO-Karolinska stroke update conference, Stockholm 11–13 November 2018. Eur Stroke J 4: 307–317.

Asopa S, Patel A, Khan OA et al. (2007). Non-bacterial thrombotic endocarditis. Eur J Cardiothorac Surg 32: 696–701.

Baddour LM, Wilson WR, Bayer AS et al. (2015). Infective endocarditis in adults: diagnosis, antimicrobial therapy, and management of complications: a scientific statement for healthcare professionals from the American Heart Association. Circulation 132: 1435–1486.

Baumgartner H, Falk V, Bax JJ et al. (2017). 2017 ESC/EACTS Guidelines for the management of valvular heart disease. Eur Heart J 38: 2739–2791.

Bin Abdulhak AA, Qazi AH, Tleyjeh IM (2018). Workup and management of native and prosthetic valve endocarditis. Curr Treat Options Cardiovasc Med 20: 73.

Cahill TJ, Prendergast BD (2016). Infective endocarditis. Lancet 387: 882–893.

Chandrashekhar Y, Westaby S, Narula J (2009). Mitral stenosis. Lancet 374: 1271–1283.

Chlumsky J, Bojar M, Svab P et al. (2006). Transoesophageal echocardiography in patients with stroke and low risk of embolic etiology. Neuroradiol J 19: 394–398.

Cho HJ, Choi HY, Kim YD et al. (2010). Transoesophageal echocardiography in patients with acute stroke with sinus rhythm and no cardiac disease history. J Neurol Neurosurg Psychiatry 81: 412–415.

Dawn B, Hasnie AM, Calzada N et al. (2006). Transesophageal echocardiography impacts management and evaluation of patients with stroke, transient ischemic attack, or peripheral embolism. Echocardiography 23: 202–207.

de Bruijn SF, Agema WR, Lammers GJ et al. (2006). Transesophageal echocardiography is superior to transthoracic echocardiography in management of patients of any age with transient ischemic attack or stroke. Stroke 37: 2531–2534.

Douen A, Pageau N, Medic S (2007). Usefulness of cardiovascular investigations in stroke management: clinical relevance and economic implications. Stroke 38: 1956–1958.

Dutta T, Karas MG, Segal AZ et al. (2006). Yield of transesophageal echocardiography for nonbacterial thrombotic endocarditis and other cardiac sources of embolism in cancer patients with cerebral ischemia. Am J Cardiol 97: 894–898.

Ellis K, Ziada KM, Vivekananthan D et al. (2006). Transthoracic echocardiographic predictors of left atrial appendage thrombus. Am J Cardiol 97: 421–425.

Fralick M, Goldberg N, Rohailla S et al. (2019). Value of routine echocardiography in the management of stroke. CMAJ 191: E853–E859.

Furlan AJ, Reisman M, Massaro J et al. (2012). Closure or medical therapy for cryptogenic stroke with patent foramen ovale. N Engl J Med 366: 991–999.

Galougahi KK, Stewart T, Choong CY et al. (2010). The utility of transoesophageal echocardiography to determine management in suspected embolic stroke. Intern Med J 40: 813–818.

Gorani D, Dilic M, Kulic M et al. (2013). Comparison of two and three dimensional transthoracic versus transesophageal echocardiography in evaluation of anatomy and pathology of left atrial apendage. Med Arch 67: 318–321.

Gupta V, Yesilbursa D, Huang WY et al. (2008). Patent foramen ovale in a large population of ischemic stroke patients: diagnosis, age distribution, gender, and race. Echocardiography 25: 217–227.

Gupta N, Lau C, Al-Dehneh A et al. (2010). Importance of performing transesophageal echocardiography in acute stroke patients older than fifty. Echocardiography 27: 1086–1092.

Habib G, Lancellotti P, Antunes MJ et al. (2015). 2015 ESC Guidelines for the management of infective endocarditis: the task force for the management of infective endocarditis of the European Society of Cardiology (ESC). Endorsed by: European Association for Cardio-Thoracic Surgery (EACTS), the European Association of Nuclear Medicine (EANM). Eur Heart J 36: 3075–3128.

Hagen PT, Scholz DG, Edwards WD (1984). Incidence and size of patent foramen ovale during the first 10 decades of life: an autopsy study of 965 normal hearts. Mayo Clin Proc 59: 17–20.

Harloff A, Handke M, Reinhard M et al. (2006). Therapeutic strategies after examination by transesophageal echocardiography in 503 patients with ischemic stroke. Stroke 37: 859–864.

Holmes M, Rathbone J, Littlewood C et al. (2014). Routine echocardiography in the management of stroke and transient ischaemic attack: a systematic review and economic evaluation. Health Technol Assess 18: 1–176.

Jacobs AK, Anderson JL, Halperin JL et al. (2014). The evolution and future of ACC/AHA clinical practice guidelines: a 30-year journey: a report of the American College of Cardiology/American Heart Association Task Force on practice guidelines. Circulation 130: 1208–1217.

Kamel H (2018). Evidence-based management of patent foramen ovale in patients with ischemic stroke. JAMA Neurol 75: 147–148.

Karakus G, Kodali V, Inamdar V et al. (2008). Comparative assessment of left atrial appendage by transesophageal and combined two- and three-dimensional transthoracic echocardiography. Echocardiography 25: 918–924.

Kent DM, Ruthazer R, Weimar C et al. (2013). An index to identify stroke-related vs incidental patent foramen ovale in cryptogenic stroke. Neurology 81: 619–625.

Kernan WN, Ovbiagele B, Black HR et al. (2014). Guidelines for the prevention of stroke in patients with stroke and transient ischemic attack: a guideline for healthcare professionals from the American Heart Association/American Stroke Association. Stroke 45: 2160–2236.

Kim SJ, Choe YH, Park SJ et al. (2012). Routine cardiac evaluation in patients with ischaemic stroke and absence of known atrial fibrillation or coronary heart disease: transthoracic echocardiography vs. multidetector cardiac computed tomography. Eur J Neurol 19: 317–323.

Lee JM, Park JJ, Jung HW et al. (2013). Left ventricular thrombus and subsequent thromboembolism, comparison of anticoagulation, surgical removal, and antiplatelet agents. J Atheroscler Thromb 20: 73–93.

Lee PH, Song JK, Kim JS et al. (2018). Cryptogenic stroke and high-risk patent foramen ovale: the DEFENSE-PFO trial. J Am Coll Cardiol 71: 2335–2342.

Leow AS, Sia CH, Tan BY et al. (2018). A meta-summary of case reports of non-vitamin K antagonist oral anticoagulant use in patients with left ventricular thrombus. J Thromb Thrombolysis 46: 68–73.

Liu J, Frishman WH (2016). Nonbacterial thrombotic endocarditis: pathogenesis, diagnosis, and management. Cardiol Rev 24: 244–247.

Marino B, Jaiswal A, Goldbarg S et al. (2016). Impact of transesophageal echocardiography on clinical management of patients over age 50 with cryptogenic stroke and normal transthoracic echocardiogram. J Hosp Med 11: 95–98.

Mas JL, Arquizan C, Lamy C et al. (2001). Recurrent cerebrovascular events associated with patent foramen ovale, atrial septal aneurysm, or both. N Engl J Med 345: 1740–1746.

Mas JL, Derumeaux G, Guillon B et al. (2017). Patent foramen ovale closure or anticoagulation vs. antiplatelets after stroke. N Engl J Med 377: 1011–1021.

McCarthy CP, Vaduganathan M, McCarthy KJ et al. (2018). Left ventricular thrombus after acute myocardial infarction: screening, prevention, and treatment. JAMA Cardiol 3: 642–649.

McIntyre WF, Spence J, Belley-Cote EP (2018). Assessing the quality of evidence supporting patent foramen ovale closure over medical therapy after cryptogenic stroke. Eur Heart J 39: 3618–3619.

McNamara RL, Lima JA, Whelton PK et al. (1997). Echocardiographic identification of cardiovascular sources of emboli to guide clinical management of stroke: a cost-effectiveness analysis. Ann Intern Med 127: 775–787.

Meenan RT, Saha S, Chou R et al. (2007). Cost-effectiveness of echocardiography to identify intracardiac thrombus among patients with first stroke or transient ischemic attack. Med Decis Making 27: 161–177.

Meier B, Kalesan B, Mattle HP et al. (2013). Percutaneous closure of patent foramen ovale in cryptogenic embolism. N Engl J Med 368: 1083–1091.

Menon BK, Coulter JI, Bal S et al. (2014). Acute ischaemic stroke or transient ischaemic attack and the need for inpatient echocardiography. Postgrad Med J 90: 434–438.

Merkler AE, Navi BB, Singer S et al. (2015). Diagnostic yield of echocardiography in cancer patients with ischemic stroke. J Neurooncol 123: 115–121.

Miles JA, Garber L, Ghosh S et al. (2018). Association of transthoracic echocardiography findings and long-term outcomes in patients undergoing workup of stroke. J Stroke Cerebrovasc Dis 27: 2943–2950.

Mojadidi MK, Winoker JS, Roberts SC et al. (2014). Two-dimensional echocardiography using second harmonic imaging for the diagnosis of intracardiac right-to-left shunt: a meta-analysis of prospective studies. Int J Cardiovasc Imaging 30: 911–923.

Mugge A, Daniel WG, Hausmann D et al. (1990). Diagnosis of left atrial appendage thrombi by transesophageal echocardiography: clinical implications and follow-up. Am J Card Imaging 4: 173–179.

Murdoch DR, Corey GR, Hoen B et al. (2009). Clinical presentation, etiology, and outcome of infective endocarditis in the 21st century: the International Collaboration on Endocarditis-Prospective Cohort Study. Arch Intern Med 169: 463–473.

Nishimura RA, Otto CM, Bonow RO et al. (2014). 2014 AHA/ACC Guideline for the management of patients with valvular heart disease: a report of the American College of Cardiology/American Heart Association Task Force on Practice Guidelines. Circulation 129: e521–e643.

Nishimura RA, Otto CM, Bonow RO et al. (2017). 2017 AHA/ACC focused update of the 2014 AHA/ACC Guideline for the management of patients with valvular heart disease: a report of the American College of Cardiology/American Heart Association Task Force on Clinical Practice Guidelines. Circulation 135: e1159–e1195.

Nkomo VT, Gardin JM, Skelton TN et al. (2006). Burden of valvular heart diseases: a population-based study. Lancet 368: 1005–1011.

O'Gara PT, Kushner FG, Ascheim DD et al. (2013). 2013 ACCF/AHA guideline for the management of ST-elevation myocardial infarction: a report of the American College of Cardiology Foundation/American Heart Association task force on practice guidelines. Circulation 127: e362–e425.

Omran H, Rang B, Schmidt H et al. (2000). Incidence of left atrial thrombi in patients in sinus rhythm and with a recent neurologic deficit. Am Heart J 140: 658–662.

Pallesen LP, Ragaller M, Kepplinger J et al. (2016). Diagnostic impact of transesophageal echocardiography in patients with acute cerebral ischemia. Echocardiography 33: 555–561.

Pepi M, Evangelista A, Nihoyannopoulos P et al. (2010). Recommendations for echocardiography use in the diagnosis and management of cardiac sources of embolism: European Association of Echocardiography (EAE) (a registered branch of the ESC). Eur J Echocardiogr 11: 461–476.

Pinede L, Duhaut P, Loire R (2001). Clinical presentation of left atrial cardiac myxoma. A series of 112 consecutive cases. Medicine (Baltimore) 80: 159–172.

Powers WJ (2018). Additional factors in considering patent foramen ovale closure to prevent recurrent ischemic stroke. JAMA Neurol 75: 895.

Powers WJ, Rabinstein AA, Ackerson T et al. (2019). Guidelines for the early management of patients with acute ischemic stroke: 2019 update to the 2018 guidelines for the early management of acute ischemic stroke: a guideline for healthcare professionals from the American Heart Association/American Stroke Association. Online data supplement 1, table LXXVII. Stroke 50: e344–e418.

Pristipino C, Sievert H, D'Ascenzo F et al. (2019). European position paper on the management of patients with patent foramen ovale. General approach and left circulation thromboembolism. Supplementary table 10. Eur Heart J 40: 3182–3195.

Raskob GE, van Es N, Verhamme P et al. (2018). Edoxaban for the treatment of cancer-associated venous thromboembolism. N Engl J Med 378: 615–624.

Sallach JA, Puwanant S, Drinko JK et al. (2009). Comprehensive left atrial appendage optimization of thrombus using surface echocardiography: the CLOTS multicenter pilot trial. J Am Soc Echocardiogr 22: 1165–1172.

Saric M, Armour AC, Arnaout MS et al. (2016). Guidelines for the use of echocardiography in the evaluation of a cardiac source of embolism. J Am Soc Echocardiogr 29: 1–42.

Saver JL, Carroll JD, Thaler DE et al. (2017). Long-term outcomes of patent foramen ovale closure or medical therapy after stroke. N Engl J Med 377: 1022–1032.

Schwartz BG, Alexander CT, Grayburn PA et al. (2018). Utility of routine transesophageal echocardiography in patients with stroke or transient ischemic attack. Proc (Bayl Univ Med Cent) 31: 401–403.

Secades S, Martin M, Corros C et al. (2013). Diagnostic yield of echocardiography in stroke: should we improve patient selection? Neurologia 28: 15–18.

Sen S, Laowatana S, Lima J et al. (2004). Risk factors for intracardiac thrombus in patients with recent ischaemic cerebrovascular events. J Neurol Neurosurg Psychiatry 75: 1421–1425.

Shantsila E, Lip GY (2016). Antiplatelet versus anticoagulation treatment for patients with heart failure in sinus rhythm. Cochrane Database Syst Rev 9: CD003333.

Sondergaard L, Kasner SE, Rhodes JF et al. (2017). Patent foramen ovale closure or antiplatelet therapy for cryptogenic stroke. N Engl J Med 377: 1033–1042.

Srichai MB, Junor C, Rodriguez LL et al. (2006). Clinical, imaging, and pathological characteristics of left ventricular thrombus: a comparison of contrast-enhanced magnetic resonance imaging, transthoracic echocardiography, and transesophageal echocardiography with surgical or pathological validation. Am Heart J 152: 75–84.

Stefanou MI, Rath D, Stadler V et al. (2018). Cardiac myxoma and cerebrovascular events: a retrospective cohort study. Front Neurol 9: 823.

Stewart MH, Gilliland Y (2018). Role of transesophageal echocardiography in patients with ischemic stroke. Prog Cardiovasc Dis 61: 456–467.

Strandberg M, Marttila RJ, Helenius H et al. (2002). Transoesophageal echocardiography in selecting patients for anticoagulation after ischaemic stroke or transient ischaemic attack. J Neurol Neurosurg Psychiatry 73: 29–33.

Vitebskiy S, Fox K, Hoit BD (2005). Routine transesophageal echocardiography for the evaluation of cerebral emboli in elderly patients. Echocardiography 22: 770–774.

Wasser K, Weber-Kruger M, Jurries F et al. (2019). The cardiac diagnostic work-up in stroke patients–a subanalysis of the find-AFRANDOMISED trial. PLoS One 14: e0216530.

Wein T, Lindsay MP, Cote R et al. (2018). Canadian stroke best practice recommendations: secondary prevention of stroke, sixth edition practice guidelines, update 2017. Int J Stroke 13: 420–443.

Whitlock RP, Sun JC, Fremes SE et al. (2012). Antithrombotic and thrombolytic therapy for valvular disease: antithrombotic therapy and prevention of thrombosis, 9th ed: American College of Chest Physicians Evidence-Based Clinical Practice Guidelines. Chest 141: e576S–e600S.

Yaghoubi E, Nemati R, Aghasadeghi K et al. (2011). The diagnostic efficiency of transesophageal compared to transthoracic echocardiographic findings from 405 patients with ischemic stroke. J Clin Neurosci 18: 1486–1489.

Yahia AM, Shaukat A, Kirmani JF et al. (2009). Prevalence and prediction of left atrial thrombus in patients with a recent cerebral ischemic event, who are in sinus rhythm: a single-center experience. J Neuroimaging 19: 323–325.

Yuan SM, Humuruola G (2015). Stroke of a cardiac myxoma origin. Rev Bras Cir Cardiovasc 30: 225–234.

Zhang L, Harrison JK, Goldstein LB (2012). Echocardiography for the detection of cardiac sources of embolism in patients with stroke or transient ischemic attack. J Stroke Cerebrovasc Dis 21: 577–582.

Handbook of Clinical Neurology, Vol. 177 (3rd series)
Heart and Neurologic Disease
J. Biller, Editor
https://doi.org/10.1016/B978-0-12-819814-8.00023-8

Chapter 32

The relationship between heart disease and cognitive impairment

KRISTIN L. MILLER*, LAURA PEDELTY, AND FERNANDO D. TESTAI

Department of Neurology and Rehabilitation, University of Illinois at Chicago, Chicago, IL, United States

Abstract

Neurodegenerative dementias, such as Alzheimer's disease, and vascular cognitive impairment were once considered unrelated processes. Emerging evidence, however, shows that both conditions often coexist and that vascular risk factors in midlife predispose to the development of cognitive decline later in older adults. In addition, recent advanced in basic science research have elucidated key underpinnings of this association. In this chapter, we review the clinical and basic science data that explain the relationship between vascular risk factors, heart disease, and cognitive decline.

INTRODUCTION

Acquired cognitive impairment (CI), including in its most severe form, dementia, and heart disease are prevalent conditions, both increasing in prevalence with aging. There are approximately 50 million cases of dementia worldwide and the number is expected to triple in the coming decade (Alzheimer's disease facts and figures, 2020). The social and economic burden are substantial and are expected to increase with the aging population. Heart disease is the leading cause of death in the developed world. There is increasing evidence that heart disease and cognitive decline are linked.

CI presents as a gradually progressive acquired deterioration in one or more cognitive domains (memory, attention, language, visuospatial skills, and executive function) due to degenerative changes or other insults to the brain. Dementia is diagnosed when deficits are severe enough to interfere with instrumental activities of daily living, employment, or recreational activities. Mild cognitive impairment (MCI) refers to a state of impaired cognition that involves a single domain, or is well enough compensated to allow independence in daily activities, employment, and recreation. MCI reflects a point on a spectrum of functional and CI and is associated with increased risk of progression to dementia, and presents as a potential target for disease-modifying interventions.

Alzheimer's disease (AD) is the most common type of dementia. Histologically, AD is characterized by neurodegeneration with amyloid plaques and neurofibrillary tangles made up of hyperphosphorylated tau. It typically presents with amnestic deficits reflecting early regional involvement, and is the most common form of dementia in the US, representing 60%–80% of cases (Alzheimer's disease facts and figures, 2020). Vascular cognitive impairment (VCI) is the second most common recognized form of dementia but mixed pathologies increasingly reported at autopsy. VCI is a protean entity, broadly defined as cognitive and functional decline in the presence of and attributable to vascular disease (Gorelick et al., 2011; Wolters et al., 2018; Sachdev et al., 2019). While strategic single strokes or substantial ischemic burden due to multiple or recurrent stroke have long been recognized as contributors to VCI, increasing evidence suggests that cardiac risk factors and coronary heart disease (CHD) may be associated with CI independent of overt stroke. CHD, congestive heart failure (CHF) and low-output states, and atrial fibrillation (AF) have all been implicated in the development of CI; cardiac risk factors, particularly at midlife, are associated with development of CI in later life, and may potentiate degenerative changes in AD. Here we explore the

*Correspondence to: Kristin Lucille Miller, MD, Department of Neurology and Rehabilitation, University of Illinois at Chicago, NPI Building, North Tower, 912 S Wood Street, Chicago, IL, (60612), United States. Tel: +1-312-413-7729, Fax: +1-312-413-8215, E-mail: kmille25@uic.edu

relationship between heart disease and dementia, including the impact of stroke, structural heart disease, shared pathologies; cardiogenic changes in cerebral perfusion and metabolism, and complex interactions between neural tissues and their vascular supply.

CORONARY HEART DISEASE

CHD, both clinical and subclinical, is associated with an increased risk of CI (Wolters et al., 2018). Cognitive decline following myocardial infarction (MI) has been described, and prospective and longitudinal population-based studies demonstrate increased risk of dementia in individuals with a history of CAD and angina or MI (Gharacholou et al., 2011; Gu et al., 2019). Surrogate markers of coronary artery disease (CAD), including EKG evidence of prior MI in individuals with a history of CHD and coronary artery calcium (CAC), have also been associated with impaired cognitive performance and dementia (Breteler et al., 1994; Newman et al., 2005; Kuller et al., 2016). While MI may predispose to brain changes through systemic as well as cardiac factors including inflammation, prothrombotic state, decreased cardiac output and arrhythmias, the association of markers of atherosclerosis and CHD, including carotid intimomedial thickening, ankle-brachial index, and CAC with CI may be mediated in part by systemic atherosclerosis-related brain pathology. Peripheral artery disease is a potent risk factor for cognitive decline in CHD, and individuals with CI related to increased CAC burden have been shown to manifest increased brain pathology, including ischemic stroke, microbleeds, white matter changes, and atrophy (Breteler et al., 1994; Kuller et al., 2016; Cannon et al., 2017).

CONGESTIVE HEART FAILURE AND LOW OUTPUT STATES

Heart failure is a well-established risk factor for CI and dementia (Breteler et al., 1994; Gorelick et al., 2011; Cannon et al., 2017). In stroke and dementia-free older patients, low cardiac index is associated with a greater risk of developing all-type dementia and AD, independent of other cardiac risk factors including AF, and incident heart failure (HF) associated with more rapid progression of CI (Jefferson et al., 2015; Hammond et al., 2018).

The brain is a highly metabolically active organ, requiring 15%–20% of the systemic blood flow and oxygen supply despite representing ~2% of total body mass. With limited intracellular energy storage, the brain relies on a constant stable cerebral blood flow (CBF) to support metabolism, maintained in healthy individuals through a range of systemic blood pressure through cerebral autoregulatory mechanisms. In HF, CBF may be altered in a number of ways. Chronically low cardiac output is associated with reduced CBF, and even in the absence of globally reduced CBF, HF patients manifest changes in cerebral autoregulation and vasoreactivity (Van Der Velpen et al., 2017a). Chemoreactivity may be altered in the setting of chronic hypercarbia. HF thus may promote diffuse ischemic brain injury, manifest as increased burden of pathological changes, including diffuse white matter changes, but also differentially affecting substrate-sensitive regions important to memory and cognition including the hippocampus (Vogels et al., 2007). Even in the absence of clinically reduced ejection fraction (EF) impaired left ventricular function is associated with vascular ischemic changes (Russo et al., 2013). HF is associated with systemic inflammation which may be exacerbated by renal hypoperfusion-induced activation of the renin-angiotensin system, with alterations in autonomic tone and cardiovascular instability. Finally, a complex cascade of endothelial dysfunction, platelet hyperreactivity, chronic inflammation, increased levels of procoagulants, and stasis contribute to hypercoagulability in patients with HF, with attendant large- and small-vessel ischemic pathology (Kim et al., 2016; Van Der Velpen et al., 2017b).

ATRIAL FIBRILLATION

AF is the most common cardiac arrhythmia, affecting approximately 9% of the population over age 65, with a projected tripling in prevalence in the coming decades (Ott et al., 1997; Miyasaka et al., 2006). AF has long been recognized as a risk factor for CI. Cross-sectional studies have documented increased prevalence of dementia of all types and impaired cognition in patients with AF compared to those without it (Ott et al., 1997; Kilander et al., 1998; Knecht et al., 2008). Prospective studies have found prevalent AF to increase the risk of dementia and have documented greater cognitive decline and higher risk of dementia in patients with incident AF (Bunch et al., 2010; Santangeli et al., 2012; Bruijn et al., 2015; Chen and Soliman, 2019). AF is associated with a four- to fivefold risk of ischemic stroke, and at least 15% of all strokes are related to AF, creating a potent substrate for multiinfarct dementia (Wolf et al., 1991). Overt or clinical strokes, however, may represent only the tip of the iceberg. A meta-analysis of 15 prospective studies evaluating cognitively normal individuals with or without AF found prevalent AF to increase the risk of cognitive decline or dementia, with a subgroup analysis suggesting that the finding was largely driven by increased risk in patients with stroke (Kwok et al., 2011). Other studies, however, document an independent effect of AF, persisting after controlling for a history of stroke and two independent meta-analyses of longitudinal studies reported a higher risk of CI or dementia in patients with AF in the absence of a history of stroke (Wolf et al., 1991; Kalantarian et al., 2014).

Imaging studies document multiple pathologies in AF. Silent cerebral infarcts (SCI), radiographic ischemic lesions occurring in the absence of a clinical history of stroke or TIA, have been described. The burden of SCI has been reported to be increased in patients with persistent as compared to paroxysmal AF. A recent cross-sectional MRI imaging study documented a substantial burden of large cortical and noncortical infarcts among patients with AF. Most of the lesions were clinically silent; infarct burden was associated with increased CI even in the setting of clinically silent infarcts without history of stroke or TIA (Gaita et al., 2013; Kalantarian et al., 2014).

These pathological findings may be driven by mechanisms other than discrete symptomatic or silent ischemic strokes. Radiographic studies also report associated diffuse white matter changes, microbleeds, and decreased global and regional brain volumes, including cortical gray matter and hippocampal atrophy, which was more strongly associated with persistent than with paroxysmal AF, and was found to correspond to deficits in learning and memory (Knecht et al., 2008; Conen et al., 2019).

AF is often associated with HF, and is associated with increased risk of CI in patients with comorbid HF (Myserlis et al., 2017). Even in the absence of clinical HF, AF is associated with disruptions in hemodynamics. Decreased atrial contribution to LV filling and beat-to-beat variability lead to reduced stroke volume and CO. Indirect evidence suggests that reduced CO and resultant hypoperfusion may underlie pathological changes including brain atrophy associated with CI. Both high (>90 bpm) and low (<50 bpm) as contrasted to moderate (50–90 bpm) ventricular rate response were associated with a sevenfold risk of dementia in patients with preexisting cognitive decline (Cacciatore et al., 2007). Physiological studies provide evidence of a role for decreased cerebral perfusion in the evolution of AF-related CI. A phase-contrast MRI study evaluating blood flow in the cervical arteries found lower CBF in individuals with AF compared to those without; furthermore, persistent AF was associated with greater decrements in total CBF than paroxysmal AF (Gardarsdottir et al., 2018). A cross-sectional TCD (transcranial Doppler) found that the presence of AF in patients with HF was associated with greater reductions in CBF velocities and increased CI, independent of severity of HF (Alosco et al., 2015).

ATRIAL CARDIOMYOPATHY

The above observations, along with the lack of a clear temporal relationship between AF and stroke, the modulation of stroke risk in AF by comorbid vascular risk factors (VRFs), the low risk of stroke in the absence of such factors (lone AF), and the failure of rhythm control to ameliorate stroke risk, have led to the evolution of the concept of "atrial cardiomyopathy" (Kamel et al., 2016). VRF, including aging, obesity, the metabolic syndrome, sleep apnea, and others may promote structural, functional, and electrical changes in the left atrium that contribute to stroke and cerebral pathology independent of arrhythmia. Atrial cardiomyopathy is characterized by fibrosis, which is associated with dilatation and hypocontractility, by endothelial dysfunction, and by hypercoagulability due to stasis, increased platelet activation, and secretion of prothrombotic factors. Inflammation and hypercoagulability in turn promote further atrial remodeling and fibrosis, as does the inflammatory state associated with stroke. Thus, there is a bidirectional association between the atrial cardiomyopathy associated with AF and stroke. Atrial cardiomyopathy may further contribute to cerebral pathology through hypoperfusion due to impaired filling and reduced atrial systolic function. The burden of atrial fibrosis on cardiac MRI imaging and decreased atrial strain on echocardiography are associated with increased stroke risk independent of arrhythmia (Daccarett et al., 2011; Costa et al., 2016). The Atherosclerosis Risk in Communities Neurocognitive Study (ARIC-NCS) documented higher incident dementia and cognitive decline in individuals with abnormal P-wave indices which are associated with associated with atrial cardiomyopathy, independent of AF and ischemic stroke (Gutierrez et al., 2019). A complex interplay of structural and functional cardiac changes, hematological factors, and comorbid cardiovascular risk factors, thus underlying the evolution of CI and dementia in AF and atrial cardiomyopathy (Fig. 32.1).

CARDIOVASCULAR RISK FACTORS AND COGNITIVE IMPAIRMENT

Epidemiologic studies evaluating the association of cardiovascular risk factors and cognitive decline are tasked with demonstrating the presence of vascular disease as well as CI using neuropsychiatric testing; this is best done with large, population-based cohort studies. While there are a great number of population-based cohort studies looking at VRFs and dementia, some of the best known include the Framingham Heart Study, ARIC Study, the Rotterdam Study, and the Honolulu-Asia Aging Study (HAAS).

The Framingham Heart Study, started in 1948, is one of the first long-term population-based cohort studies for dementia with comprehensive neuropsychological testing to examine the effects of VRFs (Mahmood et al., 2014). The ARIC Study is a prospective epidemiologic study involving four United States communities designed to

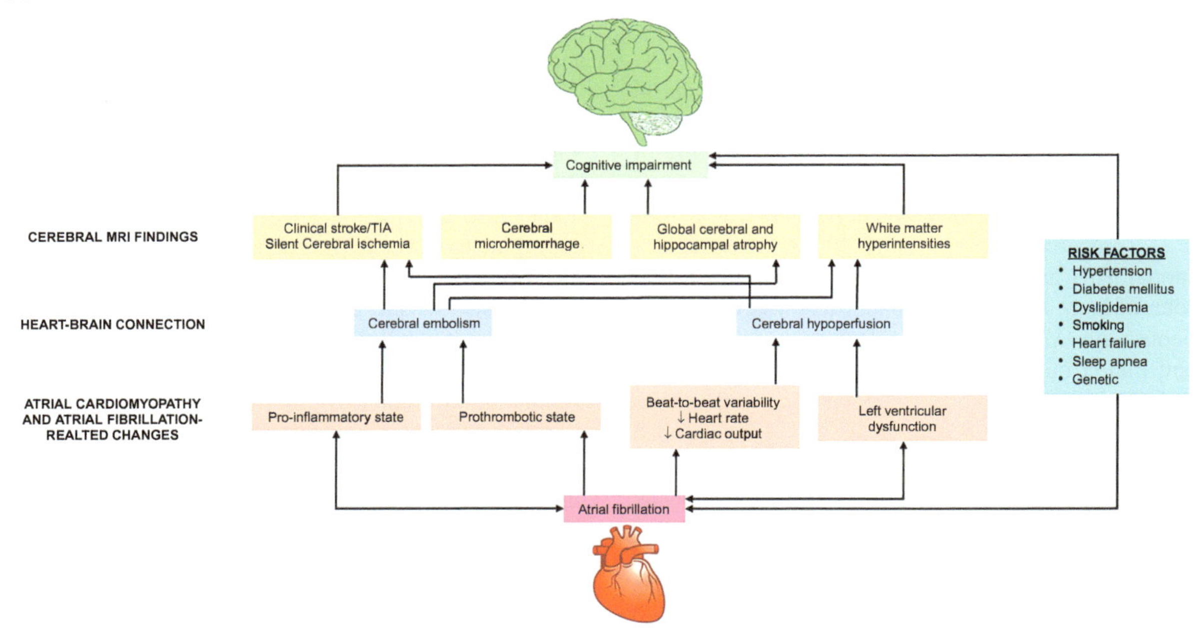

Fig. 32.1. Potential contributors to cognitive impairment in atrial fibrillation and atrial cardiomyopathy.

investigate causes of atherosclerosis and its outcomes. In a cohort followed from 1987 to 1989 through 2011–2013, 1516 cases of dementia were identified in 15,744 participants (Gottesman et al., 2017). The Rotterdam Study is a prospective cohort study of 7983 people living in Rotterdam in the Netherlands. This study has been ongoing since 1990 and participants are monitored for a variety of diseases common in the elderly, including CAD, diabetes, and dementia of various types. The prevalence of dementia increased from 1% in participants aged 55–65 years to over 40% in those aged 90 years and above (Hofman et al., 2007). The HAAS is a longitudinal cohort study investigating the prevalence, incidence, and risk factors for cognitive decline and dementia in Japanese-American men and was established in 1991. Its cohort comprises surviving participants of the Honolulu Heart Program and at baseline the prevalence of dementia was 7.6% in men aged 71–93 years (2.1% for men aged 71–74 and 33.4% for men aged 85–93) (Gelber et al., 2012). These studies as well as other important epidemiological studies have evaluated different modifiable cardiovascular risk factors and their effects on dementia (Norton et al., 2014).

Hypertension

Several studies have shown an association with hypertension, especially midlife hypertension, with dementia and CI. Launer et al. evaluated the relationship between midlife blood pressure and cognitive functioning in 3735 Japanese-American men participating in the HAAS (Launer et al., 2000). The average age was 78 years when the cognitive testing was done between 1991 and 1993. Blood pressures were categorized based on readings from prior examinations in 1965, 1968, and 1971. When controlled for age and education, for each 10 mmHg increase in midlife systolic blood pressure (SBP) there was an increase in risk for intermediate cognitive function of 7% (95% confidence interval [CI], 3%–11%) and poor cognitive function of 9% (95% CI, 3%–16%). The strength of this relationship was diminished when adjusting for stroke, coronary heart disease, and atherosclerosis. This relationship was not seen with midlife diastolic blood pressure (DBP).

In the ARIC STUDY, both prehypertension (hazard ratio [HR], 1.31; 95% CI, 1.14–1.51) and hypertension (HR, 1.39; 95% CI, 1.22–1.59) were associated with an increased risk of dementia. In a subgroup of the Framingham study with 1702 participants aged 55–88 years, blood pressure levels and chronicity of hypertension were associated with worse scores on neuropsychological testing (Elias et al., 1993; Gottesman et al., 2017).

In a subanalysis of the Longitudinal Population Study of 70-year-olds in Göteburg, Sweden, 382 nondemented subjects were followed for 15 years with repeated examinations. Participants who developed dementia by ages 79–85 had higher SBP at age 70 (mean 178 vs 164 mmHg, $P = 0.034$) than those who did not develop dementia. Similarly, higher DBP at ages 70 and 75 were associated with increased incidence of dementia later in life (Skoog et al., 1996).

In 2005, Whitmer et al. published on a retrospective cohort of 8845 study members of the Kaiser Permanente

Medical Care Program of Northern California who underwent evaluation between 1964 and 1973 when they were between the ages of 40 and 44. Approximately 8% of the cohort was later diagnosed with dementia based on medical records from 1994 to 2003. Participants with hypertension were 24% more likely to have dementia (95% CI, 1.04–1.48). This study looked at other risk factors, including smoking, total cholesterol, and diabetes at midlife; the results of these other risk factors and the associated with dementia are discussed in their respective subsections (Whitmer et al., 2005).

Diabetes mellitus

In 1997, Elias et al. published on a cohort of 187 participants of the Framingham study with noninsulin dependent diabetes mellitus (DM) and 1624 nondiabetic participants who had been followed for 28–30 years (Elias et al., 1997). They found that the diagnosis and duration of DM were associated with a greater risk of poor performance on cognitive tests, especially those of visual memory. This risk was greatest in those with both DM and hypertension. In 1999, Ott et al. published data from the Rotterdam Study, a prospective, population-based cohort study of elderly subjects, 126 of whom developed dementia. In this subjects, the presence of DM doubled the risk of dementia (relative risk [RR], 1.9; 95% CI 1.3–2.8) (Ott et al., 1999).

In 2008, Roberts et al. conducted a population-based study in Olmsted County, Minnesota of patients aged 70–89. A total of 329 participants were found to have MCI) and/or dementia compared to 1640 participants without. The frequency of DM was similar in participants with and without MCI (odds ratio (OR), 1.16; 95% CI, 0.85–1.57). However, MCI was associated with earlier onset DM (before age 65), longer duration of DM (greater than 10 years), and greater severity of DM (Roberts et al., 2008). In the aforementioned retrospective cohort study by Whitmer et al., DM was associated with a 46% increase in likelihood of being diagnosed with dementia (95% CI, 1.19–1.79) (Whitmer et al., 2005).

A systematic review by Biessels et al. (2006) identified 14 longitudinal population-based studies evaluating the incidence of dementia and diabetes These studies provide evidence that the risk of dementia is increased in patients with DM, though were limited in identifying which diabetes-related factors are at the root of this association (Biessels et al., 2006).

Dyslipidemia

In a random sample of subjects from two large population-based studies, the North Karelia Project and the Finnish Multinational Monitoring of Trends and Determinants in Cardiovascular Disease (FINMONICA) study, 6.1% of participants were diagnosed with MCI (Kivipelto et al., 2001). This study found that those subjects with MCI in late life had significantly higher serum total cholesterol (≥250.9 mg/dL) in midlife. Similarly, in the retrospective cohort study by Whitmer et al., subjects with high cholesterol (≥240 mg/dL) were 42% more likely to have dementia (95% CI, 1.22–1.66) (Whitmer et al., 2005).

Tobacco use

In 1998, as part of the population-based Rotterdam Study, Ott et al. found that former and current smokers had an increased risk of dementia compared to never-smokers (RR, 2.2; 95% CI, 1.3–3.6) (Ott et al., 1998)). In the Honolulu-Asia Aging Study, Tyas et al. studied the associated between midlife smoking and late-life dementia (Tyas et al., 2003). After adjusting for age, education, and apolipoprotein E genotype, the risk of dementia in smokers increased in a dose-dependent pattern. At medium smoking levels, the OR was 2.18 (95% CI, 1.07–4.69), and at heavy smoking levels the OR was 2.40 (95% CI, 1.16–5.17). However, at very heavy smoking levels, there was no association with dementia (OR, 1.08; 95% CI, 0.43–2.63). In the Whitmer et al. retrospective cohort study, subjects who reported ever smoking in midlife were 26% more likely to have dementia (95% CI, 1.08–1.47) (Whitmer et al., 2005).

Coronary artery and peripheral artery disease

In the Cardiovascular Health Study (CHS), Rosano et al. demonstrated an association between higher coronary artery calcification (CAC) scores, a marker of coronary atherosclerotic disease, and abnormal cognitive functioning in 409 older adults with a mean age of 79 years (Rosano et al., 2005). More recently, the Age, Gene, Environment Susceptibility-Reykjavik Study (AGES-RS) revealed a similar correlation between higher CAC scores and lower cognitive functioning as well as brain pathology on magnetic resonance imaging (MRI), including infarcts, microbleeds, and white matter lesions (Vidal et al., 2010).

Multiple risk factors

Many of these cohort studies have demonstrated a higher association of dementia when multiple risk factors are present. In 2005, Luchsinger et al. published on a longitudinal cohort of 1138 Medicare recipients in northern Manhattan followed for a mean length of 5.5 years (Luchsinger et al., 2005). Four cardiovascular risk factors (hypertension, diabetes, CAD, and current smoking) were associated with a higher risk of dementia when analyzed individually. That risk further increased with an increasing number of coexisting risk factors (Luchsinger et al., 2005). Whitmer et al. demonstrated a dose-dependent pattern of dementia with an increasing number of risk factors. When comparing to those with no cardiovascular risk factors, those who had one risk factor were 27% more likely to be diagnosed with dementia (95% CI, 1.02–1.58) and those with 2 or more risk factors were 70% more likely to be diagnosed with dementia (95% CI, 1.34–2.12) (Ott et al., 1999).

Using data from the Rush Memory and Aging Project, Song et al. assessed cardiovascular risk burden in 1588 patients using the Framingham General Cardiovascular Risk Score (FGCRS) at baseline and at follow-up (mean follow-up 5.8 years, range 1 to 21) (Song et al., 2020). Those participants in the highest tertile of FGCRS compared to the lowest tertile had faster decline in multiple domains of cognition, including global cognition. When analyzing magnetic resonance imaging data, those participants with higher FGCRS had smaller volumes in the hippocampus, gray matter, and total brain as well as greater volume of white matter hyperintensities.

LINKS BETWEEN VASCULAR RISK FACTORS AND NEURODEGENERATION

Alzheimer disease, vascular dementia, and mixed dementias

Classic neurodegenerative diseases such as AD and vascular-associated cognitive disorders, including VCI and its most severe form, vascular dementia, have historically been considered independent processes with established predisposing factors, histologic findings, and pathophysiologic mechanisms.

Most of the cases of AD are sporadic and several risk factors, including unhealthy lifestyle, pollutants, heart disease, and genetic variants, mostly APOE4, have been identified. In addition, there are familial cases of AD caused by mutations in the β-amyloid precursor protein (APP), presenilin 1 (PSEN1), and presenilin 2 (PSEN2) genes. APP is a transmembrane protein that, once cleaved by an α-secretase, results in the formation of nonamyloidogenic fragments that are efficiently eliminated from the brain. However, the sequential action of a β-secretase and by γ-secretase on APP results in the formation of the amyloidogenic fragments $A\beta_{1-40}$ and $A\beta_{1-42}$. At the histopathological level, AD is characterized by the presence of neurofibrillary tangles (NFT), hyperphosphorylated tau protein (p-tau), and amyloid-beta (Aβ), which is mainly represented by $A\beta_{1-42}$. AD patients also have an increased deposition of amyloid protein, mainly $A\beta_{1-40}$, in the wall of leptomeningeal and cortical vessels. The accumulation of amyloid on the vessel wall starts in the tunica media and adventitia but eventually involves all the layers. This process, called cerebral amyloid angiopathy (CAA), is associated with loss of vascular smooth muscle cells (VSMC) and leads to vascular disarrangement and degeneration. The development of CAA can result in convexal subarachnoid hemorrhage, cerebral microhemorrhages, cortical superficial siderosis, lobar hemorrhage, or white matter hyperintensities that are more commonly seen in the posterior regions of the brain (Alzheimer's Association, 2016; Scheltens et al., 2016; Gorelick et al., 2017). In comparison, the term VCI refers to the cognitive dysfunction resulting from vascular injury. The mechanisms of cognitive decline in these patients include cerebral hypoperfusion, infarction of areas involved in cognitive function, white matter disease, and axonal integrity. In addition, VRFs upregulate the production of proinflammatory mediators and reactive oxygen species which affect axonal function, neuronal survival and synaptic connectivity (Rosenberg et al., 2016; Skrobot et al., 2017).

There is solid evidence showing that AD and VCI often coexist. As an example, in an autopsy study, as many as a third of the individuals with diagnosis of dementia had mixed AD- and vascular-related findings (Jellinger, 2006). Also, in the US National Alzheimer Disease Coordinating Center show, out of 4629 patients with diagnosis of AD 80% had evidence of coexisting cerebrovascular pathology including lacunes, multiple microinfarcts, atherosclerosis, arteriolosclerosis, CAA, or hemorrhage (Toledo et al., 2013). Furthermore, numerous longitudinal observational studies have confirmed a strong association between midlife VRFs, including hypertension, DM, smoking, and hypercholesterolemia, with late occurrence of both AD and VCI (Testai and Gorelick, 2016; Livingston et al., 2017). These observations suggest that the pathogenic mechanisms involved in neurodegeneration and VCI are intertwined and could synergistically compromise the normal function of neural and vascular cells (Iadecola and Gorelick, 2003; Iadecola et al., 2016).

Structure and function of the neurovascular unit

To understand the complex interactions that take place between VRFs and neurodegeneration, we have to define the neurovascular unit. From the anatomical, functional and trophic standpoints, an active codependence exists between neurons, astrocytes, and vascular cells, the latter represented by endothelial cells, VSMC and pericytes. Together, these cells constitute the *neurovascular unit* (NVU) which is a functional structure that has an active role in the maintenance of the cerebral homeostasis. The concerted action of these cells regulate CBF, cellular trophism, cerebral inflammation, and blood–brain barrier (BBB) function. In the next section, we briefly review these four functions.

The blood supply to the brain is provided by the anterior, middle and posterior cerebral arteries which run through the subarachnoid space. At the brain base, these arteries give rise to small perforating branches that irrigate the basal ganglia and brainstem. As these arterioles arise directly from large vessels, they are more susceptible to the mechanical stress caused by blood flow pulsation and chronic hypertension. Through their course, the large intracranial vessels divide successively and form penetrating arteries and arterioles that dive into the brain parenchyma. The composition of the NVU changes based on the vessel segmentation. The penetrating pial arteries, still surrounded by an extension of the subarachnoid space, are formed by a monolayer of endothelial cells and one to three layers of VSMCs surrounded by the basal membrane which separates the vessel from the cerebrospinal fluid (CSF). The CSF-filled cavity delimited by the basal membrane of the vessel and the glia limitans, the latter formed by the astrocytic end-feet, is called the perivascular space. The perivascular space, or Virchow-Robin space, has an important role in neuroinflammation and the clearance of waste materials such as Aβ. As the pial arteries penetrate the parenchymal, they give rise to arterioles. The basal membranes that cover the glia limitans and the vessel come in close proximity and the perivascular space eventually disappears. At this level, the NVU is formed by the endothelial cell surrounded by a single layer of VSMC which is in close relationship with the astrocytic end-feet. In the capillaries, the VSMC is replaced by pericytes and these are covered by astrocytes. Along the course of the vessel, local neurons innervate contractile cells and astrocytes (Fig. 32.2) (Kisler et al., 2017).

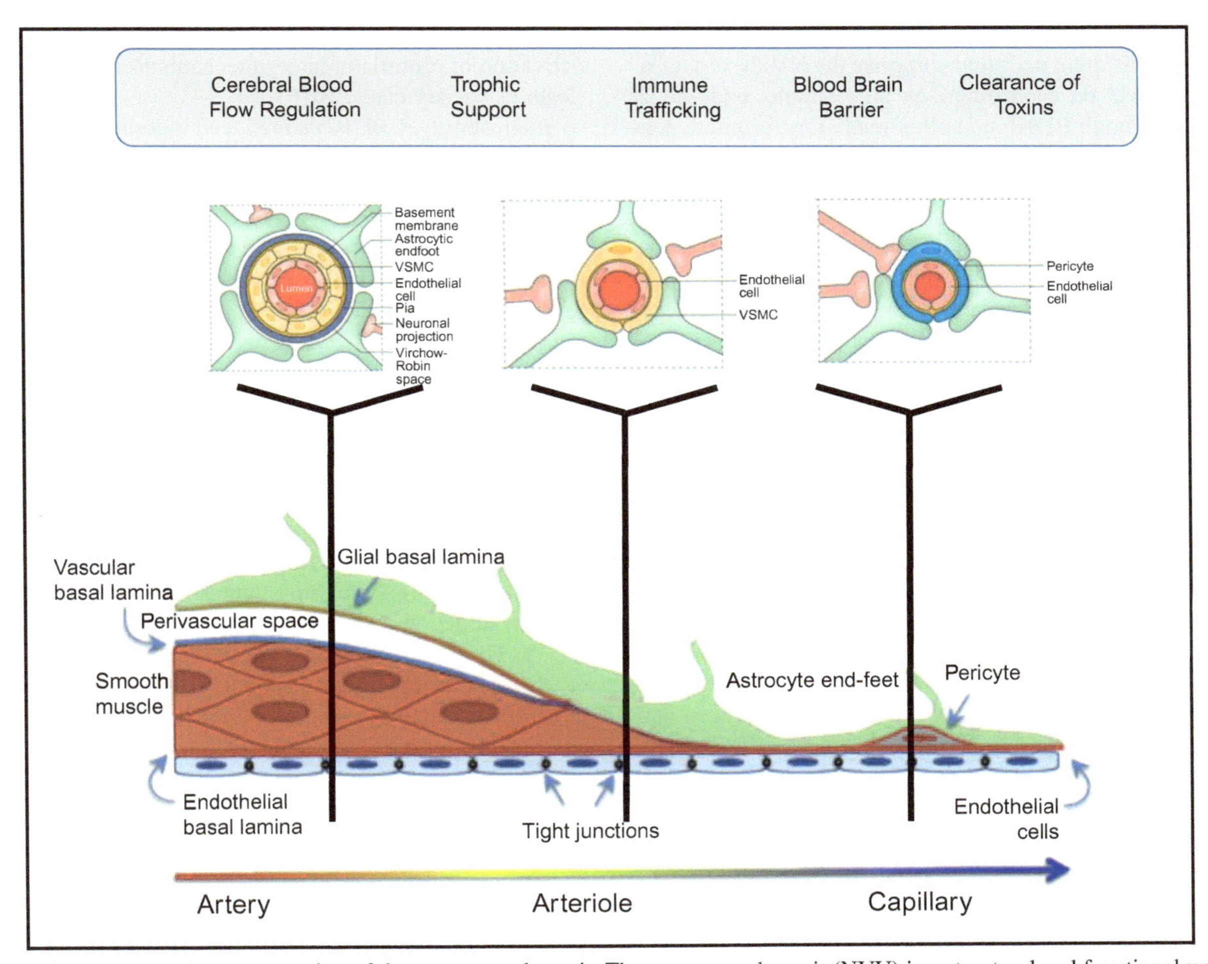

Fig. 32.2. Schematic representation of the neurovascular unit. The neurovascular unit (NVU) is a structural and functional unit comprised by neurons, astrocytic end-feet, VSMC, and endothelial cell. At the level of capillaries, the VSMC are replaced by pericytes. The CSF-filled space between the basal membrane of the vessel wall and the glia-limitans constitutes the perivascular space. The NVU regulates cerebral blood flow, immune cell trafficking, blood–brain barrier permeability, cellular trophism, and clearance of toxic substances. Modified from Kisler K, Nelson AR, Montagne A et al. (2017). Cerebral blood flow regulation and neurovascular dysfunction in Alzheimer disease. Nat Rev Neurosci 18: 419–434 with permission.

During neuronal activation, different diffusible chemicals, including nitric oxide (NO), prostanoids, adenosine, and ions, regulate VSMCs and, consequently, arteriolar tone and cerebral perfusion. At the capillary level, pericytes regulate CBF in response to local neurotransmitters, such as noradrenaline, electrical stimulation, and neuronal activity. In addition, endothelial cells have a central role in CBF control by producing potent vasodilators, such as NO or prostacyclin, or vasoconstrictors, such as endothelin-1, in response to chemical stimuli and shear stress. This synchronized process, called *neurovascular coupling*, ensures the fine adjustment of tissue perfusion in response to the changing neuronal metabolic demands (Kisler et al., 2017). In addition, through myogenic responses, smooth muscle cells constrict in response to hypertension and dilate if blood pressure decreases. This process, called *cerebrovascular autoregulation*, ensures a relatively constant CBF despite fluctuations in the blood pressure (Iadecola and Gottesman, 2019).

The different cells that comprise the NVU exert a *trophic* effect on each other. As an example, endothelial cells, through BDNF and other mediators, promote neuronal survival and neuroblast recruitment. In addition, through the activation of the Akt/PI3K pathway, they support oligodendrocyte proliferation and survival. Astrocytes have angiogenic and neurogenic properties that facilitate neuro-repair after brain injury. This trophic interdependence correlates with the microvascular rarefaction and endothelial atrophy that occur after neuronal or glial damage (Iadecola and Gottesman, 2019). At the capillary level, endothelial cells, along with the mural vascular cells and astrocyte end-feet, from the blood–brain barrier (BBB). The BBB has a fundamental role in the transport of nutrients to the brain and the removal of toxic material. Cerebral endothelial cells are tightly packed and have an elevated expression of adherent and tight junction molecules, such as cadherins, claudins, occludin, zonula occludens (ZO) proteins, and others. These cells form a layer with elevated transendothelial electrical resistance and low paracellular and transcellular permeability. Small molecules, such as oxygen and carbon dioxide, and lipophilic substances diffuse freely across the BBB. In comparison, large molecules, including glucose, lactate, and essential aminoacids, enter the brain using specialized membrane transporters. Aβ, in particular, is transported across the BBB to the periphery by the low-density lipoprotein receptor-related protein 1 (LRP1) (Sweeney et al., 2016).

Cross-talk between vascular risk factors and neurodegeneration

Several observational and preclinical studies have shown that VRFs affect the abnormal cerebrovascular responses. This process has been extensively studied in association with HTN and DM. In a large longitudinal study, for example, CBF was reduced in individuals with uncontrolled or untreated HTN, but not in those with controlled HTN (Muller et al., 2010). Though the study of the neurovascular function in humans is often confounded by the multiplicity of factors that can influence it, the evidence gathered in animal models suggests that the reduced CBF seen in association with VRFs is the result of endothelial dysfunction and neurovascular uncoupling. In the model of hypertension, for example, circulating angiotensin II causes vascular dysfunction and CBF dysregulation by activating prooxidative and inflammatory mechanisms. In the case of diabetes, advanced glycation end products (AGEs) increase the endothelial activation of matrix metalloproteinase-9 which disrupts BDNF-mediated neurotrophic signaling pathways supporting the potential role of trophic uncoupling. In addition, in models of diabetic retinopathy, hyperglycemia downregulates PDGFR resulting in the activation of proinflammatory mechanisms and pericyte death (Sweeney et al., 2016).

Interestingly, CBF is also reduced in individuals with AD and this seems to precede the development of brain atrophy and neurocognitive decline. Moreover, there is evidence suggesting that CBF decreases even before the deposition of Aβ takes place. The experimental data demonstrates that Aβ has a detrimental effect on the different components of the NVU. In addition to its known neurotoxic effect, Aβ causes pericyte loss and leads to the development of CAA. Also, autopsy studies and data obtained in animal models show that cerebral arteries of AD patients have VSMCs with a hypercontractile morphology and downregulation of endothelial-dependent and independent arterial relaxation mechanisms. Furthermore, Aβ is associated with an increased expression of proinflammatory cytokines and endogenous vasoconstrictors, such as endothelin-1, which result in microvascular dysfunction and further reduce CBF (Kisler et al., 2017). These observations support the notion that Aβ can cause vascular pathology. VRFs, in turn, close the loop by enhancing neurodegenerative processes. For example, hypertensive animals have enhanced tau phosphorylation and Aβ accumulation. Several pathogenic mechanisms seem to contribute to this phenomenon. Decreased NO and prostacyclin bioavailability, as occurs in VRF-associated endothelial dysfunction, increase the production of APP and β-secretase activity. In addition, angiotensin II promotes tau phosphorylation, increases β- and γ-secretase activity as well as Aβ deposition, and attenuates CBF responses independent of the effect on blood pressure. Similarly, animals with DM have an upregulation of APP processing and Aβ production. Furthermore, AGEs and its receptor (RAGE), both hallmarks of diabetes, facilitate the influx of peripheral

Aβ across the BBB to be brain. Aβ is a substrate of different proteases, including insulin-degrading enzyme; thus, hyperinsulinemia, as typically seen in type 2 diabetes, can reduce the clearance of Aβ. These observations explain, at least in part, the association observed between VRF and AD-type pathology (Testai and Gorelick, 2016). Similarly, in AD models, oligemia and ischemia–reperfusion increase the production of hypoxia-inducible factor 1a (HIF1a) which mediates the expression of β-secretase and facilitates the amyloidogenic processing of APP. In addition, through the upregulation of the mitogen-activated protein kinase (MAPK) pathway, hypoxia phosphorylates and inhibits the Aβ degrading enzyme, neprilysin (Zlokovic, 2011).

The BBB has a relevant role in the genesis and progression of CI (Tarasoff-Conway et al., 2015). Endothelial degeneration leads to the loss of adherent and tight junction molecules and results in BBB breakdown. As a consequence, there is extravasation of blood products to the brain parenchyma, including RBCs, Fe^{2+}, and thrombin, which increase the production of ROS and inflammatory mediators, including TNF-α, interleukin-1β, and MMP. These mediators are toxic to neurons, glial cells, and the endothelium, cause neuronal cell death and perpetuate the degradation of the BBB (Kisler et al., 2017). Additionally, the extravasation of albumin to the parenchyma results in edema which compromises the microvascular circulation and decreases tissue perfusion. Furthermore, plasmin has been associated with neuronal toxicity and fibrinogen with axonal retraction. LRP1 participates of the efflux of Aβ across of the BBB and APOE4, the most important genetic risk factor for AD, inhibits blocks this process. These results indicate that BBB dysfunction and genetic variants associated with AD jointly compromise the clearance of Aβ across the BBB (Tarasoff-Conway et al., 2015; Testai and Gorelick, 2016). The role of APOE4 in the development of dementia, however, continues to evolve and emerging evidence suggests that this mutation may also induce neurovascular dysregulation, BBB breakdown and loss of white matter integrity (Koizumi et al., 2018; Ishii and Iadecola, 2020).

In summary, the available evidence indicates that AD and VCI, once thought to be independent processes, have interconnected pathophysiologic mechanisms. These observations have been summarized in the *two hit hypothesis* (Fig. 32.3). This postulates that environmental, genetic, and vascular factors have a detrimental effect on the NVU and result in the activation of proinflammatory and prooxidative pathways that increase BBB permeability and compromise CBF (Hit 1). Consequently, there is a dysregulation of processes involved in the production and clearance of Aβ (Hit 2). The inflammatory and prooxidative environment, along with the deposition of Aβ, have a cytotoxic effect on neural, glial and vascular cells that perpetuates the NVU damage and leads to CI.

INTERVENTIONAL TRIALS

The compelling evidence of association between traditional VRFs and neurocognitive impairment suggests that treatment of modifiable risk factors may help prevent development of dementia in a number of cases. However, the randomized trials investigating VRF modification and its effect on cognitive decline have shown conflicting results.

Hypertension

In the Systolic Hypertension in Europe (SYST-EUR) phase 1 study, active treatment of hypertension to a goal SBP less than 150 (compared to untreated controls) demonstrated a reduction in incidence of dementia by 50% (Forette et al., 1998). However, as the trial was ended early due to a significant benefit in stroke there were only 32 incident cases of dementia. After these results, the controls from the original trial were actively treated and followed in SYST-EUR trial phase 2 (Gasowski et al., 1999). In this study, a reduction in SBP by at least 20 mmHg decreased the incidence of dementia at 2 years by 55% ($P < 0.001$). In the Perindopril Protection Against Recurrent Stroke Study (PROGRESS) the use of perindopril with or without indapamide was associated with a reduced rate of cognitive decline by 19% over a 4 year follow-up period (Applegate et al., 1994; Tzourio et al., 2003; Diener et al., 2008; Peters et al., 2008; Anderson et al., 2011; Bosch et al., 2019).

However, several other studies, including the Ongoing Telmisartan Alone and in Combination with Ramipril Global Endpoint Trial (ONTARGET), Telmisartan Randomized Assessment Study in ACE Intolerant Subjects with Cardiovascular Disease (TRASCEND), Prevention Regimen for Effectively Avoiding Second Strokes (PROFESS), Systolic Hypertension in the Elderly Program (SHEP), Heart Outcomes Prevention Evaluation-3 (HOPE-3), and Hypertension in the Very Elderly Trial cognitive function assessment (HYVET-COG), showed no effect of blood pressure lowering intervention on cognitive decline (Applegate et al., 1994; Diener et al., 2008; Peters et al., 2008; Anderson et al., 2011; Bosch et al., 2019). In 2019, the Systolic Blood Pressure Intervention Trial (SPRINT) MIND, designed to evaluate the effects of intensive SBP control (SBP $<$120 mmHg) compared with standard SBP treatment (SBP $<$140 mmHg) and its effect of the rate of probable dementia and MCI, demonstrated that over a median follow-up period of 5.11 years, there was no significant difference between the intensive treatment group compared to standard treatment in the occurrence

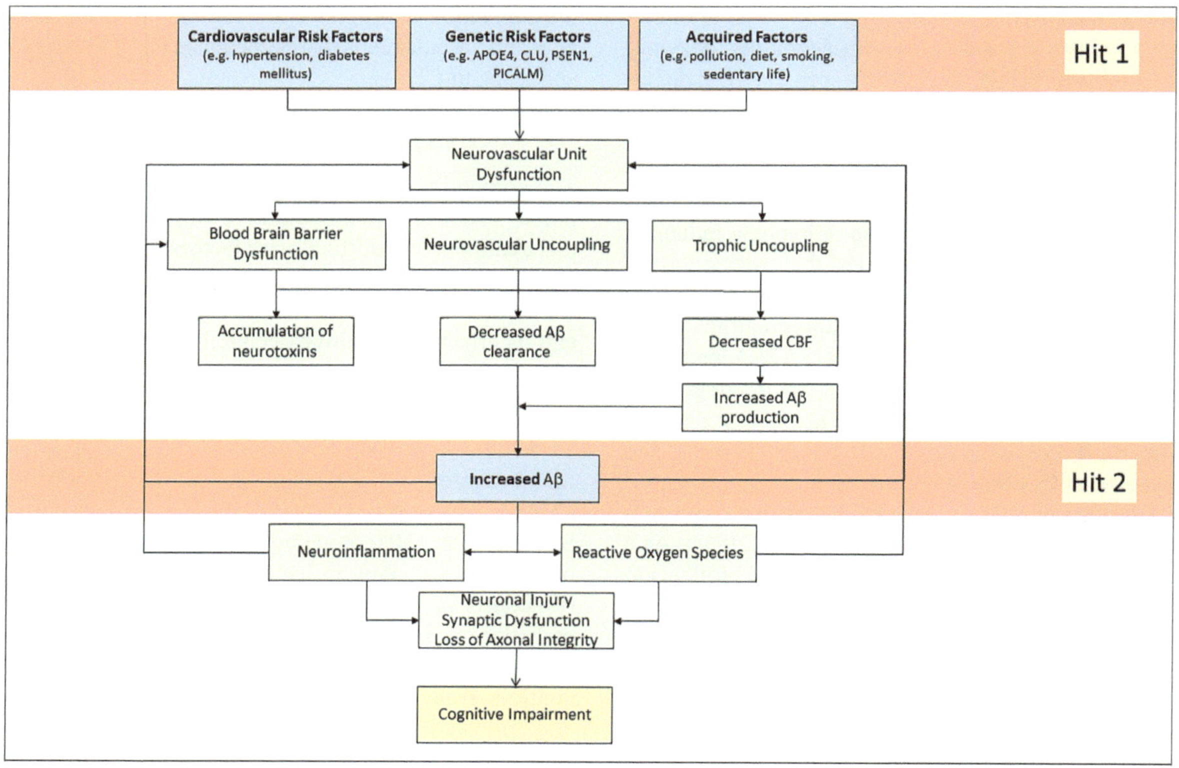

Fig. 32.3. Two hit hypothesis. The chronic exposure to acquired, genetic, and vascular factors causes NVU injury and results in blood–brain barrier (BBB) dysfunction, neurovascular uncoupling, and reduced cellular trophism (Hit 1). BBB dysfunction impairs the clearance of waste substances, including Aβ, and allows the influx of blood products to the brain, such as thrombin, plasmin, RBC, and Fe^{2+}. These upregulate the production of proinflammatory mediators and reactive oxygen species that further contribute to BBB breakdown and neurovascular uncoupling resulting in cerebral blood flow (CBF) dysregulation. Hypoperfusion and VRFs enhance the amyloidogenic processing of APP. In addition, BBB dysfunction compromises the clearance of Aβ. Together, these processes facilitate the deposition of Aβ which further affects the normal functioning of the neurovascular unit (NVU). Consequently, there is neuronal degeneration, synaptic dysfunction, and loss of axonal integrity, which manifest with cognitive impairment.

of probable dementia (7.2 vs 8.6 per 100,000 person-years; HR, 0.83; 95% CI, 0.67–1.04 (Williamson et al., 2019). There was a significant difference in the intensive treatment group in secondary cognitive outcomes of occurrence of MCI (14.6 vs 18.3 cases per 100,000 person-years; HR, 0.81; 95% CI, 0.69–0.95) and composite outcome of probable dementia or MCI (20.2 vs 24.1 cases per 100,000 person-years; HR, 0.85; 95% CI, 0.74–0.97). The results of this study are confounded by early termination due to benefit in composite of cardiovascular events and fewer than expected cases of dementia.

A recent meta-analysis performed by Hughes et al. included 12 randomized controlled trials comparing blood pressure lowering treatment with a control group and the effect on incidence of dementia or composite of dementia and CI. The meta-analysis included 92,135 participants across 12 trials, including SPRINT MIND, HYVET-COG, and SYST-EUR, and found that blood pressure lowering with antihypertensive agents compared to control was significantly associated with a lower risk of dementia or CI (OR, 0.93; 95% CI, 0.88–0.98) (Hughes et al., 2020).

Diabetes mellitus

While there are several studies investigating the effect of diabetic interventions on cognitive decline, they have failed to demonstrate benefit. In the Action to Control Cardiovascular Risk in Diabetes—Memory in Diabetes Study (ACCORD-MIND), 2977 patients with type 2 DM were randomized to aggressive diabetic control (HgbA1c goal <6.0%) or standard of care (HgbA1c goal 7.0%–7.9%). Neurocognitive testing was done at baseline and at 20- and 40-months and brain MRI was done at baseline and at 40 months post-randomization. While aggressive glycemic control was associated with decreased brain atrophy on MRI, there were no differences in cognitive outcomes between the two groups (Launer et al., 2011). In the Outcome Reduction with Initial Glargine Intervention (ORIGIN) trial, 10,320 participants over the age of 50 with were randomized to titrated insulin glargine targeting a fasting plasma glucose concentration under 95 mg/dL or standard care and to either omega 3 fatty acids or placebo. In the median follow up period of 6.2 years, insulin-mediated glycemic control had a neutral effect on cognitive function compared to standard of care (Cukierman-Yaffe et al., 2014). A meta-analysis including more than 24,000 individuals with type 2 DM randomized to intensive or standard glycemic control showed no effect of treatment on cognitive decline (Tuligenga, 2015). A Cochrane database systematic review concluded that there is limited evidence that any specific treatment can prevent cognitive decline in type 2 DM (Areosa Sastre et al., 2017).

Dyslipidemia

The Heart Outcomes Prevention Evaluation-3 (HOPE-3) was a double-blind, randomized, placebo-controlled trial randomizing participants without known cardiovascular disease to candesartan plus hydrochlorothiazide or placebo and to rosuvastatin or placebo (Gorelick et al., 2017; Bosch et al., 2019). A total of 1626 participants aged 70 years and older completed baseline and study end cognitive assessments and were followed for a median of 5.7 years. There was no significant difference in the scores of baseline and study end cognitive assessments in either the antihypertensive and/or statin vs placebo groups, despite reduced SBP by 6.0 mmHg in the candesartan/hydrochlorothiazide group and low-density lipoprotein cholesterol level by 24.8 mg/dL in the rosuvastatin group.

Multidomain interventions

Given the often multiple risk factors at play in patients with cardiovascular disease and the varied pathogenic mechanisms leading to CI, studies focusing on single interventions may be insufficient to demonstrate an effect on, and ultimately preservation of, cognitive function. Several studies have attempted to demonstrate the effect of multidomain interventions on cognitive function and preservation. In the Prevention of Dementia by Intensive Vascular Care (PreDIVA) study, 3526 individuals aged 70–78 years were randomized to usual risk factor care versus nurse-led multidomain intervention. In the median follow up period of 6.7 years, the rate of dementia did not differ between the two arms (Van Charante et al., 2016). In the Multidomain Alzheimer Preventive Trial (MAPT), 1680 subjects aged over 60 years were randomized to usual care versus multidomain intervention and with omega 3 polyunsaturated fatty acids or placebo. At 3 years, there were no significant differences in cognition decline between any of the three intervention groups and the placebo group (Andrieu et al., 2017). In the Finnish Geriatric Intervention Study to Prevent Cognitive Impairment and Disability (FINGER) study, 2645 individuals aged 60–77 years were randomized to multidomain intervention (diet, exercise, cognitive training, and vascular risk monitoring) or standard of care and underwent a comprehensive neuropsychological test battery (NTB). At 2 years, there was a reduction in the rate of cognitive decline in favor of the multidomain intervention with a between-group difference in the change of NTB total score per year of 0.022 (95% CI, 0.002–0.042; $P = 0.030$) (Ngandu et al., 2015). While the FINGER study did show a benefit in the multidomain intervention, the degree of benefit was modest. All these studies were limited by relatively short follow-up periods with small changes in VRF control between baseline and the follow-up period.

CONCLUSIONS

The evidence obtained in longitudinal population studies and basic science research support the notion that VRFs participate of the pathogenesis of heart disease and cognitive decline of the vascular and neurodegenerative types. Clinical trials designed to prevent the development of cognitive deterioration, however, have produced mixed results. Most of these studies are confounded by the multiplicity of factors that affect cognitive vitality, short follow up, lack of consensus on the definition of CI, and attrition, among other factors. Future randomized studies will likely need to target younger at-risk populations in order to halt or prevent the initiation of neurodegenerative changes. In addition, the use of state-of-the-art imaging modalities, such as diffusion tenor MR, positron emission tomography and others that provide information about microstructural integrity and physiologic function may identify individuals at risk of developing CI before this becomes clinically apparent (Maillard et al., 2012, 2015, 2016). Moreover, emerging evidence support the relevant role of healthy lifestyle choices in the preservation of cognitive health (Lourida et al., 2019). Thus, the early implementation of broad multidomain interventions and reliable early biomarkers of dementia and new imaging techniques should be incorporated into further studies.

References

Alosco ML, Spitznagel MB, Sweet LH et al. (2015). Atrial fibrillation exacerbates cognitive dysfunction and cerebral perfusion in heart failure. Pacing Clin Electrophysiol PACE 38 (2): 178–186.

Alzheimer's disease facts and figures (2020). Alzheimers Dementia 2020 (16): 391–460.

Alzheimer's Association (2016). 2016 Alzheimer's disease facts and figures. Alzheimers Dement 12 (4): 459–509.

Anderson C, Teo K, Gao P et al. (2011). Renin-angiotensin system blockade and cognitive function in patients at high risk of cardiovascular disease: analysis of data from the ONTARGET and TRANSCEND studies. Lancet Neurol 10 (1): 43–53. https://www.ncbi.nlm.nih.gov/pubmed/20980201. https://doi.org/10.1016/S1474-4422(10)70250-7.

Andrieu S, Guyonnet S, Coley N et al. (2017). Effect of long-term omega 3 polyunsaturated fatty acid supplementation with or without multidomain intervention on cognitive function in elderly adults with memory complaints (MAPT): a randomised, placebo-controlled trial. Lancet Neurol 16 (5): 377–389.

Applegate WB, Pressel S, Wittes J et al. (1994). Impact of the treatment of isolated systolic hypertension on behavioral variables. Results from the systolic hypertension in the elderly program. Arch Intern Med 154 (19): 2154–2160.

Areosa Sastre A, Vernooij RW, Gonzalez-Colaco Harmand M et al. (2017). Effect of the treatment of Type 2 diabetes mellitus on the development of cognitive impairment and dementia. Cochrane Database Syst Rev 6: CD003804.

Biessels GJ, Staekenborg S, Brunner E et al. (2006). Risk of dementia in diabetes mellitus: a systematic review. Lancet Neurol 5 (1): 64–74.

Bosch J, O'Donnell M, Swaminathan B et al. (2019). Effects of blood pressure and lipid lowering on cognition: results from the HOPE-3 study. Neurology 92 (13): e1435–e1446. https://www.ncbi.nlm.nih.gov/pubmed/30814321. https://doi.org/10.1212/WNL.0000000000007174.

Breteler MM, Claus JJ, Grobbee DE et al. (1994). Cardiovascular disease and distribution of cognitive function in elderly people: the Rotterdam Study. BMJ (Clinical research ed) 308 (6944): 1604–1608.

Bruijn RFD, Heeringa J, Wolters FJ et al. (2015). Association between atrial fibrillation and dementia in the general population. JAMA Neurol 72 (11): 1288–1294.

Bunch TJ, Weiss JP, Crandall BG et al. (2010). Atrial fibrillation is independently associated with senile, vascular, and Alzheimer's dementia. Heart Rhythm 7 (4): 433–437.

Cacciatore F, Mazzella F, Abete P et al. (2007). Mortality and heart rate in the elderly: role of cognitive impairment. Exp Aging Res 33 (2): 127–144.

Cannon JA, Moffitt P, Perez-Moreno AC et al. (2017). Cognitive impairment and heart failure: systematic review and meta-analysis. J Card Fail 23 (6): 464–475.

Chen LY, Soliman EZ (2019). P wave indices-advancing our understanding of atrial fibrillation-related cardiovascular outcomes. Front cardiovasc Med 6: 53.

Conen D, Rodondi N, Muller A et al. (2019). Relationships of overt and silent brain lesions with cognitive function in patients with atrial fibrillation. J Am Coll Cardiol 73 (9): 989–999.

Costa C, Gonzalez-Alujas T, Valente F et al. (2016). Left atrial strain: a new predictor of thrombotic risk and successful electrical cardioversion. Echo Rse Pract 3 (2): 45–52.

Cukierman-Yaffe T, Bosch J, Diaz R et al. (2014). Effects of basal insulin glargine and omega-3 fatty acid on cognitive decline and probable cognitive impairment in people with dysglycaemia: a substudy of the ORIGIN trial. Lancet Diabetes Endocrinol 2 (7): 562–572.

Daccarett M, Badger TJ, Akoum N et al. (2011). Association of left atrial fibrosis detected by delayed-enhancement magnetic resonance imaging and the risk of stroke in patients with atrial fibrillation. J Am Coll Cardiol 57 (7): 831–838.

Diener H, Sacco RL, Yusuf S et al. (2008). Effects of aspirin plus extended-release dipyridamole versus clopidogrel and telmisartan on disability and cognitive function after recurrent stroke in patients with ischaemic stroke in the Prevention Regimen for Effectively Avoiding Second Strokes (PROFESS) trial: a double-blind, active and placebo-controlled study. Lancet Neurol 7 (10): 875–884. https://www.ncbi.nlm.nih.gov/pubmed/18757238. https://doi.org/10.1016/S1474-4422(08)70198-4.

Elias MF, Wolf PA, D'agostino RB et al. (1993). Untreated blood pressure level is inversely related to cognitive functioning: the framingham study. Am J Epidemiol 138 (6): 353–364.

Elias PK, Elias MF, D'agostino RB et al. (1997). NIDDM and blood pressure as risk factors for poor cognitive performance: the framingham study. Diabetes Care 20 (9): 1388–1395.

Forette F, Seux ML, Staessen JA et al. (1998). Prevention of dementia in randomised double-blind placebo-controlled Systolic Hypertension in Europe (Syst-Eur) trial. Lancet (London, England) 352 (9137): 1347–1351.

Gaita F, Corsinovi L, Anselmino M et al. (2013). Prevalence of silent cerebral ischemia in paroxysmal and persistent atrial fibrillation and correlation with cognitive function. J Am Coll Cardiol 62 (21): 1990–1997.

Gardarsdottir M, Sigurdsson S, Aspelund T et al. (2018). Atrial fibrillation is associated with decreased total cerebral blood flow and brain perfusion. Europace 20 (8): 1252–1258.

Gasowski J, Staessen JA, Celis H et al. (1999). Systolic Hypertension in Europe (Syst-Eur) trial phase 2: objectives, protocol, and initial progress. Systolic Hypertension in Europe Investigators. J Hum Hypertens 13 (2): 135.

Gelber RP, Launer LJ, White LR (2012). The Honolulu-Asia aging study: epidemiologic and neuropathologic research on cognitive impairment. Curr Alzheimer Res 9 (6): 664–672.

Gharacholou SM, Reid KJ, Arnold SV et al. (2011). Cognitive impairment and outcomes in older adult survivors of acute myocardial infarction: findings from the translational research investigating underlying disparities in acute myocardial infarction patients' health status registry. Am Heart J 162 (5): 860–869.e1.

Gorelick PB, Scuteri A, Black SE et al. (2011). Vascular contributions to cognitive impairment and dementia: a statement for healthcare professionals from the american heart association/american stroke association. Stroke 42 (9): 2672–2713.

Gorelick PB, Furie KL, Iadecola C et al. (2017). Defining optimal brain health in adults: a presidential advisory from the American Heart Association/American Stroke Association. Stroke 48 (10): e284–e303.

Gottesman RF, Schneider ALC, Zhou Y et al. (2017). Association between midlife vascular risk factors and estimated brain amyloid deposition. JAMA 317 (14): 1443–1450.

Gu SZ, Beska B, Chan D et al. (2019). Cognitive decline in older patients with non- ST elevation acute coronary syndrome. J Am Heart Assoc 8 (4): e011218.

Gutierrez A, Norby FL, Maheshwari A et al. (2019). Association of abnormal P-wave indices with dementia and cognitive decline over 25 years: ARIC-NCS (The atherosclerosis risk in communities neurocognitive study). J Am Heart Assoc 8 (24): e014553.

Hammond CA, Blades NJ, Chaudhry SI et al. (2018). Long-term cognitive decline after newly diagnosed heart failure: longitudinal analysis in the CHS (Cardiovascular Health Study). Circ Heart Fail 11 (3): e004476.

Hofman A, Breteler MMB, Duijn CMV et al. (2007). The Rotterdam study: objectives and design update. Eur J Epidemiol 22 (11): 819–829.

Hughes D, Judge C, Murphy R et al. (2020). Association of blood pressure lowering with incident dementia or cognitive impairment: a systematic review and meta-analysis. JAMA 323 (19): 1934–1944.

Iadecola C, Gorelick PB (2003). Converging pathogenic mechanisms in vascular and neurodegenerative dementia. Stroke 34 (2): 335–337.

Iadecola C, Gottesman RF (2019). Neurovascular and cognitive dysfunction in hypertension. Circ Res 124 (7): 1025–1044.

Iadecola C, Yaffe K, Biller J et al. (2016). Impact of hypertension on cognitive function: a scientific statement from the American heart association. Hypertension (Dallas, Tex: 1979) 68 (6): e67–e94.

Ishii M, Iadecola C (2020). Risk factor for Alzheimer's disease breaks the blood-brain barrier. Nature 581: 31–32.

Jefferson AL, Beiser AS, Himali JJ et al. (2015). Low cardiac index is associated with incident dementia and Alzheimer disease: the Framingham Heart Study. Circulation 131 (15): 1333–1339.

Jellinger KA (2006). Clinicopathological analysis of dementia disorders in the elderly–an update. J Alzheimers Dis JAD 9 (3 Suppl): 61–70.

Kalantarian S, Ay H, Gollub RL et al. (2014). Association between atrial fibrillation and silent cerebral infarctions: a systematic review and meta-analysis. Ann Intern Med 161 (9): 650–658.

Kamel H, Okin PM, Elkind MS et al. (2016). Atrial fibrillation and mechanisms of stroke: time for a new model. Stroke 47 (3): 895–900.

Kilander L, Andren B, Nyman H et al. (1998). Atrial fibrillation is an independent determinant of low cognitive function: a cross-sectional study in elderly men. Stroke 29 (9): 1816–1820.

Kim JH, Shah P, Tantry US et al. (2016). Coagulation abnormalities in heart failure: pathophysiology and therapeutic implications. Curr Heart Fail Rep 13 (6): 319–328.

Kisler K, Nelson AR, Montagne A et al. (2017). Cerebral blood flow regulation and neurovascular dysfunction in Alzheimer disease. Nat Rev Neurosci 18 (7): 419–434.

Kivipelto M, Helkala EL, Hanninen T et al. (2001). Midlife vascular risk factors and late-life mild cognitive impairment: a population-based study. Neurology 56 (12): 1683–1689.

Knecht S, Oelschlager C, Duning T et al. (2008). Atrial fibrillation in stroke-free patients is associated with memory impairment and hippocampal atrophy. Eur Heart J 29 (17): 2125–2132.

Koizumi K, Hattori Y, Ahn SJ et al. (2018). Apoepsilon4 disrupts neurovascular regulation and undermines white matter integrity and cognitive function. Nat Commun 9 (1): 3816–018–06301-2.

Kuller LH, Lopez OL, Mackey RH et al. (2016). Subclinical Cardiovascular Disease and Death, Dementia, and Coronary Heart Disease in Patients 80+ Years. J Am Coll Cardiol 67 (9): 1013–1022.

Kwok CS, Loke YK, Hale R et al. (2011). Atrial fibrillation and incidence of dementia: a systematic review and meta-analysis. Neurology 76 (10): 914–922.

Launer LJ, Ross GW, Petrovitch H et al. (2000). Midlife blood pressure and dementia: the Honolulu-Asia aging study. Neurobiol Aging 21 (1): 49–55.

Launer LJ, Miller ME, Williamson JD et al. (2011). Effects of intensive glucose lowering on brain structure and function in people with type 2 diabetes (ACCORD MIND): a randomised open-label substudy. Lancet Neurol 10 (11): 969–977.

Livingston G, Sommerlad A, Orgeta V et al. (2017). Dementia prevention, intervention, and care. Lancet (London, England) 390 (10113): 2673–2734.

Lourida I, Hannon E, Littlejohns TJ et al. (2019). Association of lifestyle and genetic risk with incidence of Dementia. JAMA 322 (5): 430–437.

Luchsinger JA, Reitz C, Honig LS et al. (2005). Aggregation of vascular risk factors and risk of incident Alzheimer disease. Neurology 65 (4): 545–551.

Mahmood SSM, Levy D, Vasan RS et al. (2014). The Framingham Heart Study and the epidemiology of cardiovascular disease: a historical perspective. Lancet 383 (9921): 999–1008.

Maillard P, Seshadri S, Beiser A et al. (2012). Effects of systolic blood pressure on white-matter integrity in young adults in the Framingham Heart Study: a cross-sectional study. Lancet Neurol 11 (12): 1039–1047.

Maillard P, Carmichael OT, Reed B et al. (2015). Co-occurrence of vascular risk factors and late-life white matter integrity changes.

Maillard P, Mitchell GF, Himali JJ et al. (2016). Effects of arterial stiffness on brain integrity in young adults from the framingham heart study. Stroke 47 (4): 1030–1036.

Miyasaka Y, Barnes ME, Gersh BJ et al. (2006). Secular trends in incidence of atrial fibrillation in Olmsted County, Minnesota, 1980 to 2000, and implications on the projections for future prevalence. Circulation 114 (2): 119–125.

Muller M, Van Der Graaf Y, Visseren FL et al. (2010). Blood pressure, cerebral blood flow, and brain volumes. The SMART-MR study. J Hypertens 28 (7): 1498–1505.

Myserlis PG, Malli A, Kalaitzoglou DK et al. (2017). Atrial fibrillation and cognitive function in patients with heart failure: a systematic review and meta-analysis. Heart Fail Rev 22 (1): 1–11.

Newman AB, Fitzpatrick AL, Lopez O et al. (2005). Dementia and Alzheimer's disease incidence in relationship to cardiovascular disease in the Cardiovascular Health Study cohort. J Am Geriatr Soc 53 (7): 1101–1107.

Ngandu T, Lehtisalo J, Solomon A et al. (2015). A 2 year multidomain intervention of diet, exercise, cognitive training, and vascular risk monitoring versus control to prevent cognitive decline in at-risk elderly people (FINGER): a randomised controlled trial. Lancet (London, England) 385 (9984): 2255–2263.

Norton S, Matthews FE, Barnes DE et al. (2014). Potential for primary prevention of Alzheimer's disease: an analysis of population-based data. Lancet Neurol 13 (8): 788–794.

Ott A, Breteler MM, Bruyne MCD et al. (1997). Atrial fibrillation and dementia in a population-based study. The Rotterdam Study. Stroke 28 (2): 316–321.

Ott A, Slooter AJ, Hofman A et al. (1998). Smoking and risk of dementia and Alzheimer's disease in a population-based cohort study: the Rotterdam Study. Lancet 351 (9119): 1840–1843. https://www.ncbi.nlm.nih.gov/pubmed/9652667. https://doi.org/10.1016/s0140-6736(97)07541-7.

Ott A, Stolk RP, Harskamp FV et al. (1999). Diabetes mellitus and the risk of dementia: the Rotterdam Study. Neurology 53 (9): 1937.

Peters R, Beckett N, Forette F et al. (2008). Incident dementia and blood pressure lowering in the hypertension in the very elderly trial cognitive function assessment (HYVET-COG): a double-blind, placebo controlled trial. Lancet Neurol 7 (8): 683–689. https://doi.org/10.1016/S1474-4422(08)70143-1.

Roberts RO, Geda YE, Knopman DS et al. (2008). Association of duration and severity of diabetes mellitus with mild cognitive impairment. Arch Neurol (Chicago) 65 (8): 1066–1073.

Rosano C, Naydeck B, Kuller LH et al. (2005). Coronary artery calcium: associations with brain magnetic resonance imaging abnormalities and cognitive status. J Am Geriatr Soc 53 (4): 609–615.

Rosenberg GA, Wallin A, Wardlaw JM et al. (2016). Consensus statement for diagnosis of subcortical small vessel disease. J Cereb Blood Flow Metab 36 (1): 6–25.

Russo C, Jin Z, Homma S et al. (2013). Subclinical left ventricular dysfunction and silent cerebrovascular disease: the Cardiovascular Abnormalities and Brain Lesions (CABL) study. Circulation 128 (10): 1105–1111.

Sachdev PS, Lipnicki DM, Crawford JD et al. (2019). The vascular behavioral and cognitive disorders criteria for vascular cognitive disorders: a validation study. Eur J Neurol 26 (9): 1161–1167.

Santangeli P, Biase LD, Bai R et al. (2012). Atrial fibrillation and the risk of incident dementia: a meta-analysis. Heart Rhythm 9 (11): 1761–1768.

Scheltens P, Blennow K, Breteler MM et al. (2016). Alzheimer's disease. Lancet (London, England) 388 (10043): 505–517.

Skoog I, Lernfelt B, Landahl S et al. (1996). 15-year longitudinal study of blood pressure and dementia. Lancet 347 (9009): 1141–1145.

Skrobot OA, O'Brien J, Black S et al. (2017). The vascular impairment of cognition classification consensus study. Alzheimers Dement 13 (6): 624–633.

Song R, Xu H, Dintica CS et al. (2020). Associations between cardiovascular risk, structural brain changes, and cognitive decline. J Am Coll Cardiol 75 (20): 2525–2534.

Sweeney MD, Ayyadurai S, Zlokovic BV (2016). Pericytes of the neurovascular unit: key functions and signaling pathways. Nat Neurosci 19 (6): 771–783.

Tarasoff-Conway JM, Carare RO, Osorio RS et al. (2015). Clearance systems in the brain-implications for Alzheimer disease. Nat Rev Neurol 11 (8): 457–470.

Testai FD, Gorelick PB (2016). Vascular cognitive impairment and alzheimer disease: are these disorders linked to hyeprtension and other cardiovascular risk factors? In: V Aiyagari, PB Gorelick (Eds.), Hypertension and stroke, Second edn Humana Press, Switzerland, pp. 261–284.

Toledo JB, Arnold SE, Raible K et al. (2013). Contribution of cerebrovascular disease in autopsy confirmed neurodegenerative disease cases in the National Alzheimer's Coordinating Centre. Brain 136 (Pt 9): 2697–2706.

Tuligenga RH (2015). Intensive glycaemic control and cognitive decline in patients with type 2 diabetes: a meta-analysis. Endocr Connect 4 (2): R16–R24.

Tyas SL, White LR, Petrovitch H et al. (2003). Mid-life smoking and late-life dementia: the Honolulu-Asia Aging Study. Neurobiol Aging 24 (4): 589–596.

Tzourio C, Anderson C, Chapman N et al. (2003). Effects of blood pressure lowering with perindopril and indapamide therapy on dementia and cognitive decline in patients with cerebrovascular disease. Arch Intern Med (1960) 163 (9): 1069–1075.

Van Charante EPM, Richard E, Eurelings LS et al. (2016). Effectiveness of a 6-year multidomain vascular care intervention to prevent dementia (preDIVA): a cluster-randomised controlled trial. Lancet (British edition) 388 (10046): 797–805.

Van Der Velpen IF, Yancy CW, Sorond FA et al. (2017a). Impaired cardiac function and cognitive brain aging. Can J Cardiol 33 (12): 1587–1596.

Van Der Velpen IF, Feleus S, Bertens AS et al. (2017b). Hemodynamic and serum cardiac markers and risk of cognitive impairment and dementia. Alzheimers Dement 13 (4): 441–453.

Vidal JS, Sigurdsson S, Jonsdottir MK et al. (2010). Coronary artery calcium, brain function and structure: the AGES-Reykjavik Study. Stroke 41 (5): 891–897.

Vogels RL, Van Der Flier WM, Van Harten B et al. (2007). Brain magnetic resonance imaging abnormalities in patients with heart failure. Eur J Heart Fail 9 (10): 1003–1009.

Whitmer RA, Sidney S, Selby J et al. (2005). Midlife cardiovascular risk factors and risk of dementia in late life. Neurology 64 (2): 277–281.

Williamson JD, Pajewski NM, Auchus AP et al. (2019). Effect of intensive vs standard blood pressure control on probable dementia: a randomized clinical trial. JAMA 321 (6): 553–561.

Wolf PA, Abbott RD, Kannel WB (1991). Atrial fibrillation as an independent risk factor for stroke: the Framingham study. Stroke 22: 983–988.

Wolters FJ, Segufa RA, Darweesh SKL et al. (2018). Coronary heart disease, heart failure, and the risk of dementia: a systematic review and meta-analysis. Alzheimers Dement 14 (11): 1493–1504.

Zlokovic BV (2011). Neurovascular pathways to neurodegeneration in Alzheimer's disease and other disorders. Nat Rev Neurosci 12 (12): 723–738.

Handbook of Clinical Neurology, Vol. 177 (3rd series)
Heart and Neurologic Disease
J. Biller, Editor
https://doi.org/10.1016/B978-0-12-819814-8.00026-3

Chapter 33

Anxiety and psychological management of heart disease and heart surgery

PATRICK RIORDAN[1]* AND MATTHEW DAVIS[2]

[1]*Department of Neurology, Loyola University Chicago, Stritch School of Medicine, Maywood, IL, United States*
[2]*Mental Health Service Line, Edward Hines Jr. VA Hospital, Hines, IL, United States*

Abstract

Anxiety is associated with many forms and facets of heart disease, and, by extension, neurologic manifestations of heart disease. Despite its seeming self-evidence, anxiety is challenging to consistently define, measure, and operationalize in the context of medical research. Various diagnostic nosologies have been defined and refined over time, but anxiety is also a universal human experience that may be "normal" in many circumstances, particularly in the face of major medical issues. For these and other reasons, the research on anxiety and heart disease is mixed, incomplete, and often characterized by challenging questions of causality. Nonetheless, a broad body of literature has established clear connections between anxiety and vascular risk factors, cardiac disease, and cardiac surgery. These relationships are often intuitive, with research suggesting, for example, that chronic activation of the sympathetic nervous system is associated with increased risk of heart disease. However, they are sometimes complexly reciprocal or even surprising (e.g., with high-anxiety individuals found to have better outcomes in some cardiac conditions by virtue of seeking evaluation and treatment earlier). This chapter reviews the construct of anxiety and its complexities, its associations with heart disease, and the established treatments for anxiety, concluding with questions about anxiety, heart disease, and their optimal management that still need to be answered.

INTRODUCTION

Early in his pioneering 1848 work, "On Disorders of the Cerebral Circulation: and on the Connection between Affections of the Brain and Diseases of the Heart," Sir George Burrows references Ovid: "Facies non omnibus una, nec diversa tamen, (qualem decet esse sororum)"—"Their faces looked not all alike, nor yet unlike: as sisters' would" (Burrows, 1846). In the intervening century and decades, scientific understanding of the complex, reciprocal relationships between the vascular and nervous systems, and the neurologic manifestations of diseases of the heart and vascular system, has advanced exponentially. While there are undoubtedly discoveries yet to be made in the field of neurocardiology, a large body of scientific evidence linking diseases of the heart to risk of stroke (Hart et al., 2007), dementia (Sabia et al., 2019), and other neurologic conditions (Slade et al., 1989) has been amassed. Concurrently, tremendous advances have been made in the methods for assessment of vascular and neurologic health, allowing researchers and clinicians to collect and work with a wealth of objective, reliable data. But returning to Ovid, heart disease and its neurologic manifestations have a third sister, one more ephemeral and with less distinct features, but inextricably related and no less important.

Anxiety is strongly associated with heart disease (Tully et al., 2016). To the layman, "anxiety" is likely a more recognizable and easily understood condition than many of the cardiac diseases to which it has been

*Correspondence to: Patrick Riordan Ph.D. Neurophysiology, 2160 S First Ave, Maywood, 60153, United States. Tel: +1-708-216-3539, Fax: +1-708-216-4629, E-mail: patrick.riordan@lumc.edu

linked. While any researcher or physician can also see this same popular conception of "anxiety", its shape and parameters seemingly obvious from a distance, those who try to look more closely at it, to record its features, to compare its manifestations across one patient and another, quickly find themselves departing the objective realm of electrocardiograms, prothrombin times, and magnetic resonance images.

While strides have been made in operationalizing the assessment of anxiety and the diagnosis of its disorders, the fact remains that assessment and diagnosis of anxiety is inherently indirect, the symptoms themselves are often transient or situational, and our understanding of its physiology remains tenuous. Patient self-reports of anxiety symptoms are filtered through complex and idiosyncratic algorithms of self-image, sociocultural conditioning, and emotional self-awareness. Additionally, anxiety may manifest generally as an overarching "trait" or sporadically or eccentrically as a transient "state"—readers of this volume are likely familiar with patients whose blood pressure runs unexpectedly high at their doctor's office, or who confidently deny any experience of anxiety but present with implausible neurologic symptoms that seem rooted in underlying psychological distress. Even the most seemingly objective, quantitative physiologically-based attempts to measure anxiety—some of which excel in the right context—are difficult to deploy and often fail to capture the construct, with a recent international expert consensus statement concluding that there are no reliable physiological biomarkers for anxiety (Bandelow et al., 2017). A generally "low-anxiety" person may uncharacteristically develop palm sweat and subtle tremulousness as electrodes and cuffs are placed, while a generally "high-anxiety" individual may find the quietude of the lab and the passivity of their participation in procedures reassuring and relaxing. Borrowing again from antiquity—all we may ever really "see" of anxiety are its shadows on the cave wall; however, these shadows undoubtedly represent something real, and something that looms large in the study of cardiac disease and its neurologic manifestations. Despite these caveats and ambiguities—caveats which remain worth keeping in mind throughout this chapter—the existence of important, complex interrelationships between anxiety, cardiovascular disease, and cerebrovascular disease, and, by extension, opportunities to improve medical outcomes through better management of anxiety, are undeniable.

OVERVIEW OF ANXIETY AND HEART DISEASE

Anxiety disorders are the most common class of mental disorders in the United States (Kessler et al., 2005). Non-U.S. prevalence estimates vary across countries and cultures, but anxiety disorders are extremely common worldwide (Baxter et al., 2013). Similarly, heart disease is the leading cause of death in the United States (Heron and Anderson, 2017) and worldwide (Nowbar et al., 2019). While the causes of heart disease are many, there is growing recognition of the role anxiety can play in heart disease and the need for further research in this area, with, e.g., the American Heart Association formally recommending further research on the relationship between anxiety and acute coronary syndrome (Lichtman et al., 2014). Importantly, high rates of comorbidity between anxiety and other mental health conditions such as depression are documented (Lamers, 2011) and have raised questions about the unique role of each with various cardiac conditions. There has been increasing recognition of this issue, however, with more recent studies of anxiety and heart disease routinely controlling for depression and other mood disorders (Roest et al., 2010; Scherrer et al., 2010).

Although a firm causal relationship has not been established, there is considerable evidence to suggest that anxiety increases the likelihood of various forms of cardiovascular disease, particularly coronary artery disease (CAD). Rates of many common anxiety disorders are much higher in cardiovascular disease patients than in the general population, with metanalytic data indicating approximately 16% rates of any anxiety disorder in coronary heart disease patients (Tully et al., 2014). Other meta-analytic research has demonstrated a 26% increase in risk for coronary heart disease in initially healthy individuals with elevated anxiety levels (Roest et al., 2010). The same study also identified a 48% increased risk of cardiac death in individuals with anxiety. Elevated risk of cardiovascular disease has also been observed within various specific anxiety disorders, including generalized anxiety disorder (GAD) (Tully et al., 2014; Easton et al., 2016), posttraumatic stress disorder (Coughlin, 2011), and panic disorders (Walters et al., 2008). Notably, some of these studies are characterized by important caveats and unexpected findings. In one panic disorder study (Walters et al., 2008), for example, panic was associated with a slightly reduced rate of CHD mortality, a finding thought to be attributable to earlier pursuit of medical evaluation/treatment in this population. Other research has reinforced this finding, showing that myocardial infarction (MI) patients with GAD seek medical treatment more quickly than MI patients without GAD (Fang et al., 2018).

There is considerable agreement that there is a lack of well-controlled experimental evidence for a causal relationship between anxiety and cardiovascular disease, but some researchers (Batelaan et al., 2016) argue that application of Hill criteria (Hill, 1965) to available meta-analytic evidence supports presumption of causality.

However, others (Tully et al., 2015) note the possibility of reverse causality (i.e., anxiety symptoms as a result of cardiovascular disease) cannot be ruled out. Although the question of causation may not be completely closed, a large, well-controlled, prospective study of young Swedish military conscripts ($N = 49{,}321$) provided clear evidence that participants with an ICD-based, expert diagnosis of anxiety disorder (but not those with a depressive disorder) were at significantly higher risk of subsequent development coronary heart disease (Hazard Ratio: 2.17) and MI (Hazard Ratio: 2.51) at a 37-year follow-up (Janszky et al., 2010). In addition to CAD and general cerebrovascular disease, various studies have also examined the relationships between anxiety and other specific vascular conditions.

Myocardial infarction

Numerous studies have demonstrated links between anxiety and MI. In a retrospective cohort study of Veterans Administration patients without heart disease at study onset, various anxiety disorders (unspecified, panic disorder, and posttraumatic stress disorder (PTSD)) were all associated with substantially increased risk of incident MI in patients with and without comorbid depression (Scherrer et al., 2010). A 10-year longitudinal study of individuals with a history of MI found a nearly twofold increase in the risk of adverse outcomes amongst individuals with a diagnosis of GAD, even when accounting for demographics, cardiac disease severity, and depression (Roest et al., 2012). Other studies have identified an association between anxiety and increased risk of both cardiac events and increased healthcare usage in post-MI patients (Strik et al., 2003). Anxiety (but not depression) is also associated with reduced parasympathetic modulation of heart rate in post-MI patients (Martens et al., 2008). In terms of reverse causality, rates of anxiety (and depression) symptoms following MI are elevated relative to the general population and have been shown to persist for substantial periods of time in many patients (Lane et al., 2002).

Arrhythmia

Perhaps more than any other cardiovascular condition, research on the association between anxiety and arrhythmia has often focused on anxiety as a product as opposed to cause of medical illness. Patients with paroxysmal atrial fibrillation (AF) experience elevated rates of agoraphobia and attack anxiety, with these secondary symptoms demonstrated to be strong predictors of patient quality of life (Suzuki and Kasanuki, 2004). Some research has even suggested that anxiety and other psychological symptoms associated with AF have a larger impact on health-related quality of life than the physical symptoms of AF itself (Perret-Guillaume et al., 2010). Other research has shown that patient attitudes and perceived understanding of AF strongly influence the extent to which patients experience negative emotions related to the condition (Tully et al., 2011a). A systematic review of anxiety and depression in patients with AF indicated that the relationships between these conditions are complex and bidirectional, with some evidence suggesting that anxiety is a risk factor for both the development and the perpetuation of AF (Patel et al., 2013). Other studies have shown that psychological "tension" predicted higher likelihood of developing AF in males over a 10-year study period (Eaker et al., 2005). Anxiety has also been shown to be a risk factor for the development of AF after cardiac surgery (Tully et al., 2011b). Additionally, various physiological mechanisms by which anxiety could promote or exacerbate AF (e.g., increased sympathetic nervous systems activation, elevated levels of hs-CRP) have been proposed (Patel et al., 2013). Other research has challenged these findings, with, e.g., a meta-analytic study of anxiety as a risk factor for various forms of cardiovascular disease failing to find an association between anxiety and AF (Emdin et al., 2015), although the authors noted limited power and wide confidence intervals in this data.

Other cardiac conditions

Multiple other cardiovascular conditions have been linked to anxiety. A number of studies have demonstrated elevated rates of mitral valve prolapse in individuals with panic disorder relative to controls (Tural and Iosifescu, 2019), as well as a broader link between mitral valve disorders and anxiety states in general (Dubey et al., 2016). Various forms of cardiomyopathy also have associations with anxiety. Anxiety but not depression is higher in patients treated for takotsubo stress cardiomyopathy relative to those seeking treatment for acute coronary syndrome (Goh et al., 2016). Significantly elevated rates of anxiety disorders have been observed in patients with hypertrophic cardiomyopathy (Morgan et al., 2008) and in dilated cardiomyopathy (Guan et al., 2014). Higher rates of anxiety disorders are also observed in patients with congenital heart disease, although rates within this population are generally lower than in individuals with acquired heart disease (Jackson et al., 2018). Evidence of an association between anxiety and heart failure is mixed. A systematic review and meta-analysis of depression and anxiety as predictors of mortality in heart failure patients found evidence of a significant effect of depression ($HR = 1.57$), but minimal evidence of anxiety impacting mortality ($HR = 1.02$, 95% CI 1.00–1.04), although a limited number of relevant studies was noted (Sokoreli et al., 2016).

However, other reviews have linked anxiety to heart failure in terms of development, progression, and associated mortality (Celano et al., 2018).

Heart surgery and anxiety

In addition to heart disease itself, anxiety has also been linked to heart surgery as both a moderator of surgical outcomes and a potential by-product. Elevated rates of anxiety symptoms have been reported in approximately a third of precoronary artery bypass graft (CABG) patients (Tully et al., 2008), have been linked to surgical wait times (Koivula et al., 2002), and predict a number of poor outcomes following CABG surgery (Joseph et al., 2015). Preoperative anxiety has been shown to strongly predict hospital readmission following CABG, even controlling for relevant medical variables (Tully et al., 2008). Presurgical anxiety is also associated with increased odds of in-hospital complications like stroke and MI (Tully et al., 2011b), post-surgical AF (Tully et al., 2011a), and increased long-term mortality (Tully et al., 2008), with the latter study noting that this effect was unique to anxiety and not observed with preoperative depression. Furthermore, preoperative anxiety is associated with persisting mental health symptoms post-operatively (Krannich et al., 2007; Oxlad and Wade, 2008).

ANXIETY AND VASCULAR RISK FACTORS

Hypertension

In addition to its documented relationships to various forms of heart disease, anxiety is also associated with a variety of vascular risk factors. Evidence of a link between hypertension and anxiety has been mixed. A large cross-sectional study of older adults found evidence of an association between hypertension and symptoms of depression, but not anxiety (Maatouk et al., 2016). In contrast, baseline anxiety has been shown to significantly increase the risk of 1-year incident hypertension (Bacon et al., 2014) and lifetime incidence of hypertension (Stein et al., 2014). Similarly, cross-sectional studies have demonstrated positive, bidirectional relationships between hypertension and anxiety (Player and Peterson, 2011). Another study (Ogedegbe, 2008) used ambulatory blood pressure monitors to evaluate the "white coat hypertension" phenomenon, finding that 9% of their sample exhibited physician—specific hypertensive readings. Amongst these participants, they found elevated rates of "state anxiety" (but not "trait anxiety") relative to the rest of their sample, leading them to interpret "white coat" hypertension as a conditioned response.

Diabetes

Numerous studies and several systematic reviews and meta-analyses have established links between anxiety and diabetes. Relative to population estimates, significantly elevated rates of GAD and subsyndromal/nonspecific anxiety symptoms have been found in people with diabetes (Grigsby et al., 2002; Smith et al., 2013). A large population-based, U.S. study controlling for a wide range of demographic and lifestyle variables estimated an approximately 20% higher prevalence of lifetime anxiety disorder in people with diabetes versus those without, with particularly pronounced associations observed amongst young adults and in certain ethnic groups (Li et al., 2008). Although the directionality of the relationships between anxiety and diabetes are not well established, a prospective population-based study found that anxiety and depression were significant predictors of type II diabetes development, independent of various lifestyle and socioeconomic risk factors, over a 10 year study period (Engum, 2007).

Cholesterol and diet

Numerous studies have found elevated levels of cholesterol in patients with various anxiety disorders, including GAD, panic disorder, obsessive compulsive disorder (OCD), and PTSD, whereas research with depressed and suicidal patients has typically found lower than average cholesterol levels (Papakostas et al., 2004). More broadly, a range of links have been discovered between anxiety and diet. In animal studies, high-fat maternal diets in the perinatal period have been linked to alterations in gene expression in the amygdala and hippocampus, increased anxiety behaviors, and prolonged corticosterone elevations in response to stress (Sasaki et al., 2013). Similarly, initiation of and withdrawal from a high-fat diet in adult mice has been shown to produce a cascade of responses in brain reward circuitry and a concomitant increase in anxiety behaviors, with withdrawal from a high-fat diet increasing basal corticosterone levels and food motivated behaviors (Sharma et al., 2013). Human research has also demonstrated elevated rates of anxiety disorders amongst people with obesity (Gariepy et al., 2010). Although the directionality of this relationship has been questioned, hypercaloric diets in childhood and adolescence have been theorized to impact neuroendocrine development and risk of developing anxiety later in life (Baker et al., 2017). Within studies of adults identified as at risk for cardiovascular disease, levels of anxiety (as well as depression) have been linked to a range of unhealthy behaviors, including physical inactivity, unhealthy diets, and smoking (Bonnet et al., 2005). Anxiety and other mood symptoms have also been associated with reduced dietary adherence within specific groups of cardiac patients (Luyster et al., 2009). Various human and animal studies have also demonstrated beneficial effects of probiotic supplementation on anxiety and general mood (Luna and Foster, 2015).

Tobacco and substance use

Substantially elevated rates of tobacco, alcohol, and other substance use are well-documented amongst individuals with anxiety disorders. Relative to individuals without any history of psychiatric illness, highly elevated rates of current and lifetime tobacco use have been observed amongst individuals with anxiety disorders (Ziedonis et al., 2008). The directionality of these associations is complex and incompletely understood (Fluharty et al., 2016), with limited evidence suggesting that anxiety predicts tobacco initiation in adolescents, and adult research indicating that smoking may alleviate transient "state" anxiety but that long-term cessation may reduce overall anxiety (Morissette et al., 2007). Similarly high rates of comorbidity are observed across anxiety disorders and alcohol use disorders (AUDs). Rates of AUD have been observed to vary across specific anxiety disorders, with especially pronounced rates of AUD typically observed in panic disorder and social phobia populations, but with two to three times higher rates of AUD estimated across anxiety disorders in general (Smith and Randall, 2012). As with tobacco research, the directionality of the relationships between anxiety and alcohol use are complex and difficult to firmly establish, with many studies showing positive associations but some suggesting that certain forms of anxiety may be protective against alcohol use (Battista et al., 2010). Similarly, high levels of comorbidity between anxiety and use of cannabis and other substances have been documented, but the relationships between anxiety and substance use disorders are notoriously complex and remain incompletely understood, with regional variability in availability and legality of substances often limiting the generalizability of individual studies (Crippa et al., 2009; Lai et al., 2015). Nonetheless, meta-analytic research has estimated that the likelihood of illicit drug use is nearly triple amongst individuals diagnosed with an anxiety disorder (Lai et al., 2015).

PHYSIOLOGY OF ANXIETY

Genetics

Although no reliable predictors of who will develop anxiety disorders or diagnostic biomarkers for their presence have been established, anxiety disorders have numerous cardiac, neurologic, and other physical correlates. Family and twin research has demonstrated that anxiety disorders are moderately heritable and that certain gene variants are associated with increased risk of certain anxiety disorders, but no reliable genetic foci for the heritability of these disorders have been identified (Bandelow et al., 2016; Smoller, 2016). Typical age of onset for the various anxiety disorders varies substantially, with, e.g., many phobias tending to develop in childhood or adolescence, but GAD tending to develop in adulthood (de Lijster et al., 2017). Nonetheless, childhood abuse and neglect are strong predictors of later development of an anxiety disorder (Li et al., 2016). Disruptions of HPA-axis and neuroendocrine development are thought to account for these relationships, with emerging evidence also supporting ongoing potential for epigenetic modulation of neuroendocrine function as a result of stress exposure later in life (Dirven et al., 2017).

Imaging and neurotransmitters

Despite extensive study, increasingly sophisticated imaging techniques, and some suggestive findings, to date structural and functional neuroimaging research have not identified reliable relationships between anxiety disorders and underlying neuroanatomy or neurophysiology (Bandelow et al., 2016). Broad associations between anxiety disorders and electrophysiological readings have been reported, most notably basal instability in cortical arousal on qEEG in multiple anxiety disorders (Clark et al., 2009); however, electrophysiological findings related to anxiety remain nonspecific, vulnerable to confounds, and inadequate for diagnostic purposes (Bandelow et al., 2017). Similarly, various neurochemical correlates of anxiety disorders have been identified, although reliable relationships between particular anxiety disorders and pathognomonic neurotransmitter profiles have not. As would be expected based on the efficacy of SSRIs in the treatment of anxiety disorders, studies have demonstrated atypical function of the serotonergic system in multiple anxiety disorder including panic disorder with agoraphobia, GAD, OCD, and PTSD (Bandelow et al., 2017). Relatedly, strong links have been established between anxiety and the gamma aminobutyric acid system, which benzodiazepine drugs act upon. There is also limited evidence of atypical dopaminergic system function and specific anxiety disorders, especially OCD (Bandelow et al., 2017). In contrast, the noradrenergic system is closely associated with various anxiety disorders, especially panic disorder, and has been shown to have a complex modulatory effect on anxiety levels depending on the chronicity and predictability of situational stressors (Goddard et al., 2010). General links between anxiety and various neuropeptides, including cholecystokinin, central neurokinins, atrial natriuretic peptides, and oxytocin, have also been identified (Bandelow et al., 2017).

Neuroendocrine

Cortisol levels and hypothalamic–pituitary axis function have also been extensively studied in various anxiety disorders and anxiety in general. Findings regarding basal cortisol levels in various anxiety disorders have been mixed, with reduced cortisol levels in PTSD representing the only relatively consistent finding (Bandelow et al., 2017). However, atypical cortisol responses in response to stressors or anticipated stressors have been observed in some studies, particularly with phobias. Nonetheless, a lack of measurable cortisol response differences between control and anxiety disorder subjects has frequently been reported as well (Bandelow et al., 2017). There is also limited evidence linking anxiety disorders to immune responses and inflammatory medical conditions, including cardiovascular disease.

Cardiovascular

A number of cardiovascular variables have also been studied as potential biomarkers for anxiety. Heart rate variability (HRV) has been extensively studied as a potential biomarker for anxiety, with impaired vagal nerve activity theorized to compromise regulation of inflammatory processes and resting HRV (Chalmers et al., 2014). Meta-analytic data have demonstrated a significant association between most anxiety disorders and reduced HRV, with the overall effect size characterized as small-to-moderate (Chalmers et al., 2014). Less directly, mentally and physically stressful tasks have been demonstrated to produce blood pressure elevation, transient myocardial ischemia, and left ventricular dysfunction amongst patients with chronic CAD (Deanfield et al., 1984; Specchia et al., 1984; Bairey et al., 1990). Similarly, the phenomenon of Takotsubo cardiomyopathy is characterized by transient, nonischemic cardiomyopathy induced by extreme emotional stress (Ako et al., 2006), with an acute rise in catecholamine secretion thought to contribute to disruption of normal cardiac function (Tavazzi et al., 2017).

NEUROLOGIC MANIFESTATIONS OF ANXIETY AND HEART DISEASE

Not surprisingly, in addition to cardiovascular disease, anxiety has also been extensively linked to cerebrovascular disease and other associated neurologic comorbidities. Cross-sectional stroke registry data have shown high prevalence rates for anxiety and other mood disorders in both stroke and TIA patients (Broomfield et al., 2014). Additionally, in a large, prospective study, elevated levels of anxiety symptoms were associated with significantly increased risk of incident stroke (HR = 1.14), even after controlling for depression, demographics, and cardiovascular risk factors (Lambiase et al., 2014). Changes in cognitive functioning and increased risk of dementia have also been addressed in various studies of anxiety and cardiac disease/surgery. Anxiety has been identified as a significant risk factor for a future dementia diagnosis, both generally (Burton et al., 2013) and specifically for vascular dementia (Becker et al., 2018). Conversely, elevated rates of anxiety have been reported in dementia patients, particularly in vascular dementia (Seignourel et al., 2008), as well as a range of other neurologic disorders like Parkinson's disease, migraine, MS, and seizure disorder (Davies et al., 2001).

Outside of dementia, increased risk for measurable changes in cognitive functioning has been linked to preoperative anxiety in cardiac patients (Andrew et al., 2000), although this phenomenon has not been observed in other studies (Tsushima et al., 2005; Stroobant and Vingerhoets, 2008). Additionally, anxiety disorder populations exhibit elevated rates of neuropsychological impairment on tests of mental capacities like episodic memory and executive functioning (Airaksinen et al., 2005) and treatment for anxiety is associated with improvements in cognitive functioning, even in older adult populations (Butters et al., 2011). Similarly, deficits in cognitive functioning have been reported in a range of heart disease and heart surgery patient populations, including CAD and heart failure (Roberts et al., 2010; Almeida et al., 2012), cardiac arrest (Prohl et al., 2009), and AF without stroke (Silva et al., 2019). Meta-analytic data have also demonstrated a relationship between cognitive test performance and composite cardiovascular risk scores (DeRight et al., 2015). Relationships between cognitive outcomes and cardiac surgeries are complex, incompletely understood, and influenced by numerous procedure, condition, and patient specific variables (Eggermont et al., 2012; Fink et al., 2015). For example, evidence regarding changes in cognitive functioning following CABG is mixed (Fink et al., 2015). However, postoperative reductions in cognitive functioning have been reported with various populations and procedures, such as heart transplant recipients (Strauss et al., 1992), patients undergoing catheter ablation for AF (Schwarz et al., 2010), and in both infant (Snookes et al., 2010) and childhood (Wray and Sensky, 2001) surgery for congenital heart disease. A number of studies have also evaluated anxiety and other mood disorder as potential predictors of delirium following cardiac surgery, with some finding a relationship between preoperative depression and delirium, but no evidence of preoperative anxiety predicting delirium (Detroyer et al., 2008; Tully et al., 2010).

ASSESSMENT AND DIAGNOSIS OF ANXIETY

Variability in nosology

Although the preceding literature makes a compelling case for the importance of identifying and managing "anxiety" in reducing risk of cardiovascular disease and improving outcomes in its treatment, assessment and diagnosis of anxiety is less straightforward than it may first seem. There has been and will continue to be considerable variability in the ways anxiety has been conceptualized, assessed, and diagnosed in existing research, raising concomitant questions about how to best apply this knowledge base to clinical practice.

At a fundamental level, the nosology and diagnostic criteria for anxiety disorders have varied by time and region. The Diagnostic and Statistical Manual of Mental Disorders, or DSM-V (APA, 2013), is generally considered the authoritative reference on anxiety and other psychiatric disorders in the United States, while the World Health Organization's International Classification of Mental and Behavioral Disorders (ICD-10: Mental and Behavioral Disorders; WHO, 1993) is used more worldwide. While there is considerable overlap between these two sources, specific diagnostic categories and diagnostic criteria sometimes differ. For example, both systems include a diagnosis of "GAD" to refer to a pervasive condition of excessive and irrational worry; however, the specific criteria for diagnosis in each system—many of which, notably, can overlap with symptoms of heart disease—differ appreciably. More broadly, both classification systems include diagnostic entities for common manifestations of anxiety such as specific phobias, social phobia, agoraphobia, and panic disorders. However, the terminology and exact criteria again differ. Even within diagnostic systems, the specific criteria for the same disorders have varied over time. "GAD" did not exist as a diagnostic entity in the DSM system until the third edition (having previously been subsumed under the category of "anxiety neurosis" alongside symptoms of "panic disorder"), and the criteria for its diagnosis have changed slightly with each successive edition of the manual (Crocq, 2017). Similarly, changes are planned for the ICD criteria for GAD in the transition from ICD-10 to ICD-11 (Crocq, 2017). The extent to which specific conditions are classified as "anxiety disorders" has also varied over time. In prior DSM editions PTSD was classified as an "anxiety disorder"; in the most recent edition, however, it retains many overlapping symptoms with other anxiety disorders but is classified under a separate "Trauma and Stressor-Related Disorders" heading. Conversely, several disorders previously classified in other sections, "Separation Anxiety Disorder" and "Selective Mutism," are now classified under the anxiety disorders heading. While these changes and differences are often subtle, they still introduce variability in what may or may not have been classified as an "anxiety disorder" in the accumulated body of research on anxiety and heart disease.

Common anxiety disorders

Despite this variability, a number of core diagnostic categories constitute the majority of diagnosed anxiety conditions and share primary features across diagnostic systems. "GAD" is characterized by pervasive, exaggerated, and chronic worry in everyday life. "Specific phobias" refer to intense and/or avoidant reactions to specific objects or situations. "Social anxiety disorder" is defined by intense fear or avoidance of social situations, either generally or in specific circumstances (e.g., public speaking). "Agoraphobia" refers to intense anxiety surrounding specific situations where escape or control may be limited, as well as avoidance behaviors associated with this anxiety. "Panic disorder" is frequently associated with agoraphobia but can occur with or without it, and is characterized by intense attacks of fear and discomfort, as well as a physical symptoms such as elevated heart rate, nausea, breathing difficulty, etc. "Separation anxiety disorder" is characterized by intense distress resulting from separation from a person or place, and can occur in both adults and children. "Posttraumatic stress disorder" and "obsessive–compulsive disorder" are no longer classified as "anxiety disorders" within the DSM diagnostic system, but still share many features with anxiety disorders and have often been included in research on anxiety disorders and heart disease. PTSD is characterized by a range of symptoms associated with exposure to a traumatic event, such as reexperiencing, hypervigilance, avoidance, and distorted beliefs, while OCD is characterized by chronic, uncontrollable preoccupations and compulsive behaviors or rituals designed to relieve associated distress. Prevalence rates of the various anxiety disorders vary substantially across large epidemiological studies in the United States and Europe (Bandelow and Michaelis, 2015), highlighting the significance of variations in diagnostic criteria, and across ethnic/cultural groups (Asnaani et al., 2010) and countries (Bijl et al., 1998); in general, however, large epidemiological studies in the United States and Europe have found the highest rates of specific phobias, followed by social anxiety disorder, GAD, and panic disorder (Bandelow and Michaelis, 2015). Additionally, while not yet codified in formal mental health diagnostic systems, the construct of "Cardiac Anxiety" has been proposed, a measure of it has been developed (Cardiac Anxiety Questionnaire, CAQ)

(Eifert et al., 2000), and it is increasingly being incorporated into research (Marker et al., 2008).

Assessment of anxiety

A range of options are possible for assessment of anxiety in clinical and research contexts, but there are important considerations and caveats with each. A number of brief screening measures are available and typically have the benefit of little to no cost, rapid administration, and ease of completion. However, susceptibility to normal, transient anxiety symptoms associated with illness and overlap between symptoms of anxiety and symptoms of cardiac disease have been identified as significant issues with screening measures (Celano et al., 2013). Anxiety symptoms also frequently overlap with other mood and psychiatric disorders, with one study identifying an association between cardiovascular disease and combined anxiety/depression, but not anxiety alone (Seldenrijk et al., 2015). Furthermore, studies with cardiac populations (Easton et al., 2016) and other medical populations (Hitchon et al., 2020) have highlighted diagnostic performance issues with anxiety screening measures, especially relative to depression screening.

With these caveats in mind, a meta-analysis of anxiety assessment measures with heart failure patients (Easton et al., 2016) recommended use of the GAD-7 (Spitzer et al., 2006) and Hospital Anxiety and Depression Scale (HADS) (Zigmund, 1983) based on their ability to discriminate between anxiety and depression symptoms, their minimization of somatic symptoms, and their inclusion of well-defined cut-off scores. Other studies have recommended a two-step approach to assessing for anxiety and depression in cardiac disease patients (Celano et al., 2013), beginning with two-item screening measures and expanding to more detailed measures in the event of positive initial screenings. Referral to mental health specialists can be useful for more sophisticated clinical evaluations and administration of detailed, diagnostic symptom inventories (e.g., SCID (First et al., 2015), SCAN (Wing et al., 1990)) that can help mitigate the issues of symptom overlap between anxiety and heart disease. However, these diagnostic inventories generally require extensive training and administration time, and issues with mental health provider access and wait times are common barriers to their routine use (Easton et al., 2016).

There are also important questions of whether to assess anxiety as a continuous construct or within a binary diagnostic framework, and how much to focus on anxiety as a general patient "trait" versus a context specific "state," as distinction highlighted in the "State–Trait Anxiety Inventory" (Spielberger et al., 1970) and considered in many studies of anxiety. Ultimately, decision making about the need for and the approach to anxiety assessment in clinical practice is highly context dependent. For example, in the context of evaluating anxiety as a risk factor for mortality and possible target for intervention in CAD patients, experts recommend assessing anxiety during periods of clinical stability and not around the time of an acute cardiac episode, and using a diagnostic threshold as opposed to a continuous symptom rating approach (Celano et al., 2015). In contrast, acute screening models have been validated for ruling out panic disorder in patients presenting to ED with chest pains (Fleet and Beitman, 1997).

PREVENTION AND TREATMENT OF ANXIETY DISORDERS

Evidence for treatment

As we have shown thus far in this chapter, connections between anxiety and cardiovascular health are well established. Yet despite this evidence, anxiety, as well as related psychological impacts, is often overlooked, undiagnosed, and left untreated (Pedersen and Andersen, 2018). Further, the study of interventions to target anxiety in cardiac risk populations remains a wide hole in the research literature (Richards et al., 2017; Farquhar et al., 2018). This dearth of research occurs, despite more attention given to depression, stress, and trauma in cardiology studies (Cohen et al., 2015).

A limited number of studies can provide some narrow understanding of the benefits of treating anxiety post cardiac conditions; however, significantly more research is needed to further understand the impacts of treatment. In a Cochran report that synthesized the effects of past research on treatment of anxiety for coronary heart disease (CHD) patients, Richards et al. (2017) found evidence that treatment decreased mortality attributed to CHD, as well as improved psychological symptoms. However, risk of further cardiovascular procedures, nonfatal MI, and total mortality were not affected by anxiety treatment (Richards et al., 2017).

A similar systematic review of the literature by Farquhar and colleagues also found consistent results, showing one-third of studies documenting reductions in anxiety symptoms following treatment, compared to controls (Farquhar et al., 2018). These authors discussed two significant limitations in the research. First, the methodology and measures of anxiety used varied, with many studies not examining anxiety outcomes post-treatment of CHD or presented as a secondary outcome (Farquhar et al., 2018). Second, few studies specifically targeted patients who were reporting high anxiety or diagnosed with an anxiety disorder (Farquhar et al., 2018). This second point was also supported by Pedersen and Andersen (2018), who reported that no randomized control trial (RCT) could be found that combined screening of anxiety with subsequent treatment in a cardiac population.

Additionally, we do not yet understand which treatments would be most effective, when they should be delivered, and which patients would be most impacted. This issue is exacerbated by the few studies published being highly variable in their definition of treatment (Richards et al., 2017; Farquhar et al., 2018). Much of this research has focused on nontraditional treatment of anxiety. For example, in Farquhar and colleagues review of 119 studies, only one study examined the effects of SSRI medications, despite SSRIs being a well-established treatment of anxiety (Baldwin et al., 2009), safe in cardiac populations (Funk and Bostwick, 2013), and highly studied for treating depression in cardiac populations (Farquhar et al., 2018).

Given these limitations, the following treatment recommendations are based on related research. While we do not know the effectiveness of anxiety treatment in cardiology populations specifically, we do know much on the impacts of anxiety treatment in general. We also know more about the impacts of treatment on related psychological concerns (e.g., depression), which has been shown in the literature to respond similarly to anxiety in response to treatment in general populations. Furthermore, we know about many lifestyle and wellness treatments that have been found to impact both cardiovascular risk factors, as well as anxiety. Finally, we know that different patients respond to different interventions, given individual preference. From this, we have developed the following spectrum of treatment recommendations that can be tailored to both the individual patient, as well as the resources and skills of the treating provider.

PREVENTION AND TREATMENT

Motivational interviewing

Although the connections between physical and psychological symptoms are well established, the acceptance of this fact for a patient, let alone acknowledgement that treatment would be beneficial, can be a challenge for the treating provider. Experts have noted that not only is acceptance of treatment difficult, but also once treatment is accepted, we have added challenges of providing adequate treatment and then supporting maintenance of treatment by the patient (Wolever et al., 2007). This is why we are advocating that the best place to start with a patient who screens positive for anxiety is to engage in brief motivational interviewing (MI).

Motivational interviewing is a counseling technique developed by Miller and Rollnick with strong evidence in a variety of health and psychological conditions (Miller and Rollnick, 2013). For example, MI has been found to be as effective as longer term CBT and more effective than nontreatment controls for improving diet and increasing exercise (Burke et al., 2003; Morton et al., 2015), weight loss (Barnes and Ivezaj, 2015), reducing cholesterol, blood pressure, improving glycemic control (Cummings et al., 2009), improving activities of daily living after stroke (Cheng et al., 2015), medication adherence (Palacio et al., 2016), and smoking cessation (Lindson-Hawley et al., 2015).

While motivational interviewing can be used as a stand-alone treatment, it can also be thought of as a supplemental treatment to enhance the effectiveness and acceptance of additional treatments. At its core, motivational interviewing is a conversational strategy employed by the provider. It helps the provider engage a patient in such a way as to pull out and enhance their internal motivations for change and reduce ambivalence. Thus, a provider wanting to encourage treatment of anxiety can engage patients in open-ended questioning, reflecting the patient's motivations, thoughts about treatment, their current experiences with anxiety, and ambivalence toward change, helping them identify their own rationale for engaging in treatment. Effective use of simple MI-consistent strategies can work to encourage a referral to specialty providers for psychotherapy or biofeedback, acceptance of a prescription of a psychotropic, and/or consideration of lifestyle behaviors that can improve anxiety and cardiac health.

A provider wanting to incorporate MI-consistent methods into their standard practice would benefit from three basic strategies: (1) Use of open-ended questions before education. For example, rather than immediately discussing the benefits of anxiety treatment, the provider can first ask, "What thoughts have you had about treatment for worry and anxiety?" (2) Use of reflective listening to a patient concern. For example, rather than immediately following a patient's response with another question or education, the provider can repeat back a portion of what the patient stated. This helps the patient feel heard and increases the relationship alliance. Finally, (3) Recognition of patient autonomy. For example, rather than a provider detailing a treatment plan to the patient, an MI-consistent provider will provide their recommendations, usually including multiple options, which the patient can choose from or choose to decline. A statement such as, "Given what we have talked about, I recommend you pursue treatment for both your heart and general quality of life. There are several options. Therapy might be what I would choose, but you have other options as well. What do you think is best for you?" These simple conversation strategies can go a long way to increase a patient's likelihood of treatment engagement and adherence. There is considerable evidence for the efficacy of these basic strategies in healthcare settings (Lundahl and Burke, 2009) and the interested reader is referred to Rollnick et al.'s (2008) *Motivational Interviewing in Health Care: Helping Patients Change Behavior* for more information.

Lifestyle behavior

In our proposed model of treatment recommendations for anxiety in heart disease (Fig. 33.1), lifestyle behaviors can be viewed as a secondary goal or a "plan b." In truth, however, we feel that lifestyle behavior changes are important for both prevention of cardiovascular disease and risk reduction in existing disease, particularly in patients experiencing anxiety. As we have shown, there are complex, reciprocal relationships between modifiable vascular risk factors (e.g., diet, tobacco, substance use), anxiety, and cardiovascular disease. Therefore, healthy lifestyle promotion can rightly be viewed as a prevention and intervention strategy for cardiac disease, as well as anxiety itself.

Interventions as simple as providing education on cardiac health behaviors have been found to reduce anxiety, though these studies have typically only examined health anxiety (Farquhar et al., 2018). Other interventions, focused on increasing physical exercise in cardiology patients with anxiety, found greater improvements in reported anxiety compared to control groups (Farquhar et al., 2018). Further interventions, focused on general anxiety, have found improvements from stress management and mindfulness techniques (Khoury et al., 2015). It is clear that taking time to educate patients on strategies to live a healthier lifestyle can lead to improvements in both heart health and anxiety. Incorporating these interventions in an MI consistent way, as described earlier, can enhance both their acceptance and effectiveness.

Medication management

For many, medication management of anxiety can be the most tolerable treatment approach, as it requires

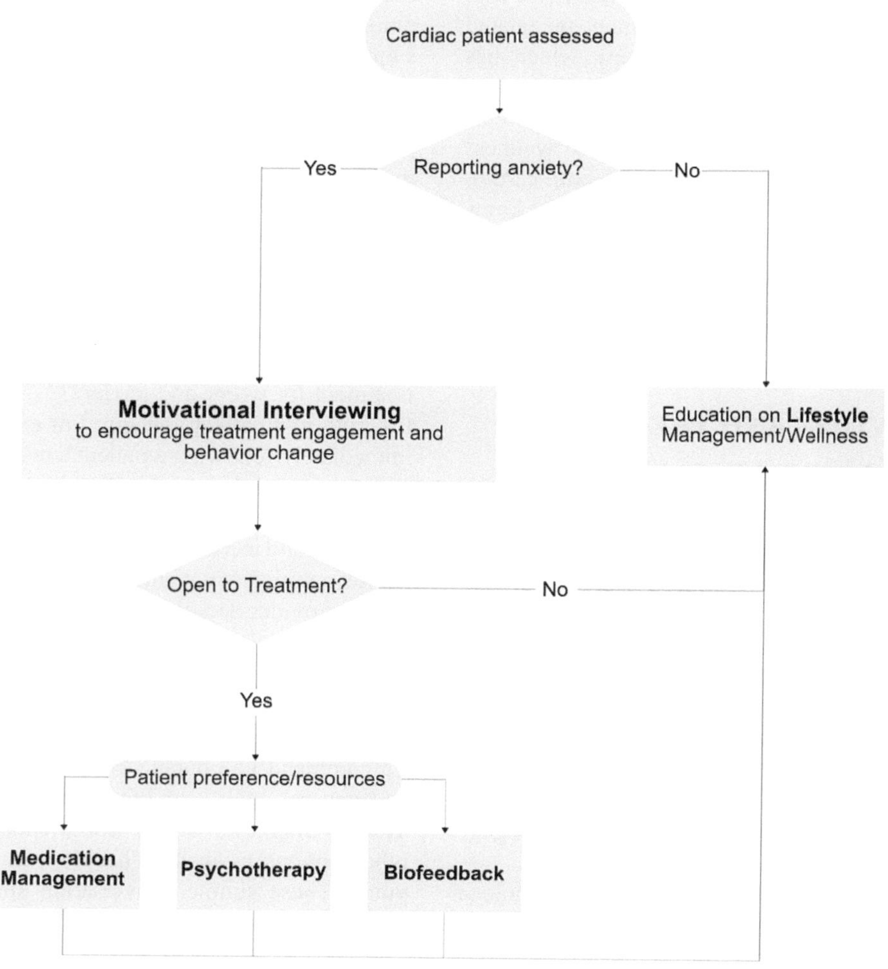

Fig. 33.1. Flow chart for treatment decision making.

minimal time commitment and behavior change. Selective serotonin reuptake inhibitors (SSRIs) and Serotonin-norepinephrine reuptake inhibitors (SNRIs) are frequently used to treat anxiety (Janeway, 2009; Baldwin et al., 2011) and have been found to be effective for patients with cardiac conditions (Clarke and Currie, 2009). SSRIs have also been found to have minimal cardiac risk (Funk and Bostwick, 2013). While there is increasing recognition of side effect and dependence concerns with benzodiazepines (Seldenrijk et al., 2017) and there has been a general move away from long-term use, their efficacy as a fast-acting agent for anxiety reduction is well established, and consideration of short-term use in specific cardiac care contexts may still be warranted. Symptom severity and complexity should be considered when determining the appropriate prescriber for medication management of anxiety. For mild to moderate anxiety, it may be appropriate for the cardiac provider to serve as a stand-alone prescriber. For moderate concerns, the addition of a care manager, a supportive nonprescriber (e.g., nurse, counselor, or psychotherapist) who will monitor anxiety symptoms and consult with the cardiac provider, may be warranted. For moderate to severe concerns it may be most appropriate to refer to a psychiatrist for further evaluation and management of prescriptions.

Psychotherapy

Cognitive behavioral therapy (CBT) is one of the most widely studied psychotherapy treatments for a variety of anxiety diagnoses. Additional evidence-based psychotherapy interventions exist, but few are as widely studied in RCTs. CBT has also been specifically tested in RCTs examining depression treatment in cardiac patients, finding patients receiving CBT in individual or group treatment had greater improvement in depression, compared to treatment as usual controls (Berkman et al., 2003). A meta-analysis also found support for use of CBT in depressed and anxious patients with cardiovascular disease in terms of both long-term mood symptom reduction and improved quality of life ratings (Reavell et al., 2018). They did not find evidence of CBT affecting cardiovascular outcomes, but noted limitations in availability of mortality and long-term cardiac outcome data in existing studies.

CBT works to help the patient recognize the cognitions, behaviors, and emotional responses that perpetuate their anxiety (Clark and Beck, 2011). In the treatment, the patient will work to identify new behaviors that reduce anxiety, restructure anxious thoughts into more realistic thinking, and utilize coping strategies to reduce anxious emotions. Although obviously impractical to implement in cardiac care settings, as little as 5 h of CBT psychotherapy has been shown to be effective in cardiac populations (Reavell et al., 2018) and it is recommended that patients presenting to cardiac clinic who are motivated to complete CBT or alternative evidence-based psychotherapy be referred to a qualified mental health provider.

Biofeedback

Research has found that relaxation techniques tailored to increase heart rate variability (HRV) have improved both cardiac risk and anxiety in patients. A meta-analysis examining impacts of HRV biofeedback on self-reported anxiety and stress symptoms found a substantial decline in symptoms for patients receiving HRV biofeedback (Goessl et al., 2017). Several studies have also shown that HRV biofeedback is linked to reduced readmissions and emergency visits, compared to controls amongst CAD patients (Yu et al., 2018b), reduced blood pressure in prehypertensive individuals (Lin et al., 2012), and reduced risk of CVD (Thayer et al., 2010).

HRV Biofeedback is a specialty treatment referral option for cardiology providers. HRV, the variation amongst the intervals between heartbeats, is different from heart rate, which is an average number of heart beats per minute. HRV is a good measure of heart health, reflecting autonomic balance, blood pressure, vascular tone, and other heart health indicators (Shaffer and Ginsberg, 2017). HRV biofeedback teaches techniques to increase HRV with immediate feedback of physiological response. A typical example of HRV biofeedback teaches breathing principles with graphical output of HRV, measured via blood volume pulse, for patients to review and monitor, while breathing at their resonant frequency or personalize ideal breath rate (Yu et al., 2018a).

CONCLUSIONS

Anxiety is extensively linked to vascular risk factors, heart disease, cardiac surgery outcomes, and associated neurologic manifestations. Although anxiety is a challenging construct to operationalize in research and to reliably assess in clinical practice, the accumulated literature clearly establishes anxiety as an important consideration and treatment target in the prevention and management of cardiac illness. In spite of the expansive and outstanding research that has already been done in this area, important questions remain. Improved understanding of the directionality and causality of associations between anxiety and heart disease is needed to clarify treatment emphases, cross-cultural studies are necessary to better understand how region and culture influence manifestations of anxiety and its optimal assessment and treatment, and cross-cutting and translational research efforts between mental health, cardiology, and other disciplines are necessary to establish clinically viable, efficacious methods for the assessment and treatment of anxiety in a range of clinical cardiac contexts. While these challenges are not insignificant, the literature reviewed in this chapter leaves little doubt that there is extraordinary potential for improvement in patients' cardiac, neurologic, and general physical health outcomes with increased attention paid to this important mental health variable.

References

Airaksinen E, Larsson M, Forsell Y (2005). Neuropsychological functions in anxiety disorders in population-based samples: evidence of episodic memory dysfunction. J Psychiatr Res 39: 207–214.

Ako J, Sudhir K, Farouque HO et al. (2006). Transient left ventricular dysfunction under severe stress: brain-heart relationship revisited. Am J Med 119: 10–17.

Almeida OP, Garrido GJ, Beer C et al. (2012). Cognitive and brain changes associated with ischaemic heart disease and heart failure. Eur Heart J 33: 1769–1776.

Andrew MJ, Baker RA, Kneebone AC et al. (2000). Mood state as a predictor of neuropsychological deficits following cardiac surgery. J Psychosom Res 48: 537–546.

AP Association (2013). Diagnostic and statistical manual of mental disorders (DSM-5®), American Psychiatric Pub.

Asnaani A, Richey JA, Dimaite R et al. (2010). A cross-ethnic comparison of lifetime prevalence rates of anxiety disorders. J Nerv Ment Dis 198: 551.

Bacon SL, Campbell TS, Arsenault A et al. (2014). The impact of mood and anxiety disorders on incident hypertension at one year. Int J Hypertension 2014 (953094): 7. https://doi.org/10.1155/2014/953094.

Bairey CN, de Yang L, Berman DS et al. (1990). Comparison of physiologic ejection fraction responses to activities of daily living: implications for clinical testing. J Am Coll Cardiol 16: 847–854.

Baker KD, Loughman A, Spencer SJ et al. (2017). The impact of obesity and hypercaloric diet consumption on anxiety and emotional behavior across the lifespan. Neurosci Biobehav Rev 83: 173–182.

Baldwin DS, Ajel KI, Garner M (2009). Pharmacological treatment of generalized anxiety disorder. Behavioral neurobiology of anxiety and its treatment, Springer.

Baldwin DS, Waldman S, Allgulander C (2011). Evidence-based pharmacological treatment of generalized anxiety disorder. Int J Neuropsychopharmacol 14: 697–710.

Bandelow B, Michaelis S (2015). Epidemiology of anxiety disorders in the 21st century. Dialogues Clin Neurosci 17: 327.

Bandelow B, Baldwin D, Abelli M et al. (2016). Biological markers for anxiety disorders, OCD and PTSD–a consensus statement. Part I: neuroimaging and genetics. World J Biol Psychiatry 17: 321–365.

Bandelow B, Baldwin D, Abelli M et al. (2017). Biological markers for anxiety disorders, OCD and PTSD: a consensus statement. Part II: neurochemistry, neurophysiology and neurocognition. World J Biol Psychiatry 18: 162–214.

Barnes R, Ivezaj V (2015). A systematic review of motivational interviewing for weight loss among adults in primary care. Obes Rev 16: 304–318.

Batelaan NM, Seldenrijk A, Bot M et al. (2016). Anxiety and new onset of cardiovascular disease: critical review and meta-analysis. Br J Psychiatry 208: 223–231.

Battista SR, Stewart SH, Ham LS (2010). A critical review of laboratory-based studies examining the relationships of social anxiety and alcohol intake. Curr Drug Abuse Rev 3: 3–22.

Baxter A, Scott K, Vos T et al. (2013). Global prevalence of anxiety disorders: a systematic review and meta-regression. Psychol Med 43: 897–910.

Becker E, Rios CLO, Lahmann C et al. (2018). Anxiety as a risk factor of Alzheimer's disease and vascular dementia. Br J Psychiatry 213: 654–660.

Berkman LF, Blumenthal J, Burg M et al. (2003). Effects of treating depression and low perceived social support on clinical events after myocardial infarction: the Enhancing Recovery in Coronary Heart Disease Patients (ENRICHD) Randomized Trial. JAMA 289: 3106–3116.

Bijl RV, Ravelli A, Van Zessen G (1998). Prevalence of psychiatric disorder in the general population: results of The Netherlands Mental Health Survey and Incidence Study (NEMESIS). Soc Psychiatry Psychiatr Epidemiol 33: 587–595.

Bonnet F, Irving K, Terra J-L et al. (2005). Anxiety and depression are associated with unhealthy lifestyle in patients at risk of cardiovascular disease. Atherosclerosis 178: 339–344.

Broomfield NM, Quinn TJ, Abdul-Rahim AH et al. (2014). Depression and anxiety symptoms post-stroke/TIA: prevalence and associations in cross-sectional data from a regional stroke registry. BMC Neurol 14: 198.

Burke BL, Arkowitz H, Menchola M (2003). The efficacy of motivational interviewing: a meta-analysis of controlled clinical trials. J Consult Clin Psychol 71: 843.

Burrows G (1846). On disorders of the cerebral circulation, and on the connection between affections of the brain and diseases of the heart. Med Chir Rev 4: 34–48.

Burton C, Campbell P, Jordan K et al. (2013). The association of anxiety and depression with future dementia diagnosis: a case-control study in primary care. Fam Pract 30: 25–30.

Butters MA, Bhalla RK, Andreescu C et al. (2011). Changes in neuropsychological functioning following treatment for late-life generalised anxiety disorder. Br J Psychiatry 199: 211–218.

Celano CM, Suarez L, Mastromauro C et al. (2013). Feasibility and utility of screening for depression and anxiety disorders in patients with cardiovascular disease. Circ Cardiovasc Qual Outcomes 6: 498–504.

Celano CM, Millstein RA, Bedoya CA et al. (2015). Association between anxiety and mortality in patients with coronary artery disease: a meta-analysis. Am Heart J 170: 1105–1115.

Celano CM, Villegas AC, Albanese AM et al. (2018). Depression and anxiety in heart failure: a review. Harv Rev Psychiatry 26: 175.

Chalmers JA, Quintana DS, Abbott MJ et al. (2014). Anxiety disorders are associated with reduced heart rate variability: a meta-analysis. Front Psych 5: 80.

Cheng D, Qu Z, Huang J et al. (2015). Motivational interviewing for improving recovery after stroke. Cochrane Database Syst Rev 2015: CD011398.

Clark DA, Beck AT (2011). The anxiety and worry workbook: the cognitive behavioral solution, Guilford Press.

Clark CR, Galletly CA, Ash DJ et al. (2009). Evidence-based medicine evaluation of electrophysiological studies of the anxiety disorders. Clin EEG Neurosci 40: 84–112.

Clarke DM, Currie KC (2009). Depression, anxiety and their relationship with chronic diseases: a review of the epidemiology, risk and treatment evidence. Med J Australia 190: S54–S60.

Cohen BE, Edmondson D, Kronish IM (2015). State of the art review: depression, stress, anxiety, and cardiovascular disease. Am J Hypertens 28: 1295–1302.

Coughlin SS (2011). Post-traumatic stress disorder and cardiovascular disease. Open Cardiovasc Med J 5: 164.

Crippa JA, Zuardi AW, Martín-Santos R et al. (2009). Cannabis and anxiety: a critical review of the evidence. Hum Psychopharmacol Clin Exp 24: 515–523.

Crocq M-A (2017). The history of generalized anxiety disorder as a diagnostic category. Dialogues Clin Neurosci 19: 107.

Cummings SM, Cooper RL, Cassie KM (2009). Motivational interviewing to affect behavioral change in older adults. Res Social Work Pract 19: 195–204.

Davies RD, Gabbert SL, Riggs PD (2001). Anxiety disorders in neurologic illness. Curr Treat Options Neurol 3: 333–346.

de Lijster JM, Dierckx B, Utens EM et al. (2017). The age of onset of anxiety disorders: a meta-analysis. Canadian J Psychiatry Revue Canadienne de Psychiatrie 62: 237.

Deanfield J, Kensett M, Wilson R et al. (1984). Silent myocardial ischaemia due to mental stress. Lancet 324: 1001–1005.

DeRight J, Jorgensen RS, Cabral MJ (2015). Composite cardiovascular risk scores and neuropsychological functioning: a meta-analytic review. Ann Behav Med 49: 344–357.

Detroyer E, Dobbels F, Verfaillie E et al. (2008). Is preoperative anxiety and depression associated with onset of delirium after cardiac surgery in older patients? A prospective cohort study. J Am Geriatr Soc 56: 2278–2284.

Dirven B, Homberg J, Kozicz T et al. (2017). Epigenetic programming of the neuroendocrine stress response by adult life stress. J Mol Endocrinol 59: R11–R31.

Dubey NK, Syed-Abdul S, Nguyen PA et al. (2016). Association between anxiety state and mitral valve disorders: a Taiwanese population-wide observational study. Comput Methods Programs Biomed 132: 57–61.

Eaker ED, Sullivan LM, Kelly-Hayes M et al. (2005). Tension and anxiety and the prediction of the 10-year incidence of coronary heart disease, atrial fibrillation, and total mortality: the Framingham Offspring Study. Psychosom Med 67: 692–696.

Easton K, Coventry P, Lovell K et al. (2016). Prevalence and measurement of anxiety in samples of patients with heart failure: meta-analysis. J Cardiovasc Nurs 31: 367.

Eggermont LH, De Boer K, Muller M et al. (2012). Cardiac disease and cognitive impairment: a systematic review. Heart 98: 1334–1340.

Eifert GH, Thompson RN, Zvolensky MJ et al. (2000). The cardiac anxiety questionnaire: development and preliminary validity. Behav Res Ther 38: 1039–1053.

Emdin C, Odutayo A, Hsiao A et al. (2015). Association of cardiovascular trial registration with positive study findings: epidemiological Study of Randomized Trials (ESORT). JAMA Intern Med 175: 304–307.

Engum A (2007). The role of depression and anxiety in onset of diabetes in a large population-based study. J Psychosom Res 62: 31–38.

Fang X, Spieler D, Albarqouni L et al. (2018). Impact of generalized anxiety disorder (GAD) on prehospital delay of acute myocardial infarction patients. Findings from the multicenter MEDEA study. Clin Res Cardiol 107: 471–478.

Farquhar JM, Stonerock GL, Blumenthal JA (2018). Treatment of anxiety in patients with coronary heart disease: a systematic review. Psychosomatics 59: 318–332.

Fink HA, Hemmy LS, MacDonald R et al. (2015). Intermediate-and long-term cognitive outcomes after cardiovascular procedures in older adults: a systematic review. Ann Intern Med 163: 107–117.

First M, Williams J, Karg R et al. (2015). Structured clinical interview for DSM-5—Research version (SCID-5 for DSM-5, research version; SCID-5-RV), American Psychiatric Association, Arlington, VA, pp. 1–94.

Fleet RP, Beitman BD (1997). Unexplained chest pain: when is it panic disorder? Clin Cardiol 20: 187–194.

Fluharty M, Taylor AE, Grabski M et al. (2016). The association of cigarette smoking with depression and anxiety: a systematic review. Nicotine Tob Res 19: 3–13.

Funk KA, Bostwick JR (2013). A comparison of the risk of QT prolongation among SSRIs. Ann Pharmacother 47: 1330–1341.

Gariepy G, Nitka D, Schmitz N (2010). The association between obesity and anxiety disorders in the population: a systematic review and meta-analysis. Int J Obes (Lond) 34: 407–419.

Goddard AW, Ball SG, Martinez J et al. (2010). Current perspectives of the roles of the central norepinephrine system in anxiety and depression. Depress Anxiety 27: 339–350.

Goessl VC, Curtiss JE, Hofmann SG (2017). The effect of heart rate variability biofeedback training on stress and anxiety: a meta-analysis. Psychol Med 47: 2578–2586.

Goh AC, Wong S, Zaroff JG et al. (2016). Comparing anxiety and depression in patients with Takotsubo stress cardiomyopathy to those with acute coronary syndrome. J Cardiopulm Rehabil Prev 36: 106–111.

Grigsby AB, Anderson RJ, Freedland KE et al. (2002). Prevalence of anxiety in adults with diabetes: a systematic review. J Psychosom Res 53: 1053–1060.

Guan S, Fang X, Hu X (2014). Factors influencing the anxiety and depression of patients with dilated cardiomyopathy. Int J Clin Exp Med 7: 5691.

Hart RG, Pearce LA, Aguilar MI (2007). Meta-analysis: antithrombotic therapy to prevent stroke in patients who have nonvalvular atrial fibrillation. Ann Intern Med 146: 857–867.

Heron M, Anderson RN (2017). Changes in the leading cause of death: Recent patterns in heart disease and cancer mortality, National Center for Health Statistics. Data Brief.

Hill AB (1965). The environment and disease: association or causation?, Sage Publications.

Hitchon CA, Zhang L, Peschken CA et al. (2020). The validity and reliability of screening measures for depression and anxiety disorders in rheumatoid arthritis. Arthritis Care Res 72: 1130–1139.

Jackson JL, Gerardo GM, Monti JD et al. (2018). Executive function and internalizing symptoms in adolescents and young adults with congenital heart disease: the role of coping. J Pediatr Psychol 43: 906–915.

Janeway D (2009). An integrated approach to the diagnosis and treatment of anxiety within the practice of cardiology. Cardiol Rev 17: 36–43.

Janszky I, Ahnve S, Lundberg I et al. (2010). Early-onset depression, anxiety, and risk of subsequent coronary heart disease: 37-year follow-up of 49,321 young Swedish men. J Am Coll Cardiol 56: 31–37.

Joseph HK, Whitcomb J, Taylor W (2015). Effect of anxiety on individuals and caregivers after coronary artery bypass grafting surgery: a review of the literature. Dimens Crit Care Nurs 34: 285–288.

Kessler RC, Berglund P, Demler O et al. (2005). Lifetime prevalence and age-of-onset distributions of DSM-IV disorders in the National Comorbidity Survey Replication. Arch Gen Psychiatry 62: 593–602.

Khoury B, Sharma M, Rush SE et al. (2015). Mindfulness-based stress reduction for healthy individuals: A meta-analysis. J Psychosom Res 78: 519–528.

Koivula M, Tarkka M-T, Tarkka M et al. (2002). Fear and anxiety in patients at different time-points in the coronary artery bypass process. Int J Nurs Stud 39: 811–822.

Krannich J-HA, Weyers P, Lueger S et al. (2007). Presence of depression and anxiety before and after coronary artery bypass graft surgery and their relationship to age. BMC Psychiatry 7: 47.

Lai HMX, Cleary M, Sitharthan T et al. (2015). Prevalence of comorbid substance use, anxiety and mood disorders in epidemiological surveys, 1990–2014: a systematic review and meta-analysis. Drug Alcohol Depend 154: 1–13.

Lambiase MJ, Kubzansky LD, Thurston RC (2014). Prospective study of anxiety and incident stroke. Stroke 45: 438–443.

Lamers F (2011). van OP, Comijs HC, Smit JH, Spinhoven P, van Balkom AJ et al. Comorbidity patterns of anxiety and depressive disorders in a large cohort study: the Netherlands Study of Depression and Anxiety (NESDA). J Clin Psychiatry 72: 341–348.

Lane D, Carroll D, Ring C et al. (2002). The prevalence and persistence of depression and anxiety following myocardial infarction. Br J Health Psychol 7: 11–21.

Li C, Barker L, Ford E et al. (2008). Diabetes and anxiety in US adults: findings from the 2006 Behavioral Risk Factor Surveillance System. Diabet Med 25: 878–881.

Li M, D'arcy C, Meng X (2016). Maltreatment in childhood substantially increases the risk of adult depression and anxiety in prospective cohort studies: systematic review, meta-analysis, and proportional attributable fractions. Psychol Med 46: 717–730.

Lichtman JH, Froelicher ES, Blumenthal JA et al. (2014). Depression as a risk factor for poor prognosis among patients with acute coronary syndrome: systematic review and recommendations: a scientific statement from the American Heart Association. Circulation 129: 1350–1369.

Lin G, Xiang Q, Fu X et al. (2012). Heart rate variability biofeedback decreases blood pressure in prehypertensive subjects by improving autonomic function and baroreflex. J Altern Complement Med 18: 143–152.

Lindson-Hawley N, Thompson TP, Begh R (2015). Motivational interviewing for smoking cessation. Cochrane Database Syst Rev CD006936. https://doi.org/10.1002/14651858.CD006936.pub3.

Luna RA, Foster JA (2015). Gut brain axis: diet microbiota interactions and implications for modulation of anxiety and depression. Curr Opin Biotechnol 32: 35–41.

Lundahl B, Burke BL (2009). The effectiveness and applicability of motivational interviewing: a practice-friendly review of four meta-analyses. J Clin Psychol 65: 1232–1245.

Luyster FS, Hughes JW, Gunstad J (2009). Depression and anxiety symptoms are associated with reduced dietary adherence in heart failure patients treated with an implantable cardioverter defibrillator. J Cardiovasc Nurs 24: 10–17.

Maatouk I, Herzog W, Böhlen F et al. (2016). Association of hypertension with depression and generalized anxiety symptoms in a large population-based sample of older adults. J Hypertens 34: 1711–1720.

Marker CD, Carmin CN, Ownby RL (2008). Cardiac anxiety in people with and without coronary atherosclerosis. Depress Anxiety 25: 824–831.

Martens E, Nyklíček I, Szabo B et al. (2008). Depression and anxiety as predictors of heart rate variability after myocardial infarction. Psychol Med 38: 375–383.

Miller WR, Rollnick S (2013). Applications of motivational interviewing, Motivational interviewing: Helping people change. Guilford Press New York, NY, US.

Morgan JF, O'Donoghue AC, McKenna WJ et al. (2008). Psychiatric disorders in hypertrophic cardiomyopathy. Gen Hosp Psychiatry 30: 49–54.

Morissette SB, Tull MT, Gulliver SB et al. (2007). Anxiety, anxiety disorders, tobacco use, and nicotine: a critical review of interrelationships. Psychol Bull 133: 245.

Morton K, Beauchamp M, Prothero A et al. (2015). The effectiveness of motivational interviewing for health behaviour change in primary care settings: a systematic review. Health Psychol Rev 9: 205–223.

Nowbar AN, Gitto M, Howard JP et al. (2019). Mortality from ischemic heart disease: analysis of data from the world health organization and coronary artery disease risk factors from NCD risk factor collaboration. Circ Cardiovasc Qual Outcomes 12: e005375.

Ogedegbe G (2008). White-coat effect: unraveling its mechanisms. Am J Hypertens 21: 135.

Oxlad M, Wade TD (2008). Longitudinal risk factors for adverse psychological functioning six months after coronary artery bypass graft surgery. J Health Psychol 13: 79–92.

Palacio A, Garay D, Langer B et al. (2016). Motivational interviewing improves medication adherence: a systematic review and meta-analysis. J Gen Intern Med 31: 929–940.

Papakostas GI, Öngür D, Iosifescu DV et al. (2004). Cholesterol in mood and anxiety disorders: review of the literature and new hypotheses. Eur Neuropsychopharmacol 14: 135–142.

Patel D, Mc Conkey ND, Sohaney R et al. (2013). A systematic review of depression and anxiety in patients with atrial fibrillation: the mind-heart link. Cardiovasc Psychiatry Neurol 2013.

Pedersen SS, Andersen CM (2018). Minding the heart: why are we still not closer to treating depression and anxiety in clinical cardiology practice?, SAGE Publications Sage UK, London, England.

Perret-Guillaume C, Briancon S, Wahl D et al. (2010). Quality of Life in elderly inpatients with atrial fibrillation as compared with controlled subjects. J Nutr Health Aging 14: 161–166.

Player MS, Peterson LE (2011). Anxiety disorders, hypertension, and cardiovascular risk: a review. Int J Psychiatry Med 41: 365–377.

Prohl J, Bodenburg S, Rustenbach SJ (2009). Early prediction of long-term cognitive impairment after cardiac arrest. J Int Neuropsychol Soc 15: 344–353.

Reavell J, Hopkinson M, Clarkesmith D et al. (2018). Effectiveness of cognitive behavioral therapy for depression and anxiety in patients with cardiovascular disease: a systematic review and meta-analysis. Psychosom Med 80: 742–753.

Richards SH, Anderson L, Jenkinson CE et al. (2017). Psychological interventions for coronary heart disease. Cochrane Database Syst Rev 4: CD002902.

Roberts RO, Knopman DS, Geda YE et al. (2010). Coronary heart disease is associated with non-amnestic mild cognitive impairment. Neurobiol Aging 31: 1894–1902.

Roest AM, Martens EJ, de Jonge P et al. (2010). Anxiety and risk of incident coronary heart disease: a meta-analysis. J Am Coll Cardiol 56: 38–46.

Roest AM, Zuidersma M, de Jonge P (2012). Myocardial infarction and generalised anxiety disorder: 10-year follow-up. Br J Psychiatry 200: 324–329.

Rollnick S, Miller WR, Butler C (2008). Motivational interviewing in health care: helping patients change behavior, Guilford Press.

Sabia S, Fayosse A, Dumurgier J et al. (2019). Association of ideal cardiovascular health at age 50 with incidence of dementia: 25 year follow-up of Whitehall II cohort study. BMJ 366: l4414.

Sasaki A, De Vega W, St-Cyr S et al. (2013). Perinatal high fat diet alters glucocorticoid signaling and anxiety behavior in adulthood. Neuroscience 240: 1–12.

Scherrer JF, Chrusciel T, Zeringue A et al. (2010). Anxiety disorders increase risk for incident myocardial infarction in depressed and nondepressed Veterans Administration patients. Am Heart J 159: 772–779.

Schwarz N, Kuniss M, Nedelmann M et al. (2010). Neuropsychological decline after catheter ablation of atrial fibrillation. Heart Rhythm 7: 1761–1767.

Seignourel PJ, Kunik ME, Snow L et al. (2008). Anxiety in dementia: a critical review. Clin Psychol Rev 28: 1071–1082.

Seldenrijk A, Vogelzangs N, Batelaan NM et al. (2015). Depression, anxiety and 6-year risk of cardiovascular disease. J Psychosom Res 78: 123–129.

Seldenrijk A, Vis R, Henstra M et al. (2017). Systematic review of the side effects of benzodiazepines. Ned Tijdschr Geneeskd 161: D1052.

Shaffer F, Ginsberg J (2017). An overview of heart rate variability metrics and norms. Front Public Health 5: 258.

Sharma S, Fernandes M, Fulton S (2013). Adaptations in brain reward circuitry underlie palatable food cravings and anxiety induced by high-fat diet withdrawal. Int J Obes (Lond) 37: 1183–1191.

Silva RMFLD, Miranda CM, Liu T et al. (2019). Atrial fibrillation and risk of dementia: epidemiology, mechanisms, and effect of anticoagulation Front Neurosci 13: 18.

Slade Jr WR, McNeal AC, Tse P (1989). Neurologic complications of cardiovascular diseases. J Natl Med Assoc 81: 193.

Smith JP, Randall CL (2012). Anxiety and alcohol use disorders: comorbidity and treatment considerations. Alcohol Res.

Smith KJ, Béland M, Clyde M et al. (2013). Association of diabetes with anxiety: a systematic review and meta-analysis. J Psychosom Res 74: 89–99.

Smoller JW (2016). The genetics of stress-related disorders: PTSD, depression, and anxiety disorders. Neuropsychopharmacology 41: 297–319.

Snookes SH, Gunn JK, Eldridge BJ et al. (2010). A systematic review of motor and cognitive outcomes after early surgery for congenital heart disease. Pediatrics 125: e818–e827.

Sokoreli I, De Vries J, Pauws S et al. (2016). Depression and anxiety as predictors of mortality among heart failure patients: systematic review and meta-analysis. Heart Fail Rev 21: 49–63.

Specchia G, de Servi S, Falcone C et al. (1984). Mental arithmetic stress testing in patients with coronary artery disease. Am Heart J 108: 56–63.

Spielberger CD, Gorsuch R, Lushene R et al. (1970). Manual for the state-trait inventory. In: Consulting psychologists, Palo Alto, California.

Spitzer RL, Kroenke K, Williams JB et al. (2006). A brief measure for assessing generalized anxiety disorder: the GAD-7. Arch Intern Med 166: 1092–1097.

Stein DJ, Aguilar-Gaxiola S, Alonso J et al. (2014). Associations between mental disorders and subsequent onset of hypertension. Gen Hosp Psychiatry 36: 142–149.

Strauss B, Thormann T, Strenge H et al. (1992). Psychosocial, neuropsychological and neurological status in a sample of heart transplant recipients. Qual Life Res 1: 119–128.

Strik JJ, Denollet J, Lousberg R et al. (2003). Comparing symptoms of depression and anxiety as predictors of cardiac events and increased health care consumption after myocardial infarction. J Am Coll Cardiol 42: 1801–1807.

Stroobant N, Vingerhoets G (2008). Depression, anxiety, and neuropsychological performance in coronary artery bypass graft patients: a follow-up study. Psychosomatics 49: 326–331.

Suzuki S-i, Kasanuki H (2004). The influences of psychosocial aspects and anxiety symptoms on quality of life of patients with arrhythmia: investigation in paroxysmal atrial fibrillation. Int J Behav Med 11: 104–109.

Tavazzi G, Zanierato M, Via G et al. (2017). Are neurogenic stress cardiomyopathy and takotsubo different syndromes with common pathways?: Etiopathological insights on dysfunctional hearts. JACC: Heart Failure 5: 940–942.

Thayer JF, Yamamoto SS, Brosschot JF (2010). The relationship of autonomic imbalance, heart rate variability and cardiovascular disease risk factors. Int J Cardiol 141: 122–131.

Tsushima WT, Johnson DB, Lee JD et al. (2005). Depression, anxiety and neuropsychological test scores of candidates for coronary artery bypass graft surgery. Arch Clin Neuropsychol 20: 667–673.

Tully PJ, Baker RA, Knight JL (2008). Anxiety and depression as risk factors for mortality after coronary artery bypass surgery. J Psychosom Res 64: 285–290.

Tully PJ, Baker RA, Winefield HR et al. (2010). Depression, anxiety disorders and Type D personality as risk factors for delirium after cardiac surgery. Aust N Z J PsyEchiatry 44: 1005–1011.

Tully PJ, Bennetts JS, Baker RA et al. (2011a). Anxiety, depression, and stress as risk factors for atrial fibrillation after cardiac surgery. Heart Lung 40: 4–11.

Tully PJ, Pedersen SS, Winefield HR et al. (2011b). Cardiac morbidity risk and depression and anxiety: a disorder, symptom and trait analysis among cardiac surgery patients. Psychol Health Med 16: 333–345.

Tully PJ, Cosh SM, Baumeister H (2014). The anxious heart in whose mind? A systematic review and meta-regression of factors associated with anxiety disorder diagnosis, treatment and morbidity risk in coronary heart disease. J Psychosom Res 77: 439–448.

Tully P, Turnbull D, Beltrame J et al. (2015). Panic disorder and incident coronary heart disease: a systematic review and meta-regression in 1 131 612 persons and 58 111 cardiac events. Psychol Med 45: 2909–2920.

Tully PJ, Harrison NJ, Cheung P et al. (2016). Anxiety and cardiovascular disease risk: a review. Curr Cardiol Rep 18: 120.

Tural U, Iosifescu DV (2019). The prevalence of mitral valve prolapse in panic disorder: a meta-analysis. Psychosomatics 60: 393–401.

Walters K, Rait G, Petersen I et al. (2008). Panic disorder and risk of new onset coronary heart disease, acute myocardial infarction, and cardiac mortality: cohort study using the general practice research database. Eur Heart J 29: 2981–2988.

WH Organization (1993). The ICD-10 classification of mental and behavioural disorders: diagnostic criteria for research, World Health Organization.

Wing JK, Babor T, Brugha T et al. (1990). SCAN: schedules fonr clinical assessment in neuropsychiatry. Arch Gen Psychiatry 47: 589–593.

Wolever R, Ladden L, Davis J et al. (2007). EMPOWER: Mindful maintenance therapist manual. Unpublished treatment manual for NIH funded grants 5U01 AT004159 and 5: U01.

Wray J, Sensky T (2001). Congenital heart disease and cardiac surgery in childhood: effects on cognitive function and academic ability. Heart 85: 687–691.

Yu B, Funk M, Hu J et al. (2018a). Biofeedback for everyday stress management: a systematic review. Front ICT 5: 23.

Yu L-C, Lin I-M, Fan S-Y et al. (2018b). One-year cardiovascular prognosis of the randomized, controlled, short-term heart rate variability biofeedback among patients with coronary artery disease. Int J Behav Med 25: 271–282.

Ziedonis D, Hitsman B, Beckham JC et al. (2008). Tobacco use and cessation in psychiatric disorders: National Institute of Mental Health report, Society for Research on Nicotine and Tobacco.

Zigmund A (1983). The hospital anxiety and depression scale. Acta Psychiatr Scand 67: 361–370.

Index

NB: Page numbers in *italics* refer to figures and tables.

D

E

I

N

O

P

T